T5-CAO-665

Energy Sources: Conservation and Renewables

(APS, Washington, DC, 1985)

AIP Conference Proceedings
Series Editor: Rita G. Lerner
Number 135

Energy Sources: Conservation and Renewables

(APS, Washington, DC, 1985)

Edited by
David Hafemeister
California Polytechnic State University
Henry Kelly
Office of Technology Assessment
Barbara Levi
Princeton University

American Institute of Physics
New York 1985

Copy fees: The code at the bottom of the first page of each article in this volume gives the fee for each copy of the article made beyond the free copying permitted under the 1978 US Copyright Law. (See also the statement following "Copyright" below.) This fee can be paid to the American Institute of Physics through the Copyright Clearance Center, Inc., 21 Congress Street, Salem, MA 01970.

Copyright © 1985 American Institute of Physics

Individual readers of this volume and non-profit libraries, acting for them, are permitted to make fair use of the material in it, such as copying an article for use in teaching or research. Permission is granted to quote from this volume in scientific work with the customary acknowledgment of the source. To reprint a figure, table or other excerpt requires the consent of one of the original authors and notification to AIP. Republication or systematic or multiple reproduction of any material in this volume is permitted only under license from AIP. Address inquiries to Series Editor, AIP Conference Proceedings, AIP, 335 E. 45th St., New York, N.Y. 10017.

L.C. Catalog Card No. 85-73019
ISBN 0-88318-334-X
DOE CONF-8504174

Printed in the United States of America

Pstack
SCIMON

ENERGY SOURCES: CONSERVATION AND RENEWABLES

TABLE OF CONTENTS

I. TECHNICAL PROGRESS AND POLICY OPTIONS (Chapters 1–5)

II. ENERGY AND BUILDINGS (Chapters 6–11)

III. WINDOWS, LIGHTING, APPLIANCES AND HVAC (Chapters 13–17)

IV. INDUSTRIAL AND AUTOMOTIVE (Chapters 18–19)

V. ELECTRICITY AND RENEWABLES (Chapters 20–28)

Section VI. APPENDICES A–J

PREFACE

A decade has passed since the oil embargo of 1973-74. The physics community initially responded to the problem of the "energy crisis" by conducting a summer study at Princeton University on THE EFFICIENT USE OF ENERGY during the summer of 1974. The American Institute of Physics book containing this study (AIP 25, 1975) has been the largest selling AIP Conference Series book; its impact went far beyond the physics community. Many of the technical ideas discussed by physicists in AIP 25 were untested concepts at the time; some of these ideas later became the focus of research, development and, ultimately, commercialization. This effort a decade ago helped shape some of the questions and discussion of how the U.S. should respond to the issue of energy use and planning.

The oil embargo of 1973-74, and the sharp rise in the price of imported petroleum from $2.50/barrel in 1972 to $30/barrel in 1980 forced the world to think more seriously about the fuels that drive our economic engine. Prior to 1973, the era of cheap energy had propelled the industrial revolution and helped develop exuhuberant consumer lifestyles. The government responded to the energy crisis in many ways: incentives for more production of energy from many different sources, incentives to encourage reduced consumption of energy by enhanced end-use efficiency, a strategic petroleum reserve to give the U.S. protection from sudden disruptions in imports, efficiency labels for appliances and automobiles, mileage standards for automobiles, and so forth. As we look back on the results a decade later, it is clear that the new sources of energy did not produce very much in the decade after the oil embargo. What is also clear is that conservation (enhanced end-use efficiency) made the lion's share contribution to our present state of relative well-being.

First, THE GOOD NEWS: Because of progress on energy, the U.S. has improved its status in financial, political and environmental matters. The U.S. is importing about 40% less petroleum than in the peak year of 1977; a drop from 8.8 million barrels per day (Mb/d) in 1977 to 5.4 Mb/d in 1984. This reduction in imports alone has saved the U.S. about $40 billion per year. The total savings of energy from all sources is about $150 billion per year when compared to projections of energy consumption of more than 100 quads/year for 1985. The "lock-step" relation between GNP and energy has been unlocked; as the GNP increased by 30% since 1974, total energy consumption has remained relatively constant. Primarily because of conservation, our national security has been enhanced since we have dramatically reduced our dependence on imports from OPEC, a reduction from 6.2 Mb/d in 1977 to 2.0 Mb/d in 1984.

Now, THE BAD NEWS: The euphoria of the results above should be tempered for a variety of reasons. In February, 1985, the Department of Interior slashed its estimates for offshore oil and natural gas resources by about a factor of two; from 27 to 12 billion barrels of oil, and from 163 to 91 trillion cubic feet of gas. In spite of the $250 billon investment to discover and develop new petroleum wells, the U.S. oil reserves have declined by 13% in 6 years (0.7 billion barrels/year = 1.9 Mb/d), and the U.S. natural gas reserves have remained essentially constant. The U.S. still spends about $60 billion per year to import oil, about 25% of the present deficit. There is continued concern about the CO_2 greenhouse effect since the CO_2 content of the atmosphere has risen from 330 ppm to 345 ppm in the last decade. And, acid rain threatens lakes and forests. Utilities are hesitant to build new power stations since both coal and nuclear have drawbacks.

The chapters in this book, ENERGY SOURCES: CONSERVATION AND RENEWABLES, were initially presented at a conference held at the Congressional Office of Technology Assessment in Washington, D.C. in April, 1985. This book is an appropriate complement to its visionary predecessor since it presents the results of a decade of research since 1975. The technical progress has been very encouraging in such areas as buildings, appliances, lighting, windows, indoor air quality measurements, passive solar, off-peak cooling, smart electrical meters, photovoltaics and wind energy, automobiles, industrial conservation and so forth. The ideas discussed in these chapters are too numerous to summarize, but together they give a clear signal that there are soundly based technical options which can continue to reduce America's energy consumption in the future.

Lastly, one of us (DH) would like to thank Art Rosenfeld and the members of the Energy Efficient Buildings Program at the Lawrence Berkeley Laboratory for their kind hospitality during the summer of 1985.

David Hafemeister
San Luis Obispo and
Berkeley, CA

Henry Kelly
Washington, DC

Barbara Levi
Princeton, NJ

August 30, 1985

CHAPTER 1

REFLECTIONS ON FIFTEEN YEARS OF ENERGY POLICY

John H. Gibbons*
Office of Technology Assessment
Washington, D.C. 20510

ABSTRACT. *The events of the 1970's — both the "energy crises" and the measures taken to alleviate them — changed our ways of thinking about energy. We now look at energy consumption as a largely substitutable means to various ends, not a goal or a measure of progress in and of itself. Energy demand growth has dropped markedly, even as the economy has grown. But there are many issues yet to be resolved if the United States is to have a comprehensive, rational energy strategy. This paper tackles four of them: Is there a place for continued government economic and regulatory intervention in the energy marketplace? What should be the federal role in energy research and development? What are our prospects for new discoveries in domestic oil and gas? What is the future of nuclear power in the United States? The author believes that the best way to solve our energy problems is to gauge, and then reflect in our energy policy, the true costs and benefits of energy production and consumption. He concludes that conservation investments have proven to be so rewarding that energy efficiency should be receiving a major amount of attention from energy policy makers for reasons of economic efficiency and in order to minimize the impact of future crises.*

Time flies. Reflecting on only the last ten years of energy policy -- my original intent -- no longer even takes us back to the Arab oil embargo of 1973, the crisis that precipitated much of our current thinking about energy policy. Getting a handle on our present energy situation requires an understanding of energy use and attitudes at least as far back as 1970. A snapshot of that year reveals that:

1. Both average and marginal energy prices had been level or declining for the previous fifty years;

2. Since 1960, the United States had experienced a steady rise in both energy consumption and oil imports;

3. Planners held grand notions of what one could accomplish when prices went even lower. Slogans for the future, such as "energy too cheap to meter" were still taken seriously. There were proposals for outdoor, as well as indoor, air conditioning, and technological visions of using thermonuclear fusion devices to atomize wastes and make them reusable. The Federal Power

---

* While the author drew upon OTA material in the preparation of this paper, the opinions given and the conclusions made are personal and do not necessarily represent those of the Office of Technology Assessment.

0094-243X/85/1350001-14$3.00 Copyright 1985 American Institute of Physics

Commission (FPC) was predicting a 3.4% annual growth in total U.S. energy demand out to 1990, and expected a concurrent 6.7% annual rate of increase in electricity consumption.[1]

4. Nuclear power was hailed as the fuel of the future. It was widely expected that electricity, continuing the downward cost spiral brought on by the advent of cheap nuclear power and abundant coal, would overtake oil as America's primary energy source sometime in the late 20th century; and

5. It was (then as now) difficult to coordinate federal energy policy because pieces of it were vested in various entities. The Atomic Energy Commission controlled nuclear power and energy research, while the Department of the Interior was in charge of fossil fuels and the Environmental Protection Agency wrestled with pollution from energy. Public and private utilities, mostly regulated state-by-state, controlled the distribution of electricity, and private cartels wielded enormous influence over the supply and price of oil.

A sense of a potentially gathering storm accompanied the general optimism, however. Events of the late 1960's highlighted many of the external costs of energy production and consumption. Many of the worsening problems of air and water pollution were directly attributable to use of energy. Apocryphal scenarios envisioned in Limits to Growth,[2] in contrast to those presented by the FPC, warned of the dangers of continuing, unmodified growth in energy consumption. The growth in the power of OPEC and increasing tensions in the Middle East contributed to a growing wariness of import dependence. High projected electric demand growth exacerbated worries not only about oil imports and the environment, but also about the sheer amount of capital investment that was implied. Gas curtailments began to show up along with spot shortages of heating oil and gasoline. And, most surprising to many in the energy field, there were growing doubts about the nuclear option.

As the ramifications of energy consumption were considered, some concluded that what might be needed was a basic transformation in the way we used energy. Easily foreseeable troubles -- a turn around in cost trends, or import constraints -- would require well planned responses. Four major (nonexclusive) options were generally presented as means to avoid increased dependence on energy, particularly oil imports. The United States could:

1. Increase exploration, development, and production of domestic conventional energy;

2. Quickly enter into large-scale development and commercial production of synthetic liquids and gas from coal and shale;

3. Rely more heavily on nuclear power, and even accelerate its already rapid development;

4. Accelerate a shift to more efficient use of energy by a variety of conservation technologies and policies.

I and many others became advocates for conservation (using energy wisely and carefully) or what some prefer to call energy productivity --

decreasing the amount of energy required to sustain a given level of goods and services. Extensive analysis had quickly proved the existence of many technical and economic opportunities to save energy. Efficiency of energy use had increased, especially in industry, for the previous 30 years, despite the fact that energy prices had been declining in real terms all the while. Conservation would have positive impacts on environmental quality and would be a generally distributed, incremental activity that would not prove overly burdensome to any one segment of society. And conservation activities would provide employment opportunities where the people are -- at the point of consumption.

I pursued this choice through several routes: through my work at Oak Ridge National Laboratory (systems and engineering analysis of efficient energy use); in setting up the first federal energy conservation office at the Department of the Interior and, later, at the Federal Energy Administration (FEA); at the Energy, Environment, and Natural Resources Center at the University of Tennessee; as chairman of the demand and conservation panel of the National Academy of Sciences CONAES study; on the Energy Research Advisory Board of the Department of Energy; as a member of the Board of the Tennessee Energy Authority; and finally at the Congressional Office of Technology Assessment (OTA). I recall with considerable pleasure my successful argument with my boss at FEA that we should provide half of the funds to sponsor the 1974 summer study at Princeton on energy conservation opportunities. In a book (Energy: The Conservation Revolution)[3] I coauthored with Bill Chandler, we laid out the ingredients for a comprehensive energy policy that had conservation at its core. We envisaged an energy future from these perspectives:

1. Consider the production and use of energy as means to certain ends, not as goals in and of themselves. Remember always that, given time and the capital for adjustment, energy is a largely substitutable input in the provision of most goods and services;

2. Application of technical ingenuity and institutional innovation can greatly facilitate energy options;

3. Energy decisions, like other investment decisions, should be made using clear signals of comparative total long-run costs, marginal costs, and cost trends;

4. It is important to correct distorted or inadequate market signals with policy instruments; otherwise external costs can increase and resources can be squandered. This correction includes internalizing in energy price, to the extent possible, the national security, human health, and environmental costs attributable to energy;

5. Investment in both energy supply and utilization research and development is an appropriate activity for both the public and private sectors, since costs and benefits accrue to both sectors;

6. There are other, generally more productive, ways to assist underprivileged citizens with their energy needs than subsidizing energy's price to them; and

7. In a world characterized by tightly integrated economies, we need to increase our cognizance of world energy resource conditions and needs, with special regard for international security as well as concern for the special needs of poor nations.

The energy situation today reflects the adoption of some of these ideas. Oil price deregulation is virtually complete, and natural gas is headed in that direction as "old" gas declines in importance. Legislation has been adopted to influence both demand (e.g., the Energy Policy and Conservation Act (EPCA), Corporate Average Fuel Efficiency Standards (CAFE)) and supply (e.g., the Synthetic Fuels Corporation (SFC), Public Utilities Regulatory Policy Act, (PURPA), and the Strategic Petroleum Reserve (SPR)). Thanks to conservation successes, imports are comfortably low (though rising), a condition that, along with the Strategic Petroleum Reserve, gives us better ability to manage energy shocks over the next several years. Our enormous bow wave of overinvestment in electric capacity is slowly ebbing, and cogeneration is making our electric system more diversified and flexible. Energy efficiency is widely increasing as energy-efficient capital stock is replaced throughout the economy, but these improvements are naturally slackening in the face of recently falling energy prices.

The seeds planted in the mid-70's have begun to sprout. Energy demand growth slowed markedly. Total industrial consumption fell 15% between 1974 and 1983, a result of both increased efficiency and product switching. Most importantly, energy use and GNP growth officially divorced, as the E/GNP ratio (energy use per unit of gross national product) fell 22% between 1973 and 1983. But now some energy prices in the United States have peaked and fallen (since oil prices are in $U.S., prices have not similarly declined for many other countries). That is good news for the United States, in as much as it contributes to lowering of inflation and a resurgence of industry, but it also contributes to a complacency about our long term energy situation that could become dangerous. Because of lower prices and domestic oil and gas discoveries far below what was once estimated, there are projections that the United States may, by the year 2000, import as much as 10 million barrels per day of oil -- in excess of our 1977 high.[4] And there have been some casualties along the way, particularly the collapse of nuclear orders and the dearth of both 1) energy research and development expenditures; and, very recently, 2) oil and gas exploration activities. The problems presented by a meager information base, the lack of sustained support of systems analysis, and the cultural, institutional, and political barriers to widespread acceptance of conservation technologies were also grossly underestimated when we first began to extol the virtues of energy efficiency.

There are many issues to be resolved before the United States can claim a coherent energy strategy. I would like to highlight just a few:

- Is there a place for continued government economic and regulatory intervention in the energy marketplace (e.g., price controls, emissions regulations)?

- What should be the federal role in energy research and development?

- What are the prospects for new discoveries in domestic oil and gas?

• What is the future of nuclear power in the United States?

As you will see from my comments, I believe a critical question, which encompasses all of these, concerns our ultimate ability to gauge, and then reflect in our energy strategies, the true costs of energy production and use. Key elements of that ability are committed leadership and access to information.

GOVERNMENT INTERVENTION IN THE ENERGY MARKETPLACE

The first question I would like to address is, Is there a place for government intervention in the energy marketplace?

Price decontrol was a crucial first step in internalizing the costs of energy consumption. The federal government's move out of that marketplace was a big step in the right direction, but not the whole answer. For example, market prices for energy do not reflect: 1) the environmental costs of energy production and use; 2) the costs of defending Mid-East oil production and shipping lanes; 3) the costs of purchasing and storing oil to meet emergencies; or 4) the cost to our allies and the impoverished Third World countries of U.S. competition for petroleum in the world market.

Electricity production consumes one-third of the primary energy used in the United States, yet the price the consumer pays does not reflect the marginal cost of providing that energy; therefore market signals for more efficient use are suboptimal.* Though all but "old" natural gas prices are decontrolled, natural gas transmission and distribution could remain controlled for many years to come.

Energy cannot be treated as a commodity only, for it involves political, social, national security, and environmental implications that must be addressed. Governments have historically intervened in energy markets in order to: 1) control and regulate a monopoly; 2) provide for the health and safety of citizens; and 3) provide for national security.

I was first drawn to energy policy when I began to notice the effects of strip mining on the mountains to the north and west of Oak Ridge National Laboratory. Now it is possible to document the fact that air pollutants are destroying parts of the beautiful Smokey Mountains. There is growing concern about the uncertain but potentially large future costs of acid rain, NO_x, and CO_2 buildup in the atmosphere. All of these problems are linked to energy -- utilities, industry, transportation -- and we should be striving for integrated, least-cost solutions. The government has, in effect, entered the marketplace by enacting legislation (the Clean Air Act, for instance) to protect the health of its citizens and their environment, but is this approach the best possible solution to the problem?

---

* There is growing evidence that in this period of overcapacity, with prospects for new kinds of generating technology and life extension of existing plants, electricity prices over the next several years will decline. Therefore marginal and average costs may be converging.

Enforcement of the Clean Air Act has resulted in greatly decreased sulfur emissions from electric power plants -- a great benefit to society. If instead of desulfurizing flue gas we used the same resources to cut energy consumption through conservation technologies, we would help to alleviate a lot of other problems in addition. Similarly, investment in research into advanced combustion technologies might reduce both sulfur and nitrogen emissions. In fact, generation technologies currently in the major demonstration phase of development, such as integrated gasification combined cycle and atmospheric fluidized bed combustion, may be able to produce electricity with lower costs and lower SO_x and NO_x emissions than can conventional coal technologies with scrubbers.[5]

A comparison of the relative costs of applying the Clean Air Act versus enacting appliance efficiency standards is a case in point. Recent research indicates that energy efficiency standards for water heaters, refrigerators, and room and central air conditioners could avoid installed capacity of between 40,000-100,000 megawatts of electricity by the year 2005. Savings at the low end would eliminate the need to burn over 100 million tons of coal each year and provide a 12-20 percent reduction in current annual SO_2 emissions. Installing flue gas scrubbers to achieve a similar reduction in sulfur would cost $5-10 billion; appliance efficiency increases could cost less than a fraction of that amount. Appliance efficiency standards would also enable the reduction of nitrogen oxide and carbon dioxide emissions at no extra cost.[6]

The leap in complexity (i.e., the number of actors) in shifting from utility investment strategies and coal industry initiatives to act on air pollution regulations, to the myriad capital investments in higher energy efficiency and associated regulatory requirements and consumer decisions is enormous. Only a fraction of economically attractive conservation options have been exercised to date. But that leap in seeming complexity should not be sufficient to deter us if the economic and other rewards are attractive. Consumers generally act collectively, and state or federal rulemaking procedures are actions by bodies similar to corporate boards. In other words, what may seem to be the independent decisions of millions of people can actually follow from the actions of only a few people (regulators, marketers, etc.). And, best of all, we have already seen considerable change in consumer habits, despite the complexity of decisions.

The fact that within the Congressional arena, decisions concerning appliance efficiency standards are made entirely separately from decisions to enact emissions standards, means that changing that decisionmaking system to effect a more integrated approach would be extremely difficult. Other governments have intervened in industrial energy policy with very satisfactory results. In the 1970's, the Japanese invested $50-150 million per year in energy efficiency in the steel industry. They concentrated on refitting primary steel plants for higher overall productivity, including such energy-related features as continuous casting and heat recovery. The result was energy demand reduced to 40 percent below the world average per ton of steel produced. Partly as a result of these investments, the Japanese steel industry has remained competitive.[7]

One of the best ways to extend oil supplies and improve air quality would be to improve automobile fuel economy. Yet 1985 is the last year in which American automakers are required to raise gasoline efficiency. Existing legislation requires only that auto manufacturers reach a new fleet sales average of 27.5 miles per gallon (though new American cars are currently only at 22 mpg). The Secretary of Transportation can raise the standard (or lower it to 26 mpg), but either the House or the Senate can override such an action. Legislation that forced American cars to become as efficient as Japanese cars would ultimately (at full fleet turn-over) reduce world oil demand by 5 percent at only a fraction of the cost of enhancing liquid fuels supplies with synfuels or alcohol. Automobile fuel efficiency was headed toward that level of efficiency -- 30 miles per gallon -- when gasoline prices rose precipitously in 1980. But the marketplace alone cannot ensure fuel efficient cars, for it is only one consideration in the consumer's decision. Because there are several major areas of public interest in cutting our dependence on oil imports and in using energy resources more efficiently, it is a legitimate if not essential area for government intervention in the cause of national security.

The advance of technology is often as important to adoption of resource conservation measures as is government intervention in the marketplace. It was the rise of "high-tech" mini-mills dependent upon scrap steel that pulled the junk cars off the hills of Appalachia and back into the marketplace. But the intervention of government -- uniquely equipped to devise integrated measures for resource conservation beyond the capability scope of individual industries or consumers -- is often required to bring that technology to fruition.

THE FEDERAL ROLE IN ENERGY R&D

The idea that government can play a role in forcing technology brings us to the next issue I would like to address, What should be the federal role in energy R&D?

The current Administration has put forward some very reasonable criteria for public investment in energy R&D.(8) They feel that the government should become involved only when the investment has:

- High risk (including technical risks, economic risks, and the risks associated with acceptance of new technologies);

- Potentially high pay off -- either financially or through better understanding (this is a fairly recent addition to the criteria for federal involvement -- past administrations having been criticized for emphasis on short-term pay-off);

- Long term to fruition (beyond the normal interest and ability of the private sector to fund because of the payback period); and

- Generic qualities (will benefit more than one firm and have wide spread applicability).

At the same time, however, the historic over-emphasis of supply technologies at the expense of conservation and renewables continues, and R&D spending in energy has generally been in retreat. Federal support for R&D in energy conservation has survived in the last several years only by Congressional action. In the face of large non-market costs in the energy sector, and great and uncontrollable uncertainties about long-term future energy prices, it is difficult for the private sector to justify extensive investment in R&D for energy efficiency substantially greater than is economic at today's energy market prices. Thus federal investment in the energy sector should be made to ensure reliability and minimum cost of the needs and amenities that energy helps provide. Public benefits of generic R&D in energy productivity include the following:

- Energy savings can lighten the heavy burden of balance of payments, and new, world markets will open to eventual private sector producers of new energy efficient technologies;

- Existing supplies of energy resources will last longer, providing more time for developing successors, just as a strategic oil reserve would last longer, increasing our resilience to short term interruptions;

- The health and environmental impacts of energy conservation technologies can be understood and dealt with prior to wide-spread implementation; and

- Less developed nations will have a better chance for economic growth if energy efficiency in the major energy consuming countries can hold down the rate of energy price increases by reducing demand pressure.[9]

There have been a variety of problems with federal R&D work in energy conservation, for example the early emphasis on near-term pay-offs. <u>Still, there is a role for conservation research and development in expanding the boundaries of efficiency that economically match a given energy price.</u> Long-term research also provides improved insight into the nature and trends of our energy consuming society -- the choices we make, how and why we adopt or reject technology. Similarly, research yields improved understanding of the social and economic implications of energy conservation: employment, social equity, freedom of choice, resilience in the face of emergencies, international security, environment and health, and urgency for new supply development. One likely outcome of such continuing analysis would be an ability to project energy demand futures with greater confidence. That result alone would justify an enormous research investment. Thus we do need continued federal involvement in energy R&D (and, fortunately, it is not likely to require extravagant sums).

Among the most notable results of public and private R&D related to energy conservation have been high temperature metals, electronic controls, and new chemical processing techniques. Also very important are the increases in understanding of building energy use, lighting, coated glass, and integrated energy systems in buildings. I emphasize the importance of conservation R&D in this discussion because of historic lack of funding in this area, but there also have been some important and notable advances in energy supply R&D. Research into uranium separation technologies and reactor fuel performance has gone a long way toward eliminating our fears of a shortage of uranium and keeping the nuclear option viable. Oil and gas exploration technologies have made remarkable advances in recent years. Advances in flue-gas scrubbing, gasification, and combustion have made it safer to burn coal. Fusion, solar-direct conversion, and biomass conversion are now remarkably advanced technically, compared to a decade ago, but still are far from being economically feasible.

PROSPECTS FOR DOMESTIC OIL AND GAS DISCOVERIES

Advances in supply technologies are very important since energy demand will continue to grow as our economy grows, though not in lock step. And recent disappointing discovery rates of petroleum manifest that it may be the less conventional technologies on which we must rely, sooner than we thought. This leads to my next question, What are our prospects for major new discoveries in domestic oil and gas?

A recent OTA report, Oil and Gas Technologies for the Arctic and Deepwater,[10] found that despite the large oil and gas price increases of the 1970's, with accompanying major increases in exploration, domestic energy production remained virtually level over the past decade. Between 1980-1984, the United States spent $250 billion on domestic petroleum exploration and development, yet oil production remained almost level. The slight increase in domestic oil production since 1980 is due entirely to production from the Prudhoe Bay Field on Alaska's North Slope, and if its contribution is removed, domestic oil production declined more than 18 percent between 1974 and 1983. Domestic oil and gas reserves have declined even more rapidly than production, despite enormous increases in resource exploration and development since 1973, and particularly since 1980. According to the Department of Energy (DOE), proven reserves of economically recoverable oil dropped from 47 billion barrels in 1970 to 35 billion barrels in 1984. And the Minerals Management Services (Department of the Interior) recently revised resource estimates for offshore oil and gas. The new estimate of undiscovered oil resources in the OCS (outer continental shelf) lease sale planning areas are 55 percent lower than the U.S. Geological Survey estimates published in 1981; natural gas resources are 44 percent lower.

A major oil price rise in 1979 and cumulative conservation efforts begun earlier led to declining imports and a record oil import low of 4.9 million barrels per day in 1983. However, in 1984 oil imports increased about 7 percent over 1983, accounting for about one-third of U.S. petroleum requirements. Oil import levels have increased as growth in domestic demand has outpaced domestic oil production.

The DOE and Gas Research Institute (GRI) energy forecasts indicate a continuing decline in the production of domestic oil and natural gas to the year 2000. In both forecasts, oil and gas imports are expected to increase substantially, to between 7.1 and 7.5 million barrels of oil per day and 2.8 and 3.8 trillion cubic feet (Tcf) of natural gas per day. There are indications, however, that even the DOE and GRI projections may be overly optimistic and that imports may reach higher levels. Continued low energy prices may lead to greater fuel usage, reduced conservation efforts, lower exploration efforts, and limited replacement of oil by alternative fuels. There are also great uncertainties about future natural gas supplies.[11]

DOE and GRI projections of year 2000 domestic production of oil are 8.1 million barrels per day and 9.2 million barrels per day, respectively. Studies by OTA and the Congressional Research Service (CRS) forecast even greater declines. OTA, in World Petroleum Availability: 1980-2000, projected that domestic oil and natural gas liquids production would decline to between 4 and

7 million barrels per day by 2000. CRS estimated a decline in production to 7.3-8.5 million barrels per day. These production levels indicate that oil imports may range from 7 to as high as 10 million barrels per day in 2000, contributing to high trade deficits and decreases in energy and economic security. Discovery rates continue to be disappointing and exploration is fading.

Recent energy forecasts underline the importance of the oil and gas resources of the U.S. Outer Continental Shelf and the Exclusive Economic Zone. An increasing percentage of our domestic oil production must come from oil reserves as yet undiscovered, yet widespread exploration and development of the lower 48 states make large field discoveries in onshore areas of the United States, outside Alaska, somewhat doubtful. Offshore resources, particularly those of the unexplored deepwater and Arctic frontier regions, offer the best hope for limiting future U.S. energy import dependence.

THE FUTURE OF NUCLEAR POWER

Meeting the growth in energy demand is a serious issue for the United States for many reasons. Because it is certain that electric demand will grow, and clear that old, coal-fired capacity will need to be replaced, another important energy supply question is, What is the future of the nuclear power industry in the United States?

OTA's study, Nuclear Power in an Age of Uncertainty,(12) found that without significant changes in the technology, management, and level of public acceptance, nuclear power in the United States is unlikely to be expanded in this century beyond the reactors already under construction. The financial risks presented by uncertainties in electric demand growth, very high capital costs, operating problems, increasing regulatory requirement, and growing public opposition appear prohibitive. Yet the outstanding success of some nuclear plant operations, both here and in other countries, has made it clear that some of these risks are not inherent to nuclear power. There are good reasons why it could be highly desirable to have a nuclear option in the future if present problems can be overcome. If, as electricity demand grows, we cannot overcome the environmental problems presented by coal or the economic problems presented by renewables, then nuclear may be very important ... our ace in the hole. Another reason to hang on to the nuclear option is that it can provide important diversity to our electric system -- preventing overwhelming dependence on coal. The diversity, some say, is worth at least a 20% penalty in apparent cost.

In the future, we should be able to avoid many of the problems of the present generation of nuclear plants. Technological improvements, while insufficient by themselves, can be very important. One approach would be to focus research and development on further improving light water reactors, with designs representing an optimal balance of costs, safety, and operability. Features that might be applied to any reactor technology include smaller sizes and standardized designs. Smaller nuclear plants would provide greater flexibility in utility planning -- especially in times of uncertain demand growth -- and less extreme economic consequences from an unscheduled outage. We might also pursue R&D on alternative reactors with inherently safe characteristics, rather than relying on active, engineered systems to protect against accidents. Several concepts appear promising, including the High

Temperature Gas-cooled Reactor (HTGR), the PIUS (Process Inherent Ultimately Safe) reactors, and heavy water reactors. Such R&D should also be directed toward design and developing smaller reactors such as the modular HTGR.

Improvements in areas outside the technology, per se, must start with management of existing reactors. A high level commitment to excellence in design, construction, and operation is essential, and must be followed by improved training programs, tightened procedures, and heightened awareness of opportunities for improved safety and reliability. Serious consideration should be given to bars against licensing or certifying demonstrably unqualified companies, since one bad company can give the whole industry a bad name.

OTA found that existing legislation is sufficient for the task of regulating the nuclear industry but must be enforced with more consistency. Encouraging preapproved standardized designs and developing procedures and the requisite analytical tools for evaluating proposed safety backfits would help make licensing more efficient without sacrificing safety.

THE POTENTIAL OF CONSERVATION

As my comments on these questions indicate, I believe that investigation and recognition of the full costs and benefits of energy use are essential to a successful energy policy. And I believe that conservation technologies offer among the best cost/benefit ratios available to energy planners today.

Conservation investments are potentially so rewarding that they deserve a large amount of attention from energy policy makers. For instance, OTA's study, Industrial Energy Use,[13] found that between 1980 and 2000, investments in new processes, changes in product mix, and technological innovation can lead to dramatically improved industrial productivity and energy efficiency, enough that the rate of industrial production can grow three times faster than the rate of energy use needed for that production.

U.S. buildings and appliances could be much more efficient, without economic sacrifice. OTA's study, Energy Efficiency in Buildings in Cities,[14] found that about 7 Quads per year of energy savings is technically and economically possible by the year 2000 if certain institutional barriers to investment are overcome. The EPRI Journal[15] reported that over half the electricity used for lighting in the U.S. -- 420 billion kWh/year -- could be saved through energy efficient strategies without imposing any hardships on productivity, safety, or aesthetics. This figure represents 35 percent more electricity than the total current hydro and nuclear output in the United States.

There are any number of conservation investments that seem too good to pass up. These investments are not just in energy saving equipment, but also in the way we actually supply goods and services: silicon chips v. vacuum tubes; planes v. passenger ships.[16] But we are slow to capture these opportunities, a symptom of the complacency brought forth by the current lull in energy price increases or supply shortages. Complacency may be easy and appealing now, but could prove to be a very costly strategy later on. We might do well, as they say, "to make hay while the sun shines!"

You who are here today are an important part of today's energy leadership. And meetings such as today's are important for keeping energy in the public consciousness. At the very least, you are the main sources of information for today's decisionmakers, and information is key to good policy. The points that I hope you take away from this talk are:

1. Energy demand price elasticity is real and large. Though there may be some short term (low) elasticity problems, in the long term, energy is an extraordinarily cost-sensitive, substitutable factor in production of goods and services. The long run energy supply price elasticity is uncertain, but not as large as has been thought in the past.

2. The Greek letter "tau," used to signify time, is very important in energy policy. Although large changes (in use of existing equipment) can happen remarkably fast, time constants in most energy conservation opportunities are measured in decades because the "tau" of utilization is related to the turnover of capital stock.

3. Movement toward capital intensive energy supplies such as nuclear power creates special problems in times when capital formation is a problem. Additionally, nuclear power, in its present form, is an unforgiving technology with inescapable overtones for social concerns.

4. Conservation, aided by price decontrol, was the dominant actor in the past decade in improving the energy position of the United States. We have learned that consumers will respond to prices while mostly dismissing notions of the "spiritual benefits or costs" of conservation.

5. People delight in procrastination, but it is important that we address foreseeable energy problems before they become crises.

6. We are wrestling with "frontier culture" attitudes when we talk about energy. Talking about conservation, in particular, is strangely characterized as admitting to limits that we would prefer did not exist. These have been and will continue to be difficult hurdles to overcome, but the lessons we learn in energy will be usable in other contexts as well.

Market pricing and the Strategic Petroleum Reserve offer us some protection from ourselves. Yet many of the issues that the "energy crises" raised -- the finiteness of fossil fuel supplies, the necessity of energy supplies for development, the appropriate role for government in energy development -- still require attention. To quote Antonio Gramsci, "The crisis consists precisely in the fact that the old is dying and the new cannot yet be born; in this interregnum a great variety of morbid symptoms appears."[17]

Our symptoms are not yet morbid, but they are serious. And the consequences -- including acid rain, CO_2, species extinction, water quality degradation, human dislocations, capital shortages, debt -- of ignoring conservation opportunities would largely be a result of our failure to internalize the full costs of energy.

When the next big energy crisis arises, as it surely will if our oil imports rise as much as some predict, unless we use the intervening time to develop alternatives, it may happen that our only feasible response would be military intervention. So let's not squander the apparent lull in the "energy crisis" and begin to repeat the mistakes of the 1970's when we pushed projects of questionable merit out of desperation. Let's concentrate instead on using energy more efficiently, cutting down on the wasteful external costs of energy use, and assuring long-term successors to cheap oil and gas supplies, proceeding at a realistic pace and aimed at systems that are reliable, resilient, simple, and more forgiving.

Acknowledgment: I am particularly grateful to Holly Gwin for assisting me in the preparation of this material.

NOTES

(1) Report by the Federal Power Commission, The 1970 National Power Survey, Part I, U.S. Government Printing Office, Washington, D.C., 1971.

(2) Meadows, D.H., Meadows, D.L., Randers, J., and Behrens, W.W., III, Limits to Growth, Potomac Associates, Washington, D.C., 1972.

(3) Gibbons, John H., and Chandler, William U., Energy: The Conservation Revolution, Plenum Press, New York, 1981.

(4) World Petroleum Availability: 1980 - 2000 (Washington, D.C.: U.S. Congress, Office of Technology Assessment, OTA-TM-E-5, October 1980). Congressional Research Service, Domestic Crude Oil Production Projected to the Year 2000 on the Basis of Resource Capability (July 1984).

(5) New Electric Power Technologies: Problems and Prospects for the 1990's (Washington, D.C: U.S. Congress, Office of Technology Assessment, OTA-E-246, July 1985).

(6) Postel, Sandra, Air Pollution, Acid Rain, and the Future of Forests, Worldwatch Paper #58, March 1984. Geller, Howard S., "Efficient Residential Appliances and Space Conditioning Equipment: Savings Potential and Cost Effectiveness as of 1984," in American Council for and Energy-Efficient Economy, Doing Better, Vol. E. (Proceedings from the Panel on Appliances and Equipment; Washington, D.C.: 1984).

(7) Tsuchiya, Haruki, (Research Institute for Systems Technology, Tokyo, Japan) "Case Study on Japan," presented at the Global Workshop on End-Use Energy Strategies, Sao Paolo, Brazil, June 4-15, 1984.

(8) Energy Conservation and the Federal Government: Research, Development, and Management, A Report of the Energy Research Advisory Board to the United States Department of Energy (January 1983).

(9) John Sawhill, editor, Energy Conservation and Public Policy, Prentice-Hall, Inc., 1979.

(10) Oil and Gas Technologies for the Arctic and Deepwater (Washington, D.C: U.S. Congress, Office of Technology Assessment, OTA-O-270, May 1985).

(11) U.S. Natural Gas Availability: Gas Supply Through the Year 2000 (Washington, D.C.: U.S. Congress, Office of Technology Assessment, OTA-E-245, February 1985).

(12) Nuclear Power in an Age of Uncertainty (Washington, D.C: U.S. Congress, Office of Technology Assessment, OTA-E-216, February 1984).

(13) Industrial Energy Use (Washington, D.C.: U.S. Congress, Office of Technology Assessment, OTA-E-198, June 1983).

(14) Energy Efficiency in Buildings in Cities (Washington, D.C: U.S. Congress, Office of Technology Assessment, OTA-E-168, March 1982.

(15) Lihach, Nadine, and Partusiello, Stephen, "Evolution in Lighting," EPRI Journal, June 1984.

(16) Rose, David, "Conservation in the Context of Global and U.S. Energy," East-West Center, 1985.

(17) Gramsci, Antonio, Prison Notebooks, as cited in Daniel Martin, by John Fowles (1977).

CHAPTER 2

THE PHYSICIST'S ROLE IN USING ENERGY EFFICIENTLY: REFLECTIONS ON THE 1974 AMERICAN PHYSICAL SOCIETY SUMMER STUDY AND ON THE TASK AHEAD

R. H. Socolow
Center for Energy and Environmental Studies
Princeton University, Princeton, N. J. 08544

We physicists who worked together on the 1974 Summer Study of the American Physical Society entitled "The Efficient Use of Energy"[1] believed we were doing something important, in questioning two strongly held beliefs. One concerned how energy relates to well-being, and the other concerned how physicists relate to energy. The first affirmed that only by ever greater use of energy could greater societal well-being be achieved. The second affirmed that it was appropriate for physicists to work on problems of energy supply but it was inappropriate for us to work on problems of energy use. The shared goal of the participants in the 1974 APS Summer Study was to overturn both of these beliefs -- by creating counterexamples. In the first instance our counterexamples would be analyses which demonstrated the emptiness of the connection between various aspects of well-being (personal mobility, light to read by) and the level of energy use required to achieve them. In the second instance our counterexamples would be ourselves.

In this brief essay, I would like to track these two arguments over the decade since the APS Summer Study, and to address five questions: 1) What kind of evidence supported the majority position in 1974? 2) What was the source of our confidence that the majority position was misguided? 3) To what extent have the events of the past decade vindicated our disbelief? 4) What have turned out, so far, to be the major shortfalls in our own analysis in 1974 in the two areas? 5) And, finally, what new "wisdom" is in evidence concerning energy and well being and concerning physicists and energy which physicists ought to be challenging at this time?

THE WISDOM OF 1974

In 1974 it was easy to believe that improvements in societal well-being required continued expansion in the rate of use of energy in the economy. One had to be impressed with the strength of the data throughout the period after World War II, over many countries, which showed two essentially unrelated aggregate variables moving together in eerie synchrony. The two variables were annual gross national product and annual use of commercial energy. The gross national product is a carefully constructed measure of overall economic activity measured in units of inflation-corrected dollars. (To be sure, it is a far from perfect measure of societal well-being; in fact, in the same mid-1970s period, there was a flurry of activity among economists to explore other measures. However, no alternative measure gained enough adherents to become a serious challenger to GNP, as the measure of choice in non-technical

0094-243X/85/1350015-18$3.00 Copyright 1985 American Institute of Physics

discussions of societal welfare.) Use of commercial energy is a sum over the uses of coal, oil, gas, nuclear energy, and hydropower -- measured in physical units like exajoules, and involving physical measurements of the energy released in combustion to permit uses of different fossil fuels to be put into common units. One plot of these data is seen in Figure 1, which shows the variation in the ratio of energy use to real (i.e. inflation-corrected) GNP over one hundred and thirty years of United States history.

The graph tells a number of stories. First, during the period from 1850 to 1900, coal is substituting for wood, and calculations which do and do not include wood look very different. Today's less developed countries are recapitulating this portion of the graph in the current period: both a declining share of energy from wood and a downward trend in overall energy use per unit of GNP are in evidence today in much of the world. However, countries today have an option not available a century ago: conversion technologies

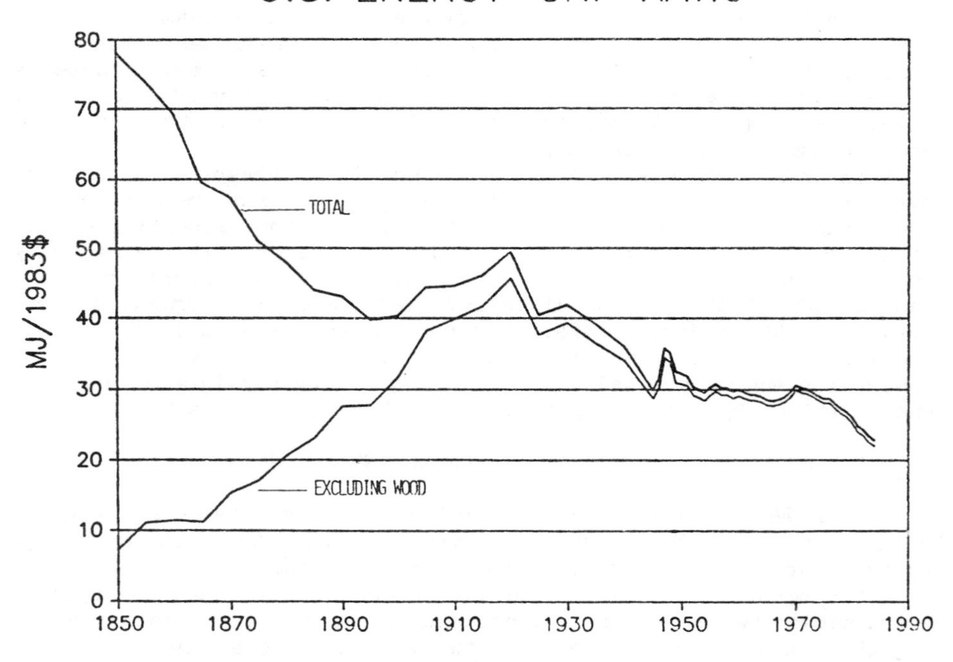

Fig. 1. Energy divided by GNP, 1850-1984, for the United States. "Energy" is primary consumption, with both hydropower and nuclear power treated as if the equivalent electricity had been produced by fossil fuels in average plants. GNP is adjusted to 1983 dollars using canonical GNP deflators. Energy data: wood and non-wood (1850-1948)[2]; non-wood (1949-1983)[3] and (1984)[4]; wood (1949-1984).[5] GNP data: (1850-1948),[2] (1949-1981),[6] and (1982-1984).[7]

permitting wood (and crop wastes) to be transformed with high efficiency into liquids and gases, to provide cooking and transportation, again at high efficiency.[8] It may well turn out that today's developing countries will deploy these biomass-conversion technologies widely, in which case the fractional share of biomass will not fall to anything like the 3 percent value which currently characterizes the United States.

A second story told by Figure 1 is the 23 percent rise from 1900 to 1920 in energy use per unit of GNP, followed by a 40 percent fall from 1920 to 1945. The increasing intensity of energy use in an early stage of industrialization and the subsequent prevalence of energy efficiency is a poorly understood phenomenon, worthy of further study. The third story told by Figure 1, however, is the one which had mesmerized many energy analysts in 1974: the constancy of the ratio from 1949 through 1973. For 17 of these 25 years the ratio had been within two percent of its average value of 30 MJ/1983 dollar (28,000 Btu/1983$), while real GNP grew by a factor of 2.5.

A quarter century of steadiness is a long time, and data from 1949 to 1974 looked roughly similar in many other countries and also in subregions of countries. Such high correlations of two variables don't happen all that often in the social sciences. Moreover, all of us are predisposed to believe that the recent historical past is a useful guide to the near-term future. Thus it did not seem all that surprising that intelligent people could believe with considerable conviction that this value of 30 MJ/1983$ had become an immutable fact of the U.S. economy.

Physics, on the other hand, leads one to reason differently. It seeks models which preserve only the essentials of a problem; here, it forces one to ask: "What is energy being used _for_?" The APS report places strong emphasis on distinguishing _tasks_ from _devices_, and then on using the language of thermodynamics to restate a given task in terms of flows of work and heat and entropy. Such an inquiry establishes thermodynamic minima for the performance of tasks, against which the actual energy use can be compared. Where the discrepancies are large, one concludes, tentatively, that mankind has not yet been particularly clever about providing the best devices to accomplish the task, and one predicts, again tentatively, that large energy savings will be possible, _with no loss of amenity_, as technology evolves. Far from seeming like a guide for prediction, the constancy, over 25 years, in the ratio of energy use to GNP struck the Summer Study physicists as a testament to the sustained inattention to energy use on the part of the technological community.

Such sustained inattention was easy to document and was bound up in the second belief with which the Summer Study physicists had to contend. Although the book about quiche hadn't been written yet, a strong message in that period was that real men don't study how to use less energy. The "real" work of physicists, it was claimed, included realizing the promise of the breeder reactor and of controlled fusion, as well as helping to coax more oil out of an underground reservoir, converting a higher fraction of the carbon atoms in coal and shale into low-molecular-weight hydrocarbons, and turning a higher fraction of the energy stream from solar photons into electric power. Energy demand, we were told, was "exogenous"; this fancy word meant not only that energy demand was a variable which could not be determined from within a model of national energy accounts but also that it was an issue which lay outside the arena of discussions about energy by serious people.

One memorable event for me was a presentation about breeder reactors in 1972 at which it was asserted that the doubling time for fissile material from a commercial breeder system would have to be less than ten years, since the doubling time for use of electric power was well established to be ten years. (Indeed, the data for more than five previous decades did show this doubling time.) Such a design criterion, for a system which was being planned for the second decade of the 21st century, implied that the economic environment in which the breeder reactor would begin to function would show sixteen times more use of electricity than the world in which we as an audience were living. I personally can date my own commitment to work full time on energy use to that event.

REASONS TO DOUBT THE WISDOM

By 1974, environmentalism and economics had provided two quite separate reasons to doubt the claim that energy and GNP were destined to be coupled together. Environmentalism is much like special relativity, subsuming a previous world view within a more inclusive one. The analog of Newtonian mechanics is what Kenneth Boulding called the cowboy economy, and the analog of relativistic mechanics is the spaceship economy. At low levels of energy use, one does not have to consider planetary constraints, while at higher levels (as at velocities close to the speed of light) these constraints dominate reality. Since these constraints eventually affect prices, economics, from this perspective, is an elaboration of one aspect of environmentalism.

The list of planetary constraints is a long one. Many of these constraints are summarized in an interesting way in dimensionless ratios, N/D, where N (the numerator) is an effect of man and D (the denominator) is an effect of nature. Such a formulation calls attention to vulnerable subsystems of nature, such as the stratosphere, the arctic icepack, and the unbuffered mountain lakes, which are easily overwhelmed by deposition of, respectively, nitrogen oxides, albedo-changing particulates, and sulfuric acid -- each pollutant arising from man's technology and potentially appearing in orders of magnitude greater quantities than what nature generates without our help. Table 1 shows a few N/D ratios importantly related to energy. We see that the direct thermal impact of man's use of energy has a much smaller fractional effect on the earth's thermal balance, and thus on climate, than the indirect impact resulting from the annual fractional increase of carbon atoms in the atmosphere resulting from the combustion of fossil fuels. The carbon dioxide increase in the atmosphere is the

Achilles heel of a multihundred-year future based on coal, oil shale, and tar sands.

TABLE 1

ECOLOGY -- THE SUBVERSIVE SCIENCE

N/D arguments

N measures an activity of man

D measures a background in nature

N	D	N/D
thermal output (10 TW)	solar input	0.0001
annual oil production	"oil" in the ground	0.01 - 0.02
annual CO_2 from burning fossil fuels	CO_2 in atmosphere	0.01
annual plutonium from reactors	Pu on the earth, 1980	0.1
land for 1 TW from biomass plantations (10 tonnes/ha)	arable land (10^{-1} of surface)	0.04

Implication: Stewardship of the Planet

Table 1 also shows what appear to be the Achilles heel of a future based on nuclear fission: the build-up of plutonium on the earth's surface, where none had existed in 1940 and 500 tonnes exist today. At an average rate of about 50 tonnes/year about as much will be produced in the next decade from civilian nuclear power plants as was produced in the past four decades by the world's civilian and military programs combined. Plutonium is a material, in any likely isotopic mixture, capable of eliciting many kinds of malevolent behavior from imperfect people -- it is reasonable to wish that there be as few tonnes generated as possible, at least in the current epoch of disarray among nations. Perhaps some day international relations will have developed to the point where the world is safe for abundant nuclear power.

Table 1 shows, as well, what is probably the Achilles heel of the solar energy future: a requirement for land which may equal what is required for agriculture. At an annual yield of 10 tonnes dry biomass per hectare (and 15 GJ/t energy content), the output of a biomass plantation is 0.5 W/m^2, roughly 0.2 percent of the incident sunlight. Perhaps advances in bioengineering and ecology, or in the competitive solar technology of photovoltaics, will significantly increase this overall efficiency of conversion. But at present it does appear that no road to a long-term energy future is free of severe environmental constraints.

Such an analysis of the human predicament, based on a concatenation of environmental constraints, motivated the first wave of researchers who began studying how energy is used. But the sudden prominence of another constraint, the finiteness of world oil, gave the subject an urgency -- and a second, larger wave of researchers -- which greatly accelerated the rate at which insights were gained. The 1974 APS Summer Study had been conceived some months before the first of the world's oil price shocks, but the Study occurred in its wake, giving great impetus to our work.

The first oil price shock provided new insights into the apparent constancy of the ratio of GNP to energy use. Might two economic forces have been accidentally cancelling over the previous twenty-five years -- a falling price of energy and a steady stream of inventions of energy-saving technologies? Accidental cancellations of large terms with opposite signs are never as aesthetically pleasing as genuine constants; nonetheless, the cancellation hypothesis began to look plausible, as responses to the rising oil price began to appear. Countless consumers of energy whose decisions were routinely assisted by optimization procedures in which the current and expected future prices of energy were input parameters, began reducing their consumption of energy almost immediately.

Suddenly, saving energy became important. For physicists to ignore the challenge on the grounds that their real work lay elsewhere began to seem irresponsible, at least to a few. Indeed, the argument that physicists interested in energy should concentrate exclusively on improving the prospects for new energy sources acquired a self-serving character, especially when it came from communities (the breeder reactor community, for one) where the prospect of reductions in future energy demand diminished the urgency of big programs.

Physicists who insisted on taking a look at the problem of saving energy quickly found that the insights of their discipline were missing from some of the most widely used tools of analysis -- for example, from the "spaghetti diagram" (see Figure 2) which exhibited the transformation of all energy inputs into definite combinations of "useful" and "rejected" energy. On close inspection, Residential and Commercial end uses were claimed to convert a high percentage of energy inputs into useful energy (in Figure 2, 13.0/17.8, or 73 percent), whereas for Electrical Generation the corresponding percentage was small (in Figure 2, 7.0/22.4, or 33 percent). Were America's weakly engineered home furnaces and water heaters really, on average, more than twice as successful devices as her highly engineered power plants?

The resolution of this conundrum emerged from the Second Law of Thermodynamics. Spaghetti diagrams, and energy accounting generally, systematically ignored any reference to the quality of

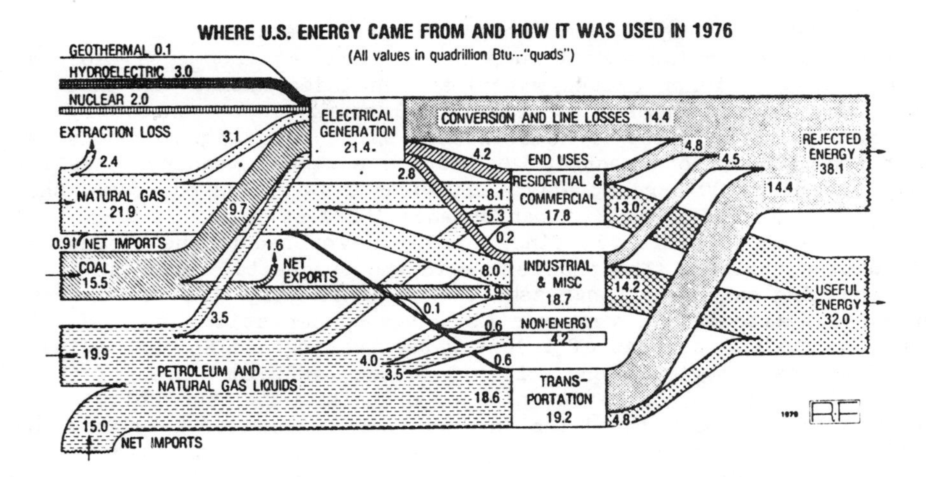

Fig. 2. Conventional rendition of the energy flows in the U.S. economy ("spaghetti diagram"), reproduced from Ref. 19, p. 47. Distinction between rejected and useful energy is oblivious to thermodynamics (see text).

energy, a horrendous omission. Why "horrendous"? This needs several sentences to explain:

The assertion that 33 percent of the chemical energy delivered to an electric power plant becomes useful electricity, while 67 percent becomes rejected heat, is, of course, a reflection of the fact that the electricity is being generated in a thermal (usually Rankine) cycle, where the complete conversion of heat to work is not possible. Both the chemical energy and the electricity are high quality forms of energy.

When chemical energy or electricity is delivered to residential and commercial buildings, on the other hand, the dominant conversion device is a furnace, boiler, or water heater, and the output is not a high quality form of energy but rather <u>heat</u> at a temperature close to the temperature of the environment. Traditionally, one rates the heating system by an efficiency equal to the fraction of the incident chemical or electrical energy which is transferred to a fluid (water or air) as the fluid flows past a heat exchanger, and these efficiencies generally turn out to be from 60 to 100 percent (at the low end for older furnaces, and at the high end for electric resistive heating). This measure of efficiency is deeply imbedded in the rivalry between gas heat and oil heat, and it is an excellent measure of the quality of furnaces, boilers, and water heaters.

However, another way to accomplish space heating or water heating is by running a heat pump between some environmental source (e.g., outdoor air or ground water) and an indoor sink. In that case, one unit of chemical or electrical energy input can easily bring, say, three units of heat indoors while removing two units of heat from the outdoor source. By the conventional measure, this is like having a 300 percent efficient furnace -- or, as one says, the system has a coefficient of performance (COP) of 3.

All this is perfectly fine, and has never led to the misdesign of heating systems. The problem is in the spaghetti! Many an unsuspecting banker or congressman, seeking to gauge the relative promise of alternative ventures in energy conservation, was led to conclude that 73 percent efficient water heating and space heating could scarcely be improved, while 33 percent efficient electricity production had to be ripe for a major breakthrough. The ubiquitous spaghetti diagrams pointed the investor and legislator away from the distinction between heat and work, by means of which it would have been possible to comprehend a heat pump four or more times more effective at using fuel or electricity to heat air or water.

The APS Study reminded its audience that thermodynamics led naturally to the concept of "available work" and, thence, to a "Second Law Efficiency" by means of which all tasks (including, in particular, space and water heating and electricity generation) could be evaluated, without distorting comparisons across tasks. "We strongly recommend," the Report said (p. 5), "that this formulation, or a similar one, be widely adopted by the scientific and technical community as a standard from which all tasks should be measured, and against which all devices should be evaluated."

The APS Study, moreover, estimated the Second Law Efficiency by which each of the major energy-using tasks is accomplished in modern society. Figure 3 shows the Study's conclusions, in the form of a heat wheel: it is seen that the provision of heat near environmental temperatures is done least well and the provision of electricity is done best.

The main point I want to make here is that physicists saw quickly that they could be useful in discipling the discussion of using energy efficiently.[10] Moreover, we perceived that even our own special contribution, The Second Law of Thermodynamics, could well stand some deep thought to allow its generalization to problems occurring in finite time, instead of infinitely slowly and reversibly. (In the past decade such work has begun; see.[11] With such evidence of their value and such intimations that there was even some good physics to do, all doubts vanished about the appropriateness of physicists in the enterprise.

DEGREE OF VINDICATION

The United States used one percent less energy in 1984 than in 1973, yet the GNP grew 31 percent (in constant dollars) in that period. A similar decoupling of energy from GNP occurred in Japan, Australia, New Zealand, Canada, and the countries of Western Europe.

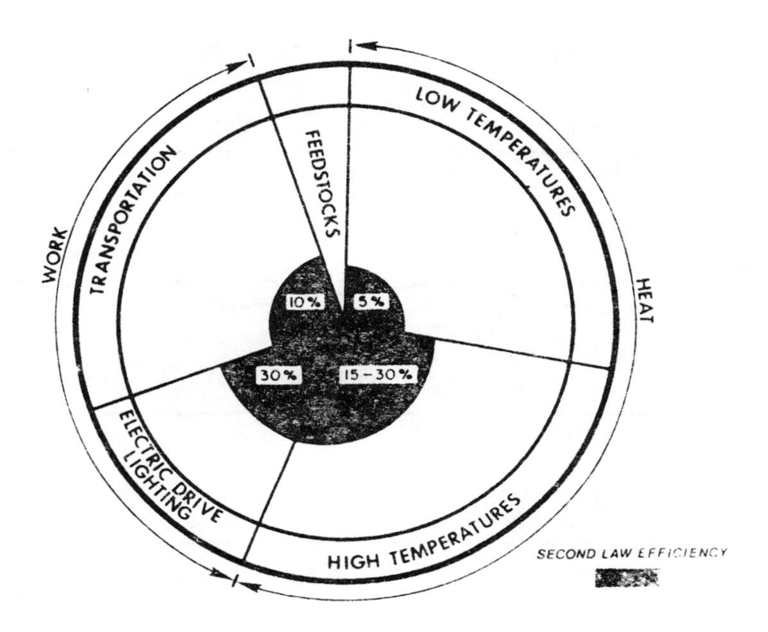

Fig. 3. Thermodynamics of Energy Use, by Task (United States). Based on numbers in Table 2.8, Ref. 1.

Only very imperfectly can modelers tease apart the relative roles of the rising price of energy and of non-price effects, like increased attention to the objective of efficiency, in bringing about the decoupling that has been observed. But most modelers agree that comparable roles for price and for non-price effects appear to be necessary to explain the data. (By non-price effects one means adjustments which are required to fit data when elasticities are confined to values based on historical precedents.) The answer matters, for if the entire cause of the decoupling were a price response, and if the price rises of the past decade were to turn out to be a transition from one trajectory of gradually falling price to another, then a recoupling of energy and GNP might be expected.

Interestingly, in the Soviet Union and in Eastern Europe a decoupling of energy from GNP has not yet occurred. Figure 4 compares trends in the ratio of energy to GNP in both the Soviet Union and the United States, revealing a remarkable dance of the ratios, in which the United States ratio falls when the Soviet Union ratio rises, and vice versa. In particular, between 1975 and 1982 (a pair of years for which all four variables are cited in a single source,[12] annual GNP grew 16.3 percent in the Soviet Union and 20.6 percent in the United States, while annual energy use grew 26.0

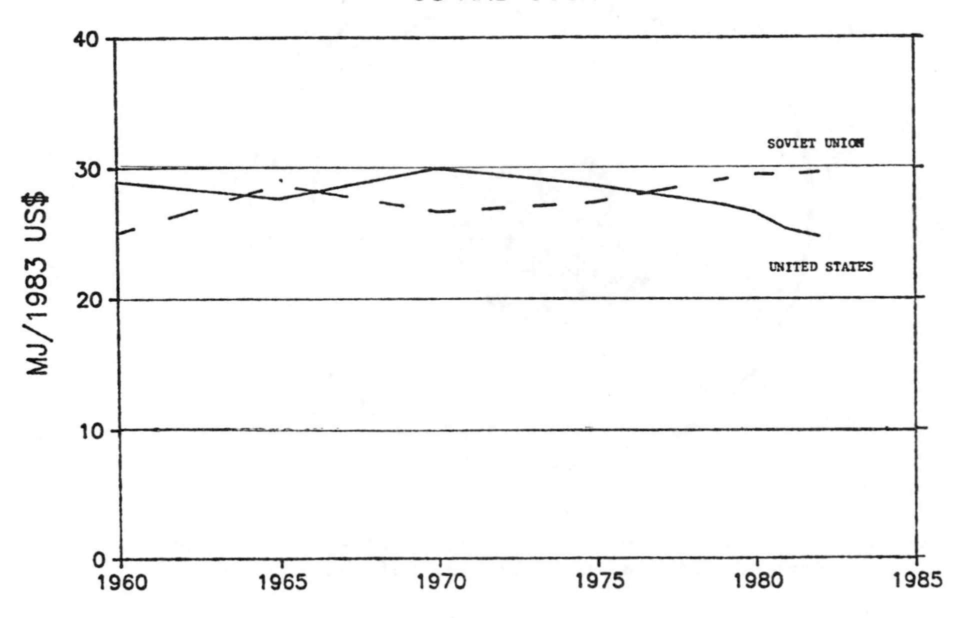

Fig. 4. Energy divided by GNP, 1960-1983, for the United States and the Soviet Union. Energy does not include wood. All data are from Ref. 12, which constructs the Soviet GNP and estimates a conversion factor from rubles to dollars. U.S. values of energy use are not quite identical to those in Fig. 1, as a result of an undocumented difference in accounting convention between Ref. 3 and Ref. 12.

percent in the Soviet Union and 3.9 percent in the United States. The E/GNP ratio, then, increased 8 percent over these seven years in the Soviet Union but decreased 14 percent in the United States. (The very careful reader will discover that the E/GNP ratios for the United States shown in Figures 1 and 4 are not identical: the conventions by which total energy is calculated apparently differ in the two sources -- in such a way, in fact, that E/GNP falls a little more slowly with time in Figure 4 than in Figure 1.) Perhaps the steady increase of E/GNP in the Soviet Union and Eastern Europe since 1970 is a reflection of domestic economies much better shielded from higher world energy prices, and perhaps the increase is also a reflection of an inactivity among technologists, who have been less strongly drawn to the tasks of using energy efficiently. My own guess is that a decoupling of energy from GNP will occur in Eastern Europe and the Soviet Union in the next decade, and one reason I think so is that the Soviet physics community, in particular, is only now enlarging the scope of its engagement with energy issues to include energy efficiency. In fact, energy efficiency may turn out to be an appropriate field for broad-scale East-West collaboration.

The perspective of physics continues to elicit the observation that there is enormous room for being more clever at providing greater amenities with fewer resources. Indeed, not merely a constant level of energy use but more like a halving of the level of energy use in thirty years appears entirely feasible in the world's industrialized countries.[13] My own view is that such a rate of reduction is not only feasible but likely, because technological change for the next several decades will be strongly driven by what I have called "molecular control".[14] Building on the breakthroughs in quantum mechanics which tamed the atom in the first forty years of this century, technologies of molecular control will displace sledgehammers by scalpels, in one task of the economy after another, with the inevitable consequence that inputs of natural resources will be sharply reduced.

As for the achievements of the insurgency which sought to achieve a less unequal balance between physicists working on energy supply and physicists working on efficiency at the end use, these have been much more spotty. One not terrible measure is the budget of the U.S. Department of Energy for research on energy conservation, as a fraction of its budget for research on fission, fusion, fossil fuels, and solar energy. This ratio rose in the Department's early years, but has fallen since 1981. Such a measure ignores end-use research elsewhere in the government and in private industry and also ignores the possibility that the fraction of the research enterprise carried out by physicists has not been constant across time and across programs; although neither omission can be defended, this measure is one of many which show that the level of involvement of American physicists in end-use issues grew rapidly for a while, then peaked, and now has subsided somewhat.

The current enterprise has considerable staying power. Every national laboratory of the U.S. Department of Energy now has an end-use research program, and one of these, the Lawrence Berkeley Laboratory, has in its Applied Physics Division a constellation of experienced researchers and a vigor of coordinated programs which rivals anything the supply research community can muster.

The past decade's transformation of end-use efficiency from a set of sleepy backwaters of engineering to a set of exciting enterprises has been called by my colleague, Robert Williams, a Quiet Revolution in end use efficiency. There is now a weekly publication called Energy Users News, whose pages testify to continuing innovation in devices and systems. A biannual meeting at Santa Cruz captures in its discussions and in its Proceedings[15] an increasingly vigorous set of activities in the subspecialty of energy conservation in buildings. And groups like the American Society of Heating, Refrigeration and Air Conditioning Engineers (ASHRAE) have been gradually transformed by newcomers into intellectually vigorous professional societies.

In the universities the story is not as impressive, though there are a few success stories. One such is my own group at Princeton, which specializes in field research in buildings[16] and has just received a kind of recognition which perhaps can become an exemplar for other university-based research in the country: We have been asked by the State of New Jersey to configure a large part of our research in a New Jersey Energy Conservation Laboratory (NJECL), funded directly by the State's gas and electric utilities and a few of its larger corporations. Since our energy end-use research has always had a regional focus, this new institutional arrangement will not require major revisions in our choices of projects, while it gives us much improved access to many of those to whom we most need to communicate our research results. All fifty States have gas and electric utilities, a state energy office of some form, and strong university research groups with at least embryonic interests in end-use efficiency; I see no barrier to forty-nine more ECL's coming into existence -- and there would be a role for physicists in each of them.

SHORTFALLS IN OUR ANALYSES

When one reads the APS Study with a decade of hindsight, its omissions are more conspicuous, fortunately, than its outright errors. Foremost among these omissions is the general absence of discussion about the adverse side effects which might accompany energy conservation strategies. It was only one or two years after the 1974 Summer Study that researchers in end-use efficiency lost a particular naivete about these side effects. We had not imagined that conservation strategies could possibly have negative consequences for safety or environmental quality comparable to those associated with the supply technologies conservation was intended to displace. But we came to understand that there is nothing which cannot be done stupidly, and that it was up to us to be at least as assiduous in identifying the negative externalities of energy conservation as we were asking proponents of coal, fission, and solar energy to be of their conversion technologies.[17]

One good example is indoor air quality, which in first approximation is made worse as outdoor air flow through a building is reduced, for those pollutants whose sources are largely indoors. The 1974 APS Study scarcely mentions this problem. Yet within two years researchers on energy use in buildings had inserted considerations of indoor air quality into the country's research agenda, and the single objective of saving energy cost-effectively was replaced by a program of greater sophistication with multiple objectives. It has since become clear that the application of scientific thinking to the performance of buildings will have benefits along both dimensions, and that control of sources of pollution, provision of controlled air exchange through heat exchangers, monitoring with advanced sensors, and improved understanding of health effects are all going to emerge from this research along with improvements in the efficiency with which energy is used. Similarly, one has come to expect that each year's cars will be both more fuel-efficient and more crashworthy.

The APS Study also failed to anticipate the range of opportunities which the microprocessor would open up for the accurate control of processes. One generic area of application which we did anticipate was the problem of part-load, whether in an auto engine or a furnace or a factory process, and we were correct in believing that devices could be developed at reasonable cost which had much flatter performance curves over the range of conditions in which they would actually operate. Today's variable-speed-drive fans and compressors, with solid state controls, and today's microprocessor-driven carburetors, are among many examples.

Of the many other areas of naivete and ignorance one could discover in the 1974 APS Study, one which is still with us is the issue of durability of conservation investments. Conservation researchers acquiesced in thousands of analyses which purport to establish the payback period (or internal rate of return, or cost of saved energy) for conservation investments (like clock thermostats and puttying around windows), in which an input to the calculations is the lifetime (in years) over which the savings could be expected -- without insisting that such lifetimes be measured, not guessed. We still know pitifully little about the determinants of durability of hardware, and even less about the determinants of durability of attitudes and behavior.

Shortfalls of our analysis of the physics profession (the other column vector, so to speak, in the structure of this essay) were perhaps most glaring insofar as they concerned physics education. "One realization we shared during our study," says the APS book (p. 23), "is the importance of classical physics, notably areas such as fluid mechanics and classical physical and chemical thermodynamics. Despite the power and scope of these subjects (and despite their practical importance), they have been neglected in our educational enterprise. Academic physicists, while witnessing with satisfaction the adoption and adaptation of these subjects within other disciplines, have counseled their students, openly or subtly, to put their efforts elsewhere. This lack of breadth and diversity is a disservice to our science." To restore this "breadth and diversity" apparently will require much more of an assault than what is engendered by a concern for energy efficiency and environmental quality, for freshman physics textbooks have scarcely changed in the past decade. In particular, heat pumps continue to get short shrift. I am still waiting for a diagram like Figure 5 to appear in the chapter where Carnot cycles are discussed.

Inherent in the impedance mismatch between physics and societal concerns is the striving for generality in the one and the need for specificity in the other. Even the concept of available work in thermodynamics (mentioned above) requires a descent from universality, because the concept, which expresses the maximum amount of work which can be done by a system enclosed within a much larger system with fixed temperature and pressure, only becomes worth the trouble to introduce when one agrees to consider processes being carried out on or near the surface of this particular Earth,

Tasks Modeled by Heat Pumps

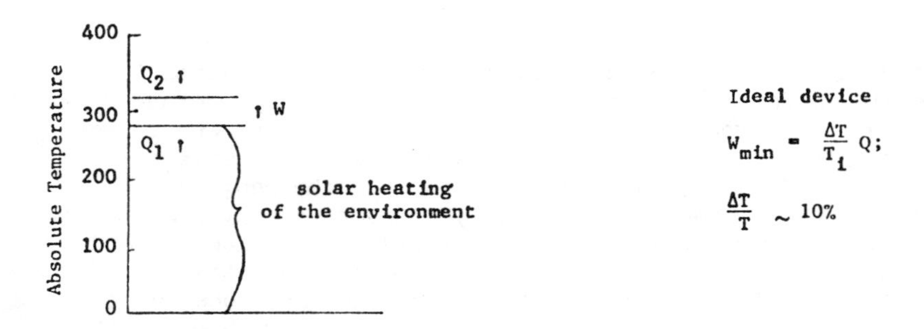

Many important tasks of everyday life require the transfer of heat from a lower to a higher temperature, where the absolute temperatures before and after differ by only ten percent or so. Examples are space heating, water heating, air-conditioning and refrigeration. When actual systems accomplishing these tasks (such as furnaces and water heaters) are compared with ideal heat pumps, very large opportunities for energy savings are revealed.

Fig. 5. A suggested Figure for the next edition of elementary physics textbooks.

with its particular environmental temperature near 300K and its particular surface pressure near 100,000 Pascals. Earth is fascinating to many of us, but it is, after all, just a special case: its radiative balance (equally warmed by the sun and cooled by outer space), its atmospheric concentrations, its ocean salinity, its magnetic field, are all just particular numbers on a continuum of numbers to which theory applies with indifference.

A related, even more implacable source of discordance in environmental science occurs when the subject at issue is human beings, rather than inanimate earth. Of the many disconnected fields of psychophysics, one subspecialty related to vision appears to have insinuated itself far more deeply into canonical physics, than have any subspecialties related to hearing or the other senses. If our own group was a good sample, most physicists learn to calculate light fluxes in lumens/m^2, or lumens/ft^2 (otherwise known as footcandles), without having any idea that the lumen is defined by means of a weighting function over wavelengths which is a measure of relative sensitivity of the cones in the human eye (see [1], pp. 73-74). Getting this straight was one of the pleasures of the summer's work; failing to get the lumen introduced properly in physics books (as one of a class of useful hybrid units drawing on both physics and psychology) is one of the many shortfalls in implementation.

THE WISDOM OF 1985

Today's popular beliefs about energy and economics and about priorities for the attention of real physicists are not the same ones as a decade ago but they are at least as worthy of disputation. The heir to the belief that prosperity requires endless increases in the rate of use of energy is the belief that the three quarters of the world's population who live in the "less developed" countries of this planet cannot achieve a level of amenity comparable to the current level of amenity in the industrialized countries without overwhelming the planet's life support systems; in short, the developing countries are believed to be undevelopable. And the heir to the belief that real physicists interested in energy should work on energy supply, not efficient use, is the belief that real physicists interested in national security should work on weapons systems, not civilian concerns. It seems that there will never be a shortage of nefarious conventional wisdom.

Those physicists who have been looking at energy use in developing countries have been learning that thermodynamic analysis of tasks, detailed modeling of devices, and multiobjective modeling of performance are all at least as applicable in the setting of a poor village as elsewhere. The cooking stove, which is the world's most important conversion device for harnessing the chemical energy in wood, is yielding its secrets to careful studies,[18] and it is becoming clear that more efficient cooking technology will permit the freeing up of enormous energy resources for redeployment in other tasks.

Moreover, when the full set of tasks requiring important amounts of energy are examined from the point of view that they should be performed with equal technical cleverness everywhere in the world, it turns out that the total rate of use of energy by the world's seven or so billion people in 2020 need not be any greater than the current rate of use, while substantial economic development occurs everywhere. This first cut at a planet-wide analysis, undertaken by Jose Goldemberg (Brazil), Thomas Johansson (Sweden), Amulya Reddy (India) and Robert Williams (USA), demonstrates that a world in which there is a uniform distribution of 80-mpg cars, 500 kWh/year refrigerators, superinsulated houses, variable-speed-drive fans and compressors, microprocessor-controlled rolling mills for sheet steel, *etc.*, is a livable, sustainable world.[19] Their version of energy use in 2020, and its comparison with more traditional analyses, is summarized in Figure 6. Thirty-five years out, according to Goldemberg, Johansson, Reddy, and Williams (GJRW), the developing world can be using 66 percent of the world's energy, double its 32 percent share today, if a vigorous rate of introduction of energy-efficient technology is sustained in the industrialized countries. Not visible in Figure 6 but a main conclusion of their work, the current rate of per capita consumption in the developing countries, about 1.0 kilowatts, is consistent with

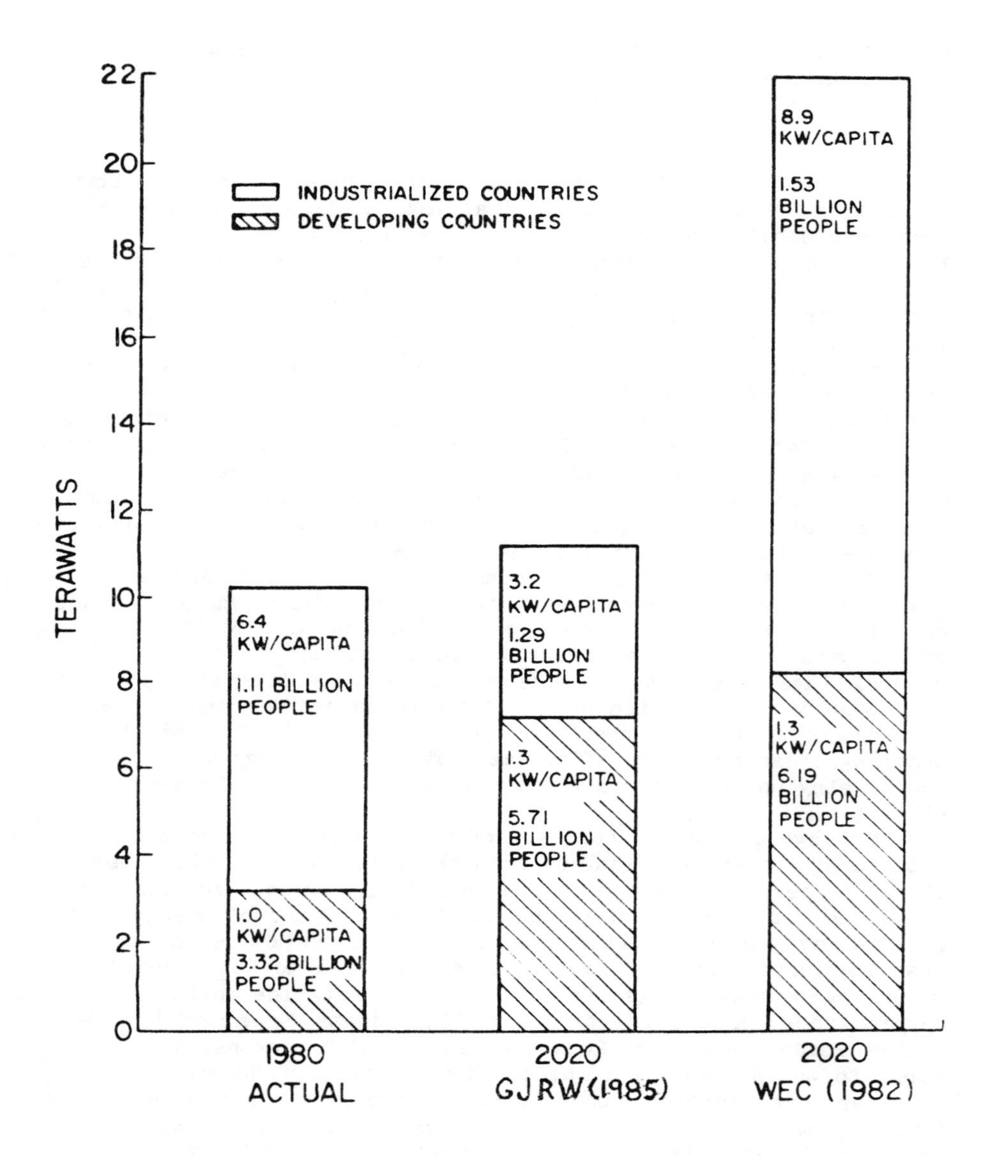

Fig. 6. World primary energy use, including wood, in TW (10^{12} watt-years/year), and its disaggregation into industrialized and developing countries. Figure adapted from Ref. 19, which is labeled GJRW (1985). WEC is the average of a high-energy and a low-energy scenario for 2020 in Ref. 20.

every standard of living from the present one to one equivalent to the level of amenities enjoyed in Western Europe, so that the increase in total consumption of energy in a developing country which gives priority to efficient use of resources need grow only as fast as its population.

As for the argument that national security equals weapons systems, it can best be undermined by creative work in other areas which demonstrably improves the society. If the United States and the Soviet Union lose ground to the rest of the world by engaging an ever increasing share of their most creative people in the enterprise of modern arms, cause and effect will be difficult to prove. What those of us appalled at such present trends can probably best do is to continue to make as vivid as possible several alternative ways of working as scientists on problems of concern to society. One such path, and an endlessly fascinating one, remains the career of physicist engaged in designing a resource-respectful world.

REFERENCES

1. Efficient Use of Energy, K. Ford, G. Rochlin, A. Rosenfeld, M. Ross, R. Socolow, eds. American Institute of Physics Conference Proceedings No. 25, AIP, New York, 1975. See also "Efficient Use of Energy," Physics Today, August 1975, pp. 23-33.
2. Bureau of the Census, Historical Statistics of the United States: Colonial Times to 1970, Part II, data series M-83, 90, 92, Dept. of Commerce, Washington, D.C., 1975.
3. Energy Information Administration, Annual Energy Review 1983, U.S. Department of Energy, Washington, D.C., April 1984.
4. Energy Information Administration, Monthly Energy Review, December 1984, U.S. Department of Energy, Washington, D.C., March 1985.
5. Energy Information Administration, "Estimates of Wood Energy Consumption, 1980-1983," U.S. Department of Energy, Washington, D.C., November 1984.
6. Bureau of Economic Analysis, "Gross National Product by Sector or Industry of Origin, 1947-1983" computer printout, U.S. Department of Commerce, Washington, D.C., 1984.
7. Bureau of the Census, Statistical Abstract of the United States 1985, U.S. Department of Commerce, Washington, D.C., December 1984.
8. R.H. Williams, "Potential Roles for Bioenergy in an Energy-Efficient World," Center for Energy and Environmental Studies Report R-183, Princeton University, Princeton, N.J., February 1985.
9. S.H. Schurr, J. Darmstadter, H. Perry, W. Ramsey, and M. Russell, Energy in America's Future: The Choices Before U.S., Resources for the Future, Johns Hopkins University Press, Baltimore, 1979.
10. R.H. Socolow, "The Coming Age of Conservation" Annual Reviews of Energy 1977, 2, 239-289, 1977.

11. B. Andresen, P. Salamon, R.S. Berry, "Thermodynamics in Finite Time," Physics Today 37 (9) 62, 1984.
12. Central Intelligence Agency, Handbook of Economic Statistics 1983, U.S. Directorate of Intelligence, Washington, D.C., September 1984.
13. R.H. Williams, "A Low Energy Future for the United States," Center for Energy and Environmental Studies Report R-186, Princeton University, Princeton, N.J., February 1985. See also E.D Larson, E.D., R.H. Williams, and A. Bienkowski, "Material Consumption Patterns and Industrial Energy Demand in Industrialized Countries," Center for Energy and Environmental Studies Report R-174, Princeton University, Princeton, N.J., December 1984; and T.B. Johansson, and R.H. Williams, "An End-Use Energy Strategy for Industrialized Countries," Center for Energy and Environmental Studies Report R-185, Princeton University, Princeton, N.J., February 1985.
14. R.H. Socolow, "Resource-Efficient High Technology, Basic Human Needs, and the Convergence of North and South," Center for Energy and Environmental Studies Report R-157, Princeton University, Princeton, N.J., April 1982.
15. Proceedings of the ACEEE 1984 Summer Study on Energy Efficiency in Buildings, Santa Cruz, California, 1984, 10 volumes. Available from the American Council for an Energy-Efficient Economy, 1001 Connecticut Ave., N.W., Washington, D.C. 20036.
16. R.H. Socolow, "Field Studies of Energy Savings in Buildings: a Tour of a Fourteen-Year Research Program at Princeton University," Center for Energy and Environmental Studies Report R-191, Princeton University, Princeton, N.J. To be published in Energy: The International Journal as part of the Proceedings of the Soviet- American Symposium on Energy Conservation, Moscow, June 1985.
17. R.H. Socolow, "Four Anxieties about a Vigorous National Conservation Program: Discussion Paper" Annals of the New York Academy of Sciences, Vol. 324, 1979, pp. 28-30.
18. H. Geller, S. Baldwin, G. Dutt, and N. H. Ravindranath, "Improved Woodburning Cookstoves: Signs of Success," Center for Energy and Environmental Studies Report R-184, February 1985.
19. J. Goldemberg, T. Johansson, A.K.N. Reddy, and R.H. Williams, "An End-Use Oriented Global Energy Strategy," Center for Energy and Environmental Studies Report R- 179, Princeton University, Princeton, N.J., March 1985.
20. World Energy Conference, 1983. Energy 2000-2020: World Prospects and Regional Stresses, Frisch, J.-R., ed. London: Graham & Trotman.

CHAPTER 3

THE ECONOMICS OF ENERGY CONSERVATION IN DEVELOPING COUNTRIES: A CASE STUDY FOR THE ELECTRICAL SECTOR IN BRAZIL

Jose Goldemberg
Companhia Energetica de Sao Paulo (CESP)
Sao Paulo, Sao Paulo, Brazil

Robert H. Williams
Center for Energy and Environmental Studies, Princeton University
Princeton, N.J 08544

ABSTRACT

A wide range of high efficiency, energy-using technologies have become commercially available in recent years, in North America, Western Europe, and Japan. Contrary to the widely held view that these technologies are relevant mainly to the rich, already-industrialized countries, we show that from an economic perspective, energy efficiency improvements often make as much or even more sense for capital-poor, developing countries.

We illustrate the relevance to developing countries of more energy-efficient end-use technology, with an analysis of the economics of energy-efficient refrigerators and lightbulbs in the context of the electrical system of Brazil, from both the consumer's perspective and that of society. We show that the required extra investments in energy efficiency generate attractive returns in electricity savings for the consumer. Moreover, for the country as a whole, investments in energy efficiency can lead to net savings of scarce capital resources, by reducing the need for new electrical generating capacity. Because electricity in Brazil is largely based on low-cost hydro-electric power, showing the importance of energy efficiency improvements in this situation is an "acid-test" for the relevance of energy efficiency to developing countries more generally.

Capturing the economic benefits of energy efficiency improvements probably requires that utilities be transformed from being energy supply companies into companies that market energy services, by facilitating investments on the "customer's side of the meter" as well as in new supplies. Some utilities in industrialized countries are already beginning to shift their activities in this direction. An even more active utility role may be desirable in developing countries, because there most of the population is poor, and the poor tend to be far more first-cost sensitive, and thus resistant to making investments in energy efficiency improvement, than higher income consumers.

INTRODUCTION

As a response to the energy price increases of the 1970s, many energy-efficient end-use technologies have been commercialized for

0094-243X/85/1350033-19$3.00 Copyright 1985 American Institute of Physics

the domestic sector, for commercial buildings, for transportation, and for industry, in various parts of the world.[1]

To date most of these new products are available largely in industrialized countries with market-based economies. Their relevance to developing countries is an issue of considerable controversy. Three questions have been raised concerning such technologies:

- First, how can one justify emphasizing energy conservation in countries so poor that they have little to save?
- Second, even if there were significant energy conservation opportunities in poor countries, wouldn't the pursuit of these opportunities imply technological dependency on industrialized countries, since much of the needed energy-efficient technology is not now manufactured in developing countries?
- Third, in light of the extra investment usually required to obtain improvements in energy-efficiency, isn't more energy-efficient end-use technology inappropriate for developing countries, where capital is so scarce?

In this paper we address these questions -- the first two briefly and the last in some depth, with a detailed discussion of the economics of energy efficiency in the context of the Brazilian electrical sector. We believe that our analysis will lead the reader to understand why investments in energy-efficiency are not only relevant to developing countries but essential to bringing about rapid development.

What is there to save in poor countries? While it is true that the rich industrialized countries can save far more energy than developing countries, it does not follow that there is little energy to save in developing countries.

The elites, which typically make up 1/10 of the population but receive one third to one half of all income and account for 9/10 of all commercial energy use in developing countries, have energy-wasting habits very similar to those of citizens of industrialized countries -- and often there is even greater waste in the developing country situation.

Consider refrigerators, which are now present in about 60% of all Brazilian households.[2] Two-door refrigerator/freezers, which are becoming popular among the elites, accounted for about 1/3 of all new refrigerators sold in Brazil in 1983.[2] The new Brazilian units of this type, though smaller (340 to 420 liters) than typical units in the US (about 500 liters), consume between 1310 and 1660 kWh per year[2], much more than than the average for new units in the US -- some 1150 kWh in 1983.[3] Moreover, the most efficient model available in the US, introduced commercially in March 1985, was a 490 liter unit requiring just 750 kWh per year.[4]

Even poor households dependent largely on "non-commercial" energy (fuel which is not purchased in market transactions and which accounts for nearly half of all energy use in developing countries[5]) and having few if any modern amenities tend to use energy very inefficiently. Figure 1 shows the per capita energy use for cooking, via wood stoves in developing countries and via modern energy

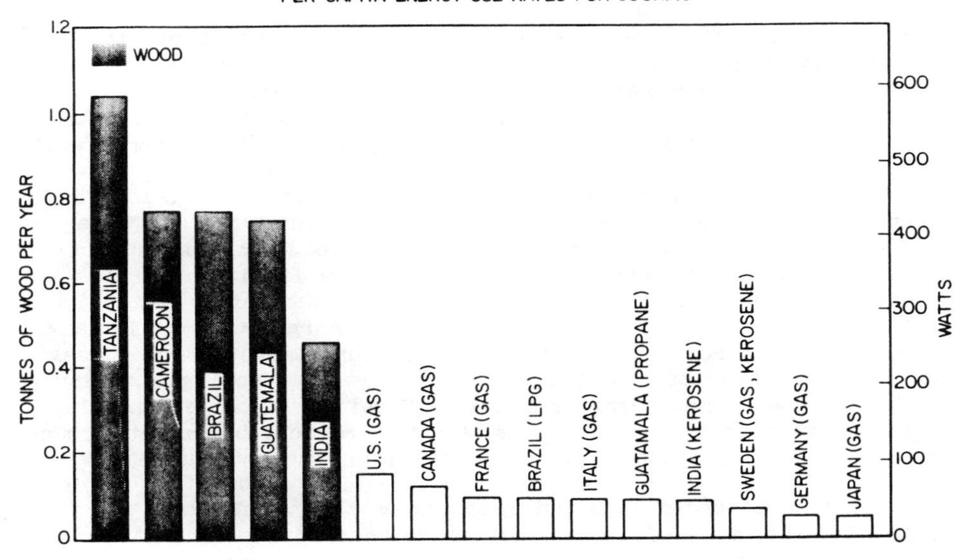

Fig. 1. Per capita energy use rates for cooking in various parts of the world with both modern energy carriers (natural gas, LPG, kero-ene) and wood[16]. For both wood stoves and stoves involving high quality energy carriers the per capita energy use rate for cooking is expressed in Watts. The wood consumption rate is also given in tonnes of dry wood per year. Assuming 1 tonne = 18 GJ, 1 tonne per year = 570 Watts.

carriers (LPG, kerosene, and natural gas) in both developing and industrialized countries. Over a wide range of conditions it is seen that the poor who depend on wood for cooking consume 3 to 10 times as much energy for cooking as those who have access to modern energy carriers. The high level of energy use for cooking via wood arises from the low energy efficiencies of wood stoves (typically 10% or less) and the lack of controllability of the heat of most wood stoves. For the urban poor who buy wood or charcoal for cooking, this inefficiency implies a large expense. For the rural poor who are not active participants in the market economy, it implies a great deal of time committed to wood gathering (especially by women and children) that could otherwise be occupied more productively -- an amount of time which is increasing in many parts of the world, as the process of deforestation makes wood supplies scarcer.

<u>Where will efficient end-use technology come from?</u> To the extent

that energy-efficient end-use devices are not now available in developing countries, these devices must either be imported, or a local manufacturing capability must be established.

Those developing countries with little industrial infrastructure may have to rely on imported technologies, at least for a while. But these countries would have to import the conventional, less-efficient technologies anyway. For these countries the issue is whether the increased foreign exchange expenditures for the more efficient end-use technologies can be justified. In this instance it is important to calculate the foreign exchange implications of choosing the more efficient technologies by considering the entire system of improved end-use technology plus energy supply. In many cases the extra foreign exchange required for a more efficient device will be more than offset by the resulting reduced foreign exchange requirements for new energy supplies. Clearly, to the extent that net foreign exchange requirements would be reduced by importing more efficient end-use technologies, a developing country would be better off.

In more advanced developing countries the manufacturing capability for many efficient end-use devices could be developed in just a few years, if manufacturers believed there were sufficient markets. Strong evidence that Brazilian manufacturers could do this for electricity-using devices (refrigerators, efficient lighting technologies, heat pumps, motors, motor control devices) is indicated in a recent study by Geller;[2] indeed Brazil is already manufacturing heat pump water heaters, sodium vapor lamps for street lighting, motor control devices, and some lighting control devices. Moreover, it is ironic that the most efficient refrigerator/freezer available in the US achieves its efficiency advantage in large part from the use of a compressor imported from Brazil.[4] The Brazilian manufacturer involved exports a high efficiency line and markets a less efficient product domestically.

What about the problem of capital scarcity? It is certainly true that consumers must usually pay more for the purchase of energy-efficient devices. But the difficulty the consumer has in obtaining capital in a developing country should be distinguished from the problem of the overall limited supply of capital. As we shall show by example in this paper and has been shown in detail elsewhere,[1,2] investments in energy efficiency improvement as an alternative to investments in the equivalent amount of energy supply often lead to major reductions in overall capital requirements by society for providing a given level of energy services. This prospect provides a strong motivation for finding ways to make capital more readily available for individual consumer investments in energy-efficiency improvements.

Scope of the Paper: In what follows we discuss in some detail the economics of energy efficiency in the context of the electrical sector of Brazil. While our discussion focuses on two technologies (more efficient refrigerators and compact fluorescent light bulbs) in the Brazilian context, our methodology is applicable to a wide range of technologies in market-based economies, for developing and industrialized countries alike.

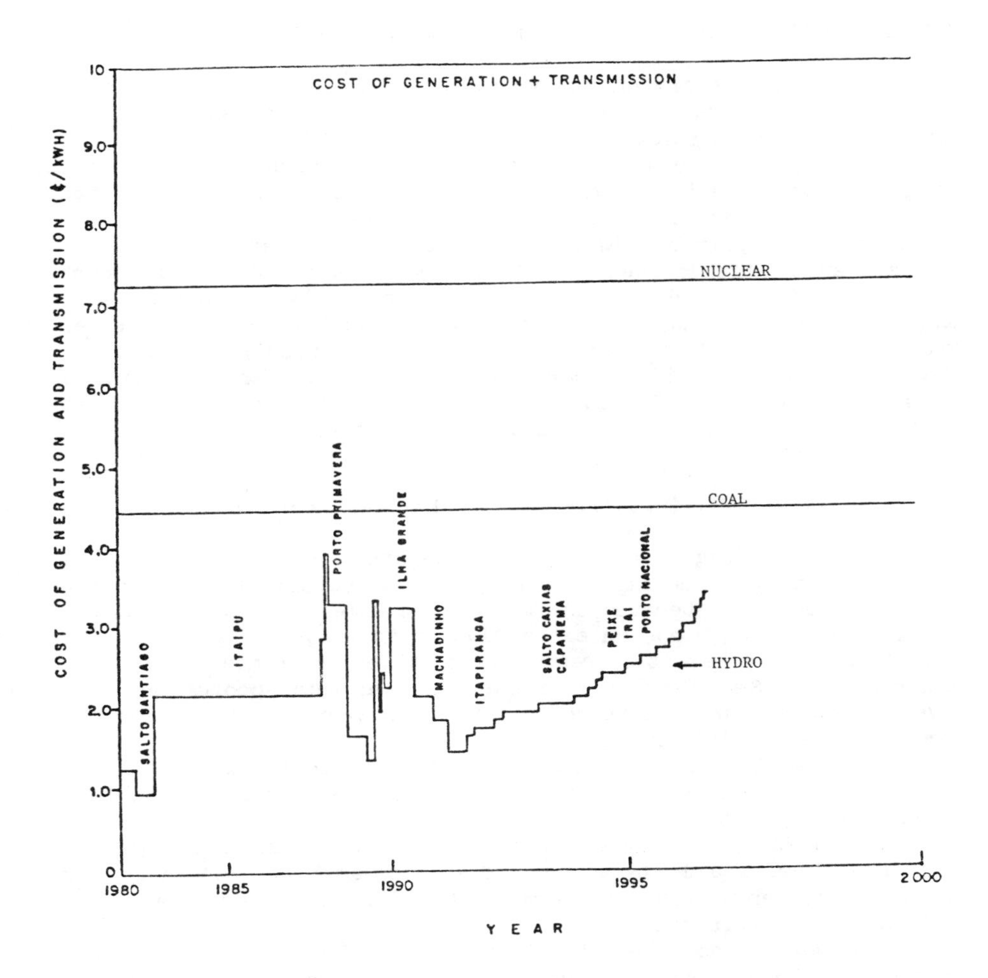

Fig. 2. Cost estimates for the generation and transmission of electricity from new hydroelectric facilities in Southeastern Brazil -- for plants coming on line, under construction and planned. Costs for electricity from coal and nuclear thermal plants that could be built in Brazil are shown for comparison. Source: Companhia Energetica de Sao Paulo

In fact, showing the relevance of conservation to the electrical sector of Brazil is an "acid test" for the relevance of energy conservation generally for developing countries, since Brazil, unlike most industrialized market economies, is blessed with considerable hydro-electric resources that are much less costly than thermal power sources based on coal or nuclear energy (see Figure 2) -- a blessing which would seem to suggest that conservation is really not all that important for the Brazilian electrical sector.

REFRIGERATORS AND LIGHT BULBS: THE CONSUMER'S PERSPECTIVE

While energy efficiency improvements usually add to the first cost of an energy-using device, this cost is often offset by the savings in operating cost.

The simplest economic measure relating these quantities is the "simple payback" period -- the ratio of the extra first cost to the expected savings in the first year. The concept is illustrated by comparison of a pair of 510 liter (18 cubic foot) refrigerator/freezers available in the US in 1982/83 which were identical except for the first cost and the energy-conserving design features. The more efficient one had a $100 higher price tag but used 460 kWh per year less electricity.[6] Even at the low residential Brazilian electricity price of $0.047 per kWh* the more efficient refrigerator had a simple payback of about 5 years. Since US refrigerators last 19 years on average,[6] the consumer would thus gain a "pure profit" after the fifth year of some

$$(19 - 5) \times \$0.047/\text{kWh} \times (460 \text{ kWh per year}) = \$303 \quad (1)$$

during the remaining life of the refrigerator.

This simple calculation overstates the economic benefit of the more efficient unit because it ignores the time value of money.

If the consumer has an extra $100 to invest, he could alternatively invest it in real estate, stocks or bonds, a savings account, etc. Some measure is needed to take into account such investment opportunities forgone if the consumer were to invest in a more efficient refrigerator.

Sometimes the consumer won't have $100 to invest without borrowing -- in which case the interest costs of borrowing should be taken into account.

Such problems having to do with the "time value of money" can be handled systematically by using a "discount rate" to express dollar values at different times in terms of the dollar values at a particular point in time.

In terms of the discount rate "i", the relative merits of two alternative investment opportunities which are identical in every way except for first cost and energy efficiency can be assessed via calculations of the lifecyle costs involved. In the case of the two US refrigerator/freezers described above, the lifecycle costs are:

$$LCC_a = I_a + \sum_{n=1}^{19} E_a \times P_n \times (1 + i)^{-n}, \quad (2)$$

----

* This was the rate in 1983 for consumption in the range 31 to 200 kWh per month. For consumption in the range 201 to 500 kWh it was $0.061 per kWh, and for consumption over 500 kWh per month it was $0.076 per kWh. For comparison, the average residential electricity price in the US was $0.072 per kWh in 1983.

$$LCC_b = (I_a + 100) + \sum_{n=1}^{19} (E_a - 460\text{ kWh}) \times P_n \times (1 + i)^{-n}, \quad (3)$$

where I_a is the initial cost of the conventional unit, E_a is its annual electricity consumption, and P_n is the electricity price in the year "n". For the case where the electricity price is constant ($P_n = P$), the sums are easily performed so that:

$$LCC_a = I_a + E_a \times P / CRF(i, 19), \quad (4)$$

and

$$LCC_b = LCC_a + \$100 - [460 \times P / CRF(i,19)], \quad (5)$$

where CRF is a so-called capital recovery factor:

$$CRF(i,N) = i / [1 - (1 + i)^{-N}]. \quad (6)$$

One way to use the lifecycle cost concept would be to specify the discount rate and compare the resulting lifecycle costs. The option with the least lifecycle cost would be the most cost-effective. While a real (inflation-corrected) discount rate of the order of 10% (comparable to the long term average rate of return on private corporate investment in many industrialized countries) is often used for planning purposes by governments in comparing alternative long term investment options, the appropriate discount rate is not at all well defined for most consumers. Riskier investment opportunities should be discounted more than less risky ones. Different individuals and institutions have different kinds of investment opportunities available to them and can borrow at different rates. And real (inflation-corrected) interest rates can vary over time, according to how tight the supply of capital is.

One use of the lifecycle cost concept that does not require the specification of a discount rate in comparing two alternative options is in "internal rate of return" analyses. With this methodology it is assumed that the consumer is indifferent to the two alternatives being compared, the lifecycle costs are equated, and the discount rate is solved for. The solution is the annual internal rate of return on the extra investment, associated with choosing the energy-saving option. An energy efficiency improvement would be attractive as an investment opportunity if the estimated rate of return were in excess of the opportunity cost of capital or the market rate of interest. If there were a choice of several alternative technologies for making efficiency improvements, the alternatives could be ranked by the rate of return as investment opportunities.

For the refrigerator/freezer example the real internal rate of return (obtained by solving the equation $LCC_a = LCC_b$ for "i"), would be 21%. This is a tax-free return on the investment that is much higher than the "opportunity cost of capital" or the "market rate of interest" for most consumers.

As favorable as this result is for the more efficient units, this calculation probably underestimates the benefits of the energy savings two ways. First it is based on the assumption of constant prices. But in Brazil electricity prices can be expected to rise, as more and more costly hydro sources have to be exploited (see Figure 2). The effect of a constant annual rate of increase "r" in the electricity price can easily be incorporated into the analysis, by setting:

$$P_n = P_o (1 + r)^n, \tag{7}$$

so that in (4) and (5) "i" is replaced by

$$j = (i - r) / (1 + r). \tag{8}$$

If the electricity price escalation rate were 5% per year, the internal rate of return for the above example would increase to 27%.

The second problem with the above analysis is that the increased price of the more efficient refrigerator/freezer probably overstates the cost of making improvements of this magnitude. Design studies have shown that comparable savings could be achieved with relatively modest improvements (more insulation and more efficient compressor) for an extra cost of only about $20.[7] Moreover, the price of the most efficient unit now available in the US (the 490 liter unit described earlier consuming 750 kWh per year, having more insulation and a more efficient compressor than conventional units) is no higher than that of the previous model having similar features.[4]

Consider next the compact fluorescent bulb, now commercially available from a few manufacturers in Europe, Japan, and the US. Compact fluorescents are small bulbs that fit into ordinary incandescent sockets, last 5 to 6 times as long as incandescents, but require only 1/4 to 1/3 as much electricity for the same light output and approximately the same light color as incandescents.

A 13 Watt compact fluorescent lightbulb is available in Japan as an alternative to a 40 Watt incandescent bulb. The compact fluorescent is expected to last 6000 hours or six times as long as the incandescent. Even though the compact fluorescent costs $9.20 (the price in Japan), compared to $0.50 for the ordinary incandescent (the price in Brazil), the more efficient bulb would often be cost-effective.

The calculation of the internal rate of return is slightly more complicated than in the case of the refrigerator, as six incandescents must be purchased during the life of the fluorescent, and the lifetime depends on "H", the number of hours of use per day:

$$LCC_a = \$0.50 \times [1 + \sum_{n=1}^{5} 1 \times (1 + i)^{(-1000n/365H)}] + \sum_{n=1}^{6000/365H} E_a \times P_n \times (1 + i)^{-n}, \tag{9}$$

and

$$LCC_b = \$9.20 + \sum_{n=1}^{6000/365H} E_b \times P_n \times (1 + i)^{-n}. \tag{10}$$

For the case where the electricity price is constant and H = 5 hours per day, the internal rate of return obtained from equating the lifecycle costs is 8%, 20%, and 31%, respectively for the three residential electricity rates in Brazil -- $0.047, $0.061, and $0.076 per kWh. At the two higher electrical rates, appropriate for more affluent households, the rates of return are obviously quite good.

In reality buyers often do not opt for the more efficient appliances, given the choice. For a variety of reasons (lack of information about the energy savings or disbelief of claims of energy savings, scarcity of capital, the belief that someone else* will capture the energy savings benefits, higher priority given to features other than energy-savings, etc.) buyers tend to shy away from appliance purchases that involve added first costs.[8-12] This first cost sensitivity can be expressed in terms of an implicit discount rate for actual appliance purchases, determined by (a) equating the lifecycle costs for a choice between items which differ only in initial price and energy efficiency and (b) solving for the discount rate. If the buyer chooses the less efficient option having the lower initial price, his implicit discount rate is at least as large as the solution of this equation. High implicit discount rates provide a measure of imperfections in "energy conservation markets."

Generally speaking, the implicit discount rate is not readily measurable, because it is usually difficult to isolate the influence of efficiency from other factors influencing buyer decisions. Most analyses that have been done, however, suggest that the typical buyer behaves in "energy conservation markets" as though his discount rate were far in excess of market interest rates.

One study provides an unequivocal demonstration of this phenomenon.[8] In this analysis by Meier and Whittier, sales data were examined for a large number of two refrigerator models sold in the US by a major retailer between 1977 and 1979. The two models differed only in initial price ($40) and annual energy use (410 kWh per year), and retail clerks were instructed to explain to customers the energy saving advantages of the most costly model. The study showed that 2/5 of the consumers behaved as if there discount rates were above 60%, and 3/5 above 35%.

Another study by Hausman[9] showing high implicit discount rates for room air conditioners in the US goes further to show that the implicit discount rate tends to be much higher for poor people than for those in high income brackets (see Table I). This is hardly a surprising result. With pressing immediate needs to be satisfied, poor people rarely have surplus dollars available to invest in the future, and most do not have good enough credit to borrow. Since a major fraction of the population is poor in developing countries, consumer reluctance to invest in energy efficiency improvement can be expected to be greater in developing countries than in industrialized

---

* The purchaser concerned that "someone else" would get the benefits could be a builder who installs refrigerators in the houses he sells, the landlord who provides refrigerators for his renters, or the home-owner who has to sell an appliance before it wears out, when he moves.

Table I. Hausman's estimate of the discount rate implicit in consumer purchases of room air conditioners in the US, by income class[9]

Income Class	Implicit Discount Rate
up to $6000	89%
$6000 to $10000	39%
$10000 to $15000	27%
$15000 to $25000	17%
$25000 to $35000	8.9%
$35000 to $50000	5.1%

countries, where the poor make up only a minor fraction of the population.

REFRIGERATORS AND LIGHTBULBS: THE SOCIETAL PERSPECTIVE

Many would argue that the general reluctance to invest in energy efficiency improvements is not only natural but appropriate for developing countries, which are struggling to provide the most basic amenities to their populations. The conventional wisdom is that in the early stages of development it is more appropriate to make available at low cost refrigerators, light bulbs, and other appliances to as many of the population as quickly as possible, and worry about energy efficiency only later, after achieving a reasonable amount of affluence.

The problem with this line of thinking is that it ignores the enormous costs to society as a whole associated with neglecting opportunities for energy efficiency improvement and continuing to pursue a "business as usual" supply expansion approach to the energy problem.

At the aggregate level the emphasis on energy supply expansion cost developing countries $25 billion in foreign exchange in 1982 -- over 1/3 of the foreign exchange required for all kinds of investment. And in 1983 the World Bank estimated that investments averaging some $130 billion a year (in 1982 dollars) would be required between 1982 and 1992, to achieve a targeted 44% increase in per capita commercial energy use in developing countries in the period 1980 - 1995; half of this investment would have to come from foreign exchange earnings, requiring an average annual increase of 15% in real foreign exchange allocations to energy supply expansion in this period.[13]

The indicators of the high costs of ignoring opportunities for energy savings are perhaps even more impressive at the "micro" level. For example, while a typical new refrigerator in Brazil costs the consumer about $200, the cost to the utility for the investment in new baseload electrical supplies needed to run one of these "electricity guzzling" refrigerators could cost up to 1 1/3 times as much.[14] Moreover, in the case of lighting, for each dollar the consumer spends on incandescent bulbs, the utility must spend about ten dollars on new peaking electrical supplies![15]

To illustrate how the costs to society compare for investments in energy efficiency versus investments in supply expansion, we consider once more the refrigerator/freezer and lighting examples described earlier.

In the case of the refrigerator/freezer, each more efficient unit saving 460 kWh per year would obviate the need for

$$460 \text{ kWh}/8760 \text{ hours} = 0.0525 \text{ kW} \qquad (11)$$

of new baseload power delivered to the residential customer at a cost (for a refrigerator/freezer costing $100 more) of:

$$\$100 / 0.0525 = \$1904 \text{ per kW}. \qquad (12)$$

In making a comparison to new hydro-electric power supplies, this is not quite the apppropriate comparison to make, as the expected life of the refrigerator is only 19 years, whereas that of the hydro facility is expected to be some 50 years. In a 50 year period some 2.63 refrigerators would have to be purchased. Using the cost of all these refrigerators in the calculation brings the total "cost of saving power" to $5007 per kW. But this is not quite correct either, because all these refrigerators don't have to be purchased initially. The correct way to do the calculation is to discount the future purchases at the appropriate rate. Figure 3 shows the cost of saving power as a function of the discount rate. For a 10% rate the cost is about $2260 per kW or about 20% more than the "naive" estimate above for a single refrigerator. This cost of saving power should be compared to the corresponding cost of producing power from new supplies.

Consider first generation. Each kW of baseload power delivered to a residential consumer in Brazil requires 1.25 kW of power at the power generating plant, to compensate for transmission and distribution losses. Moreover, because the available hydroelectric capacity varies with the season (wet or dry) and from year to year, some 1.8 kW of installed capacity is needed to support each kW of "firm baseload power," bringing the total required installed capacity to 2.25 kW for each kW of baseload residential demand.

The total cost of new hydro capacity, installed "overnight," for plants that would be ordered today, is about $1200 per kW. Taking into account interest charges during construction (assuming a 6 year construction period and a 10% interest rate) the cost would rise to $1500 per kW installed. The corresponding cost per kW of baseload residential demand is $3375 per kW. Figure 3 shows the total cost of generation per kW of baseload residential demand as a function of the interest rate. Note that for a discount rate greater than 5% the cost of new generation is greater than that for the corresponding cost of saving power by investing in more efficient refrigerators.

If new generating capacity could be deferred because of investments in energy efficiency improvements, there would also be some savings associated with transmission lines that would not have to be built. Transmission lines lasting 30 years can be built for a construction cost of about $700 per kW of residential demand.[2] The total cost (generation plus 50 years worth of transmission capacity) per kW of of baseload residential demand is also shown in Figure 3. The total of new construction that could be avoided if improvements were made in residential refrigerators as an

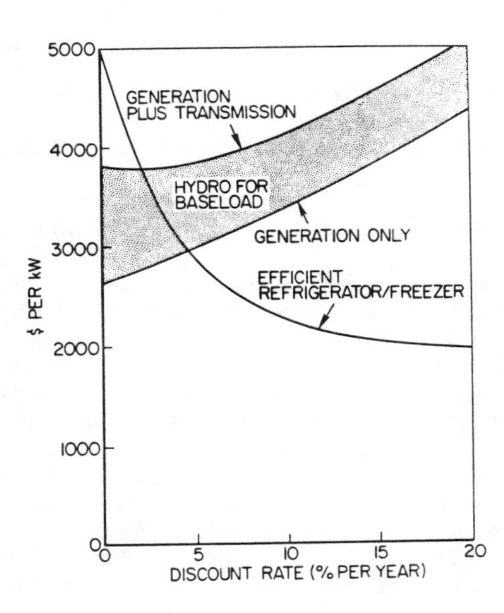

Fig. 3. The discounted present value of baseload electricity produced via a hydroelectric power system or saved via the installation of efficient refrigerators, in dollars per kW measured at the household.[17] All capital investments over the estimated 50 year life of the hydroelectric power plant are included.

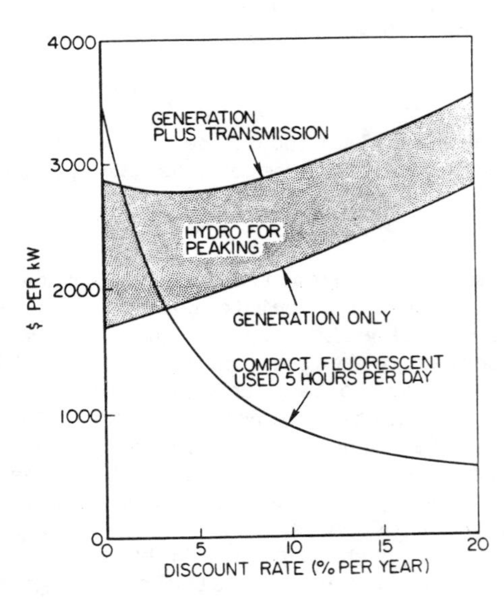

Fig. 4. The discounted present value of peaking electricity produced via a hydroelectric power system or saved via the installation of efficient compact fluorescent lightbulbs, in dollars per kW measured at the household.[18] All capital investments over the estimated 50 year life of the hydroelectric power power plant are included.

alternative to supply expansion would lie somewhere in the shaded region shown in Figure 3. Note that at a 10% discount rate the cost of supply expansion would be 1.5- to 1.8- times the cost of the corresponding investment in refrigerator efficiency improvement.

Figure 4 shows plots for the costs of peaking power saved and produced, for the case of compact fluorescent light bulbs replacing incandescent bulbs. Replacing a 40 Watt incandescent with a 13 Watt compact fluorescent would save 27 Watts of peak demand in Brazil, where the system peak occurs in the early evening. Again because of transmission and distribution losses the peaking demand reduction at the power plant would be 34 Watts, and with a 16% reserve margin the corresponding savings in installed hydro capacity would be 39 Watts. The curves for the costs of producing and saving peaking power over a 50 year lifecycle, as a function of the discount rate, are similar to those described above for the refrigerator/baseload power example (see Figure 4). Note that for a 10% discount rate the supply expansion cost per kW is 2.4- to 3.2- times the cost of saving a kW by investing in more efficient bulbs.

The opportunities for capital savings via energy efficiency improvements are not limited to refrigerators and lightbulbs. In his analysis of the entire Brazilian electrical sector,[2] Geller estimated that for a (discounted) total investment of some $4 billion in more efficient end-use technology [for more efficient refrigerators, street lighting, lighting in commercial buildings, and motors, and the deployment of variable speed motor drives], it would be feasible to defer construction of some 21 GW (e) of new electrical supply capacity, corresponding to a discounted capital savings for new supplies of some $19 billion in the period 1986 to 2000. The resulting capital savings could be used to speed up the development process via investments in other areas.

RECONCILING CONSUMER AND SOCIETAL PERSPECTIVES ON ENERGY EFFICIENCY

We have shown that because appliance buyers are "first-cost sensitive" they are generally reluctant to make investments in energy efficiency improvements. This reluctance is probably inherently greater for developing countries than for industrialized countries, because in developing countries the majority of the population is poor and preoccupied with immediate needs.

Investment capital is also scarce for poor countries, which like poor individuals, have pressing immediate needs to satisfy. But it does not follow that poor countries should view investments in energy efficiency as a burdensome demand on scarce capital. On the contrary, investments in energy efficiency can often lead to a net reduction in the total capital required by society to provide a given level of energy services. Moreover, from the societal perspective the economic benefits of investments in energy efficiency often look more attractive as capital resources get scarcer. This phenomenon can be quantified by comparing the economics of energy efficiency investment from the individual and societal perspectives.

Whereas the individual sees expenditures on energy supply as operating costs (e.g., paid on the monthly utility bill), and expenditures on energy efficiency improvments as capital investments, the expenditures seem to be ordered the other way around from the societal perspective. New supply for the utility requires up front expenditures for new facilities that may last 50 years (e.g., new hydro-electric facilities). The expected lifetimes of most investments in energy efficiency are typically much shorter than this and usually in the range of the two examples discussed here (3 up to 20). Because the investments for conservation equivalent to new supply expansion are spread out over time, they look from the societal perspective more like operating costs, which decline rapidly with increasing discount rate (see Figures 3 and 4). In sharp contrast (see also Figures 3 and 4), the investment required for energy supply expansion tends to increase with the discount rate, because of interest charges accumulated during the long construction period*. The net result is that the benefits of conservation investments tend to increase with the discount rate. We are led therefore to a rather surprising result: investments in energy efficiency improvement will often make even more sense in capital-short developing countries than in industrialized countries. Ironically, from a strictly economic point of view, rich countries can afford to waste more!

Enormous societial benefits would result if the consumer's interests on energy efficiency could be brought in line with the societal interest. This might be accomplished if the utility were to help promote consumer investments in energy efficiency improvements. To illustrate the possibilities consider once again the compact fluorescent bulb. Costing nearly 20 times as much as an incandescent, it is unlikely that many consumers would buy these bulbs outright. But suppose the utility were to finance the investment, allowing the consumer to pay off the loan on his monthly utility bill -- thus effectively converting the high first cost into an operating cost. Specifically, suppose the utility were to purchase many such bulbs from the manufacturer and make make them available to customers on "the installment plan," via loans of duration N months and interest rate "i/12" per month, where "i" is the annual rate. The required constant monthly payment M would be the solution of the equation:

$$\$9.2 = M \times \sum_{n=1}^{N} 1 \times (1 + i/12)^{-n} = M \,/\, CRF\,(i/12,\, N) \qquad (13)$$

Suppose that the term of the loan is equal to the expected life of the investment: 39.5 months, for a bulb used 5 hours a day. The required monthly payment, for a 10% annual interest rate would be:

$$CRF(i/12,\, N) \times \$9.2 = CRF\,(0.00833,\, 39.5) \times \$9.2 = \$0.27/\text{ month.} \qquad (14)$$

With such loan payments, the consumer's monthly utility bill would be less than if he were to keep using ordinary incandescents, at

---

* The cost of transmission plus generation shown in Figures 3 and 4 first declines with the discount rate, reaches a minimum, and then begins to rise. The initial decline arises because of the discounting of the cost of the replacement transmission facility installed after 30 years.

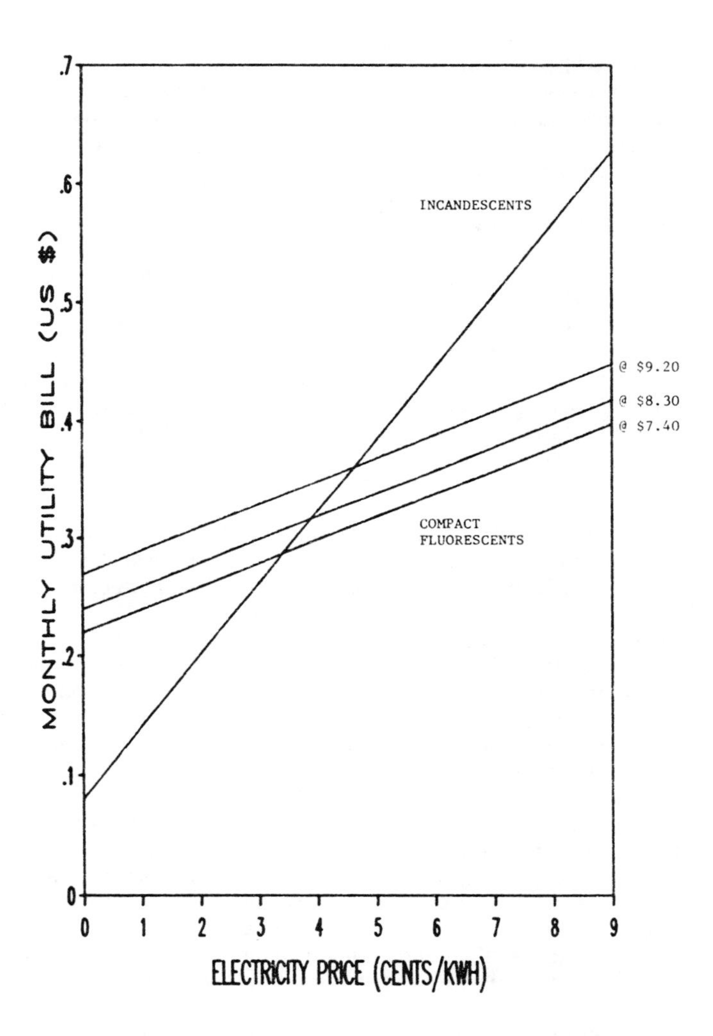

Fig. 5. The monthly utility bill (in US dollars) versus electricity price, for lighting services provided by both incandescent and compact fluorescent light bulbs.[19]

electricity prices in excess of about 4 1/2 cents per kWh -- i.e., at all residential electricity prices in Brazil (see Figure 5). The situation could actually turn out to be better for the consumer than this calculation suggests, if the utility were to purchase the bulbs at a bulk rate below the retail price and pass on some of these savings to the consumer (see Figure 5).

Utility involvement in investments on the "customer's side of the meter" is not a new idea. Thomas Edison's original concept of the utility was to sell, not electricity, but lighting. Moreover, private utilities in Brazil were involved with consumer investments some 50 years ago, when the American Foreign Power Company and Light and Power Company Ltd. "loaned" or simply gave away electric cooking stoves as a strategy to promote electricity markets. And today, some electric and gas utilities in the US and other industrialized market economies are beginning to promote consumer investments in energy efficiency improvements.

But institutional changes are usually needed to enable energy companies to engage in these activities. Today these companies in many developing countries are sectorially organized as electricity, oil, or coal producers or vendors. In some cases "energy companies," such as the Companhia Energetica de Sao Paulo (CESP) in Sao Paulo, Brazil (of which one of the authors is President), have been organized to provide a variety of energy carriers. Further institutional innovations are needed to give energy companies the authority to finance consumer investments in energy efficiency improvements or otherwise facilitate such investments.

CONCLUSION

The benefits resulting from investments in energy efficiency improvements in electricity-driven technologies can be large. For the individual consumer an emphasis on energy efficiency means lower energy bills. For society as a whole it means a slowing of the trend to ever more costly supply sources (see Figure 2) and a freeing up of capital resources for other needs such as sanitation, housing, and and schools. Moreover, we have argued that conservation investments will often make more sense for capital-poor developing countries than for industrialized countries, because the benefits of conservation in comparison to investments in supply expansion tend to increase with the discount rate (see Figures 3 and 4).

But to realize the cost-effective savings potential would require conversion of electric utilities from being purveyors of electricity supplies into being purveyors of the energy services (lighting, refrigeration, motive power, etc.). As energy service companies, utilities would allocate capital resources between supply expansion and end-use efficiency improvement in a manner that would minimize the cost of providing desired energy services.

In this new role of promoting energy conservation, the utility effort must be greater in dealing with the poor than with the rich, as the empirical evidence suggests that the poor are far less willing and able to invest in conservation than the rich (see Table I).

It will not be easy for utilities to change their role in these ways. But the enormous benefits that would result from such conversions would make the needed efforts worthwhile.

NOTES AND REFERENCES

1. T.B. Johansson and R.H. Williams, "An End-Use Oriented Energy Strategy for Industrialized Countries," PU/CEES Report No. 185, Center for Energy and Environmental Studies, Princeton University, Princeton, New Jersey, February 1985.
2. Howard S. Geller, "The Potential for Electricity Conservation in Brazil," Companhia Energetica de Sao Paulo, Sao Paulo, Brazil, April 1985.
3. The Association of Home Appliance Manufacturers, Chicago, Ill.
4. Personal communication from Howard S. Geller, March 1985.
5. Jose Goldemberg, Thomas B. Johanssson, Amulya K.N. Reddy, and Robert H. Williams, "An End-Use Oriented Global Energy Strategy," _Annual Review of Energy_, vol. 10, pp. 613-688, 1985.
6. R.H. Williams, G.S. Dutt, and H.S. Geller, "Future Energy Savings in US Housing," _Annual Review of Energy_, vol. 8, pp. 269-332, 1983.
7. US Department of Energy, "Consumer Products Efficiency Standards Engineering Analysis Document," DOE/CE-0030, March 1982.
8. A.K. Meier and J. Whittier, "Consumer Discount Rates Implied by Purchases of Energy-Efficient Refrigerators," _Energy, the International Journal_, vol. 8, No. 12, pp. 957-962, 1983.
9. J.R. Hausman, "Individual Discount Rates and the Purchase and Utilization of Energy-Using Durables," _The Bell Journal of Economics_, vol. 10 (Spring), pp. 33-54, 1979.
10. Harry Chernoff, "Individual Purchase Criteria for Energy-Related Durables: the Misuse of Life Cycle Cost," _The Energy Journal_, vol. 4, No. 4, pp. 81-86, 1983.
11. D. Gately, "Individual Discount Rates and the Purchase and Utilization of Energy-Using Durables: Comment," _The Bell Journal of Economics_, vol. 11 (Spring), 373-374, 1980.
12. H. Ruderman, M.D. Levine, and J.E. McMahon, "Energy Efficiency Choice in the Purchase of Residential Appliances," _Proceedings from the Panel on Perspectives on Individual Behavior (Volume F), in Doing Better: Setting an Agenda for the Second Decade, the ACEEE 1984 Summer Study on Energy Efficiency in Buildings_, American Council for an Energy-Efficient Economy, Washington DC, August 1984.
13. The World Bank, _The Energy Transition in Developing Countries_, Washington, DC, 1983.

14. In Brazil new single door refrigerators (which dominate sales) consume essentially baseload electricity at an average rate of about 75 Watts, last about 15 years, and cost about $200. Thus 3 1/3 units would have to be purchased during the 50 year expected life of a new hydro-electric facility. The present value of such purchases would be:

$$\$200 \times [1 + (1 + i)^{-15} + (1 + i)^{-30} + (1 + i)^{-45} - 0.667/(1 + i)^{50}],$$

where i is the discount rate. For a 10% discount rate the

present value of all these purchases is $230.

Following the procedure outlined in note 17 it is found that 2.25 kW of installed new hydro-electric capacity costing $3375 is required (for a 10% discount rate) for each extra kW of residential baseload electrical demand that would be needed in Brazil in the 1990s. An additional $750 would be required for the associated 50 year supply of transmission equipment. Thus the total cost to the utility that can be associated with the purchase of a new refrigerator is between

$$0.075 \times \$3375 = \$250$$

and

$$0.075 \times (\$3375 + \$750) = \$310.$$

15. A typical 40 Watt incandescent lightbulb costs $0.50 and, when used in the early evening, contributes to the peak demand for electrcity in the Brazilian context. Assuming such bulbs are used 5 hours a day and last 1000 hours, some 91 bulbs would have to be purchased during the 50 year expected life of a new hydro-electric facility. The present value of such purchases would be:

$$\$0.50 \times \{1 + \sum_{n=1}^{90} 1 \times (1 + i)^{[-1000n/(365\times5)]}\}.$$

For a 10% discount rate the present value of all these purchases is $9.7.

Following the procedure outlined in note 18 it is found that 1.45 kW of installed new hydro-electric capacity costing $2175 is required (for a 10% discount rate) for each extra kW of residential baseload electrical demand that would be needed in Brazil in the 1990s. An additional $750 would be required for the associated 50 year supply of transmission equipment. Thus the total cost to the utility that can be associated with the purchase of a new 40 Watt light bulb is between

$$0.040 \times \$2175 = \$87$$

and

$$0.040 \times (\$2175 + \$750) = \$117.$$

16. R.H. Williams, "Potential Roles for Bio-Energy in an Energy-Efficient World," PU/CEES Report No. 183, Center for Energy and Environmental Studies, Princeton University, Princeton, NJ 08544, February 1985

17. In Figure 3 it is assumed that the "overnight" construction cost of the hydro-electric facility is $1170 per kW of installed capacity, that the plant is constructed over a 6 year period, and that the plant is paid for with 6 equal payments over this construction period.

To provide 1 kW of firm baseload demand requires 1.8 kW of installed hydro-electric capacity. The total installed capacity required to provide 1 kW of residential baseload demand is 2.25

kW, when allowance is made for 20% transmission and distribution (T&D) losses.

It is assumed that transmission facilities lasting 30 years cost $710 per kW. Some of this transmission cost could be avoided if new generating capacity could be deferred via investments in more efficient refrigerators.

It is assumed that efficient refrigerators costing $100 more than conventional refrigerators and lasting 19 years save 460 kWh per year, corresponding to a savings rate of 0.0525 kW.

18. In Figure 4 it is assumed that the "overnight" construction cost of the hydro-electric facility is $1170 per kW of installed capacity, that the plant is constructed over a 6 year period, and that the plant is paid for with 6 equal payments over this construction period.

To provide 1 kW of peaking demand requires 1.16 kW of installed hydroelectric capacity, to allow for a 16% reserve margin. The total installed capacity required to provide 1 kW of peak residential demand is 1.45 kW, when allowance is made for 20% T&D losses.

It is assumed that transmission facilities lasting 30 years cost $710 per kW. Some of this transmission cost could be avoided if new generating capacity could be deferred via investments in more efficient lightbulbs.

It is assumed that 13 Watt compact fluorescent lightbulbs lasting 6000 hours and costing $9.20 each replace 40 Watt incandescent bulbs lasting 1000 hours and costing $0.50 each. It is assumed that the lightbulbs are used 5 hours per day, including the early evening hours, so that they contribute to the peak electricity demand.

19. The lightbulbs compared in Figure 5 are (i) a Brazilian 40 Watt incandescent providing 480 lumens of light, costing $0.50, and having an estimated life of 1000 hours, and (ii) a 13 Watt Japanese compact fluorescent (which can be screwed into an ordinary incandescent socket) providing 500 lumens of light, and having an estimated life of 6000 hours.

The lightbulbs are assumed to be used 5 hours a day and paid for "on a utility installment plan," in which the customer pays for the bulbs via a loan @ 10% interest and loan term equal to the life of the compact fluorescent bulb (3.3 years). During this period 6 incandescent bulbs would be needed.

Compact fluorescent bulb prices 10 and 20 percent below the retail price of $9.20 are considered here as alternatives to the base case. Such lower prices might arise if the utility gets a volume discount for a large purchase from the lightbulb manufacturer.

CHAPTER 4 ENGINEERING/ECONOMIC END-USE ENERGY MODELS

Daniel M. Hamblin and Teresa A. Vineyard
Oak Ridge National Laboratory
Oak Ridge, Tennessee 37830

SECTION I. INTRODUCTION: THE BROAD PERSPECTIVE OF END-USE MODELING

Engineering/economic end-use energy models are forecasting devices for predicting energy use and policy impacts over a 20- to 30-year period. The models are capable of highly articulated structure which permits energy and policy impacts prediction by building type, energy service equipment type, fuel type, household income stratum, etc. The energy forecast outcome of end-use models is an accounting composite of predictions from the building blocks comprising the model structures. Equation 1 is an example of the fundamental energy forecasting equation for a residential end-use model:

$$\text{Energy Consumption} = \begin{pmatrix}\text{stock and}\\ \text{capacity of}\\ \text{equipment}\end{pmatrix} \cdot \begin{pmatrix}\text{efficiency of}\\ \text{equipment (and}\\ \text{of shell for space}\\ \text{conditioning)}\end{pmatrix} \cdot \begin{pmatrix}\text{intensity}\\ \text{of}\\ \text{utilization}\end{pmatrix} \cdot \begin{pmatrix}\text{base period}\\ \text{equipment}\\ \text{energy use}\end{pmatrix}$$

$$\text{from} \quad \begin{bmatrix}\text{building stock}\\ \text{and fuel-and-}\\ \text{equipment choice}\end{bmatrix} \quad \begin{bmatrix}\text{technology}\\ \text{choice}\end{bmatrix} \quad [\text{usage}] \qquad (1)$$

or equivalently

$$Q_t^{ikl} = \left(HT_t^l \cdot HS_t^l \cdot C_t^{ikl}\right)\left(TI_t^{ikl} \cdot \bar{h}_t^{ikl}\right)\left(U_t^{ikl}\right) \cdot EU^{ikl}$$

where the building stock and fuel-and-equipment choice components are

HT	number of houses
HS	average house size (e.g., sq. ft. floor space)
C	fuel-and-equipment saturation market share;

the technology choice components are,

TI	building stock average thermal performance (space conditioning end uses only)
$\bar{h}$	average equipment efficiency relative to base period energy use performance--which bears an inverse relationship to conventional efficiency ratings that indicate output per unit of input;

0094-243X/85/1350052-29$3.00 Copyright 1985 American Institute of Physics

the usage component is,

U | short-run usage intensity;

and the base value for adjustment in energy consumption is,

EU | base period equipment energy use (e.g., Btu/year)

for which subscripts and superscripts are

t | year of interest

i | fuel type

k | end use

l | building type.

The input of the physics profession in the practice of end-use modeling has been in three areas primarily. First, fundamental physics relationships or laws underlie the characterization of technical efficiency. For example, in the case of space conditioning, a numerical estimate of a particular period's new house joint technical and economically efficient solution for shell thermal performance is included in the TI in Equation (1) (representing total stock); and likewise, a jointly efficient solution for new house space heating (or cooling) equipment is included in $\overline{h}$. The second input to the practice of end-use modeling has been the active role which professional physicists have played in the design of advanced energy-efficient technologies. The third role has been in defining resource conservation policy strategies.

The objective of our paper is to provide a glimpse of the current technology of engineering/economic end-use energy models. In providing this glimpse, we have found it desirable to articulate three subsidiary objectives. The first objective, served by this Introduction, is to describe end-use modeling within the broader context of an analytical framework capable of producing statistically sound and valid forecasts. The second objective, supported by following Sections III and VI, is to highlight those aspects of the end-use modeling problem which are associated with technology and technology characterization. The third objective, also served by Section VI, is to describe results of a policy application. We hope that the latter two objectives provide insights to the physics community concerning how and how well their inputs to the end-use modeling problem are employed.

Better end-use models are justifiably more complex because they bring more information to the decision process. However, there can be and has been a significant lag between development and application of technique, and development and application of data to support technique. End-use models have typically been described in terms of the selling characteristics of the "internal structure." This code definition and the associated typical forecast documentation of input data, code, and output are overly restrictive, and overly tolerant of the use of anecdotal performance evidence and internal calibration to replicate historical consumption patterns. The appropriate description of end-use modeling includes information development within the definition of the end-use modeling construct. We believe that this is an important point of departure for a forum on physics and society, because consideration of the broad theoretical and analytical underpinning requires concomitant consideration of fundamental concepts and modeling paradigms originating from various disciplines--Statistics, Physics, Engineering, Economics, and Operations Research.

Figure 1 is a flow schematic of the end-use model analytical framework, broadly considered. The analytical framework originates in a commmon ground of design of statistical experiments, and terminates in the end-use model and its output. In between are parallel tracks of data development. The left track deals with the physics and engineering of technical change. Its objective is to produce--for each building, end use, equipment type and fuel--a locus of technically efficient combinations of capital and energy capable of providing a given (and constant or iso-) level of energy service amenity (e.g., 72°F ambient space heat). In practice, it is exceedingly rare for the technical process analysis to be driven by a formal experimental design. Figure 2 illustrates a sample outcome of this analysis for the case of natural gas space heating equipment in the residential sector. A locus of technically efficient combinations referred to as an isoamenity curve or isoquant (equal quantity) is a graphical and numerical construct used by economists to relate technical efficiency to economic efficiency. In

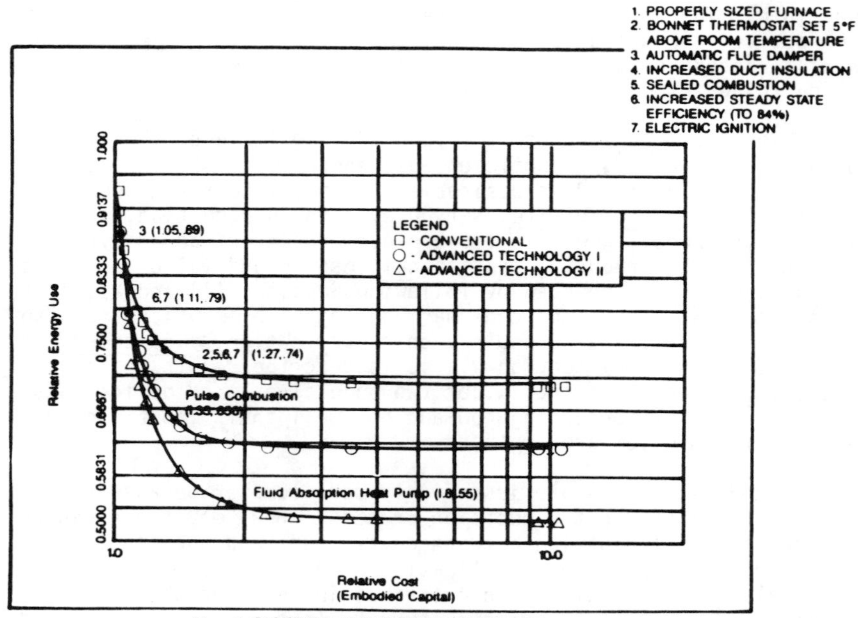

Fig. 2 GAS SPACE HEATING ISOAMENITY CURVES
Baseline Unit Energy Consumption = 84.9 × 10^6 Btu/yr
Cost = $1310 (1975-$)

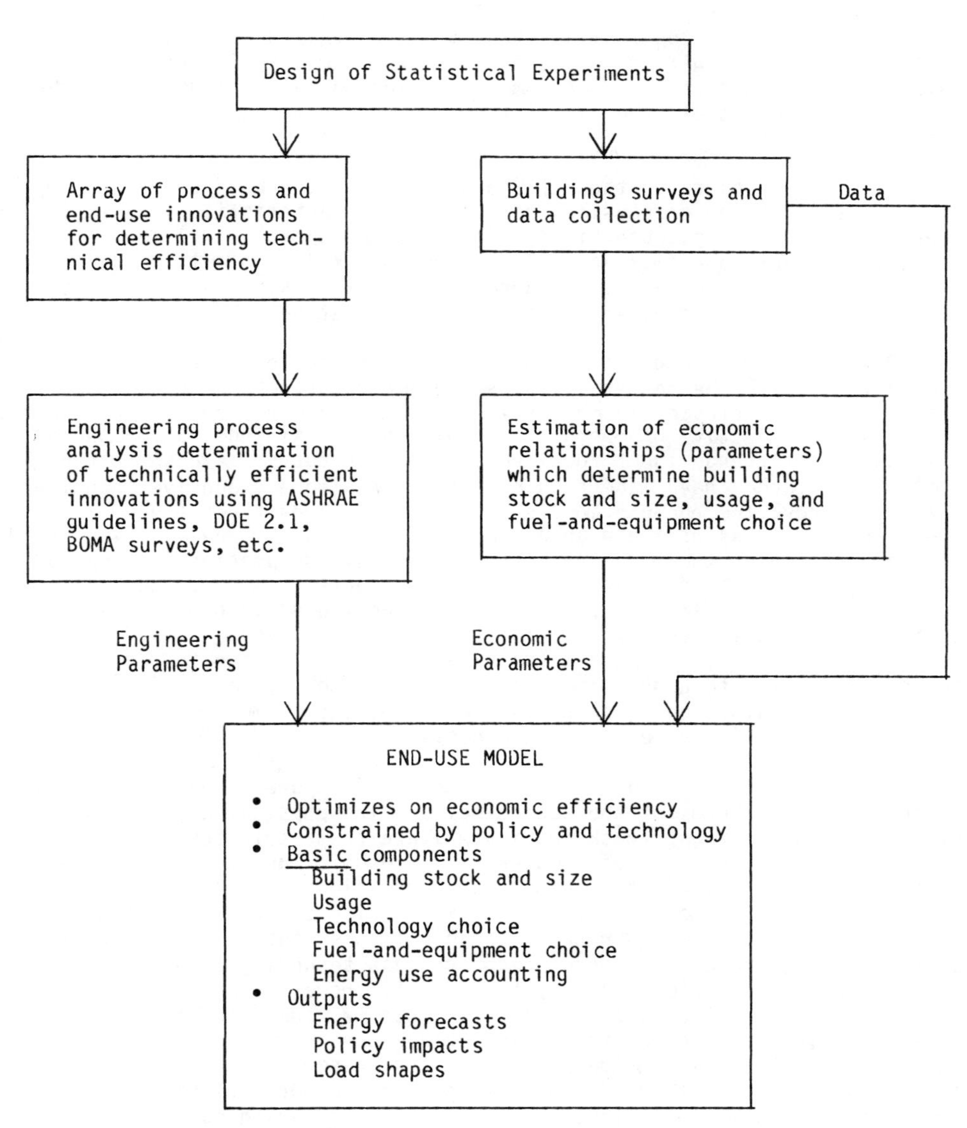

Figure 1. End-Use Model Analytical Framework

this instance, constant or equal quantity entails provision of the same ambient temperature by each option. A technically efficient option has the lowest real cost for a given quantity of energy consumption (required for isoamenity output), or the lowest amount of energy consumed for a given real cost. A companion isoamenity curve depicts technically efficient shell thermal performance options (e.g., wall insulation).

The topmost block on the left-hand technical track is concerned with the translation of physical performance characteristics into engineering criteria (e.g., Coefficient of Performance or C.O.P.) for measuring technical efficiency. The second block is a modeling subconstruct in which energy performance and capital costs are jointly considered in defining frontiers of technical innovations.

Physicists make a distinction between "1st Law" efficiency and "2nd Law" efficiency. The latter embodies the important notion that a Btu of hot energy is intrinsically more valuable than a Btu of cool energy; hence, a derivative objective of energy conservation in buildings is to lower the quality of energy input required for amenity output. To the extent that market forces have made high quality energy more expensive than low quality energy, the end-use models will depict a "positive" economic preference for less of the high and more of the low. However, given the central purpose of the models--to predict energy consumption (or energy "sales"), technical efficiency is defined in terms of minimum cost of delivered amenity. If, as is the case with active solar heating systems, low quality energy systems (solar) are dominated by (higher cost than) high quality systems (fossil-fueled), it is the "normative" purpose of policy to address market failures which may be implicit in this relative valuation. The policy responses may be modeled as constraints on private technical choices.

It might also be said of the left track of Figure 1 that it is concerned with attributes of the technology and its physical environment. Of concern in the right track in Figure 1 are the attributes of the decisionmaker and his or her demographic, economic, and social environment. The right track sets the boundary conditions for the end-use model determination of economic efficiency. The notion of economic efficiency has to do with obtaining a rate of technical substitution of capital for energy (in the choice of an end-use or process technology) which is equal to the real price ratio of energy to capital. However, because capital is durable and provides services over a lifetime, the price relationship between capital and energy must also include consideration of the decisionmaker's marginal rate of time preference for present consumption over future consumption. This rate of time preference, or implicit discount rate, is an economic efficiency boundary condition which may be different from (favor more present consumption than) a socially optimum rate of time preference. The higher the implicit discount rate, the less valuable is the decisionmaker's perceived benefit from energy conservation capital investments. Determining economic parameters also properly emanates from a statistical design construct for conducting sector-specific commercial and residential surveys. These surveys and associated consumption history and price data provide baseline characteristics directly to the end-use energy model. Moreover, they are employed for estimating cause and effect in economic behavior. The estimates take the form of economic parameters which may determine building stock and size, intensity of utilization, and the relative attractiveness of different fuel-and-equipment combinations available for producing a common energy service amenity.

We now enter the end-use model with engineering parameters from the left track, and economic parameters and baseline data from the right track. The end-use model looks to the future. In doing so, it combines the vision of technological change embodied in the left track process analysis determination of technical efficiency with information on economically efficient choice contained in the right track economic parameters. The end-use model pursues a dynamic simulation of behavior by optimizing on economic efficiency constrained by social policy and technological possibilities. This means that as energy becomes scarce relative to capital, the end-use model depicts resource conservation by substituting technology capital for energy, whenever the present discounted cost of owning and operating energy service equipment and processes can be reduced by doing so. The optimizing simulation output typically includes detailed energy forecasts, conservation policy impacts, and not so typically, electricity load profiles.

This is the Gestalt of engineering/economic end-use energy modeling. In following sections, we shall discuss components and sub-components of the Gestalt. We preface our attention to the building blocks with a Section II which provides a cursory overview of the governing concepts and purpose by discipline associated with accomplishing the end-use modeling objectives. Section III is a methodological exposition of engineering parameters development. Section IV is a parallel methodological exposition of economic parameter development. Section V provides a cursory conceptual structural overview of the end-use model. Section VI focuses upon the technology choice sub-component of the end-use model and the treatment of policy therein. The paper closes with concluding remarks and suggestions for further development.

SECTION II. THE INTERDISCIPLINARY NATURE OF END-USE MODELS

The objective of this Section is to disclose the interdisciplinary nature of the underlying tools and concepts which support the end-use analytical framework. Figure 3 is a "Family Tree" of end-use modeling--which begins with discipline and objectives, and proceeds through tools and concepts either directly to the end-use model analytical framework, or to another discipline and objective.

The statistical objective is the faithful replication of population characteristics, both with respect to characteristics and attributes of decisionmakers, and with respect to the cost and performance of energy service technologies. The statistician's tools are experimental design

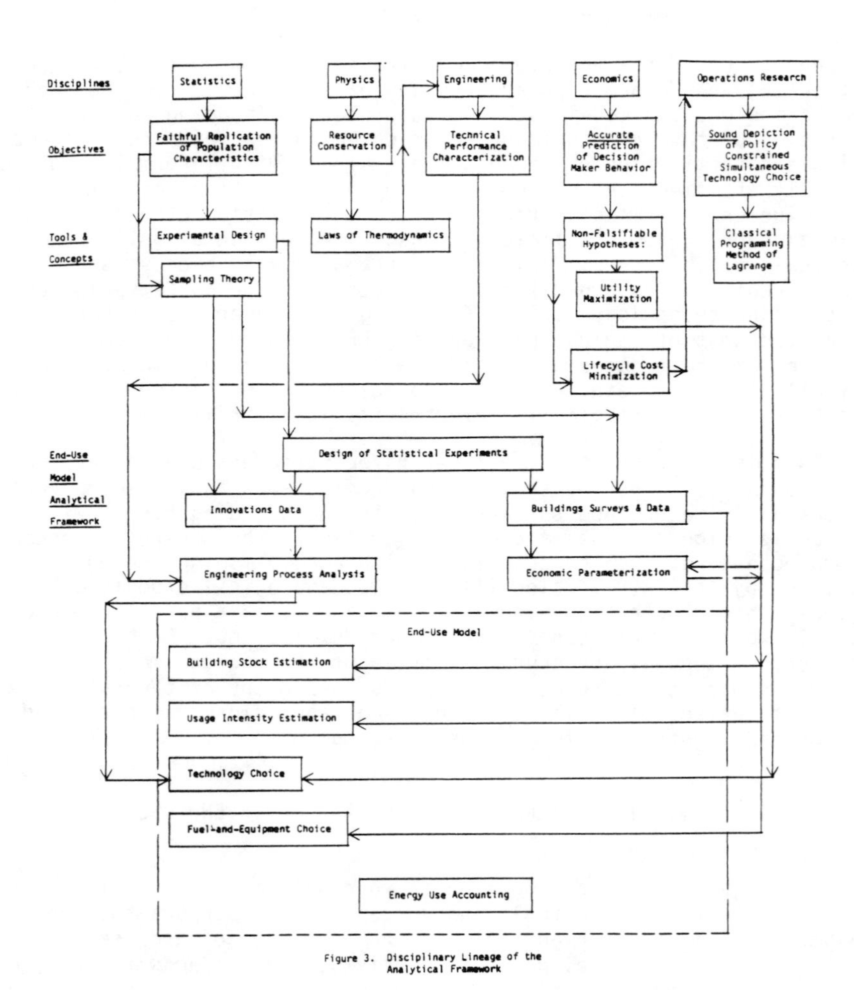

Figure 3. Disciplinary Lineage of the Analytical Framework

and sampling theory. A snapshot of how these tools should be exercised is illustrated in what we choose to call the "building performance analytical approach" for sound determination of engineering parameters. This approach embodies six steps:

1. Define building systems which interact physically and/or economically. For example, commercial building systems which interact are the thermal envelope, the heating equipment, the distribution system, the ventilation fan motors, the cooling equipment, and the lighting equipment.
2. Identify data sources from which building performance information can be developed. These data sources include surveys, audits, load simulation analyses, conservation program evaluations, and monitored buildings.
3. Develop "robust" methods for producing (as required) end-use energy use characteristics from aggregate building energy use characteristics.
4. Design a statistical experiment which includes a sampling construct in order to utilize data sources from number 2 above to develop system-specific incremental performance data for the number 1 processes and end-uses.
5. Develop capital costs and durability data for all options identified in step 4.
6. Combine steps 4 and 5 to determine, for each system delineated under step 1, the "frontier" of technically efficient options.

The success of this approach is fundamentally linked to the soundness of the statistical method employed in steps 3 and 4. In actual practice, the poor "unadulterated" performance of most end-use models can be attributed to the fact that physics and engineering have been targeted to the production of anecdotal evidence in lieu of attempting the more comprehensive task of producing statistically sound evidence.

Our statement of the Physics objective is resource conservation. That is, the laws of thermodynamics lay out the physical parametrics governing the substitution of less scarce for scarce resources. The governing rules of resource conservation are the first and second laws of thermodynamics.

The thermodynamic tool kit is made operational through Engineering analysis. The objective of this analysis is to characterize the technical performance of energy service processes and equipment in such a way that technical efficiency can be usefully described for forecasting.

The Economics objective for end-use modeling is accurate prediction of societal decisions with respect to technology and fuel choice, and intensity of use. These predictions have typically been based upon two underlying non-falsifiable hypotheses--that decisionmakers maximize the utility or satisfaction obtained from the provision of energy service amenities, or that decisionmakers minimize the life-cycle cost of producing a given level of energy service amenity. Figure 3 illustrates the linkage of utility maximization to economic parameterization, building stock estimation, usage intensity estimation, and fuel-and-equipment choice. This occurs because utility maximization is a useful (but not necessary[1]) construct for fabricating demand curves. The simple hypothesis that consumers maximize the utility provided by the negative of discomfort and a composite other good, subject to expenditure constraints, has been used to underpin the residential sector development of models of fuel-and-equipment choice and household electricity usage.[2] On the other hand, life-cycle cost minimization is the driving hypothesis underlying the determination of economic efficiency, given technical efficiency along the isoamenity curve. The nonfalsifiability of economic hypotheses means that whether decisionmakers are economically efficient actors (in pursuit of private optima or social optima) cannot be determined from the end-use modeling construct. Rather, the quality of the prediction is indicative of the strength of these hypotheses for explaining behavior.

Life-cycle cost minimization calls upon the Operations Research discipline for accomplishing the sound depiction of policy constrained simultaneous and sequential technology choice analysis. A builder confronts a simultaneous choice of process and end-uses--for the provision of space heating and cooling amenity--which differs from individual equipment replacement choices confronted by the building occupant or manager. The Operations Research tool employed for constrained life-cycle cost minimization is that of classical programming utilizing the Method of Lagrange. The Lagrange technique provides shadow prices which are economic valuations of the difference between private market unconstrained optima in the choice of technologies and programmatically determined social optima.

SECTION III. DEVELOPING ENGINEERING PARAMETERS

The physical characteristics of a building structure and the primary energy-consuming equipment contained therein must be transformed into measurable performance factors or indices which serve as inputs to the end-use energy model. These inputs are in the form of parameters defining technology curves for each end-use and for each type of building structure. A technology curve or isoamenity curve is the convex (to the origin) envelope of technical options, located with coordinates in which energy use intensity determines the ordinate and relative (or incremental) real capital cost determines the abscissa. Energy use intensity is the ratio of energy required by a technical option to produce a given amenity level and energy required by a base-period technology to produce the same amenity. The residential gas space heating curves, depicted in Figure 2, are examples of technology curves. At Oak Ridge National Laboratory (ORNL), the following three-parameter formulas are used to fit technical options to curves:

$$h_{j,k} = a_{j,k} + \left(\frac{b_{j,k} + 1}{b_{j,k} + C_{j,k}} \right)^{\alpha_{j,k}} \left(1 - a_{j,k} \right) \qquad (2)$$

$$T_{j,k} = ta_{j,k} + \left(\frac{tb_{j,k}}{tb_{j,k} + C_{sh}}\right)^{t\alpha_{j,k}} \left(1-ta_{j,k}\right) \quad (3)$$

where

$a_{j,k}$ or $ta_{j,k}$	the high efficiency (lowest h- or T- value) asymptote parameter representing lowest attainable relative per unit fuel-j-in-end-use-k or process energy use, given the modeled technical options
$b_{j,k}, \alpha_{j,k}$ or $tb_{j,k}, t\alpha_{j,k}$	shape parameters of the curve
$h_{j,k}$ or $T_{j,k}$	the coordinate for efficiency
$C_{j,k}$ or C_{sh}	the coordinate for capital cost

The formula (2) depiction of a technology curve for end-use equipment includes capital cost as an index relative to a base year value. The formula (3) depiction of a technology curve for a process such as shell thermal performance or distribution system performance (in the commercial sector) includes real incremental capital costs associated with a process improvement. It should be noted that the left-hand-side efficiency values are not end-use model solutions. These solutions are the result of taking derivatives of (optimizing) life-cycle-cost equations which have the curve formulas explicitly embedded in them. Solving these "first-order conditions" produces efficiency points (or sets of points) which satisfy the above formula(s), and hence, lie on the technology curves.

The energy performance of heating equipment is typically described in terms of its efficiency. However, there are several acceptable efficiency definitions used by the furnace industry--steady-state efficiency, utilization efficiency, and annual fuel utilization efficiency. Each is discussed briefly below.

1. Steady-State Efficiency. Traditional American National Standards Institute (ANSI) definition based on fuel plus electrical input; assumes jacket losses are useful heat.
2. Utilization Efficiency. Based on fuel input only; accounts for cyclic effects, assumed combustion-induced infiltration, and seasonal pilot input.
3. Annual Fuel Utilization Efficiency (AFUE). Same as item 2, but includes annual pilot input.

All of the above factors are limited to describing the performance of the furnace rather than the entire heating system since none of the factors account for system effects, such as the interaction of the furnace with the heated structure, venting and distribution system losses, and the effect of controls. However, Federal law has dictated that manufacturers of furnaces use AFUE to rate efficiency. For any model furnace, the AFUE rating can be found at the point of sale on Fact Sheets required by the Federal Trade Commission. Similar energy efficiency ratings based on test procedures prescribed by the Department of Energy (DOE) are required for other appliances such as air conditioners and refrigerators.

Other sources which have been used to develop equipment technology curves include load simulation models such as the DOE2.1 model. DOE2.1 calculates the heating, cooling, and water heating loads for buildings. Therefore, DOE2.1 can be used to calculate the loads for various architectural, mechanical, and architectural/mechanical options to a baseline or reference house. Another source of data for developing technology curves for equipment is actual consumption data. These data are available only if a house has been submetered to determine the consumption of individual equipment components in the house. Due to the equipment expense required to submeter a house, field studies provide little data of this type. Consequently, the AFUE and other energy efficiency ratings required by the Federal Trade Commission provide the most available and consistent source of rating the performance of different equipment.

Often a building standard is specified in terms of maximum permissible overall thermal transmittance; i.e. U values for walls, roof/ceiling assemblies, and floors. Determining compliance with the standard requires the translation of the performance standard into prescriptive measures. For example, the maximum allowable U value for the roof/ceiling must be translated into the minimum amount of attic insulation needed to meet the standard. The building structure is then characterized in a load simulation model such DOE2.1 to determine process coordinates for the structural thermal performance/cost tradeoff curves used in the end-use model. The cost of structural modifications is often best estimated by an architectural/engineering firm having expertise in commercial and residential construction costing.

Technologies which are under development and therefore not available on the market must also be represented on the technology curves. The end-use model has the capability to substitute an advanced technology curve for an existing curve in a specified future year. The advanced technology curve includes new technology with the existing equipment options. Information about the performance and cost of advanced technologies are estimates obtained from the principal investigators of the research work. Because private manufacturers of equipment are very reluctant to discuss future products, the performance and cost of advanced technologies represented in the end-use energy models are usually derived from research activities being conducted at DOE National Laboratories. Examples of advanced technologies include daylighting techniques, glass innovations which enhance light transmittance and thermal insulation properties, Stirling-engine-driven heat pumps, high efficiency fluid absorption heat pumps, refrigeration systems with nonazeotropic refrigerant mixtures, and internal combustion engine-driven heat pumps.

SECTION IV. DEVELOPING ECONOMIC PARAMETERS

The "econometric" technique for estimating economic parameters is a way of depicting cause and effect among economic and demographic variables. Once the cause-and-effect equations are estimated, they are employed (along with predicted values of exogenous-to-the-model explanatory variables such as household income, energy prices, etc.) in end-use model simulation of future energy consumption. Economic relationships requiring estimation are developed for building stock, usage intensity, and fuel-and-equipment choice.

Residential and commercial end-use models employ an area measure of the breadth or extensity of energy use in buildings. In general, the objective of floor space estimation in the commercial sector is to predict the number of square feet of floor space served for each building type (e.g., small office buildings); whereas the objective of floor space estimation in the residential sector is to predict the number of households by building type (e.g., multifamily) and average residence size in square feet. The absence of building size information in predicting cause and effect in the choice of commercial process and equipment efficiency and fuel has been a significant impediment to reasonable accuracy.

In the commercial sector, building stock estimation equations have tended to be simple and _ad hoc_. The two approaches followed are to either estimate an investment demand equation for new floor space plus depreciation, or to estimate a gross stock equation. Explanatory variables commonly employed in estimation are population and per capita income, or population and commercial-building-type-subsector employment.

Residential building stock parametrics typically emanate from a more sophisticated estimation endeavor. The approach followed at ORNL was to estimate parameters for a primarily demographic age-cohort housing submodel. Parameters were estimated for equations which determine headship rates and consequent number of households, net household formation, number of housing units retired, and number of housing units constructed. The demographic and economic determinants include population, income, age-group specific marital status, average price per square foot of new homes, average number of people per household, average dwelling size, and region-specific dummy variables. The approach which has enjoyed more recent acceptance is to estimate building stock parameters as part of a demographic submodel attached to a national or regional economic model. While the determinants of household formation may not differ significantly between the two approaches, the advantage of the latter is that building stock estimation parameters are produced within a logically consistent framework for predicting exogenous variables (e.g., income) employed in end-use model forecasting.

Within an econometric estimation framework, fuel-and-equipment choice and usage intensity in the residential and commercial sectors have been treated as a joint modeling problem. Usage intensity refers to the short-run decisionmaker response to changes in economic variables such as the price per unit of delivered amenity and household income. Both fuel choice and usage elasticities in the commercial sector have been estimated from a multinomial logit specification originally developed by Baughman and Joskow.[3] The Baughman and Joskow framework permits estimates of market shares of the three major fuels used for commercial sector space heating on the basis of fuel prices, climatic variables, and natural gas availability. The fuel share equations are linked to a third flow adjustment equation which estimates total energy consumption as a function of an energy price index, the consumer price index, income, and other variables. The linkage established is that weights for the energy price index are the fuel shares given in the fuel-choice model. From this framework, fuel-specific short-run own-price elasticities may be computed. These estimated elasticities are utilized to simulate the short-run utilization response in the commercial end-use models.

In commercial models which employ an econometric fuel-choice component, the Baughman-Joskow formulation provides estimates of choice coefficients. The synthetic fuel-choice coefficients are computed to represent the long run effects of fuel price upon adjustments in capital stock, net of the short run usage intensity effect. However, the Baughman-Joskow framework has not worked well for fuel choice. The estimated fuel-choice coefficients tend to produce simulations of new investment decisions quite unresponsive to changes in relative prices. One problem with the use of the Baughman-Joskow for determining market penetrations of commercial space heating is that the interpretation of the specification has been incorrect. It has been argued that one could derive the relative price coefficient for space heating system choice by subtracting the short-run utilization impact coefficient from the long-run capital-stock-adjustment-and-utilization impact coefficient.[4] What this imputation actually provides is the coefficient for predicting the change in equipment <u>stock</u> saturation, given relative price changes. One advantage of econometric estimation techniques is that sluggishness in the occurrence of effect given causal factors can be depicted by the inclusion of a constant term, the inclusion of so-called choice-specific dummy variables, and the magnitude of explanatory variables coefficients. The commercial sector alternative to the Baughman-Joskow mechanism for fuel choice has been a "no econometrics" micro-simulation technique which projects fuel choice and fuel switching on the basis of energy service optimal choice set life-cycle cost solutions. This procedure has tended to produce fuel shares much too responsive to relative price changes. A significant impediment to sound estimation of fuel-choice decisions in the commercial sector has been the absence of survey data on attributes of commercial building decisionmakers.

In the residential sector, a joint structure for estimating the fuel choice and usage has been developed by Dubin and McFadden.[5] Choice parameters for fuel-and-equipment selection are estimated in a nested logit qualitative discrete choice framework in which choices for some end uses may be conditioned or influenced by choices for other end uses. For example, in the Pacific Northwest version of the ORNL Residential Reference House Energy Demand (RRHED) Model, the parameter estimates for water heating were conditioned by the space heating system in the house, and the parameter estimates for cooling and clothes drying were conditioned by the space heating fuel and gas availability to the house.[6] The probabilistic choice parameters are estimated by limited information maximum likelihood, and the space heating system choice equation includes an inclusive value instrumental variable parameter, for which solution on the unit interval indicates logical consistency with the hypothesis of utility maximization. The error structure of the representative nested logit equation takes the form of the extreme value distribution. Equation (4) below depicts the probability of space heating system choice involving fuel i, and equipment type im.

$$P(i,im) = \frac{\exp(U_{i,im} + \theta \cdot I_{(i,im)})}{\sum_{j=1}^{3} \sum_{im=1}^{IM} \exp(U_{j,im} + \theta \cdot I_{(j,im)})} \tag{4}$$

where,

$U_{i,im}$ — space heating utility for heating system choice fuel i (electric, natural gas, or heating oil), type im (central forced air, non-central, hydronic, or heat pump).

θ — the inclusive value coefficient

$I_{(i,im)}$ — the inclusive value instrumental explanatory variable determined from the utility or satisfaction of water heating choices

$$IM = \begin{cases} 4 & \text{for } i = 1 \text{ electric space heat denoting choice among central forced air, non-central, hydronic, and heat pump} \\ 3 & \text{for non-electric fuels } i \text{ denoting choice among central forced air, non-central, and hydronic} \end{cases}$$

Equation (4) interpreted in terms of its physics analogue is the probability that the utilty or satisfaction from occupying state (i,im) is at least as great as the utility or satisfaction from occupying other states represented by alternative heating system choices.

The embedded equation for space heating utility ($U_{i,im}$) may take many forms. Equation (5)is a representative specification which was estimated for the Pacific Northwest.[7]

$$U_{i,im} = \beta_0 \cdot OC + \beta_1 \cdot (OC \cdot Y) + \beta_2 \cdot C + \beta_3 \cdot HP + \beta_4 \cdot GFA + \beta_5 \cdot GNC + \beta_6 \cdot ONC \tag{5}$$

where additionally

$\beta_0, \beta_1, \beta_2, \beta_3, \beta_4, \beta_5$, and β_6 — coefficients to be estimated

OC — operating cost

C capital cost

Y household income

and for capturing inertial effects and intrinsic effects...

HP $\begin{cases} 1 & \text{simulated utility is for heat pump} \\ 0 & \text{otherwise} \end{cases}$

GFA $\begin{cases} 1 & \text{simulated utility is for gas forced air} \\ 0 & \text{otherwise} \end{cases}$

GNC $\begin{cases} 1 & \text{simulated utility is for gas non-central} \\ 0 & \text{otherwise} \end{cases}$

ONC $\begin{cases} 1 & \text{simulated utility is for oil non-central} \\ 0 & \text{otherwise} \end{cases}$

The household-specific choice probabilities are used as endogenous instruments in the second step estimation of short-run household usage intensity. The second step usage intensity equation typically estimates whole-house energy use as a function of the end-use appliance stock available in the house, energy price, and attributes of the household such as number of persons, income, etc. An example of the whole-house energy usage equation is as follows:

$$W = \gamma_0 + \sum_{\substack{k=1 \\ k \neq EXO}}^{K} \gamma_k \cdot CN_k + \sum_{\substack{k=1 \\ k \neq END}}^{K} \gamma_k \cdot d_k + \sum_{h=K+1}^{K+H} \gamma_h \cdot HC_{h-K} + \sum_{j=K+H+1}^{K+H+J} \gamma_j \cdot P_{j-(K+H)} + \gamma_Y \cdot Y \quad (6)$$

where,

W household-specific electricity consumption

CN first-step choice probabilities for equipment "ENDogenous" to the estimation procedure (e.g., space heating, water heating, cooking, clothes drying probabilities estimated by Berkovec, Hausman, and Rust[8] framework)

d 0,1 dummy variable indicating the absence or presence of equipment "EXOgenous" to the estimation procedure (e.g., room or central air conditioning, refrigerators, freezers, etc.)

HC — household characteristics such as dwelling type (e.g., 0 0 1 dummies indicate mobile home), number of rooms, number of occupants

P — energy price vector containing one element (average or marginal price) or two elements (marginal price and "block-rate premium") according to the rate structure characterization selected

Y — household income

k, h, j — end-use, household characteristics, and energy price component subscripts, respectively

$\gamma_0 - \gamma_{K+H+J}, \gamma_Y$ — coefficients which represent (Btu or KWh) usage associated with particular end uses, or changes in (Btu or KWh) usage associated with changes in particular demographic and economic variables

Price can also be endogenous or predetermined in the specification. For example, for electricity usage, the usage equation is first estimated on the basis of end-use/ appliance holdings and non-price characteristics, the resultant energy use is applied to the household's electricity rate schedule, and price is "endogenously" imputed from the rate schedule.[9] A third endogenous component to an electricity usage equation can be the presence of wood heat in the house. This requires the estimation of a wood-heat choice equation and the use of the choice probability as an instrument in the usage equation.[10] The estimated usage parameters may be used to impute short-run responses to price and income effects, and with some perhaps heroic imputation, to disaggregate the usage intensity response to responses across specific end-uses. However, a significant limitation of this approach has been that estimations performed to date have been limited to electricity usage. Given this shortcoming, the ORNL RRHED Model also employs fuel-specific usage elasticities from a log-linear specification developed by Eric Hirst and others.[11]

SECTION V: THE END-USE MODEL STRUCTURAL OVERVIEW

Sections III and IV described development of parameters for the end-use model. As is depicted in Figure 3 above, the Section IV economic parameters are employed directly for building stock, usage intensity, and fuel-and-equipment choice simulation. The Section III engineering parameters determine the technical alternatives available for technology choice.

The engineering/economic models are stock-adjustment in the sense that certain parameters (e.g., appliance efficiencies) of the model change with respect to the corresponding values in the base year. The primary output of the models is total energy consumption classified by fuel, building type, and end use. The models also calculate estimates of new equipment penetration, equipment efficiencies, structure thermal performance, usage factors, fuel expenditures, equipment costs, and cost for improving thermal performance of new and existing building units.

End-use models typically consist of four structural components--building stock, usage, technology choice, and fuel-and-equipment choice--and an accounting equation relating these components. As an example, Table 1 depicts that structural articulation of the ORNL Residential Reference House Energy Demand Model, and Figure 4 is a flow schematic depicting interaction of the conceptual structural components for that model. A cursory overview of component functions for the building sector end-use models is provided below with emphasis placed upon the description of the technology choice component because this is the optimizing segment of the model to be discussed in the following section.

The objective of the building stock component is to determine the amount of sector floor space. The simulation of floor space employs parameters described in the preceding section. Therefore, in the commercial sector, simulation depends upon outside the end-use model forecast inputs of employment, population, and/or per capita income--depending upon the forecasting specification chosen. In the residential sector, outside-the-model forecasts are required for population, household income, and a number of demographic/socioeconomic factors and family characteristics. The sector computation of energy consumption depicted in Equation (1) above includes the number of houses HT and average house size HS predicted by the building stock component.

In the commercial sector, usage determination employs a model of decisionmaker responses to changes in fuel prices and efficiency of the available stock of energy-using equipment. In the residential sector, usage determination employs a model of consumer responses to changes in income, fuel prices, and efficiency of the available stock of energy-using equipment and shell. In the residential sector, responses are modeled for both the short and long term. For both sectors, the short term response, constrained by previous efficiency and equipment choices, is manifested in changes in day-to-day usage patterns (i.e., changes in thermostat settings). In the residential sector, the long run response reflects the expected level of utilization in shell and equipment choice (i.e., a customer expecting high usage will choose more efficient equipment). The sector computation of energy consumption as depicted in Equation (1) includes the short-term energy usage intensity response U.

Of course, for space conditioning end-uses, significant variations in year-to-year usage patterns are attributable to annual weather variations for which long term forecasts are of questionable validity. The end-use models implicitly assume that weather patterns do not vary from year to year by predicting usage variation based upon changes in end-use operating costs which emanate from weather normalized consumption data in the base year.

Table 1. Structural Articulation of the ORNL Residential Reference House Energy Demand Model

Fuel Type:	Electricity, Gas, Oil, Other
Housing Type:	Single-Family, Multi-Family, Mobile Homes
End Use:	Space Heating, Air Conditioning, Water Heating, Cooking, Clothes Drying, Refrigeration, Food Freezing, Lighting, Other
Housing States:	New Structures Added During Forecasting Period, Existing Structures Built Prior to Base Period
Year:	Flexible Base Period - 2005

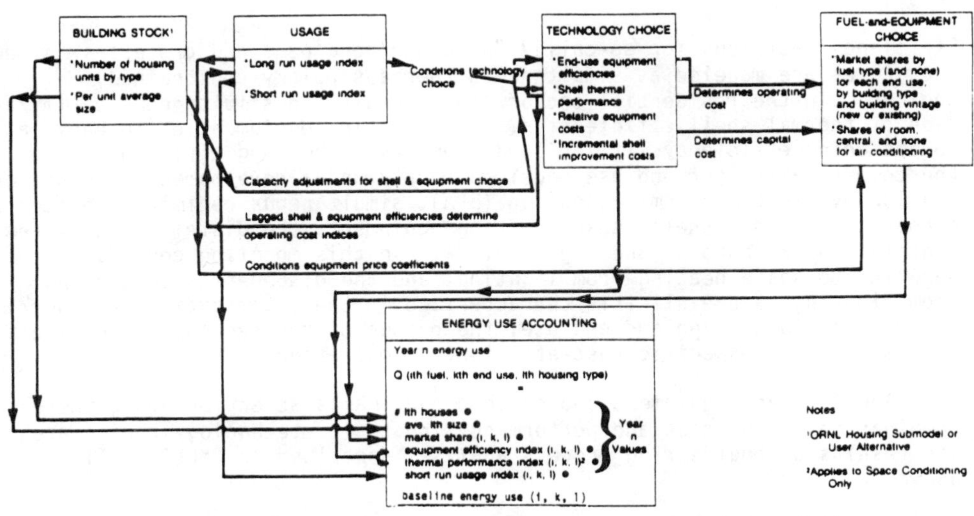

Fig. 4 Interaction of model components in the ORNL Residential Reference House Energy Demand Model

The technology choice component involves a constant dollar value optimization process for selecting appliance efficiency and thermal performance of the shell. Year-to-year solution values for efficiency and thermal performance are included in the Equation (1) $\bar{h}$ and TI, respectively. In commercial model development work now under way, the optimization process is extended to consideration of the efficiency of distribution systems. The kernel of this submodel is the "technology curve" which depicts the relationship between equipment or structure energy use or heat loss and capital costs for a given amenity level (see Figure 2 and formulas (2) and (3) above). The curves represent the technical frontier by defining options that provide the largest efficiency gains per unit increase in capital cost. The fundamental technology choice decision paradigm (depicting selection of energy service equipment and/or processes which are jointly technically and economically efficient) is that the average decisionmaker choice in a forecasting period is that point on a technology curve that minimizes the life-cycle cost (LCC) (in discounted present value) of owning and operating or inhabiting the energy-using equipment or shell. This average choice represents a weighted average of individual choices involving lower and higher equipment or shell efficiencies and respectively higher and lower individual discount rates. The LCC is given by:

LCC = initial cost + present value operating costs.

Components of this formulation are explicitly depicted in Equation (7) below.

Efficiency decisions for so-called "attached" end uses and processes in new structures are modeled as simultaneous (perhaps policy constrained) choices. In the residential sector, this results in simultaneous optimization of thermal shell efficiencies, space heating equipment efficiency, air conditioning efficiency (if included), and water heating efficiency. In the commercial-sector end-use model currently under development at ORNL for the Bonneville Power Administration (BPA), simultaneous optimization occurs across the thermal shell, space heating equipment, the distribution system, ventilation fan motors, and lighting. Within this modeling construct, the benefits to space heating from lighting, and the disbenefits to cooling from lighting, are explicitly characterized in the optimization framework. For each building type and end use, the output of the technology component is a set of fuel-specific cost-efficiency combinations.

The fuel-and-equipment choice component makes selection among these combinations. The cost and performance inputs for technology choice are the results of engineering process analyses described in Section III above.

The fuel-and-equipment choice component determines new equipment market shares. These market shares are folded into an equipment saturation C value depicted in Equation (1) above for the residential sector. Market shares are typically simulated as functions of capital and operating costs and, in the residential sector, per capita income. The functional forms for these simulations have been described in Section IV above on estimating economic parameters. The market shares estimated determine fuel selection in new structures and fuel switching in existing structures. Additionally, the discount rates implicitly associated with capital and operating cost coefficients, and, in the residential sector, a multiple of the income coefficient and time-variant-and building-type-specific household income, are utilized in the present value formulations of the technology choice component.[12]

The primary output of the end-use models--energy use by fuel, end-use, and building type--is determined from accounting equations which incorporate the outputs of the four conceptual structural components. In the above commentary, each of these outputs has been referenced to the fundamental end-use model energy accounting equation as depicted in Equation (1) for the residential sector.

SECTION VI: TECHNOLOGY CHOICE AND POLICY ANALYSIS

Technology choice is the energy/economic end-use modeling component which optimizes on economic efficiency by equating the rate of technical substitution of end-use or process capital for energy on a year-to-year basis to the ratio of the real price of energy to capital. The addition of "optimizing" to simulation provides two benefits. First, the operating cost advantages of advanced technologies can be explicitly accounted for in the choice sets available to future decisionmakers. Second, the market distorting impacts of policy which constrains decision-maker behavior may be depicted within the optimization paradigm which also underlies the estimation of fuel-and-equipment parameters input into the models.

Another significant point to make about the technology choice component is that it puts to use the technical efficiency information developed under the disciplinary domains of physics and engineering. The purpose of technology choice is to determine economically efficient processes and end uses, given portfolios of technically efficient alternatives, to make the consequent operating and capital costs available to the fuel-and-equipment choice decision, and to pass the solution efficiencies to energy use accounting for predicting the final energy consumption numbers. These flows of technology choice information are explicitly depicted in Figure 4 above. The economic efficiency solutions can represent an unconstrained private optimum, or constrained social optima resulting from energy policy such as prescriptive or aggregate performance standards.

Economists in particular confront a significant set of alternatives in determining how to specify the technology choice optimization problem. A salient objective of modeling is to simplify reality. This objective can be satisfied by tailoring the estimation process around technical alternatives embodied in the specification of an economist's "production function." The functional form which has been used for commercial sector engineering/economic end-use models is the Cobb-Douglas.[13] The Cobb-

Douglas, which has been traditionally used for aggregate (national or regional level) economic analysis, has tractable econometric estimation properties such as linearity in the logarithms. Outside the strict domain of energy analysis, optimizing-simulation models have employed flexible functional forms such as the transcendental logarithmic function used in the Dynamic General Equilibrium Model (DGEM).[14] The production function approach has two significant drawbacks. First, economists' production functions can have weird characteristics in the light of actual technologies. For example, the unitary-elasticity-of-technical-substitution property of the Cobb-Douglas entails that altering parameters to reflect technical change tends to rotate the isoquant rather than change its shape. The translog is known to exhibit erratic properties when employed for simulation of technological possibilities outside the range of those available for estimation.[15] A physicist or mechanical engineer (such as Eric Hirst) who was not familiar with the notation of production functions and isoquants might have attempted to determine economically efficient solutions directly from a "technology curve" which best fit the technical alternatives. In fact, Hirst[16] used the three-parameter specifications depicted in formulas (2) and (3) above. These "technology curves" are capable of a wide variety of shapes, and therefore, capable of producing outcomes consistent with a number of production functions and production correspondences. The second significant drawback of economists' production functions is that it is difficult to relate the efficiency solution outcome to a particular technology with particular performance characteristics. The efficiency solution emanating from optimizations employing the Hirst functional form can be directly referenced to their location on the technology isoquant, and thereby compared to the actual technical options depicted. The increased arithmetic complexity inherent in minimizing a multivariate life-cycle cost function with explicit "technology curve" isoquants constraining the technical possibilities available may be justified in more accurate and sensible results than are produced by utilizing production functions as an intermediate step.

In the remainder of the Section, we walk through the technology optimization problem in its most interesting form--the decision on economic efficiencies for end uses and processes in new structures. In doing so, we shall highlight four aspects of the analytical exercise:

1. the number and nature of technical blueprints which technology choice optimization makes available to fuel-and-equipment choice decisionmaking;

2. the objective function and constraints for technology choice optimization;

3. the meaning of first order conditions and solution equations; and,

4. the meaning of economic valuation in the policy outputs.

At the point of new structure fuel-and-equipment choice simulation, the building sector end-use models have life-cycle-cost minimizing "blueprints" for simultaneous choice combinations of energy service processes and attached equipment. A blueprint is a simultaneous economic efficiency solution set for a particular configuration of process, and fuels assigned to end uses. As a first example, one residential sector configuration includes the building envelope, a natural gas hydronic (steam/hot water) space heating system, electric room air conditioning, and gas water heating. As a second example, one commercial sector new structure configuration includes the thermal envelope, gas central space heating, a variable air volume distribution system, electric ventilation fan motors, electric reciprocating compressor cooling equipment, and lighting (which provides benefits to heating and disbenefits to cooling). For each commercial or residential configuration, an economic optimum array of efficiency solutions and underlying combination of techniques is determined. Hence, each solution set is equivalent to a blueprint of technically efficient processes and end-uses which also satisfy the criterion of constant $ value economic efficiency, given relative prices and decisionmaker implicit discount rates.

The optimization of the LCC is with respect to the life of the longest lived asset, typically assumed to be the shell. Equipment is assumed to have a fixed life and is replaced over the life of the longest lived asset. The decision is made assuming premature retirement of surviving equipment at the end of the longest lived asset life.

The least complex specification of the technology choice problem is for the residential sector. The problem may be stated as follows: Minimize:

$$LCC = S\left[C_{sh} + \sum_{k=1}^{3}\sum_{n=0}^{n_k}\left(\frac{C_{j,k}}{(1+r)^{nL_k}}\right) + \sum_{t=0}^{L_s-1}\sum_{k=1}^{3}\left(\frac{P_j^t \cdot h_{j,k}^t \cdot T_{j,k}^t}{(1+r)^t}\right)\right] \tag{7}$$

Subject to:

$$S\left[\sum_{K=1}^{2}\left(w_j \cdot h_{j,k}^0 \cdot T_{j,k}^0\right) + w_j \cdot h_{j,3}^0\right] = C_{max} \tag{8}$$

where

the unknowns are

C_{sh} = Capital cost for particular combinations of shell modifications.

$C_{j,k}$ = Capital cost of equipment using fuel j for end use k (e.g., cost of oil furnace).

$h^t_{j,k}$ = energy consumed by equipment using fuel j, for end use k to produce a given amenity level in year t (related to efficiency).

$T^t_{j,k}$ = thermal performance (annual heat loss or gain) of shell given fuel i and end use k. Relative to base year value (i.e., $T^0_{j,k}$ = 1).

and the parameters are

j = j(mm,k) = fuel assignment to end use k in configuration mm

k = a "simplified" designation for end use; separate optimizations are performed for space-heating equipment types (e.g., gas non-central)

t = year of shell life

n = number of equipment replacements over the life of the shell

n_k = maximum number of equipment replacements (= greatest integer less than $\frac{L_s}{L_k}$)

L_s = life of shell

L_k = equipment life for end use k

r = implicit discount rate

p^t_j = price of fuel j in year t

S = index of new housing size relative to base year

w_j = weighting factor reflecting relative economic value of each fuel

C_{max} = limit on energy consumption implied by a building energy performance standard, or some combination of end-use and shell prescriptive standards.

Several assumptions and consequential results further specify the problem. They include assuming that equipment replacements perceived in optimizing (7) are of the same efficiency as original, technological degradation of efficiency and thermal performance do not influence choices made, the contribution to air conditioning energy use intensity from shell improvements is a constant fraction of the contribution to space heating energy use intensity from shell improvements, water heating is independent of thermal performance, and end-use energy consumption (efficiency) and thermal performance solutions are located on the technology curves as specified in formulas (2) and (3) above. Also, for heat pumps, it is assumed that cooling efficiencies bear a fixed and proportionate relationship to heating efficiencies.

The technology choice problem can be solved by Lagrangian optimization. Under this regime, the Lagrange multiplier is the shadow price for the last unit of energy conservation (expressed in \$/million Btu) and may be considered as an addition to the expected market price of energy.

The commercial sector optimization problem is directly analogous to that stated for the residential sector. However, there are at least two complications. In addition to the shell considered as a process, the distribution system is considered as a process capable of technical efficiency improvement. Therefore the numerator of the operating costs term of Equation (7) above would in the commercial sector include an additional variable, say $D^t_{j,k}$, which represents the distribution system performance relative to a base year value. The second complication is that lighting is assumed to provide lighting amenity, heating benefits for reducing space heating load and cost, and cooling disbenefits for increasing cooling load and cost. Water heating is not included in the commercial sector joint optimization.

A unique feature of the above specification for residential sector technology choice is that any combination of prescriptive standards can be dealt with without adding explicit constraints to the objective function. This is done through a sequential solution framework (using a variant of Everett's method[17]).

For the purpose of illustration, we here highlight only the residential sector optimization problem, and verbally describe two solution equations for residential shell thermal integrity. These equations are informative because they focus on the derivations of a normative "socially optimum" thermal performance, given a frontier of technically efficient possibilities, and a "positive" determination of privately optimum thermal performance. One solution equation depicts the thermal performance selected along the technology curve or technology isoquant, given a Lagrange multiplier representing the stringency of policy. This equation embodies not only the technical characteristics of thermal performance, but also the technical characteristics of space heating equipment and air conditioning equipment simultaneously considered. The other thermal performance solution equation represents the normative determination of thermal performance from the policy constraint. In the case of an aggregate performance requirement (e.g., a design energy budget), these equations are systematically iterated on Lagrange multiplier and thermal performance values until solution convergence is achieved.

One of the several justifications for energy performance standards in the case of commercial or residential electric space heat has been the difference between the private and social cost of the energy resource, which has resulted in violation of the second law of thermal dynamics and resource conservation. For predicting policy impacts, the difference between the private costs of an energy resource and the social costs implied by a conservation measure is represented by the Lagrange multiplier shadow price obtained from the classical programming solution. The solution social cost (private costs plus Lagrange multiplier) for policy-constrained consumed energy may be spread over the base of energy savings to produce a social cost per unit of conserved energy. This economic valuation of resource conservation can be compared directly with the economic valuation of power generation alternatives. Table 2 presents a representative set of social cost comparisons between a residential Model Conservation Standard (MCS) for new single family houses in the Pacific Northwest and a number of power generation alternatives costed at the Bonneville Power Administration (BPA) official 3% real social rate of discount.[18] The standard was designed to reduce energy consumption for space heating in new single family homes to approximately one third of its level under current building codes. The comparisons depicted in Table 2 indicate that the MCS would not be considered privately cost-effective relative to both conventional and non-conventional alternatives available for the BPA least cost mix.

However, it should be recalled that private willingness to undertake residential energy conservation actions has been dampened in the Northwest by implicit and explicit subsidies, as well as by average cost pricing. Moreover, if private conservation initiative is typified by response to "catastrophic" energy price increases or by "threshold" reaction to energy price levels, then the conservation willingness revealed by the historically determined private implicit discount rates may be biased downward. An alternative marginal social cost computation which may be argued to define the sympathetic boundary condition for MCS cost effectiveness is obtained by imposing the social (BPA 3%) rate of discount upon the builder or homeowner. Table 3 provides a single family electric space heat snapshot of MCS marginal social cost at a 3% real rate.

Table 3. Standard-Induced Marginal Social Cost Per Unit of Conserved Electricity Valued at the BPA 3% Real Social Rate of Discount (in 1984 ¢/kWh)

	Public Rate Pool			
	1986	1990	1995	2000
Central Forced Air	2.7	2.9	2.9	2.8
Non-Central	2.9	3.2	3.2	3.1
Hydronic	2.9	3.2	3.1	3.1
Heat Pump [b]	8.0	8.8	8.9	8.9

[b]The implication of footnote "a" of Table 2 below is that heat pump cost of conserved energy should be disregarded.

Tables 2 and 3 considered together indicate that the MCS is indeed cost-effective relative to many power generation alternatives--when viewed from the policy maker's perspective. The difference between private (Table 2) and public (Table 3) perception of MCS cost is attributable to different

Table 2. SOCIAL COST COMPARISONS
(In 1984 Value Units)

STANDARD-INDUCED MARGINAL SOCIAL COST PER UNIT OF CONSERVED ENERGY PUBLIC RATE POOL

	1986	1990	1995	2000
Electricity In ¢/kWh				
1	12.6	13.6	13.0	12.4
2	13.8	15.3	14.7	13.8
3	13.9	15.1	14.5	13.6
4	37.7	45.2	43.9	43.2
Natural Gas In $/THERM				
1	2.80	2.72	2.57	2.31
2	3.43	3.72	3.86	3.59
3	2.80	3.05	3.16	3.07
Heating Oil In $/GALLON				
1	3.85	4.14	4.01	4.73
2	4.64	5.27	5.06	4.81
3	3.79	4.27	4.38	4.98
Other Fuels In $/MMBTU Primary				
5	23.98	27.23	28.41	31.99

MARGINAL SOCIAL COSTS IN CENTS PER DELIVERED kWh of POWER GENERATION ALTERNATIVES VALUED AT THE BPA 3% REAL SOCIAL RATE OF DISCOUNT

		1985	1990	1995	2000
Conventional					
HYDRO	1	-	2.0	2.1	2.3
	2	-	3.0	3.3	3.6
	3	-	3.4	3.7	4.1
	4	-	3.7	4.1	4.4
	5	-	4.4	4.8	5.2
	6	-	5.5	6.0	6.5
Coal		-	6.3	6.5	6.7
Combustion Turbine		-	7.8	8.5	9.1
Non-Conventional					
Industrial Cogeneration	1	3.4	3.7	3.9	4.0
	2	4.8	5.1	5.4	5.5
	3	6.1	6.5	6.9	7.0
	4	-	9.3	9.8	10.0
Geothermal		-	7.9	7.3	6.6
Biomass		7.6	7.8	8.0	8.0
Wind		6.8	6.3	5.6	4.4
BPA Least Cost Mix		1.6	3.0	3.5	4.4

Key: 1. Central Forced Air 2. Non-Central 3. Hydronic 4. Heat Pump 5. Generic

a Predicted heat pump performance under the MCS is on the basis of implementation of shell measures identical to those prescribed for electric resistance options. In fact, MCS compliance can be satisfied through a prescriptive mix of heat pump and shell (which is less tight than for resistance heat). Because no cost and performance data were developed for this option, it was not considered in the demand analysis. As a consequence, the heat pump results overstate the severity of the MCS for that option. Because of the very low cooling loads in the Pacific Northwest and the low historical penetration of heat pumps, it was not deemed cost effective to develop the additional cost and performance data.

time horizons for receipt of benefits. The new home purchaser may have reasonable expectations of 20 to 30 years service from new central station power provision. However, the length of expected benefit from programmatic conservation is dependent upon the expected time of home residence, and the expected return on the programmatic investment upon resale. Of course, the dilemma is that the policy rationale--significantly lower private than social resource cost--ensures expected resale returns (implied by positive shadow prices) which will be negative. The different private and public valuation of marginal social cost as well as the consequent market distorting impacts (e.g., utility customer loss) could be mitigated were the MCS packaged and financed as an investment annuity whose rate of return is linked to the social cost advantage of the MCS vis-a-vis construction of new electric power generation.

SECTION VII: CONCLUSIONS

Although end-use modeling has entered an era of "model simplification for the sake of personal computer-compatibility," we may appear to have been merciless and uncompromising in our penchant for complexity. Our goal has been to identify the modeling construct which appropriates in a logically consistent and useful manner the technical information which physics and engineering bring to the traditionally economic forecasting and policy analysis tools. It is our contention that rigorous adherence to the modeling construct from the design of statistical experiments as a control for both technical and economic data development to the final structural requirements for policy constrained technology characterization in the models makes end-use modeling a scientifically useful endeavor. End-use modeling in simplified and anecdotal form may be adequate for political purposes, but it does not serve scientific purposes.

In our discussion, we initially introduced an underlying theoretical construct for viewing the end-use modeling exercise as one component in a larger analytical framework. We then described the analytical framework as derivative from tools and concepts developed to satisfy objectives which were meaningful within the operating context of different professional disciplines--Statistics, Physics, Engineering, Economics, and Operations Research. We next described how engineering and economic parameters are developed for input into the end-use models. Then, a cursory structural overview of the end-use model was presented. Finally, because the lineage of technology choice traces backwards through the engineering and physics disciplines, we highlighted technology choice and policy analysis.

We believe there are, at least, two requirements to make end-use modeling scientifically thoroughbred. The development of technical efficiency parameters, particularly in the commercial sector, has borne little relationship to sound statistical design. Until this is accomplished, there is no way of knowing what you've got when you've got it. Second, the choice of fuel and equipment in commercial sector end-use models simply does not embody sufficient information concerning decisionmaker attributes (and also technology attributes) to make sound and accurate predictions. We know of no commercial sector model being exercised in the field in which the fuel choice component is not encumbered with chewing gum and baling wire.

Finally, there is one area in which most extant models fail to account benefits and costs from "2nd Law" low quality energy sources. This is in the modeling of internal gains. As mentioned above, "internal" benefits and costs of lighting in the commercial sector are explicitly accounted for in the HVAC-choice decision problem. Kenton R. Corum has developed a variant of the ORNL residential end-use model which considers internal gains from appliances in evaluating the efficacy of efficiency standards.[19] If social policy becomes increasingly directed towards internalizing the social cost of high quality energy consumption for low quality purposes, the modeling of internal gains (in the tighter building stock) will become increasingly important for accurate prediction and policy analysis.

REFERENCES

1. Gary S. Becker, Economic Theory, New York: Alfred A. Knopf, Inc., 1971, pp. 14-23.

2. J. A. Hausman, "Individual Discount Rates and the Purchase and Utilization of Energy-Using Durables", The Bell Journal of Economics, Vol. 10, No. 1, Spring 1979, pp. 35-40.

3. Martin L. Baughman and Paul L. Joskow, "Energy Consumption and Fuel Choice by Residential and Commercial Consumers in the United States," Energy Systems and Policy, Vol. 1, No. 4, 1976, pp. 305-323.

4. Jerry R. Jackson et al., The Commercial Demand for Energy: A Disaggregated Approach, Oak Ridge National Laboratory, ORNL/CON-15, April 1978, pp. 31-32.

5. J. A. Dubin and D. McFadden, "An Econometric Analysis of Residential Electric Appliance Holdings and Consumption," Econometrica, 52, 1984, pp. 345-367.

6. J. Berkovec, J. Hausman, and J. Rust, "Heating System and Appliance Choice," Studies in Energy and the American Economy, Energy Laboratory, Massachusetts Institute of Technology, MIT-EL 83-004WP, January 1983.

7. Ibid., 30-31.

8. Ibid.

9. J. A. Hausman, M. Kinnucan, and D. McFadden, "A Two-Level Electricity Demand Model," Journal of Econometrics, 10, 1979, pp. 263-289.

10. J. A. Hausman and J. Mackie-Mason, "Price and Weather Effects in Electricity Demand," First Draft Report to the Bonneville Power Administration, May 1984, pp. 12-15.

11. E. Hirst and J. Carney, The ORNL Engineering-Economic Model of Residential Energy Use, Oak Ridge National Laboratory, ORNL/CON-24, July 1978, p. 31.

12. J. A. Hausman, Op. Cit., pp. 50-54.

13. Jerry R. Jackson and Robert B. Lann, Development of Fuel Choice and Floor Space Models for BPA's Commercial Model, Atlanta: Georgia Institute of Technology Economic Development Laboratory, Final Report prepared for Synergic Resources Corporation, February 1983.

14. Kenneth C. Hoffman and Dale W. Jorgenson, "Economic and Technological Models for Evaluation of Energy Policy," The Bell Journal of Economics, Vol. 8, No. 2, Autumn 1977, pp. 444-466.

15. David B. Reister and James A. Edmonds, "Energy Demand Models Based Upon the Translog and CES Functions," Energy, Vol. 6, No. 9, 1981, pp. 917-926.

16. Hirst and Carney, Op. Cit., pp. 17-25.

17. D. T. Phillips, A. Ravindran, and J. V. Solberg, Operations Research Principles and Practice, New York: John Wiley and Sons, 1976, pp. 510-513.

18. BPA Assessment and Evaluation Branch, Division of Power Resources Planning, Generating Resource Supply Curves, September 15, 1983, p. 5.

19. Kenton R. Corum, "Interaction of Appliance Efficiency and Space Conditioning Loads: Application to Residential Energy Demand Projections," Proceedings of ACEEE 1984 Summer Study on Energy Efficiency in Buildings, Santa Cruz, August 1984.

CHAPTER 5

SOCIAL PSYCHOLOGICAL ASPECTS OF ENERGY CONSERVATION

Elliot Aronson
University of California, Santa Cruz, CA 95064

Suzanne Yates
Lehman College, SUNY, NY 10468

ABSTRACT

Although some increases in the adoption of energy-efficient practices have been noted, only a small fraction of the potential savings are being realized, perhaps because human behavior is too complex for existing economic models. The rational-economic model is able to predict behavior in many situations, but it has limitations. To design effective public policy, the social, cognitive, and personal forces, that in addition to the economic realities define the situation, must be understood. This chapter examines one aspect of current energy conservation policy, the home energy audit program mandated by the Residential Conservation Service, and attempts to show how existing social psychological research might be beneficially applied.

The Residential Conservation Service (RCS) program was created by the 1978 National Energy Conservation Policy Act. The program requires major gas and electric utility companies to offer customers a variety of conservation services, including information about conservation practices and programs, lists of contractors, financial institutions, and suppliers who can help retrofit houses, and home energy audits. The goal of the low-cost home energy audits is to provide information about energy efficiency which is uniquely tailored to meet the needs of individual households. The guidelines require that "useful, reliable, and accurate" information be provided to customers in all socio-economic subgroups by auditors who have been trained to act as effective communicators.

At the present time, little is known about the degree to which the home energy audits have actually helped residents to become energy efficient. According to one review[1] of 35 utilities across the nation systematic efforts to evaluate the programs are rare-only three utilities examined changes in fuel consumption and only two used control groups in order to determine whether or not the audited residents were more likely to invest in energy saving products. According to the review, one program in Seattle, Washington found that participants were more likely than nonparticipants to take the following three actions: insulating and/or reducing thermostat setting on hot water heaters (30% vs 20%), installing wall and/or ceiling insulation (9% vs 6%), and installing storm doors and/or windows (20% vs 14%). The other program using a control group, one in Minnesota, reported that the only action participants were more

0094-243X/85/1350081-11$3.00 Copyright 1985 American Institute of Physics

likely to take as a result of the audit was to install weather-stripping and caulking (55% vs 45%). One California utility reported that the audits had no measurable impact on fuel consumption, while the program in Seattle reported energy savings of close to 9%. An evaluation of the Tennessee Valley Authority program reported energy savings of 20%--but the evaluation limited its sample to include only those audited households who obtained loans to install attic insulation. Finally, the review finds that nearly all of these programs have participation rates of less than 5% of the potential population.

While the review does find evidence to suggest that home energy audit programs do have a positive effect, it is clear that they are not yet producing the major increases in residential energy efficiency which are possible. Some, viewing these initial reports, may be tempted to conclude that the RCS program is inefficient and not worth continuing. Others might argue that the program simply requires a little more time or a better advertising system to prove its worth. We believe that the RCS home energy audit program represents an extremely promising strategy and deserves continued support. But we would argue that it is not simply time, money, or advertising which is needed to make the program a success. The RCS mandate calls for auditors to provide customers with useful, reliable, and accurate information. The key word in this sentence is "useful" and we would interpret this term social psychologically. First, no matter how accurate or reliable the information is, if it is not presented in a form which the consumer can easily understand and will readily attend to, then it will fail to have any impact. But even beyond this, an intervention involving human interaction is useful social psychologically if strategies are employed which maximize the intended impact of that interaction on the behavior of the individual. Social psychologists can make important contributions to this project in two major areas. First, we can make recommendations for modifying the auditor-resident interaction in order to maximize the impact of the exchange on the resident's subsequent behavior. And secondly, we have some knowledge about factors which can expedite the diffusion of innovation throughout a community.

Many of our recommendations and suggestions are based upon recent research on the way humans process information. Psychological research has demonstrated that, in many cases, people do not assign weight to information in strict accordance with its economic properties[2]. Certain situational factors, as well as limitations of human cognitive functioning, can override the individual's economic concerns. The same information can be presented in many different ways, each of which will produce a different response. If the energy audit is going to produce a major change in behavior then the structuring of the presentations must be based on our knowledge of what it is that people attend to and why.

VIVID AND PERSONAL INFORMATION

Several studies have shown that people tend to weight information in proportion to its vividness. Statistical data summaries and impersonal informational sources are less vivid than face-to-face interactions and case studies. Accordingly, impersonal data summaries, while accurate and efficient, have shown to have less impact than more "concrete" information even when the more vivid information is less representative. Here's a rather vivid example[3] of what we mean:

> Let us suppose that you wish to buy a new car and have decided that on grounds of economy and longevity you want to purchase one of those solid, stalwart, middle-class Swedish cars--either a Volvo or a Saab. As a prudent and sensible buyer, you go to Consumer Reports, which informs you that the consensus of the readership is that the Volvo has the better repair record. Armed with this information, you decide to go and strike a bargain with the Volvo dealer before the week is out. In the interim, however, you go to a cocktail party where you annouce this intention to an acquaintance. He reacts with disbelief and alarm: "A Volvo! you've got to be kidding. My brother-in-law had a Volvo. First, that fancy fuel injection computer thing went out. 250 bucks. Next he started having trouble with the rear end. Had to replace it. Then the transmission and the clutch. Finally sold it three years later for junk."

Now, rationally, your acquaintance's story should have no more and no less impact than any one of the other several hundred individual repair records summarized in the Consumer Reports. Assuming the Consumer Reports article was based on information from 900 car owners, then your acquaintance's story just changes the data base available to you from 900 to 901. But if you are like most of the people who served as subjects in this kind of research, the eyewitness will weigh much more heavily in your imagination than the hundreds of unknown persons summarized in the survey. Indeed, it might be decisive.

By acknowledging this fact of human behavior, we can incorporate practices into the home energy program that mitigate against any unintended consequences of tendencies that are not economically rational. One major recommendation, therefore, is to train auditors to present information not only completely and accurately but also in the most vivid and personal manner possible.

Specifically, the auditors should not simply leave the consumer with a computer summary of the potential savings associated with various retrofits. As indicated, presentations of statistical data alone are ineffective at inducing attitude or behavior change. The auditor might use a copy of the customer's own utility bill rather than using a fictitious example to illustrate a point. Whenever possible, auditors should provide normative information that acquaints residents with how much neighbors have been able to save through retrofitting. Similarly, the auditor might use a "smoke stick" to convince residents about the value of weather stripping and caulking. The smoke stick, invented by the Princeton House Doctor Program, works like an atomizer-instead of containing a liquid, however, it is filled with a very fine, colored powder. The powder when sprayed, flows on air currents and resembles smoke. When the smoke stick is operated below a window that has not been properly sealed, the "smoke" rushes along the path of the draft, clearly indicating when one is heating the out-of-doors in winter or drawing in heat from the outside in summer. Telling people that they are losing a certain percentage of home heat through the cracks around the windows is reasonable, but demonstrating the point by allowing the customer to watch the smoke pour out under doors and over window sills is far more compelling.

The fact that people are frequently swayed more by the report of a single individual than they are by a comprehensive data summary reflects, in part, people's tendency to overweight low probabilities and underweight large ones. It follows from this line of research that after describing the average cost-benefit ratios associated with various conservation practices, it would also be beneficial to provide each consumer with the concrete example of a "super conserver." Super conservers are families who save more energy and money than average. Thus, after describing the average cost-benefit ratios associated with various conservation practices, the auditor might say something such as:

> Of course, these estimates are averages; your savings might be greater or less than average. But, I'd like to tell you about one local success story so you can get a sense of just how effective these practices can be. The [Smiths], who live in this town, installed weather stripping. They save about [the average percentage saved by this super-conserving family] every month.

Identifying super conservers would provide a vivid and dramatic exception to the rule, which though low in probability would have more than its share of impact on the consumer.

FRAMING INFORMATION: LOSS VERSUS GAIN

People respond more seriously to a loss than they do to a gain. That is, the amount of joy someone experiences when winning $100 is not equal to the consternation suffered when losing the same amount. As a result, individuals are more willing to take a risk if it is to avoid or minimize a loss than they are if the purpose of the gamble is to increase their fortunes. We would expect, therefore, that people who fear a loss are more open to innovation, whereas those who are concerned with protecting their savings are more inclined to eschew risk or change. Accordingly, informational campaigns stressing the amount of money and energy that can be saved by investing in alternative energy sources and conservation devices encourage people to define this as a gain or win situation. Thus labeling may function as a deterrent to the aceptance of the new technology or behavioral practice. The typical campaign strategy with its great emphasis on savings inadvertently may be discouraging people from changing their energy use habits.

Based on this research, auditors should focus on showing residents how much they are losing every month by not investing in alternative energy sources and conservation measures. Once the loss becomes salient, people will take action. People may not go out of their way to save money but it appears that they are willing to act to avoid losing it.

INTEGRATING COMPLEX INFORMATION

Determining whether an investment will be cost-effective is a complex task, requiring the resident to evaluate a myriad of factors. It has been shown that people often have trouble integrating quantitative information. To simplify the choice between alternatives, people tend to disregard components that the alternatives share and focus on the components that distinguish between them. Such a decision-making strategy can produce very inconsistent responses because different decompositions of the same problem often lead to different preferences. One implication of this finding is that people may have trouble integrating tax credit information or the effect of rising utility prices into their assessment of whether an item is actually cost-effective.

There is considerable evidence to support the contention that the average citizen does not properly integrate all information relevant to any given decision. People are aware of their energy bills and attempt to calculate cost-benefit analyses when they are considering installing insulation, adopting conservation habits, and the like. Unfortunately, consumers tend to use a simplified strategy which, though intuitively appealing, produces systematic errors in quantification. For example, when calculating the energy savings due to added insulation, most homeowners fail to take into account rising fuel costs. This results in a systematic underestimate of the impact of such actions. Even if the consumer is motivated to seek out and consider the available information, this research makes clear that there is little reason to expect that the information will be used in the way it was intended.

Yates[4] has provided an energy-related field test for two of the principles outlined above: the contention that we are more likely to take an action to avoid a loss than we are to achieve a gain, and the finding that people have trouble integrating quantitative information properly. In the study, homeowners were asked to evaluate some cost-benefit information about either a solar water heater or an insulating blanket for the hot water heater. The presentations were designed to focus on either potential savings if the investment was made, or current losses as a result of failing to take action. In addition, the statements either explicitly incorporated the impact of available tax credits into the presentation, or presented this information separately and left it up to the homeowner to integrate the two. Subjects evaluated the worth of the product on a number of dimensions and indicated their intentions to install the device in question within the coming year.

In general, the findings confirm the contention that homeowners will find energy efficient technology more attractive if they are given price information which incorporates the effects of tax credits and which encourages them to consider the negative consequences of non-action.

SOCIAL DIFFUSION THEORY AND THE MODELING BEHAVIOR OF CONSUMERS

The acceptance of innovation rarely, if ever, occurs simply as a function of appeals through mass media. While mass media have been effective in inducing people to choose one brand over other (almost identical) brands of products such as toothpaste and beer, getting people to adopt new and different behaviors is a much more difficult process. In these situations, the example of others is often far more effective than advertising, appeals to patriotism, and the like. If mass media are effective at all, it is in conjunction with "social diffusion." That is, people are most likely to accept an innovation because they come into contact with others who have successfully adopted it. It is our contention that most people will be more influenced by the report of a single neighbor who contacted an auditor and subsequently reduced her/his energy consumption than they will by exposure to the data of summary charts or by watching a TV appeal, no matter how slick, humorous or interesting. TV commercials are quite effective at inducing people to buy one brand of toothpaste rather than another. They are strikingly _less effective at inducing people to begin_ using toothpaste in the first place. Social diffusion effects have been observed in sitations as diverse as inducing people to use shopping carts at supermarkets and persuading farmers about the benefits of using improved agricultural practices[3].

These findings underscore the importance of modeling. Recently, Aronson and O'Leary[5] conducted a small demonstration project on the importance of modeling for energy conservation behavior. They observed shower-taking behavior in a university athletic field house. Introducing an obtrusive prompt into the men's shower room (a large sign in the middle of the room instructing people to turn off the water while soaping up) resulted in gain from 6% (control condition) to 19% in the requested behavior. When the researchers employed one student to serve as an appropriate model by turning off the water and soaping up whenever someone came in to use the facility, the number of people turning off the water to soap up climbed to 49%. When two people simultaneously showered with the proper behavior, it rose to 67%. Similarly, Winett recently reported how he and his colleagues have been exploring ways of using videotapes of models in group settings. Based on a number of behavioral-communication strategies, these videos differ from traditional TV ads in several important ways. They are carefully designed to use characters and locales which correspond to a specific target population; they show models who must learn to plan and readjust their lifestyles; the actions taken are clearly depicted and reinforced; and explicit care is taken to defuse counter-beliefs and attitudes. These videos work because they show people what others are actually doing, how they are doing it, and what the effect of such actions are. According to one study by Winett, et al, homeowners who watched videos about efficient electricity use in the summer decreased the amount of electricity they used for cooling by 35% and the amount of electricity they used overall by 16%.

"Social diffusion" can be viewed as modeling on a broad scale--i.e., innovation spreading through exemplary behavior. How can we accelerate social diffusion? If social diffusion theory is correct, two ideas follow: 1) If left to occur naturally, we should not expect the adoption of retrofit technology to occur in a smooth linear fashion. Rather, few effects will be apparent for several years, followed by more rapid diffusion as increasing numbers of people know people who have adopted the new technology. 2) We can intervene in order to help accelerate the early "dead" period by cultivating visible and credible models. For example, the RCS program could place retrofit technology into the homes of selected people at little or no cost. The people selected would be the so-called "sociometric stars"--the individuals within a community who are acquainted with and respected by a great many people from different walks of life. The more embedded the adopted is in cross-cutting networks of people and the more respected he or she is, the more quickly will the innovation spread.

Actively encouraging residents to provide these demonstrations would have many positive results. Reviews of peer tutoring programs and laboratory research provide clear support for the fact that we learn something better if we learn it in order to teach someone else. We expect that residents who provide demonstrations to their neighbors will asssimilate the information offered by the auditor more fully. Moreover, evidence indicates that peers are frequently able to teach their classmates more effectively than specially trained experts. Applied to the energy situation, this example underscores the importance of the diffusion model approach. Clearly, the most credible and effective communicator a utility company can hope to employ is the person next door.

THE THEORY OF COGNITIVE DISSONANCE

Leon Festinger's theory of cognitive dissonance is central to the areas of persuasion and attitude change. As a cognitive theory it argues that humans are active, organizing beings and that people's behavior, attitudes, and definitions of themselves are influenced by their actions and the situations surrounding them. In its most basic form the theory states that if an individual simultaneously holds two cognitions that are psychologically inconsistent, he/she will experience discomfort. Consequently, the individual will strive to reduce the inconsistency (dissonance) by changing one or both cognitions to make them more consonant or by adding a third cognition which will render the original cognitions less inconsistent with one another. Later research clarified this basic formulation by adding that in order for dissonance to be aroused in most situations, the person must feel a connection between his or her own behavior and its consequences. Thus, to the extent that people act in the absence of coercion, publicly commit themselves to act in front of others, or invest time, money, or personal prestige in an activity, they come to see themselves as believers in that sort of activity and develop a personal interest in it.

This model suggests that larger, ongoing commitments can be obtained from people by first soliciting a smaller commitment. Such a process is commonly called the foot-in-the-door technique. The percentage of people who would agree to an unsightly sign being put on their front lawn urging people to "Drive Carefully" increased dramatically (from 17% to 55%) if they had first been given the opportunity to sign a petition favoring safe driving. Similarly, the foot-in-the-door technique has been used successfully to encourage recycling behavior. Residents were asked to commit themselves to one, two, or three minor actions: complete a survey about recycling behavior, save cans for one week, and/or send a post card to the city council urging an increased recycling program. As the number of requested commitments increased so did the recycling behavior. The effects were still observed ten months later.

What does this have to do with how the RCS audit can be most effectively conducted? Consider the case of Mr. X who is initially mildly interested in conservation but is not ready to retrofit his home. Suppose he agrees to have an energy audit done in order to see if he can easily reduce his utilities bill. The audit should be designed to encourage Mr. X to take an active role in the process. The auditor could begin by inviting Mr. X to accompany her on her rounds so she can explain what she was looking for and why. Then as the audit progressed, the auditor could enlist Mr. X's aid in a number of ways--he could hold one end of a tape measure, or read a meter, or work the smoke stick, for example. These behaviors are not of consequence in and of themselves, but they would provide Mr. X with a basis for thinking of himself as someone who is actively interested in energy conseration. "If I'm not interested in energy conservation, why am I spending two hours actively helping this auditor track down heat leaks in my home?" The beginning of a new self-perception would increase the likelihood that Mr. X would take some action such as weatherstripping his windows, as a result of the audit.

Moreover, people are particularly prone to increase their commitment to a cause which they have attempted to persuade another to adopt. If Mr. X had been left with a smoke stick, he might tell his neighbors about the RCS program and give them a small demonstration. If Mr. X talks to his neighbor, Ms. Y, about the benefits of weatherstripping, he is likely to start thinking of himself as someone who is concerned about residential heat loss. People who are concrned about heat loss tend to be interested in wall and ceiling insulation, storm windows, drapes, and air leaks as well. As a result, Mr. X will probably be more inclined to close and open drapes at the appropriate times, consider installing insulation, and so on, than he was before he had the energy audit done, weatherstripped his windows, and spoke with Ms. Y.

The operation of the "foot-in-the-door" phenomenon is not immediately obvious to non-social psychologists. For example, at the 1980 Santa Cruz Conference on Energy Efficiency, an engineer noted that most people are more interested in solar than in insulation--even though solar is typically less cost effective-because solar is "sexier" than insulation. One of the tasks for social psychologists is to unpack the term "sexy." For example, we suspect that one of its meanings (in this context) is "visible, positive, and dramatic." One cannot easily point to a wall and announce to one's neighbors, "See that wall? Inside there is some terrific insulation." It is far more dramatic to have solar collectors on one's roof as a demonstration of one's smartness and patriotism. Moreover, to turn on a tap and get hot water that is heated by the sun is visible and dramatic proof of one's cleverness. Interestingly, one of the participants at the Conference suggested that his "irrational" behavior could be remedied by withdrawing tax credits for solarizing unless the home had first been insulated and weatherstripped." This approach has a certain intuitive wisdom to it; it's akin to telling a child that he can't get dessert unless he first eats his brüssel sprouts. It is believed that such a recommendation is social psychologically naive. We cannot legislate how people become interested in energy conservation; neither can we dictate how they express their interest. Withdrawing tax credits from people who want to put in solar water heaters before they install insulation is likely to discourage people from doing anything at all. Conversely, providing people with some encouragement for engaging in whatever energy efficient behavior they choose to become interested in should plant the seeds for the continuation of energy efficient behavior.

THE IMPORTANCE OF CHOICE AND CONTROL

Social psychologists have learned that the feeling of choice and control is an important determinant of happiness and behavior. The control need not involve major or dramatic events. On the contrary, allowing people to make choices in simple situations can have quite powerful results. For example, patients in a nursing home were specifically encouraged to make decisions about their daily routines. Patients decided how they wanted the furniture in their rooms arranged, where they wanted to meet with visitors, what kind of plants they wanted in their rooms, and when they wanted to view the weekly movie. On the surface, these choices might appear trivial; i.e., the content of these decisions in and of themselves would not appear to have exerted a dramatic effect on the quality of life in nursing homes. But the process of making decisions does have a powerful impact on people. The results of this study are clear. Patients who exercised some control over their lives became much more active and alert. In general, they expressed a greater sense of personal well-being than a comparison group of patients in the same home. Eighteen months later, nurses continued to rate the residents who had been specifically encouraged to exercise control as being more active, more social, and more vigorous than other patients. Doctors rated them as being in better health. Finally, and most striking of all was the difference in mortality rate. Whereas 30% of the control group had died during the intervening months, only 15% of the residents in the personal control group had died.

Work done by social psychologists[6] at the Center for Energy and Environmental Studies at Princeton University provides a good example of the application of this principle to energy conservation. The Princeton group considered people's resistance to installing automatic day-night thermostats. They believed that people didn't feel as if they had enough control over the temperature setttings once the program was operative. Accordingly, the thermostat was redesigned to allow residents to override the system temporarily. This simple modification made the automatic thermostats much more attractive to the residents because it gave them a real feeling of control by enabling them to adjust the system to their particular needs whenever they deemed it essential. It should be noted that this override mechanism would serve the psychological purpose even if the homeowner never actually utilized it.

In keeping with this phenomenon, we would recommend that the RCS program be structured so as to allow the customer as much decision control as is feasible. The customer should be encouraged to choose to accompany the auditor rather than be given the impression that he has no choice. If the customer is more interested in hearing about the house's potential for active solar than she/he is about the house's potential for weatherstripping, this should not be discouraged. Active solar can serve as a foot in the door. That is, if people become both actively and freely involved in <u>any</u> aspect of an energy awareness program, then their own commitment will inevitably continue to develop and expand.

Finally, it should be noted that we all have difficulty overcoming inertia, even when we are convinced that taking a particular action would be beneficial. Accordingly, the homeowners should be given every opportunity to commit themselves to take action within a specific time frame, <u>AND</u> ideally the auditor should inform the homeowner that she/he will return at the end of that time frame to offer further assistance. This recommendation is based upon some classic experiments by Kurt Lewin and his associates. In these experiments, it was found that if individuals were induced to agree publicly to prepare, serve and eat unappealing intestinal meats, the probability of their following through was far higher than a group that simply heard a lecture on the nutritional benefits of such meats. This was especially true if the individuals were led to believe that their behavior would be monitored. This is analogous to Pallack and Cummings[7] finding that people who volunteered to attempt to save energy were far more effective if they were informed that their names and intentions would be written up in a newspaper article. Their increased effectiveness continued for at least a year after they were informed that the article would not be published.

IMPLICATIONS FOR POLICY

The policy implications of our discussion are self-evident. With these implications in mind, we note with dismay, that the Department of Energy has announced its intention to cut back on the home audit program. Specifically, to save money, it is no longer required that the auditor return to the home to present recommendations in person--rather, they can be mailed in. Undoubtedly, this will reduce the cost of the audit. Unfortunately, such a policy decision is naive social psychologically. If our reasoning in this paper has merit, such a "minor" change in the program will almost certainly destroy its effectiveness. For further references and discussion on the topic of social psychology of energy conseration, see reference 8.

REFERENCES

1. E., Hirst, L. Berry, and J. Soderstrom, Energy 6 (7) 621 (1980).

2. R. M. Hogarth (Ed.), QUESTIONS FRAMING AND RESPONSE CONSISTENCY, Jossey-Bass, San Francisco, 1982.

3. R. E. Nisbett, E. Borgida, R. Crandall, and H. Reed, in J. S. Carroll and J. W. Payne (Eds.), Cognition and Social Behavior 2, 227 (1976).

4. S. M. Yates, USING PROSPECT THEORY TO CREATE PERSUASIVE COMMUNICATIONS ABOUT SOLAR WATER HEATERS AND INSULATION, Unpublished doctoral dissertation, University of California at Santa Cruz, 1982.

5. E. Aronson and M. O'Leary, Jour. Environmental Systems 12, 219 (1983).

6. L. J. Becker, C. Seligman, and J. M. Darley, PSYCHOLOGICAL STRATEGIES TO REDUCE ENERGY CONSUMPTION, The Center for Energy and Environmental Studies, Report PU/CEES 90, Princeton University, 1979.

7. M. S. Pallack and W. Cummings, Personality and Social Psychology Bulletin 2, 27 (1976).

8. P. Stern and E. Aronson, ENERGY USE: THE HUMAN DIMENSION, Freeman, New York, 1984. S. Yates and E. Aronson. Amer. Psychologist 38, 435 (1983).

CHAPTER 6

RESIDENTIAL ENERGY EFFICIENCY: PROGRESS SINCE 1973 AND FUTURE POTENTIAL

Arthur H. Rosenfeld
Department of Physics
University of California, Berkeley,
Energy-Efficient Buildings Program
Lawrence Berkeley Laboratory
Berkeley, CA 94720

ABSTRACT

Today's 85 million U.S. homes use $100 billion of fuel and electricity ($1150/home). If their energy intensity (resource energy/ft^2) were still frozen at 1973 levels, they would use 18% more. With well-insulated houses, need for space heat is vanishing. Superinsulated Saskatchewan homes spend annually only $270 for space heat, $150 for water heat, and $400 for appliances, yet they cost only $2000 $\pm$ $1000 more than conventional new homes.

The concept of Cost of Conserved Energy (CCE) is used to rank conservation technologies for existing and new homes and appliances, and to develop supply curves of conserved energy and a least cost scenario. Calculations are calibrated with the BECA and other data bases. By limiting investments in efficiency to those whose CCE is less than current fuel and electricity prices, the potential residential plus commercial energy use in 2000 AD drops to half of that estimated by DOE, and the number of power plants needed drops by 200.

For the whole buildings sector, potential savings by 2000 are 8 Mbod (worth $50B/year), at an average CCE of $10/barrel.

I. INTRODUCTION

In 1984, U.S. buildings used about $165 B (billion) of energy (38% of the U.S. total costs) of which about half was "wasted." By "wasted" I don't want to invoke the first or second laws of thermodynamics, I only mean that if for the next 20 years we were to follow a "least-cost" investment scenario (optimizing our investment in efficient use vs. new supply), the buildings sector would emerge using only abut half as much energy as is projected today by most economists and policymakers.

Although its energy use is huge and wasteful, the buildings industry is badly fragmented and supports very little research and development. Since 1973, many physicists have switched their research from more traditional fields to building science and are proud of their contributions to spectacular gains in efficiency. I think there is a need for even more of us to be doing such rewarding research and development. Right now,

0094-243X/85/1350092-30$3.00 Copyright 1985 American Institute of Physics

under the Reagan Administration, federal and state support has dropped sharply, but I still assert that any field which has a potential annual savings of $85 B [see Note a] is bound to support increasing R&D. In other words, conservation is not a transient slogan; it has grown to be a profession, it will be with us henceforth.

Table I and *Fig. 1* show the importance of buildings in the U.S. economy. In 1984, our buildings sector used $165 Billion worth of energy, mainly (60%) as electricity. In fact, of the total annual U.S. electricity sales of $135 Billion, most ($100 B or 75%) went to the equipment and appliances in buildings.

In 1984, 236 million Americans spent, per capita, $1800 for energy, of which $700 went into buildings and their appliances and equipment. The average home pays $1150 of annual bills for 2.8 people. Based on Table I we can make the following remarks about the building sector. Of the national costs for buildings; $165 B, 60% goes to residences and 40% to non-residential (called "commercial") buildings; 60% of the $165 B for buildings goes for electricity (accounting for 75% of all the $135 B of electric revenues). It's not a bad approximation to say that the past and the future of the electric industry depends on trends in the buildings sector. Thus in *Fig. 3* (below), you'll note that our least-cost scenario[1] frees 200 standard plants to serve more productive uses or even not be built. This 1980 estimate was based on whatever technology was already on the market; it did not count on any of the dramatic improvements in lighting or daylighting discussed elsewhere in this book.

Table I. 1984 U.S. Energy Expenses

	Fuel ($B)	Electricity ($B)	Total ($B)
Buildings Sector	65	100	165
Residential	(45)	(55)	(100)
Commercial	(20)	(45)	(65)
Industry	70	35	105
Transport	160	0	160
Total	295	135	430

To get a better feeling for the cost of energy in buildings, we note that the U.S. has 85 M occupied dwellings, with a total floorspace of about 110 B ft^2, and another 50 B ft^2 of non-residential ("commercial") space. So every square foot costs about $1/year in energy services, with residential space costing $0.90 and commercial space costing $1.30, while the energy for new office space costs $0.50 to $1.00.

Because of rising prices and enhanced awareness during the period 1970-84, the energy/GNP ratio for the entire economy dropped to 73.5% of its former value. If our energy efficiency were still frozen at 1970 levels,

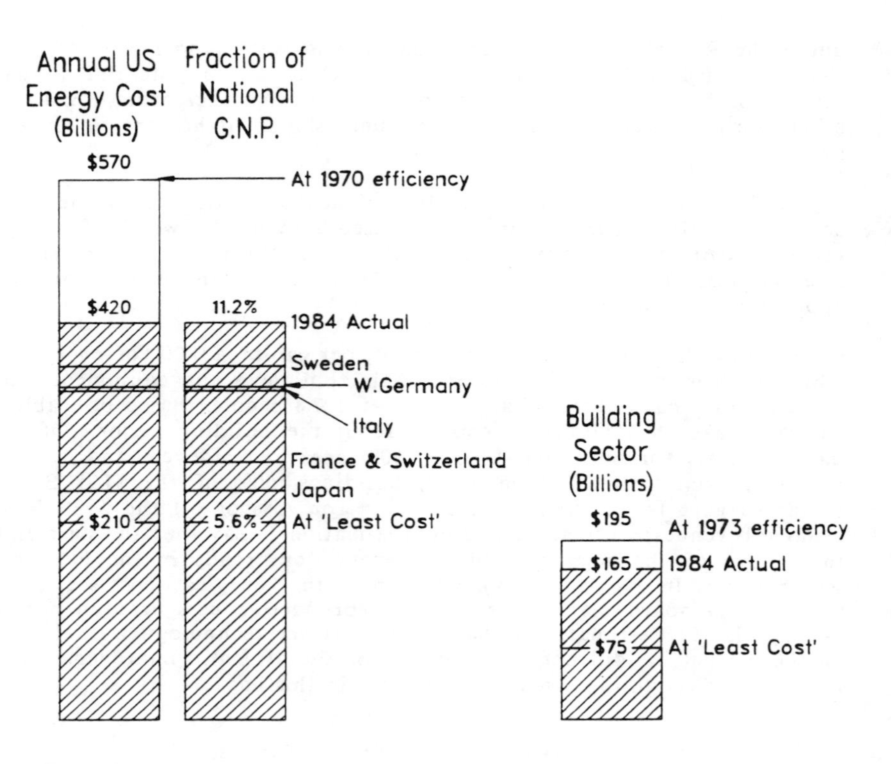

Fig. 1. Annual U.S. Energy Cost, from Table 1.
(a) Energy use per $ of GNP (in constant $) has dropped to 74% from 1970 to 1984; if efficiencies had stayed frozen at 1970 values, our $420 B annual costs today would instead be $420 B/0.74 = $570 B. On right bar, "Fraction of National G.N.P.", are lines (from "Btu Plot" figure 4) representing 1984 fractions for European countries and Japan. These lines extended left to the U.S. bar show what the 1984 U.S. economy would pay for energy at foreign efficiences. Source: DOE/EIA-0376 (85).

we would today be paying $420 B/0.735 = $570 billion annually, i.e., we are actually saving $150 billion, ($570 B - $420 B) each year, a very significant sum which is comparable to our highly publicized national deficit.

In buildings, the percentage savings are comparable. In the last decade, we have built 27% more homes and added 32% to our commercial floor space; yet primary energy use in buildings is up only 10%; so the energy/ft^2 is down to 85% (1.10/1.29) of its former value, and we are actually saving $30 B/year.

That brings us to trends in homes and to *Fig. 2*, we just mentioned an energy bill per home of $1200; the left part of Fig. 2 shows how it is distributed between space heat (50%), appliances (35%), and water heat (15%). (To get the total cost of $1200 from the costs/ft^2 of Table I and Fig. 2, remember that the average existing single- family home has a floor area of 1320 ft^2.)

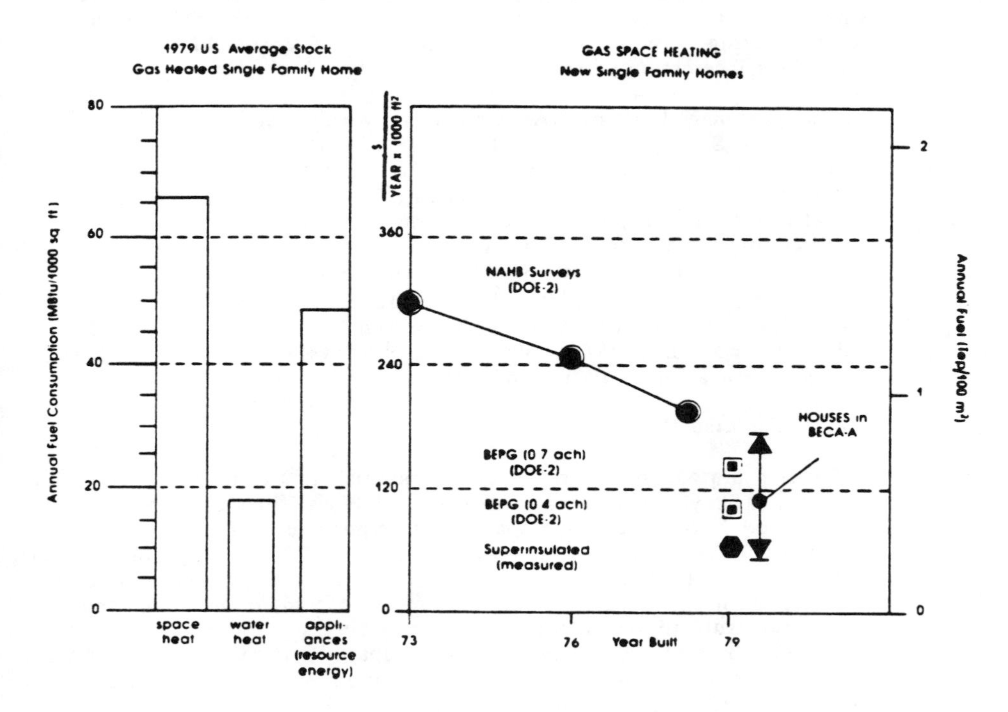

Fig. 2. Energy Use in New and Existing Gas Heated Single Family Houses.
The bar graph shows average space heat and appliance energy use for the 1979 stock of gas heated single family homes. Space heat and hot water use were calculated from NIECS utility billing data (Meyers, 1982). Appliance use is based on unit consumption and appliance saturations used in the ORNL model and includes electric appliances, such as refrigerators and lighting (air-conditioners are excluded), with electricity counted in resource energy units, using 1 kWh = 11,500 Btu. The points labeled "NAHB" are DOE-2 computer simulations of space heating in homes built by builders surveyed by the National Association of Home Builders in 1973, 1976, and 1979. The simulations were normalized to the Washington D.C. climate, which has approximately the same number of degree-days as average new building stock. Because of the non-random nature of the NAHB survey, results cannot be extrapolated to all new homes. Furthermore, The assumptions used in the simulation may not accurately represent lifestyle or building characteristics, however, they serve here as an example of energy use in new homes now on the market. "BEPG" represents proposed federal energy guidelines for practice that more closely approaches minimum life-cycle costs, using the same assumptions about thermostat settings, furnace efficiency, and free heat as the NAHB points. "Superinsulated" is the average of the 15 best-performing superinsulated houses of 30 for which detailed data were available in Ribot et al., 1982. It represents measured energy use, normalized to average degree-days for new buildings, using assumptions comparable to the NAHB and BEPG points. Source: Rosenfeld/Wagner (1983)-Labels. XBL 856-2816

The right part of Fig. 2 deals with new homes; it shows the impact of rising fuel prices--builders learn to build and sell more efficient homes, and the heating needs of new homes fell 25% in 6 years. Plotted in 1979, we see four interesting cases:

1. average homes built by NAHB members, which were heated for $250/year per 1320 ft^2 home.
2. Building Energy Performance Guideline (BEPG) "optimized" home, but without mechanical ventilation ("mv.") (unwise), $170/year.
3. BEPG optimized home, with mv., and heat recuperation ("hr.") (wiser), $120/year. (Mv. and hr. are discussed by Fisk elsewhere in this book.)
4. superinsulated homes, again with mv. and hr., heated for $100/year (for 2300 heating ^{o}C-days), which we shall now discuss (wisest).

o Superinsulated Homes

Superinsulated homes are becoming popular in the northern U.S. and in Canada. They typically have at least "R-20" wall insulation and "R-40" ceilings, and have an average space heat requirement of approximately 5 kW for a ΔT (outdoor - indoor) of 30°C [Note b]. The heating system needs such a small capacity (10 kW for the coldest days) that it is often combined with the domestic water heater whose rating is also about 10 kW. Superinsulated Saskatchewan homes, using natural gas heat, have typical heat bills of around $250--small compared with $550 for hot water plus appliances (which, in fact, provide much of the yearly heat). In Saskatoon recently, one-quarter to one-half of new homes are superinsulated (the fraction varies along with changes in the Canadian incentive programs and the economy). These homes take advantage of passive solar gain by mildly concentrating their windows towards the south, but they need not have large windows, so they look more "conventional" than "passive."

Before we leave this topic, I should try to explain as best I can why superinsulated homes fall below the BEPG "economic optimum" point in Fig. 2. Part of the explanation is a difference in the definition of "optimum." The square labelled BEPG (0.4 ach) was calculated using the DOE-2 computer program, but the economics failed to include the dollar savings available as the furnace is downsized or eliminated. Builders of superinsulated homes, of course, consider (indeed, aim for) these savings. A second part of the explanation is that occupants of superinsulated homes probably operate them very carefully and efficiently. A third part is that these homes may indeed be slightly over-optimally insulated and glazed, but it doesn't appear to be a very serious over-investment; the homes typically cost $2000 ± $1000 above conventional practice (with a very long high cost tail). See Fig. 14 below.

o Integrated Appliances

In the winter there are two recuperable leaks of heat from the home--the exhaust air and the used ("grey") water. In the colder parts of the U.S., it pays to recuperate at least one of these two. A superinsulated house has a very short heating season (only the few months when the outside temperature is below about 55°F). Above 55°F outdoors, one still has to keep the windows closed, but the appliances and the sun supply enough heat, and one needs no auxiliary space heat. In that case, one should start putting the excess heat from the refrigerator and the exhaust air into the hot water tank. By summertime, one should add the waste heat from the air conditioner. So, of course, by the turn of the century we can expect to see many integrated appliances, combined in a central utility core and controlled by a microprocessor.

o More Efficient Appliances

If we have good information, labelling, incentive, and loan programs, Americans will pay more attention to life-cycle cost when purchasing appliances. In that case, the overall[2] potential saved operating expense is about 40%, but refrigerators, freezers, and lighting can each drop about 50%. In Section II, Fig. 7 will show a complete "supply curve" for electrical appliances.

o More Efficient Lighting

With the introduction of high-frequency ballasts for fluorescent lamps (see Berman's chapter), their energy use in homes will drop about 25% (and by 40% in offices, where they can cheaply capture the added savings from daylighting). With the introduction of small screw-in fluorescent bulbs to replace incandescents, the residential lighting bill will decrease to one half. In the next 10-15 years, as these two remarkable devices replace today's ballasts and lamps, they will together save about 200 BkWh, worth $15 Billion/year and corresponding to the output of 40 standard 1000-MW power plants (200 BkWh is 60% of the 325 BkWh sales in 1984 by our entire stock of nuclear plants).

o Halving the Energy for Heating Water

Even without integrated appliances, the energy needed to heat hot water can decrease to about 60% as appliances are redesigned to use less hot water, people learn about cold-water laundry soap, and hot water heaters use solar preheat or heat pumps. [Ref. 1, Fig. 1.36]

o Indoor Air Quality

Before the 1973 oil embargo, we were beginning to be concerned with smog and soot and outdoor air quality in general, but never dreamed that indoor air quality was an even more pressing environmental problem. Nobody pointed out that indoor air is mainly outdoor air with some added pollutants, or pointed out that we spend most of our time indoors. Starting about 1974, and before inaugurating programs to "tighten" homes, i.e., reduce their infiltration rate below the typical 3/4 to 1 "ach" (air changes per hour), building scientists did have the wisdom to measure indoor air quality, so as to determine a "safe" number of ach. We then learned two things which may appear contradictory the first time you hear them.

1) It is safe to reduce the infiltration rate about in half the homes in the U.S.

2) Radon in U.S. homes causes about 10,000 lung cancers per year (within an uncertainty factor of 2-3). Over one-half of these cancers are caused in a relatively small number (about 10%) of homes. Clearly in these homes the need is not to reduce the infiltration rate but to remove the source of radon. The unit of radioactivity for radon (1 pCi/litre) can be equated with the risk of smoking about 1/2 cigarette/day (when the windows are closed, mainly during the winter). About 2500 of 10,000 total lung cancers per year may come from the worst few percent (about 3%) of our homes. Therefore, clearly, in many parts of the U.S., new homes will have to be monitored for radon and other indoor pollutants. [See Sextro's Chapter.]

o Load Levelling

Today, because of air conditioning, most U.S. utilities experience their peak power demand on hot afternoons. The cheapest (non-hydro) new peak power is generated by a gas turbine, which costs about $1000/kW of capacity, and burns expensive kerosene. (The $1000 includes transmission and distribution.) Thus a 100-W lamp burning on a summer afternoon requires a utility investment of $100! Or an uninsulated, uncovered water bed (150 Watt average) requires an investment of $150, even though insulation and a quilt will cut its losses to 50 W and cost far less. And an electric hot water heater (diversified afternoon load of 350 W) costs a utility $350. These examples explain why homes must soon have time-of-day meters and why these smart meters must control appliances. A brilliant example of this is the British Credit and Load Management System ("CALMS"), which listens by radio to a new price of electricity every 5 minutes, as broadcast by a BBC sideband. For $200 of hardware, CALMS turns the home into nine different "interruptible" circuits, controls appliances, performs as a clock thermostat, and does other clever things [see chapter by Bulleit and Peddie].

o Home Energy Ratings and Labels

One of the reasons that homes have not responded to the energy crisis as fast as autos or commercial buildings is that homes have not had "mile per gallon" stickers, and (unlike the buyer of a car or an office building) the purchaser of a home is usually unable to predict his energy bills. Today we know enough to rate homes to an accuracy of $50-100/yr.[2], and labels are being introduced in the U.S. and Western Europe. The impact of ratings is amplified because U.S. wholesale lenders ("Freddie Mac" and "Fannie Mae") are now willing to offer bigger and better loans on energy-efficient homes.

o Existing Homes and Commercial Buildings

In the discussion above, I have tried to give an impression of the new "turn-of-the-century home" and to show that there is room for physical and engineering innovation. Many of the improved appliances and controls will, of course, also be installed in existing homes. As for new commercial buildings, the changes are even more striking--many modern office blocks in Sweden get through the winter entirely on heat from the lights and occupants; in fact, they heat up during a winter day and use their thermal mass to coast over nights and weekends. Modern office buildings in warm climates like Reno, Nevada, can store enough "coolth" from the night air to get along without conventional air conditioning, and even in soggier climates, air conditioning (which amounts to about 40% of all our peak power) can (with the help of thermal storage) be shifted ahead to the previous night, when power is cheaper and cooling towers are more efficient.

In this introduction, I have tried to interest you enough to induce you to put up with Section II, which discusses Least Cost Studies and the cost of conserved energy; this section has more economics and methodology and less physics. Section III will discuss results from the residential sector.

II. LEAST COST STUDIES AND CONSERVATION "SUPPLY CURVES"

The Potential, in Buildings, for Saving 8 Mbod by 2000 AD, at a Cost of $10 Per Conserved Barrel

I will focus on the conclusions of the Buildings Panel from the SERI Solar/Conservation Study[1] (of which I was chairman). In Section A., I will summarize the conclusions; in Section B., I will define "cost of conserved energy," "supply curves of conserved energy," and then "conservation potentials". I will discuss commercial buildings in another chapter in this book.

A. Summary of Results

To whet your appetite, before stopping to define the method, I present *Fig. 3* which shows the potential for both fuel and electric use for the buildings sector to drop to half of conventional wisdom, by 2000 AD--a savings of 8 Mbod [Note c].

Let's discuss first Fig. 3(a): "Fuel". Two "Base Cases" were shown; they are the 1978 and 1979 medium- price, medium-growth projections from DOE's 1978 Annual Report to the Congress. Dropping faster than the base case is our Potential, made up of a decreasing white bottom area (existing buildings) and a small but growing shaded top (new buildings). The white bottom falls mainly because existing buildings can be retrofit; in addition, 20% of them will be demolished by 2000 AD. The grey wedge is small because new homes tend increasingly to be superinsulated and so to use very little heat, and new commercial buildings to use none. The "Low Renewable" line assumes that most homes install solar domestic hot water by 2000, and most new homes gain some passive solar heat. The "High Renewable" goes a bit further out the supply curve for solar options than we did for conventional options, only because we had more data for solar products.

Figure 3(b): Electricity. Unlike fuel use, where the base case is falling, conventional wisdom forecasts annual electric growth at 2.3% (compounding to 60% by 2000 AD). By contrast, the SERI study saw a potential <u>drop</u> to 3/4 of present use (at about 1% per year) despite an 80% growth in GNP by 2000. The

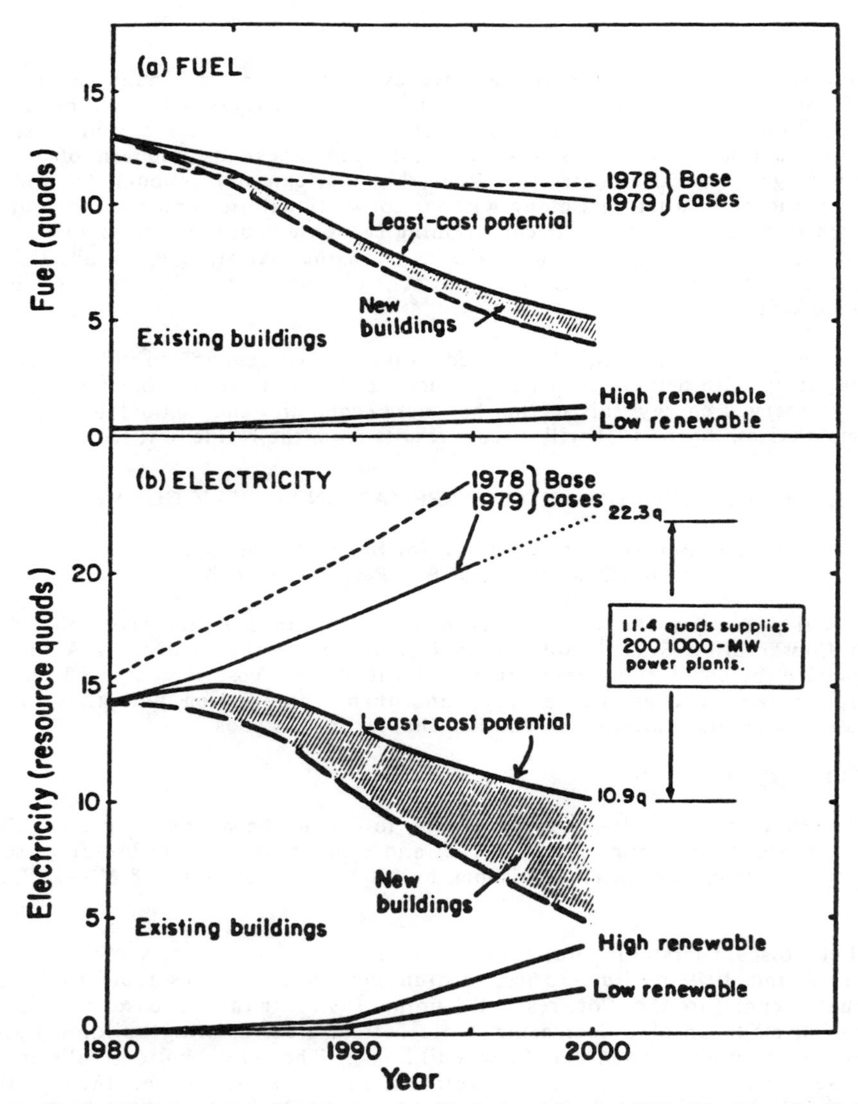

Fig. 3. Potential for saving half of fuel and half of electricity by 2000 AD, from the Building Chapter of SERI Solar/Conservation Study [A New Prosperity (1981)]. Base cases are from 1978 and 1979 Annual Reports to the Congress of DOE, which project fuel declining 6.7%/year with electricity growing 2-3%/year. White area is use by existing buildings, which decline because of retrofit or replacement of old buildings; shaded area is use by new buildings, which need little fuel for heating, but use proportionally more electricity. For details see figure 13 in Rosenfeld-Hafemeister Commercial Building chapter. XBL 807-1450

Low Renewable potential is mainly from wind in rural areas; the High(ly unlikely) Renewable line includes some photovoltaics cells on roofs. Daylighting of commercial buildings is included, but is not counted as "renewable." (We classified it as a savings resulting from better design and controls.) The Potential savings is 200 of the 400 standard 1000-MW power plants serving the sector and, the potential wind generation could conceivably replace 30 more. If we take the cost of a new 1000 MW plant to be $1.5-2B, then 200 avoided plants saves $300-400 B, which roughly covers the capital necessary for our entire Least-Cost Scenario, even the non-electric investments that end up saving 5 quads/year of fuel.

These investments will be discussed below, but it may be appropriate to summarize them here. We conclude that we should invest about $2000 in each existing and new dwelling unit (100 million units by 2000), for a total of $200 B in residences. We should also invest about $2/ft.sq. of existing and new commercial space (50 billion ft.sq. by 2000) or $100 billion more. Finally, we would invest about $1250/home in more efficient appliances (furnaces, heat pumps, air conditioners, heat exchangers, water heaters, refrigerators, freezers, low-flow shower heads, etc.). The appliance investment of $125 billion is surprisingly large. The total investment is $425 billion, and it will save, in 2000 AD, about 16 quads. The average cost of conserving these 16 annual quads would be about $10/barrel of oil equivalent.

"Advancing the Market"

Of this 16 quads of annual potential savings by 2000, probably about half will inevitably be captured by action of the marketplace as energy prices rise. However, government and utilities can speed the process by sponsoring applied research, education, training of house- doctors and retrofit contractors, monitoring and evaluation of retrofit and new buildings, energy labels for appliances, homes, and commercial space. More controversial are tax credits for conservation, and performance standards for appliances and buildings. I am against most tax credits, but an argument for tax credits and standards is that they help correct a 10-to-1 imbalance in federal subsidies; annually new supply receives about $50 B, efficiency investments receive only about $1 B, but the gasoline tax ($6 B) favors conservation.[3]

o Comparison with Western Europe and Japan

This buildings summary has described a potential drop to half the energy use typically projected today. Roughly the same factor of one half applies to all sectors of the whole SERI Study, which gives a potential U.S. use dropping to 60-65 in 1983 annual quads by 2000, versus 74 today, and 100 projected by DOE for 2000.

It is of interest, then, to compare our potentials with what is already going on in Western Europe and Japan, where oil has been imported for a much longer time than in the U.S. and consequently has been more expensive and used more carefully.

Figure 4 allows us to compare the U.S. and Canada with these other countries. it is a scatter plot of energy use (per capita) versus income (per capita). Each country is a snake, with its tail at 1973 and its head at 1983. We can draw three inferences:

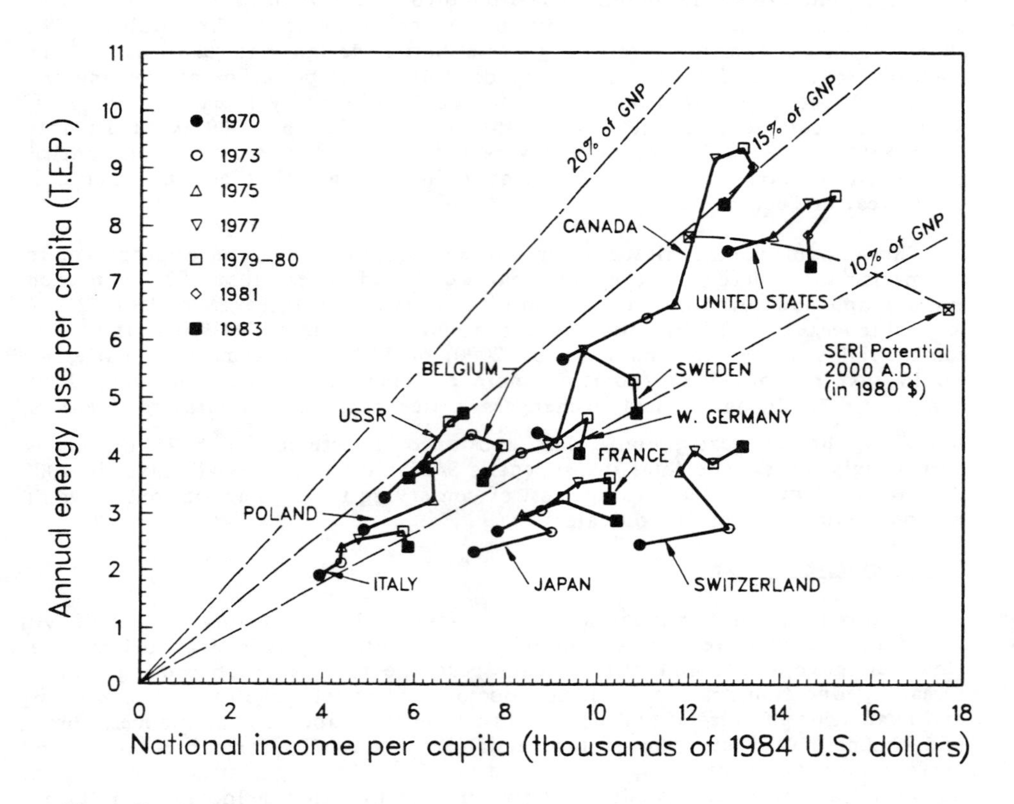

Fig. 4. Resource Energy Use vs. GDP (both per capita) For 11 Industrial Countries.
Each country is a sequence of 7 or fewer points joined by straight lines. The conversion from local GDP to dollars depends only on the 1984 exchange rate; earlier points are plotted using individual national deflators. The energy data comes from the OECD/IEA volume "Energy Balances". In the case of income data, there are two different series (before and after 1980), so we scaled the incomes to match at 1980. We convert electricity to resource (primary) energy using the national heat rate (e.g., U.S. efficiency = 35%), except Japan, which uses a nominal efficiency of 35.1%, and 3 "hydro" countries, which use an OECD average efficiency of 37%. For the lines labeled 10%, 15%, and 20% of GNP, we use an average 1984 price of resource energy of $5.66/MBtu or about $226 per TEP. Conversion: 1 TEP (Tonne equivalent of petroleum) = 40 MBtu. Source for price: DOE/EIA 0376 (1983).

1. For the same income/capita, the rest of the industrial world has for a long time used only about half as much energy per capita as used by Canada, the U.S., and the USSR.

2. As energy prices rise and as more efficient cars, buildings, and industrial processes appear, the energy use per dollar of Gross Domestic Product (GNP corrected for exports and imports) is falling for all the countries plotted except the USSR, Poland, and Switzerland. The drop since 1980 is particularly steep. The buildings sector follows a similar trend; see Section III.

B. Supply Curves of, and the Cost of, Conserved Energy

This section follows closely Chapter 2 of our recent book[4].

o Defining Conservation

Consumers want the the services that energy provides, not energy itself. Furnaces burn gas to provide heat, air conditioners use electricity to cool and dry the air, and motors use electricity to provide mechanical drive. The amount of energy used for a particular service depends on the efficiency of the service mechanisms and the level of service demanded. One approach to energy conservation is to accept lower levels of service (turning down the thermostat, for example). Our approach, however, favors simple, economic measures that improve efficiency and save large amounts of energy without changing the service.

Trade-offs between energy efficiency and capital costs are common. *Figure 5* summarizes the progress in energy conservation for refrigerators. The California standard has progressively been improved from 1900 kWh/year in 1977, to 1500 kWh/year in 1979, to 1000 kWh/year in 1987, and to 700 kWh/year in 1993. The

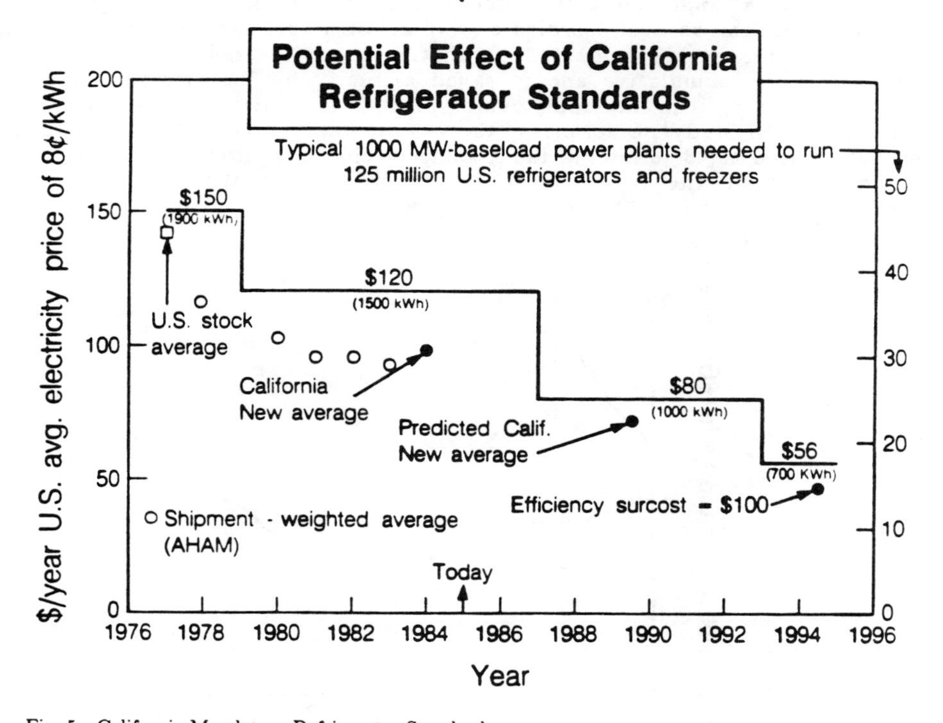

Fig. 5. California Mandatory Refrigerator Standards.
The standards reduce the average energy intensity from 1900 kWh/year in 1977 to 700 kWh/year by 1993. The additional cost of $100 retail for the 1993 standard will be paid back in one year (left-hand scale). The number of 1000-MW base-loaded power plants needed to run the nation's refrigerators and freezers is displayed on the right-hand scale. XCG 858-9882

additional cost to comply with the 1993 standard will have a payback period of one year since there will be a one- time cost of $100 to save $100/year for each of the remaining 19 years of the life of the refrigerator. Figure 5 also indicates the U.S. stock averages and the yearly production averages. the right-hand margin lists the number of 1000 MW-baseload power plants required to operate the 125 million U.S. refrigerators and freezers; the improved refrigerators will save the need to operate about 30 power plants when compared with the 1977 U.S. stock average.

o A Supply Curve of Conserved Energy

A supply curve for any energy source ranks the various reserves of that energy in order of increasing cost and shows how large each reserve is. *Figure 6* depicts supply curves for two grades of coal. A supply curve of conserved energy is the analog of a supply curve for reserves of gas, coal, or other tangible energy resources--the curve slopes upward since more conserved energy becomes available only at increasing costs. The reserves of conserved energy can be tapped by a sequence of conservation measures, each having its own size and cost.

To develop a supply curve of conserved energy, two values must be found for each measure. The vertical coordinate (y-value) of a conservation measure is the unit cost of the energy conserved by that measure; the horizontal coordinate (x-value) is the cumulative energy saved annually by that measure and all measures preceding it in the supply curve. *Figure 7* is an actual supply curve, Fig. 1.40 of the SERI Study[1]. Determining the y- value (unit cost) requires engineering and economic data; determining the x-value (savings) requires research into the characteristics of the energy-using stock. We discuss these two types of investigations in detail in the next two sections.

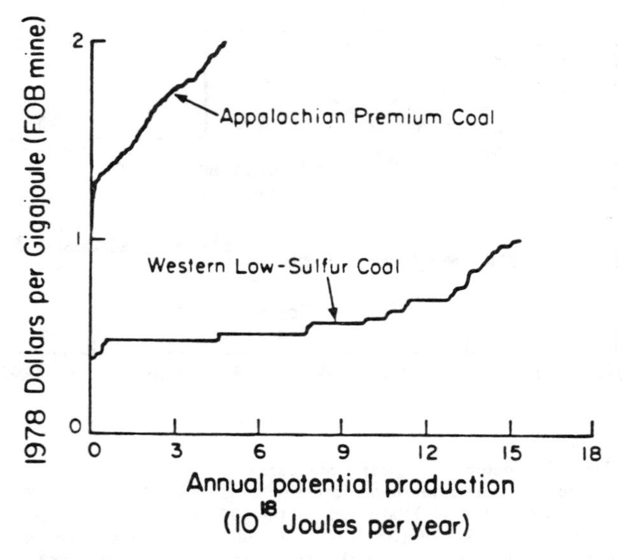

Fig. 6. Supply curves for two grades of coal. The reserves of Western coal are cheaper and three times as large as the reserves of the Appalachian coal. Source: EIA 1978.

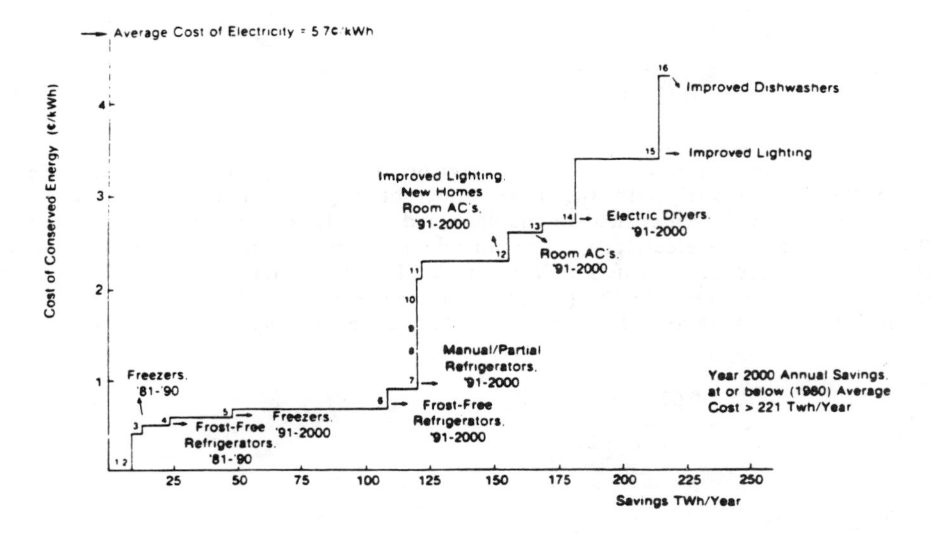

Fig. 7. Year 2000 Supply Curve of Conserved Energy (in Twh/year) for Electric Appliances. **Year 2000 baseline annual use for this sector is 581 Twh, where "baseline" assumes continuation of 1980 average unit energy consumption for existing stock or new additions in that year. Unit cost of conserved energy (in constant 1980 $) assumes that all increased costs are amortized over the useful life of the measure, using a 3% (real-dollar) interest rate. Potential savings in 2000, at or below 1980 average cost of $0.057/kWh is 221 Twh, or 38% of the year 2000 baseline.** Source: SERI Solar/Conservation Study, Brick House Press, 1981.

o The Cost of Conserved Energy

To establish the unit cost of the conserved energy CCE, such as cents per kWh, the annualized investment I in conservation (for materials and labor) is divided by the annual energy savings E:

$$CCE = \frac{\text{Annualized Cost}}{\text{Annual Energy Savings}}. \qquad (1)$$

Since investment (I) actually occurs just once, it must be annualized by multiplying it by the "capital recovery rate," CRR,

$$CRR = \frac{d}{1 - (1+d)^{-n}}, \qquad (2)$$

where "n" is the number of years over which the investment is written off, or amortized, and "d" is the discount rate. The unit cost is thus determined by the formula:

$$\text{CCE} = \frac{I \cdot \text{CRR}}{E} = \frac{I \cdot d}{E\,[1 - (1+d)^{-n}]} . \qquad (3)$$

Let us take an example. A consumer wishes to buy a new refrigerator. We compare the 1977 model with the 1993 model of Fig. 5. The high-efficiency model (offering services identical to the standard model) costs \$100 more but uses 1200 kWh per year less electricity. The consumer wants to recover his investment in 20 years (the useful life of a refrigerator). The consumer has a "real" (after inflation) discount rate of 5% (at an inflation rate of 5%, this would be equivalent to borrowing at 10%). The cost of conserved energy in this case is:

$$\text{CCE} = \frac{\$100}{1200 \text{ kWh/yr}} \times \frac{0.05}{1 - (1.05)^{-20}}$$

$$= \frac{\$100 \times 0.08}{1200 \text{ kWh}} = 0.7¢/\text{kWh}$$

For the case of a shorter-lived appliance (10 years), the CRR rises from .08 to .13, giving CCE = 1.1¢/kWh.

Compared with paying the utility 7.5¢/kWh (1984 U.S. residential average), it is very profitable to pay only 1¢/kWh saved by purchasing an efficient refrigerator where one's old one needs replacing.

Furthermore, the cost of the <u>conserved</u> electricity will stay the same for the 20-year life of the refrigerator. In contrast, the <u>real</u> price of electricity will most likely "escalate," that is, exceed general inflation. Note that the <u>cost</u> of the conserved electricity is independent of the <u>price</u> of electricity. The payback period in avoiding the 7¢/kWh electricity with the \$100 investment is 1.1 years.

Calculating the cost of energy "supplied" by a conservation measure thus involves four variables:

1. Investment (initial) of the conservation measure.
2. Annual energy savings expected from the measure.
3. Amortization period of the investment.
4. Discount rate of the investor.

These variables are analogous to the criteria for investment in the supply sector:

1. Cost of extraction facility.
2. Rate of extraction.
3. Depreciation of facility (and possible depletion of the reserve).
4. Discount rate of the firm.

In the rest of Chapter 2 of Meier et al.[4], we discuss each of these four variables in detail, but our procedures are summarized on page 14 of the SERI Study.

1. For investment cost, we took contractor costs or retail prices of appliances.

2. For annual savings, we took empirical data where available; if we had to use calculations, we scaled them to agree with measured results.

3. For amortization, we took the shorter of the physical life of a measure, or 20 years.

4. For real discount, rate we distinguished between:

   a) 3% real for residential investment, corresponding to the historic real interest rate on mortgages.
   b) 10% real for commercial buildings, where energy investments have to compete with investments for more production and profit.

o "Conservation" or "Least Cost" Potential

We define the potential from the consumer's point of view, i.e., where the supply curve crosses the consumers price for fuel or electricity.

Please note that although we define the residential (commercial) potential corresponding to a 3% (10%) real discount rate, the reader can easily select any rate he chooses. Thus, after we calculate our "grand supply curves" (Fig. 16 below), we recalculate them for 3%, 10%, 30%, and 40%. We see that an increase in the perceived consumer discount rate from 3% to 30% loses roughly half the potential savings. We look on this not only as a sensitivity analysis, but a nice way to estimate the potential savings attributable to information programs and labels, which effectively remove uncertainty and risk, and lower the consumer's discount rate towards the 3% and 10% values available for home mortgages or commercial borrowing.

o Time Perspective for a Conservation Potential

A potential energy savings is, of course, a function of one's time horizon. Since we are interested in changes in efficiency, not behavioral changes, nothing can be done overnight. For the SERI Study, we chose a 20-year perspective. In 20 years most appliances will have worn out at least once and been replaced with more efficient models. By 2000, 20 million of the 1980 count of 80 million dwelling units will have vanished, the remaining 60 million will have been retrofit, and 40 million new ones will be built. Unlike homes, which have a mean life of about 100 years, commercial buildings last only 40 years. So, by 2000, 20 B ft^2 of our current 50 B ft^2 of commercial floor space will have vanished, but the total will have risen to 62 B ft^2 , i.e., 32 B new ft^2 will have been added.

o Why Use "Resource" Energy Instead of "Site" Energy?
i.e., What is the energy value of 1 kWh of electricity?

In many of our figures, electricity and fuel both enter--and we want to express them to be of comparable dollar value. Restated: we are not interested in conserving energy per se, only in reducing the cost of the services that it supplies. But fuel and electric prices are unstable, so people like to work in units of energy. If we convert to Btu 1.0 kWh of electricity as if it were electric resistance heat (1 kWh liberates 3415 Btu of heat within the building), we find that it costs about $20/MBtu, while gas and oil cost only about $7/MBtu. But if we convert electricity according to the heat used back at the power plant to generate 1 kWh (11,600 Btu burned per kWh generated, transmitted, distributed, and sold), then electricity costs $7/Btu of "resource" or "primary" energy (the same as the price of oil). Hence, from our point of view, it makes sense to use resource energy when comparing electricity and fuel.

o More Examples of Supply Curves and the Cost of Conserved Energy

To give some physical examples, we introduce another pair of figures. *Figure 8* is an energy-cost curve for retrofitting an existing home, and *Fig. 9* is the same set of calculations replotted and reordered as a supply curve of conserved resource energy.

TABLE II. COST OF CONSERVED GASOLINE, assuming a car is driven 10,000 miles/year and lasts 10 years.

YEAR	MILES PER GAL	EXTRA FIRST COST $\Delta\$$	ANNUAL GAL USED	ANNUAL GAL SAVED	ANNUAL LOAN PAYMENT (\$)*	COST OF CONSERVED GASOLINE (¢/GAL)
1975 Fleet	14	0	700	0	0	--
1985 Standard	27.5	1000	350	350	160	45¢
1995 Proposed	40	1500	250	450	240	53¢
1995 Import (VW RV 2000)	65	2000	150	550	320	58¢

I. CRUDEST ARGUMENT:

$$\frac{\text{INCREASE FIRST COST}}{\text{TEN YEAR GAS SAVED}} = \frac{\$1000}{350\text{ GAL}} = \frac{28¢}{\text{GAL}}$$

*II. USING 10% "REAL" INTEREST RATE, (ie., in constant $):

A 10% bank loan, repaid in 10 constant annual payments, costs $160/y

$$\frac{\text{ANNUAL COST}}{\text{ANNUAL SAVINGS}} = \frac{\$160}{350\text{ GALLONS}} = \frac{45¢}{\text{GAL}}$$

- This calculation has nothing to do with the price of gasoline, i.e.

Current Price of Gasoline - $1.30/GAL
Price of Synfuels - $2-3/GAL

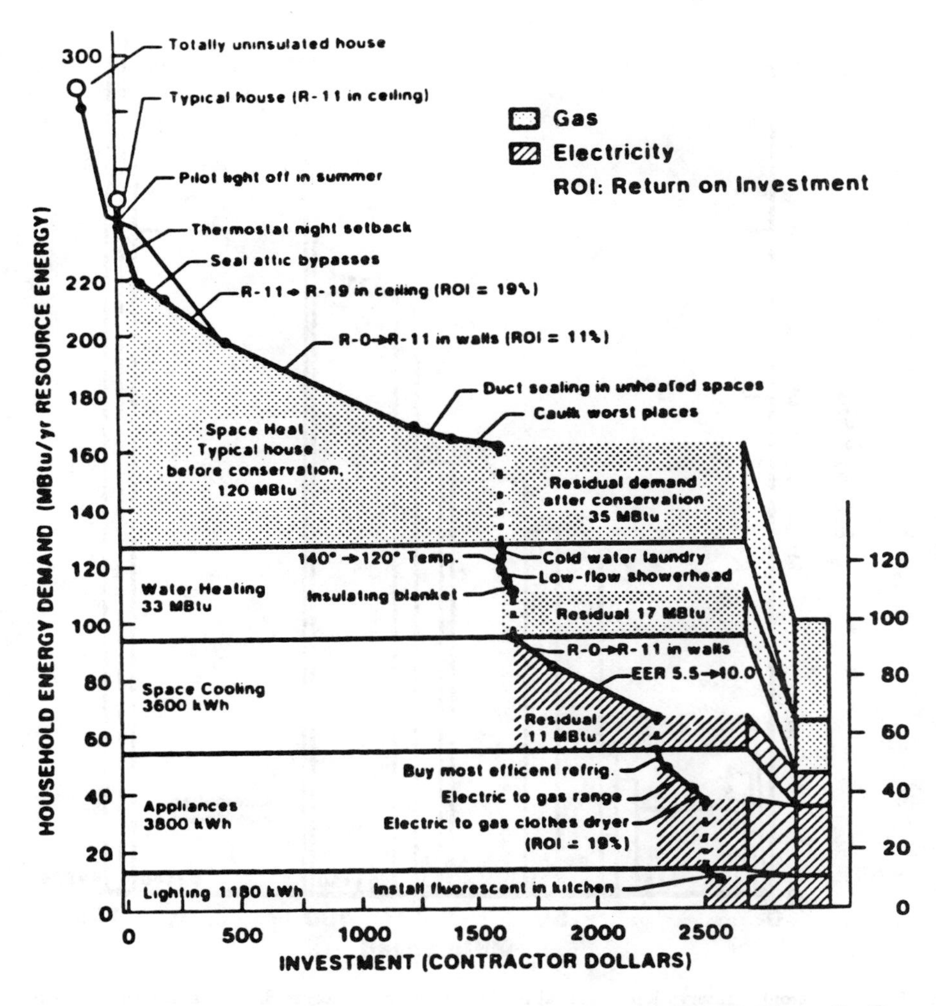

Fig. 8. Retrofit Conservation Potential in a Northern California Single Family Home, Gas Heat. (1200 sq.ft., 3000 Heating Degree-Days) XBL 812-7946 (See figure 9).

Table II switches temporarily from buildings to cars. It compares the 1975 "gas-guzzler" with the 1985 "social drinker." It shows that even if we include the cost of the catalytic converter (which is added to abate pollution, not to save energy), the cost of conserving a gallon is still 45¢, much cheaper than buying a gallon for $1.30. Again this illustrates the nice feature of using the cost of conserved energy--it makes it trivial to tell whether to invest in efficiency or in new supply.

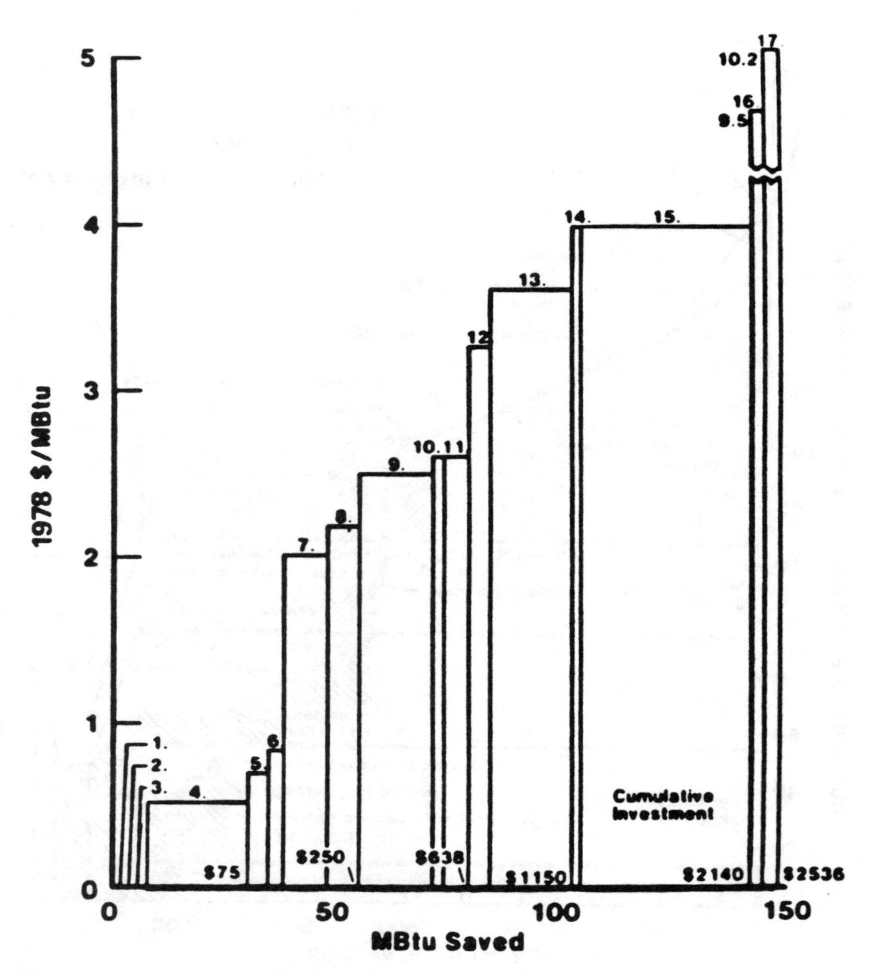

Fig. 9. Retrofit Conservation Potential in the Northern California Single Family Home, replotted as a conservation supply curve. Cost of conserved energy in $/MBtu of resource energy, calculated using 10% cost of money, plus depreciation, with all installation done by a contractor. Total annual energy consumption of the house before retrofit was 250 MBtu/year.

1. Turn furn. pilot off in summer.
2. Reduce hot water temperature, 140 to 120 degree-F.
3. Cold laundry rinse.
4. Thermostat night setback to 60 degree-F.
5. Buy most efficient refrig.
6. Install low-flow shower head.
7. Furnace tune-up (biennial).
8. Change from electric to gas range.
9. Increase ceiling insulation, from R-11 to R-19.
10. Install fluorescent lighting in kitchen.
11. Change from electric to gas clothes dryer.
12. Seal attic bypasses.
13. Change to high-efficiency air conditioner, EER 5.5 to 10.0.
14. Install water heater insulating blanket.
15. Insulate walls, R-0 to R-11.
16. Seal and insulate ducts in unheated spaces.
17. Caulk building shell (in worst places).

XBL 812-7947

At LBL (Lawrence Berkeley Laboratory), we have for many years been pursuing the concepts of Cost of Conserved Energy, and Conservation Supply Curves, with considerable success. The California Energy Commission and the Northwest Power Planning Council have adopted them, but unfortunately most energy planners still do not treat efficiency and supply investments symmetrically; thus our tax credits and deductions, and other subsidies, contribute annually about $50 B to investments in supply and less than $1 B to investments in efficiency[3] .

o Simple Payback Period

The concept of cost of conserved energy CCE is useful to policymakers for two reasons: (1) Conservation and production can be discussed together on the same footing since the costs of supply and demand are in the same units such as ¢/kWh or $/barrel. (2) Since CCE is calculated without reference to the utility price P, or guesses as to the future price of energy, one can calculate CCE once and for all and not have to repeat it for every utility and cost scenario.

On the other hand, the consumer usually is concerned with the return on his investment which can be expressed in terms of the simple payback period, the time to recover his investment in the conservation technology, or

$$SPT = I/P\ e. \quad (4)$$

By dividing Eq. 4 by CCE (Eq. 3), we obtain

$$SPT = (1/CRR)\ (CCE/P). \quad (5)$$

For the first example of the refrigerator, SPT = (1/0.08) (0.7/7.5) = 1.1 years.

III. RESULTS FOR THE RESIDENTIAL SECTOR

In the SERI report[1] you will find eight residential supply curves; "eight" because there are two types of energy (fuel and electricity), times four subsectors (homes new and existing, hot water, and appliances). We used computer simulation as little as possible and always normalized their results to empirical data. In this section I want to discuss some of the data, because they add reality and contribute confidence to the results. Although the discussion refers to the 1980 SERI study, all the data below are more modern.

At LBL, our Buildings Energy Data Group publishes a series of review articles called BECA (Buildings Energy Use Data and Critical Analysis), and most of the following figures are from BECA.

o Retrofit of Existing Homes

Figure 10 is a scatter plot of 47 different retrofit projects or experiments, each experimental point involving typically 10-20 homes. To be cost effective, each point must fall above the sloping lines representing the current prices of electricity, gas, and oil. Restated in terms of our "cost of conserved energy"--a point on the line will have a cost of energy just equal to the price of energy. You can see that overwhelmingly the retrofits are cost effective and lead to a 35-40% savings for an average cost of $1370. Note also the CSA/NBS open circles

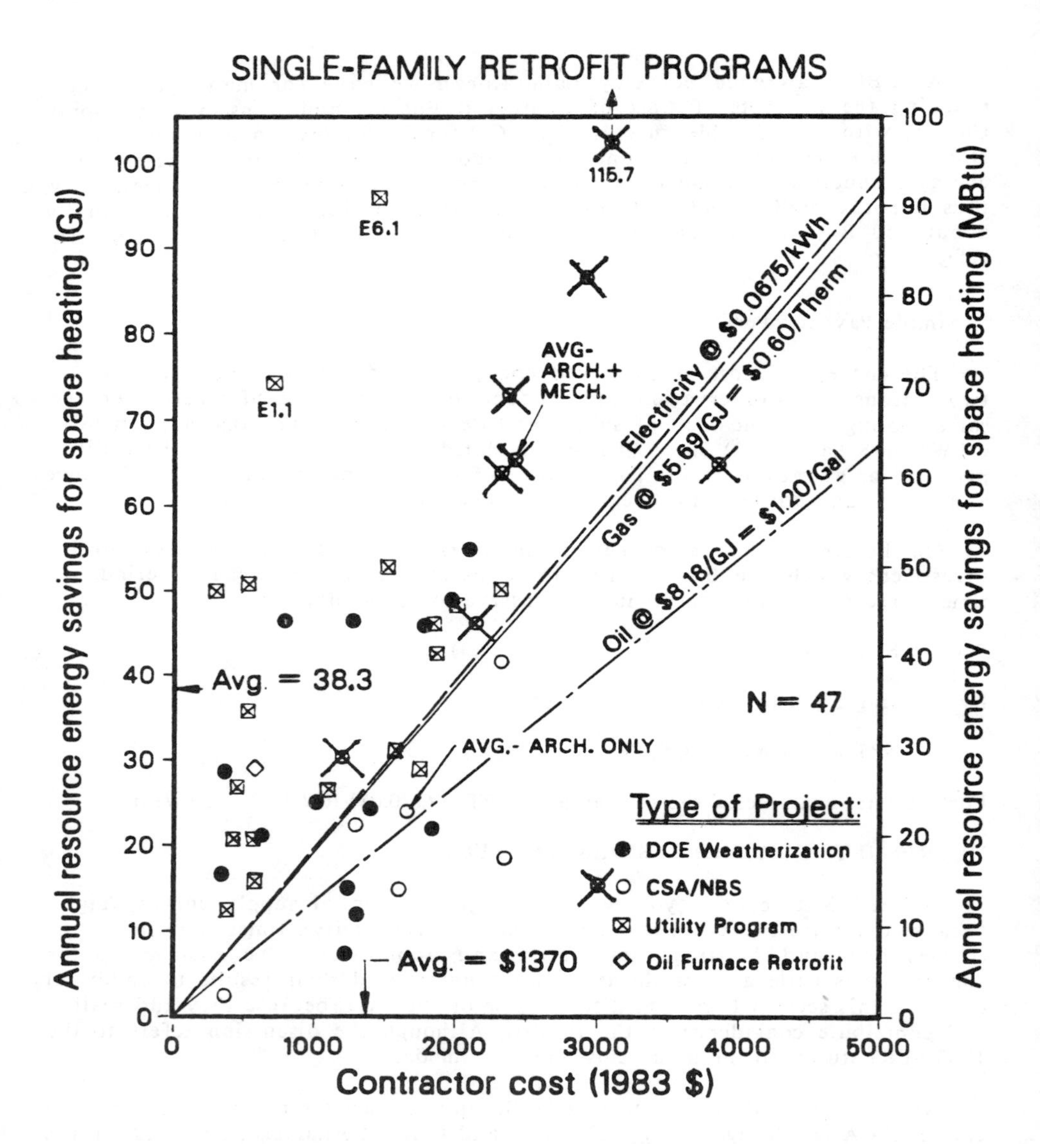

Fig. 10. Annual space heat energy savings are plotted against the first-cost of the retrofit investment for utility-sponsored or low-income weatherization programs. Average space heat savings are 36.3 million Btu (MBtu). The 47 data points represent results from over 50,000 homes. The sloping reference lines show the minimum energy savings that must be achieved, for each level of investment, if the retrofit is to be cost-effective compared to national average residential prices for fuel and electricity. The future stream of energy purchases for 15 years, assuming constant energy prices (in 1983 $), is converted to a single present-value, using a 7% real discount rate, in order to compare it with the "one-time" conservation investment. Roughly 75% of the data points lie above their respective reference line. Electricity is measured in resource units of 11,500 Btu per kWh. Source: BECA-B (Oct. 1983). XCG 839-7233

showing that "architectural" measures (insulation and other repairs to the shell of the home) are less effective alone than the big X's which are CSA results when shell retrofits were combined with "mechanical" retrofits, i.e., repairs and tune-up of the heating system.

How well do measured saving agree with predictions made by home auditors? From project to project, there is great variation in the ratio of saved/predicted, but averaged over hundreds of homes, we find that 2/3-to-3/4 of the predicted savings are actually achieved[5]. This shortfall (and the fluctuations) can be explained by many factors: poor auditing, inadequate quality control, and finally "respending" (i.e., the occupant plugs leaks and then decides he can now afford to

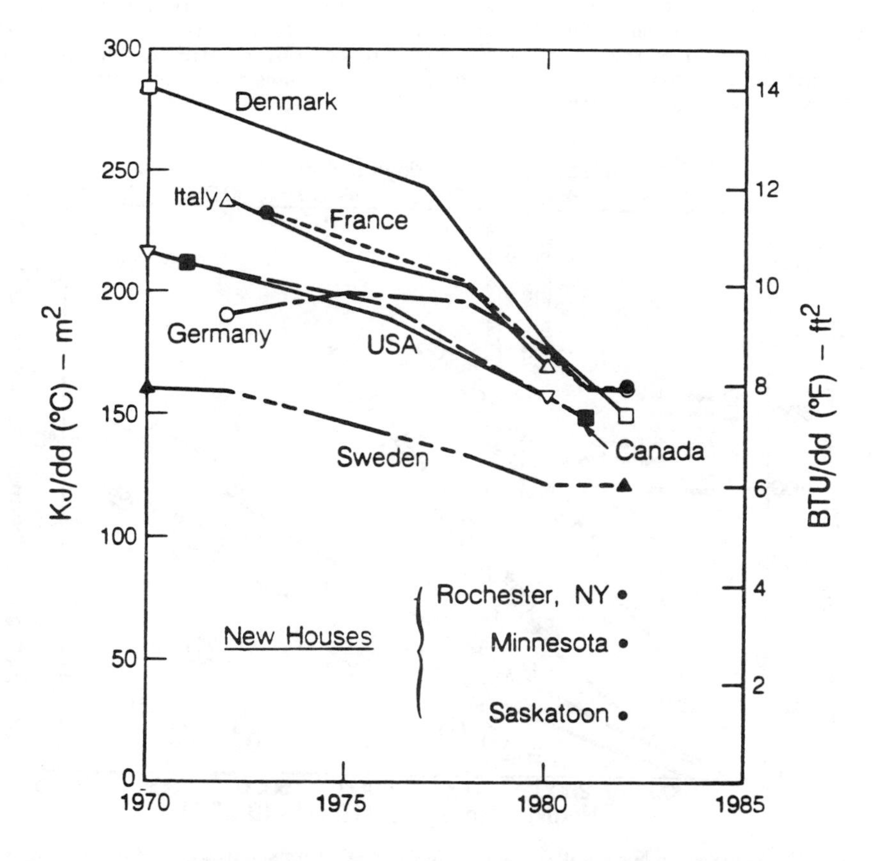

Fig. 11. Residential space heating *load* (useful furnace output) per degree-day for existing homes. To convert to English units use 1 kJ/m^2-dd(°C) = 0.049 Btu/ft^2-dd(°F). The new houses data are from a BECA survey of 175 new homes by J. Busch, A. Meier, T. Nagpal, American Council Energy Efficient Economy, 1984 and LBL-17883. Source: L. Schipper, A. Ketoff, A. Kahane, Annual Review Energy 10 (1985). LBL-19448.

heat additional rooms). Unfortunately, there has been very little monitoring before-and-after retrofit and little post-retrofit inspection, so it is difficult to sort out these factors and "tune up" the audit/retrofit industry.

The trends in international residential space heating, tracked by Schipper, Ketoff, and Kahane[6], are plotted in *Fig. 11.* United States space heat per m^2 has dropped 25% in 5 years, but sadly most of the conservation has come from lifestyle changes and not from retrofitting. For commercial buildings as a whole, fuel use/ft^2 has dropped 4%/y and electrical use has grown 0.7%/y.

o New Homes

Figure 12 compares the heat needs of various U.S. homes. It presents the same data as already plotted as a time-series in the right-hand part of Fig. 2, but this time the x-axis is the climate, measured in "heating degree days," and the units are heat (i.e., furnace output, not fuel as in Fig. 2). The higher dashed line represents the "Stock" of Fig. 2; the lower dashed line is the improved 1979 home-building practice; the solid BEPG lines are computer-optimized homes; both have 0.7 air changes per hour, but the lower line assumes heat recuperation from the exhaust air with an "effectiveness" of 2/3.

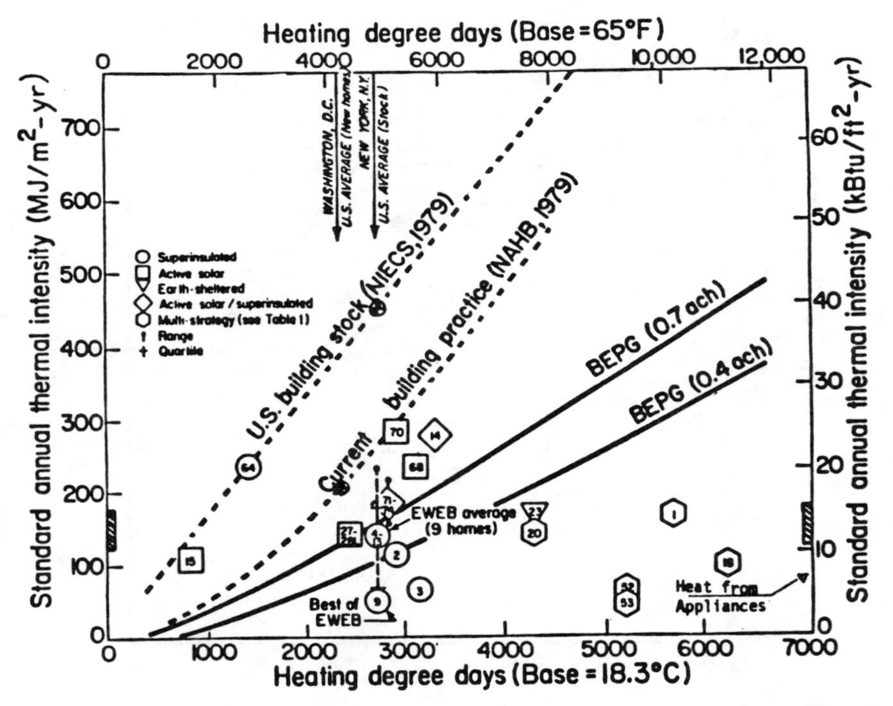

Fig. 12. Thirty-seven home scatter plot of "standardized" thermal intensity vs. climate. The various comparison curves are defined in the text. The average thermal intensity per degree-day for our 37 homes is 50 kJ/(m^2-°C-day), or half of the current building practice. Shaded bars represent typical annual water heating use - comparable to space heating in the low-energy homes shown. 5 months of winter heat from appliances is approx. 6.5 kBtu/sq.ft. Source: BECA-A (1983).

On Fig. 12, note the many superinsulated and passive solar homes with thermal intensities of 10-20 kBtu/ft^2 , which is even below the theoretical BEPG (Building Energy Performance Guidelines), for reasons already discussed in Section I. Figure 12 displays only those points in the BECA-A collection where there were enough measurements to correct each house to standard occupancy conditions; but *Fig. 13* displays many more points where internal temperatures and/or appliance usage were not measured--but where it is still clear that 20 kBtu/ft^2 is adequate even in Saskatchewan. For a 1500 ft^2 home, this translates to 30 MBtu/winter; for a modern gas furnace with an efficiency of 90%, the annual heating bill is then down to $200.

Figure 14 shows that most of these low-energy homes are cost effective. As in Fig. 10, a point on one of the sloping lines (i.e., gas at $5.60/MBtu and a 3% real interest rate) has a cost of conserved energy just equal to the price of the gas. We see that many points lie above these "indifference lines." The key (circles for superinsulation, triangles for passive solar,...) shows the strategy used. It has been very hard to find good data for active solar homes. We started with many leads, but had to throw out most of them for one of three reasons:

1. Poor thermal data,
2. Poor cost data,
3. Not measuring the "auxiliary" thermal output of a wood stove.

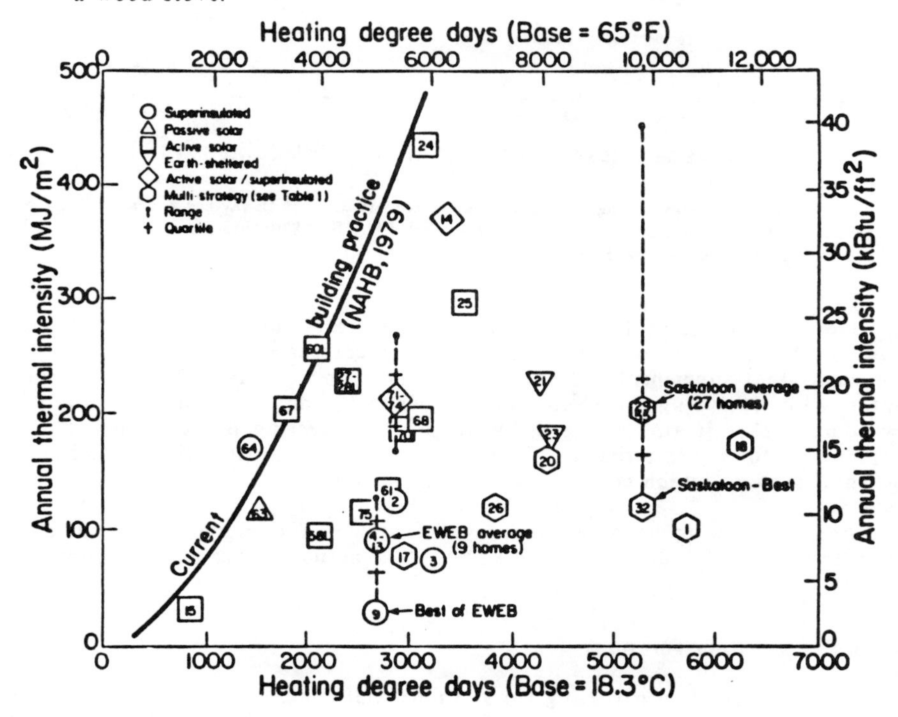

Fig. 13. Scatter plot of annual heating load/m^2 vs. climate for 28 points representing 128 sub-metered energy-efficient new homes. The solid curve is NAHB's 1979 survey of U.S. building practice, taken from Fig. 12. Source: BECA-A (1983).

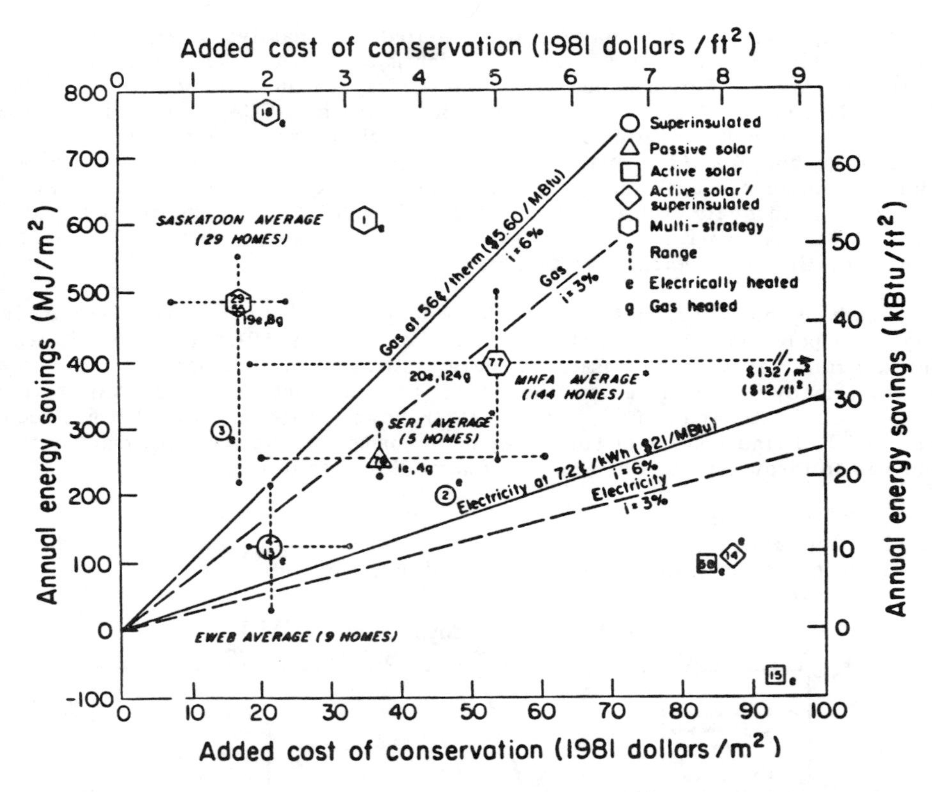

Fig. 14. Annual energy savings vs. added cost of conservation. Both savings and added cost are relative to 1979 current building practice curve labeled NAHB, 1979; thus a point at 0,0 would be a typical 1979 home. Source: BECA-A.

Several active solar houses did pass our criteria, but they are all economic failures. We are still eagerly looking for successes, but the basic economic problem is that superinsulated and passive homes need only $200 of fuel for the winter, all of it during a few cold months when the days are short and the sun is low. This makes it almost impossible to justify investing many thousand dollars in solar collectors. Note that this is not an argument against active solar domestic hot water systems which collect heat all year round.

I hope by now to have convinced you that we have enough empirical data to calibrate our supply curves for new homes. The actual curves can be found in the SERI Study[1] .

o Appliances (Refrigerators)

I'll try to give you a feeling for the potential for increasing the efficiency of appliances by discussing a single figure, 15, on refrigerators [Rosenfeld and Goldstein, 1978].

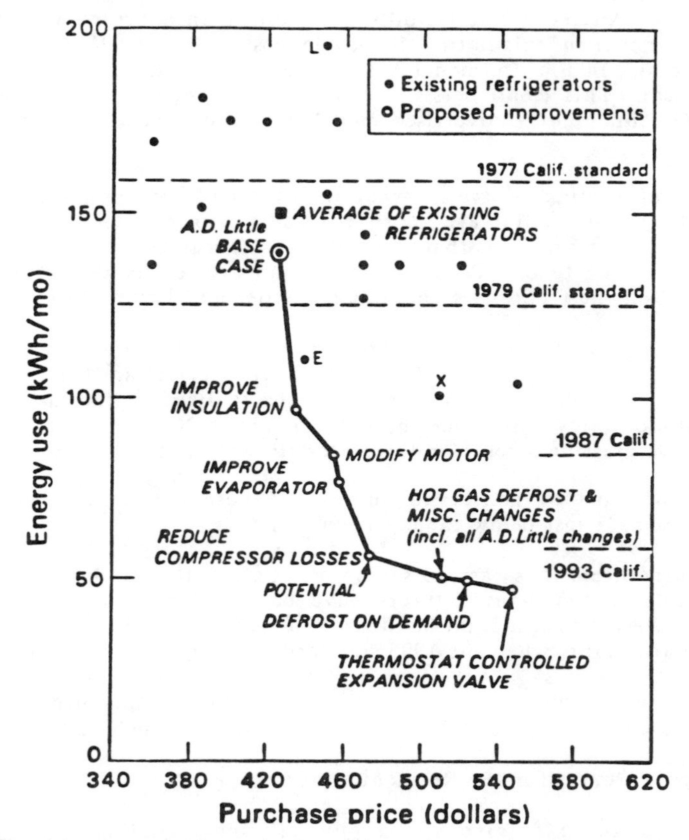

Fig. 15. Electricity use vs. purchase price for existing and proposed refrigerators. The closed circles in the upper half of the figure represent 16-17.5 cu. ft. top-freezer, automatic defrost models sold in California in 1976. The open circles joined by a heavy line are improved design steps proposed by A.D. Little (May 1977). All U.S. refrigerators plus freezers in 1980 used about 140 BkWh, so the vertical scale can also be read in BkWh, for the U.S. The potential savings of 85 BkWh is the output of 17 1000-MW baseload power plants. See figure 5 for the evolution of California refrigerator standards. XBL 7712-11464, from LBL 6865 (CA Policy Seminar)

Figure 15 dates back to 1976, before there were any appliance efficiency labels. The solid dots in the upper half of the figure represents 16-to-17.5 ft^3 automatic defrost top-freezer models for sale in California in '76. The first thing to point out is that there was almost no correlation between efficiency and price. By buying model E (for Economical) instead of model L (for Lemon), a homeowner could save 1000 kWh/year, worth (in 1976) $50/year, i.e., $1000 over the 20-year life of the refrigerator. It was data such as these that convinced the California Energy Commission that appliance standards would cost the consumer nothing and yet (for refrigerators and freezers alone) would save California 1-2.5 power plants over 20 years.

With the advent of appliance efficiency labels, there is now some mild correlation between efficiency and price, but the wise comparison shopper can still save hundreds of dollars of electric bills by investing an hour on the phone asking about prices and then calculating life-cycle costs. One further remark about appliance pricing: In Fig. 15, model X is the most efficient, but by no means the most economical. This trend persists today for most appliances and equipment. The new, efficient, highly- advertised model is almost never the most economical to buy.

So much for pricing questions--what improvements are practical? The open circles joined by a heavy line are improved design steps proposed in 1977 by A.D. Little. We should point out that several manufacturers have since produced models corresponding to the "Potential" point and are selling them for about the price indicated on the figure [see chapters by Geller and Levine in this book].

Noted on Fig. 15 (which dates from 1977) are the subsequent California mandatory refrigerator standards. They show that as prices and energy awareness both rise, engineering calculations can indeed be realized in the market place, if helped by public policy. However the delay is about 20 years, plus an added 20 years for all the existing stock of refrigerators to wear out and be replaced.

The 1¢/kWh cost of conserved energy of these better refrigerators was calculated in Sect. II just below the equations for CCE.

For manufacturers it is easy to satisfy the California standards, so whenever California updates, the manufacturers have been following for the whole U.S. Hence, in Fig. 5 we added a right-hand scale of power plant savings for the entire U.S.--30 standard plants when the 1993 standards propagate into the stock.

It should be no surprise that refrigerator improvements are a significant contribution to conservation supply curves for appliances, such as Fig. 7.

o Grand Supply Curves for the Residential Sector

I have tried above to give you a physical understanding for the sorts of options that go into the eight supply curves for the residential sector. The individual supply curves are in A New Prosperity, and the grand ensemble is displayed in *Fig. 16.* Each curve is calculated four times, for four different discount rates, ranging from 3% real (the low interest that corresponds historically to a home mortgage, where there is good information, security for the bank, and minimal risk to the lender) to 30% real (more characteristic of the appliance market, with its poorer information and security).

If we had good information programs (labels, fact sheets, buying guides) and perhaps standards, we could use the supply curves corresponding to 3% or 10% discount rates, and read off a potential for saving 500 or 550 BkWh (the output of about 100 plants). As the "implicit discount rate" rises to 20% we see that about 40% of the potential savings are lost, but then a discount rate rise from 20% to 30% does not make a lot more difference. This family of curves is a good way to display both the physical and information potential of conservation.

The lower figure tells the same story for fuel. We see that for a 3% real discount rate and the December '80 price of fuel oil (which is about the December '83 price of natural gas), there is a potential savings of about 6 quads out of the 9 needed for the base case. If the implicit discount rate rises to 20%, again about 40% of the potential is lost.

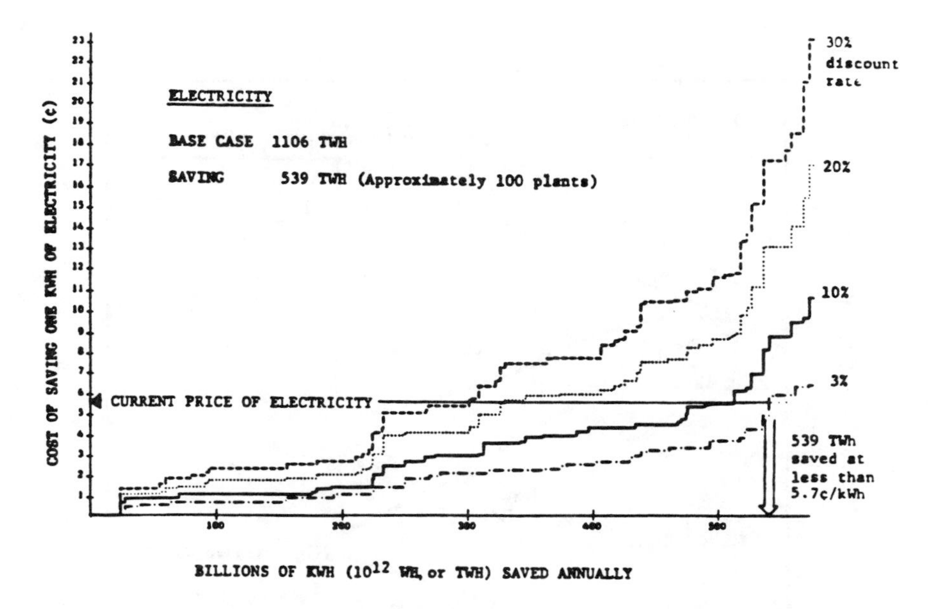

Fig. 16(a). Supply curves of conserved electricity for the U.S. residential sector (retrofit of stock, plus 20 years of construction) in the year 2000. Four curves are plotted, each corresponding to a different real discount rate. LBL-11300

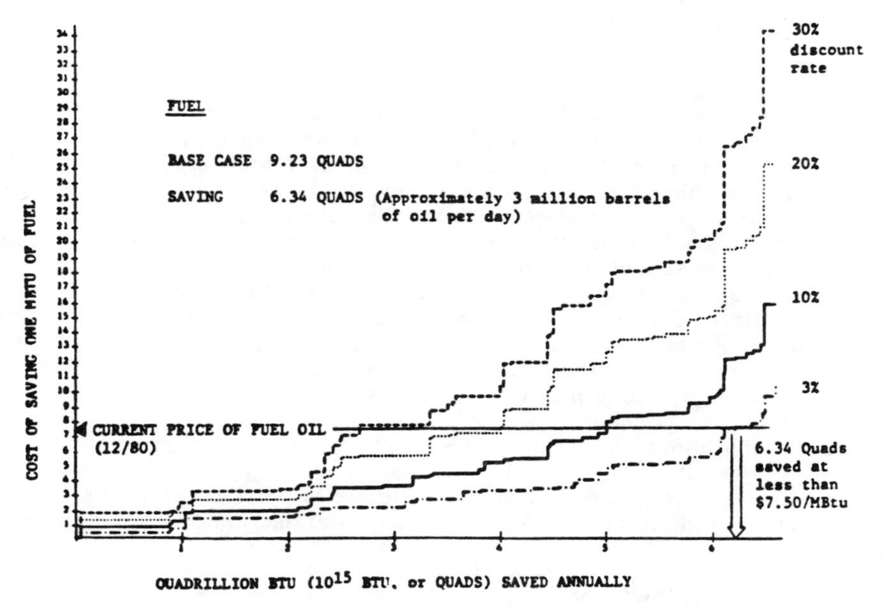

Fig. 16(b). Supply curves of conserved fuel for the U.S. residential sector (retrofit of stock plus 20 years of construction) in the year 2000. Four curves are plotted, each corresponding to a different real discount rate. Source: LBL/SERI Buildings Study. LBL-11300 (April 1981)

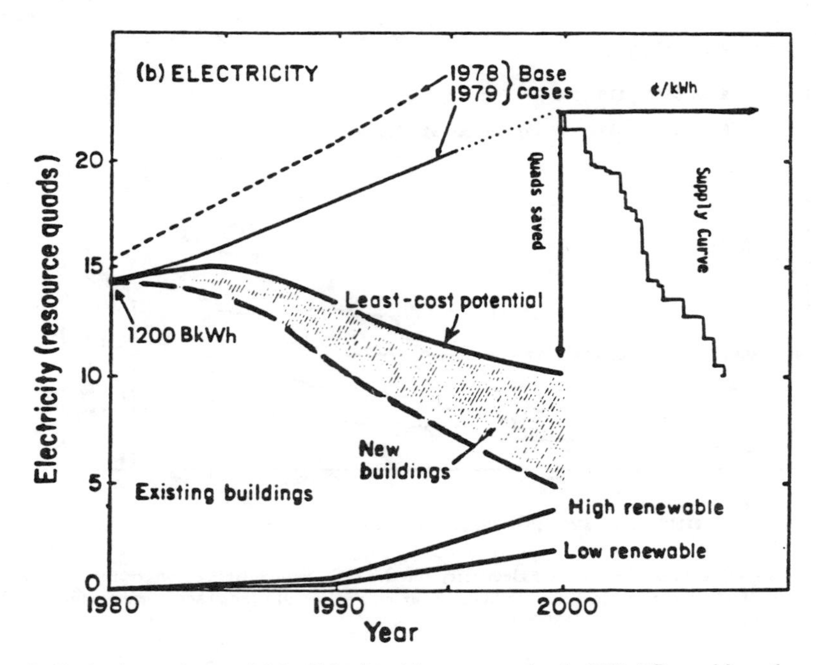

Fig. 17. Projection and potential for U.S. electricity consumption in 2000 A.D., and how they are joined by a supply curve of conserved energy, shown in figure 16. Source: SERI Report.

We want to close the loop, for the residential sector, by relating the supply curves of Fig. 16 to the time series of Fig. 3. (For the moment, pretend that the grand supply curves already include the commercial building options to be discussed in a later chapter.) *Figure 17* shows how the x-axis (energy saved) of the grand supply curve points downwards from the base case, for the appropriate year. (Remember that the supply curves of Fig. 16 apply to 2000 AD.) The "potential" then corresponds to the savings at which the supply curves cross today's price of energy. Remember, for new homes we used a 3% real discount rate (in constant dollars); for retrofit and appliances we used a 10% real discount rate. For the commercial sector the facts are generally the same, except that we shall use a discount rate of 10% real. But if the reader prefers other discount rates, he can now easily replot his own time series.

D. Commercial Buildings

Conservation potentials for commercial buildings are just as spectacular as for residences, and are discussed by Rosenfeld and Hafemeister elsewhere in this book.

ACKNOWLEDGMENTS

I want to thank Professor David Hafemeister for many interesting discussions, and both Hafemeister and Rosemary Riley for help with preparing the paper.

REFERENCES

1. A NEW PROSPERITY: BUILDING A SUSTAINABLE ENERGY FUTURE, Brick House Publishing, Andover, MA, 1981.

2. Y.J. Huang, J.B. Dickinson, A.H. Rosenfeld and B.S. Wagner, American Council for an Energy-Efficient Economy B, 135 (1984), LBL-18669.

3. H.R. Heede, Rocky Mountain Institute, Testimony to House Subcommittee on Energy Conservation and Power, June, 1985. We cite Heede because it is current and fairly complete, and shows that subsidies to supply are much larger than to investment in efficiency. But Heede counts investment tax credit when it applies to supply, but not, for example, when it helps reduce the cost of retooling to produce more efficient cars. Nor does Heede include 6% federal and 4% state gasoline tax, which adds to $10 B/year and stimulates efficiency. Also, he omits sales of electricity below market prices by Federal Power Authorities, which is another $6 B/year subsidy to the supply side.

4. A. Meier, J. Wright, and A. Rosenfeld, SUPPLYING ENERGY THROUGH GREATER EFFICIENCY, University of California Press, Berkeley, 1983.

5. E. Hirst, D. White, R. Goeltz and M. McKinstry, Energy and Buildings 8, 83 (1985).

6. L. Schipper, A. Ketoff and A. Kahane, Annual Reviews of Energy 10 (1985).

NOTES

a. $85 B is just half of $165 B (1984 $) used by today's stock. By 2000 AD our 85 million homes will grow to 110 million, our 50 billion sq. ft. of commercial floor space will grow to 70 billion, and our 25 quads of resource energy for buildings is forecast to grow to 32 (see Fig. 3), costing $200 B even at today's prices. So, a 50% savings swells to $100 B/year.

b. For a compilation of superinsulated homes, see BECA-A, ref. 2. The R-value is the thermal Resistance measured in English units, i.e., $[\text{Btu ft}^{-2}\ \text{hr}^{-1}\ {}^{o}\text{F}^{-1}]^{-1}$. R-1 converted to SI (Systeme Internationale) is $[0.176\ \text{Watt m}^{-2}\ \text{K}^{-1}]^{-1}$.

c. Mbod = Million barrels of oil equivalent per day.
1 Mbod = 2.12 x 10^{15} Btu/year = 2.12 annual "quads."

d. For an average 1985 U.S. household using gas for heating both space and water, average U.S. annual electric use is: a/c, 1200 kWh (includes homes with no a/c); refrigerator + freezer, 1800; lighting, 1000; misc., 700; Total, 4700. In addition, electric cooking uses 1000 kWh, but the electric saturation of cooking is only 60%, and drying uses 1000 (sat. = 50%). Source: J. McMahon, LBL Residential Model.

CHAPTER 7

UNDERSTANDING HEAT LOSS IN HOUSES

Gautam S. Dutt
Center for Energy and Environmental Studies
Princeton University, Princeton, NJ 08544

ABSTRACT

Heat loss from houses obeys the standard physics of conductive, convective, and radiative heat transfer. However, in a real house two- and three-dimensional conduction can greatly increase heat loss relative to a simple one-dimensional model, and subtle convective phenomena can produce additional heat loss anomalies. Radiative processes can interact with conduction and convection with surprising results. Recent developments in modeling these heat transfer processes are presented in this paper. Steady state formulations are adequate for understanding elusive heat loss paths. Non-steady-state models permit the measurement of house heat loss from short term data and permit the characterization of transient heat flow processes. The applications of several steady and unsteady heat loss models to energy conservation problems are reviewed in this paper.

INTRODUCTION

In 1980 houses consumed 8.2 exajoules of energy; the space heating component of 5.0 EJ has a substantial conservation potential.[1] Understanding heat loss is essential to realizing these savings. Early models of house heat loss used simple, one-dimensional, steady-state models and while they worked reasonably well, the omissions are important if we are to realize the full potential for energy savings in houses. In this paper recent experimental and theoretical work on heat loss from houses is reviewed. Many of the concepts presented are also applicable to the analysis of heat gain and space cooling energy needs.

In the next section, steady-state heat loss is considered in all its mechanisms including conduction, convection, radiation, and air infiltration. The effects of two-and three-dimensional conduction on house heat loss are considered. Theoretical and measurement aspects of convection particularly relevant to house heat loss -- convective loops and air infiltration -- are reviewed. Improved techniques for reducing energy use in houses using the new methods are presented.

The following section describes two approaches to modeling the transient thermal behavior of houses on the basis of small amounts of time-varying data. One model permits the determination of equivalent steady state heat loss rate. The other permits the characterization of the transient response of houses.

0094-243X/85/1350122-26$3.00 Copyright 1985 American Institute of Physics

THE STEADY STATE

The principle of conservation of energy applied to a house states that the rate of energy stored in the structure must equal the net energy input to the structure. The energy balance may be represented by:

$$C \frac{dT}{dt} = \text{Heat gain rate} - \text{heat loss rate} \qquad (1)$$

where C is the thermal capacitance of the house structure at temperature T. If the energy balance is averaged over a long time, the time derivative term vanishes and Eq. 1 reduces to a steady-state form.

In the simplest steady-state formulation, the house's heat loss rate is assumed to be proportional to the difference between a constant indoor temperature (T_i) and the outside temperature (T_o). The constant of proportionality is called the lossiness, L.

$$\text{Heat loss rate} = L\,(T_i - T_o) \qquad (2)$$

Heat loss from the house is balanced by energy input from the heating system and from other sources termed intrinsic heat. Fuel consumed by the furnace (by which is meant any space heating system) at a rate F, is delivered into the house with efficiency e. Letting I denote the average intrinsic heat, which comes from the sun, people in the house, lights and appliances, we can write the energy balance as:

$$e\,F + I = L\,(T_i - T_o) \qquad (3)$$

Solving for the furnace energy consumption rate, we obtain:

$$F = (\,L\,(T_i - T_o) - I)\,/\,e \qquad (4)$$

Eq. 4 may be restated as:

$$F = L\,(\,T_r - T_o)\,/\,e \qquad (5)$$

where $T_r = T_i - I / L$ is called the reference temperature. Physically, because of intrinsic heat, furnace energy input is not needed typically until the outside temperature falls below this reference temperature, T_r. T_r is lower than the interior temperature by an amount that increases with increased intrinsic heat and/or reduced lossiness. The term L / e is termed the heating slope, represented as β, and Eq. 5 becomes:

$$F = \beta\,(T_r - T_o)^+ \qquad (6)$$

The + in the above equation indicates that furnace heat is needed and the equation is valid only when ($T_r - T_o$) is positive; otherwise, F = 0.

Often, the space heating fuel is also used for other, non-space conditioning needs -- such as cooking or water heating -- which are assumed to consume a constant amount of fuel termed the base level, α.

The consumption of heating fuel is then given by:

$$F = \alpha + \beta(T_r - T_o)^+ \tag{7}$$

This simple, steady-state formulation has turned out to be very practical for using energy consumption billing data and commonly available outside temperature data to calculate a weather-normalized energy consumption index for a house and for measuring energy savings through energy conserving retrofits. Additional description is given in the accompanying chapter by Fels.[2] Retrofits reducing non-heating energy use lead to reduced α. Heating energy use reduction results in a lower β or lower T_r. Since β is the ratio of the house lossiness and the furnace efficiency, reducing β involves some combination of lowering the rate of heat loss from the house (lossiness) and increasing the furnace efficiency. T_r may be lowered by reducing the interior temperature spatially by zoned thermostats, or temporally by timed-setback thermostats. T_r may also be reduced by increasing the intrinsic heat, for instance, by increasing the solar energy input.

One-dimensional heat flow

We will focus here on understanding house lossiness and ways of reducing it. Consider heat loss through a house window (Fig. 1). Heat is transmitted to the interior surface of the window glass by convection from the room air and by radiation from the surfaces in the room. The heat is then conducted through the glass to the exterior surface, from which it dissipates to the outside by convection and radiation.

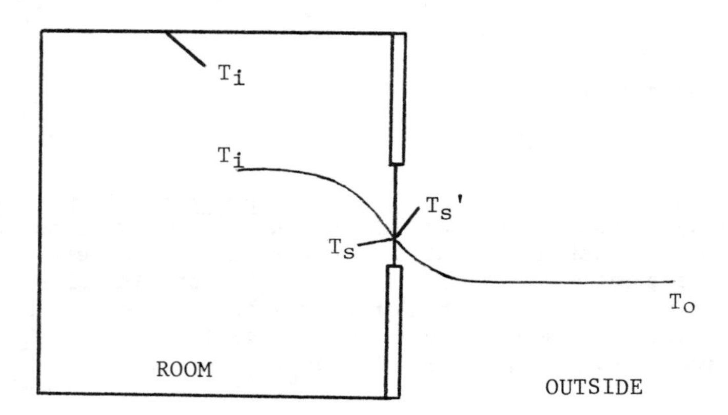

Figure 1

Heat loss from a room through a single-pane window

Consider first the heat transfer processes to the interior surface of the window, at a temperature T_s. In general the interior air and room surface temperatures will be comparable (= T_i, say).

The convective heat transfer rate is modeled as the product of a convective heat transfer coefficient and the appropriate temperature differential:[3]

$$Q_{conv} = h_{ci} (T_i - T_s) \qquad (8)$$

where Q_{conv} is the convective heat transfer rate per unit area of window, and h_{ci} is the convective heat transfer coefficient for the window interior. The radiative heat transfer rate per unit window area is:

$$Q_{rad} = \varepsilon \sigma (T_i^4 - T_s^4) \qquad (9)$$

where ε is the emissivity of the interior surfaces (assumed constant) and σ is the Stefan-Boltzmann constant (56.687 nW m^{-2} $^oK^{-4}$). Here the temperatures are expressed on the absolute scale (^oK). For small differences between T_i and T_s relative to their absolute value, Eq. 9 may be factored as:

$$Q_{rad} = 4 \varepsilon \sigma [(T_i^2 + T_s^2) (T_i + T_s)] (T_i - T_s) \qquad (10)$$

which may be written as:

$$Q_{rad} = h_{ri} (T_i - T_s) \qquad (11)$$

where h_{ri} is the interior radiative heat transfer coefficient and is approximately constant. At typical room temperatures, the radiation to the window is in the infrared range where glass is largely opaque. For visible radiation to the window (e.g. sunlight), one would have to consider radiant energy transmission through the window as well.

The sum of convective and radiative heat transfer from the room interior to the window surface facing the interior is given by:

$$Q_i = (h_{ci} + h_{ri}) (T_i - T_s) = h_i (T_i - T_s) \qquad (12)$$

where h_i is the total (convective plus radiative) heat transfer coefficient at the window interior.

Conduction through the window pane is modeled using Fourier's Law as the product of the temperature difference between the interior and exterior surface of the window and the thermal conductance, k, of the glass:

$$Q_k = k (T_s - T_{s'}) \qquad (12a)$$

where $T_{s'}$ is the exterior surface temperature of the window. Finally, the convective-radiative heat transfer from the window exterior to the outside is modeled in a manner similar to that described above for heat transfer from the room to the interior surface of the window, using a total heat transfer coefficient, h_o, for the exterior surface.

An equation in which heat transfer rate is proportional to a constant times a temperature difference is analogous to Ohm's Law for the electrical current between two points separated by a resistance R and a potential V (Ohm's Law):

$$I = (1/R)\ V \tag{13}$$

In this analogy, the heat flow rate is the current, the temperature difference is the potential, and the heat transfer coefficient is the electrical conductance or the inverse of the electrical resistance R. The three heat transfer steps for heat loss through the window are analogous to three thermal "resistances" in series between the interior and outdoor temperatures. The three resistances may be added in the thermal case as well. The heat transfer from inside to outside is then given by:

$$Q\ (I\ \text{to}\ O) = \frac{1}{\frac{1}{h_o} + \frac{1}{h_i} + \frac{1}{k}}\ (T_i - T_o) \tag{14}$$

A single-paned window is an example of a single element boundary of the house. For a multi-element wall or ceiling there will be additional resistances in series which must be included.

<u>Two- and three-dimensional effects -- thermal bridges</u>

In the discussion above, we have considered heat transfer through the wall or window to be one dimensional. Clearly this would not be the case for a corner, even in a single-element wall. To calculate the conductive heat transfer from the inside surface of a corner at temperature T_s to the outside surface at temperature T_s' (Fig. 2), we look at the two-dimensional heat transfer equation:

$$\nabla^2 T = 0 \tag{15}$$

For certain geometries, the solution to Eq. 15 is facilitated by a technique known as <u>conformal mapping</u>.[4] By this method, a difficult geometry such as that in Fig. 2 is mapped to a geometry in which the solution to Eq. 15 is straightforward. The appropriate transformation for our problem is given by:[4]

$$\frac{dz}{dt} = \frac{G\ (t - 1)^{1/2}}{t\ (t + a)^{1/2}} \tag{16}$$

The corner in the z-plane is transformed to the real axis on the t-plane. The points A, B, C, D, E, and F in the physical wall (z-plane) transform to $-\infty$, $-a$, 0, 1, $+\infty$ in the t-plane.

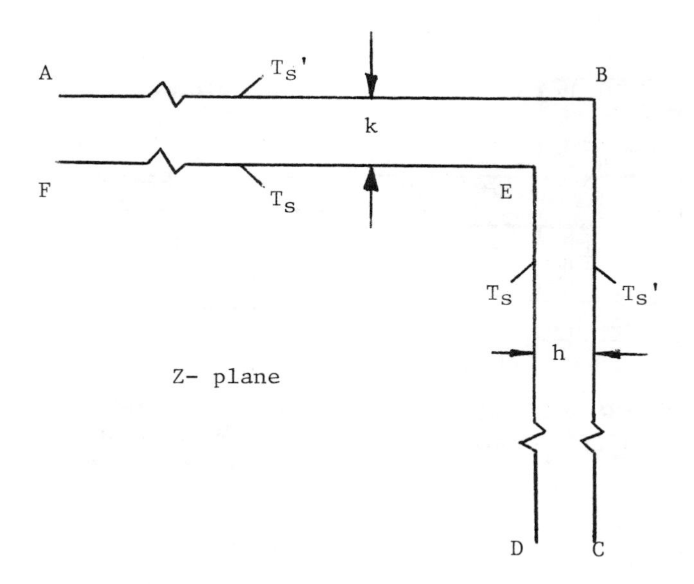

Figure 2

Calculating corner heat loss
by conformal mapping
(Source: ref. 4)

To integrate, let
$$\xi = \left(\frac{t - 1}{t + a}\right)^{1/2} \tag{17}$$

so that
$$z = -\frac{2G}{a^{1/2}} \tan^{-1}(\xi a^{1/2}) + G \ln\left(\frac{1 + \xi}{1 - \xi}\right)$$

where the constant of integration has been made zero by choosing z = 0 when t = 1, i.e., E to be the origin in the z plane.

Conformal mapping permits analytical solutions to relatively few two-dimensional heat conduction situations. Two-dimensional conduction is typically calculated numerically using finite-element techniques.[5] Consider a masonry wall supported on a concrete floor. Figures 3a and 3b show cross sections of the floor-wall joint and wall-interior-partition joint.[6] If the wall is insulated on the inside, then the concrete floor and the partition are gaps in the insulation and act as thermal bridges. Figure 3c shows the effective conductance of a 4 m by 3.2 m wall with different thicknesses of insulation. At low levels of insulation, the effective conductance is close to the nominal value, calculated assuming one-dimensional conduction. However, as the insulation is increased on the inside, the thermal conductance soon becomes over twice that of the nominal case, due to gaps and other factors. Insulation placed on the outside of the wall leaves fewer gaps and its effective conductance is closer to the nominal value. The concrete floor and interior

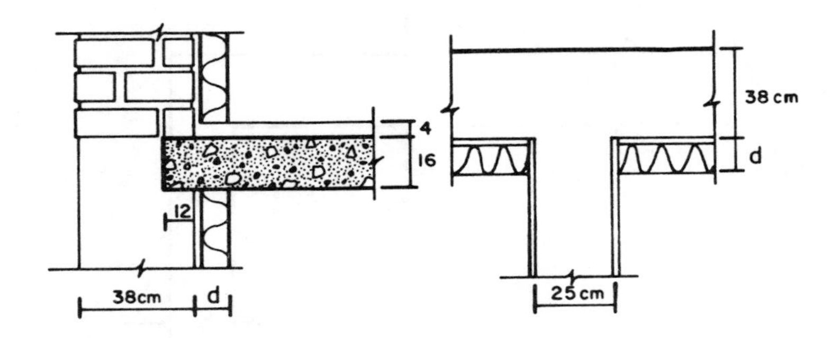

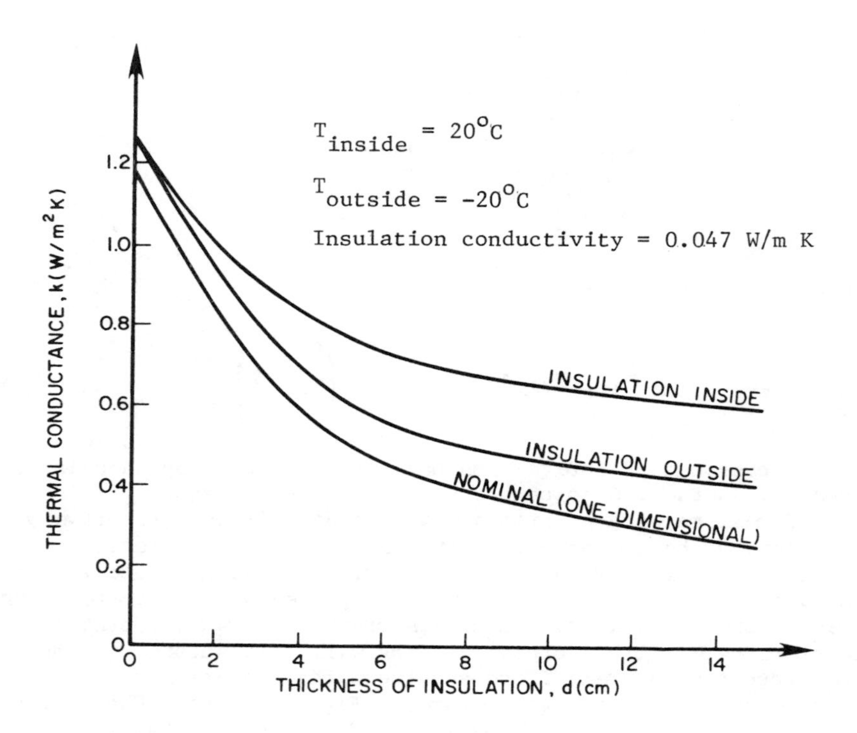

c) Effective conductance with insulation inside and outside, compared to nominal values.

Figure 3

Effect of thermal bridges on conductance
(Source: ref.6)

partition thus permit a heat loss path of higher conductance than the insulation, and belong to a class of heat loss paths known as thermal bridges.

Figures 4a and 4b illustrate thermal bridges in wood-frame walls created by the framing elements, which have much higher thermal conductance than insulating materials.

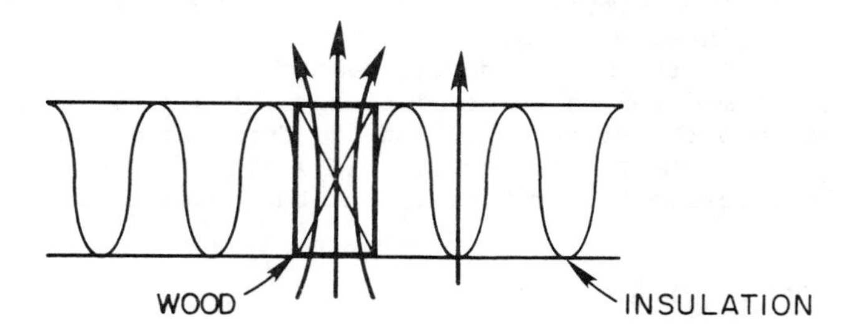

a. STUD IN AN INSULATED WOOD FRAME WALL

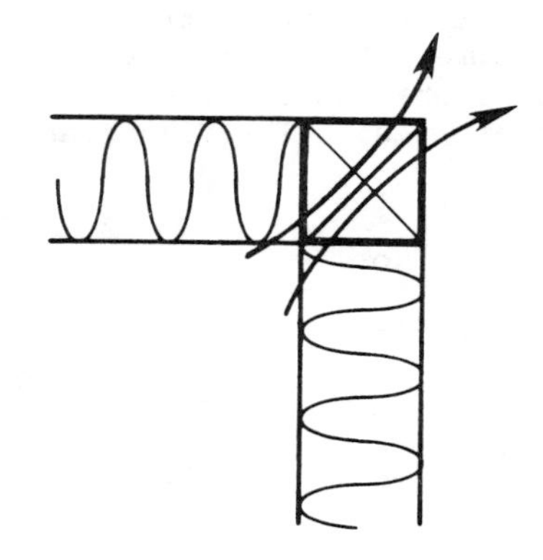

b. CORNER ELEMENT IN AN INSULATED WALL

Figure 4

Thermal bridges in a wood-frame wall

Three-dimensional heat conduction is also encountered in houses. Important cases involve heat transfer from the house to the ground and to the outside through basements, crawl spaces, and slabs. When houses are better insulated above grade, heat transfer through the foundation becomes more important. Only recently have Mitalas[7] and Kusuda[8] developed procedures for calculating basement and slab heat loss in houses, respectively. In addition to three-dimensional conduction, the involvement of the ground with a large heat capacity implies that the steady state heat loss assumption is no longer valid and transient heat flow modelling is required.

<u>Bypass heat losses and convective loops</u>

As shown above, two- and three-dimensional heat conduction effects cause heat loss from houses to be significantly different from the one-dimensional models on which most house heat loss calculations are based. Princeton researchers have observed another major departure from simple modeling, in measurements of attic temperature in a number of townhouses at Twin Rivers instrumented in the mid 1970s. In the simple model, heat flow from the living space to the attic is modeled as two conductances in series -- one from the living space into the attic (W_{ia}) and one from the attic to the outside (W_{ao}). The attic temperature (T_a) can then be expressed in terms of the interior temperature T_i and the outside temperature T_o by the equation:

$$W_{ia} (T_i - T_a) = W_{ao} (T_a - T_o) \quad (17)$$

Eq. 17 may be expected to hold during late night periods when the effects of solar radiation on the roof, not taken into account here, are absent. Since the Twin Rivers houses had insulation on the attic floor and no insulation on the roof, Eq. 17 implies that the attic temperature would be much closer to the temperature outside than to the house interior temperature. Princeton's data from many instrumented townhouses indicated that attics were much warmer than predicted by Eq. 17, even at night. The discrepancy is quantified below:[9,10]

$$\text{Expected:} \quad \frac{T_i - T_a}{T_a - T_o} = \frac{W_{ao}}{W_{ia}} = 8.9;$$

$$\text{Measured:} \quad \frac{T_i - T_a}{T_a - T_o} = 1.0, \text{ on average;}$$

At night, the attic temperatures were typically halfway between the inside and the outside. The anomaly was eventually identified as the result of parallel heat loss paths of considerable conductance bypassing the attic floor insulation. The heat transfer through the attic was better modeled by a three-resistance model[10] (Fig. 5). The bypass conductance could be determined from measurements of night-time attic temperature.[9] For the Twin Rivers townhouses, the bypass conductance implied that actual nighttime heat losses were <u>five times</u> those predicted on the basis of the traditional models which only considered conduction in and out of the attic and attic ventilation. These results were startling and in the following winter (1977-78) a larger survey was conducted to determine whether the attic bypass heat loss phenomenon was present in other houses as well. This survey used low cost maximum-minimum thermometers to record the night-time minimum temperatures in the living space, attic and outside.[11] The survey demonstrated that attic heat losses were 2.9

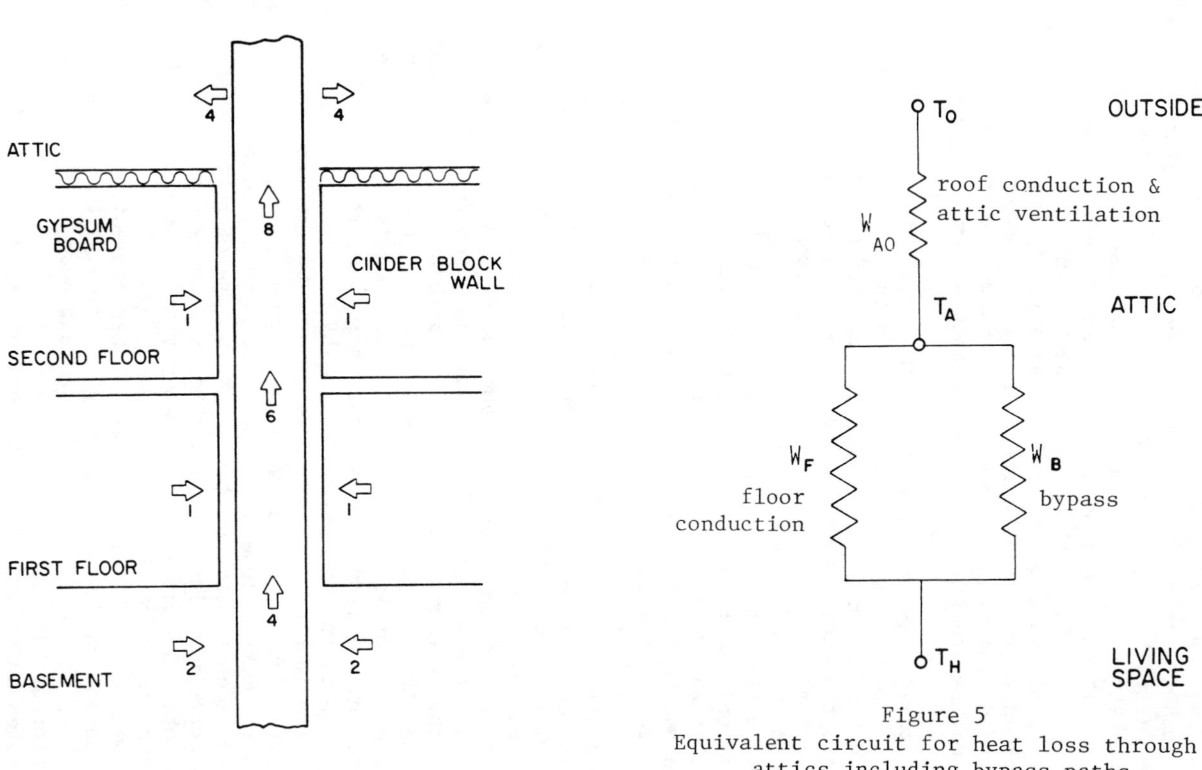

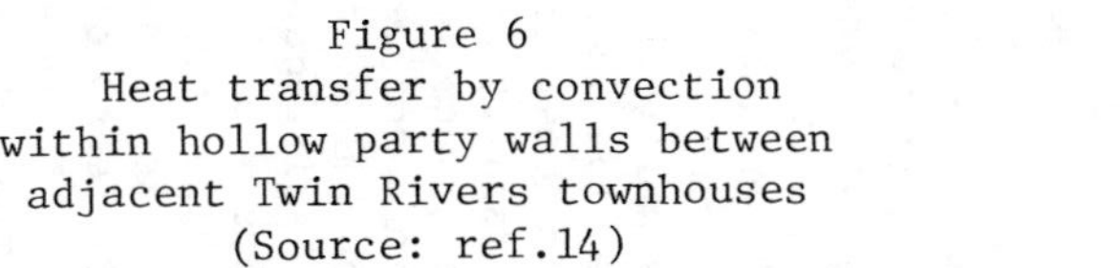

Figure 6
Heat transfer by convection
within hollow party walls between
adjacent Twin Rivers townhouses
(Source: ref.14)

Figure 5
Equivalent circuit for heat loss through
attics including bypass paths
(Source: ref.12)

to 7.5 times as much as predicted by the traditional two-conductance model.[12]

What constitutes these bypass heat losses? A number of possibilities were investigated in the Twin Rivers houses.[13] Heat flux measurements through the attic floor insulation indicated that the bypass heat losses were not the result of poorly performing insulation. Similarly, heat transfer from a metallic furnace flue that passed through the attic also could not add enough heat to account for the attic overheating. Although, air leakage from living space to attic accounted for a significant amount of heat transfer, even after these air leakage paths were sealed and the measured house-to-attic air flow had dropped to very small values, the attic temperature discrepancy remained. The major culprit, in fact, was eventually shown to be the hollow masonry "party" walls that separate adjacent townhouses.

Twin Rivers townhouses are separated from their neighbors by hollow cinder block party walls that extend from the basement floor to the roof. The two-dimensional conduction "fin effect" of these walls jutting into the attic was first suspected as a potential heat loss path; however, simple heat conduction calculations quickly revealed that conduction could not account for the extent of heat transfer implied by the attic temperature discrepancy.[13] Eventually, virtually at wits end, we hypothesized that convection within the hollow cinder block walls accounted for the bulk of the heat transfer. There was no theory against which we could judge our hypothesis. A relatively cold spell in May of 1977 allowed us to test our hypothesis. We rushed to Twin Rivers and covered both the party walls in an attic with fiberglass insulation. If the party walls were indeed responsible then this would considerably reduce the amount of heat contributed to the attic through them. Our hypothesis proved to be correct -- the attic was much colder on subsequent nights, in fact scarcely warmer than the outside.[13] Thus heat was being conducted into the air in the party wall cavities from the living space and the basement, convected up through the cavities, and transferred to the attic by conduction (see Fig. 6). The numbers in the figure represent relative magnitudes of heat transfer within the block wall, as determined from an empirical three-zone model.[10] The convection is driven by an instability caused by air density differences: the air in the block cavities between two attics is considerably colder than the air between two living spaces; the colder, denser air located above the warmer, lighter air below is unstable and sets up convection patterns. This type of heat transfer is now called a <u>convective loop</u>. A quantitative theory for calculating convective loop heat losses has yet to be developed and remains a challenge to researchers.

Since the Twin Rivers experience, many other convective loop heat loss paths have been discovered. They often exist in lowered ("soffited") ceilings above bathtubs, kitchen cabinets, stairwells as well as in interior partitions.

Thermal bridges and convective loops may be rapidly located using infrared thermography.[14] In this technique, an optical image is created from infrared radiation emitted from various surfaces in the house. The infrared radiation intensity depends on the surface temperature and emissivity. On surfaces of uniform emissivity, variations in radiation levels can be used to determine surface temperature distributions. Patterns on visual image of the infrared radiation field seen on a CRT, for example, are extremely useful in identifying the type of heat loss -- insulation voids, thermal bridges, or convective loops. It was infrared thermography that first showed, in Twin Rivers townhouses, that more heat was being lost through some of the interior walls than any of the exterior walls.[15] This was due, of course, to the party wall convective loops discussed above.

Air infiltration

Air leakage from living space into attic is only one of several pathways by which air flows into and out of a house. These air flows -- called infiltration and exfiltration -- provide ventilation and account for a significant amount of energy use. The air leaking into the house must be heated before it leaks out again. The heat loss associated with air leakage is:

$$Q_{ai} = AI \cdot V \cdot \rho C_p (T_i - T_o) \qquad (18)$$

where AI is the air infiltration rate defined as the number of house volumes of air entering (or leaving) the house each hour, ρ is the air density, C_p is its specific heat, and V the house volume. The air infiltration rate depends on the distribution of leaks and pressure over the building envelope. The leakage sites are cracks and openings; the pressure distribution is induced by wind and by air density differences (resulting from temperature differences). The pressure distribution imposed by density differences, which depend on the inside-outside temperature difference, is called the stack effect. The wind-induced pressure differences depend on the wind speed, its orientation, the obstructions around the house, as well as the distribution of leaks. The wind and stack effects interact in such a way as to reinforce or cancel each other, depending on wind direction and the location of the leaks.[16] Sinden has shown that the effects of the wind and stack effects are subadditive, i.e., the combined effect is always smaller than the sum of the two effects taken separately:[16]

$$AI(W, \Delta T) \leq AI(W, 0) + AI(0, \Delta T) \qquad (19)$$

where $AI(W, \Delta T)$ is the air infiltration induced by a wind W and a temperature difference ΔT.

In the Sherman air infiltration model, developed at Lawrence Berkeley Laboratory, the stack and wind effects add in quadrature:[17]

$$AI(W, \Delta T)^2 = AI(0, \Delta T)^2 + AI(W, 0)^2 \qquad (20)$$

which is consistent with, and a special case of, the Sinden inequality. Sherman has developed a procedure for calculating the air infiltration rate of a house based on the wind speed, the inside-outside temperature difference, a measure of the leakiness of the house called the equivalent leakage area, and certain parameters characterizing the wind environment of the house. The model additionally assumes that the leaks are evenly distributed with respect to the compass directions. In order to calculate the air infiltration of a house its equivalent leakage area must be known; this can be readily determined using a house pressurization technique originally developed in Sweden in the mid 1970s. In the USA, houses are now most often pressurized using a portable device called a blower door.[14]

A blower door is a high flow fan that can be attached to an exterior doorway of a house. With all other intentional openings in the house closed, the fan is used to slightly pressurize or depressurize the house. Pressures typically attained -- in the range of 10 to 75 pascals -- are much larger than the pressure differences normally experienced by the house under the stack or wind effect. Thus the fan imposes a uniform pressure difference across the house envelope. If the fan is calibrated, then both the induced pressure and the corresponding flow through the fan may be determined. The flow through the fan equals the flow through the leaks in the house envelope, and a plot of the leakage flow against pressure is called the pressure-leakage profile. The profile usually takes the shape shown in Fig. 7, and the relationship between flow and leakage may be expressed as:

$$Q = C\,(\Delta P)^n \qquad (21)$$

The parameters C and n which characterize the flow can be determined by regression of the data. The index n usually falls between 0.5 and 1.0. These two values correspond to fluid mechanic "limits" -- 0.5 corresponds to orifice or turbulent flow while 1.0 corresponds to laminar flow. The air flow through the numerous leakage sites in a house must be some combination of laminar and orifice flow.

The air leakage of a house as determined by a blower door may be expressed as the flow rate at a particular interior-exterior pressure difference, usually 50 pascals. For use in his model for the air infiltration rate, Sherman calculates an equivalent leakage area from the pressurization data as follows:

$$ELA = \frac{Q\ (4\ Pa)}{(2\ \Delta P/\rho)^{1/2}} \qquad (22)$$

where ELA is the equivalent leakage area and Q (4 Pa) is the leakage flow at an induced pressure difference of 4 pascals, and ρ is the air density.

Using ELA measured from house pressurization, the Sherman model may be used to estimate the air infiltration of a house under the normally occurring stack and wind effects. The air infiltration is constantly changing in response to changes in driving forces. The overall energy use due to air infiltration is only sensitive to the seasonally-

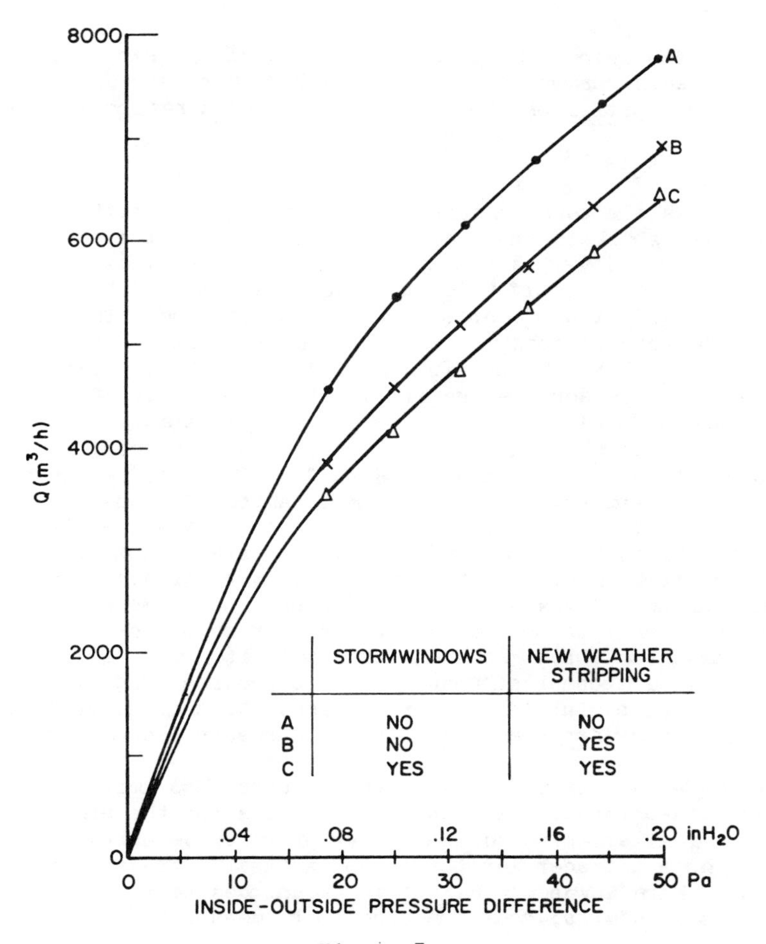

Figure 7
Pressure-leakage profile of a house (Source: ref.18)
These data illustrate that window leakage typically accounts for a small part of the whole house air leakage

averaged infiltration rate. A simplified version of the Sherman model can be used to calculate a seasonal average value of the air infiltration rate.[19] Much cruder empirical relationships can also be used to estimate seasonal average values of the air infiltration rate from pressurization data. According to one empirical relationship:[20]

$$AI \text{ (seasonal average)} = Q \text{ (50 Pa)} / 20 \tag{23}$$

The air infiltration rate can also be measured directly using a variety of tracer gas techniques.[21] In the most popular method, a a few cm^3 of a harmless tracer gas (sulfur hexafluoride) is mixed with the house air, and the tracer gas concentration is monitored over time. The indoor tracer concentration decays exponentially as indoor air leaks out and is replaced with outdoor air containing negligible quantities of the tracer. The concentration is given by:

$$C(t) = C_o e^{-kt} \tag{24}$$

where C (t) is the concentration at time t, C_o the initial concentration and k is the constant characterizing the exponential decay. The quantity k is the air infiltration rate, expressed in house volumes per hour if t is in hours, and can be determined from log-linear regression of concentration against time. The air infiltration rate commonly carries the units of air changes per hour (ACH) even though the house air volume is not actually replaced.

The winter season average air infiltration rates of most houses fall between 0.3 and 1.0 ACH. However air infiltration depends on wind and temperature difference and is continually changing; for the same house it can vary from 0.1 ACH to 1.5 ACH at different times. This large variability has serious implications for the majority of houses that depend on natural air leakage to provide ventilation: a part of the time, they are inadequately ventilated and for a significant portion of time the leakage is much higher than needed for ventilation and wastes energy. Making a house more air tight reduces the energy waste but extends the period of inadequate ventilation. For these reasons, new houses in Sweden are required to have controlled mechanical ventilation to provide adequate ventilation while minimizing energy waste. Controlled ventilation will be necessary for energy-efficient houses in Canada and the U.S. as well.[22]

The wide variability in air infiltration also implies that extensive measurements are needed to characterize the air leakage of a house. An alternative to the tracer dilution technique uses a passive constant tracer gas source and a passive tracer gas sampler.[23] More advanced tracer gas techniques permit air flows among zones in a multi-zone building to be measured.[24-26]

Finding and fixing energy flaws in houses

Air leakage sites can be readily located using a blower door, which exaggerates the leaks, together with a smoke gun or infrared thermography.[14] Cold outside air leaking into a house chills surfaces adjacent to the leaks and its imprint can be detected in an infrared radiation pattern. A calibrated blower door can be used to quantify the severity of air leakage, determine the energy conservation potential through house tightening, and make sure that the house is not too tight. If house tightening is carried out while the blower door remains installed, the effectiveness of the work can be quantified. At least 300 blower doors nationally are being used by contractors providing an air infiltration reduction service that we call shell tightening.

In the above discussion, we have treated air infiltration separately from thermal bridges and convective loops. Although conceptually different, they can be diagnosed simultaneously using a combination of house pressurization and infrared thermography.[14,27] Using additional instrumentation such as digital thermometers, furnace combustion analyzers, etc. one can carry out a comprehensive instrumented energy analysis covering the house heat losses, and the space and water heating systems. Instrumented analyses may be combined with on-the-spot retrofits to bring about immediate savings. The combination of instrumented energy analysis with on-the-spot retrofits is the basis for the house doctor approach.[28] Even though house doctors focus on obscure defects that require the use of the diagnostic equipment, the energy savings are significant. Additional savings are realized through conventional retrofits such as the addition of insulation or storm windows. The costs and savings from house doctoring and followup retrofits were first determined in the 138-house Modular Retrofit Experiment.[29] For an average investment of $1315 (in 1980), the combined house-doctor and contractor retrofits reduced the average space heat component of energy use by 30%, even though most of the houses already had attic and wall insulation and storm windows. Although the entire retrofit package shows an attractive return on investment, the house doctor retrofits were found to be more cost-effective than the conventional retrofits.

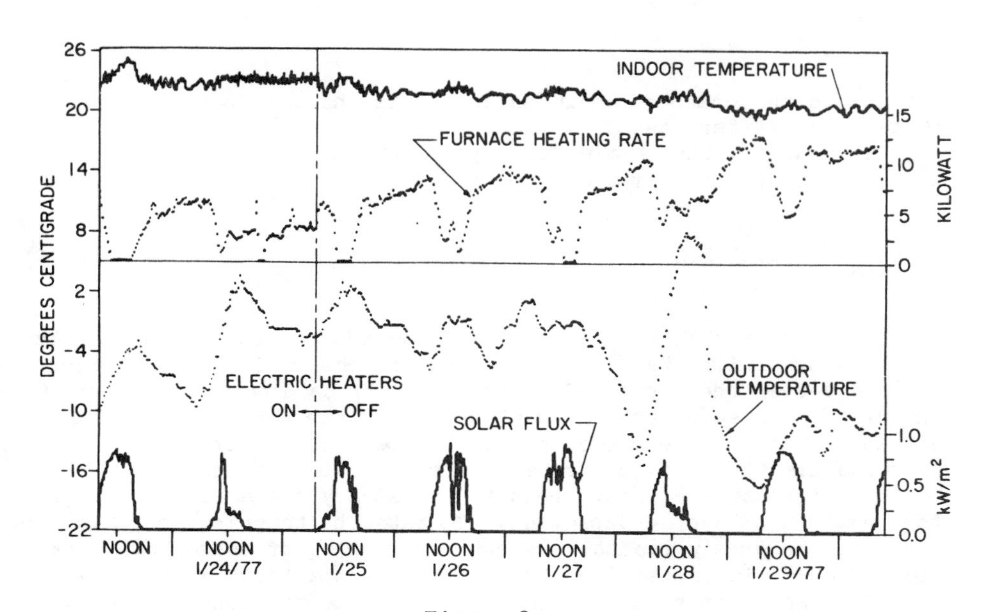

Figure 8
Indoor and outdoor temperature, furnace heating rate and solar flux on south-facing wall (Source: ref.30)

THE UNSTEADY STATE

Measuring "steady-state" lossiness from short term data

In the discussion above, we noted how the overall house thermal performance can be determined from energy billing data spanning a year or more. We also discussed diagnostic techniques by which components of the house may be evaluated. Often it is desirable to evaluate the thermal performance of the whole house with short term data of days to weeks. We do not always have the luxury to wait a year to evaluate the success of a retrofit project, especially if we need to evaluate several retrofits in sequence. In the short term, the steady-state heat balance assumption is often no longer valid. Fig. 8 shows indoor and outdoor temperature, solar flux and energy input to the heating system for a test house.[30]

To understand the transient phenomena better, consider a simple opaque chamber on a square base 2.5 m wide and 3.5 m tall, built to be extremely airtight, and relatively well insulated.

If the chamber were subjected to steady outdoor temperatures while its indoor temperature was also kept constant, the heat loss rate would be given by:

$$Q = L \Delta T \qquad (25)$$

where L is the chamber lossiness and ΔT is the inside-outside temperature difference, $T_i - T_o$.

In reality, the chamber is exposed to time varying weather and so we define an instantaneous lossiness as:

$$L_{inst} = \frac{Q_{inst}}{\Delta T_{inst}} \qquad (26)$$

In the steady state, L_{inst} and L would be identical, but in reality L_{inst} changes with time. Fig. 9 shows the change in instantaneous lossiness over the course of a night. L_{inst} is low at the start of the night because of the residual effect of the warmer temperature and solar heat of the preceding daytime period. In order to characterize the heat loss rate of the structure we define a "material" lossiness, equivalent to the steady state lossiness L, that would not be time dependent. To take the transient effects into account, the purely resistive electrical analog of house heat loss (Eq. 12 - 14) is replaced by the resistive-capacitive model shown in Fig. 10. Here the structure is represented by two conductances U_i and U_o and a capacitance C. Heat flows from an interior node (T_i) at a rate Q_i and flows into an outer node (T_o) at a rate Q_o. The central node is at a temperature T_c which is also the temperature of the thermal "capacitance" C.

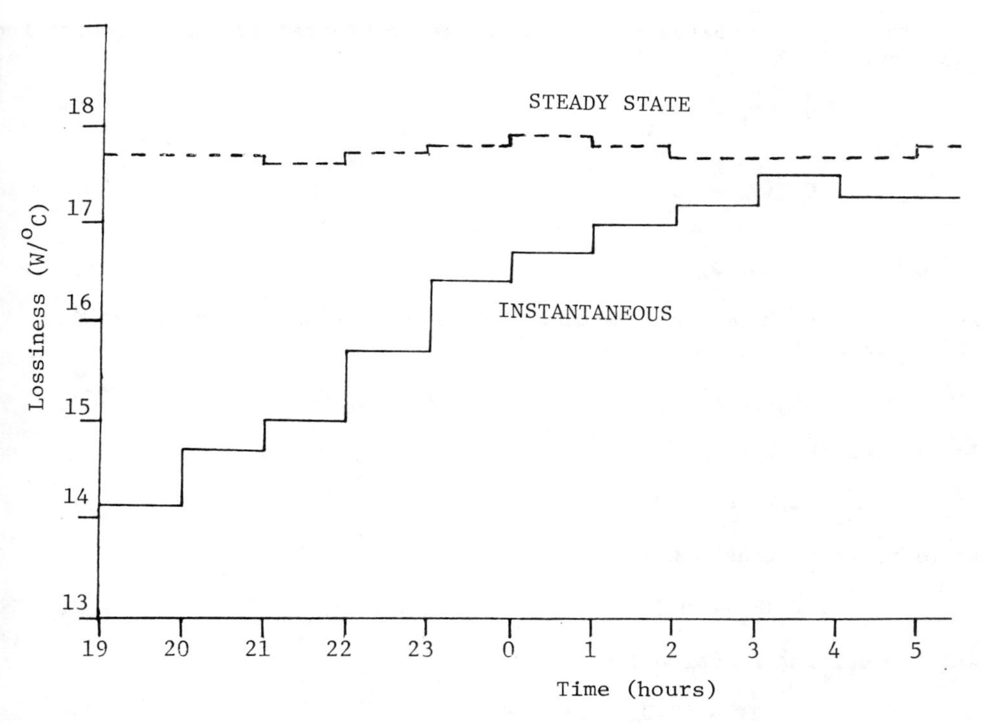

Figure 9
Instantaneous and steady-state lossiness
(Source: ref.31)

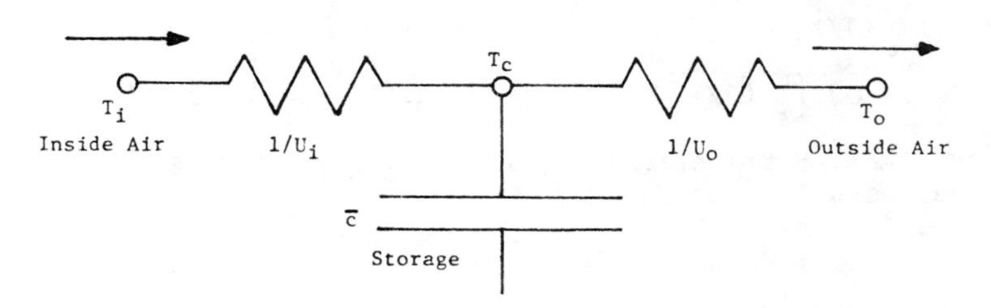

Figure 10
One-capacitance heat flow model of test chamber
(Source: ref.31)

The heat flow rates in the system are described by the following thr equations:[31]

$$Q_i = U_i (T_i - T_c) \tag{27}$$

$$Q_o = U_o (T_c - T_o) \tag{28}$$

$$C \dot{T}_c = Q_i - Q_o \tag{29}$$

where the dot above the T denotes a time derivative. Eliminating T_c and Q_o yields:

$$\dot{Q}_i + (U_i + U_o) Q_i/C = U_i \dot{T}_i + (T_i - T_o) U_i U_o/C \tag{30}$$

Eq. 30 has the form

$$\dot{Q}_i + Q_i/\tau = f(t) \tag{31}$$

where the time constant τ is:

$$\tau = C/(U_i + U_o) \tag{32}$$

and the driving function f(t) is:

$$f(t) = U_i \dot{T}_i + (U_i U_o/C)(T_i - T_o) \tag{33}$$

The solution of Eq. 31 is:

$$Q_i(t) = U_i \int_{-\infty}^{t} e^{-[(t-t')/\tau]} \dot{T}_i(t')\, dt' + (U_i U_o/C) \int_{-\infty}^{t} e^{-[(t-t')/\tau]} [T_i(t') - T_o(t')]\, dt' \tag{34}$$

or, after integrating the first term by parts,

$$Q_i(t) = U_i \int_{-\infty}^{t} (1/\tau) e^{-[(t-t')/\tau]} [T_i(t) - T_i(t')]\, dt' + (U_i U_o \tau/C) \int_{-\infty}^{t} (1/\tau) e^{[(t-t')/\tau]} [T_i(t') - T_o(t')]\, dt' \tag{35}$$

The coefficient of the second integral, $U_i U_o \tau/C$ equals the steady-state lossiness, L, (see Eq. 32)

$$L = U_i U_o/(U_i + U_o),$$

Thus the lossiness L is determined by Eq. 35 as:

$$L = \frac{Q_i(t) - U_i \int_{-\infty}^{t} (1/\tau) e^{-[(t-t')/\tau]} [T_i(t) - T_i(t')]\, dt'}{\int_{-\infty}^{t} (1/\tau) e^{-[(t-t')/\tau]} T_i(t')\, dt' - \int_{-\infty}^{t} (1/\tau) e^{-[(t-t')/\tau]} T_o(t')\, dt'} \tag{36}$$

Eq. 36 can be simplified by noting that:

$$\int_{-\infty}^{t} (1/\tau)\, e^{-[(t-t')/\tau]}\, dt' = 1 \tag{37}$$

Thus,

$$L = \frac{Q_i(t) - U_i[T_i(t) - \int_{-\infty}^{t} (1/\tau) e^{-[(t-t')/\tau]}\, T_i(t')\, dt']}{\int_{-\infty}^{t} (1/\tau) e^{[(t-t')/\tau]} T_i(t')\, dt' - \int_{-\infty}^{t} (1/\tau) e^{-[(t-t')/\tau]} T_o(t')\, dt'} \tag{38}$$

$$L = \frac{Q_i(t) - U_i\ [T_i(t) - T^*{}_i(t)]}{T^*{}_i\ (t) - T^*{}_o\ (t)} \tag{39}$$

where T^* is the <u>trailing temperature</u> defined by:

$$T^*(t) = \int_{-\infty}^{t} (1/\tau)\, e^{-[(t-t')/\tau]} T(t')\, dt' \tag{40}$$

The trailing temperature is a weighted average of all past temperatures. Since, temperatures are weighted less and less as one moves further into the past, the trailing temperature is not greatly influenced by the actual temperature more than a short time back. In practice, with data recorded at discrete time intervals, the integral defining T^* is replaced by a summation which, for sufficiently small time steps Δ, reduces to:

$$T^*(t) = [(\sum_{k=0}^{\infty} \gamma^k\ T(t-k\Delta))] \,/\, (\sum_{k=0}^{\infty} \gamma^k) \tag{41}$$

where $\gamma = e^{-\Delta/\tau}$. (42)

Eq. 41 defines the trailing temperature at time t in terms of the actual temperature history of T(t) and the weighting factor γ related to the time constant of the structure through Eq. 42. The calculation of trailing temperature is greatly facilitated by noting that Eq. 41 is a solution of the following recursive relationship:

$$T^*(t + \Delta) = \gamma\, T^*(t) + (1 - \gamma)\, T(t + \Delta) \tag{43}$$

Use of this recursive relationship to calculate T^* from actual temperatures T requires that we know only one trailing temperature, at the start of the interval of interest. Since we are interested in a number of night-time hours, we record data starting a few hours earlier, and use these data to estimate the "initial" value of the trailing temperature. The steady-state lossiness can then be estimated from Eq. 39, for various assumed values of γ. The optimum value of γ is determined as the value which leads to a relatively constant lossiness through the night (see steady state lossiness plot in Fig. 9).

L was determined using the trailing temperatures approach with many nights of data. The observed night-to-night variation of L between 16.4 and 18.7 W/°C was much larger than could be explained by experimental error. We noticed a pattern with wind speed and cloudiness. On calm nights, the lossiness was higher when it was clear corresponding to greater lossiness by radiation to the night sky. On clear nights the lossiness decreased with wind speed. This unusual phenomenon was believed to be because, on clear nights, the outside air temperature is higher than the equivalent temperature of the night sky. Thus an unheated surface can become much colder than the ambient air. Our test chamber is so well insulated and its heat loss rate is so small that it behaves like an unheated object in that its surface temperature falls below the outside air temperature. In such a case, an increase in wind speed increases the rate of heat transfer from the warmer ambient air to the cooler surface of the chamber. The result is that the lossiness is lower on clear windy nights. On cloudy nights the surface cannot become much colder than the ambient air and the wind speed has little effect on the surface temperature or the lossiness. We call this set of phenomena the <u>infrared-wind interaction</u>. Persily confirmed our initial hypotheses by making simultaneous measurements of lossiness and infrared radiation from the night sky.[31] He developed a heat transfer model that represented the lossiness as the product of two terms:

$$L = L_T S_{f,a} \tag{44}$$

where L_T characterizes the lossiness of the chamber from the inside to the outside surface and is independent of the sky radiation or wind speed, and $S_{f,a}$ is a "sky factor" which corrects for sky radiation and wind. The full formulation was presented by Persily.[31]

The infrared-wind interaction would not be seen in a conventional house because a) increased wind speed would generally increase infiltration and lossiness which would more than offset any lossiness reduction from reduced radiation to the night sky; b) heat loss from a less insulated structure would keep the outside surface temperature higher than the ambient air temperature so that heat would flow from the surface to the ambient air irrespective of the sky conditions. In either case, increased wind speed will also increase heat loss and lossiness. Our experiments in the test chamber illustrate how the behavior of well insulated and air tight houses may differ from conventional houses.

<u>Characterizing the dynamic behavior of houses</u>

Persily's method of trailing temperature attempts to find the steady-state equivalent rates of heat loss from short term data. It does not completely characterize the transient behavior of the house. Various researchers have developed models for transient heat flow. Prominent among them are Stephenson and Mitalas.[32,33] Sonderegger has developed a method, using the ubiquitous electrical analogy and drawing from control theory techniques, that enables a set of <u>equivalent thermal parameters</u> for a house to be determined from a limited amount of data (as little as a few days). The Sonderegger method includes estimates of the lossiness (L), rates of heat transfer from interior air to the thermal mass (L_s) and to a thermal "clamp" such as the ground (L_c). In addition, Sonderegger estimated an equivalent thermal capacitance (C) and a solar energy input factor expressed as an equivalent south facing solar aperture (A).[30] These terms are determined by regression analysis using the following equation:

$$T_i(t+1) = a\ T_i + b_1\ [T_o(t) + T_o(t+1)]/2 + b_2\ [T_o(t+1) - T_o(t)] + c_1\ [S(t)+S(t+1)]/2 + c_2\ [S(t+1)-S(t)] + d \qquad (45)$$

where t, t + 1 are successive points in time for data collection,
T_i is the average interior temperature,
S is the solar flux on the south facade,
a, b_1, b_2, c_1, c_2, and d are constant coefficients estimated from regression using the above equation.

Sonderegger regressed two days of data using Eq. 45 and used the regression coefficients to estimate the equivalent thermal parameters from the following relationships:[30]

$$L_c = (1 - a - b_1) / d_1 \qquad (46)$$

$$L = b_1/d_1 \qquad (47)$$

$$A = c_1/d_1 \qquad (48)$$

$$C = [(1+a)/2 - (1-a)\ c_2/c_1]\ \Delta t/d_1 \qquad (49)$$

where $d_1 = [d - (1-a_1-b_1)\ T_c] / E$ (50)

E is the energy input to the house from all electric appliances and electric resistance heaters and people in the house together with the latent load of the humidifier and the plants, and
Δt is the time interval between successive data in the regression.

The parameter L_s can be determined from the regression coefficients in two ways:

$$L_s = [(1+a)\ b_1/(2b_2) - (1 - a)] / d_1 \qquad (51)$$

or, $$L_s = [(1+a)\ c_1/(2c_2) - (1 - a)] / d_1 \qquad (52)$$

Regression estimates made using data from the first 52 hours were used to predict variations in interior temperature over the following 42 hour period: the agreement with measured temperatures is good (see Fig. 11).

In the example above, the house was heated by electric heaters that put out heat at a steady rate, and the variation in interior temperature was modeled. Sonderegger also developed a similar model for a "thermostated" house. In this situation, the parameters C and L_s relating to thermal capacitance and the rate of heat transfer between house air and the storage cannot be determined. Instead, one can determine the overall efficiency of the space heating system by heating the house alternately with the conventional furnace and by a combination of the furnace and distributed electric resistance heaters (electric co-heating). The furnace efficiency is determined from the furnace fuel savings when the auxiliary electric heating is present.[34] This furnace efficiency determination includes the effects of stack, off-cycle and heat distribution losses. This method provides the only known way of measuring the efficiency of actual heat delivered, not only for central heating systems but of fireplaces and wood stoves as well.[35]

Sonderegger's method of equivalent thermal parameters is also useful for determining the solar energy input to a house, and for calculating diurnal temperature fluctuations in un-thermostated spaces having large exposures to solar radiation, such as greenhouses and atria.

One important application of the unsteady models discussed above is the modeling of cooling. In mild periods, intrinsic heat can maintain interior temperatures substantially higher than that outside. Thus, cooling is often necessary even when the outside temperature is below the desired inside temperature. Transient models are much better able to predict cooling energy needs during these periods. Air conditioning also represents a substantial demand on electric utilities who must meet it using expensive, peak generating capacity. Transient heat transfer models are essential to predicting and managing cooling energy demand.

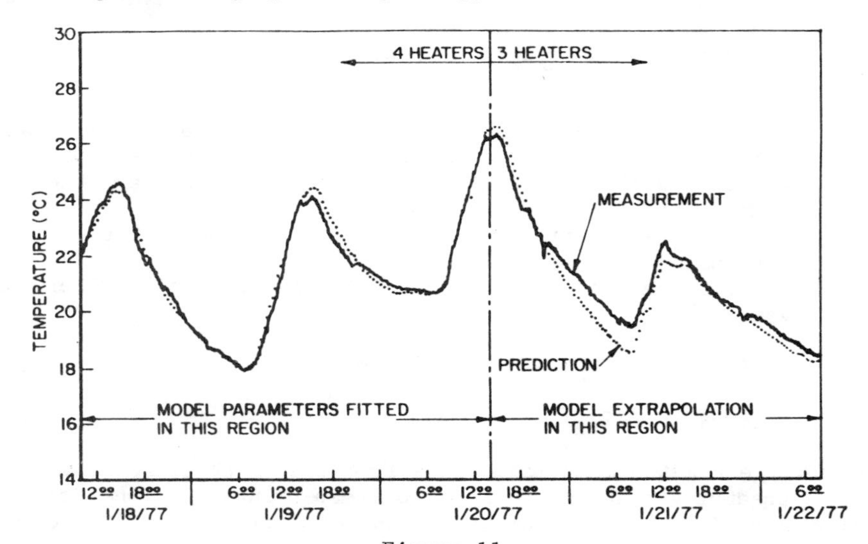

Figure 11
Measurement and prediction of indoor temperature based on equivalent thermal parameters determined by regression
(Source: ref. 30)

CONCLUSIONS

In this paper, we have examined the basic physics underlying heat transfer in houses. We have outlined procedures for modeling one-dimensional heat transfer through multi-component building elements, multi-dimensional conduction through corners and foundations, and wind- and temperature-induced air flows in houses. Our understanding has already been applied to develop instrumentation for diagnosing heat loss anomalies in existing houses, to develop techniques for eliminating these heat loss paths, and to improve new house design. The house-doctor instrumented energy analysis and retrofit procedure has demonstrated substantial and cost-effective energy savings even in houses already considered to be well insulated. Avoiding the thermal defects in new houses, combined with higher levels of insulation and airtightness, has led to the concept of superinsulation.[22] Superinsulated houses are heated largely by intrinsic heat from the sun, lights, appliances, and people. Their heating energy consumption is only 10 to 20 percent that of conventional housing. At the same time, we have improved modeling and understanding of ventilation, indoor air quality, and moisture problems. The result is that houses can be made energy efficient while reducing moisture problems and indoor air pollution hazards.[22,28] Using technology that was commercially available and cost effective in 1982, it is technically possible to reduce space heating energy use in US houses by almost a factor of two by the year 2000 compared to its 1980 rate.[1]

Transient heat transfer modeling has led to the development of techniques for characterizing the effects of thermal storage, measuring house heat loss from short term data, estimating solar gains and heating system efficiency, and peak energy requirements.

While the direct application of our theories has often led to useful insights, it is when the physical theories fail to explain what we observe that we have an opportunity for scientific detective work. Such was the case in Twin Rivers, for example, where convective loops were the mechanism for substantial heat loss through interior walls.

The sampling of physical models and mathematical methods presented here suggests that the analysis of heat loss from buildings is not just a problem for architects and builders. Although the area of heat loss from houses has been well explored in recent years, many challenges remain in the areas of cooling energy and peak power requirements and optimization of building design for both heating and cooling. Multifamily and other large buildings have also been neglected. The need to extend our physical theories to better explain phenomena related to energy use in buildings remains a challenge for applied physicists.

ACKNOWLEDGEMENTS

Thanks go to David Hafemeister for inspiring me to undertake this effort, and to Barbara Levi, Sam Baldwin, and John DeCicco, whose critical review and many valuable suggestions helped me immensely.

REFERENCES

1. R.H. Williams, G.S. Dutt, H.S. Geller, Ann. Rev. Energy 8, 269 (1983).
2. M.F. Fels, "Energy Scorekeeping," in this volume.
3. M. N. Ozisik, Basic Heat Transfer (McGraw-Hill, 1977), p. 8.
4. H.S. Carslaw and J.C. Jaeger, Conduction of Heat in Solids (Clarendon Press, Oxford, 1973), p. 445.
5. M.N. Ozisik, op. cit., Chapter 5.
6. A-C. Andersson, "Internal Additional Insulation," Lund Inst. of Tech. Div. of Building Tech, Rept. TVBH-1001 (Lund, Sweden, 1979), p.62.
7. G.P. Mitalas, ASHRAE Transactions Vol. 89, Pt. 1B, p.420 (1983).
8. T. Kusuda, and J.W. Bean, ASHRAE Transactions Vol. 90, Pt. 1B, p.611 (1984).
9. T.H. Woteki, G.S. Dutt, J. Beyea, Energy 3, 657 (1978)
10. J. Beyea, G.S. Dutt, T.H. Woteki, Energy and Buildings 1, 261 (1977)
11. G.S. Dutt, J. Beyea, "Hidden Heat Losses in Attics -- Their Detection and Reduction," Princeton Univ. Center for Energy and Environ. Studies (PU/CEES) Report 77 (1979).
12. G.S. Dutt, J. Beyea, F.W. Sinden, "Attic Heat Loss and Conservation Policy," Proc. of Am. Soc. Mech. Engrs Energy Technology Conference (1978).
13. G.S. Dutt, J. Beyea, "Attic Thermal Performance -- A Study of Townhouses at Twin Rivers," PU/CEES Rept. 53 (1977).
14. D.T. Harrje, G.S. Dutt, J. Beyea, ASHRAE Transactions Vol. 85, Pt.2, p.521 (1979).
15. R.H. Socolow (Ed.), Saving Energy in the Home -- Princeton's Experiments at Twin Rivers (Ballinger, Cambridge, MA, 1978).
16. F.W. Sinden, Energy and Buildings 1, 275 (1977/78).
17. M.H. Sherman, "Air Infiltration in Buildings," Ph.D. Thesis, Physics Dept., University of California - Berkeley, Lawrence Berkeley Laboratory Report No. 10712, 1980.
18. D.T. Harrje, A.K. Blomsterberg, A.K. Persily, "Reduction in Air Infiltration Due to Window and Door Retrofits in an Older Home," PU/CEES Rept. 85 (1979).
19. D.T. Grimsrud, M.H. Sherman, R.C. Sonderegger, "Calculating Infiltration: Implications for a Construction Quality Standard," Proc. ASHRAE/DOE-ORNL Conf. Thermal Performance of the Exterior Envelopes of Buildings II (ASHRAE, Atlanta, 1983).
20. G.S. Dutt, D.I. Jacobson, R.H. Socolow, "Air Leakage Reduction and the Modular Retrofit Experiment," PU/CEES Rept. 113 (1983).
21. D.T. Harrje, K.J. Gadsby, G.T. Linteris, ASHRAE Transactions, Vol. 88, Pt.1, p.1373 (1982).
22. J.D.N. Nisson, G.S. Dutt, The Superinsulated Home Book (John Wiley, NY, 1985).
23. R.N. Dietz, E.A. Cote, Environment Int. 8, 419 (1982).
24. F.W. Sinden, Buildings and Environment, 13, 21 (1978).
25. R.N. Dietz, T.W. D'Ottavio, R.W. Goodrich, "Multizone Infiltration Measurements in Homes and Buildings Using a Passive Perfluorocarbon Tracer Method," to appear in ASHRAE Trans. Vol.

91, Pt. 2 (1985).

26. D.T. Harrje, G.S. Dutt, D.L. Bohac, K.J. Gadsby, "Documenting Air Movements and INfiltration in Multi-cell Buildings Using Various Tracer Techniques," to appear in ASHRAE Trans. Vol. 91, Pt. 2 (1985).
27. D.T. Harrje, G.S. Dutt, K.J. Gadsby, Proc. Soc. Photo-Opt. Instr. Engrs 254, 50 (1981).
28. G.S. Dutt, "House Doctor Visits -- Optimizing Energy Conservation Without Side Effects," Proc. Int. Energy Agency Conf. — New Energy Conservation Technologies and Their Commercialization (Springer Verlag, Berlin, 1981), p.444.
29. G.S. Dutt, M.L. Lavine, B.G. Levi, R.H. Socolow, "The Modular Retrofit Experiment -- Exploring the House Doctor Concept," PU/CEES Rept. 130 (1982).
30. R.C. Sonderegger, "Dynamic Models of House Heating Based on Equivalent Thermal Parameters," Ph. D. Thesis, Aerospace and Mechanical Sciences Dept., Princeton University (1977).
31. A.K. Persily, "Understanding Air Infiltration in Homes," Ph. D. Thesis, Mechanical and Aerospace Engineering Dept., Princeton University (1982), Appendix D.
32. D.G. Stephenson, G.P. Mitalas, ASHRAE Transactions Vol. 77, Pt. 2, p.117 (1971).
33. G.P. Mitalas, ASHRAE Transactions Vol. 74, Pt.2, p.182 (1968).
34. R.C. Sonderegger, P.E. Condon, M.P. Modera, ASHRAE Transactions, Vol. 86, Pt.1, p.394 (1980).
35. M.P. Modera, R.C. Sonderegger, "Determination of In-Situ Performance of Fireplaces," Lawrence Berkeley Laboratory Rept. 10701 (1980).

CHAPTER 8

ENERGY CONSERVATION IN LARGE BUILDINGS

A. Rosenfeld and D. Hafemeister
Energy Efficient Buildings Program
Lawrence Berkeley Laboratory
Berkeley, CA 94720

ABSTRACT

As energy prices rise, newly energy aware designers use better tools and technology to create energy efficient buildings. Thus the U.S. office stock (average age 20 years) uses 250 kBTU/ft^2 of resource energy, but the guzzler of 1972 uses 500 (up x 2), and the 1986 ASHRAE standards call for 100-125 (less than 25% of their 1972 ancestors). Surprisingly, the first real cost of these efficient buildings has not risen since 1972. Scaling laws are used to calculate heat gains and losses of buildings to obtain the ΔT(free) which can be as large as 15-30°C (30-60°F) for large buildings. The net thermal demand and thermal time constants are determined for the Swedish Thermodeck buildings which need essentially no heat in the winter and no chillers in summer. The BECA and other data bases for large buildings are discussed. Off-peak cooling for large buildings is analyzed in terms of saving peak-electrical power. By downsizing chillers and using cheaper, off-peak power, cost-effective thermal storage in new commercial buildings can reduce U.S. peak power demands by 10-20 GW in 15 years. A further potential of about 40 GW is available from adopting partial thermal storage and more efficient air conditioners in existing buildings.

I. SCALING LAWS FOR BUILDINGS.

As one might expect, big commercial buildings have quite different energy characteristicts from small buildings, or residences. In large buildings the main source of heat gain is internal (equipment, people, lighting, solar, etc.). In small buildings the main heat gains and losses are external, the heat/coolth from the outside climate passing through the envelope, or shell, of the building. Let's roughly examine this transition from small to big by considering some scaling laws for energy gains and losses. Our building will be a cube of length L and of volume L^3.

The rate of winter heat loss from our building is proportional to its surface area, or $L^2\Delta T$, where ΔT is the inside-outside temperature difference. If the thermal conductivity of the building envelope (and fresh air) is KL^2, then $\dot{Q}(\text{loss}) = KL^2\Delta T$. On the other hand, the internal heat gains in our builidng are proportional to the floor space of the building which is proportional to the volume of a multistory building, or L^3, or $\dot{Q}(\text{gain}) = GL^3$. We ignore a smaller term SL^2 for solar gain in winter. Without space heat or

0094-243X/85/1350148-21$3.00 Copyright 1985 American Institute of Physics

airconditioning the gains and losses are equal, or

$$\dot{Q}(\text{gain}) = GL^3 = \dot{Q}(\text{loss}) = KL^2 \Delta T(\text{free}), \quad (1)$$

and the building floats above the ambient temperature by an amount

$$\Delta T(\text{free}) = (G/K)\ L. \quad (2)$$

Obviously the thermostat will not call for heat until T(ambient) drops ΔT(free) below the comfort temperature T(thermostat). This temperature when the furnace turns comes on (ignoring thermal mass) is called the "balance point" of a building, when T(ambient) = T(thermostat) - ΔT(free). At the balance point, the internal heat gains are exactly balanced by the heat losses without auxilary space heat and the occupants are at the thermostat temperature.

As we scale up the size of the building, $\dot{Q}$(gain) raises ΔT(free). For a "free heat" of 15°C (30°F), the length L must be about 15(K/G) = 10 m for the example in Sec. II. Even in winter, the internal heat gains in a large building can overwhelm the loss of heat through the walls, overheating the building. In summer the air-conditioning used to remove the excess heat from the buildings causes most U.S. utilities to experience their peak demand in the afternoon. On the other hand, the internal gains can be beneficial since they are sufficient to heat a large building or a superinsulated small building. In the next section we will equate the gains to the losses, using the appropriate numerical parameters and determine the amount of "free heat" available in a building.

II. FREE HEAT, ΔT(free), FOR BUILDINGS

The average (sensible) power of a person[1] is 75-100 watts (350 BTU/h). In a large building the density of people is such that they provide a heat intensity of about 11 W/m^2 (1 W/ft^2). The lighting and equipment gains can be about three times (or more) this amount, or 33 W/m^2 (3 W/ft^2).** Since the internal and solar gains can vary widely, we shall use a range of values for the internal gain of 66 ± 22 W/m^2 (6 ± 2 W/ft^2). The floor area of a building is $nL^2 = L^3/H$ where n is the number of floors in the building and H is the interfloor height of about 3 m (10 ft). The internal gain of the occupied building in SI units (watts, mks) is:

$$\dot{Q}(\text{gain}) = (66 \pm 22)(nL^2) = (22 \pm 7)L^3. \quad (3)$$

The steady state loss rate from a building is

$$\dot{Q}(\text{loss}) = \sum_i U_i A_i \Delta T + \rho \dot{V} c \Delta T \quad (4)$$

where A_i is the area of each envelope component, U = 1/R where U is the conductance and R is the thermal resistance, ρ is the density of air, $\dot{V}$ is the flow of incoming air (m /s), and c is the specific heat of air. The metric R values are obtained from the

** See Fig. 13 for a breakout of electricity and fuel use.

English values with

$$R(m^2\ K/W) = R(hr\ ft^2\ {}^\circ F/BTU)/5.69\ . \tag{5}$$

The following SI (English) parameters represent a medium level of energy tightness for high-rise office buildings (one version of the 1985 California standards, see Fig. 13 plotted near the bottom.)

Ceilings: R-2.62 (R-14.9)
Walls: R-1.14 (R-6.5)
Single Glazing: R-0.158 (R-0.9) 30% of wall area
Basement (about 50% of ceiling loss)
Infiltration/Ventilation (about 30% of total UAdT)

The loss rate from the cubic structure is

$$\dot{Q}(loss) = 1.3\{\dot{Q}(ceiling/basement) + \dot{Q}(70\%\ walls) + \dot{Q}(windows)\} \tag{7}$$

$$\dot{Q}(loss) = 1.3L^2\Delta T(1.5/2.62 + 0.7(4)/1.14 + 0.3(4)/0.158) \tag{8}$$
$$= 13.8\ L^2\Delta T.$$

Equating the losses (Eq. 8) to the internal gains (Eq. 3), we obtain:

$$\Delta T(free) = (1.6 \pm 0.5)\ L \quad (L(m),\ T(^\circ C)) \tag{9}$$
$$\Delta T(free) = (0.9 \pm 0.3)\ L \quad (L(ft),\ T(^\circ F)) \tag{10}$$

The "free temperature rise" $\Delta T(free)$ for our balanced (occupied, unheated) new office building of 10 m (33 ft) on a side is 16 ± 5 °C (29 ± 10°F). If the thermostat was set at 20°C, the furnace would turn on at the balance point of 4°C (20°C - 16°C). These values of free heat would be 30°C (60°F) by doubling the product of internal gains and the net thermal resistance. A large building (or a superinsulated building) can have a balance point close to the average winter ambient temperature. Of course, this example is pedagogical in nature, but the basic physics is correct; large office buildings have useful free heat in winter, and too much heat in summer (and often in winter) that necessitates either air conditioning or thermal storage. Because the internal loads dominate in large buildings, the annual energy intensity (kWh/m^2, BTU/ft^2) of large buildings does not depend very much on the climate. Proper controls can minimize heating and cooling by ventilation, thermal storage, and heat recovery systems, so that in actual practice large buildings can consume less energy/area than small buildings.

Houses have 1/5 to 1/10 the intensity of internal heat, perhaps 1 kW for a typical house of 120 m^2 (1300 ft^2), or less than 1 W/ft^2, compared with 6 W//ft^2 for an office.** Houses also can lose their internal energy more easily since they have a larger surface to volume ratio, thus the energy intensity of a house is much more dependent on its climate than for a large building. These physical facts require that houses have considerably higher insulation standards (Table I) than big buildings in order to have balance

** Electricity use in houses and office buildings are compared in Fig. 13.

points similar to that of large buildings. A conventional house has 3-6°C (5-10°F) of "free heat," but a superinsulated house can have 15° C (25°F) or more.

TABLE I. California thermal resistance standards in SI (English) units for high rise office buildings (1987) and residences (1985). The R values for walls depend on their heat capacity.

	HIGH RISE OFFICE BUILDINGS	RESIDENCES
CEILINGS:	R-2.62 (R-14.9)	R-5.27 (R-30)
WALLS:	R-1.14 (R-6.5)	R-3.34 (R-19)
GLAZING	SINGLE R-0.16 (R-0.9)	DOUBLE R-0.26 (R-1.5)

III. HEAT AND COOLTH STORAGE IN HOLLOW-CORE CONCRETE SLABS.

Concrete floor/ceiling slabs have a large heat capacity (100 $Wh/m^2 {}^\circ C$), but for accoustical reasons this is normally poorly coupled to the room air. In the Swedish "Thermodeck" system[2], the supply air is distributed via hollow cores in the floor slabs as shown in Fig. 1. These cores are already extruded in slabs to reduce weight/thickness, but are normally not exploited for energy conservation. In this way, the concrete mass is made available for the storage of heat. Even though Stockholm (3580°C-day, 6444°F-day) is colder than Chicago, the Thermodeck office buildings annually use only about 4 kWh/ft^2 for electric resistance heating, so little that it does not pay to hook up to the Stockholm district heating system.

Modern Swedish buildings have small internal gains and are relatively small by American standards since every office must have a window, but they are so well insulated that their temperature floats upwards during a typical occupied winter day. The net winter heat gain in a modern Swedish building is about 15 W/m^2 for the 8 occupied hours. Figure 2 (curve "a") shows that in a normal office, with an insulated suspended ceiling, this 15 W/m^2 will raise the temperature to an unacceptable level within an hour or so, making it impossible to continue storing the heat gain (free energy) in the structure. But with Thermodeck (curve "b"), the full 8 hour gain can be stored with a temperature rise of 1-2°C, which is readily acceptable to the occupants. During the winter, this stored heat is used to compensate for night/weekend heat losses.

During the summer, daytime heat gain is again stored in the pre-cooled slabs. In Stockholm, the outdoor air temperature seldom exceeds 30°C (86°F), and the minimum temperature at night is usually 18-20°C, so the slabs can be cooled by circulated night air (and thus made ready for the next morning) without the need of air conditioning. In roughly half of the U.S., nights are not cool enough to pre-cool the building, and cheap off-peak air conditioning would still be required, but the concrete's heat capacity will still handle the daytime load. Only enough peak air conditioning is needed to dry outside ventilation air. This peak can be made negligible with a water-permeable heat exchanger.

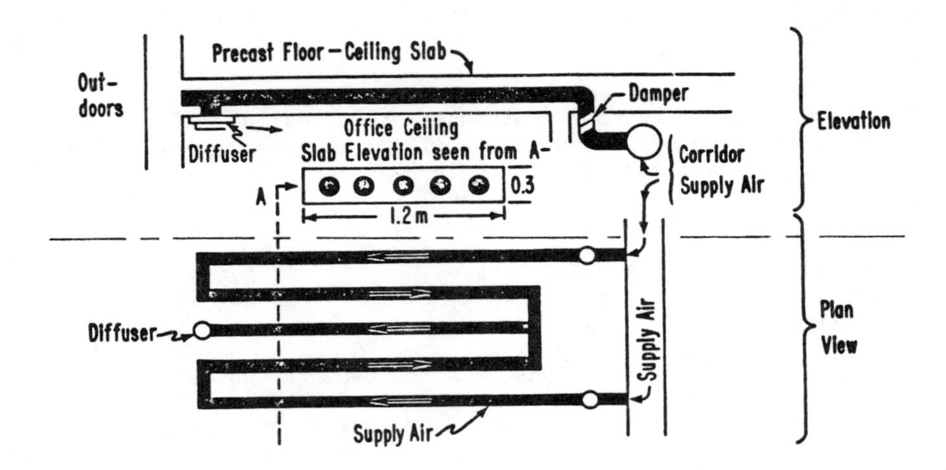

Fig. 1. Forced convection and increased thermal surface area enhance the thermal storage of a Swedish Thermodeck Office Building. Each 10 m^2 office module has two slabs (1.2 m x 4.2 m). Source: LBL- 8913. XBL7910-13105

Fig. 2 shows some computer simulations of heating cycles in the Thermodeck building. These buildings have a thermal relaxation time similar to an RC circuit (Appendix F). The choice of τ = RC for a building is critical for energy management. From Fig. 2 (curve "a", no hollow cores) we see that a typical office has $\tau \cong 5$ hours, but when the mass of the concrete is coupled to the room, τ is raised to about 100 hours, and enough heat can be stored to carry the space through unoccupied hours, and even weekends of 60 hours.

Let us estimate the heat gains and losses for a Thermodeck building to confirm these energy management concepts. A single-occupant Thermodeck office is 2.4 m wide by 4.2 m deep by 2.7 m high, or 10 m^2 in area and 27 m^3 in volume. We will assume a cold day in Stockholm of -8°C (18°F) for a temperature difference between inside and outside of ΔT = 22 - (-8) = 30°C (54°F).

HEAT GAINS per 10 m^2 office when occupied:

1. 1 person/10 m^2 = 100 W (sensible heat only)
2. Lights and machines = 300 W
3. Solar Gain (small in winter) through 1.5 m^2 = 30 W

TOTAL GAIN 8 OCCUPIED HOURS = 430 W/10 m^2

HEAT LOSSES per 10 m^2 office: (losses are negative gains)

1. Wall (U)(A)(ΔT) = (0.25)(5)(30) = -38 W
2. Window (U)(A)(ΔT) = (2)(1.5)(30) = -90 W
3. Outside Air = -200 W (occupied), -50 W (unoccupied.)

TOTAL LOSS = -330 W (occupied), -180 W (unoccupied)

GAINS-LOSSES: Occupied = +100 W, Unoccupied = -180 W/10 m^2

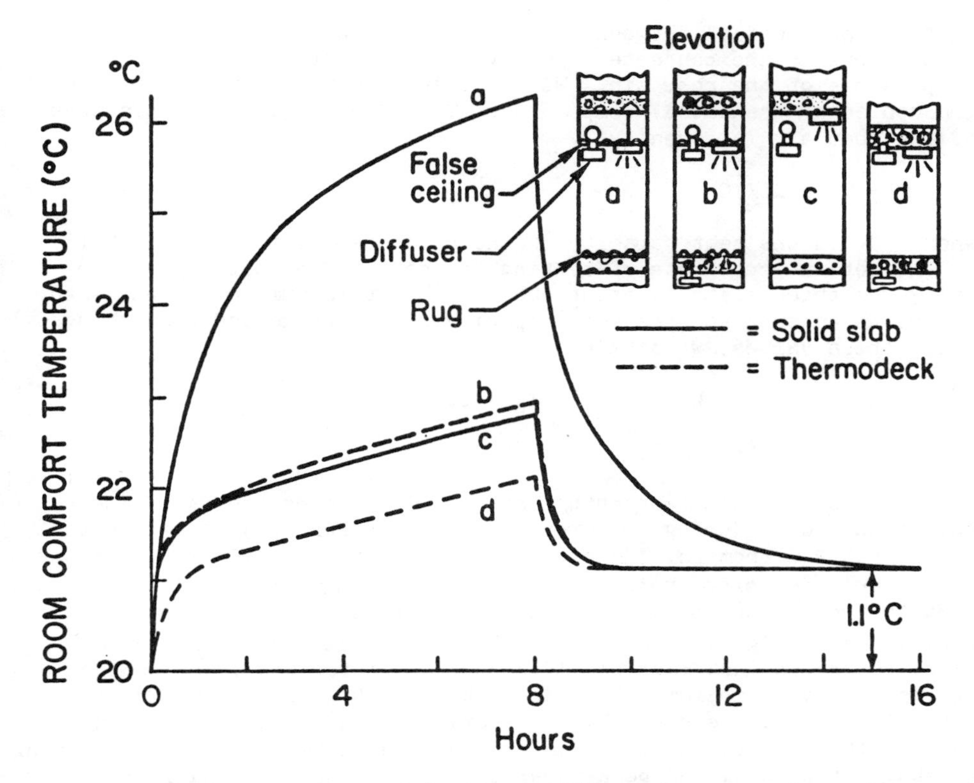

Fig. 2. Response/relaxation curves calculated by the BRIS computer program for equal rooms with two different slabs, each with a heat capacity of 100 Wh/m^2-°C. The surroundings are assumed symmetric on all sides (as in an office in the core of a building). Lighting (15 W/m^2, 50% radiation) is turned on for the first 8 hours of each run. The cases are as follows:

a. 20-cm thick solid concrete slab, with rug, insulated, suspended ceiling, and plenum. Resistances assumed were: rug (0.1 m^2-°C/W) ; insulated false ceiling (0.5); plenum (0.17).
b. Same as a., but slab is 30-cm thick Thermodeck.
c. 20-cm thick concrete slab, but bare -- no rugs, suspended ceilings, plenum.
d. Same as c., but slab is 30-cm thick Thermodeck.

Source: LBL-8913. XBL7910-13104

The heat loss from heating cold, outside air is the largest loss for the Thermodeck building. As in all office buildings, to control indoor contaminants, 20 m^3/hr of outside air is mixed with the 120 m^3 of air recirculated to each office, thus, changing the building air every 1.3 hours. During unoccupied hours, fans are off, but natural infilatration is about 5 m^3/hr per office. During the 8 hour work day, this outside air corresponds to a 200 W heat loss, and 50 W during the unoccupied hours. If additional heat is needed, air-to-air heat exchangers could be used to recover about 70% of the heat in the exhaust air stream.

Thus far, we have treated the curves of Fig. 2 as exponentials, but now we want to calculate their numerical slope. Because of the good thermal contact between the hollow cores and the room air, the temperature of the concrete is not very different from the temperature of the room air. We start at time t = 0, with an offset (precooled or preheated) temperature T_o. Then T, the temperature of the room air is given by

$$T = T_o + \dot{Q}t/ C \tag{11}$$

where C is the heat capacity of the concrete slabs, and $\dot{Q}$ (W/m^2) is the net internal rate of heating the room. The heat capacity of the 30 cm thick slabs is about 100 $Wh/m^2 {}^oC$; this number is increased by 20% to account for the heat capacity of the walls and furnishings. Using these values, we obtain:

$$\text{Occupied } (W = 10 \text{ W/m }), \quad T = T_o + 0.1t \tag{12}$$

$$\text{Unoccupied } (-18 \text{ W/m}^2), \quad T = T_1 - 0.2t \tag{13}$$

Eq. 12 gives a small temperature rise of 1°C during the day. The temperature drop during the evening (with the fan off) is closer to 1° C (and not 2°C from Eq. 13) since the rooms are are allowed to become quite cool, reducing their thermal losses through the envelope. These results agree with the data of Fig. 3 for heating in the winter and Fig. 4 for cooling in the summer. In the US the storage of summer night coolth is much more significant than winter heat. As can be seen in Fig. 13 for a medium office in Washington DC, annual cooling per ft^2 costs 35¢; heating costs only 5¢. During the deep cooling season, one can run the chillers at night to precool the slabs. There is no saving of kWh, but by avoiding peak power charges one saves annually $50-100/ kW shifted. A slab does not quite have the heat capacity to keep an American office cool all day, but can be aided with a small water or ice storage system, or with phase change material, tuned to about 21°C, canned and loaded loosely into the cores. In mid-season, nights are cool enough to precool without running the chiller, thus saving kWh.

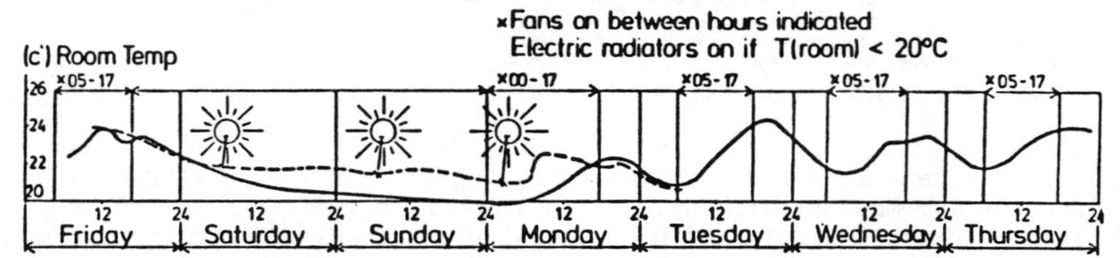

Fig. 3. Winter. During a winter week, the outdoor temperature varied between -2 and -10°C. On Friday afternoon, when the internal gains end and the fans and radiators are turned off, the indoor temperature starts to fall from 24°C. By about Monday morning, 20°C is reached. Fans are turned on (the ventilating air system runs with 100% recirculation) and the air is heated one or a few °C depending on the outdoor temperature. At 8:00 Monday morning, the temperature level is still about 20°C. Each weekday the occupied offices climb 2-3°C in temperature, and empty rooms remain about 20°C. Each night the indoor temperature falls 1-2°C. By Friday afternoon, the cycle is complete. Source: LBL-8913. XBL7910-13107

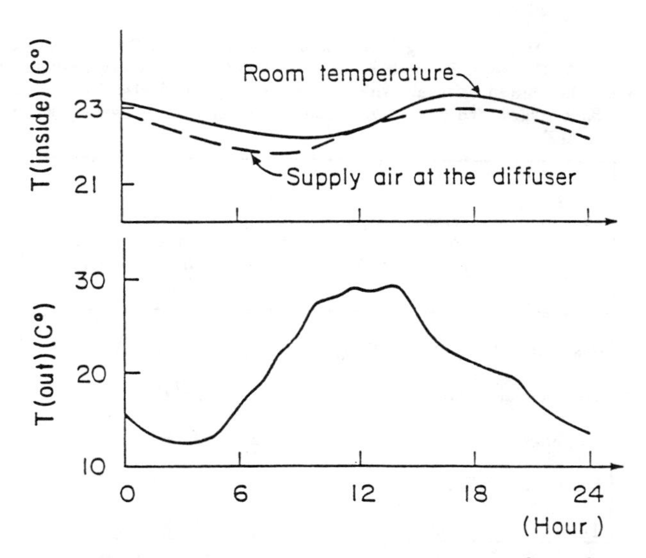

Fig. 4. Summer. During the day, the outdoor air temperature varies between 12°C and 30°C and the air going into the system has about the same temperature. The supply air, after passing through the slab cores, has a much smaller variation (22.5 +/- 0.5°C) and the room temperature during office hours increases from 22°C to 23°C. Source: LBL-8913. XBL7910-13106

IV. THERMAL STORAGE CAN REDUCE PEAK POWER DEMANDS

A. The Potential for Savings of Peak Power (kW). Since internal heat gains dominate in large buildings, air conditioning must be used to make these buildings both comfortable and useable. Primarily because of air conditioning, the nation's power grids have a severe peak power problem, the peak demand on hot afternoons can often be 2 or 3 times the demand at night.** And as more air conditioning is installed, the utilities demand problem worsens. Table II contains some estimates of peak cooling and possible displacements of this cooling by using cost-effective thermal storage for large buildings. The fraction of new, single-family homes installing air conditioning has dramatically risen from 25% in 1966 to 70% in 1983, increasing the peak demand of electricity by about 2 GW/year. Presently 58% of U.S. homes[3] are air conditioned. The high growth rate for new commercial buildings (annually 5% = 2.5 B ft^2) causes peak demand growth of about 1.6 GW/year. Table II shows that residential and commercial air conditioning each account for 80 GW, totalling to 160 GW (32% of peak summer power demand of 500 GW). The potential savings in peak power (kW) are very large; the adoption of off-peak cooling with thermal storage on new commercial buildings would avoid the need of about 10-20 standard 1 GW plants in the next 15 years, with a furthr potential savings of about 40 GW by adopting partial thermal storage and more efficient air conditioners in existing buildings.

** For power profiles, see Rosenfeld's introduction to Peddie/Bulleit.

Table II. Peak power demand for cooling U.S. buildings extrapolated 20-fold from So. Calif. Edison's 1985 summer peak of 13 GW, to U.S. peak capacity of 500 GW. The a/c peak includes both chiller-and-pumps (which can be shifted off peak with thermal storage) and also fans (which cannot). Of the residential 80 GW, about 10% is fan power; for small commercial, fan power runs around 20%, and for large commercial up to 30%.

	A. Sector Peak (GW)	B. a/c Peak (GW)	C=B/A a/c Fraction of Peak	D. '77-'82 Annual Growth (GW)	E=C*D Annual a/c Growth (GW)
Residential	(170)	(80)	(47%)	(4)	(2.0)
Commercial	(195)	(80)	(40%)	(4)	(1.6)
Buildings	365	160	44%	8	3.6
Industrial	135	–	0%	~0	0.0
Total	500	160	32%	8	3.6

Comments:

Column C - These fractions apply only to SCE, but we assume that they apply to the U.S. We should, of course, use a weighted average of the peak fractions for about 10 utilities.

Column D - We have no U.S.-wide annual data on peak demand (GW) disaggregated by sector, but annual sales (BkWh) by sector are readily available, so to estimate GW growth, we use BkWh growth and assume that the GW/BkWh does not change. This ratio, for example in 1982, was 2086 BkWh/418 GW peak demand = 5000 hours equivalent production per peak watt.

Column E - For the Total, E is simply not equal to C times D.

B. Off-Peak Cooling with Thermal Storage. In order to gauge the potential for saving peak power, one should examine the disagregation of peak power demands in large buildings. Fig. 5 displays the peak power components in the summer for a large office building in Madison, Wisconsin, as calculated with DOE.2 (Appendix E). Nearly 2 W/ft^2, fully one-third of the peak demand of 5 W/ft^2 is used to run the chillers that could be operated in the off peak hours. Many new commercial buildings store "coolth" in chilled water or ice during the unoccupied hours. This approach allows the downsizing of the chillers by 50-60%. The block diagrams in Fig. 6 compare:

(top) Conventional Cooling on Demand; chillers run 8 hours per day, no thermal storage.

(left) Partial Storage; small size chiller (40% of conventional) runs the entire day, storing 2/3 of the coolth during the unoccupied hours for later use during peak demand.

(right) Demand Limited Storage; medium size chiller (50% conventional) runs only during the unoccupied, 2/3 of the day, and the thermal storage is about 50% larger than for partial storage.

The economics[4] for the transition to off-peak cooling are very favorable. The price of off-peak electricity is as much as 6 ¢/kWh cheaper than the peak price, and the demand charges for power during peak hours can be as large as $9/kW-month. Thus the annual savings

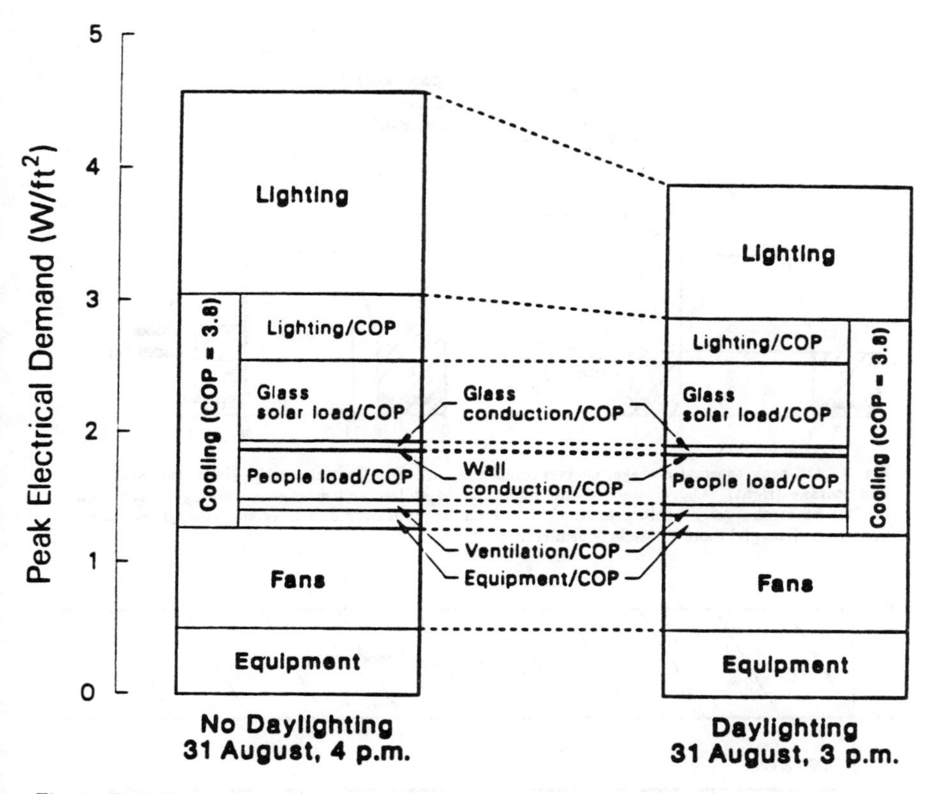

Fig. 5. Peak Power demand per square foot for a large office building, as simulated for Madison, WI weather by the DOE.2 program. Each floor has 10,000 ft^2 of core (cooled only) plus 6000 ft^2 of perimeter floor area.

If built according to ASHRAE Standard 90-75, its peak demand would be 6.7 W/ft^2; if it satisfies Standard 90-E (revised, 1985), with daylighting it would use only 4 W/ft^2. Thus its peak demand is down 40%, its yearly energy use is down by 40%, yet its first cost is also slightly down, mainly because of savings by downsizing the air conditioning.

Note that about half the peak demand goes to running chillers. With *thermal storage* this 2 W/ft^2 can be moved entirely off peak for a first cost of about \$0.50/W, which is only half of the utility's cost for new peak capacity. The residual peak demand is then down to about 2 W/ft^2. Alternatively, and cheaper, the chilling can be partially (60%) moved off peak for only \$(0.00 to 0.25)/W. XCG 855-223, 1985

by shifting 1 kW of chilling off-peak is \$30-100. The combined savings from reduced electrical bills and from downsizing the chillers by 50-60% provides a strong economic incentive to use off-peak cooling with thermal storage in new and existing buildings.

In 1977, Stanford University realized that its daytime cooling requirements were going to rise from 5 MW (5000 tons of air conditioning) during the peak hours to about 8 MW by 1986. The additional 3 MW of chillers and cooling towers were going to cost about \$1.5 million, but Stanford found out that for the the same

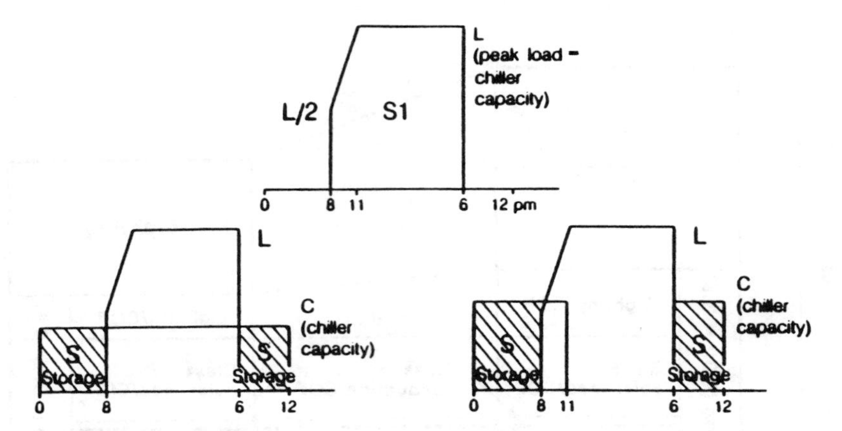

Fig. 6. Off-peak thermal storage. Conventional system (top); Partial storage (left); Demand limited storage (right). Calculation of the capacities of chiller and storage according to the load profile. S1 is the daily cooling load, C is the chiller capacity, and S is the storage capacity. Source: A. Rosenfeld and O. de la Moriniere.

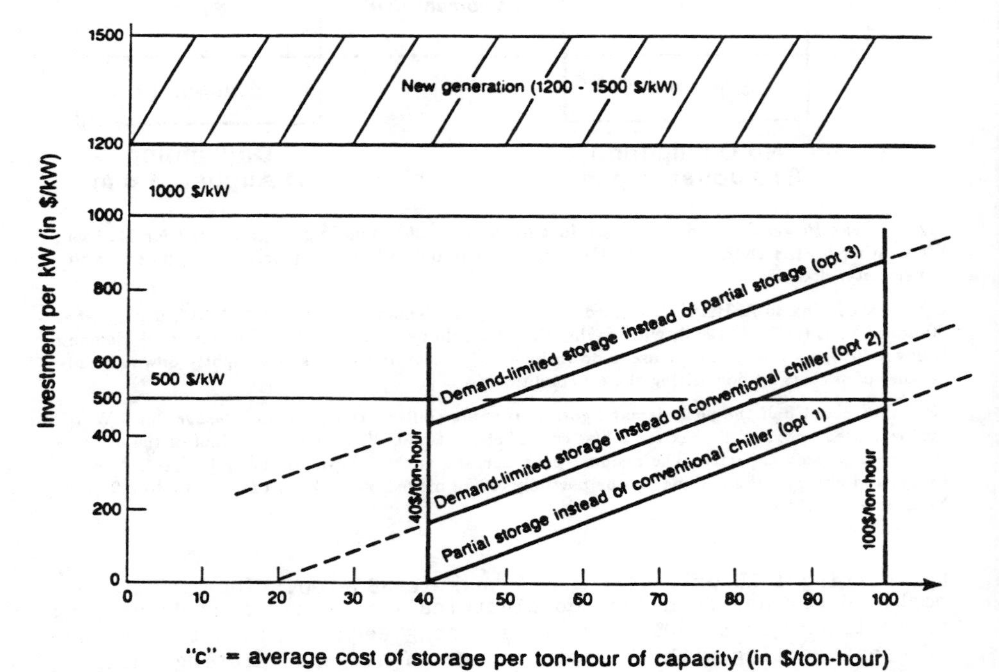

Fig. 7. Investment to save one peak kW using cool storage. For an office building, the cost range of installed storage is from $40 to $100 per ton-hour. At $40, the saving from downsizing the chiller pays for the storage. To go from partial to demand-limited storage costs about $500/avoided peak kW, still much cheaper than the indicated cost of new generation. This figure does not include the cost savings from moving electricity charges off-peak. Source: A. Rosenfeld and O. de la Moriniere, ASHRAE Transactions, HI-85-15, #4 (Hawaii Meeting).

price it could build a 4-million gallon insulated tank for cold water storage, and connect it to the present chillers. In this way Stanford could meet its 8 MW afternoon load by running its present capacity at night, saving all peak power charges. Thus Stanford, at no increase in first cost, saved operating and peak power costs, and shaved 5-8 MW in its peak load, which saves $300,000 to $500,000/year.

The investment necessary to save 1 peak kW with off-peak cooling are considered in Fig. 7. For the case of partial storage, an optimistic cost of $40/ton-hour would take no additional investment, and would save the utility about $1200-1500. For a pessimistic cost of $100/ton-hour, there is a finite first cost of $500/peak kW avoided. To go from this most lucrative option (partial storage) to full demand-limited storage is more expensive; the incremental increase in first cost is $450/kW, and the payback time is about 7 years. This is attractive to a utility, which otherwise must pay off the expensive new plants over 30 years, but it is not as attractive (without incentives) to most builders.

C. An Example. In order to gain a quantitative understanding for these large savings, let us examine the partial storage system of a single[5] facility, the headquarters of the Alabama Power Company in Birmingham. The five large ice cells contain 550 tonnes of ice to cool the 1.2 million ft^2 building, or 0.46 kg/ft^2 (an equivalent layer 5 mm thick per ft^2 of floor). The latent heat/ft^2 is

$$Q = (0.46 \text{ kg})(3.4 \times 10^5 \text{ J/kg}) = 1.6 \times 10^5 \text{ J/ft}^2 . \quad (14)$$

The electrical power to make the ice during the 16 off-peak hours is

$$P = Q/(COP)(\Delta t) = (1.6 \times 10^5)/(2.5)(16 \text{ hr}) = 1.1 \text{ W/ft}^2. \quad (15)$$

This gives a total of 1.3 MW for the entire building, which is less than 1/2 of the 2.8 MW required without thermal storage. From this, we can determine the average heating intensity during a summer day (solar, internal, envelope). Since the coolth stored in the ice is only about 2/3 of the cooling requirement, the daily gain is about 2.4×10^5 J/ft^2 which corresponds to a heating intensity of about 10 W/ft^2 in the day.

V. DATA ON COMMERCIAL BUILDINGS

Commercial buildings use a considerable amount of energy, about one-seveth of the U.S. total annual consumption of energy. The commercial sector builds at the rate of 5%/year of which about half it to replace old buildings, leaving 2.5%/year net growth. In spite of these high growth and replacement rates, the commercial sector has a considerable longevity because commercial buildings last 50 years, with the result that about 2/3 of the projected floorspace for the year 2000 is already in place. The average annual cost for energy in a commercial building is about $1.20/$ft^2$, or about 1.5%/year of the total capital cost of a typical new building of $75/$ft^2$. Over the lifetime of a building, the cost of energy for the building approaches the cost of constructing the building.

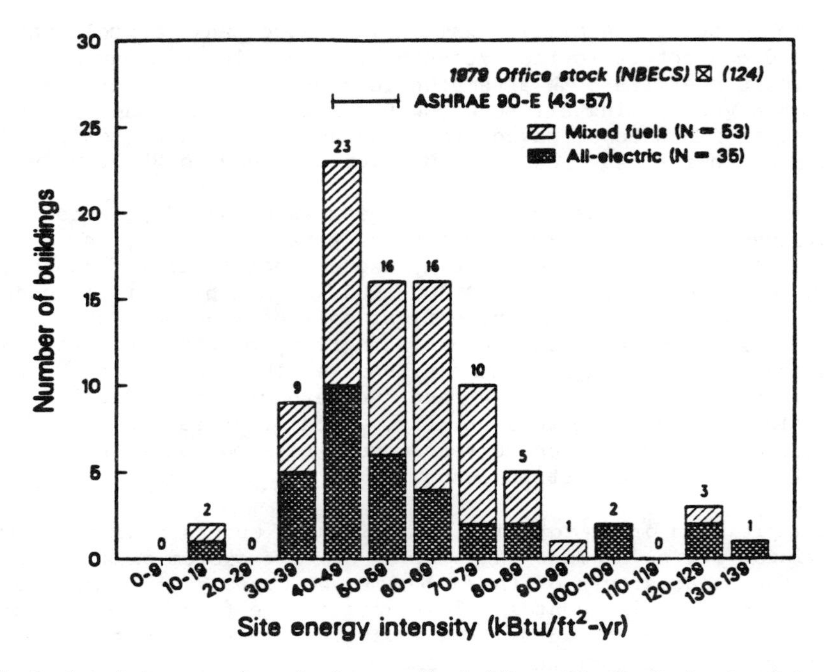

Fig. 8. Actual site energy intensity for new office buildings. The distribution for all-electric and mixed-fuel buildings are similar. Over 60% use a site intensity of 40-70 kBtu/ft^2-yr. The average U.S. office stock (EIA, 1981) and the proposed ASHRAE 90-E values for large offices are included for reference. Source: LBL BECA-CN. A compilation of current standards and data can be found in figures 12 and 13. XCG 851-48

In order to quantify progress in reducing energy use in the commercial sector, the BECA (Buildings Energy-Use Compilation and Analysis) project of the Building Energy Data group at LBL has compiled data bases on existing, retrofitted, and new commercial buildings. From these compilations of actual, measured data, the BECA group has estimated the cost-effectiveness of various retrofit measures. Since most of the energy consumed in new large buildings is electrical energy, the intensity of energy used on site is approximately 1/3 of the intensity of energy resources used. Some of the results from BECA are as follows: The data set for the new[6] commercial buildings (Fig. 8) is a selected set mainly comprised of buildings that have energy efficient designs. Most of these new buildings use a site energy intensity of 40-70 kBTU/ft^2-yr (resource intensity of 125-220 kBTU/ft^2-yr). The large office median site intensity is 59 kBTU/ft^2-yr (resource intensity of 185), while small office buildings use a median site intensity of 47 kBTU/ft^2-yr (resource of 148). The data on commercial buildings is disaggregated among building types in Fig. 9. The average intensities for both large and small buildings are well below the intensities of the existing U.S. building stock (resource intensity of 264 kBTU/ft^2-yr), but slightly higher than the simulations for buildings designed to the proposed ASRHAE standards (90-E).

Summary of Energy Performance in Comparison to Baseline (NBECS) and Standards (ASHRAE 90-E)

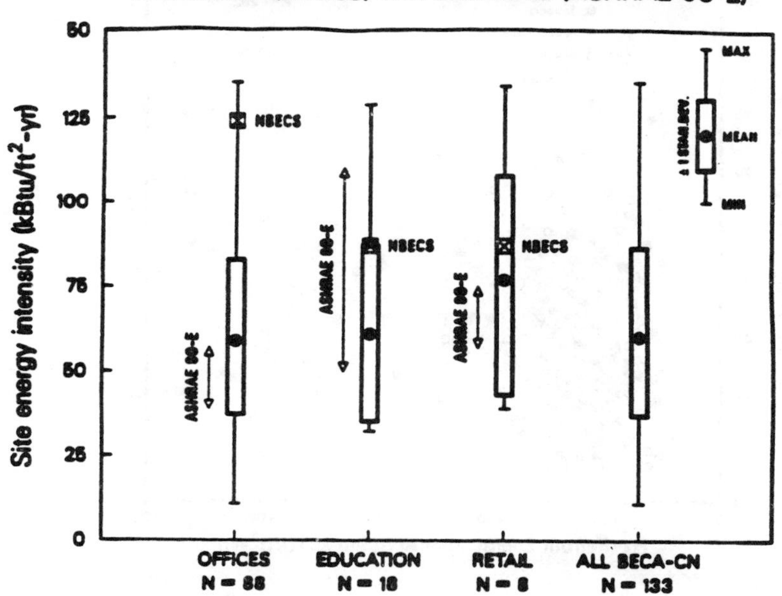

Fig. 9. Summary of energy performance by building type for BECA-CN compared to 1979 average U.S. stock (EIA, 1981) and proposed ASHRAE Standard 90-E. The minimum, maximum, mean, and standard deviation are presented for each of the four BECA-CN categories of buildings. No NBECS average or Standard 90-E data are presented for the fourth category because of the wide variety of building types in the total data base. For all three building types the BECA-CN mean is clearly below the U.S. average stock, but in only one case is it within the range of the standard. The high value for one DOE-2.1 calculation on educational buildings was caused by high ventilation rates and high use of hot water (40% of total). Source: BECA-CN. XCG 851-24

A variety of measures can be used to retrofit existing buildings to save energy by improving operation and maintenance, HVAC (heating, ventilation, and air conditioning) systems, lighting, building envelopes, windows and doors, and so forth. The BECA-CR data set[7] shown in Fig. 10 shows that building owners and managers are biased towards retrofit measures which had a short payback period. This compilation shows that about 10 to 40% of a building's annual energy use can be saved by cost-effective measures. The median cost of the energy saved was about $0.90/MBTU with a payback period of about 1 year (using a discount rate of 7% and an amortization of 10 years).

VI. COMPARISON OF ELECTRIC GROWTH IN TEXAS AND CALIFORNIA

While Texas is still a "laissez-faire" state, California practices vigorous conservation with multi-tier increasing residential gas and electric rates, mandatory standards for appliances and new buildings, zero- and low-interest loans, rebates for efficient appliances, home energy ratings, etc., and in 1985

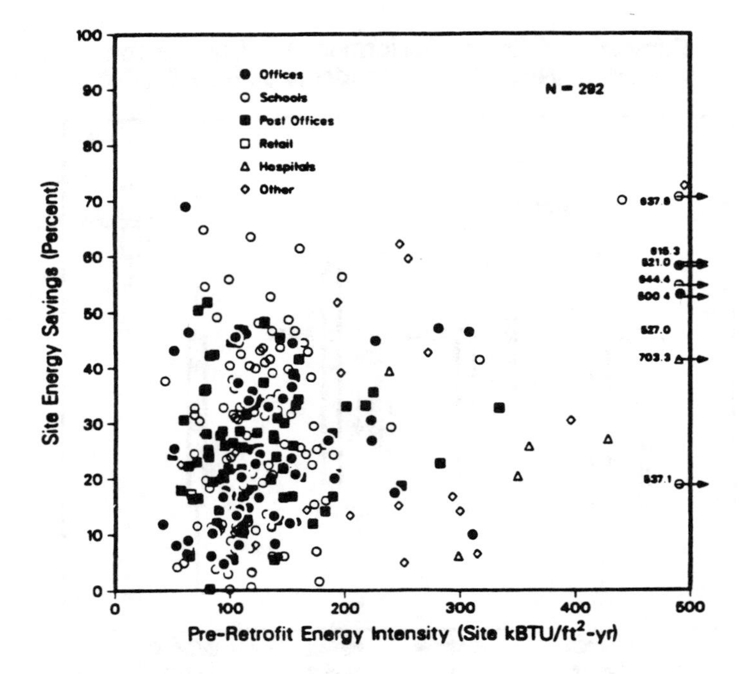

Fig. 10. Percent site energy savings vs. site energy intensity (kBtu/ft^2-yr). There does not appear to be any correlation between the pre-retrofit energy intensity and the percentage savings realized. Both high and low energy users achieved a wide range of percentage savings. Source: BECA-CR. XCG 805-13075

California completed its millionth residential audit. An example[8] of the success of this policy is the drop in the median capacity of air conditioning units sold; from 4 "tons" in 1977 to 3 "tons" in 1955.

A comparison of the growth in the electricity (kWh) for Texas and California in Fig. 11 suggests that California's conservation tools are very cost effective. The 1985 population of Texas is 16 million (growing at 2.8%/year); California has a population of 26 million (58% larger and growing at 1.7%/year). As shown in Fig. 11, Texas electricity use crossed that of California in 1978-79, and since then Texas has required 1.3 nominal 1-GW plants every year, while California has needed only 1 plant in 5 years.

We won't make a big point of the 1978 difference in kWh use per capita (Texas used 70% more than California). A defiant Texan could cite a high need for air-conditioning and electricity-intensive industry. But once we have corrected for, or ignored, the higher use per capita, we do think that the difference in growth rate is significant: annually 4.3% for Texas, and 0.9% for California. If we correct for the 1.1% higher population growth of Texas, the diference is still 2.1%/year.

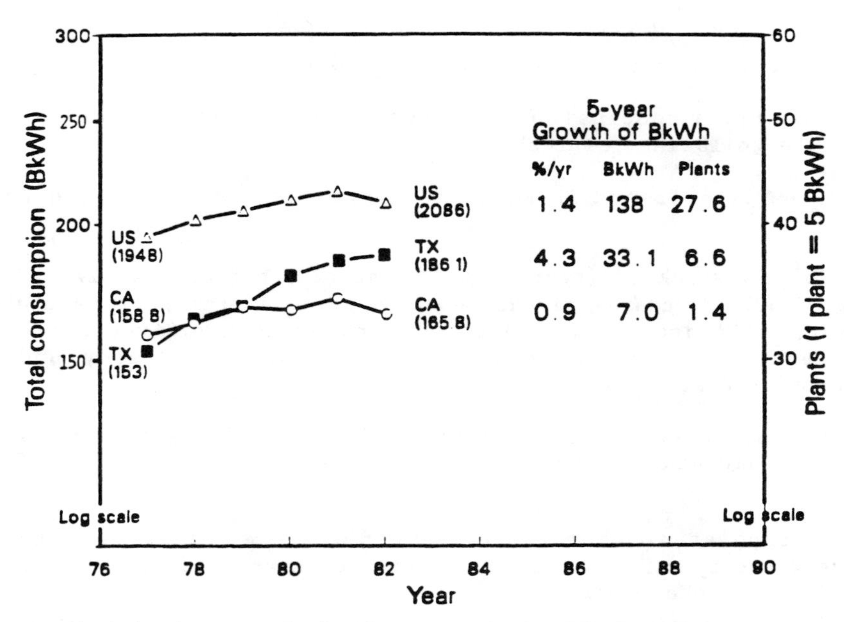

Fig. 11. Total electric consumption by all customers for CA, TX, and U.S. The y-axis is logarithmic and the U.S. is shifted down one decade. BkWh are converted to 1000 MW (1 BW or 1GW) using 1 plant = 5 BkWh/year. Source: Consumption-Electric Power Annual (EIA 0348(82)) p. 167. GW-Annual Energy Review (EIA 0384(83))p. 195 and 201. XCG 846-13083A

What are the economics to California of being able every year to avoid the construction and operation of one nominal 1 GW power plant? Let us take a 10-year perspective. If we focus only on the first cost of 10 plants, then we defer the investment of $10-20 billion, but that is an under-estimate of the full story by about a factor of two. To make a better estimate we note that the cost of new electricity is at least 10 ¢/kWh, and from Fig. 11 we note that after 10 years California has saved about 50 BkWh/year, worth $5 billion/year in the 10th year. The total electric bill saved over 10 years is then about $25 billion.

In a forthcoming study by the University of Texas and LBL[9] (ELECTRICITY CONSERVATION IN TEXAS BUILDINGS, 1985), we discuss the Figure 11 difference in terms of the price of electricity, or growth in the individual sectors. But we find that over the five year period (1977-82), Caifornia has added the same population (2.1 million), more square feet of commercial buildings, and twice as much "industrial value added," all for the one equivalent plant, compared with Texas' need for 6.6 plants. As to price effects, in the buildings sector both states had average prices of 7 ¢/kWh, but the Texas industrial rate was indeed cheaper: 4 ¢/kWh instead of 6 ¢/kWh for California.

This discussion is surely not rigorous, but we find it suggestive that California's conservatin tools are effective and cost effective.

VII. Trends: Saving 2 Alaskas and 70 Power Plants

In this brief conclusion we present two summary figures which point to the following remarkable facts.

Figure 12. Trends in Resource Energy Use (per year and per ft^2).

1. Today's stock of (typically 20-year old) offices use 270 kBtu (costing $1.30). Standards already enacted in California, or in draft by ASHRAE, will drop this 270 to 100 or 130 kBtu. Given further improvements in lighting, controls, and storage, already under development, 100 kBtu should become routine.

2. Because of savings by downsizing air conditioning and windows, new office buildings cost no more than the 1973 models, which use 500 kBtu.

3. Extrapolated to the whole 50 B ft^2 of commercial space, this future decrease by a factor of 2.7 in resource energy corresponds to a saving of 2.2 Alaska pipelines.

Figure 13. Separates the data of Fig. 12 into fuel (whose use is vanishing) and electricity.

4. Per year and per ft^2, electric use is dropping from 17.5 kWh to 11.5 (both numbers within a range of $\pm$2.5 kWh). The California mandatory standard dropped a factor of 2 from 18 to 9 kWh in 10 years (see the CA line joining these two points low in Fig. 13). Given the further improvements under development it seems realistic to extrapolate this factor of 2 to the U.S.

5. Extrapolating again to the whole 50 B ft^2 sector, this gain of a factor of 2 will avoid the need to build 70 power plants.

We now return to Fig. 12 for some additional comments.

The sharp rise in resource energy use from 1950 to the OPEC embargo is explained by the low prices of energy, accompanied by buildings with acres of single-glazing, acres of lights, and oversized HVAC (heating, ventilation, and air conditioning) systems, which cooled and then reheated the same air, ignored the availability of cool outside air, failed to use free heat from the core to heat the perimeter, and, although they ran at part load most of the time, were not designed with much consideration of part-load efficiency. Consequently, after the Embargo, it was easy to improve the design of these buildings and cut their annual energy intensity from 500 to 200 Btu/ft^2, with no increase in first cost.

The line starting in 1975 is the ASHRAE standard, calculated using the DOE-2 program for prototypes. Real buidings under-perform by 10-20%, with 25% of the buildings using 1.5 times the design energy--see Figs. 8 and 9.

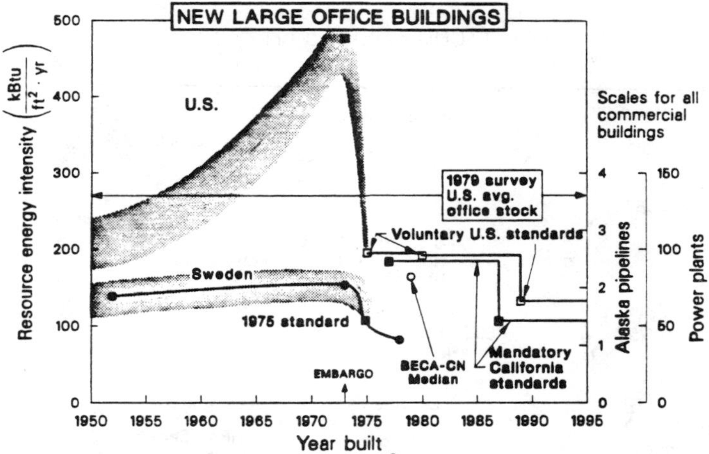

Fig. 12. Trends in annual energy intensity (use per ft^2) of new office buildings. Electricity is counted in resource energy units of 11,600 Btu/kWh.

Dots represent data from real buildings. Squares are computer simulations from prototypes. Thus, the U.S. sequence is represented with Zipatone and is a crude measure of New York City office buildings by Charles W. Lawrence, Public Utilities Specialist for the city of New York (1973).[a] The 1973 (pre-embargo) square is a simulation by A.D. Little for FEA; the later squares are simulations of buildings conforming to the indicated standards.

Interpretation of right-hand scales for all commercial buildings, using data from 1979 NBECS (Non-residential Building Energy Consumption Survey, DOE/EIA-0138).

U.S. Commercial buildings in 1979 used 3.4 q (quads) of fuel and 613 BkWh of electricity (equivalent to 7.1 q), for a total of 10.5 q.

By 1985, 6 years later, with growth of 2.6%/year, use is probably up 13% to 3.8 q fuel + 690 BkWh; total 12 q.

In 1984, the Alaska pipeline carried 1.73 Mbod, equivalent to 3.5 q. Hence, commercial buildings (12 q) use the resource output of 3.5 pipelines.

A typical 1000-MW baseload power plant generates 5 BkWh each year, so commercial buildings need the output of 690/5 = 137 standard plants.

In 1979, according to NBECS, the average U.S. office building used 270 $kBtu/ft^2$ of resource energy. The right-hand scales are then adjusted to that stock office energy intensity (270 $kBtu/ft^2$) corresponds to 3.5 pipelines and to 137 power plants. Next we assume that efficiency trends in offices reflect the same percentage trends for all commercial buildings. Thus, if the 1973 office building (up to 500 $kBtu/ft^2$) had gained permanent acceptance, our present floorspace would need the equivalent of 6.5 Alaska pipelines, and 250 power plants.

Significant further improvements in lighting, controls, and thermal storage are already in the pipeline, so it seems plausible that office energy intensity will drop to the 1987 CA standard of 100 $kBtu/ft^2$, i.e. drop a factor of 2.7 compared with office stock, and that the whole sector will follow this trend. For the present floorspace, resource energy use will then drop from 3.5 to 1.3 pipelines, but one has to wait many years to achieve equilibrium.

For electricity, figure 13 and the same reasoning show that without thermal storage we can only expect to save a factor of 2.0. For the present 50 B ft^2 of buildings, power plants needed will then drop from 140 to 70.

The *first cost* of these new energy-efficient offices is still falling, mainly because of savings from downsizing chillers.[b] With *thermal storage*, another 40% of the peak power demand could be displaced off-peak. To compare first cost of 1972 prototype with ASHRAE 90-75 see reference (a.). To compare 90-75 with later version see reference (b.).

a. A.D. Little, FEA Conservation Paper 43 B (1976).

b. ASHRAE Special Project 41, Vol. III. DOE/NBB 51/6(1983).

XCG 853-111 D

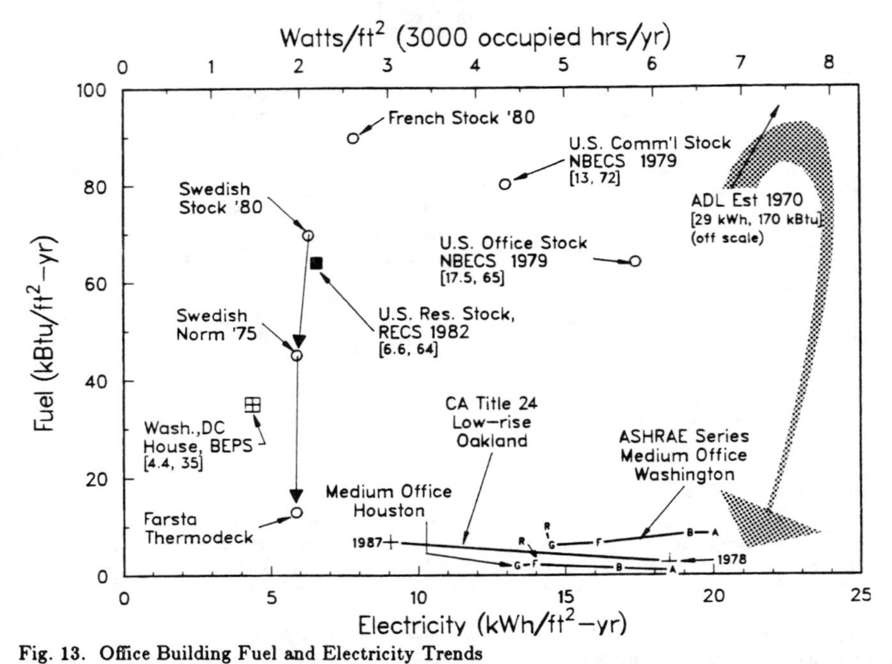

Fig. 13. Office Building Fuel and Electricity Trends
Some of the data of figure 11 are replotted with the electricity separated from *fuel* (mainly gas for heating). Commercial buildings have so much *free* heat from equipment and people that they are now need almost no space heat, even in Sweden. So modern office buildings are becoming almost entirely electric. Thus the sequences labeled A,B,F,G,R (representing *modern office building* prototypes conforming to the ASHRAE Standard 90 Series) *are almost lost at the bottom of the figure.* Similarly for the 2-point sequence representing the California Title 24 mandatory standard. Figures 7 and 8 show that 124 real buildings used 10-20% more energy called for by standards, and several used twice as much energy.

Data on the stock of existing buildings come from NBECS [Non-Residential Building Energy Conservation Survey, DOE/EIA-0318(79)] and RECS [Residential Energy Conservation Survey, DOE/EIA-0321(81)].

To compare office trends with residential trends, note the dark sequence representing U.S. residential stock and the + for a prototype BEPS home [Building Energy Performance Standards -- Federal Register 44, p. 68170 (Nov. 28, 1979), or the LBL Affordable House Data Base – DOE/SF/98-1, 1983]. BEPS specified only the cooling and heating loads per ft^2 (1 kWh, 20 Btu) for Washington, D.C.); to plot a real BEPS house to compare with a RECS house, we have added gas for domestic hot water (15 kBtu/ft^2). For a new 1700 ft^2 U.S. single-family home using gas for heating both space and water, average U.S. annual electricity use is 4.4 kWh/ft^2-yr calculated as follows: a/c 2100 kWh (includes homes with no a/c); refrigerator + freezer 1400; lighting 1200; cooking 900; drying 800; misc. 900; Total 7300 kWh. Source: J. McMahon, LBL Residential Model.

Key to symbols: Open circles are measurements, +'s and letters A,B,F,G,R are calculations based on prototypes. The letters A through R are the notation of ASHRAE Special Project 41, published as DOE/NBB-0051/6.

A Standard 90-75 (1975)
B Standard 90A-1980
F SPC 41 (90E)
G SPC 41 (90E with daylighting)
R Draft Standard 90R (will appear in 1986 as 90.1)

Note: For the Medium Office in Houston, F and R are coincident.

The ASHRAE standards are targeted towards least life-cycle cost, but real-world considerations cause them to fall somewhat short, as is illustrated by the fact the the "real" (inflation-corrected) first cost of several building types (large offices, retail, hotels) has not yet begun to rise.

Next we return to Fig. 13, in which the same energy trends are separated into fuel and electricity.

As we saw in Sections I, II and III, large buildings have a large ΔT(free), and so can have "balance" temperatures at or below freezing. Hence the need for space heat is vanishing. This is easily seen for the Swedish sequence at the left, but can be missed for the U.S. because the inefficient 1973 building falls vertically off scale by a factor of 1.7, and could be missed. Thus the discontinuity in resource energy use of Fig. 12 becomes even more striking in fuel use alone (Fig. 13).

Even by keeping the scale large enough to show the U.S. stock of existing buildings, the ASHRAE Standard 90 series fall almost on the x-axis, as does the California mandatory Title 24 sequence.

The ASHRAE voluntary standard has gone through the sequence of Standard 90-75(1975) [plotted with symbol A], 90-A(1980) [B], and soon 90.1 (for commercial buildings) [plotted as R]) and 90.2 (for residences). In preparing these standards, there was a major engineering/economic study knows as ASHRAE Special Project 21, cited in the figure caption 12. Some intermediate calculations are presented in the series [symbols F and G, as explained in the caption].

Residential squares are presented for comparison purposes, particularly to show the differences in internal load (kWh).

We hope that with these comments and the detailed captions, the reader can easily verify all five of the conclusions stated at the beginning of this section.

Shifting the Summer Peak

To complete a discussion of trends, we must recall thermal storage for load management.

1. Thermal mass, as in Thermodeck (Sect. III) can shift 50-75% of chilling off peak and requires only about 0.1 W/ft^2 of fan power. But precast slabs are currently used in only a few percent of U.S. buildings.

2. 21°C PCM's, i.e., phase change materials tuned to change at room temperature will <u>eventually</u> become cheap enough not only to handle the summer peak, but to lock the building at the comfort temperature, say 23°C, all year. The amount of material to do this

would be relatively small; for the partial storage mode (Sect. IV) it would take for each floor a layer of 6 cm for chilled water, 0.6 cm for ice, and 1.5 cm for phase-change polyalcohols.

An attractive combination of 1. and 2. is to load hollow cores in concrete with 21°C PCM's.

3. Water and ice storage (Sect. IV) cost about the same as thermodeck, but need about 1 W/ft^2 of fan power. We should strive to develop a PCM which is more attractive than water/ice; for example it could freeze at 10°C and contract as it freezes, so as to tear itself off of freezer coils.

4. To maximize thermal capacity/watt of fanpower, we should plan to use a combination of the technologies above.

5. The potential for summer peak shaving is summarized in Table II.

ACKNOWLEDGMENT

We want to thank Rosemary Riley for her help with the figures.

REFERENCES

1. For the same level of activity, additional latent heat is 95-130 W, but this affects cooling loads only. ASHRAE HANDBOOK: 1985 FUNDAMENTALS, Am. Soc. Heating, Refrig. Air-cond. Engineers, Atlanta, GA, p 26.21.

2. L. Anderson, K. Bernander, E. Isfalt, A. Rosenfeld, STORAGE OF HEAT AND COOLTH IN HOLLOW CORE CONCRETE SLABS. LBL-8913, 1979.

3. ENERGY CONS RVATION INDICATORS 1983, and ANNUAL ENERGY REVIEW 1984, Energy Information Agency, Washington, D.C.

4. A. Rosenfeld and O. de la Moriniere, ASHRAE Trans. HI-85-15, no. 4, 1985. LBL-19448.

5. R. Reardon and K. Penuel, ASHRAE Jour. 27, 24 (May, 1985).

6. M. Piette, L. Wall, and B. Gardiner, BECA-CN, LBL-19413, submitted to ASHRAE Journal, 1985.

7. B. Gardiner, M. Piette, and J. Harris, BECA-CR, LBL-17881, American Council Energy Eff. Econ. D, 31 (1984).

8. M. Messenger, California Energy Commission, private communication.

9. H. Akbari, M. Baughman, B. Hunn, A. Rosenfeld, and S. Silver. ELECTRICAL ENERGY CONSERVATION AND PEAK DEMAND REDUCTION IN TEXAS BUILDINGS, Center for Energy Studies, U. Texas and LBL, in process.

CHAPTER 9

ENERGY CONSERVATION SCOREKEEPING: THE PRISM METHOD*

Margaret F. Fels

Center for Energy and Environmental Studies
Princeton University, Princeton, N. J. 08544

ABSTRACT

The Princeton Scorekeeping Method, or "PRISM", is a statistical procedure which uses available billing and weather data to produce accurate estimates of saved energy. Derived from simple physical principles, the method provides physical descriptors as well as a weather-normalized index of consumption for each house analyzed. While similar in some respects to weather-adjustment techniques used in-house by utilities and others, this work differs in its long-range objective: a standardized approach which is equally applicable to all fuels, which can be used for a wide range of climates and building types, and which offers accurate diagnostics. This chapter summarizes the scorekeeping methodology -- its motivation, its current status, and needed areas of expansion.

INTRODUCTION

Several chapters in this book describe the considerable progress that has been made over the past ten years in discovering and understanding various methods for conserving energy in the home. Many of these measures have been implemented, some in large-scale programs. As in most other areas of science, however, an analytical prediction must be put to experimental test. The energy consumption of a residential or commercial building can depend on many variables, and it is not an easy matter to determine the effect on energy use produced by a change in one or a few of those variables. Thus, a vital part of energy conservation work is the proper assessment of savings that result from measures initiated. We have given the term of "scorekeeping" to this task of measuring the value and magnitude of the energy saved by the measures implemented. In this chapter we will describe the particular methodology that has been developed for this task at Princeton University - PRISM, or the Princeton Scorekeeping Method.

Years of research in housing conservation have convinced us that serious scorekeeping is essential to the success of all conservation ventures. Without it, the importance of conservation cannot be effectively communicated to homeowners, the best programs cannot be distinguished from ineffective ones, and the credibility of conservation is being threatened.

Many utilities in the U.S. have undertaken extensive retrofit assistance programs for their customers, not only because of the

* This chapter is taken from the introductory paper to the Scorekeeping Issue of "Energy and Buildings" 9, 1985/86.

0094-243X/85/1350169-15$3.00 Copyright 1985 American Institute of Physics

federal Residential Conservation Service (RCS), which mandates nearly free energy audits for customers, but also because of a growing commitment to energy conservation as a utility investment strategy. RCS audits have reached some two million homes. In addition, the Low-Income Weatherization Program, federally funded but managed at the community level, is reaching many additional homes, not only with an audit but with extensive, often costly, retrofits as well.

Only rarely do these programs include an accurate evaluation of how much energy is actually being saved by the actions taken. The program's yardstick of success is often the number of participants with no regard for the number of kilowatt-hours of electricity, barrels of oil, or cubic feet of gas saved. Estimates of savings are often based on engineering models typically without calibration to real-world experience. Such estimates, though useful for planning purposes, are notoriously higher than the actual savings realized, in part because they do not accurately take into account either human behavior or the irregularities in the complex heat flows of real buildings.

On the private side, companies which sell conservation services invariably omit feedback to the customer on how much energy -- and money -- the purchase is saving. Without records of actual savings achieved, companies deny themselves information from which they can understand -- and project -- the value of the services they sell. As a result, one may find many dissatisfied, confused customers dealing with a company unable to convey accurate estimates of the value of its own services.

APPLICATIONS OF SCOREKEEPING: PAST, PRESENT AND FUTURE

Over the past several years, the PRISM tools have been enormously valuable to our own buildings research program. One of the first applications was to the Modular Retrofit Experiment, a collaborative conservation project between Princeton and the natural gas utilities in the New Jersey area[1], which assessed the energy savings from house doctor visits and from subsequent, more extensive treatments. Some of that work is described in the preceding chapter by Dutt. PRISM has also been used for monitoring statewide conservation trends in New Jersey.[2]

Conservation researchers outside of Princeton are now showing increasing interest in the scorekeeping method. Recently, it was used for the evaluation of Wisconsin's low-income weatherization program[3]. For their evaluation of Residential Conservation Service and other utility conservation programs, the staff at Oak Ridge National Laboratory are using PRISM as stage one of their two-stage evaluation approach (see, for example, their scorekeeping of Bonneville Power Administration's weatherization pilot program.[4]) The method is being used extensively in Minnesota to monitor the success of a variety of city and state programs.[5]

There is much more to be learned before PRISM will work equally well for all major fuels, over a wide range of climates and building types. While the initial emphasis of the methodology development was on gas-heated single-family houses, we are focusing our current research in three areas: 1) the inclusion of cooling for electrically heated houses -- a nasty problem because the demand for cooling is far more erratic (people-dependent) than it is for heating; 2) the treatment of "bad" houses that don't respond predictably to weather; and 3) the applicability of the approach to large multi-family buildings, to understand its limitations as well as its strengths. With the benefit of the wealth of real-world experiences embodied in ongoing scorekeeping projects such as the above, we are optimistic that these advancements in the methodology are feasible.

The long-range goal of our scorekeeping research at Princeton is to produce a standardized, easy-to-use approach which utilities, communities and others throughout the country may adopt for measuring the savings achieved by their retrofit programs.

AN OVERVIEW OF THE SCOREKEEPING METHOD

Accurate estimates of actual energy savings require only two sets of readily available data - utility bills and daily average temperatures. As depicted in Figure 1, the Princeton scorekeeping method uses utility bills from before and after the retrofit installation, together with average daily temperatures from a nearby weather station for the same periods, to determine a weather-adjusted index of consumption, called the Normalized Annual Consumption or NAC, for each period. Analogous to the EPA miles-

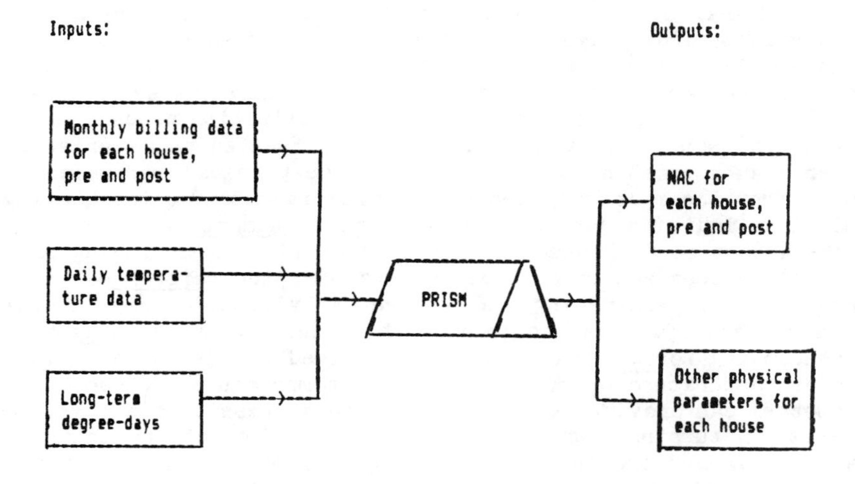

Fig. 1. Schematic diagram showing the data requirements for the Princeton scorekeeping method (PRISM) and the estimates that result from it.

per-gallon rating, the NAC index provides a measure of what energy consumption would be during a year with typical weather conditions. However, it is more accurate that the EPA ratings because it is based on field measurements.

The total energy savings are derived as the difference between the NAC in the pre- and post-periods. A conservation effect is thus neither masked by a cold winter nor exaggerated by a warm one, nor is it obscured if the time covered by billing periods in one "year" is longer than in another.

Scorekeeping experiments often require a control group. This control is a set of houses to which the treatment being studied has not been applied. Such houses help correct for the influences of occupant behavior and externalities such as energy price changes, and in effect to isolate the savings due to the program from savings that would otherwise have occurred. The same procedure applied to both the treatment and control houses gives a measure of control-adjusted savings for the treatment group. The analysis can then be updated for succeeding years, to track the durability of the savings.

Another application of PRISM is for house energy labeling, whereby an actual energy consumption index (based on energy bills) would be attached to any occupied house. Such an index could be extremely valuable to a house purchaser at time of sale, and to the homeowner or billpayer for energy conservation investment choices.

A more complete evaluation is often desired, to determine the cost-effectiveness of various approaches to conservation, for example, or to assess the effect of program participation and other explanatory variables. The energy savings estimates, along with other PRISM outputs, provide reliable input to such analyses. Thus the PRISM analysis depicted in Fig. 1 may be thought of as a standardized first stage of an evaluation, with subsequent analyses, limited by available data and shaped by the specific needs of the project being evaluated, comprising the second stage.

PRISM differs from other weather-normalization procedures in that the house's break-even temperature is treated as a variable, rather than a constant such as 65°F. Three physical parameters result from the model applied to the billing data for the heating fuel of an individual house: <u>base level consumption</u>, corresponding to the amount of fuel used per day (for appliances including water heaters) independent of outside temperature; the <u>reference temperature</u>, approximating the average daily outside temperature above which no fuel is required for heating; and the <u>heating slope</u>, corresponding to the amount of fuel required per degree drop in outside temperature below the reference temperature. These parameters can provide indications of the sources of conservation: insulating, turning down thermostats, more efficient appliance usage, etc., and thus define an "energy signature" of the house. The Normalized Annual Consumption (NAC) index is derived from these parameters applied to a long-term (say, ten-year) annual average of heating degree-days.

It turns out that NAC is extremely well determined (its standard errors are typically 3±4% of the estimate), so that savings of 6% or more may generally be considered significant. Furthermore, NAC is quite insensitive to which periods are included, or their length. We know of no more reliable index for monitoring conservation.

DETAILS OF THE SCOREKEEPING METHOD

For each house being analyzed, the PRISM method requires meter readings (or for fuel oil, delivery records) for about one year in each period. The consumption data are then corrected for the effects of weather, which of course is never the same for two different years, and also for differences in the time spanned by the two periods. From these results the NAC is calculated.

THE PHYSICAL BASIS FOR THE MODEL

We start by describing the method developed for fuels used for heating but not cooling. Generally, whether for natural gas, oil or electricity, a house's heating system is first required when the outside temperature (T) drops below a certain level (the heating reference temperature **T**), and for each additional degree drop in temperature a constant amount of heating fuel (the heating slope **B**) is required. Thus, the required heating fuel is linearly proportional to **T** - T, and the proportional constant **B** represents the house's effective heat-loss rate. In addition, the house may use a fixed amount of heating fuel (the base level **A**) which is independent of outside temperature T. As illustrated in Fig. 2

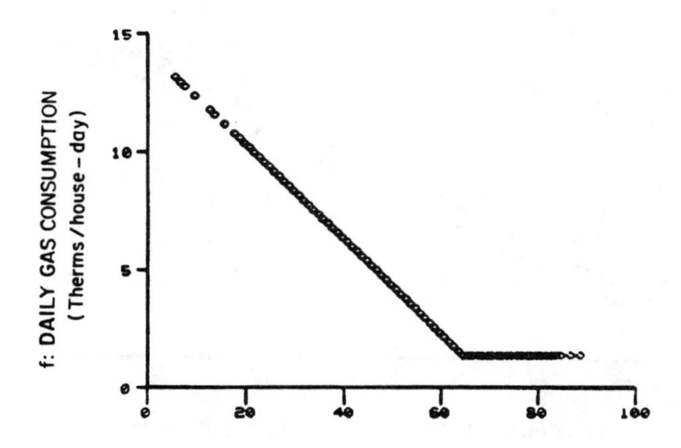

Fig. 2. Daily gas consumption as a function of outside temperature for a single idealized house.

for an idealized house, the expected fuel consumption f is given by

$$f = \mathbf{A} + \mathbf{B}\ (\mathbf{T} - T)_+ \qquad (1)$$

where the term in parentheses is the heating degree-days to base **T** and the "+" indicates zero if the term is negative.

The physical justification for assuming that both the base level **A** and the heat-loss rate B are constant has been carefully analyzed in previous work.[6] The derivation leads to a simple physical interpretation for each of the three parameters. The reference temperature **T** represents the outdoor temperature below which the heating system is required. The value of **T** is influenced primarily by the indoor temperature (thermostat setting) and, in addition, an offsetting contribution from the free heat (i.e., heat generated by appliances and occupants). The heat-loss rate **B** depends on the conductive and infiltration heat losses, while the base level **A** represents the fuel requirements of appliances (including lights, for electricity, and the water heater if fueled by the heating fuel).

If **T** is not accurately determined, or if it changes significantly over the time periods studied, the error or change in **T** will directly affect **A**, with an opposite sign. In fact, the slope **B** will be affected as well. Fig. 3 illustrates this for the idealized house by plotting f vs. h for one correct and two incorrect values of **T**. A straight-line fit through each set of points will have a different slope and intercept. Therefore, an assumed (incorrect) reference temperature, such as the value of

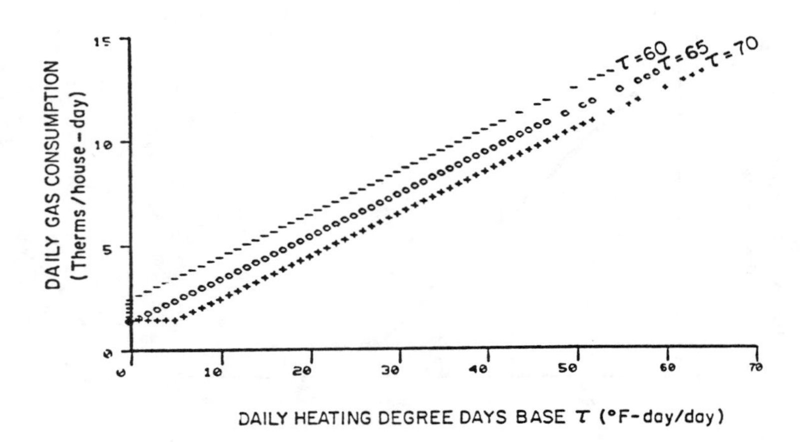

Fig. 3. Daily gas consumption as a function of degree-days to three different base temperatures for an idealized house.

65^oF so commonly used, is likely to lead to less physically meaningful values of the base level and the heat-loss rate. However, estimates of total consumption over a time period of about a year are much less sensitive to choice of **T**.

INDIVIDUAL-HOUSE ANAYLSIS

Based on this physical interpretation, the two data requirements for the analysis are actual meter readings, from which consumption is calculated, and daily average temperatures, from which heating degree-days to different reference temperatures are computed in exact correspondence to the consumption periods. The input to the procedure is then F_i and H_i where:

F_i = average daily consumption (eg, kwh/day) in time interval i

$H_i(T)$ = heating degree days per day to reference temperature **T** in time interval i.

Here, $H_i(T)$ is computed from average daily temperatures T_{ij} for the N_i days in interval i, i.e.,

$$H_i(T) = \sum_{j=1}^{N_i} (T-T_{ij})_+/N_i \qquad (2)$$

where "+" indicates that the term in parentheses is set to zero if T_{ij} is above **T**. Fig. 4 shows a plot of F_i against H_i for the 1978-79 heating year, for a sample house from the Modular Retrofit Experiment (MRE). A straight-line relationship is clearly suggested.

The set of data points $\{F_i\}$ and $\{H_i\}$ for an approximately year-long period are then fit to a linear model:

$$F_i = \mathbf{A} + \mathbf{B}\, H_i\ (T) + \mathbf{e}_i \qquad (3)$$

where $\mathbf{e}_i$ is the error term. For a guessed value of reference temperature **T**, base level and heating slope parameters **A** and B are found by standard statistical techniques (ordinary least-squares linear regression). The parameters **A** and B are calculated in this way for several different values of **T**. "Best **T**" is the one for which a plot of F_i vs. H_i (T), such as the one shown in Fig. 4, is most nearly a straight line. Formally, **T** is the value for which the mean squared error is minimized, or equivalently, for which the R^2 statistic is highest. The corresponding values of **A** and **B** are the best estimates of base level and heating slope.

In our model, the term **A** characterizes the temperature-independent component of consumption (in units/day, where units may be ccf or therms for gas, kwh for electricity or gallons for fuel oil), which is dominated by appliance and water heater usage. The parameter **B** represents the incremental amount of gas required for

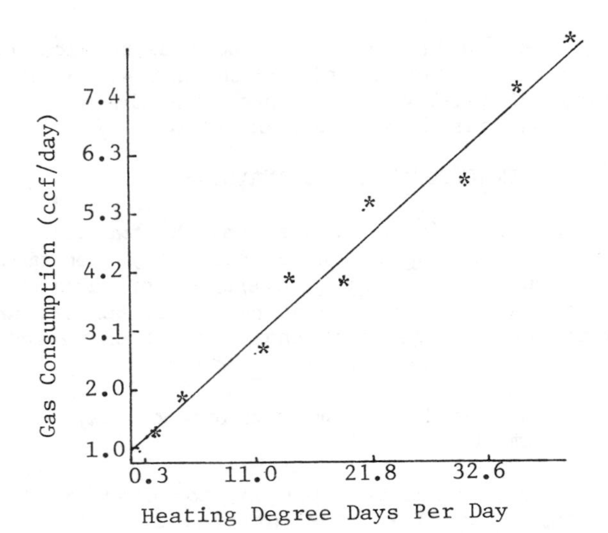

Fig. 4. Consumption data for monthly periods (asterisks) versus heating degree-days to base 68°F for a sample gas-heated house in New Jersey. The straight line represents the best fit to Eq. 3.

---

each degree drop in temperature below the reference temperature. Referring to **B** as the heating slope (in units/°F-day), the term $\mathbf{B}SH_i(\mathbf{T})$ gives an estimate of temperature-dependent demand, which is dominated by space-heating. The reference temperature **T** (in °F), which varies from house to house, represents the average outside temperature below which a house's heating system is required.

The parameters **A, B** and **T** determine the NAC, the consumption that would occur in a year with typical weather conditions:

$$NAC = 365\ \mathbf{A} + \mathbf{B}\ H_o\ (\mathbf{T}) \tag{4}$$

where $H_o(T)$ is the heating degree-days (base T) in a "typical" year. For our recent New Jersey analyses we represented a typical year by an average over the twelve years from 1970 through 1981. Weather data were collected from National Oceanic and Atmospheric Administration data for the Newark, NJ, weather station. Values of H_o for **T** = 60, 65 and 70°F are 3807, 4917 and 6181 °F-days/year.

Eq. 3 applied to the house data in Fig. 4 gives the following results for the best-**T** approach:

$\mathbf{T}$ = 68.1 (±2.7) °F

$\mathbf{A}$ = 0.90 (±0.26) ccf/day

$\mathbf{B}$ = 0.18 (±0.01) ccf/°F-day

NAC = 1324 (±27) ccf/year

R^2 = 0.985.

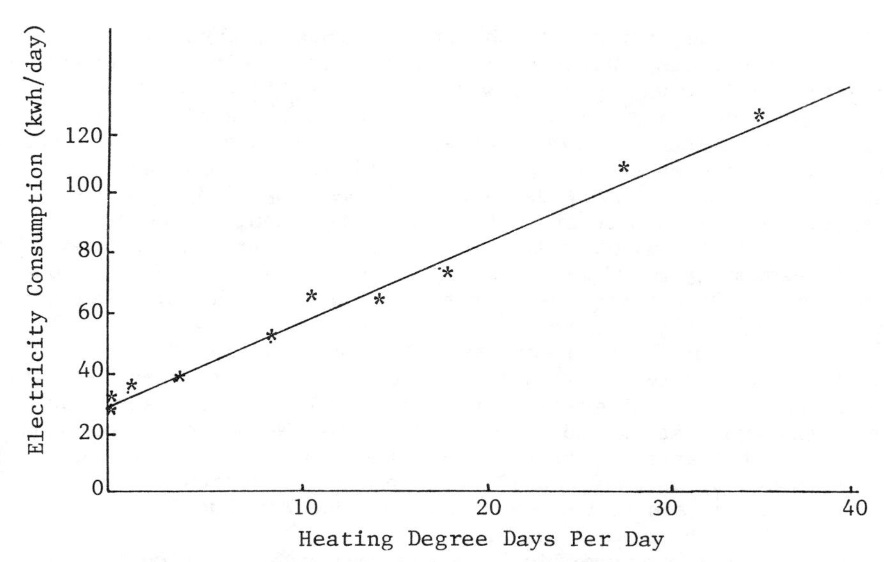

Fig. 5. Consumption data for monthly periods (asterisks) versus heating degree-days to base 59.9° F for an electrically-heated house in New Jersey. The straight line is the best fit to Eq. 3.

The numbers in parentheses represent the standard errors. The relatively small standard error for NAC as compared with **A** and B (2% vs. 12% and 6%) is typical of results from this model. The heating component $\mathbf{B}H_o(\mathbf{T})$ represents 63% of the total consumption. The R^2 statistic indicates a very good straight-line fit, corresponding to the line drawn in Fig. 4.

The methodology can be directly extended to electrically heated houses without cooling. For example, the heating-only model in Eq. 3 applied to the house data in Fig. 5 gives the following results:

$$\mathbf{T} = 59.5\ (\pm 1.6)\ ^{\circ}\mathrm{F}$$

$$\mathbf{A} = 29.0\ (\pm 1.6)\ \mathrm{kwh/day}$$

$$\mathbf{B} = 2.73\ (\pm 0.17)\ \mathrm{kwh/^{\circ}F\text{-}day}$$

$$\mathrm{NAC} = 20{,}700\ (\pm 375)\ \mathrm{kwh/year}$$

$$R^2 = 0.990.$$

Again, the NAC, with a standard error of 2%, is extremely well determined.

In general the NAC estimate provides a reliable conservation <u>index</u> from which energy savings and conservation trends may be accurately estimated. On the other hand, the three parameters **A**, B and **T** comprising the energy signature of the house, and the estimate of annual heating consumption $BH_o(T)$ derived from them, are less well determined, and their changes over time are often difficult to interpret due to the interference of physical and statistical effects. While it is tempting to attribute a change in the base level to water heater wrap or more efficient appliances, for example, or a drop in the heating-consumption estimate to added ceiling insulation or other measures to tighten the structure, such physical inferences are often not statistically valid. We feel that these parameters provide physically meaningful <u>indicators</u>, whose changes may not be statistically significant but whose behavior can often suggest the reason for a consumption change.

A frequently mentioned shortcut is to fix T at 65°F.[7] Although **A** and **B** are highly sensitive to the **T** value used, the NAC results are not, especially when the best-**T** values are fairly close to 65°F. (The median **T** value for several samples analyzed by this method has been close to 60°F.) Nevertheless, our studies indicate that **A** and **B** are considerably more meaningful when estimated for best **T** than when estimated at a fixed value. We strongly recommend that the best-**T** approach be used when the results for **A, B** and **T** are of interest, as they usually are in a conservation analysis, or when there is reason to believe that the true **T** value is quite different from the assumed value.

Such examples assume the pre-selection of estimation periods for the model. When a continuous series of consumption data is available, it is often enlightening to run a month-sliding analysis, wherein a one-year estimation "window" is slid forward one month at a time, as shown in Fig. 6. In our experience, this

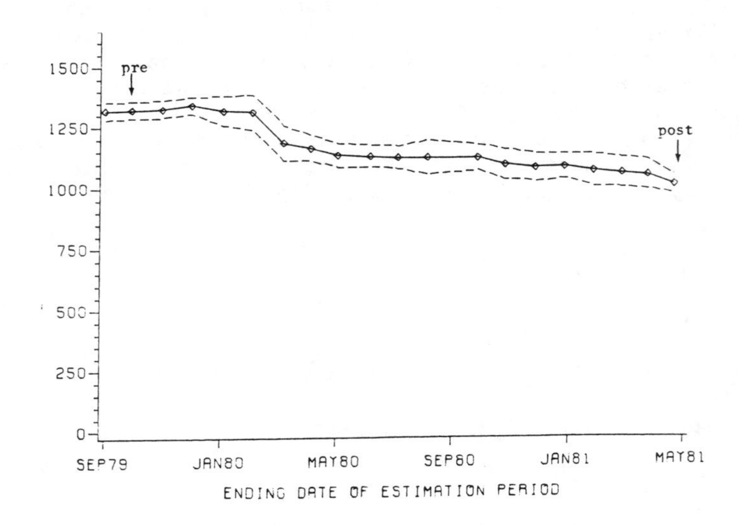

Fig. 6. Illustration of sliding-month approach. Each point is the NAC estimate that results from applying PRISM to one year of consumption data, ending on the indicated date. Dotted lines show standard estimate of the error. MRE demonstration house is used. Pre- and post-retrofit periods used for scorekeeping are indicated.

provides a powerful tool not only for the selection of final estimation periods, but also for identification of anomalies in the data, and, more generally, for monitoring gradual changes in consumption. The NAC summary in Fig. 6 illustrates the approach for the MRE demonstration house. The analogous plots for the individual parameters (not shown), which are particularly useful for flagging data anomalies in houses less well behaved than this example, demonstrate the temporary instability as the estimation window passes through the retrofit period. The drop in consumption after the retrofit is evident in the NAC plot.

THE MEASUREMENT OF SAVINGS

The NAC index provides the basic parameter for monitoring energy savings resulting from retrofit programs. Using billing and weather data for approximately year-long periods before and after (and not including) the period during which the retrofits were performed, $NAC_{pre}(T)$ and $NAC_{post}(T)$ are calculated as averages (medians or means) over houses in the treatment group, for the pre- and post-periods respectively. The raw, weather-adjusted change in energy consumption is then given by

$$S_{raw}(T) = NAC_{pre}(T) - NAC_{post}(T) . \tag{5}$$

Analogous control indices $NAC_{pre}(C)$ and $NAC_{post}(C)$ are calculated as averages over the control houses, for the same pre- and post-periods. The raw savings can then be adjusted as follows:

$$S_{adj} = NAC_{pre}(T) \, [NAC_{post}(C)/NAC_{pre}(C)] - NAC_{post}(T) \tag{6a}$$

or, in percentage terms,

$$S_{adj,\%} = [NAC_{post}(C)/NAC_{pre}(C)] - [NAC_{post}(T)/NAC_{pre}(T)] \tag{6b}$$

The raw savings for the treatment group (Eq. 5), the control savings and the savings adjusted by the control (Eq. 6) are all quantities of interest in scorekeeping.

For the MRE house in Figs. 4 and 6, the raw savings were 325 ccf/year, or 25% of pre-period consumption, with a standard error of 56 ccf/year, or 4% of pre-NAC. This house belonged to the "House Doctor" group, for which the median savings may be summarized as follows:

raw savings, treatment group: $S_{raw}(T)$ = 200 ccf/year, or 15% of pre-NAC

raw savings, control group: $S_{raw}(C)$ = 133 ccf/year, or 10%

control-adjusted savings: S_{adj} = 80 ccf/year, or 8%.

Thus the savings are highly sensitive to whether they are adjusted by a control. In the MRE, the control adjustment substantially deflated this experiment's raw savings.

INCLUSION OF ELECTRIC COOLING

The PRISM methodology has been applied extensively to gas- and oil-heated houses, and electrically heated houses without cooling. For all fuel types, R^2-values are typically 0.97 or better, and the accuracy of the estimates given here is typical of the individual houses studied.

If electricity is used for cooling but not heating, a model analogous to Eq. 3 applies, with $H_i(\mathbf{T})$ replaced by cooling degree-days $C_i(\mathbf{T}_c)$ computed to a cooling reference temperature $\mathbf{T}_c$, and with $\mathbf{B}$ replaced by the cooling rate $\mathbf{B}_c$.[8] If the house is electrically heated and cooled, the model becomes:

$$F_i = \mathbf{A} + \mathbf{B}_h H_i(\mathbf{T}_h) + \mathbf{B}_c C_i(\mathbf{T}_c) + \mathbf{e}_i \, . \tag{7}$$

The corresponding weather-normalized index is given by

$$\text{NAC} = 365\mathbf{A} + \mathbf{B}_h \, H_o(\mathbf{T}_h) + \mathbf{B}_c \, C_o(\mathbf{T}_c) \tag{8}$$

where cooling degree-days C_o are computed for the same normalization period establishing H_o.

Even in a heating-dominated climate, summer consumption not uncommonly tracks cooling degree-days, as the data in Fig. 7

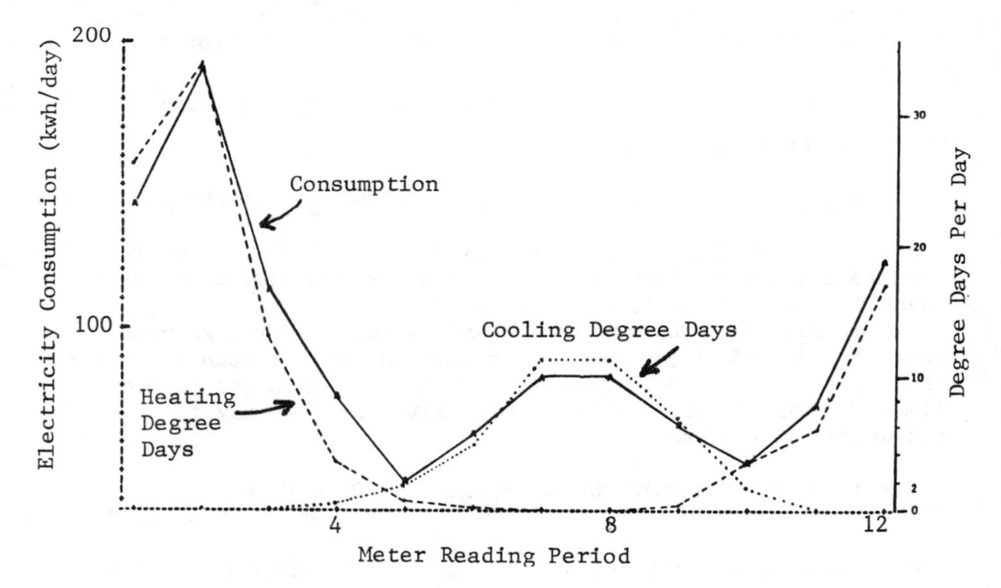

Fig. 7. Superposition of degree-day data on consumption data, as a function of meter-reading period, for a sample electrically heated and cooled house in New Jersey. Heating (---) and cooling (...) degree days are calculated to bases of 61°F and 67°F, respectively.

illustrate. The results of applying the heating-plus-cooling model in Eq. 7 to the data are shown with the figure. Once again, the NAC estimate, with a relative standard error of 2%, is extremely well determined.

As might be expected, not all houses behave as predictably as this example does. In a current project for EPRI, we have been exploring several modifications of the above equation, such as holding T_c at 70°, or estimating summer consumption in excess of base level. Our experience to date suggests that the R^2-values for houses heated and cooled by electricity will be somewhat lower than they are for a heating-only fuel, but generally high enough for an accurate measurement of savings.

EXTENSION TO UTILITY AGGREGATES

The above methodology is designed to be applied to individual-house billing data for large numbers of houses, in utility conservation programs, for example, or retrofit projects such as MRE. An analogous approach has been demonstrated to work well for utility aggregate sales of natural gas to gas-heating customers. To account for the billing lag, a simple function of this month's and last month's heating degree-day, AH_i, replaces H_i in Eq. 3. For example, the very simple form $AH_i \equiv (H_i + H_{i-1})/2$ gives reliable results in New Jersey (e.g., $R^2 > 0.99$ for each year since 1970).

As for the single-house example in the previous section, the error bars for the aggregate NAC, at approximately ±3% of NAC for single years, are considerably narrower than the bars associated with the individual parameters. The narrow bounds mean that small changes in typical consumption can be identified. Even using a stringent test that the 95% confidence intervals not overlap, a drop in aggregate NAC of 6% between two years can be judged significant. This sensitivity to small changes makes the NAC parameter a valuable conservation index for monitoring purposes.

Utility aggregates can be used in scorekeeping as substitutes for control groups, whose selection and monitoring may be costly. In all of the MRE locations, very similar estimates of percent savings were obtained from the control samples and from the corresponding utility aggregates for the same time periods. For our MRE demonstration house, the utility aggregate savings was 12%, vs. 11% for the corresponding control sample.

Another use of the aggregate approach is for monitoring conservation trends. Figure 8 summarizes the results from using utility sales data from New Jersey's four natural gas utilities aggregated to the state level (i.e., to a total of almost a million customers). After weather normalization, per-customer consumption showed a decline of 26% since the peak level before the oil embargo.[2] Our analysis of the separation between the base-level and heating components suggests that most of the conservation was due to lowering of thermostat settings, with perhaps a surprisingly small fraction due to structural retrofitting of the houses. Thus, important policy implications can emerge from the methodology

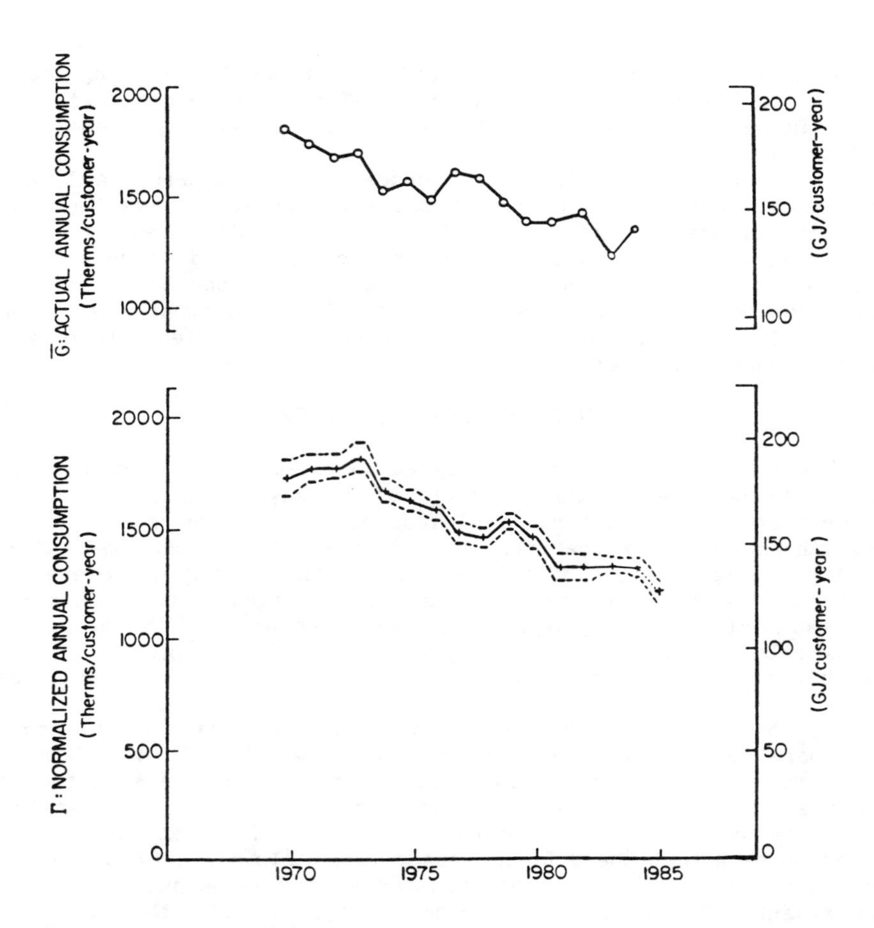

Fig. 8. Actual annual consumption compared to normalized annual consumption (NAC) as a function of time (1970-1985) for New Jersey residential gas-heating aggregate. Dotted lines show limits of 95% confidence level. Estimate for 1985 is preliminary.

applied at the aggregate as well as at the house level.

ACKNOWLEDGEMENTS

Over several years the contributors to the scorekeeping methodology have been many: Yoav Benjamin, Miriam Goldberg, Cliff Hurvich, Michael Lavine, Lawrence Mayer, James Rachlin, Thomas Schrader, Robert Socolow, Daniel Stram and Thomas Woteki. Funding, also, has been from numerous sources: the Electric Power Research Institute (Research Project RP 2034-4), the Ford Foundation State Environmental Management Program, the U. S. Department of Energy (Contract No. DE-AC02-77CS20062), and the New Jersey Energy Conservation Laboratory.

REFERENCES

1. G.S. Dutt, M.L. Lavine, B.G. Levi, R.H. Socolow, "The Modular Retrofit Experiment: Design, Scorekeeping and Evaluation", Energy and Buildings 9, 1985/86 (Scorekeeping issue); See also, "The Modular Retrofit Experiment: Exploring the House Doctor Concept", PU/CEES #130, Princeton, Univ., June 1982.

2. M. F. Fels, M. L. Goldberg, "Using Billing and Weather Data to Separate Thermostat from Insulation Effects," Energy 9, #5, 439 (1984).

3. M. L. Goldberg, A. Jaworski, I. Tallis, "Evaluation of Wisconsin's Low-Income Weatherization Program", Final Report and Appendices of Final Report to Wisconsin Departmentof Administration, October 1984.

4. E. Hirst, "Electricity Savings One, Two and Three Years after Participation in the BPA Residential Weatherization Pilot Program", Energy and Buildings 9, 1985/86.

5. M. Hewett, T. Dunsworth, T. Miller and M. Koehler, "Measured vs. Predicted Savings from Single Retrofits: A Sample Study", Energy and Buildings 9, 1985/86.

6. T. F. Schrader, "A Two-Parameter Model for Assessing the Determinants of Residential Space Heating", PU/CEES # 69, Princeton Univ., June 1978.

7. M. F. Fels, J. N. Rachlin, R.H. Socolow, "Seasonality of Non-heating Consumption and its Effect on PRISM Results", Energy and Buildings 9, 1985/86; also, "Seasonality of Non-heating Consumption: A Study Based on Submeter Data", PU/CEES #166, July 1984.

8. D. Stram, M. F. Fels, "The Applicability of PRISM to Electric Heating and Cooling," Energy and Buildings 9, 1985/86.

CHAPTER 10

PASSIVE SOLAR HEATING

David E. Claridge
Robert J. Mowris
University of Colorado, Boulder, CO 80309

ABSTRACT

Buildings have been designed to use solar gains for winter heating for several millenia, but the quantitative basis for passive solar design has only been developed in the last decade. A simplified lumped capacitance model is used to provide insight into the physics of passive building behavior. Three passive design methods are described: the Solar Load Ratio (SLR) method based on correlations to simulation results; the Gordon-Zarmi closed form analytical model; and the "unutilizability" model of Monsen and Klein. Model predictions are compared with measured results; agreement is good if measured building characteristics are used. Numerous passive houses use less than 2 Btu/ft^2-DD for auxiliary heating and consensus is developing that modest levels of passive glazing combined with superinsulation techniques can provide the best feature of both approaches.

HISTORICAL INTRODUCTION

Just over a decade ago, passive solar heating didn't have a name and the concept had a bad name. Windows were considered a major villain in building energy performance; they increased heating costs and cooling costs. They were an aesthetic necessity, but it was thought that window area should be stringently reduced or completely eliminated in energy efficient buildings.

The 1974 APS Summer Study investigated ways to increase the thermal resistance of windows[1] and meet aesthetic needs. Berman and Claridge[2] showed that over an entire heating season, south-facing windows provide net heat gain to a house. This was true for climates as diverse as Texas, New York, Northern Michigan and the cloudy Pacific Northwest. East or west-facing windows showed marginal gains in some cases. To the authors' knowledge, this was the first quantitative evaluation of the potential for passive solar heating in diverse climates.

However, passive heating has been used for millenia. Butti and Perlin[3] have written a fascinating and comprehensive history of the use of the sun for applications as diverse as greenhouses and steam generation. The brief passive heating history here is largely based on their work.

The ancient Greeks and Romans used passive solar principles on occasion. An addition to the Greek city of Olynthus built about 500 B.C. was carefully laid out to provide solar access for every house and north walls had few, if any, north windows. The Romans began using transparent glass for windows in the first century A.D. and

0094-243X/85/1350184-25$3.00 Copyright 1985 American Institute of Physics

subsequently, the popular public baths nearly always had large windows facing south - partly because heat from the sun was considered healthier than "artificial" heat.

EARLY NORTH AMERICAN USE OF PASSIVE SOLAR

The Indians of the American Southwest were the first to extensively use passive solar heating and cooling in the United States. The best known of these are the Mesa Verde cliff dwellings in southwestern Colorado which incorporated southern exposure, massive construction, and the use of overhangs to temper ambient temperatures during winter and summer. Other community dwellings built by the Anasazi Indians over 800 years ago show similar sensitivity to the sun's seasonal movement. For example, the "sky city" of Acoma (New Mexico) is similar to the Greek Olynthus; most doors and windows open to the south and full exposure to the winter sun was provided for every dwelling.

While several early American housing styles such as the Colonial "saltbox", the Spanish Colonial adobe, and southern plantation houses showed considerable sensitivity to climate, use of passive solar heating became a lost art during the 19th century. Near the end of the century, Bruce Price, Samuel Atkinson and other architects studied the effects of orientation on winter and summer temperatures in houses. Atkinson built a small "sun house" in which winter temperatures above 100 F were often observed on subfreezing winter days. But this work was quickly forgotten.

The development of double pane glass during the 1930s improved the performance of windows and spurred a revival of interest in passive heating. An experiment comparing the heating cost of two test structures identical except for a clear south window in one and a heat absorbing window in the other showed the structure with the clear glass window had nine percent lower heating requirements. This highlighted the importance of winter solar gain in buildings for the engineering community, since the test was conducted by the American Society of Heating and Ventilating Engineers.

Architect Frank Keck noted the importance of solar gain and gradually began designing houses with ever larger amounts of south windows until one had windows covering the entire south side. This house was built for Chicago real estate developer Howard Sloan, who recognized the sales potential and provided public tours of the house. In 1941 he built a small development with half the houses using this "solar" design. These houses proved very popular, and by 1946, a number of prefabricated home builders offered "solar" designs. However, there was no analytical basis for predicting the performance of these houses and the empirical basis was sketchy. Engineers were divided over the benefits of south glass. Customers liked the large windows and "solar" houses with large expanses of east, west or even north glass were built with disastrous results. Low priced fossil fuels and a decline in the wartime conservation ethic further eroded interest in solar houses.

By the 1970s, most of the lessons learned earlier had been long forgotten and passive heating had to be rediscovered and relearned by a new generation. This time, the computational tools available and the interest spurred by the oil embargo and skyrocketing energy prices resulted in the development of a sound theoretical and empirical basis for the performance of "solar" houses. The term "passive solar" was coined to differentiate solar heated houses which use the windows as collectors and massive construction for thermal storage from those which use roof-mounted collectors and water tanks or rock beds for storage.

TYPES OF PASSIVE SOLAR SYSTEMS

Three basic types of passive heating systems have emerged. These differ in configuration and storage mass placement and are commonly called (1) direct gain, (2) thermal storage wall and (3) sunspace systems.

The simplest of the three is the direct gain system. Interior spaces are heated directly by solar radiation transmitted through south-facing window area. Direct gain systems may produce uncomfortably high interior temperatures on sunny days, particularly during spring and fall. Massive floors and partition walls are often used in direct gain houses to reduce temperature swings, and overheating. These massive components act as thermal storage which releases heat at night and reduces the amount of "auxiliary" or

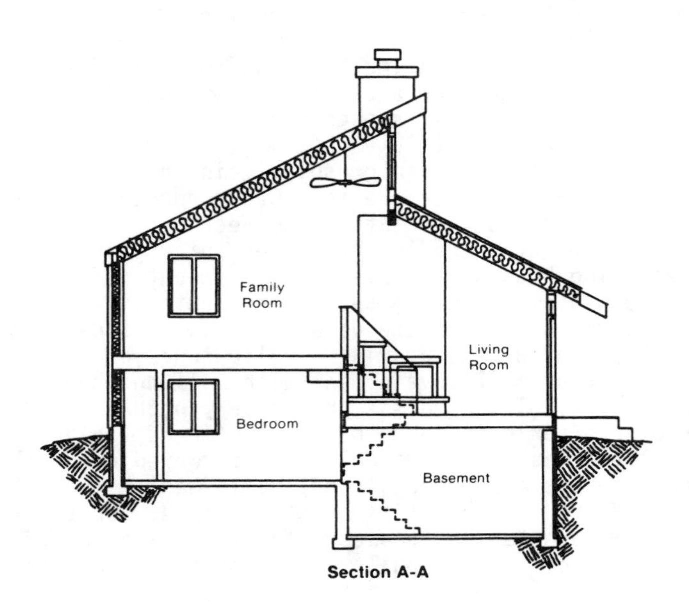

Fig. 1. Direct Gain House - SERI Class B Site DML (from ref. 4).

furnace heating required. Auxiliary heating can be further reduced by using movable insulation at night on the south glass area. Overhangs, shades or awnings are used to reduce solar gain during summer months.

A cross-sectional view of a house incorporating several typical direct gain features is shown in Figure 1. It has large south-facing windows in the living room and clerestory windows which light the brick veneer wall on the north side of the family room. It also has a concrete floor slab and uses a ceiling fan to prevent stratification in the second level. Overhangs are present on both the clerestory and the first level south windows. This house, located near Denver, CO, was monitored in the Solar Energy Research Institute (SERI) residential Class B monitoring program[4,5] and is used in several calculational examples of this chapter.

The thermal storage wall, or Trombe wall, is an indirect gain system. The wall is generally built of concrete or filled masonry blocks, but water-filled containers are sometimes used. The thermal storage wall absorbs solar radiation, heats up, and then releases the heat by radiation, conduction, and convection into the interior space. Thermal storage walls are typically designed so heat takes 8 to 12 hours to diffuse from the outside surface to the inside surface. They are sometimes vented to provide immediate heating by gravity-induced air circulation. Approximately 50% of solar heat

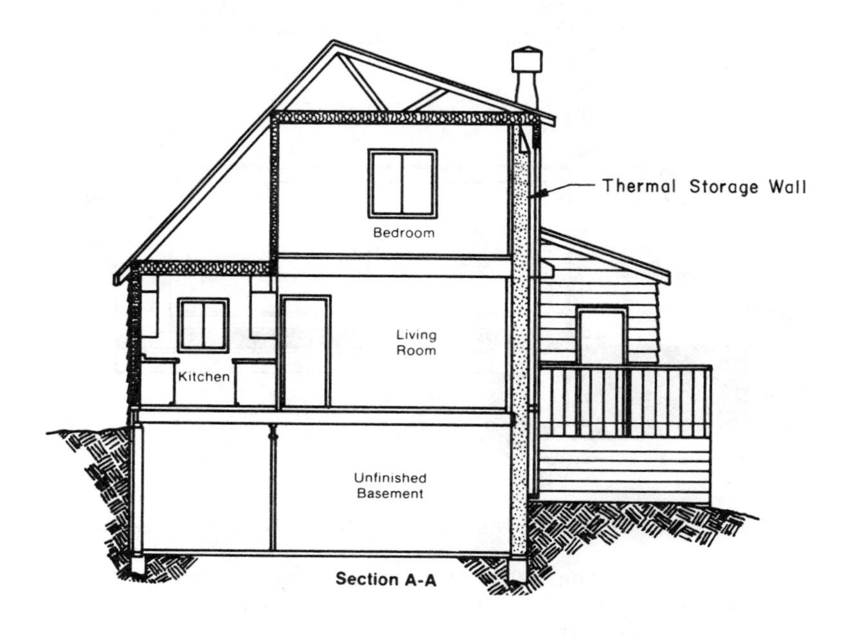

Fig. 2. Thermal storage wall - SERI Class B site DMK (from ref. 4).

produced by a south-facing wall is transferred into the circulating air.

A cross-sectional view of a house with an unvented thermal storage wall is shown in Fig. 2. The foot-thick concrete storage wall is double glazed and extends from the middle of the basement level, through the entire first level, and for half of the glazing width, through the second level. Reflective curtains and a mechanical ventilator reduce overheating problems in the summer. Not visible in the cross-section are several direct-gain windows in the south wall.

The third system, the sunspace, is an indirect gain system. The sunspace is heated directly and functions as a direct gain space which experiences wide temperature variations. Excess heat is transferred into the interior living space through a thermal storage wall, through windows or vents, by a fan, or a combination of these methods. In some cases the common wall is an insulated wall with windows, and thermal storage is in the floor of the sunspace. If the sunspace is used as a greenhouse for plants, movable insulation is sometimes used to prevent excessive temperature drops at night.

Fig. 3 shows a cross-sectional view of the SERI site NEM, located in Northwood, NH. The site uses the attached sunspace as a greenhouse. A thermal storage wall separates the greenhouse from the interior space and provides indirect gain.

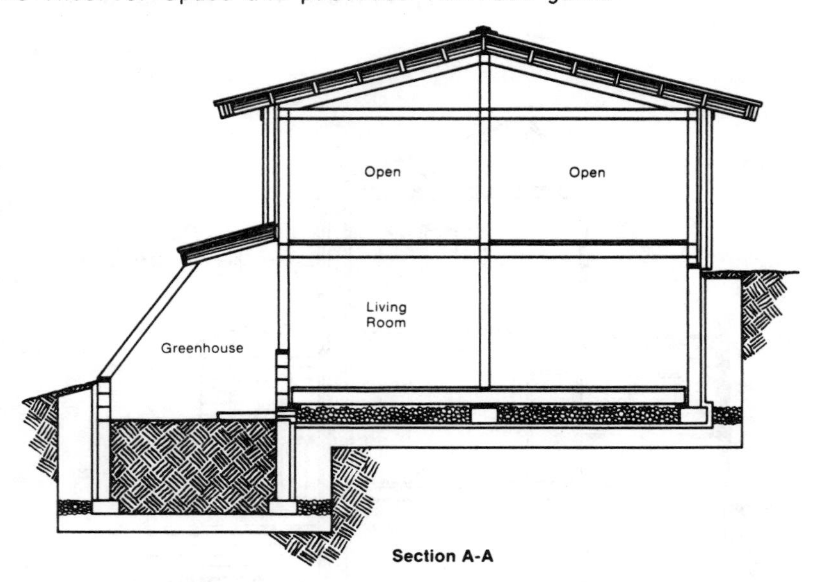

Fig. 3. Sunspace - SERI Class B site NEM (from ref. 4).

Many different sunspace designs are used in both residential and commercial buildings. They are often used as retrofits to existing houses providing energy savings and added space while opening up a new dimension to the interior living space.

WINDOW HEAT BALANCE

The fundamental requirement for a passive solar house is net heat gain from windows averaged over a heating season. We begin by examining the seasonal heat balance of typical south-facing windows. We will make simplifying assumptions to examine this heat balance. Assume that a building requires heat continuously during the heating season. Then the net heat flow, Ewindow, through a unit area of window during the heating season is simply

$$Ewindow = \int_{\text{Heating season}} [Qsolar(t) - U \cdot (Ti(t)-To(t)]dt \qquad (1)$$

where Qs is the solar flux transmitted through the window and absorbed inside, U is the overall thermal loss coefficient of the window and Ti and To are the inside and outside temperature, respectively. The length of the heating season varies, depending on the climate and the thermal properties of the house. For typical heating seasons and double glazed windows, Table I shows Ewindow for different window orientations and four different climates as given in reference 2. Note that positive values indicate the window reduces the heat required from a furnace.

Table I Net heating season window heat flow in $kBtu/ft^2$

	North	East-West	South
Dallas-Ft. Worth	1	44	109
Seattle	-14	21	56
New York City	-25	14	71
Sault Ste. Marie	-62	5	79

North glazing is a net loser except in Dallas, but even north glazing gets enough sunlight during the winter to offset a majority of the losses except in extremely cold climates. The east-west and south glazings are net gainers in all four climates. These values consider cloud cover and glass transmission, but actual performance is generally worse than shown above because of shading. Window frames, shades, drapes, overhangs, etc. often cut the solar gain by a third or more, but good passive design will avoid this shading.

We have assumed that all solar gain reduces the heating requirements of the house. How good is this assumption? This assumption is investigated with three simple models below. The house whose thermal characteristics are given in Table II will be examined for a typical January day in Denver, CO using each model. This house was monitored in the SERI Class B program.

Table II Thermal properties and January weather for SERI house DML

BLC	- Building Loss Coefficient	439 Btu/hr-F	231 W/C
C	- effective thermal Capacity	4991 Btu/F	9471 kJ/C
Cg	- thermal Capacity per unit area of south glass	31 Btu/ft^2-F	633 kJ/m^2-C
Cr	- thermal capacity per unit area of radiantly coupled storage	10 Btu/ft^2-F	204 kJ/m^2-C
Ag	- Area of south Glazing	161 ft^2	15 m^2
$\bar{\tau}$	- average Transmission of glazing	0.68	
Qi	- Internal gain rate	60 kBtu/day	63.3 MJ/day
Ts	- Thermostat Set Temperature	65 F	18.3 C
$\overline{To}$	- January average Outside Temperature - 1982	33 F	0.6 C
ΔT	- internal temperature swing	10 F	5.6 C
Iv	- Insolation on Vertical south surface - January, 1982	1523 Btu/ft^2-day	17.3 MJ/m^2-day

INFINITE THERMAL CAPACITANCE HOUSE

The house heating energy, Eaux, can be expressed as

$$Eaux = \int[BLC\cdot(Ti(t)-To(t)) - Qg(t)]dt - Estor \qquad (2)$$

where Estor is the net increase in the heat stored in the house mass during the period of integration and Qg is the heat from lights, appliances, occupants and solar gains. For the infinite mass house, the gains all reduce heating requirements as long as the limits of the integral in eq. 2 are set so Estor=0.

The heat required on a typical January day is

$$\begin{aligned} Eaux &= 439\cdot(65 - 33)\cdot24 - 60{,}000 - 0.68\cdot161\cdot1523 \\ &= 337{,}150 \text{ Btu/day} - 226{,}700 \text{ Btu/day} \\ &= 100{,}400 \text{ Btu/day} \end{aligned}$$

when values from Table II are used in eq. 2. The gains for this house offset over 2/3 of the losses and all serve to reduce Eaux since they are less than the losses.

MASSLESS HOUSE

A massless house illustrates the other extreme. Following the approach of Hafemeister[6], we assume that solar gains occur over 10 hours during the middle of the day with the gains approximated by

$$Qs(t) = \begin{cases} 0 & 0 < t < 7 \\ Qsm \cdot \sin(\pi/10 \cdot (t - 7)) & 7 < t < 17 \\ 0 & 17 < t < 24 \end{cases} \quad (3)$$

with Qsm = 26191 chosen to give the 166,700 Btu/day solar gain calculated above. A typical diurnal temperature swing in Denver is 26 F, so the ambient temperature To(t) can be approximated as

$$\begin{aligned} To(t) &= \bar{T} - Ta \cdot \cos(2\pi/24 \cdot (t - 3)) \\ &= 33 - 13 \cdot \cos(\pi/12 \cdot (t - 3)) \end{aligned} \quad (4)$$

where time is measured in hours from midnight. The furnace heat Eaux will be given by eq. 2 with Estor=0, but now the interior temperature will be given by

$$Ti(t) = \begin{cases} Ts & \text{when } BLC \cdot (Ts - To(t)) - Qg(t) > 0 \\ To(t) + Qg(t)/BLC & \text{when } BLC \cdot (Ts - To(t)) - Qg < 0 \end{cases} \quad (5)$$

where Ts is the thermostat set temperature. Note that for the massless house, the interior temperature increases when the gain level becomes larger than needed to maintain the set temperature. The integrand in eq. 2 never becomes negative.

The noon-time gains will raise the house temperature by (Qi+Qsm)/BLC = 65.4 F, raising the house temperature to 107.9 F! Evaluating Eaux, we find Eaux = 206,000 Btu and the furnace was off from t = 8.55 hours to t = 16.27 hours. If there had been no gains, Eaux would be 337,152 Btu and if there had been no solar gains (but Qi = 2500 Btu/hr), Eaux would be 277,152 Btu, so we can say that all of internal gains were <u>utilized</u> to reduce the required furnace heating, but only 71,150 Btu or 42.7 percent of the solar gains were <u>utilizable</u>.

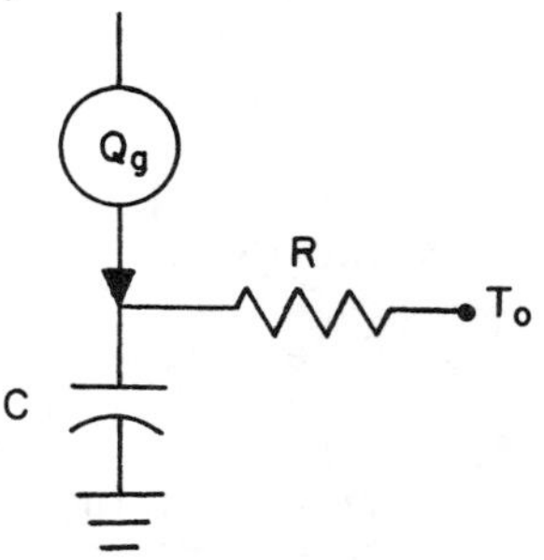

Fig. 4. RC house model.

SIMPLE RC MODEL

The simplest representation for the mass in the house assumes that the effective mass of the house is purely capacities as shown in Fig. 4. This oversimplifies the problem but permits a rough quantitative assessment of building behavior.

Consider SERI DML house used in the earlier example and assume that it has a capacitance C. Heat balance on the room node requires that the interior temperature Ti(t) be the solution of

$$Qaux(t) + Qg(t) + [To(t)-Ti(t)]/R - C \cdot dTi/dt = 0 \quad (6)$$

where the furnace output Qaux(t) is given by

$$Qaux\ (t) = \begin{cases} [Ts - To(t)]/R - Qg(t) & \text{if } Ti < Ts \\ 0 & \text{if } Ti(t) > Ts \end{cases} \quad (7)$$

Hence the room temperature Ti is constant at Ts unless it is heated above Ts by the sun or internal gains. This is normal house operation (without night set-back) except the small temperature oscillations associated with furnace cycling and control are ignored.

When Qg increases to a value Qg(to) > [Ti(to) - To(to)]/R, Qaux = 0 and Ti is the solution of

$$C \cdot dTi/dt + Ti(t)/R = Qg(t) + To(t)/R. \quad (8)$$

The solution for Ti(t) is easily found (for the temperature and gain functions described earlier) to be

$$Ti(t) = \overline{To} + R \cdot Qi + \Delta T(t) + (Ts - \overline{To} - R \cdot Qi - \Delta T(t_1)) * \exp((t_1 - t)/R \cdot C) \quad (9)$$

where

$$\Delta T(t) = Qsm/(C \cdot (a^2+1/(RC)^2) \cdot (\sin(a \cdot (t-\theta))/RC - a \cdot \cos(a \cdot (t-\theta))) + Ta/(RC \cdot (B^2+1/(RC)^2) \cdot (\cos(B \cdot (t-\phi))/RC + B \cdot \sin(B(t-\phi))) \quad (10)$$

where $a = \pi/10$, $\theta = 7$, $B = \pi/12$, $\phi = 3$ and $t_1 = 8.546$ hours is the time when Ti becomes larger than Ts. In the evening after the sun goes down, set $t_1 = 17$ and Qsm = 0 until Ti reaches 65 F.

The auxiliary heating Eaux can now be easily evaluated using eq. 2 with Estor=0 and integration limits from 0 to 8.546 and from 22 to 24 (since the furnace does not come on in the evening until hour 22). Fig. 5 shows the temperatures, gains and losses for this case. The furnace provides all the heat for the house, reduced by the internal gains Qi until 7 am. Solar gain reduces the furnace heat requirement until just before 9 am, when it heats the house above 65 F. The sun continues to heat the house until it reaches 79 F about 3:30 pm. The house then stays above 65 F using stored solar heat until 10 pm. House temperatures above 65 F result in "extra" losses, and the solar gains which meet these "extra" losses are "unutilized".

Table III shows the gains, losses and utilized gains for the three cases just considered.

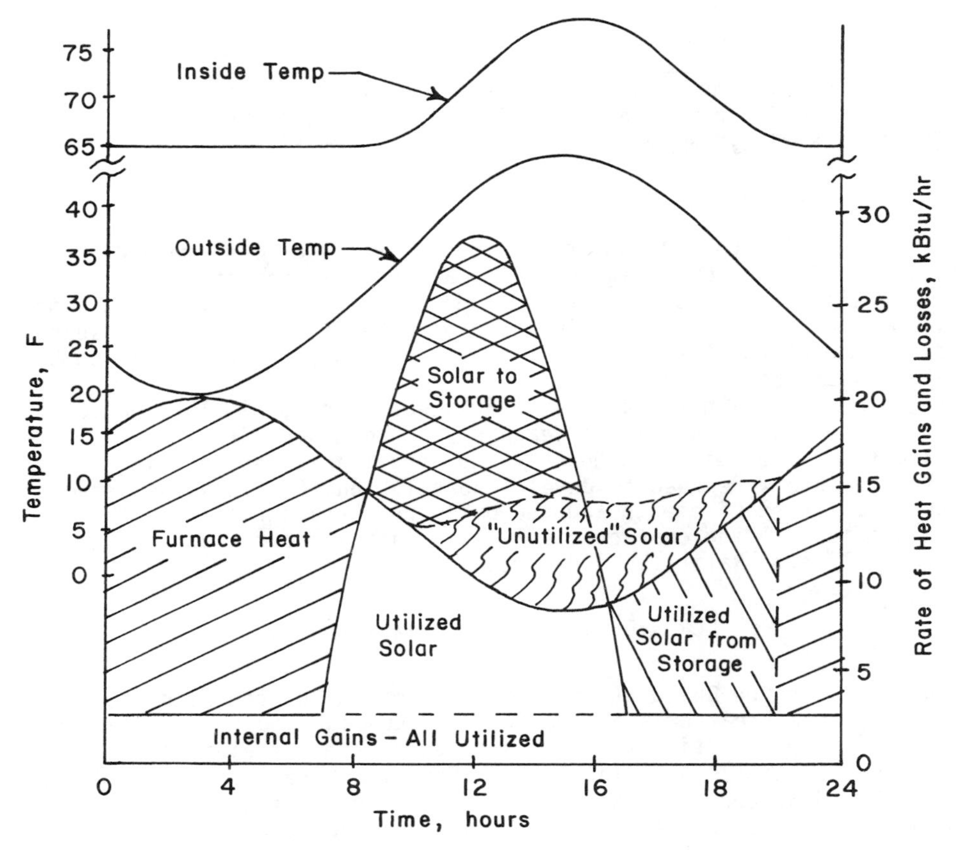

Fig. 5. Temperature Gains, Losses and Auxiliary Heating for the RC Model.

Table III Comparison of Gains, Losses, Utilized Gains and Furnace Heat Requirements for Three Mass Levels

Thermal Cap	Gains	Useful Gains	Heat Loss	Furn. Heat	Frac Sol Util.
Infinite	226,700	226,700	337,150	100,400	1.00
4991 Btu/F	226,700	172,250	392,000	154,850	0.67
Zero	226,700	132,250	431,600	204,900	0.43

PASSIVE CALCULATION METHODS

The examples just worked show the need for methods to handle passive calculations for the more complicated geometries and systems of real buildings. Over the last decade, several methods have been developed to calculate the performance of houses with substantial amounts of glazing an differing levels of interior mass. Three methods are described and a detailed example is shown for the "unutilizability" method.

SLR METHOD

The Solar Load Ratio or SLR method[7] was the first method suitable for use by designers which accurately accounted for many of the design variables and enabled a designer to calculate performance in most U.S. climates. It is based on extensive simulation research performed at the Los Alamos National Laboratory by Douglas Balcomb and associates.

The thermal network model PASOLE was used to simulate houses for every hour of a typical weather year in many different climates. The resulting monthly and annual auxiliary heating requirements were used to develop correlation relationships between the heating requirements and the building characteristics. The principle building characteristic used in these correlations are the Load Collector Ratio (LCR) defined as

$$LCR = BLC/(\text{net south glazing area})$$

where BLC is the Building Loss Coefficient calculated with the south glazing area treated as adiabatic. The LCR is used to calculate the Solar Savings Fraction for houses based on the correlations developed. Tables of Solar Savings Fraction have been prepared for over 200 U.S. locations as a function of LCR. These tables implicitly incorporate numerous assumptions about glazing characteristics, building mass, shading, ground reflectance, etc. A separate set of values is provided for different passive system types such as direct gain, vented Trombe wall, sunspace, etc. Adjustment procedures are provided to consider the effect of changes in the implicit assumptions for most of these variables.

This method can be used to predict monthly or annual heating requirements for passive houses and is the most widely used method for design calculations on passive solar houses. A simplified approach which is a correlation to the correlations of Balcomb has been developed by Kreider[8] and is called P-Chart.

GORDON-ZARMI ANALYTICAL MODEL

J.M. Gordon and Y. Zarmi have developed a relatively simple analytic method[9,10] for predicting the auxiliary heating requirements of passive solar houses. The method uses an "energy "bookkeeping" approach where the load is divided into day and night loads and solar heating fractions, SHF, are calculated for days and nights and then weighted to arrive at the total SHF. The method relies on a distribution function of the frequency of occurrence of different values of the solar load ratio, SLR. For simplicity the authors assumed the distribution function was parabolic. For many locations in the United States Cowing and Kreider[11] have shown that the distribution is more closely approximated by a higher order polynomial expression. The method may be applied to any type of passive heating system.

"UTILIZABILITY" CONCEPTS

The objective of solar thermal energy systems is to utilize or convert as much incident solar radiation as possible into useful heat. The objective of solar energy design methodologies is the prediction of the auxiliary energy requirements or conversely the utilizable solar fraction. The utilizable solar fraction is essentially a measure of the solar system's conversion efficiency. The ratio of the useful energy collected by the solar conversion system to the total incident solar radiation is the utilizability fraction. The concept of utilizability was first introduced as a design methodology by Whillier[12] for flat-plate collectors. It was later generalized by Liu and Jordan[13,14], and simplified for ease of calculation by Klein[15,16]. The design method was applied to passive solar buildings by Monsen and Klein[17]. For passive solar buildings the design method is more appropriately called "unutilizability", since the solar energy that cannot be stored must be vented hence go unutilized. This method is described in detail since it is based on the physical concepts described in the earlier examples.

"UNUTILIZABILITY": PASSIVE DIRECT GAIN METHOD

The method developed by Monsen and Klein predicts the performance of passive solar direct gain houses using the PHI, or utilizability method. For direct gain buildings the design method is more appropriately called "unutilizability" since a portion of the transmitted solar radiation that cannot be stored contributes to heating the room air temperature above the set point and must vented, hence be unutilized. The method allows for room air temperature swings and accounts for solar and internal gains in the overall heat balance. The variable-base degree-day method[18,19] can be used to calculate the estimated auxiliary energy requirement.

The derivation of the method is based on an examination of the theoretical upper and lower limits on the auxiliary energy requirements of direct gain systems. The auxiliary energy must lie between the limiting cases of zero and infinite thermal storage capacity. The solar load ratio, SLR, and a storage-to-ventilation ratio, Y, are used to calculate the solar heating fraction, SHF. Both SHF and Y are functions of PHI, the dimensionless solar statistic discussed in ref. 12. The basis of the method, and the defining equations are presented in the next section followed by an example calculation for the DML house. Details concerning the calculation of PHI may be obtained from reference 13. The treatment below follows that of ref. 17.

PHYSICAL BASIS AND DEFINING EQUATIONS OF METHOD

Energy balances on zero thermal capacity, ZTC, and infinite thermal capacity, ITC, systems will provide bounds for the finite thermal capacity auxiliary energy requirements of direct gain systems. In a ZTC system, excess solar energy which would heat the room air above the fixed set point temperature cannot be stored for later use, and must be vented. If the transmitted solar radiation is less than the required heating load, auxiliary energy from the backup system is used to keep the room air temperature at the fixed set point. A heat balance on the building will yield Qvent, the instantaneous energy removed.

$$Qvent = [It \cdot Ag - BLC \cdot (Tb - To)]^{+} \tag{11}$$

where
- It = transmitted solar radiation (includes the tansmittance absorbance product)
- Ag = area of south-facing glass
- Tb = balance point temperature
- To = ambient temperature

The plus superscript in eq. 11 indicates only positive values are considered. There is a critical radiation level Ic, at which the absorbed solar radiation is equal to the load. At this point Qvent is zero and eq. 11 yields:

$$Ic = BLC \cdot (Tb - To)/Ag \tag{12}$$

In a ZTC system, radiation levels above Ic do not contribute to offsetting the auxiliary load. Substituting eq. 12 into 11 and integrating over a month yields Event, the total energy which must be removed from the space.

$$Event = Ag \int_{month} (It - Ic)^{+} dt \tag{13}$$

An assumption in the method is that Ic is constant over the month. Eq. 13 can be expressed in terms of the monthly average utilizability, PHI, defined as the fraction of solar radiation incident on a surface that has in intensity greater than a specified critical level[12,13,16].

$$\overline{PHI} = \int_{month} (It - Ic)^{+} dt / (\overline{Ht} \cdot N) \tag{14}$$

where
- $\overline{Ht}$ = average transmitted solar radiation (includes the monthly average transmittance absorptance product)
- N = number of days in month

$\overline{PHI}$ is a radiation statistic which is a non-linear function of Ic. The monthly energy vented from a ZTC building is:

$$Event = \overline{PHI} \cdot \overline{Ht} \cdot Ag \cdot N \tag{15}$$

The monthly auxiliary energy for a ZTC building is then:

$$Eaux,z = L-(1-\overline{PHI}) \cdot \overline{Ht} \cdot Ag \cdot N \tag{16}$$

where L is the total monthly load for opaque glass.

It can be calculated using the variable-base degree-day method as follows:

$$L = 24 \cdot BLC \cdot DDvb \tag{17}$$

where DDvb are the degree-days to the balance temperature Tb and

$$Tb = Ts - Qi/BLC \tag{18}$$

During a one month period in an ITC system, all energy in excess of the heating load is absorbed by the thermal capacitance of the building. The temperature of the room air is always constant, regardless of the amount of transmitted solar radiation. The stored solar energy can be used to offset auxiliary energy requirements, but month-to-month carry over is not allowed. A monthly energy balance gives Eaux,i:

$$Eaux,i = [L-\overline{Ht} \cdot Ag \cdot N] \tag{19}$$

The two limiting cases bound the auxiliary energy used by a direct gain system. Dividing equations 7 and 8 by the total load, L, gives SHF, the solar heating fraction.

$$SHFIi = 1-Eaux,i/L = SLRi \tag{20}$$

where SLRi = SLR for infinite storage capacity

and

$$SHFz = 1-Eaux,z/L = (1 - \overline{PHI}) \cdot SLRz \tag{21}$$

The solar load ratio, SLR, is given by:

$$SLR = \overline{Ht} \cdot Ag \cdot N/L \tag{22}$$

A parameter which correlates the monthly values of SHF to monthly values of SLR for direct gain systems with finite thermal storage capacity is Y, the storage-to-ventilation ratio.

$$Y = C_g \cdot \Delta T/(\overline{PHI} \cdot \overline{Ht}) \tag{23}$$

where C_g = thermal storage capacity per square foot of south-facing glass, Ag
ΔT = allowable room temperature swing

Physically, Y is the ratio of maximum energy storage to maximum solar energy removed from the space for one month. The results of over 500 TRNSYS[20] simulations of direct gain systems were subjected to a non-linear regression analysis[14] to produce the relation for SHF in terms of SLR and Y.

$$SHF = P \cdot SLR + (1-P) \cdot (3.082 - 3.142 \cdot \overline{PHI}) \cdot (1 - \exp(-0.329 \cdot SLR)) \quad (24)$$

where P is given by:

$$P = (1 - \exp(-0.294 \cdot Y))^{0.652} \quad (25)$$

Plots of SHF as a function of SLR, Y, and $\overline{PHI}$ are contained in ref. 14. Night insulation can be handled by adjusting the BLC of the building using the following equation:

$$BLCni = fd \cdot BLCo + (1-fd) \cdot BLCi \quad (26)$$

where fd = fraction of day uninsulated
BLCo = uninsulated building load coefficient
BLCi = insulated building load coefficient

EXAMPLE CALCULATION

The example calculation is for the month of January for the DML house which was monitored as part of the 1981-82 SERI Class B performance evaluation[4]. The system parameters and necessary meteorological data are given in Table II.

a. Calculate the monthly heating load L.
Using eq. 18,

$$T_b = 65 - (2500\ \text{Btu/hr})/439\ \text{Btu/hr} \cdot \text{ft}^2 \cdot \text{F} = 59.3\ \text{F} \simeq 60\ \text{F}$$

Then DDvb = 818 F-days for January 1982 and

$$L = 24 \cdot 439 \cdot 818 = 8.62\ \text{MBtu}$$

b. Calculate the solar heating fraction SHF.

From eq. 22, $SLR = \overline{H}t \cdot Ag \cdot N/L = 0.63$

From eq. 12, $Ic = BLC \cdot (Tb - To)/(Ag \cdot (\overline{\tau\alpha})) = 105.46\ \text{Btu/hr-ft}^2$

Using the methods of ref. 16 with the critical level, $\overline{PHI} = 0.399$.

Using eq. 23 and C_g from Table II, $Y = C_g \cdot \Delta T/(\overline{PHI} \cdot \overline{Ht}) = 0.750$

From eq. 25, P = 0.348

From eq. 24, SHF = 0.423

C. Calculate the auxiliary heating required:

Eaux = L•(1-SHF) = 4.97 M Btu

This may be compared with the measured auxiliary heating load of 4.12 M Btu for the DML house during January, 1982.

The monthly results using the "unutilizability" method are shown in table IV:

Table IV Comparison of Estimated Direct Gain Performance with Measured Performance of the DML house in Denver, CO

Month	DDvb (F-days)	IV (Btu/hr ft^2)	SHF	(M Btu) Eaux	Eaux,DML
J	818	1523	0.423	4.97	4.12
F	822	1580	0.435	4.88	3.78
M	604	1320	0.470	3.37	3.04
A	355	1110	0.520	1.74	--
M	142	954	0.662	0.50	--
J	0	926	1.000	0.00	--
J	0	946	1.000	0.00	--
A	0	1066	1.000	0.00	--
S	0	1318	1.000	0.00	--
O	259	1542	0.623	1.02	--
N	497	1730	0.541	2.40	1.41
D	787	1480	0.421	4.79	3.38

The total estimated auxiliary using the "unutilizability" method was 23.7 M Btu, which is 21% higher than the total "measured" energy use of 19.56 M Btu. A discrepancy of 4.1 MBtu is rather good agreement if we note that total heat loss for the season is approximately 60 MBtu.

PREDICTED VS. MEASURED THERMAL PASSIVE THERMAL PERFORMANCE

The example of the previous section showed that the "unutilizability" method satisfactorily predicted the performance of the one house - the SERI DML house. This section compares the predictions of the SLR method, the Gordon-Zarmi method and the

"Unutilizability" method with measured results for the DML house shown in Fig. 1. These comparisons are made using measured thermal characteristics of the house. The predictions based on thermal characteristics estimated from construction drawings are also compared with measured consumption for 11 houses.

Monthly and annual predictions and measured 1981-82 auxiliary consumption for the DML house are shown in Table V. The thermal characteristics and weather data given in Table II were used as inputs for each of the three methods as required.

Table V Comparison of Predicted vs. Measured Thermal Performance of the DML house using a measured BLC = 439 Btu/hr-F (231 W/C)

Month	Auxiliary Energy MBtu/month (GJ/month)			
	SLR	"Unutilizability"	G-Z	Measured
J	4.49 (4.73)	4.97 (5.24)	4.19 (4.41)	4.12 (4.34)
F	4.74 (4.98)	4.89 (5.15)	4.90 (4.72)	3.78 (3.98)
M	3.25 (3.42)	3.37 (3.55)	2.83 (2.98)	3.04 (3.20)
A	1.73 (1.82)	1.79 (1.89)	1.77 (1.87)	--
M	0.35 (0.37)	0.50 (0.53)	0.48 (0.51)	--
J	--	--	--	--
J	--	--	--	--
A	--	--	--	--
S	--	--	--	--
O	0.56 (0.59)	1.03 (1.08)	0.45 (0.48)	--
N	1.88 (1.98)	2.40 (2.53)	1.53 (1.61)	1.41 (1.49)
D	4.27 (4.50)	4.80 (5.06)	4.09 (4.31)	3.38 (3.56)
Ann.	21.3 (22.4)	23.7 (25.0)	19.8 (20.9)	19.6 (20.6)

Note: The measured annual auxiliary is obtained by multiplying the average BPI times the 20 year average yearly degree-day value of 6016 F-day (3342 C-day) and the floor area for the DML house of 1339 ft^2 (162 m^2).

The annual total predicted is within 4 MMBtu in every case. However, the monthly predictions often differ from the measured value by over 25 percent.

Predictions made based on building design generally do not agree as well with measured auxiliary consumption. Table VI compares the predicted vs. measured performance of 11 passive solar houses that were part of the SERI Class B passive solar monitoring program in 1981-82 and 1982-83. The predictive study was performed by Claridge and Simms[20] prior to construction using the SLR method and SUNCAT 2.4, an hourly thermal network program developed at the National Center for Appropriate Technology[21]. The building parameters used as inputs for the predictive study were estimated from drawings.

Table VI Predicted vs. Measured Thermal Performance of 11 Class B Passive Solar Houses in Denver, CO

Site	System	Bldg. Perf. Index * Btu/F-day-ft^2(kJ/C-day-m^2) BPI Predicted	BPI Measured	BPI-pred / BPI-meas	BLC-pred / BLC-meas
DMA	SS/RB	1.32 (27)	1.87 (38)	0.71	0.53
DMC	DG	1.62 (33)	2.54 (52)	0.67	0.58
DMD	DG	2.35 (48)	1.32 (27)	1.78	1.22
DME	SS/DG	1.03 (21)	2.07 (42)	0.50	0.70
DMF	DG/RB	1.81 (37)	2.38 (49)	0.76	0.85
DMG	DG/TW	2.59 (53)	1.46 (30)	1.77	0.88
DMH	SS/RB	0.25 (5)	1.17 (24)	0.19	0.78
DMI	DG/SS	3.97 (81)	1.57 (32)	2.53	1.26
DMJ	DG/TW	3.18 (65)	1.95 (40)	1.63	0.88
DMK	TW/DG	1.17 (24)	0.97 (20)	1.21	0.96
DML	DG	1.32 (27)	1.87 (38)	0.71	0.70

DG = Direct Gain, SS = Sunspace, RB = Rock Bed Storage
TW = Thermal Storage Wall * Note: 1981-82 data.

The auxiliary energy is presented in terms of the building performance index (BPI), a normalized measure expressed as (energy use)/(unit floor area • heating degree-days to base 65 F or 18 C). The predicted values of BPI were not very close to the measured values. The ratio between the predicted and measured auxiliary heating is shown and varies from 0.19 to 2.53 with an average value of 1.13±0.67. An important factor in the variation is the difference between predicted and measured loss coefficient (BLC). The ratio of predicted to measured BLC ranges from 0.70 to 1.26 with an average of 0.85 and standard deviation of 0.22. With the exception of sites DMG, DMI, and DMK, the errors in estimating BLC closely follow the errors in predicting the BPI auxiliary energy requirements. Differences in gains and interior temperatures affect the comparison as well. The estimated uncertainty in measured BLC values is 10-30 percent, the higher level corresponding to high mass houses. The uncertainty in interior-exterior temperature differences is 10 percent.[11,22] The important point is that while predictions may be poor for a single house, they are quite good for the group as a whole.

SUMMARY OF THE SERI CLASS B PERFORMANCE EVALUATION

The most comprehensive study of passive performance conducted to date is the SERI Class B monitoring program. The program monitored 56 buildings of all types, located in all parts of the country from 1981 to 1983,[4,5] The thermal performance of these houses is summarized in Fig. 6.

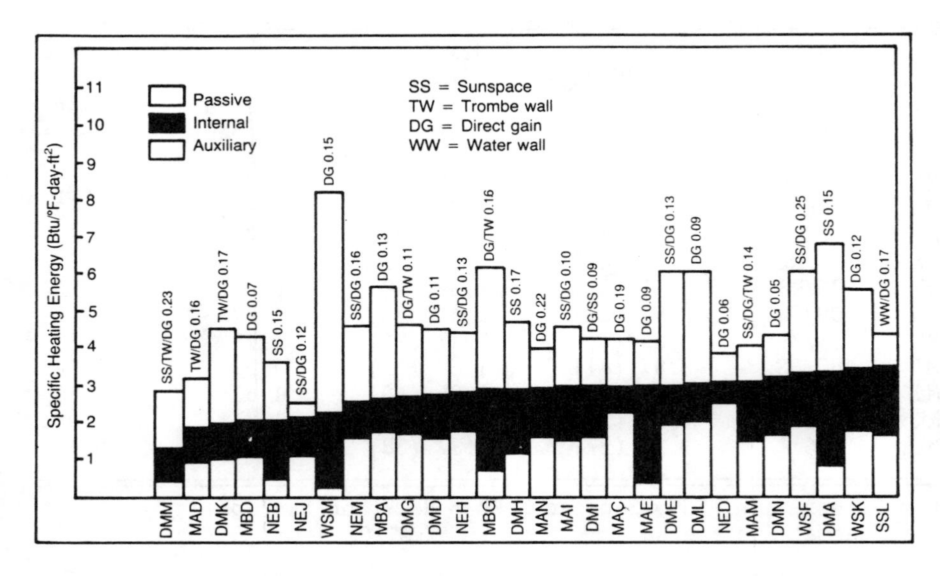

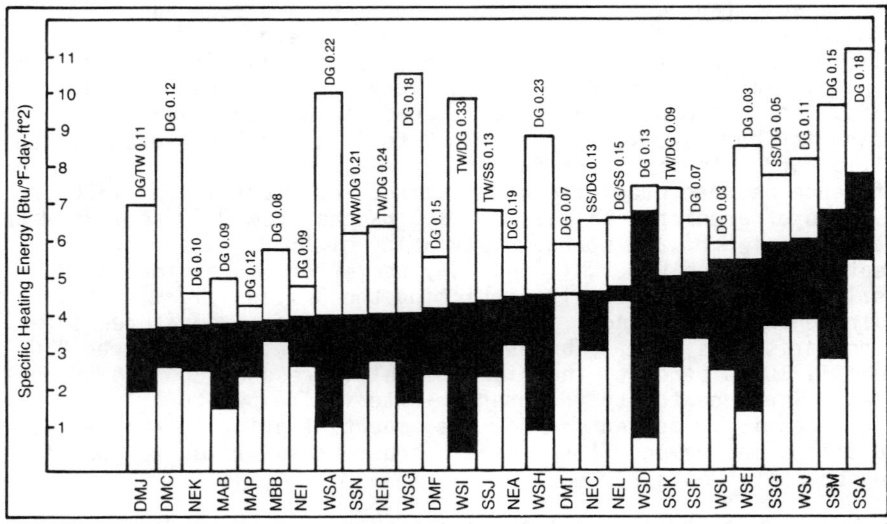

Fig. 6. Normalized heating season energy summaries for 56 passive residences (from ref. 5).

The auxiliary energy use is shown in the bottom portion of each bar, the purchased energy contributing to internal gains is the middle (dark) portion, and the top portion is the solar contribution. The solar system type or types are indicated by the letters above each bar and the decimal fraction is the ratio of south glazing area to floor area. The letters below each bar identify the house.

It can be seen from Fig. 6 that auxiliary heating by itself is sometimes a misleading measure of performance. Sites DMM and WSI both used about 0.4 Btu/ft^2-degree-day for auxiliary heating but WSI used 4.5 Btu/ft^2-dd when internal gains were included compared with 1.3 Btu/ft^2-dd at site DMM. However, most data from other sources includes only auxiliary energy, so auxiliary comparisons are discussed below.

From Fig. 6, it is clear that the monitored buildings have low auxiliary energy needs. The average BPI of 3.49 Btu/F-day-ft^2 (71.4 kJ/C-day-m^2) is well below the typical conventional building BPI of 6-12 Btu/F-day-ft^2 (120-240 kJ/C-day-m^2).

The energy saving effects of insulation and weatherization are very important. Modest solar designs on very "tight" buildings generally have lower auxiliary requirements than more ambitious solar designs on very leaky buildings. The Class B data show a definite relationship between the building heat loss coefficient (BLC) and total purchased energy. Some of the smallest purchased energy users were buildings in Mid-America and the Northeast that had moderate passive heating and very low heat loss coefficients. The Class B passive systems on average contributed 39% of the total heating load, or 55% of the net heating load (total minus internal gains). Although some buildings with large solar aperture-to-floor area ratios (Ag/Af) had low BPI's others did not, indicating that Ag/Af is not as important a design parameter as the BLC. However, the solar heating fraction (SHF) clearly increases with increasing Ag/Af.

The habits of the building occupants are critical to passive system performance when night insulation is used. Most of the buildings with poor performance had problems with operable components or occupant participation. There was no significant difference in either BPI or SHF between houses with or without night insulation. Gains lost when night insulation was left in place during the day offset the reduced uses.[23]

None of the basic passive system types (direct gain, sunspace, and thermal storage wall) demonstrated superior performance. Several of the sunspace designs used very little purchased heat, while still having low values of SHF.

Performance varied significantly by location. The SHF means for California and Denver were 55% and 40% higher, respectively, than the SHF means for the other locations. This is due to the higher solar radiation in Denver, and the large solar aperatures in the California houses. The mean BPI for California and the South were much higher than those for the colder regions. There are two reasons for this (1) the homes in colder regions had more insulation; and (2) more venting occurs in the warmer regions.

The passive solar houses are comfortable. Average indoor temperatures were 67.6 F (19.8 C), and the time averaged temperature of each site ranged from 62 to 73 F (17 to 23 C). Although many of the buildings had large interior temperature swings, serious overheating problems were infrequent. Night ventilation was sufficient to prevent summertime overheating in most northern sites, while some southern sites required air-conditioning.

The 54 buildings in the Class B study showed that passive solar design can achieve very low purchased energy requirements in all areas of the country.

PASSIVE SYSTEM COST/BENEFIT

Several methods for determining the thermal performance of passive systems have been discussed. The useful output per unit area of a passive system is clearly a decreasing function of system size, so if the algorithms for the incremental cost of passive systems can be identified, the optimum size can readily be determined. This has been attempted numerous times, but the value of the results is questionable.

There is generally no simple or unique algorithm for determining the incremental cost of a passive system. This is true for two major reasons. A passive system is incorporated into the structure of the building. It is possible to estimate the incremental cost of replacing a wall with a window, but the changes are generally more complex in good passive design. Moveable insulation may be used. Is this cost simply added, or will insulated Roman shades be used in place of conventional drapes? Interior brick may be used to add mass. Should its cost be compared to a painted frame wall or expensive wallpaper?

Cost/benefit calculations are further complicated by the way passive features change market value of a house independent of energy savings. Good passive design can sometimes completely offset any incremental cost purely on the basis of increased attractiveness to buyers. Light, airy rooms resulting from large windows or the use of clerestory windows to light north rooms; massive fireplace sections or other massive interior walls; plantings made possible by attached greenhouse sunspaces; all can dramatically affect the value and marketability of a home. The authors are aware of a passive design which was marketed at $129,000 - competitive with the builder's other designs in the same development. The passive features in this design made the model so attractive and it sold so well the price was raised to $149,000 and it continued to sell briskly. Had the construction cost gone up $20,000? Had the value of energy savings gone up $20,000? Of course not. The price was raised because the design would sell at the higher price and increase the profit margin.

Likewise, poor passive design can lower the value of a house. Some designs look garishly solar with terrible curb appeal. Large expanses of glass without adequate mass can lead to 100F temperatures - and owner dissatisfaction and lower resale values.

In spite of these difficulties, numerous attempts have been made to quantify the costs and benefits of passive systems. One approach develops an incremental cost and determines the cost of the utilized solar energy. Results of over a dozen such studies are summarized in ref. 24. The solar costs ranged from $0 to $20 per million BTU with most below $10.

Another approach compares the sale price of the passive homes with those of "comparable" non-passive homes. Born[25] found that passive homes cost from 4 to 6 percent more than conventional houses of similar design. A study of 21 speculatively built passive homes from all parts of the country[26] found that they were not more expensive on average ($/square foot) than conventional houses built in the same states. 55% sold much faster or slightly faster than average, 30% sold at the average rate and 15% sold more slowly than average.

OPTIMUM PASSIVE DESIGNS

The previous section discussed the difficulty of determining the incremental cost of passive solar designs and hence the cost of passive solar heat. However, a consensus is emerging that the optimum passive designs generally make heavy use of conservation measures - approaching or entering the superinsulated category - while using modest levels of glazing for passive gain and corresponding levels of mass.

The study of ref. 26 found that "balanced" designs which combined high levels of insulation with moderate levels of glassing (10%+ south glazing/floor area) were most often present in houses with the lowest purchased energy requirements.

The study of 144 housing units by the Minnesota Housing Finance Authority[27] found that auxiliary heating requirements below 2 Btu/ft^2-degree-day were obtained in superinsulated houses and houses with large passive apertures. However, there was more variability in the performance of the passive houses with south glass to floor area ratios above 10% and the incremental cost of the houses with smaller passive apertures was smaller. Ribot and Rosenfeld[28] showed, in a separate study of 128 houses, that the lowest cost of conserved energy was for superinsulated houses, with an average BPI of 1.9 Btu/F-day-ft^2 (39 kJ/C-day-m^2). Passive/superinsulated combined houses were close behind with an average BPI of 2.1 Btu/F-day-ft^2 (42 kJ/C-day-m^2). Others have reached similar conclusions and there is developing consensus that modest passive gain apertures in houses with thermally tight "superinsulated" construction reliably provides very low heating requirements and the aesthetic and market benefits of passive design.

CONCLUSIONS

The last decade has seen passive solar heating develop from a concept known to only a few people in this country to an approach which is now incorporated in thousands of houses. The designs have evolved rapidly, beginning as obviously "solar" designs where glass covered the entire south side of houses to more modest passive designs where the glass and interior mass are an integral part of the design and can add more value to the house than the incremental cost.

Analytical procedures have been developed which can accurately predict the performance of passive designs - provided the thermal characteristics of the house as built are the same as the design assumptions. Good passive design can readily achieve auxiliary heating requirements below 2 Btu/ft^2-degree-days.

REFERENCES

1. S.M. Berman and S.D. Silverstein, eds., "Part III - Energy Conservation and Window Systems," in Efficient Use of Energy, AIP Conference Proceedings No. 25, American Institute of Physics, New York, 1975.

2. S.M. Berman and D.E. Claridge, "The Positive Aspects of Solar Heat Gain Through Architectural Windows," in Efficient Use of Energy, AIP Conf. Proc. No. 25, AIP, New York, 1975, p. 273.

3. K. Butti and J. Perlin, A Golden Thread: 2500 Years of Solar Architecture and Technology, Van Nostrand Reinhold Co., New York, 1980.

4. Swisher, J. and Cowing, T., "Passive Solar Performance: Summary of 1981-82 Class B Results", SERI/SP-281-1847, June 1983.

5. Swisher, J., Carlisle, N. Oatman, P. Russell, K., "Passive Solar Performance: Summary of 1982-83 Class B Results", SERI/SP-271-2362, December 1984.

6. Buffa, A., Hafemeister, D. and Brown, R., Physics for Modern Architecture, Paladin House, Geneva, IL, 1983.

7. Balcombe, J. Douglas et al, Passive Solar Design Handbook, Vol. II, DOE/CS/0127/2, January 1980.

8. Kreider, J.F., "P-Chart - a Passive Solar Design Methodology," Energy Engineering, 78, 1981, p. 41.

9. Gordon, J.M. and Zarmi, Y., "Analytical Model for Passively-Heated Solar Houses-II. Users Guide", Solar Energy, Vol. 27, No. 4, pp. 343-347, 1981.

10. Gordon, J.M. and Zarmi, Y., "Analytical Model for Passively-Heated Houses-I. Theory", Solar Energy, Vol. 27, No. 4, pp. 331-342, 1981.

11. Cowing, T., "A New Passive Solar Space Heating System Design Method For The United States", Masters Thesis in Civil Engineering, University of Colorado, Boulder, Colorado, 1984.

12. Whillier, A., "Solar Energy Collection and Its Utilization For House Heating", Ph.D. Thesis in Mechanical Engineering, M.I.T., Cambridge, Massachusetts, 1953.

13. Liu, B.Y.H. and Jordan, R.C., "The Long-Term Average Performance of Flat-Plate Solar Energy Collectors", Solar Energy, 7, 53, 1963.

14. Liu, B.Y.H., and Jordan, R.C., "Availability of Solar Energy for Flat-Plate Solar Heat Collectors", Chap. V Applications of Solar Energy For Heating and Cooling of Buildings, ASHRAE GRP 170, New York, 1977.

15. Theilacker, J.C. and Klein, S.A., "Improvements In the Utilizability Relationships", Solar Energy Laboratory, University of Wisconson, Madison, WI 53706.

16. Klein, S.A., "Calculation of Flat-Plate Collector Utilizability", Solar Energy, Vol. 23, pp. 393-402, 1978.

17. Monsen, W.A. and Klein, S.A., "Prediction of Direct Gain Solar Heating System Performance", Solar Energy, Vol. 27, No. 2, pp. 143-147, 1981.

18. Kusuda, T., Sud, I. and Alereza, T., "Comparison of DOE-2 Generated Residential Design Energy Budgets with Those Calculated by the Degree Day and Bin Methods," ASHRAE Trans., 87, Part 1, 1981, p. 491.

19. Claridge, D., Jeon, H. and Bida, M., "A Comparison of Traditional Degree-Day and Variable-Base Degree-Day Predictions with Measured Consumption of 20 Houses in the Denver Area," ASHRAE Trans., Vol. 91, Part 1, 1985.

20. Claridge, D. and Simms, D., "Performance Analysis of 11 Denver Metro Passive Homes", Proceedings of Sixth National Passive Conference, September, 1981, p. 566.

21. Palmiter, L., "SUNCAT Version 2.4 User Notes".

22. Duffy, J. "Estimation of the Uncertainty in Class B Monitoring Results, Final Report", J.J. Duffy and Associates, Winchester, MA, November 1983.

23. Swisher, J. Bishop, R. and Frey, D., "Effectiveness of Moveable Window Insulation in Passive Solar Homes", prepared for the U.S. DOE under Contract No. DE-AC02-83CE8900 by NAHB Research Foundation, 1984.

24. SERI, 1981. Solar Energy Research Institute, A New Prosperity: the SERI Solar/Conservation Study. Brick House Publishing, Andover MA 1981.

25. Born, M. 1980. "TVA's Solar Homes for the Valley Program", Tennessee Valley Authority, 1980. quoted in ref. 24.

26. R.C. Bishop and D.J. Frey, "Consumer Acceptance, Marketability, and Financing of Passive Solar Homes," prepared for the U.S. Department of Energy under Contract No. DE-AC02-83CE89000 by the NAHB Research Foundation, Inc., 1984.

27. Hutchinson, M., Nelson, G. and Fagerson, M., "Measured Thermal Performance and the Cost of Conservation for a Group of Energy Efficient Minnesota Homes," in What Works: Documenting Energy Conservation in Buildings, J. Harris and C. Blumstein, eds., American Council for an Energy Efficient Economy, Washington, D.C., 1984, p. 185.

28. Ribot, J. and Rosenfeld, A., "Monitored Low-Energy Houses in North America and Europe: A compilation and Economic Analysis", Lawrence Berkeley Laboratory, LBL-14788, 1982.

CHAPTER 11

PASSIVE COOLING SYSTEMS IN RESIDENTIAL BUILDINGS

John G. Ingersoll
Hughes Aircraft Company, El Segundo, CA 90245

Baruch Givoni
University of California, Los Angeles, CA 90024

ABSTRACT

The performance of four passive cooling systems, nocturnal convective cooling, nocturnal radiative cooling, direct evaporative cooling and conductive earth-coupled cooling, is evaluated for representative environmental conditions in the temperate, hot-humid and hot-arid climatic zones of the United States. The analysis indicates that substantial portion of the cooling load of a typical energy-efficient single family residential building can be eliminated with any of these passive systems. Depending on system type and climatic zone, the building cooling load can be reduced by 1/3 to over 4/5 of its original value. The corresponding energy savings would amount to a minimum of 25 TWh/yr and could potentially exceed 50 TWh/yr, if proper passive cooling systems were to be employed throughout the country. Incorporation of passive cooling models in building energy analysis codes will be necessary to determine more precisely the potential of each system. Field testing will also be required to further evaluate this potential. Moreover, the extention of analytical modeling to include additional passive cooling systems and the research of advanced building—natural environment coupling systems and materials constitute tasks requiring further effort.

INTRODUCTION

Even though space heating constitutes by far the most significant portion of energy consumption in residential buildings in the U.S., air-conditioning is also becoming quite important as the majority of new houses are equipped with air-conditioners. Furthermore, as population moves from the northern climates to the sunbelt, a predominantly cooling climate, more and more people have the need of air-conditioning. As a result the saturation level of air-conditioners in U.S. residential buildings is expected to rise from 61 percent, of which 31 percent are central units and 30 percent room units, in 1980 to 70 percent, of which 46 percent will be central units and 24 percent room units, by the year 2000[1]. The 70 percent saturation level corresponds to an upper limit for the U.S. if one excludes areas in the country that need no air-conditioning. Moreover, the shift from room units to

0094-243X/85/1350209-20$3.00 Copyright 1985 American Institute of Physics

central units is apparent. On an energy basis, single family housing, constituting some 70 percent of all residential housing in the U.S., consumed 7.5×10^{10} kWh for air-conditioning, equal to 6.9 percent of the total residential resource energy consumption in 1980[1]. It is predicted that by the year 2000 the respective figures, asuming the same (1980) ratio of single family to all residential housing, will be 6.1×10^{10} kWh and 8.3 percent of total residential resource energy due to combination of factors, including the market penetration of more efficient air-conditioners, a 24 percent increase in housing units and relatively smaller cooling efficiency gains compared to space heating and hot water heating.[1]

On a global scale, air-conditioning is bound to also become of importance as the standard of living in developing countries rises. Need for air-conditioning in developing countries is associated with the geographic location of the majority of them, namely the tropical and subtropical regions of the earth.

Passive cooling systems constitute means by which the interior of the building can be maintained within comfort levels of temperature and humidity without the aid of an air conditioner but rather by utilizing various natural cooling sources such as ambient cool air, soil below the building, longwave radiation to the sky, water evaporation, etc.

The term passive does not exclude the use of mechanical devices such a fans and pumps. It rather excludes systems based on typical thermodynamic cycles such as Rankine, Stirling, Brayton-Joule, etc. Consequently, solar energy is not used directly for any passive cooling systems even though the availability of natural cooling sources is the indirect result of solar radiation arriving at the surface of the earth. For instance solar absorption cooling systems are not included in the discussion, as in the opinion of the authors, such systems are at present quite inefficient by consuming substantial "parasitic" energy and therefore displaying relatively low COP. On the other hand passive cooling systems included herein, even when mechanical devices such as fans and pumps are utilized, have a seasonal COP above 5 and often above 10. By comparison, the top of the line air-to-air or water-to-air air conditioners have a seasonal COP not exceeding 2.5 to 3 at the present time.[1, 2]

We might also add that passive cooling of buildings has different implications in developing and developed countries. In the former, where air-conditioning is virtually non-existent, the benefits of implementing passive cooling systems are not measured in terms of energy conservation and efficiency but rather in terms of health, comfort, and productivity. Furthermore, the lack of adequate energy supplies in most developing countries makes passive cooling systems by far the best choice to attain the goal of a healthy, comfortable, and productive indoor environment. Implementation of passive cooling systems in developed countries, on the other hand, is intimately related to improved energy efficiency and conservation with the well known consequences of lower cost to consumers, increased life span of fossil fuels and a better environmental quality.

CLASSIFICATION OF PASSIVE COOLING SYSTEMS

The various passive cooling systems are classified according to the major physical mechanism by which cooling energy is derived. Four main systems can be thus defined.

a. Nocturnal Convective Cooling. Heat from the building is discharged at night to the ambient air by natural or mechanical ventilation. Depending on the size of thermal storage of the building, a portion of the cooling load of the following day can thus be shifted in time until the next evening when the entire process is repeated.

b. Nocturnal Radiant Cooling. Heat is discharged to the sky by longwave radiation at night. The outgoing radiation can take place from a nocturnal radiator, typically attached to the roof of the building, whose temperature can be lower than that of ambient. Air passing through the radiator and entering the house can cool the latter to temperatures below ambient. Hence, a building with given thermal storage thus cooled can shift in time a larger portion of the following day's cooling load compared to a convective cooling system.

c. Evaporative Cooling. Cooling in this case is the result of latent heat absorbed during evaporation of a liquid, water in this case. This can take place in two forms:

   - Direct: Evaporation of water in the building typically with the aid of a rather simple mechanical system and distribution of cooled air throughout the building.

   - Indirect: Evaporation of water on the exterior surfaces of the building resulting in lower building envelope temperatures. In the same category we can also include the temperature depression of ambient air due to evapo-transpiration of plants and in general vegetation surrounding the building.

d. Conductive Earth-Coupled Cooling. In this case heat is transferred from the building to the soil underneath which is utilized as a heat sink. The building can be cooled either by direct contact with soil or with the assistance of air or water flow.

It should be noted that all four main systems involve "convective" heat transfer in addition to any other heat transfer mechanism employed in the system.

Within each of the above four main systems, one can enumerate a few generic designs and several adaptations applicable to specific climatic situations[3]. For example, in the case of radiant cooling we can mention two generic designs, the nocturnal roof radiator and the heavy mass concrete roof. The roof-water pond is a variant of the latter for hot-arid climates, while the desiccant dehumidification enhanced by a nocturnal radiator is a variant of the former applicable to humid climates.

In the following sections we will described the general characteristics, expected performance, and limitations of each of the nocturnal convective, nocturnal radiant, evaporative and conductive earth-coupled passive cooling systems. Examples are given for locations in each of the major climatic zones in the U.S.: temperate, hot-humid and hot-arid[4]. For simplicity the cool climatic zone is not considered separately as it is similar to the temperate zone when cooling is concerned.

NOCTURNAL CONVECTIVE COOLING

When a building is ventilated at night, its structural mass is cooled by convection from the inside. During the daytime, the cooled mass can serve as a heat sink to absorb the heat penetrating into the building as well as the building internal loads. This implies that the thermal mass of the building must be well insulated from the outdoors. Furthermore, the building must be unventilated during the daytime in the sense of open window cross-ventilation. Convective cooling is expected to be applicable mainly in climates with a large diurnal temperature range on the order of 10°C or more and with night temperature in the summer below the lower end of the comfort range, i.e., below about 20°C. Humidity may also be a limiting factor as indoor comfort should be maintained in the daytime without ventilation. Regions with vapor pressure below 17 mmHg (0.014 kg moisture/kg dry air) are therefore considered acceptable for convective cooling.

The total amount of "cold" storage is given by:

$$Q = MC\,(\bar{T}_m(\max) - \bar{T}_m(\min)) \qquad \text{(Joule)} \quad (1)$$

where M is the mass of storage in kg; C the heat capacity of the mass in J/kg°C; $\bar{T}_m(\max)$ is the average maximum mass temperature in °C; and $\bar{T}_m(\min)$ is the average minimum mass temperature. From a practical standpoint, $\bar{T}_m(\min)$ cannot be lower than about 2°C above the minimum outdoor temperature. On the other hand, $\bar{T}_m(\max)$ cannot exceed the comfort limit which can range from 24°C in hot-humid regions (vapor pressure above 22 mmHg) or 28° in hot-dry regions (vapor pressure below 10 mmHg), both under still air conditions. By increasing the indoor airspeed at daytime with a ceiling fan or the central fan, the $\bar{T}_m(\max)$ can be shifted to 28°C and 30°C for the humid and arid regions, respectively.

Thermal storage can be provided through ceilings or floors made of reinforced concrete with imbedded channels through which air flows at night from the outside and circulates during the day only within the house. Ceiling storage is preferable because of higher performance due to unobstructed radiant exchange and can be used in multistory buildings. Floor storage may be, however, the only practical option in one-story buildings.

The energy E required to power a fan moving air through such channels is given by:

$$E = W \cdot t = \frac{F \cdot \Delta p}{\eta} \cdot t \qquad \text{(Joule)} \quad (2)$$

where W is the rated fan power in W, F is the flow rate in m^3/s; Δp is the pressure drop in Pa; t is the total time of fan operation in sec, including both night and day operation; and η is the overall fan efficiency. Hence, the COP of the system will be given by:

$$(COP)_{NCC} = \frac{Q}{E} = \eta \cdot \frac{MC(\bar{T}_m(max) - \bar{T}_m(min))}{F \cdot \Delta p \cdot t} =$$

$$\frac{\dot{Q}}{W} = \frac{MC(\bar{T}_m(max) - \bar{T}_m(min))}{W \cdot t} \quad (3)$$

A typical example of diurnal indoor temperatures for a residential building in a hot-arid climate with and without nocturnal convection is given in Figure 1.[5] As a numerical example we consider a residential building with concrete slab floor of 150 m^2 in area and 0.15 m in thickness. Furthermore, we assume a flow rate of 1500 $m^3/h = 0.416\ m^3/s$ using a fan with 0.5 kW nominal power operating at an average of 16 hours per day. We have for concrete, $\rho = 2250\ kg/m^3$ and $C = 840\ J/kg°K$. Assuming three different $T_m(max) - \bar{T}_m(min)$ corresponding to typical hot-humid, hot-arid and temperate climates we calculate the power density $\dot{Q}/A$ per unit floor area available for cooling and the respective COP given in Table I. Notice that typically the average summer daily minimum temperate in humid climates in the U.S. is about 24°C, and in arid climates about 21°C. On the other hand, the average daily minimum temperature in cool and temperate climates is about 18°C but the corresponding maximum temperature does not exceed 28° to 29°C.[6]

The results of Table I point out the not so surprising result of a relatively lesser utility of nocturnal convective cooling in humid climates and the rather unexpected fact that buildings in temperate climates can benefit the most by such a passive cooling scheme. However, we should point out that the energy saved in hot-humid and hot-arid climates may exceed the energy saved in temperate climates as the cooling load in the former can be at least a few times larger than that in the latter for a building with the same construction and operation characteristics. We also note that this passive cooling system may become inoperative during hours in which the ambient temperature exceeds 28°C, 29°C, and 30°C in hot-humid, temperate and hot-arid climates, respectively. The corresponding percentages of summer time (May, June, July, August and September) that these limits are exceeded in the respective climates are roughly 30 percent, 5 percent, and 45 percent[6]. That is to say, nocturnal convective cooling may not be able to provide indoor comfort conditions during one-third to one-half of the summer time in hot-humid and hot-arid climates, but it will perform adequately in more temperate climates.

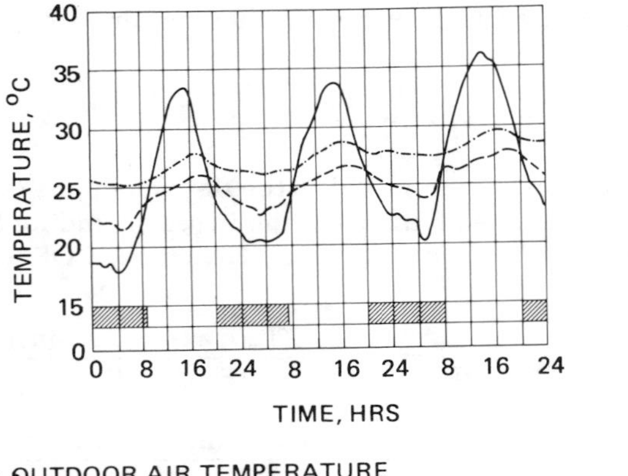

Fig. 1. Diurnal patterns of outdoor and indoor temperatures in a house with and without nocturnal ventilation during a three-day period in July 1981, Sede Boqer, Israel.

Table I. Performance of nocturnal convective cooling in hot-humid, hot-arid and temperate climates.

Climate	T_a[a] (°C)	$\overline{T}_m(max) - \overline{T}_m(min)$ (°C)	$\dot{Q}/A$ (W/m^2)	$(COP)_{NCC}$
Hot-Humid	24	3	14.8	4.4
Hot-Arid	21	7	34.5	10.3
Temperate	18	9	44.3	13.2

[a]Average summer night ambient temperature (June, July, August) in New Orleans, LA, Phoenix, AZ, and New York, NY, respectively.

NOCTURNAL RADIANT COOLING

A typical nocturnal radiant cooling system consists of a radiating surface exposed to the sky. The latter is constructed of a lightweight metallic surface such as aluminum, copper or galvanized steel 2–3 mm in thickness. This radiating surface, painted appropriately to have a high longwave emissivity and preferably a low shortwave absorptivity, comprises the upper side of a radiator channel attached to the roof of the building. Outside air is drawn from one end of the channel and discharged inside the house through the other end with the aid of a fan. Furthermore, it is assumed that the roof beneath the radiator is very well insulated so that no conductive heat transfer takes place between the house and the channel. As the temperature of the upper surface of the channel drops below ambient at night, the air entering the house is at a lower temperature than ambient and can therefore increase the cooling capacity of the building's thermal storage by increasing the range of ΔT in Eq. (1) as compared to nocturnal convective cooling.

An analytical model has been developed to predict the channel exit air temperature $T_F(L)$ as a function of environmental parameters (ambient air temperature, wind speed and humidity) as well as the geometrical and physical parameters of the radiator channel.[7]

The rate $\dot{q}$ at which cool energy can be potentially utilized over that achieved by nocturnal convection alone is:

$$\dot{q} = \rho C_P F [T_a - T_F(L)] \tag{4}$$

where L indicates the length of the radiator; $\rho = 1.24$ kg/m^3 air density; $C_p = 1.005 \times 10^3$ J/kg°C air heat capacity, and F is the air flow rate in the radiator channel.

If the required fan power is W, then the COP of the roof radiator is:

$$(COP)_{RAD} = \frac{\dot{q}}{W} = \frac{\rho C_P F (T_a - T_F (L))}{W} \tag{5}$$

The $(COP)_{NRC}$ of the nocturnal radiant cooling system is calculated from Eq. (3) with $\overline{T}_m(min)$ reduced from its previous value by $T_a - T_F(L)$.

An example of the measured temperature profile of a radiator during the night as well as the relevant weather parameters are given in Figure 2[8]. A wind protected radiator as well as a totally exposed one are included for comparison.

Assuming a roof radiator with a 0.05 m channel spacing, a 6 m width and length L = 15 m, a flow rate F = 1500 m^3/h = 0.416 m^3/s, a radiator surface thermal emissivity $\epsilon_r = 0.9$ and a fan power consumption W = 0.5 kW, we give in Table II some typical numerical results for enviromental conditions applicable to hot-humid, hot-arid and temperate climates[7].

The house is assumed to have a concrete slab, consisting of imbedded channels through which air flows, as a storage medium with the same specifications as in the nocturnal convective cooling system discussed in the previous section. We also assume that during the night hours the same fan that is drawing air through the radiator is also moving the cooled air through the concrete slab channels.

The results of Table II indicate that the radiator performance is best in hot-arid climates and about the same in hot-humid and temperate climates. However, the addition of the radiator improves substantially the performance of convective cooling in all climates, but relatively more so in humid and arid environments. As a result, the nocturnal radiant system performs as well in hot-arid and temperate climates, while its performance in hot-humid climates is also significantly better than that of a nocturnal convective cooling system. It must be noted that the dew point temperature is quite important in restraining the lowest temperature limit that the radiator and therefore the cooled air can attain. In other words, the nocturnal radiant cooling will perform the best in areas with the greatest night temperature difference between ambient and dew point. Furthermore, covering the radiator surface with "selective" materials which reflect incoming radiation from clouds while emitting strongly at the wavelengths of the atmospheric windows can ensure system performance under a cloudy sky[3].

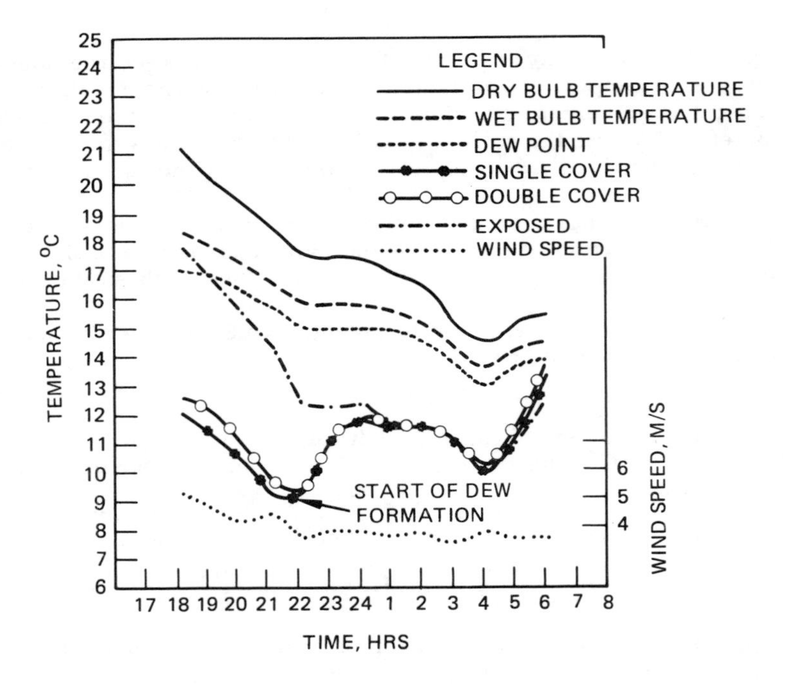

Fig. 2. Temperature patterns of exposed and wind-screened nocturnal radiators during a typical night in July 1981, Sede Boqer, Israel.

Table II. Performance of nocturnal radiant cooling system in hot-humid, hot-arid and temperate climates.

Climate	Environment[a]				Radiator				Nocturnal Radiant Cooling System		
	T_a (°C)	T_w (°C)	n^d (#)	T_{dp} (°C)	T_{RS}^b (°C)	$T_F(L)$ (°C)	$\dot{q}/S^c$ (W/m^2)	$(COP)_{RAD}$	$\Delta\bar{T}^e$ (°C)	$\dot{Q}/A^d$ (W/m^2)	$(COP)_{NRC}$
Hot-Humid	24	21	3	19	19	21	17.3	3.1	5	24.7	7.3
Hot Arid	21	13	1	6	14	17	23.0	4.1	11.0	54.2	17.3
Temperate	18	14	4	10	12.6	15	17.3	3.1	11.0	54.2	17.3

[a] Typical summer (June, July, August) night parameters for New Orleans, LA, Phoenix, AZ, and New York, NY, respectively. It is assumed that wind speed near the radiator surface is kept at 0.45 m/s with the aid of an appropriate exterior surface windscreen (e.g., egg-crate device, polyethylene screen, etc.). Other parameters include dry-bulb T_a, wet-bulb T_w, sky cloud cover n (0 = clear, 10 = completely overcast), and dew point T_{dp}.

[b] Radiator stagnation temperature, i.e., minimum temperature under prevailing environmental conditions.

[c] Power density per unit radiator surface.

[d] Power density per unit building floor area.

[e] $\Delta\bar{T} = \bar{T}_m\,(max) - \bar{T}_m\,(min)$

EVAPORATIVE COOLING

In this section, we examine only one of the several possible evaporative cooling systems, namely direct evaporative cooling. The basic reason is that indirect systems such as two-stage mechanical evaporative cooling with the aid of heat exchangers, evaporative cooling by roof ponds, shaded roof ponds, evapo-transpiration of vegetation, etc., are not well understood at the present time in terms of their performance, even though some of them may hold great cooling potential under certain conditions. In fact, development of analytical models to predict the performance of a variety of indirect evaporative cooling systems is an absolute necessity if we are eventually to utilize all natural means to reduce the cooling load of buildings.

Consequently, we consider here the direct or one-stage mechanical evaporative cooling only. In essence such a system consists of an indoor fan blowing air at a very high rate onto wetted filters or pads. The net result is the decrease of the indoor dry bulb temperature with a simultaneous increase of the moisture content or water vapor pressure of the indoor air, while the wet bulb temperature remains unchanged. This change takes place with no external heat added to the system and is called, therefore, adiabatic. The total heat content of air, sensible plus latent, is kept constant. The process is depicted schematically in the psychrometric chart shown in Figure 3 (dry-bulb and humidity ratio change from point A to point B).

The air flow rates involved can be as high as 15 to 30 building air changes per hour. Typically, the indoor air temperature, T_i, is reduced by about 80 percent of the difference between the outdoor dry bulb, T_a, and outdoor wet bulb temperatures, T_w. Building internal loads may actually reduce this 80 percent potential gain to around 60 percent. Given the fact that the summer month daytime temperature difference between dry bulb and wet bulb is 3°C to 4°C in hot-humid climates, 6°C on the average in temperate climates and 10°C or more in hot-arid climates, such a technique can have virtually no impact in hot-humid climates and a rather small one in temperate climates. Furthermore, direct evaporative cooling by elevating indoor humidity or vapor pressure can exacerbate comfort conditions in environments of high (vapor pressure 20 mmHg)[†] or even moderate natural humidity (vapor pressure 15 mmHg). Hence, direct evaporative cooling can be effectively used in hot-arid areas or temperate areas with lower natural humidity (vapor pressure of 10 mmHg or less).

We can define the amount of heat Q transferred from sensible to latent load as:

$$\dot{Q} = \rho\, C_p\, F\, (T_a - T_i) \tag{6}$$

where $T_i = T_a - 0.6\,(T_a - T_w)$; $\rho = 1.24\ \mathrm{kg/m^3}$, the air density; $C_p = 1005\ \mathrm{J/kg°C}$, the air heat capacity; and F the fan flow rate in $\mathrm{m^3/s}$. Furthermore, if Δp is pressure drop across the wetted filter or pad and η is the mechanical and motor efficiency of the fan, the coefficient of performance of this system will be:

$$(\mathrm{COP})_{\mathrm{DEC}} = \frac{\dot{Q}}{W} = \frac{\rho\, C_p\, F\, (T_a - T_i)}{W} = \frac{\eta \rho C_p (T_a - T_i)}{\Delta p} \tag{7}$$

[†]A vapor pressure of 10 mmHg corresponds to a humidity ratio of 0.008 kg moisture per kg dry air. See also Figure 3.

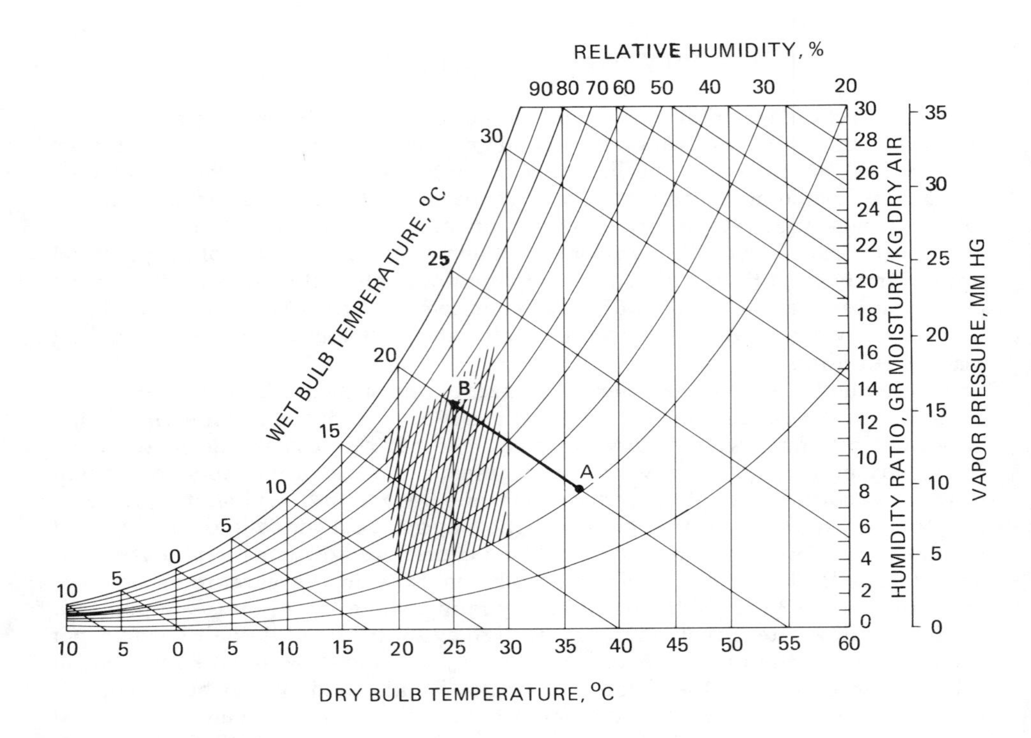

Fig. 3. Psychrometric chart indicating the effect of direct evaporative cooling (from A to B) on indoor comfort parameters. Comfort zone envelope is also shown, assuming medium human activity, light summer clothing, and air velocity up to 1.5 m/s.

As an example we assume $F = 2\ m^3/sec$, $\Delta p = 200$ Pa and $\eta = 0.40$[9]. Furthermore, the house has a floor area A of 150 m^2 and a volume of 450 m^3 — a flow rate of 1 m^3/sec corresponds to eight air changes per hour. The results of this numerical example are given in Table III.

Notice that direct evaporative cooling takes place during the day time, particularly late afternoon hours, when the building cooling load reaches its daily maximum. In this type of evaporative cooling, no thermal storage is required, although this may not be true for indirect evaporative cooling systems. We also note that direct evaporative cooling may be able to handle cooling in terms of comfort up to about 35°C ambient temperature. Given that during 20 to 30 percent of the summer time (May, June, July, August and September) the 35°C may be exceeded in hot-arid climates, we conclude that direct evaporative cooling may not be able to meet indoor comfort conditions up to 25 percent of the summer hours.

Table III. Performance of direct evaporative cooling in hot-humid, hot-arid and temperate climates.

Climate	Natural Environment[a]			Evaporative Cooling System			
	T_a (°C)	$T_a - T_w$ (°C)	Humidity Ratio (kg moisture/ kg dry air)	T_i (°C)	W_i (kg moisture/ kg dry air)	$\dot{Q}/A$ (W/m^2)	$(COP)_{DEC}$
Hot-Humid	30	4	≥0.020	$T_a - 2$	≥0.021	N/A	N/A
Hot-Arid	35	14	≤0.010	$T_a - 7$	≤0.013	58.2	8.7
Temperate	27	6	0.010 − 0.020	$T_a - 3$	≤0.015	(24.9)[b]	(3.7)[b]

[a]Typical summer (June, July, August) daytime parameters for New Orleans, LA, Phoenix, AZ, and New York, NY.

[b]Applicable only to arid temperate climates such as those western of the Rockies.

CONDUCTIVE EARTH-COUPLED COOLING

The earth under and around a building can serve in most climatic zones as a natural cooling source. In the summer the soil temperature, a few meters below the surface, is lower than the average daily temperature and, of course, the daytime ambient temperature above. Hence, the earth can serve as a heat sink. The soil underneath a typical house with floor area 150 m^2 up to a depth of 10 m and for a temperature rise of 10°C has a heat capacity of 150 m^2 × 10 m × 10°C × 2250 kg/m^3 × 1000 J/kg °C = 3.4 × 10^{10} Joule. The average cooling load of a well insulated ranch house of the same floor area is on the order of 1.5 × 10^{10} Joule/yr (of which 75 percent is sensible) in the United States, although it can be as high as 3.3 × 10^{10} Joule/yr (of which 90 percent is sensible) in a hot-arid climate such as Phoenix or as high as 4.5 × 10^{10} Joule/year (of which 65 percent is sensible) in a hot-humid climate such as Miami[10]. That is to day, the soil beneath the house alone has the potential to accommodate the maximum cooling load of the house in the hottest climates in the United States.

The heat capacity of the soil beneath and around the house can be further increased by lowering its temperature by artificial means below its natural level. Such artificial means include shading the surface to eliminate solar gains and utilizing evaporation from the same surface through vegetation or irrigation or both. Experiments in Israel and North Florida have shown that it is possible to lower the soil temperature by as many as 8° to 10°C below its natural temperature. Figure 4 gives an example of such soil temperature depression[11]. Thus, we can have access to larger than 10°C soil temperature rise from the beginning to the end of

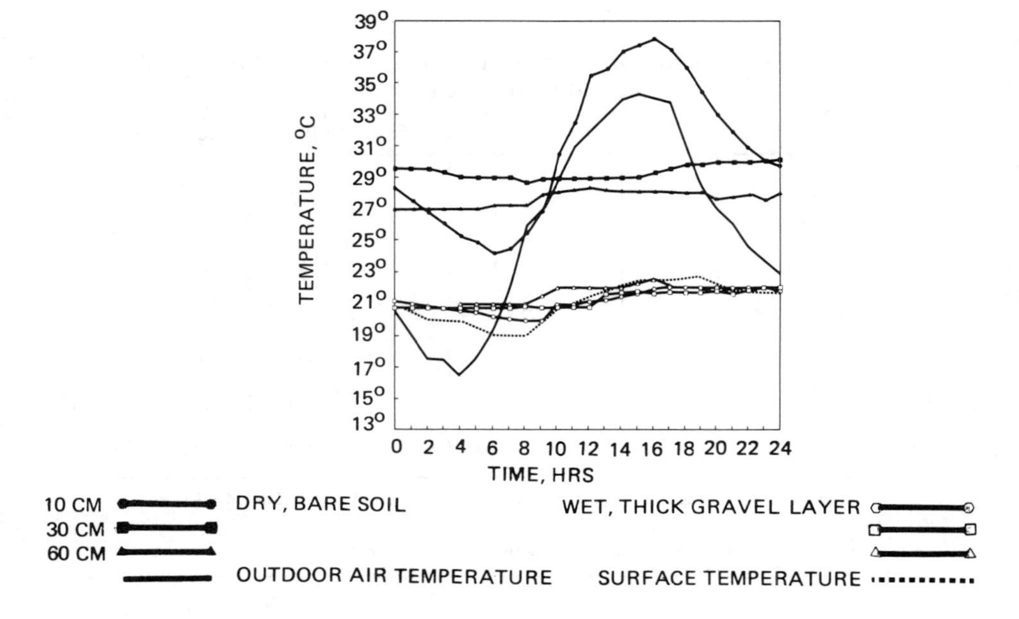

Fig. 4 Diurnal temperature pattern of exposed soil and shaded by pebbles and watered soil during a typical day in August 1981, Sede Boqer, Israel.

the summer to accommodate heat (cooling load) rejected by the house to the earth. Furthermore, we have only considered the soil underneath the house. If we include the soil up to 5 m beyond the house boundary and all around its perimeter, we can have a two to three times increase in available soil mass.

As with any other passive cooling system, one has to devise an effective and appropriate scheme whereby the building cooling load can be transferred to the soil. One such efficient and practical way of coupling the house to the soil under and around it is by the use of earth cooling tubes. Air is drawn into the house after it passes through underground tubes where it can be substantially cooled. Depending on the humidity of the entering air and the earth temperature, both latent load and sensible load are removed from the air with this system—as compared to sensible load only for the three types of cooling system discussed so far.

An analytical model has been developed by the authors that predicts the exiting from an underground tube dry bulb and wet bulb temperatures for a given set of inlet dry bulb and wet bulb temperatures, soil temperature and soil conductivity, and the tube dimensions, material properties and flow rate[12].

The rates of sensible and latent heat removal from the air are given by:

$$\dot{Q}_s = \rho \cdot v_f \left(\frac{\pi D^2}{4}\right) \cdot C_p \left[T_a - T_f (x = L)\right]$$

$$\dot{Q}_l = \rho \cdot v_f \left(\frac{\pi D}{4}\right)^2 \cdot q_\ell^0 \left[w_a - w (x = L)\right] \tag{8}$$

where $\rho = 1.214\ kg/m^3$ (air density); $C_p = 1.005 \times 10^3\ J/kg\ °C$ (air heat capacity); $q_\ell^0 = 2.46 \times 10^6\ J/kg$, the latent heat of water condensation; v_F the airspeed in the tube in m/s; D the tube inner diameter in m; T_a, w_a the tube inlet air temperature and humidity ratio; and $T_f\ (x = L)$, $w(x = L)$ the tube outlet air temperature and humidity ratio at a distance L from the inlet.

The coefficient of performance of the earth cooling tube system, consisting of N identical tubes operating in parallel, will be:

$$(COP)_{ECC} = \frac{N\,(\dot{Q}_s + \dot{Q}_l)}{W} \tag{9}$$

where W is the total fan power required to move the air through all N tubes and is given by:

$$W = \frac{N\,F\,\Delta P}{\eta} = \frac{N\left(\frac{\pi D^2}{4}\right) \cdot v_F \cdot \Delta P}{\eta} \tag{10}$$

where ΔP is the pressure drop per tube.

Based on the analytical model as well as field measurements on a number of earth cooling systems, it takes a tube length of about 20 to 30 m, with adequate flowrate ($0.1\ m^3/s$), in order for the exit air to reach within 1 to 2 degrees of the surrounding soil temperature. Typically, each tube is buried at a depth of 3 to 5 m below grade, and is made of polyethylene with a wall thickness 2 to 3 mm and a diameter of 15 to 20 cm. The system consists of a vertical inlet tube reaching the horizontal portion at the prescribed depth and connected to it through a tee type connection. The horizontal portion is slightly sloped to allow condensing water to be drained to the soil by gravity through the third leg of the tee connection. At the other end of the horizontal portion, a second vertical tube brings the cooled air into the building. Typical flow rates per tube of the preceding dimensions are on the order of 0.1 to 0.2 m^3/sec. Depending on the size of the building, two or three tubes may be used to accommodate the building cooling load. With the possible exception of hot-arid climates condensation will commence in the horizontal portion of the tube after the first 5 or 10 m. In field measurements performed in humid climates, initial humidity ratios in excess of 0.020 kg moisture/kg dry air of the inlet air were brought close to or below 0.010 kg moisture/kg dry air for the outlet air. Indeed the latent load removal capability of earth cooling tubes appears to be quite substantial and in most occasions it exceeds in magnitude the sensible load removal. In Table IV we present a numerical example, based on our analytical model, to illustrate these points. Once more we perform the anaylsis in three different climatic zones, hot-humid, hot-arid and temperate. We consider two tubes 30 m each in length with a flowrate of $0.1\ m^3/s$ in each tube. The tube diameter is 0.2 m. The pressure drop along each tube is on the order of 500 Pa, as can be readily calculated[12]. We also treat, in both hot-humid and hot-arid climates, the case of natural as well as depressed soil temperature.

The results of Table IV indicate that earth cooling tubes perform well in all climates and as one would expect perform the best in climates with high humidity. In fact in both hot-humid and temperate climates the ambient air can be brought within the comfort levels of temperature and humidity as it goes through the tubes

Table IV. Performance of earth cooling tubes in hot-humid, hot-arid, and temperate climates.

Climate	Environmental Parameters[a]				System Parameters					
	T_a (°C)	T_w (°C)	T_{dp} (°C)	T_s (°C)	$T_F(L)$ (°C)	$T_w(L)$ (°C)	$\dot{Q}_s/A$ (W/m^2)	$\dot{Q}_l/A$ (W/m^2)	$(\dot{Q}_s+\dot{Q}_l)/A$ (W/m^2)	$(COP)_{ECC}$
Hot-Humid	30	26	25	25	25.5	25	15.6	—	15.6	4.3
				20[b]	22	17.5	27.8	39.2	64.2	17.7
Hot-Arid	35	21	16	27	28	19	24.1	—	24.1	6.7
				20[b]	21	17	48.4	—	48.4	13.4
Temperate	27	21	20	15	17.5	14	32.8	20.6	54.0	14.9

[a] Typical environmental parameters for New Orleans, LA, Phoenix, AZ, and New York, NY, respectively; T_a, T_w typical daytime average values for the months of June, July and August.

[b] Lower soil temperature value indicates depression due to ground shading and water evaporation.

and enters the building. We also note that soil temperature depression around the building by both soil shading and water evaporation can dramatically increase the system performance: roughly by a factor of 2 in hot-arid climates and by a factor of 4 in hot-humid climates.

CONCLUSIONS AND RECOMMENDATIONS

The preceding examples point to at least two facts. First, passive cooling systems display a wide variability of performance both among themselves for the same set of climatic parameters as well as within each system for varying climatic parameters. Second and less apparent, passive cooling systems have the potential of substantially reducing and possibly even eliminating, depending on climatic condition, the need for air-conditioning in residential buildings, particularly single family ones, in the United States. To illustrate this last point, we give in Table V the maximum and average cooling power density required to be removed from the typical one story ranch house, 150 m^2 in floor area, in the three different climatic zones. The results are based on hourly computer simulations for the entire year, using the DOE-2.1 code[10]. The building is assumed to have internal loads, not exceeding 90 MJ per day, with 2/3 sensible and 1/3 latent, including people but excluding solar gains, and is also assumed to have an energy efficient envelope with the wall (U_w), ceiling (U_c), glazing (U_g) conductances, number of glazings (G) and average summer infiltration (I) indicated in Table V. The window area equals 10% of floor area and is evenly distributed on all four sides of the building.

It would be difficult to determine without hourly calculations for the entire cooling season the fraction of the building cooling load in a given climate that can be accommodated by a specific cooling system. For even though some passive cooling systems can accommodate the maximum cooling power density required, there may be instances during which the resulting indoor environment is outside

Table V. Maximum and average total (sensible and latent) cooling power density in hot-humid, hot-arid and temperate climates

Climate[a]	Building Envelope Characteristics					Indoor Conditions			
	U_w (W/m^2 °C)	U_c (W/m^2 °C)	G (#)	U_g (W/m^2 °C)	I (m^3/s)	T_i (°C)	W_i (kg moisture/ kg dry air)	Average Power[b] (W/m^2)	Maximum Power[c] (W/m^2)
Hot-Humid	0.30	0.19	2	2.78	0.07	25.5 ±1	≤0.014	18.0	46.7
Hot-Arid	0.30	0.19	2	2.78	0.09	25.5 ±1	≤0.014	23.1	65.5
Temperate	0.21	0.15	3	1.75	0.06	25.5 ±1	≤0.014	12.0	31.0

[a] Calculations based on climatic data from New Orleans, LA, Phoenix, AZ, and New York, NY, respectively.

[b] Average over the entire cooling season rather than the principal summer months (June, July, August).

[c] Maximum based on the day of the typical year with the highest cooling load rather than the cooling design day.

the air temperature and humidity comfort region (as shown in Figure 3). Depending on the time of occurrence of such instances and the schedule of the building occupancy, we may have to resort to a combination of a passive cooling system along with a standard air-conditioning system to ensure comfort conditions. While it is most likely that such a combination will be necessary in the extreme hot-humid and hot-arid climates, it is conceivable that only a passive cooling system, properly designed and optmized, could be adequate for many houses in temperate and mild hot-humid and hot-arid climate zones.

Of the various passive cooling systems presented herein, it appears that earth-cooling tubes can be used well in any climate, but particularly in hot-humid with soil temperature depression and temperate ones. Next, evaporative cooling and nocturnal radiant cooling are well suited for hot-arid climates and temperate climates with relatively low humidity and cloud cover. A tentative estimate of the fraction of the cooling load that can be potentially eliminated in a specific climate zone in any of the four passive cooling systems considered here is given in Table VI. These estimates are derived on the basis of the operational limits of each of the passive systems given earlier and the aforementioned DOE 2.1 hourly computer simulations.

The numbers in Table VI are only indicative of the potential of each particular passive cooling system in the specific climate zone. Detailed studies will be necessary in order to conclusively determine the true potential of these and other passive cooling systems in different climatic zones.

Although this work is intended to deal exclusively with the technological aspects of passive cooling systems, a few remarks regarding the economic aspects of such systems may also be appropriate at this point. First, we would like to point out that as a general rule the highest efficiency system does not necessarily correspond to the most economic one.[13] In other words, high efficiency can often

Table VI. Percent of building cooling load that can be eliminated with the aid of respective passive cooling system.

Climate[a]	Passive Cooling System			
	Nocturnal Convective Cooling	Nocturnal Radiant Cooling	Direct Evaporative Cooling	Earth Coupled Cooling[b]
Hot-Humid	34%	49%	—	>90%
Hot-Arid	29%	46%	58%	86%
Temperate	63%	73%	(77%)	>90%

[a] Respective building locations: New Orleans, LA, Phoenix, AZ, and New York, NY.

[b] Assumes soil temperature depression of 5°C in hot-humid and 7°C in hot-arid climate.

only be achieved by a very expensive system or installation. This is certainly true for air-conditioners and other similar equipment; it is proven to be true for solar heating systems and there is no reason to doubt that it will be valid for passive cooling systems as well. Consequently, a balance must be found between added capital expenditure and resultant monetary savings due to reduced energy consumption. Second, we note that a useful expression of the cost-effectiveness of a system is the so called figure of merit. Several figures of merit are defined in the literature. They range in complexity from the straightforward concept of simple payback to the more sophisticated ones of the cost of saved money, the internal rate of return, and the annualized life-cycle cost.[1] A modified simple payback figure of merit, recommended by some architects to evaluate solar heating systems, is defined as the ratio of the value of the energy saved by the system in 10 years to the extra cost of the system over that of a conventional one.[13] It has been further suggested that a system will be competitive if the above defined figure of merit equals or exceeds one. Even though the 10 year period appears to be somewhat arbitrary, it most probably reflects the minimum life expectancy of many solar heating systems. The passive cooling systems described herein are expected to have a minimum useful life on the order of 15 years based on their simpler design. Hence, the preceding definition can be modified accordingly in order to be used for passive cooling systems. We may also note that more sophisticated definitions of figures of merit require as input the system life expectancy, the energy price escalation rates and the net discount rate during the life of the system. Precise knowledge of these parameters is needed in order to yield valid figures of merit, although the non-availability of proper data can be circumvented, in principle at least, by performing parametric studies. Third, we believe that it is premature to attempt to estimate costs for any of the passive cooling systems due to the early stage of their development. A variety of publications can be used to obtain construction costs for items such as thermal storage slabs or flat plate collectors or below grade excavation and placing of tubing. [14, 15] However, such an approach is bound to lead to unreliable estimates for a number of reasons. The very significant local variability of construction costs and the frequent temporal cost fluctuations are two such reasons. Moreover, any passive cooling system built today will be more or less a custom design. Mass construction of standardized systems will undoubtedly reduce their cost. Yet, this will not happen until after the concept is proven to be technically viable. For these reasons, we propose to estimate instead the cost that a system should not exceed in order to be competitive. This is a relatively easier task. For instance, if we assume that a passive cooling system can save on the average 1000 kWh per year per household and if we further make the reasonable assumptions of a $0.1/kWh initial electricity residential price and a 5% net price escalation rate over a period of 15 years[16], we obtain a value of electricity saved equal to $2365 in constant dollars. Hence the cost of the particular cooling system, including anticipated maintenance costs, should not exceed this dollar figure. Incidentally, a 1000 kWh/yr savings corresponds to about 1/5 of the cooling energy required in the more extreme hot-humid and hot-arid climates in the U.S.; and to about 3/4 of cooling energy required in most temperate climates in this country.[10] This assumes the typical residential building with the characteristics given in Table V. We note from Table VI and under the preceding assumptions that $2000 to $2500 may be the most one should invest for a passive cooling system in the majority of temperate climates. However, a $7000 passive cooling system could be reasonable in extreme hot-humid and hot-arid locations if it could save 2/3 or more of the cooling energy. A final remark of a technical nature, but with significant economic implications, concerns the design of passive cooling systems so that they can be replaced or added-on without additional expenses over the cost of an original system with the same degree of performance. Situations like these would arise when the original system has to be replaced with a new one or when an existing house with no prior passive cooling is retrofitted with such a system.

The discussion so far has been applicable to the conditions prevailing in the United States and, for that matter, developed countries with cooling climates. The limitations of passive cooling systems to obtain indoor comfort conditions 100 percent of the time are irrelevant in developing countries with practically no cooling systems available to them at the present time. Consequently, passive cooling systems will be adequate for developing countries as long as such systems are compatible with the prevailing climatic conditions, i.e., either a hot-humid or a hot-arid climate. As stated in the introduction, the significance of cooling systems in developing countries is related to improving health, comfort and productivity rather than improving the efficiency with which energy is utilized. Given the limitations of energy availability as well as the scarcity of capital in most developing countries, introduction of passive cooling systems is the most prudent solution to cooling as it can improve tremendously the quality of life with a minimal expenditures of capital and the least requirements of prime energy. Incidentally, prime energy requirements can be eliminated altogether in some passive cooling systems by making use of natural wind and if degradation of system performance is of no concern. Nocturnal convective cooling and direct evaporative cooling are two such systems. In the former, we can accomplish the effect by allowing direct passage of the prevailing summer night wind through the house and onto the thermal mass. In the latter, cool but moist air can be obtained by funneling the summer daytime wind into the house through one wall of the exterior enclosed space (porch) and then into the house[17]. The wall, consisting of a 20 to 30 cm thick structure made of loose round rocks 5 cm in diameter, continuously kept wet by water dripping from the top, collected at the bottom and pumped back to the top, constitutes the wet filter. Alternatively a funnel-like structure, placed on the roof of the building, can capture and divert the wind indoors, force it to go through a structure of wetted rocks similar to the one previously mentioned but with the outside air entering at the top, exiting at the bottom, and then being distributed in the house. Of course the last scheme without the wetted rock structure has been used in the past simply as a comfort ventilation system, i.e., a system that increases the air movement into the house to improve comfort when ambient temperature and humidity fall outside the comfort range.[17]

It is evident that passive cooling holds a great promise for both developed and developing countries. However, in the last 10 years or so passive cooling has not received the same degree of attention as, for example, passive and active solar heating. The reason is not difficult to discern. Since the United States and most developed nations, where the bulk of the pertinent research and development efforts take place, are predominantly heating climates, it makes sense to deal first with the more pressing problems. This is not to detract from the fine research effort on the subject of cooling and the improvements in relevant industrial products that have been accomplished in the United Stats and elsewhere in the last decade. Suffice it to mention two examples: the development of glazing materials with apropriate properties to substantially reduce summer solar gains and the availability of air-conditioners with close to 100 percent increased coefficient of performance over that of 10 years ago.

Better understanding of the performance characteristics, applicability, and limitations of passive cooling systems can further enhance the society's ability to utilize energy more efficiently. To this end, the following research and development efforts will be necessary at a minimum:

a. Development of analytical models to better understand and predict as a function of climate the performance characteristics of alternative nocturnal radiant systems, indirect evaporative systems including the impact of vegetation, and earth-coupled systems including soil temperature depression.

b. Incorporation of such models into hourly building performance simulation codes to determine the actual potential in energy savings of specific passive cooling systems as a function of climate and indoor comfort conditions.

c. Performance of field testing of various passive cooling systems in buildings located in different climatic zones to verify and/or further refine models and simulation codes.

d. Adaptation of preceding results to the needs and realities of developing countries to ensure a continued rise in the standard of living in the Third World.

Other possibilities of passive cooling systems may also be found. The incorporation of new technologies into old ideas is an intriguing proposition. In this particular case the utilization of heat pipes to couple the building to the ground so that heat can be rejected (cooling) or absorbed (heating) more effectively is such a possibility. The building is coupled to the heat pipe(s) through a proper convective or radiative heat transfer system or even an air-to-air or water-to-air heat pump. Tests in Japan for heating rather than cooling purposes with heat pipes coupled to concrete panels have shown that such a system was able to deliver 20 to 130 W/m^2 of heat, while the soil was able to regenerate the lost heat during the summer months[18]. The coupling by heat pipes of a heat pump to the ice storage bin in the Annual Cycle Energy Storage (ACES) concept, whereby rejected coolness in the winter can be utilized in the summer is another relevant example. Another possibility is the development of a phase change material for thermal storage with a phase change temperature ideally in the range of 24 to 25°C. Such a material incorporated in the nocturnal convective and radiant cooling systems would increase the effectiveness of these systems and potentially enable them to accommodate 90% or more of the building cooling load.

Indeed, as we enter the second decade of concerted efforts toward an energy-efficient society, and looking back into the accomplishments of the first decade following the 1973 Arab oil embargo, we can be reassured that attaining our goal will be taken as far as our imagination and determination allows us.

REFERENCES

1. R. H. Williams, G. S. Dutt, and H. S. Geller, "Future Energy Savings in U.S. Housing, "Annual Review of Energy, 8, pp. 269–332, Palo Alto, CA, 1983.
2. J. G. Ingersoll and D. K. Arasteh, "Energy Efficiencies of Heat Pumps in Residential Buildings, "Energy and Buildings, 5, pp. 253–262, 1983.
3. B. Givoni, Passive Cooling of Buildings, McGraw-Hill, New York, N.Y., 1985 (in press).
4. V. Olgyay, Design with Climate, Princeton University Press, Princeton, NJ, 1973.
5. B. Givoni, measurements obtained at Institute of Desert Research, Sede Boqer, Israel, 1981.
6. Typical Reference Year Weather Summaries in DOE–2.1 User's Manual, Vol. 3.
7. J. G. Ingersoll and B. Givoni, "Modelling the Performance of Radiant Cooling," Ch. 4, Passive Cooling of Buildings, McGraw-Hill, New York, NY, 1985 (in print).
8. B. Givoni, "Cooling by Longwave Radiation," Passive Solar Journal 1, pp. 131–150, 1982.
9. ASHRAE Handbook, 1983 Equipment, "Evaporative Air-Cooling Equipment, Ch. 4, ASHRAE, Atlanta, GA, 1983.
10. J. G. Ingersoll, et al., unpublished computer simulations of the energy performance of new residential buildings, Lawrence Berkeley Laboratory, Berkeley, CA, 1982.
11. B. Givoni, "Earth Integrated Buildings—An Overview," Architectural Science Review, 24, No. 2, 1981.

12. J. G. Ingersoll and B. Givoni, "Analytical Modeling of the Performance of Earth Cooling Tubes," Ch. 7, Passive Cooling for Buildings, McGraw-Hill, 1985 (in press).
13. S. V. Szokolay, Solar Energy and Building, pp. 119–127 The Architectural Press, London, UK, 1975.
14. Building Construction Cost Data 1985, 43rd annual edition, Robert Snow Means Co., Kingtson, MA, 1984.
15. Dodge Construction Systems Costs: 1984, McGraw-Hill Co., New York, NY, 1983.
16. Energy Information Administration, Monthly Energy Review March 1985, p. 103, U.S. Department of Energy, Washington, D.C., 1985.
17. A. Konya, Design Primer for Hot Climates, pp. 56–57, The Architectural Press, London, UK, 1980.
18. P. D. Dunn and D. A. Reay, Heat Pipes, pp. 266–269, Pergamon Press, Oxford, UK, 1983.

CHAPTER 12

Richard G. Sextro, Anthony V. Nero, and David T. Grimsrud

Indoor Air Quality: Sources and Control

Building Ventilation and Indoor Air Quality Program
Lawrence Berkeley Laboratory
University of California
Berkeley, California 94720

INTRODUCTION

During recent years scientists and policymakers have paid substantial attention to airborne substances that can threaten human health or environmental quality. Studies of these pollutants have led to the enactment of measures to control them. Controls on emissions from automobiles or industries that burn gasoline, kerosene, or coal have noticeably improved the quality of the air. A growing effort is being devoted to isolation of industrial activities and waste dumps that can release complex chemicals into the air and water. And a substantial and relatively effective regulatory structure is in place to control releases of radioactivity to the general environment.

But, in terms of human health, scientists may have missed the main point: that people typically spend 80 to 90% of their time indoors. Recent research has revealed that concentrations of many pollutants can be higher indoors than out and that the factors contributing to indoor pollution are virtually unaffected by controls on outdoor air pollution. This perspective is not new just to the public or government officials. Even environmental scientists and engineers specializing in air pollution have been startled to discover that the highest personal exposures to combustion emissions occur not in urban smog but in homes with unvented combustion appliances. Concentrations of organic chemicals in homes and offices are often a hundred or a thousand times higher than they are outdoors, and airborne radioactivity in homes is more significant by far than that released from nuclear power plants.

The question arises as to how this situation came about. The reason is simply that substantial amounts of these pollutants are emitted from indoor sources into what are actually very small microenvironments, the atmospheres inside buildings. Understanding indoor air quality requires substantial efforts to characterize the sources of each pollutant class, the removal of pollutants by ventilation and air movement, and the interactions of airborne substances with each other and with the building and its contents.

In this chapter we discuss briefly the nature and origin of major classes of indoor pollutants; then discuss in some detail the nature of pollutant control. The latter discussion focuses on what is perhaps the major indoor pollutant in the developed countries of the world, radon and its progeny. For further information the reader is invited to consult the National Academy report, *Indoor Pollutants* (1981), the Spengler-Sexton paper (1983) for a general discussion of public policy, or Turiel's

0094-243X/85/1350229-18$3.00 Copyright 1985 American Institute of Physics

book (1985) for a more general introduction.

Air pollution arises indoors from two major sources: materials or appliances that are part of the building, or directly connected with it, and outdoor emissions that are carried indoors with outdoor air. The exchange of outdoor and indoor air provides ventilation of building interiors, whether by infiltration through the walls of buildings, passage through windows or doors, or forced movement by a blower and, perhaps, duct system. In a typical home, ventilation is accomplished largely by means of infiltration or open windows and serves to exchange the indoor and outdoor air about once an hour. This provides at least one limit to the buildup of pollutants emitted from indoor sources. At the same time outdoor pollutants may be carried in, but it appears that indoor pollutant concentrations are dominated by contributions from indoor sources; indoor concentrations are substantially greater than those outdoors when indoor sources are present but are less when they are not.

The indoor pollutants most closely related to outdoor pollution are emissions from the combustion of fuels such as kerosene, natural gas, and wood. To varying degrees these produce oxides of nitrogen (primarily NO and NO_2) and of carbon (CO and CO_2) as well as particles, oxides of trace substances such as sulfur, and complex organic substances (some of which are present on the particles emitted). Indoor levels of these emissions are determined by the fuel composition, the detail of the combustion process, and the degree to which the emissions are vented to the outdoors. Central home heating systems that burn oil or natural gas vent virtually all the combustion products outdoors unless there is a defect in the system. But many space heaters and gas ranges, as well as some other appliances, are not vented or merely have a local exhaust fan (which often goes unused).

As examples, a modern kerosene heater can cause indoor concentrations of nitrogen oxides approaching one part per million if used with low ventilation rates. The outdoor long-term limit for NO_2 is about 0.06 part per million. Under similar conditions indoor concentrations of CO_2, not ordinarily considered to be an environmental pollutant with health significance, can reach thousands of parts per million, which is comparable to the upper limit applied to industrial workers. Use of a gas range also noticeably affects indoor concentrations. But this has less potential for causing high concentrations than a space heater, simply because more fuel usually is burned for heating a home than for cooking.

Cigarette smoking is a form of combustion that can also influence indoor air quality, thus affecting the health, not only of the smoker, but of others nearby. It is well established that cigarettes introduce into the indoor atmosphere substantial amounts of organic material. The chemical content of airborne cigarette smoke is extremely complex, and its effect on the health of those who inhale it is not well understood. Of course, cigarette smoking itself can have significant effects (such as lung cancer and emphysema) on the health of those who smoke. Whether the same types of effects are caused to a lesser degree among those who occupy the same space with cigarette smokers may never be known with certainty.

Thus combustion in general and cigarette smoking in particular can be a rich source of airborne organics, sometimes with substantial health implications. But combustion is certainly not the only source of organics. This is particularly true indoors, where organic substances may arise from a variety of materials and products in common use and may often reach levels that far exceed those regarded to be a concern outdoors. Major classes of sources include building materials and furnishings, various types of liquid and aerosol products, specific types of machines or appliances, and the combustion sources mentioned above. Together they emit a wide range of substances with a variety of known or suspected effects on building occupants if they occur at relatively high levels. For example, many organics are known to cause nonspecific ill effects--such as eye or respiratory irritation, headaches, nausea, or general weakness--among at least a portion of the people exposed to them. In addition, some organics are known to have carcinogenic effects quite apart from those that may be attributed to cigarette smoke. However, the question of which organic substances appear in indoor air and what health effects they may have is far from fully understood. Even characterizing the wide variety of sources of these substances is a bewildering problem, since they include many of the most common features of people's daily lives.

It is not altogether surprising that building interiors should be significant causes of exposure to combustion gases and organics. But it is disconcerting to find that significant exposures to radiation take place in homes. This results from the naturally occurring radioactivity that comes from the ground. All of the Earth's crust has trace amounts of uranium. The amounts are, as expected, much higher in areas that are suitable for uranium mining, but even ordinary soil and rock contains uranium and, of more significance to indoor atmospheres, radium. When radium decays, it produces radon, a gas that is chemically inert and can therefore be transported with the air in the soil and thus reach the outdoor or indoor atmosphere. The decay products of radon are isotopes of polonium, bismuth, and lead that are chemically active. Once they are formed from radon in the air, if inhaled they can attach to the lining of the lungs, either directly or along with airborne particles to which they have become attached while in the air. This leads to radiation exposure that has been shown to be the cause of elevated rates of lung cancer among uranium miners. For that reason their exposure to radon decay products has been severely limited.

Other classes of indoor pollutants deserve attention as well. These include insulating materials containing asbetos; cooking products such as grease, water vapor, and odorants; and airborne bacteria, viruses, and fungi. As more studies are undertaken, more pollutants and sources are likely to be found. Some will be easily identified and can be remedial through standard control techniques. Others, more elusive or persistent will require development of special control techniques.

An example that emphasizes the physics of the indoor air quality problem is the control of indoor radon and radon progeny concentrations. Control of excess pollutant concentrations can be obtained in several ways. These tend to be dominated by the observation that indoor air pollution is a source dominated problem.

One of the main incentives for the recent increase in research on indoor air quality has been the awareness that energy conservation programs could have a significant effect on pollutant levels. Specifically, if buildings are "tightened" and ventilation rates reduced in order to decrease the amount of energy devoted to heating or cooling the air that comes in from the outdoors, the levels of indoor-generated pollutants could rise simply because less ventilation is being provided for removal. In homes tightening typically entails measures to reduce infiltration of outside air through walls and around windows and doors. Much concern has been voiced about such measures, but it is a fact that indoor pollutants are a problem regardless of energy conservation programs. The major determinant of whether or not a building has excessive levels is the emission rate from various indoor sources, and ordinary measures to reduce infiltration in homes, such as weather-stripping and caulking the windows, have a relatively minor effect. Concentrations of any given pollutant vary by factors of hundreds to thousands within the building stock, whereas energy-saving measures ordinarily reduce ventilation rates by only 10 to 30% and thus increase indoor concentrations by roughly the same amount.

RADON CONTROL

Radon and its immediate radioactive decay products are ubiquitous contaminants of indoor air. Radon isotopes 222 and 220 arise as part of the ^{238}U and ^{232}Th decay series, respectively. These radionuclides, and their eventual respective radium decay products ^{226}Ra and ^{224}Ra, are naturally occurring elements in the earth's crust. The relatively rapid decay of ^{220}Rn (often referred to as thoron, which has a half-life of 55 seconds compared with 3.8 days for ^{222}Rn) effectively limits the amount of this nuclide that can accumulate indoors in most situations; the average dose to the lung from ^{220}Rn progeny has been estimated to be about 25% of that from ^{222}Rn progeny (UNSCEAR 1982). Thus, while much of the discussion in this chapter is generally applicable to either ^{222}Rn or ^{220}Rn, most of the details apply to ^{222}Rn (hereinafter referred to as radon) and its progeny.

Based on a recent compilation of measured indoor radon concentrations, radon levels in detached or semi-detached housing in the United States span two to three orders of magnitude. These concentrations appear to be lognormally distributed, with a geometric mean (GM) of 33 Bq m^{-3} (0.9 $pCiL^{-1}$)* and a geometric standard deviation (GSD) of 2.8 (Nero et al. 1984). The corresponding arithmetic mean (AM) is 58 Bq m^{-3} (1.5 $pCiL^{-1}$). The National Council on Radiation Protection and Measurements (NCRP) has recently proposed a 0.04 WL guideline, equivalent to approximately 300 Bq m^{-3} (8 $pCiL^{-1}$). At this guideline value, approximately 1 to 2% of the United States housing stock -- 1 to 2 million homes -- can be expected to exceed this recommended level.

The health risks associated with radon are due to the alpha decay of two of its short-lived progeny, ^{218}Po and ^{214}Po. These polonium isotopes, and the lead and bismuth isotopes that constitute the immediate radon progeny, are shown in the radon decay chain in Figure 1. These progeny, unlike the chemically inert radon parent, are chemically active and can adhere to surfaces, such as airborne particles, room walls, and lung tissue. A number of models have been devised to estimate the lung dosimetry due to these radioactive decays. While a detailed review of these models is beyond the scope of this chapter, the resulting dosimetric calculations indicate that the alpha dose from progeny not attached to aerosols is 9 to 35 times larger than the dose estimates for progeny attached to aerosols, depending upon the modeling assumptions (James et al. 1981).

Based on lung cancer incidence among uranium miners, estimates have been made of the lung cancer incidence due to radon exposures among the general population. Although there are a number of uncertainties, the expected lung cancer incidence in the United States, based on the average radon concentrations just discussed, is between 1000 and 20,000 per year (Nero 1983). Exposures to higher radon concentrations increase the risk proportionately. This is an important health consequence, and efforts to reduce or control excessive exposures to radon and its progeny deserve attention.

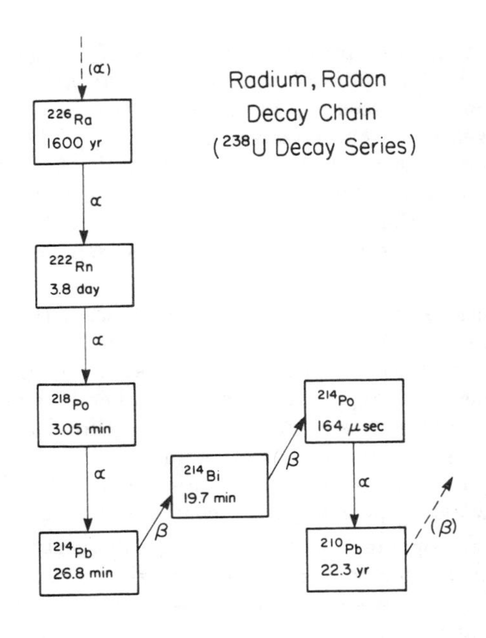

Figure 1. 11 Radon decay chain. The nuclides ^{218}Po, ^{214}Pb, and ^{214}Bi are of primary radiological concern due to inhalation and subsequent alpha decay.

CONTROL STRATEGIES

Background

Before discussing various methods of controlling or reducing radon and radon progeny concentrations indoors, it is worth reviewing the potential sources of radon in residential environments and noting the contribution each source can make to indoor concentrations. The major potential sources of radon in U.S. housing are the soil adjacent to the building substructure, domestic water supplies, and building materials. Other possible sources include natural gas and outdoor airborne radon. The average source strength for radon from each of these sources is summarized in Table 1, and each source is discussed in briefly below. The resulting *average* contribution to indoor radon concentrations can be estimated by dividing the source strengths listed in Table 1 by the air exchange rate, which is typically 0.5 to 1 hr^{-1}.

Soil. A primary source of radon indoors is the soil, where it is produced by the radioactive decay of radium found in trace quantities in all crustal materials. The average concentration of ^{226}Ra in soil samples taken in 33 states in the United States was 41 Bq kg^{-1} (1.1 pCi g^{-1}), with a range of 8.5 to 160 Bq kg^{-1} (0.2 to 4.3 pCi g^{-1}) (Myrick et al. 1983). The radon gas, which is chemically inert, is then transported through the soil and into homes via cracks and other openings in the building substructure. Molecular diffusion of radon gas from the soil through these cracks or through building materials directly is a possible mechanism for radon transport into the building, although as noted in Table 1, the magnitude of the contribution to indoor concentrations does not appear to be sufficient to account for the levels often found indoors. Another process, pressure-driven flow of soil gas, is thought to be a major mechanism for the transport of soil gas into the house (Nero and Nazaroff 1984). Pressure differentials across the building shell can arise due to wind loading and the thermally driven stack effect inside the building. These can create a slight depressurization relative to atmospheric pressure near the floor of the building shell on the order of a few pascals. This pressure differential can draw radon-bearing soil gas into the building. Thus the house, rather than being simply a passive accumulator of radon, has an active role in creating the forces responsible for a major radon entry mechanism.

Soil gas concentrations of radon range from 0.7 to 22 x 10^4 Bq m^{-3} (200 to 6000 pCi L^{-1}), with a typical concentration of 2 x 10^4 Bq m^{-3} (540 pCi L^{-1}) (Nero and Nazaroff 1984). The rate at which radon accumulates in the soil gas and the mobility of the soil gas in the soil is dependent upon a number of environmental and physical parameters. These parameters include the moisture content of the soil and other characteristics of the soil texture and structure which affect the soil permeability. Thus, while radium content of the soil is important, soil and environmental factors that affect gas flow also appear to be key components (Akerblom et al. 1984).

The cracks or other openings in the building substructure through which soil gas flow can occur may be due to the settling or aging of the building or simply a product of the construction or design practices used. The importance of this flow can be illustrated by estimating the flow needed to account for the average total entry rate for radon shown in Table 1. At the average soil gas concentration noted above, if only 0.2% of the typical building infiltration rate is due to flow through the soil, the incoming soil gas entry rate is sufficient to account for observed indoor radon levels. The radon entry rate could be much higher, due to either higher soil gas flow rates or higher radon concentrations in soil gas, or both. In many cases, the entry of radon-bearing soil gas appears to be the predominant source of radon in houses observed to have high indoor concentrations.

Building Materials. Because radium is a trace contaminant of crustal materials, it is present in all earth-based building materials. However, in the United States the emanation rate of radon from these materials is generally too small to cause elevated indoor radon concentrations. Ingersoll (1983) measured radon emanation rates for a variety of common building materials. For concrete, the emanation rate ranged from (2.6 to 19.8) x 10^{-6} Bq kg^{-1} s^{-1} (0.25 to 1.93 pCi kg^{-1} hr^{-1}), with an average of 7.7 x 10^{-6} Bq kg^{-1} s^{-1} (0.75 pCi kg^{-1} hr^{-1}). For gypsum, the average emanation rate was found to be 6.3 x 10^{-6} Bq kg^{-1} s^{-1} (0.61 pCi kg^{-1} hr^{-1}). Other materials, such as brick and rock, had lower emanation rates. The estimated source strength for indoor radon from concrete is shown in Table 1. Most of the elevated indoor radon concentrations observed in the United States are not associated with "technologically enhanced" sources. It is the control of indoor radon levels due to natural radium concentrations that is the focus of this paper.

Water. Radon dissolved in water is a potential source of indoor airborne radon, although the average transfer factor relating the resulting concentration in air to the concentration in water is approximately 10^{-4} (Nazaroff et al. 1985b) Thus, in order to produce 40 Bq m^{-3} (~1 pCi L^{-1}) in air, the radon concentration in water must be 400,000 Bq m^{-3} (~10,000 pCi L^{-1}). Surface water supplies, which provide potable water to almost half the United States population, contain very minimal concentrations of radon, averaging 1050 Bq m^{-3} (28 pCi L^{-1}).

TABLE 1

Typical Radon Source Contributions for a Single-story Residence

Source	Average Source Strength $Bq\ m^{-3}\ h^{-1}$ ($pCi\ L^{-1}\ h^{-1}$)	Reference
Outdoor air	10.0 (0.3)	Gessel 1983.
Potable water	1.0 (0.03)	a.
Concrete floor	2.3 (0.06)	b.
Soil - diffusion through floor	1.3 (0.04)	Nero and Nazaroff 1984.
Soil - uncovered soil	32.0 (0.9)	Nero and Nazaroff 1984.
Total Entry Rate:	52.0 (1.4)	c.

a. **Potable water derived from public groundwater supplies (Nazaroff et al. 1985b).**

b. Assumes half the flux from a 100 m^2, 20 cm-thick concrete floor enters the house (Ingersoll 1983).

c. Arithmetic mean indoor radon concentration (Nero et al. 1984) divided by an average ventilation rate of 0.9 hr^{-1}.

Natural Gas. Like groundwater, natural gas can accumulate radon gas from radium in the rock structures surrounding the gas formation. Surveys of radon concentrations in gas distribution lines in various locations in the United States have shown a concentration range of 37 to 3700 Bq m^{-3} (1 to 100 pCi L^{-1}), with an average of around 740 Bq m^{-3} (20 pCi L^{-1}) (Johnson et al. 1973). At typical residential gas use and air exchange rates, even for unvented gas appliances, the contribution to indoor radon concentrations from natural gas is minor, less than 4 Bq m^{-3} (0.1 pCi L^{-1}).

Source Control

As with a number of indoor air pollutants, limiting production or entry of a pollutant is often easier and more cost effective than attempting to deal with the pollutant once it has been dispersed in the indoor environment. In some cases, source exclusion or elimination is the most straightforward approach. For example, the use of water with low radon concentrations will eliminate the possibility of substantial indoor radon release from water.

Where radon source elimination is not practical, such as when the source is soil gas or where no low-radon-bearing substitutes are practical, some source control methods are available. In the following section specific radon entry points are discussed, followed by a section on source reduction techniques. A general review of indoor air quality control techniques is provided by Fisk et al. (1984).

Radon Entry. Entry points for the pressure-driven flow of radon-bearing soil gas depend on a number of factors, including the type of house substructure, the construction practices used, and the age and structural integrity of the house. Typical substructures in U.S. housing include concrete slab-on-grade, basement (partial or full), crawl-space (usually topped by a wooden floor), or some combination of these three basic designs. The first two of these substructures have similar potential entry paths, including cracks or other penetrations between the conditioned indoor space and the soil. These cracks may result from the aging and settling of the building or may be a design feature, such as the joint frequently found between the foundation walls and the floor. Penetrations for plumbing or electrical connections are also possible; often the hole surrounding the pipe or wiring is not filled or sealed.

Efforts to evaluate the differences in indoor radon concentration among houses with the various substructure types have only recently begun; thus the data are not conclusive. While homes with basements appear to have the greatest potential for high radon levels, the data collected thus far suggest that high indoor radon concentrations can also occur in houses with either a crawl space or slab-on-grade substructure. The variability in the components of the source terms for radon, such as radium content and soil permeability, may overwhelm any differences due to substructure type. In a survey of housing in the Pacific Northwest, for example, the average radon concentration in the first floor living area was 47 Bq m^{-3} (1.3 pCi L^{-1}) for 120 houses with basements, 33 Bq m^{-3} (0.9 pCi L^{-1}) for 93 houses with crawl spaces, and 43 Bq m^{-3} (1.2 pCi L^{-1}) for 7 houses having slab-on-grade construction (Thor 1984).

With regard to house substructure, clearly the potential coupling between the house substructure and the soil is largest for a basement simply on the basis of surface area alone. In many cases, concrete blocks are used for basement walls. Untreated, these can be fairly permeable to fluids, and chinking of the mortar between blocks can also occur. In addition, transport of soil gases can take place through the hollow core of the blocks. In some houses, open sumps are part of the basement construction. These sumps may be connected to a "weeping tile" system designed to remove water from beneath the basement floor and walls. This system can also serve as an effective entry pathway for soil gas when it is not occluded by water (Nazaroff et al. 1985a).

Houses built with crawl spaces appear to be less tightly coupled to the soil, although degree of coupling will depend upon features of the crawl space, such as whether the space is vented or unvented, and the number and size of penetrations between the living space and the crawl space. Many crawl spaces have open soil floors, thus radon entry into the crawl space is unimpeded. A recent study of crawl-space homes suggested that about half the radon present in the crawl space entered the home, even with the crawl space vented. When the crawl-space vents were closed, the radon concentration in both the crawl space and the living space increased (Nazaroff and Doyle 1985). In a study of twenty-two homes in the Chicago area with unvented crawl spaces almost half of the houses were found to have radon concentrations above 185 Bq m^{-3} (5 pCi L^{-1}) and about a quarter had concentrations above 370 Bq m^{-3} (10 pCi L^{-1}) (Rundo et al. 1979).

Housing built using slab-on-grade construction can also have high radon concentrations; while the surface area of the building-soil interface is smaller than a house with a basement, the coupling between the house and the soil can still be substantial. Scott and Findlay (1983) found that cracks and penetrations through the slab floors were major sources of radon entry.

Source Reduction Techniques. Reduction of radon entry from the soil into building interiors has generally involved (1) sealing specific leakage pathways, such as cracks, joints or other penetrations, (2) application of a more general surface sealant and/or (3) sub-slab or subfloor ventilation. An important element in these procedures is the identification of likely entry points for soil gas; this is especially true for remedial work.

When the inner surfaces of the building foundation are finished with floor or wall covering materials, as is virtually always the case with slab-on-grade construction, identification and access to radon entry points may be particularly difficult. The task may be less complicated for an unfinished basement (though a larger surface area may be involved). Although the effectiveness of finding and sealing these entry pathways is dependent upon a number of factors, there is growing evidence from a variety of remedial projects to indicate that significant reductions in indoor radon concentrations can result This is not always the case, however, and measurements to assess the effectiveness of remedial techniques are usually necessary.

A common entry point is a sump system connected to a sub-slab drainage system, as described earlier. Radon can enter the building if there is no water trap to isolate the incoming drain line from the interior of the house. Installing or rebuilding the sump to accommodate a trap has been shown to be effective in reducing indoor radon concentrations, often by a factor of 4 to 5.

Substructure ventilation is another technique to reduce radon entry into the living space. A common version of this is the ventilated crawl space, which can be used in conjunction with sealing of cracks and penetrations in the living space flooring and/or use of a radon barrier over the open soil floor. The effect of reducing crawl space ventilation has been examined in a few cases, and the radon concentrations in both the crawl space and the living space increased with crawl-space vents blocked. Sealing potential leakage pathways was found to reduce the flow between the crawl space and living area. (Nazaroff and Doyle 1985). In some cases, mechanical ventilation of the crawl space has also been used to reduce buildup of radon (Keith 1980).

Removal of Indoor Radon

In some cases source reduction measures may not be feasible or may not sufficiently reduce radon concentrations. Ventilation of indoor spaces is widely used for control of indoor pollutants generally, though in some cases the energy and economic costs can be substantial. Another technique for reduction of indoor concentrations is the use of a pollutant-specific removal technique; one such method that has been suggested for radon is the use of an activated charcoal adsorbent. The effects of ventilation and charcoal adsorption on radon concentrations are discussed in more detail below. The effect of ventilation on radon progeny concentrations is reviewed in the next section.

Ventilation. The effects of ventilation are generally described using a fairly straightforward, well-mixed box model to estimate steady state concentrations of the pollutant of interest. In such a mass-balance model, the average indoor concentration, C_i, is equal to the various source terms divided by the removal terms:

$$C_i = \frac{S + \lambda_V P C_O}{\lambda_V + K + \lambda_0 + \lambda_F}, \tag{1}$$

where

$S =$ source strength per unit indoor volume (Bq m^{-3} hr^{-1}),

$P =$ penetration fraction for outdoor airborne pollutants (= 1 for an inert gas such as radon),

$C_O =$ outdoor concentration,

$\lambda_V =$ air exchange rate (= ventilation rate; hr^{-1}),

$K =$ chemical or physical transformation rate (hr^{-1}),

$\lambda_0 =$ removal rate due to radioactive decay of radon (= 0.00758 hr^{-1}), and

$\lambda_F =$ pollutant removal rate due to operation of an air cleaning device (hr^{-1}).

In the case of radon, several of these parameters have a negligible effect on indoor concentrations. The outdoor airborne concentration, C_O, is usually small compared with typical indoor levels. The chemical or physical reaction constant, K, is also zero, since radon is chemically inert (except under extreme circumstances not likely to be found in a residential environment) and is not significantly adsorbed on most building surfaces. The removal rate, λ_F, due to operation of an air cleaner is also zero, for essentially the same reasons (activated charcoal filtration appears to be ineffective, as discussed in the next subsection). And finally, the radioactive decay constant for radon, 0.00758 hr^{-1}, is quite small compared with typical ventilation rates. Thus Equation 1 essentially reduces to the ratio of the source strength to the ventilation rate.

Use of this equation involves a number of simplifying assumptions, particularly the assumption of perfect mixing of the indoor air. The equation also does not account for any coupling between ventilation rate and radon source strength. As we have discussed earlier, pressure-driven flow is thought to be responsible for most of the radon entry into U.S. housing. Several recent studies have indicated that the radon entry rate is often a function of ventilation rate and that entry rates associated with air exchange can be significantly greater than diffusive transport alone. Coupling between the rates of natural infiltration and radon entry, as discussed by Nazaroff et al. (1981b), for example, may be a consequence of the fact that wind and the thermal stack effect drive both infiltration and radon entry. Studies have also been done using residential air-to-air heat exchangers, which provide more balanced ventilation flows. No increase in radon entry rates was observed, presumably because there was no net increase in building depressurization with use of the air-to-air heat exchangers, which supply incoming air mechanically to make up for the mechanically vented exhaust air (Nazaroff et al. 1981a; Offermann, et al. 1982).

The effect of ventilation rate on indoor radon concentration is shown as a short dashed line in Figure 2, where several observations can be made. In order to achieve a reduction in radon concentration equivalent to those seen from application of some source control measures, a factor of 5 to 10, for example, the ventilation rate would have to increase by the same factor (neglecting any coupling between source strength and ventilation rate). A five-to-ten-fold increase in ventilation rate is substantial.

At low initial air exchange rates, below about 0.5 hr^{-1}, such an increase may be feasible. If the initial air exchange rate is about average, from 0.6 to 1.2 hr^{-1}, a factor of 5 to 10 increase is much less practical. For example, most air to air heat exchangers used in residential applications will increase the air exchange rate 0.4 to 0.9 hr^{-1} (Fisk and Turiel 1983).

On the other hand, as illustrated by Figure 2, the indoor radon concentration rises steeply for ventilation rates below 0.5 hr^{-1}. While it is difficult to achieve ventilation rates this low on a retrofit basis, new homes can be constructed with natural ventilation rates close to 0.1 hr^{-1}. In doing so, it may be useful to provide for additional mechanical ventilation (using an air-to-air heat exchanger, for example) to bring the ventilation rate of the structure up to ~0.5 hr^{-1} if necessitated by indoor air quality problems.

Charcoal Adsorption. The adsorption of radon by activated charcoal is a well-known phenomenon, and its use for cleansing mine atmospheres of radon gas has been suggested by a number of authors. Charcoal has also been suggested for control of indoor radon, although few evaluations of its use have been made. In two recent experiments, operation of an activated charcoal filtration unit produced negligible effects on indoor radon levels (Nitschke et al. 1984; Sextro et al. 1985). Both papers report that radon progeny concentrations were reduced by use of the charcoal filtration device. As noted in Sextro et al. (1985), airborne particle concentrations were also reduced by use of the charcoal filtration unit. As discussed in greater detail in the following section, removal of particles contributes to the reduction in progeny concentration.

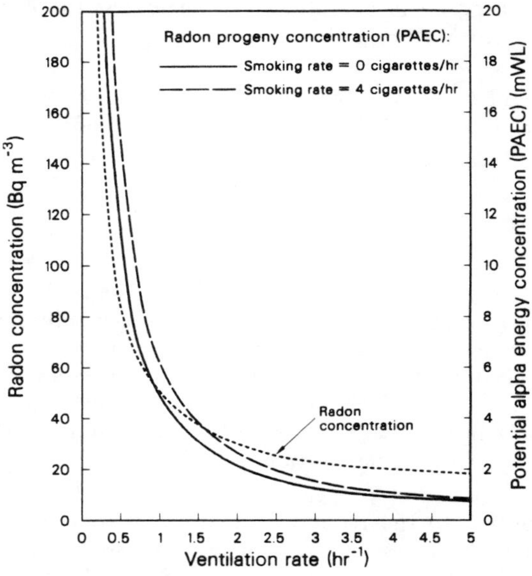

Figure 2. Radon and radon progeny concentrations versus ventilation rate. The calculated concentrations for both radon and radon progeny assume a constant radon source strength of 50 Bq m^{-3} hr^{-1}. The PAEC curves are based on an assumed constant indoor source rate of 10^{12} particles hr^{-1} (from sources other than tobacco combustion) and infiltrating outdoor air with 20,000 particles cm^{-3} and a penetration factor of 0.5. For the case with cigarette smoking, each cigarette produces 9 x 10^{12} particles.

Radon Progeny Control

Radon progeny, the radioactive products of the radioactive decay of radon, are the main source of the radiological risks of exposure to radon. Because these elements, unlike their inert radon parent, are chemically active and can therefore attach to surfaces, such as airborne particles, room surfaces, or lung tissue, control of progeny concentrations presents a different set of considerations. In this section, equations describing progeny behavior and the associated health risks are first presented, followed by discussions of the effects of radon progeny control. Finally, estimates of the relative alpha radiation dose to the lungs under various control conditions are discussed.

Background. A commonly used method of parameterizing the airborne concentration of radon progeny with respect to their alpha decay properties is the Potential Alpha Energy Concentration (PAEC), which is given by

$$\text{PAEC} = k_1A_1 + k_2A_2 + k_3A_3, \tag{2}$$

where the subscripts 1 to 3 refer to ^{218}Po, ^{214}Pb, and ^{214}Bi respectively. The coefficients, k_i, are a function of the potential alpha decay energy and the half-life of the nuclide of interest. For progeny concentrations, A_i, measured in Bq m^{-3}, the coefficients are $k_1 = 2.84 \times 10^{-5}$, $k_2 = 1.39 \times 10^{-4}$, and $k_3 = 1.03 \times 10^{-4}$, which gives the PAEC in units of working level (WL). One working level is defined as any combination of radon progeny in one liter of air such that the ultimate decay to ^{210}Pb will result in 1.3×10^5 MeV of alpha decay energy.

Another useful concept is the equilibrium factor, F, which is a measure of the degree of equilibrium established between radon and its decay products:

$$F = \frac{3700 \text{ PAEC}}{A_0}, \tag{3}$$

where A_0 is the radon concentration in Bq m^{-3}. If the radon progeny concentrations were those established solely by secular radioactive equilibrium (i.e., no other removal mechanisms exist other than radioactive decay) F would be unity. Since radon progeny are chemically active and can attach to room surfaces, the airborne concentrations are typically lower than expected from simple secular radioactive equilibrium; F is usually in the range of 0.3 to 0.7 for most indoor situations, depending upon the airborne particle concentration.

The behavior of radon progeny is illustrated in Figure 3, where the various processes contributing to the reduction of airborne concentrations are shown. The rate constant for each process is shown in parenthesis; radioactive decay as a removal process for each progeny nuclide is not indicated. While radon has two removal processes, ventilation and radioactive decay, progeny removal can occur in four ways: ventilation, air cleaning, deposition on macro surfaces, and radioactive decay. The progeny can also attach to the surfaces of indoor airborne particles, which, in turn, can be removed by ventilation, deposition, and air cleaning. As shown in Figure 3,

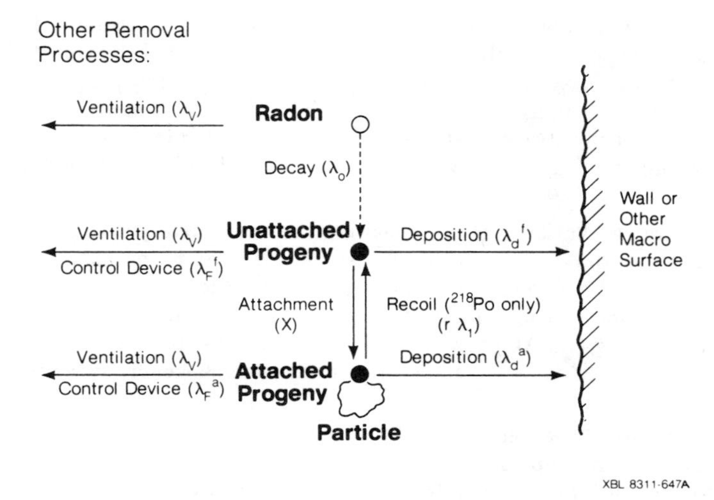

Figure 3. Schematic diagram of the removal processes (and their associated rates) affecting concentrations of radon and radon progeny. The radioactive decay pathways for radon progeny are not explicitly noted in this diagram.

the alpha decay of ^{218}Po can produce sufficient recoil momentum to detach the decay product from the particle; this detachment probability, denoted r in the figure, has been estimated to be 0.83 (most of this background discussion is based on Sextro et al. 1984 and Offermann et al. 1984 and references therein, unless otherwise noted). The rate of progeny attachment to particles, which is usually assumed to be independent of the chemical nature of the progeny, is dependent upon the particle concentration and size distribution. The mean attachment rate coefficient is 4.3 x 10^{-3} hr^{-1} x (particles $cm^{-3})^{-1}$ for particles typically found in indoor air.

In a recent series of experiments in a room-sized chamber, the number-weighted particle deposition rate was found to be 0.16 hr^{-1}; this rate is assumed to also be the average deposition rate of attached progeny. In the same set of experiments, the deposition rate of unattached progeny was estimated to be 15 hr^{-1}, almost a factor of 100 greater than for attached progeny (Offermann et al. 1984).

The various removal rates discussed thus far can be compared. At a particle concentration of 10,000 particles cm^{-3}, which is typical for homes without heavy cigarette smoking, the attachment rate for progeny to particles is 43 hr^{-1}, while the deposition rate for unattached progeny on macro surfaces is 15 hr^{-1}. Surface deposition of airborne particles (and of any progeny attached to them) is almost negligible, with a deposition rate of 0.16 hr^{-1}. At lower particle concentrations, 1000 particles cm^{-3} for example, the attachment rate to particles is 4.3 hr^{-1}, and by comparison, deposition of unattached radon progeny becomes a more important process. Similarly, at high particle concentrations, the attachment rate to particles is higher, and a greater proportion of progeny remain airborne, attached to particles.

Particle Reduction. The concentration of airborne particles is an important determinant of indoor radon progeny concentrations. Removal of particles by active air cleaning (i.e., operation of a mechanical or electrostatic filtration system) has two general effects on airborne radon progeny concentrations. First, air cleaning can remove radon progeny directly, either those attached to airborne particles which are captured by the air-cleaning system, or the unattached radon progeny which are also trapped by the air-cleaning device. Second, air cleaning also contributes to the reduction of radon progeny concentrations by reducing the particle concentration so that deposition of unattached progeny on indoor surfaces becomes a predominant removal mechanism.

There are a variety of air-cleaning devices available, including portable unducted devices that might be used for one or two rooms and devices that are installed in a forced-air space conditioning system. Although air-cleaning devices are also used in commercial and industrial applications, as part of heating, ventilating, and air conditioning (HVAC) systems, for example, this discussion focuses on residential systems since radon and radon progeny problems generally arise in residential buildings.

There are two broad categories of air-cleaning systems: mechanical fan-filters in which impaction, interception or diffusion are the major particle removal mechanisms and electrostatic filters, which rely on electrostatic forces between the particles and the collection surface.

Changes in radon progeny concentrations as a result of air cleaning show similar results. Air-cleaning devices that removed particles effectively had a commensurate effect on airborne radon progeny. The effect of particle concentration on the equilibrium factor, F, is shown in Figure 4, where the solid circles represent data from experiments with the unducted air-cleaning devices. As can been seen, the equilibrium factor decreases with decreasing particle concentration.

The calculated equilibrium factor values for unattached progeny are shown in Figure 4 as the dashed line. As can be seen, at particle concentrations below 500 particles cm^{-3}, the airborne progeny concentration is almost entirely associated with unattached progeny. The relative concentration of unattached progeny declines with increasing particle concentration. The total equilibrium factor, on the other hand, increases rapidly with particle concentration in the range of ~1000 to ~25,000 particle cm^{-3}, a concentration range typical of indoor environments. 1.

Lung Dose and Particle Concentrations. As noted earlier, the radiological effects of radon exposure are due to the alpha decay of the progeny and, based on dosimetric models, the unattached progeny produce a significantly larger lung dose than progeny attached to airborne particles. Effective air cleaning results in both particle and progeny removal, and as can be seen in Figure 4, the fraction of unattached progeny species increases with decreasing particle concentration. The lung dose expected from the resulting mixture of attached and unattached progeny can be estimated relative to the dose calculated assuming no unattached progeny (Harley and Pasternak 1972, as adapted by Jonassen 1982). Results from these calculations are shown in Figure 4, where the relative dose curves are shown as solid lines. The relative dose curves, both of which refer to the right axis, are based on two dosimetric cases; children undergoing light activity (top curve) and adults at rest (bottom curve). These two curves are reasonable representations of the limiting cases; more realistic assumptions regarding behavior patterns, breathing rate, air volume, etc., are likely to fall between these two lines. As these estimates illustrate, the radiological effects of reduced particle and progeny concentrations (without a commensurate reduction in *radon* concentration indoors) are not significantly smaller,

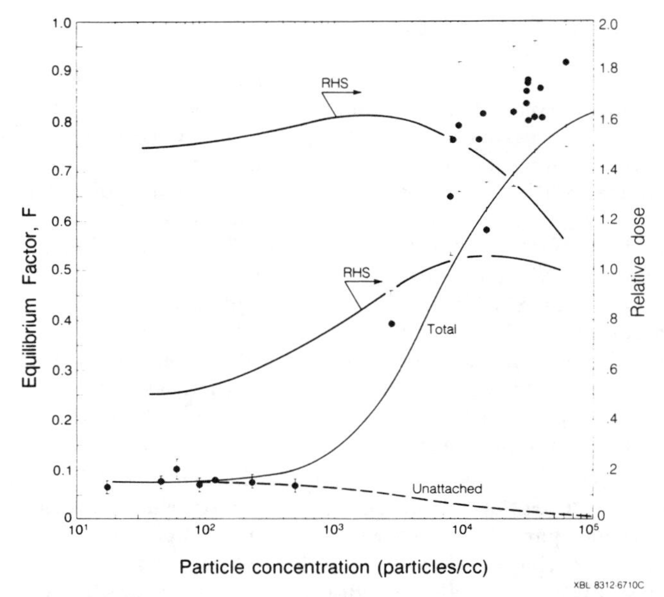

Figure 4. Equilibrium factor, F, versus particle concentration. Measured data and uncertainties are indicated by the solid circles and error bars. The solid line labeled Total represents calculated values for total airborne radon progeny, while the dashed line shows calculated values for unattached progeny. The two curves referring to the right hand scale (RHS) show the relative alpha radiation dose to lung tissue. The upper curve is based on dosimetric calculations for children undergoing light activity, while the lower curve is for adults at rest. The calculations further assume that at 10^5 particles cm^{-3} the progeny are all attached to airborne particles, and thus the relative dose is unity.

even though the *total* radon progeny concentration (at a constant radon concentration) does decrease with reduced particle concentration.

SUMMARY AND CONCLUSIONS

The nature of the indoor air quality problem has been described; methods of control or reduction of indoor radon and radon progeny concentrations have been reviewed. These techniques may be categorized as radon source reduction, radon removal, and radon progeny removal. There are a number of potential sources of radon in U.S. housing, including soil, potable water, and building materials. In most cases, it appears that flow of radon-bearing soil gas into houses, driven by a slight negative pressure differential across the building shell, is a major source of indoor radon; this pressure-driven flow appears to be the most likely source of radon that can account for the elevated radon concentrations observed in some houses. There are a number of radon source control techniques; their effectiveness will depend upon characteristics of the house substructure and the details of the specific application. While the results of such remedial measures have varied and the data base from which to generalize is small, five-to-ten-fold reductions in radon concentration have been reported.

In cases where source reduction is not possible or economically practical or is not entirely effective, concentration reduction measures can be employed, including ventilation or air cleaning. While, in principle, large changes in ventilation rate can be made, significant increases may not be economically or physically practical if the initial air exchange rate is in the 0.8 to 1.2 h^{-1} range (typical of much of the existing U.S. housing stock). In many cases, reductions in indoor radon or radon progeny concentrations by a factor of 2 to 3 are possible through increased ventilation, although unbalanced exhaust ventilation procedures that lead to additional depressurization of the building shell are not likely to produce the expected reduction in radon concentration due to coupling between the additional air exchange and infiltration of soil gas bearing radon. In very tight houses with natural ventilation rates ~0.1 to 0.2 hr^{-1} additional mechanical ventilation, as might be produced by an air-to-air heat exchanger, can be used to increase the ventilation rate of the structure to ~0.5 hr^{-1} with a minimal energy penalty.

Progeny concentration reductions may also be achieved by air cleaning, which removes progeny by filtration of the unattached or attached airborne radon progeny and also by reducing particle concentrations, thereby increasing the progeny deposition rate on indoor surfaces. However, the health risk from the alpha radioactivity of the remaining mixture of airborne radon progeny (both unattached and attached) may not be significantly reduced as a result of air cleaning.

It is clear that while an understanding of the efficacy of various radon and radon progeny control methods is emerging, substantial work remains in developing more general and systematic source control techniques. Because a large number of homes in the United States appear to exceed guideline levels for indoor radon concentrations, general indicators of the potential for high indoor radon concentrations need to be identified and investigated in order to locate geographical areas where either remedial or preventive control methods might be required.

ACKNOWLEDGMENTS

This work was supported by the Assistant Secretary for Conservation and Renewable Energy, Office of Building Energy Research and Development, Buildings Systems Division of the U.S. Department of Energy under Contract No. DE-AC03-76SF00098.

REFERENCES

Akerblom, G.; Andersson, P.; and Clavensjo, B. 1984. "Soil gas radon - a source for indoor radon daughters." *Radiation Protection Dosimetry 7,* pp. 49-54.

ASHRAE. 1981. *ASHRAE Standard 62-81,* "Ventilation for acceptable indoor air quality." American Society of Heating, Refrigerating, and Air-Conditioning Engineers, Atlanta, GA.

Fisk, W.J., and Turiel, I. 1983. "Residential air-to-air heat exchangers: Performance, energy savings, and economics." *Energy and Buildings 5,* pp. 197-211.

Fisk, W.J.; Spencer, R.K.; Grimsrud, D.T.; Offermann, F.J.; Pedersen, B.; and Sextro, R.G. 1984. "Indoor air quality control techniques: a critical review." Lawrence Berkeley Laboratory, report LBL-16493.

Gessel, T.F. 1983. "Background atmospheric ^{222}Rn concentrations outdoors and indoors: A review." *Health Physics 45,* pp. 289-302.

Harley, N.H., and Pasternak, B.S. 1972. "Alpha absorption measurements applied to lung dose from radon daughters", *Health Physics 23,* pp. 771-782.

Ingersoll, J.G. 1983. "A Survey of radionuclide contents and radon emanation rates in building materials used in the U.S." *Health Physics 45,* pp. 363-368.

James, A.C.; Jacobi, W.; and Steinhausler, F. 1981. "Respiratory tract dosimetry of radon and thoron daughters: The state-of-the-art and implications for epidemiology and radiobiology", In M. Gomez (ed.), *Radiation Hazards in Mining: Control, Measurement, and Medical Aspects,* New York: Society of Mining Engineers, pp. 42-54.

Johnson, R.H.; Bernhardt, D.E.; Nelson, N.S.; and Calley, H.W. 1973. "Assessment of potential radiological health effects from radon in natural gas." U.S. Environmental Protection Agency, report EPA 520/1-73-004, Office of Radiation Programs, Washington DC.

Jonassen, N. 1982. "Radon daughter levels in indoor air: Effects of filtration and circulation." Progress report II, Laboratory of Applied Physics I, Technical University of Denmark, Lyngby.

Keith. 1980. "Summary of Uranium City, Saskatchewan, remedial measures for radiation reduction with special attention to vent fan theory." presented at the Third Workshop on Radon and Radon Daughters, March, 1980, Port Hope, Ontario; Keith Consulting.

Myrick, T.E.; Berven, B.A.; and Haywood, F.F. 1983. "Determination of concentrations of selected radionuclides in surface soil in the U.S." *Health Physics 45,* pp. 631-642.

National Research Council 1981. *Indoor Pollutants.* National Academy Press.

NCRP. 1984. *Exposures from the Uranium Series with Emphasis on Radon and its Daughters.* National Council on Radiation Protection and Measurements, NCRP Report 77, Bethesda, MD.

Nazaroff, W.W.; Boegel, M.L.; Hollowell, C.D.; and Roseme, G.D. 1981a. "The use of mechanical ventilation with heat recovery for controlling radon and radon-daughter concentrations in houses." *Atmospheric Environment 15,* pp. 263-270.

Nazaroff, W.W., and Doyle, S.M. 1985. "Radon entry into houses having a crawl space." *Health Physics 48,* pp. 265-281.

Nazaroff, W.W.; Fuestel, H.; Nero, A.V.; Revzan, K.L.; Grimsrud, D.T.; Essling, M.A.; and Toohey, R.E. 1985a. "Radon transport into a single-family house with a basement." *Atmospheric Environment 19,* pp. 31-46.

Nazaroff, W.W.; Doyle, S.M.; Nero, A.V.; and Sextro, R.G. 1985b. "Potable water as a source of airborne radon-222 in U.S. dwellings: A review and assessment." Lawrence Berkeley Laboratory, report LBL-18514, to be submitted to *Health Physics.*

Nero, A.V. 1983. "Indoor radiation exposures from ^{222}Rn and its daughters: A view of the issue." *Health Physics 45,* pp. 277-288.

Nero, A.V., and Nazaroff, W.W. 1984. "Characterizing the source of radon indoors." *Radiation Protection Dosimetry 7,* pp. 23-39.

Nitschke, I.A.; Wadach, J.B.; Clarke, W.A.; and Traynor, G.W. 1984. "A detailed study of inexpensive radon control techniques in New York state houses." Vol 5, pp. 111-116, Swedish Council for Building Research, Stockholm.

Offermann, F.J.; Hollowell, C.D.; Nazaroff, W.W.; and Roseme, G.D. 1982. "Low infiltration housing in Rochester, New York; A study of air exchange rates and indoor air quality." *Environment International 8,* pp. 435-446.

Offermann, F.J.; Sextro, R.G.; Fisk, W.J.; Nazaroff, W.W.; Nero, A.V.; Revzan, K.L.; and Yater, J. 1984. "Control of respirable particles and radon progeny with portable air cleaners." Lawrence Berkeley Laboratory, report LBL-16659.

Porstendoerfer, J.W.; Wicke, A.; and Schraub, A. 1978. "The influence of exhalation, ventilation and deposition processes upon the concentration of radon and thoron and their decay products in room air." *Health Physics 34,* pp. 465-473.

Rundo, J.; Markun, F.; and Plondke, N.J. 1979. "Observation of high concentrations of radon in certain houses." *Health Physics 36,* pp. 729-730.

Scott, A.G., and Findlay, W.O. 1983. "Demonstration of remedial techniques against radon in houses on Florida phosphate lands." for the U.S. Environmental Protection Agency, Report EPA 520/5-83-009, American Acton, Inc., Columbia, MD.

Sextro, R.G.; Offermann, F.J.; and Nero, A.V. 1985. "Reduction of indoor particle and radon progeny concentrations with ducted air cleaning systems." Lawrence Berkeley Laboratory, report LBL-16660.

Spengler, J.D.; Sexton, K. 1983. "Indoor Air Pollution: A Public Health Perspective." *Science 221,* pp. 9-17.

Thor, P.W. 1984. "BPA radon field monitoring study." in *Doing Better: Setting An Agenda for the Second Decade. ACEEE 1984 Summer Study on Energy Efficiency in Buildings, Santa Cruz, CA,* Vol B, pp. 283-298, American Council for an Energy Efficient Economy, Washington, DC.

Turiel, I. 1985. *Indoor Air Quality and Human Health.* Stanford University Press.

UNSCEAR. 1982. *Ionizing Radiation: Sources and Biological Effects,* United Nations Scientific Committee on the Effects of Atomic Radiation, United Nations, New York, NY.

CHAPTER 13

ENERGY AND LIGHTING

Samuel Berman
Lighting Systems Research Group
Lawrence Berkeley Laboratory
Berkeley, CA 94720

ABSTRACT

Advances in research for new types of lighting with increased efficacies (lumens/watt) are discussed in the following areas: (1) high-frequency, solid-state ballasts, (2) isotopic enhancement of mercury isotopes, (3) magnetic augmentation, (4) electrodeless, ultra-high frequency, (5) tuned phosphors, (6) two-photon phosphors, (7) heat mirrors, and (8) advanced control circuits to take advantage of daylight and occupancy. As of 1985, improvements in efficacy have been accomplished on an economic basis to save energy for (1) high frequency ballasts (25%), (2) isotopic enhancement (5%), and (8) advanced control circuits (up to 50%). Most of these advances depend on a deeper understanding of the weakly ionized plasma as a radiating and diffusing medium.

INTRODUCTION

About 80 base-loaded power plants, one-fourth of U.S. electricity, are needed for lighting. We estimate that approximately 50% of the electrical energy consumed by lighting, or about 12% of total electrical energy, could be saved by gradually replacing existing lighting with energy-efficient lighting. This would amount to a yearly savings of some 220 billion kilowatt-hours of electricity, or about the equivalent of 40 based-loaded 1-GW power plants.

The objective of the Department of Energy (DOE) Lighting Program is to assist the lighting community (manufacturers, designers, and users) to achieve a more efficient lighting economy. The program, carried out Lawrence Berkeley Laboratory (LBL), exemplifies a unique partnership between a national laboratory/university complex and industry, facilitating technical advances, strengthening industry capabilities, and providing designers and the public with needed information.

PROGRAM SCOPE

To implement its objectives, the lighting program has been divided into three major categories: engineering science, electromagnetic compatibility, and impacts on human health and productivity.

The engineering science component of the program undertakes research and development projects in lamp technology that are both long-range and high-risk. These are projects in which the lighting industry has an interest but cannot pursue on its own, and from which significant benefits could accrue to both the public and industry if the technical barriers were surmounted.

The program also aims to understand the electromagnetic compatibility of high-frequency lighting with building functions including machinery, computers, and other electrical and electronic systems. The program's impacts component examines relationships between workers and the physical lighting environment to assure that energy-

0094-243X/85/1350247-11$3.00 Copyright 1985 American Institute of Physics

efficient technologies contribute to human productivity and health. These efforts are interdisciplinary, involving engineering, optometry, and medicine. To implement its Lighting Program, DOE combines the facilities and faculties of LBL with those of the University of California College of Environmental Design, School of Optometry, and School of Medicine.

TECHNICAL DESCRIPTION

Total annual costs of a lighting system are dominated by energy costs, close to 80% of the total, the remaining 20% going to lamps, fixtures, wiring, electrical components, and controls. Because lighting systems use so much energy and have much shorter lifetimes than furnaces, elevators, and other building systems, there are excellent opportunities to replace inefficient lighting systems with more efficient ones. Heightened energy concerns and consumer demands for more efficient products can hasten innovations. In this technically difficult and high-risk area, a cooperative effort between government and industry can achieve technological solutions and market penetration more quickly.

Three major components of lighting system are important for energy considerations: the lamp converts electricity into visible light and is the major user of energy in the lighting system. The electronics, switching, and controls generally use only a small fraction of the lamp energy but can have a major effect by regulating, scheduling, and switching the lamp. The fixture consumes no energy but can greatly affect illumination efficiency through illumination distribution and light capture within it.

Although in a typical residence lighting energy often represents less than 10% of net energy use, in a high-rise office building it can account for more than 50% of the building energy consumption. Table I shows how, on a national basis, lighting energy use compares with the total electrical energy use for the four main categories of consuming classes: residential, commercial, industrial, and street lighting. The category referred to as "other" includes outdoor lighting such as sports lighting, transportation lighting, or outdoor advertising. Thus lighting accounts for 25% of the national electrical energy consumption, approximately equally divided between incandescent and gas-discharge [fluorescent and high-intensity discharge (HID)] lamps.

TABLE I. NATIONAL LIGHTING ENERGY CONSUMPTION (%)

Energy use	Res.	Comm.	Ind.	Street	Other	Total
Total (BkWh)	560	400	790	14	50	1,814
Lighting (BkWh)	90	200	90	14	24	418
(%)	16	50	11	100	48	23
Incandescent (BkWh)						200
Gas-discharge (BkWh)						218

Lamps

Table II shows the use of the three principal types of lamps (incandescent, fluorescent, and HID), their annual energy consumption, number of sockets, and total number of lumen hours they provide. Annual energy is indicated in lumen hours as well as kilowatt-hours because this indicates the efficacy of energy use. Table II shows that although the energy use of incandescents is comparable to that of fluorescents, incandescents produce only one-fifth the light/watt. Clearly this represents a prime opportunity for improved efficiency. During the past few years several new lamps have appeared on the market as energy-efficient replacements for the incandescent lamp.

TABLE II. USE OF LAMPS

Lamps	Sockets	Ave. Power (W)	Ave. Light (1)	Annual Use (hr)	Annual Energy (kWh)	Annual (Lumen hr)
Incandescent	2.8×10^9	100	1,600	700	196×10^9	3.1×10^{15}
Fluorescent	1.4×10^9	50	3,300	3,000	216×10^9	14.0×10^{15}
HID	0.02×10^9	250	12,000	4,000	20×10^9	0.96×10^{15}

It is well established that no single lamp will be the unique replacement for the ubiquitous incandescent filament lamp with its "Edison" socket, for reasons including the longevity, price, color quality, weight, size, location, and heat output of new lamps. Aside from using an infrared reflective coating (a heat mirror) to reduce heat loss on a standard filament lamp, all potential replacements are gas-discharge lamps. These lamps require more sophisticated controls than the typical on-off switch. Working with the lamp and electronics industry, DOE/LBL has proved the technical and economical feasibility of high-frequency solid-state ballasts, which, together with innovations in lamp design, have accelerated the availability of a variety of energy-efficient replacements for the incandescent lamp. But this is not the whole story because, although fluorescent and HID lamps are considerably more efficient than filament incandescent lamps, their efficiencies still are far below the physical limit of about 400 lumens per watt. There are several technically feasible ideas for improving the efficiency of these lamps, but further scientific and engineering research is required before they produce marketable systems. Table III lists some promising possibilities and their target dates and Table IV lists the parameters of the lamps now available.

Controls

The revolution in microelectronics coupled with the need to reduce operating costs of lighting systems spurred the development of dynamic controls for lighting systems operable by users.

Before the days of oil embargoes, lighting systems in most commercial and industrial establishments had only on-off switches for controls and, furthermore, a single switch quite commonly controlled large banks of lighting. More appropriate lighting controls make use of daylighting and provide various light levels through scheduling or lumen depreciation. Controls can also combine general with task lighting to moderate lighting energy use. Efforts are needed to understand how a control system should work to maximize energy savings and, at the same time, provide the desired lighting services. An important component of this research is understanding the degree of light-sensitive

TABLE III. TARGETS OF OPPORTUNITY

Technology		Total Efficacy (lumens/watt)	Year in Market
Incandescent Lamps		11-18	1879
Fluorescent Lamps			
Present Fluorescent		80	1937
High-Frequency Operation		90	1980
Narrowband Phosphors		100	1983
Isotopically Enriched		110	1988
Magnetically Loaded	LBL/DOE	135	1990
Two-Photon Phosphor	Technical	200	1992
Gigahertz/Electrodeless	Initiatives	230	1994
HID Lamps			1960
Today with High-Frequency Ballast	130 lm/W (1000-W High-pressure Sodium)		
	90 lm/W (1000-W Metal Halide)		1984
	55 lm/W (1000-W Mercury Vapor)		
Electrodeless/High-Frequency	10—15% improvement, 1000 W lamps		1989
	30% improvement, low-wattage lamps		
New Gases	20 — 25% improvement		1990
Color Constant/Dimmable	20 — 25% improvement		1993

detection consistent with human needs and productivity. The potential for energy savings is large and, using the new generation of microelectronics, can be cost-effective. A successful effort would develop ties with the controls industry so that innovations in technology and strategy can be passed on to the consumer.

Fixtures

Fixtures help determine how much of the light flux emitted by a light source reaches a work surface. This metric is defined as the coefficient of utilization (CU) and is a function of the reflectivity of the material, transmissivity of the lens, geometric shape of the fixture and light source, the ambient temperature, and wall and ceiling reflectivities. Work has been completed on improving the reflectivity and transmissivity via multiple-coated films. The major problem has been in estimating the ballast factor and thermal effects of fluorescent lamp/ballast systems; mis-estimations have resulted in improper evaluation of the CU factor and, as a result, over-illumination of spaces by more than 50%. DOE, through LBL, is determining the functional dependence of the light output of fluorescent lamp/ballast systems, as well as measuring the thermal characteristics of fluorescent luminaires in environments where they typically are used. If these functional dependencies are disseminated, they can serve as design guides for meeting targeted illumination levels. This information could help designers approach the illumination target within 10%, in contrast to the +50% that is typical, and would reduce energy consumption by 20%.

ACCOMPLISHMENTS

High-Frequency Solid-State Ballasts

All gas-discharge lamps (e.g., fluorescent, mercury vapor, high- and low-pressure sodium, and metal halide) require ballasts to maintain stable electrical operation. The ballast provides the required starting voltage and limits the lamp current to a constant prescribed value. In normal operation, typical electromagnetic ballasts can undergo energy losses amounting to 25 to 35% of the overall lamp/ballast system. The advent of solid-state electronics provided the impetus for creating ballasts that experience much lower energy losses, and allowed lamps to be operated at high frequencies--in the 30,000-cycle range where the intrinsic efficiencies of the low-pressure lamps are 10 to 15% greater than 60 cycles. The Department of Energy, through LBL, worked with two small engineering firms and developed a cost-effective, high-frequency, solid-state ballast that is 20 to 25% more efficient than the typical electromagnetic ballast, raising the efficacy from 80 lumens/watt for a typical fluorescent lamp to 90 lumen/watt for the high-frequency ballast. Demonstrations conducted by LBL in occupied buildings measured savings achieved by the new ballasts and convinced the lighting community of their viability and effectiveness.

Energy-Efficient Replacements for the Incandescent Lamp

Nearly 90% of the energy applied to an incandescent filament lamp is dissipated as heat. An energy-efficient replacement for this kind of lamp is an acknowledged need and the focus of much research. Low cost, ease of use, and good color quality are all concerns that must be addressed in developing an efficient substitute.

Working with the lamp industry, DOE, through LBL, in 1979 began a program to accelerate development of a variety of energy-efficient replacements for incandescent lamps. One concept is a slimmed-down fluorescent (compact fluorescent) small enough to use in incandescent configurations. To satisfy individual preferences, the phosphor coating the lamp's inner wall can be tuned to provide a color indistinguishable from that provided by either an incandescent lamp or daylight.

Also considered as a possible replacement is an electrodeless fluorescent lamp in which the discharge is excited magnetically at very high frequencies (13.56 megacycles). Still another possibility is to cover the inside of the incandescent lamp with a very thin selective coating that transmits visible light but reflects infrared radiation. The radiation reflected back to the filament reduces its power requirement and increases efficacy. Another concept is a small metal halide lamp operated at high pressure, but with a very small arc length to provide the light output characteristic of an incandescent lamp rather than the high lumen output typical of metal halide lamps. These lamps have efficacies ranging from two to four times that of the incandescent lamp. Some can last longer than 10 years (10,000 hours), compared to less than one year (750 hours) for the incandescent. Although first costs may be higher, the longer lifetime and lower operating costs provide a lower life-cycle cost than that of an incandescent lamp.

Lighting Controls Demonstration at the World Trade Center

Many approaches have been considered for saving lighting energy in commercial buildings. Techniques such as having the lights turn off and on according to predetermined schedules, removing overhead bulbs in little-used areas, and taking advantage of available daylight have been discussed theoretically. To quantify the energy saved by introducing an energy-saving lighting management system in a real environment, LBL conducted a demonstration on the 58th floor of the World Trade Center in New York City. The tests were designed to determine:

1. which control functions have the greatest impact and why,

TABLE IV. INCANDESCENT LAMPS AND REPLACEMENTS

Lamp	Efficacy (Lumen/Watt)	Life (Hrs)	Light (Lumens)	Initial Cost ($)	Total Cost/ 10^6 Lumen-Hrs (No Labor)[a]	Cost of Conserved Energy (¢/kwh)(c)	Equivalent Electricity ($ Billions)
100 W	17.5	750	1,750	0.70[b]	3.96	-	10
100 W Long Life	14.9	2,500	1,490	0.83[b]	4.24	-	
40 W	11	1,100	440	0.70[b]	6.90	-	
20 W Circline Fl[t]	35	10,000	770	10.00[b]	3.14	0.79 (60 W)[c]	5
40 W Circline Fl	40	10,000	1,760	15.00[b]	2.35	1.01 (100 W)	4
Energy Button	7.3	37,000	440	2.70[b]	8.38	1.31 (40 W)[d]	-
Thermistor	16.1	2,000	1,610	2.70[b]	4.56	-	-
Heat Mirror Duro Test	30	2,500	1,700	5.00	3.18	2.07 (100 W)	6
Halarc-(General Electric)	40	5,000	2,200	15.00	2.86	1.75 (100 W)	4
Compact Fl (Westinghouse)	40	6,000	1,100	15.00[b]	3.77	2.81 (75 W)	4
Compact Fl (Norelco)	60	7,500	1,100	25.00[b] 15.00	4.03 2.81	4.21 (75 W) 1.87 (75 W)	3 3
Electrodeless Fl (General Electric, Litek)	55	10,000	1,750	15.00	1.94	0.83 (100 W)	3

U.S. Incandescent Lamps in 1982 used 180 BkWh, worth $10B; 180 BkWh is the output of 36 typical 1,000 MW plauts.

a - Energy cost $0.06/kWh
b - Product on market, retail price; other prices are market targets.
c - ( ) indicate wattage of incandescent from which conserved energy is calculated.
d - The CCE using a 40 W instead of a 100 W lamp with energy button.
t - Fl=Flourescent

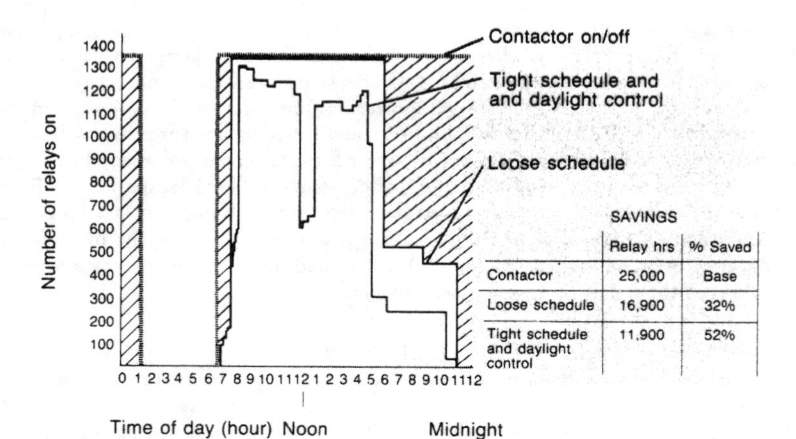

Fig. 1 Profile of the instantaneous power used during a 24-hour period at the World Trade Center before and after installation of the lighting control system. The major control strategies tested were scheduling according to occupancy and taking advantage of daylight.

2. what the economic tradeoffs are between control cost and savings,
3. how acceptable controls are to occupants, and
4. how reliable control systems are.

To provide the monitoring to evaluate a range of strategies, a relay was attached to every ballast on the 58th floor of the building. By selectively switching relays, it was possible to make each 6-lamp, 3-ballast fixture go to 1/3, 2/3, full "ON", or "OFF". The 1350 relays were operated by a computer-programmed system that allowed independent scheduling of each relay, monitoring of daylighting thresholds to provide a constant level of illumination, and manual override by occupants via their desk phones. All relay activity was stored on tape to record consumption in response to various strategies. Some of the results are shown in Fig. 1. The most significant conclusions of this demonstration were:

1. Lighting controls can have a significant and positive impact on energy consumption: a maximum reduction of 52% was achieved.
2. The payback on relay-based lighting control for new construction is extremely attractive.
3. Relay-based automatic lighting controls are acceptable to occupants, are very reliable, and provide significant energy savings, but continuous dimming is preferable.

The value of the tests was not only the 52% savings achieved, but also the insight gained into why certain strategies were successful, how strategies interact, the relationship of a strategy to the purpose of a space, occupants' responses to techniques, and the importance of providing individual overrides to ensure positive reactions to a system.

PROJECT AREAS AND CURRENT PROJECTS

The LBL lighting research program is divided into three major areas: Advanced Lamp Technology; Electromagnetic Compatibility of High-Frequency Lighting; and Impacts of New Lighting Technologies on Productivity and Health.

Advanced Lamp Technology

Today's fluorescent lamp has a luminous efficacy of approximately 80 lumens of light output per watt of electrical power input. Although this is more than four times as efficient as an incandescent lamp, greater efficacies are possible. White light can, theoretically, be produced at almost 350 lumens per watt. The technology program is working to reach an efficiency of 200 lumens per watt within the next few years.

Two loss mechanisms must be changed to allow for more efficient fluorescent lamps. First, we must reduce the self-absorption of the ultraviolet (UV) radiation within the lamp plasma before it strikes the phosphor-covered inner wall, and second, we must develop a more efficient phosphor that will convert one energetic UV photon into two visible photons. Reductions in self-absorption could provide a 30% improvement while a two-photon phosphor could double lamp efficacy.

LBL is studying two principal ways of reducing the UV self-absorption. The first method uses the naturally occurring isotopes of mercury. There are seven stable isotopes that differ according to their UV emission spectra. Because the nuclear volume of each isotope of mercury is slightly different, the interaction energy between the electrons and the nucleus will be slightly different. The resultant "isotope shift" shifts the radiations by about 10 GHz, or 5×10^{-5} eV. A small subgroup of isotopes will emit and absorb UV radiation with such a small spread in energy that a nearby subgroup of isotopes is not involved in its radiative transfer. The data of Fig. 2 show that it is then possible to make

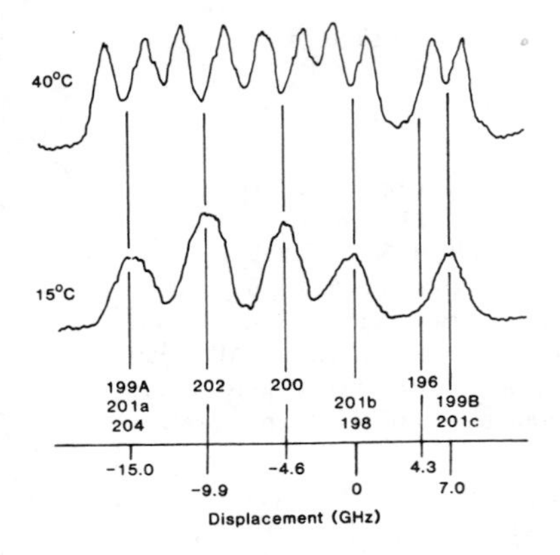

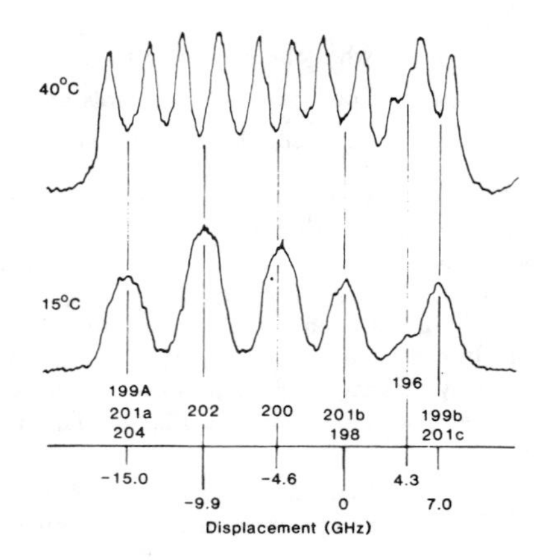

Fig. 2. (left) Fabrey-Perot-measured hyperfine structure of the 253.7 nm resonance line of mercury in a mercury-argon low-pressure discharge of the type occuring in a standard fluroescent lamp at 15° and 40°C. Note the dramatic change in line shape between 15° and 40° caused by the self-reversal phenomenon in resonance radiation. (right) Same spectra except Hg-196 was enriched to 3.9%. The efficacy of mercury-rare gas lamps can be increased by about 5% in this way. Source: J. Maya, M. Grossman, R. Lagushenko, and J. Waymouth, Science 226, 435 (1984).

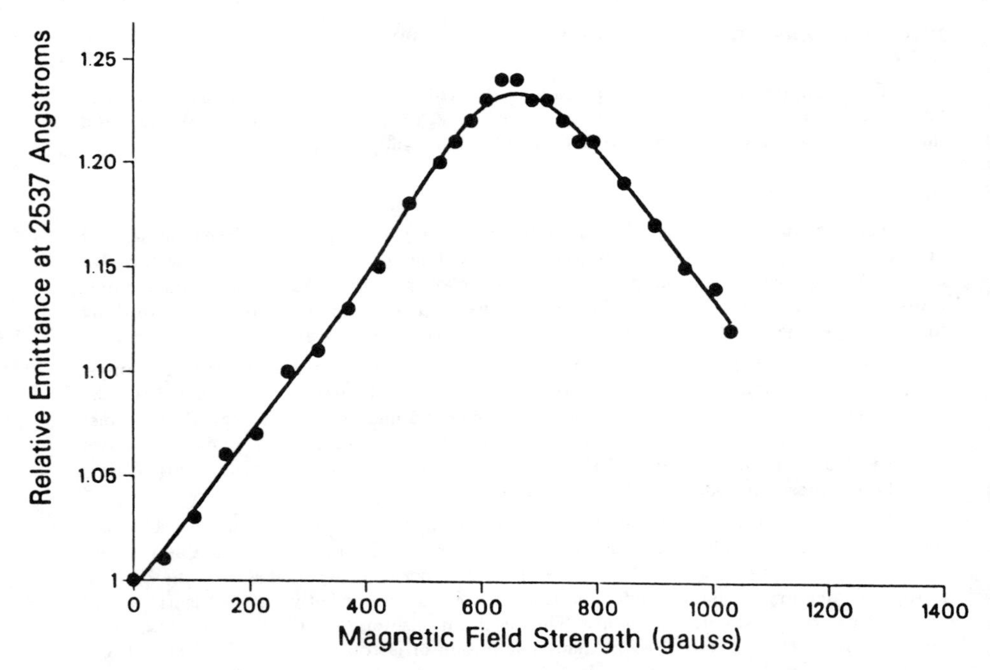

Fig. 3. Augmentation of the 2537 Å radiation from a mercury-argon fluorescent lamp as a function of applied mangetic field.

the plasma slightly optically thinner with a subsequent increase in lamp efficacy.

One possibility for isotope alteration, enrichment with mercury 196, is being pursued in a joint effort by LBL and GTE Lighting. Isotope alterations should increase efficacy by about 5% by increasing Hg^{196} from 0.14% to 2.5%. Another 5% might be gained by increasing the amount of Hg^{201}. If this proves economical, modified lamps could enter the market quickly, as lamps could simply be loaded with isotopically enriched rather than natural mercury, while other components would remain the same.

Another method for reducing UV self-absorption, recently discovered at LBL, involves an applied magnetic field having a direction parallel to the main current. Axial magnetic field strengths of about 600 gauss can increase light emission by 12% to 15% as shown in Fig. 3. LBL and major lamp industries are studying practical ways to apply this technique.

The second loss mechanism is that a lamp's phosphor converts each UV photon into at most one visible photon. Improving this conversion rate could increase the efficacy of low-pressure discharge lamps. Although the energetics are sufficient to permit the cascade conversion of a UV photon into two visible photons, this process must occur quickly and the intermediate level in the cascade must be tuned carefully to ensure that both emitted photons are in the visible spectrum. Studies on phosphor chemistry are ongoing at LBL to discover whether the two-photon phosphor is feasible. The lamp industry, long aware of this problem but not in a position to make these investigations, awaits the results.

Based on work at LBL, future fluorescent lamps should operate at high frequency and be isotopically enriched, magnetically loaded, and coated with a two-photon phosphor. Such lamps would have an efficacy of more than 200 lumens per watt, three times that of today's 60-cycle fluorescent. Other research concentrates on high-intensity discharge lamps, which can be made both more efficient and dimmable if operated without electrodes. High-frequency operation is required to excite the lamp plasma in an electrodeless mode. Operation at high frequency may permit lamps to function with just one or two metal halides and without any mercury or sodium. Electrodeless operation would also enable the use of compounds in the discharge that have desirable light output and color, but are not used today because they harm the electrodes. An electrodeless lamp, dimmable without a spectral change, would appeal to lighting designers. It would be an efficient, dimmable point source that could produce any color.

Electromagnetic Compatibility of High-Frequency Lighting

The successful development of high-frequency, solid-state ballasts provided the impetus for adopting high-frequency lighting and control systems in commercial and industrial buildings. Many benefits can be realized by operating at high frequencies: lower starting voltages, increased efficacy, ease of dimming, etc. The electronics industry is making rapid progress in developing less expensive high-power transistors and amorphous magnetic materials needed to ensure that high-frequency ballasts will be cost-effective and enter the market rapidly.

High-frequency lighting, however, can produce electromagnetic energy, which could affect sensitive machinery, computers, and electronic systems. The widespread use of these lighting systems requires studying possible effects of electromagnetic interference (EMI) on other building operations. LBL leads the effort to obtain information needed for industry and government agencies. The major technical problem is establishing reliable near-field EMI measurements for lighting systems and relating them to what the Federal Communications Commission requires for far-field radio interference. Also, a methodology must be developed to determine near-field effects when many EMI-generating systems are used.

The Federal Communications Commission and National Electrical Manufacturers Association have recommended that LBL assist in collecting data on the EMI generated by high-frequency lighting systems. The development at LBL of a standard source and measuring method for determining radiated near fields is essential to make convenient and reliable procedures available to industry. Computer models for stray EMI are being developed at LBL to further study the physical characteristics of external electromagnetic fields generated by high-frequency lighting.

Impacts of New Lighting Techniques on Productivity and Health

The idea that lighting might negatively affect health has become prevalent in the popular media in the past few years. Many factors have been implicated, but scientific data are lacking, especially to ascertain whether new energy-efficient technologies have adverse effects on human health and productivity.

Factors that may influence performance and productivity (as well as energy efficiency) may be associated with the lamp, the electronics and associated controls, the fixture, or the geometry and location of the lighting system. These lighting factors are: color variations; glare; intensity fluctuations; spectrum fields generated by the lamp, ballast, or controls; and flicker, all of which could evoke a variety of human responses (behavioral, psychophysical, physiological, or biochemical).

Our research aims to assure that new energy-efficient lighting technologies do not adversely affect human health and productivity. We are investigating whether any aspect of the new technologies can produce responses in humans. If so, the effects will be characterized and necessary changes to lighting technologies identified. Although subjective responses of workers provide some information, such responses are generally muddied by a mix of sociological factors and motivations; the investigations carried out by DOE/LBL use non-subjective responses to establish cause and effect and assure repeatability.

In this first phase of the program, lamps to be evaluated include incandescent, cool-white fluorescent, full-spectrum fluorescent, high-pressure sodium, and metal halide. Effects of exposure to various lighting conditions will be asssessed by monitoring autonomic responses. Parameters to be monitored include heart rate, respiration rate, galvanic skin response, muscle strength, exercise tolerance, facial expressions, and pupillary responses. Behavioral measures to be used include memory (Wechsler Memory Scale and Sternberg's Memory Scanning Time), cognitive function (Mental Arithmetic), time estimation, and simple reaction time. Other behavioral tasks probably will be implemented.

Data-gathering and subject control are supervised by trained medical personnel. A national technical advisory committee oversees and reviews the project. First results of this effort surround the effects of visible spectrum and low-frequency radiation on human muscle strength. Results, partially described below, indicate that subjective psychological factors are the likely cause of reported effects.

Ten subjects who stated they were bothered by fluorescent lighting were studied using maximal strength of shoulder forward flexion. Fifty trials, each a randomized sequence of paired exposures to 400-footcandle incandescent and cool-white fluorescent light, were measured in each subject to counter-balance effects of fatigue. Over all subjects the mean difference in maximal muscle strength for the two lighting conditions was less than 0.2% of the usual strength. The standard distribution was 13 times the mean difference. A paired-t test showed that our results were likely due to chance ($p = 0.85$). Similarly, there was no statistical difference in the responses under the two lighting conditions ($p = 0.39$). Power analysis showed that 90% of the time we would have detected a difference in muscle strength of 4%. Since the effects of suggestion on maximal muscle strength have been reported to be about 10%, the effects of the lighting, as tested, are likely to be obscured by psychological effects.

Besides direct effects of lamps on humans, LBL is examining how lamp characteristics--in particular compound flicker--can affect workers using visual display terminals. Compound flicker occurs in the workplace when there are two independent sources flickering at different rates that, when combined, produce a "beat". For example, light from a video display terminal whose refresh rate differs slightly from its nominal 60 Hz will combine with standard fluorescent lighting flickering at 120 Hz to form a low-frequency beat. Experiments are being performed to ascertain the effect of beats in general, and the beat between VDTs and ambient lighting in particular, on the visual system.

Present results show that beats formed by independently modulated fluorescent luminaires cause subjects to exhibit frequency-specific declines in temporal contrast sensitivity and entrained pupillary oscillations when the flicker rates of the luminaires are below critical fusion frequency (CFF). No evidence is found of these effects when the flicker rates are above CFF, or when a VDT is viewed under flickering ambient illumination. High-frequency operation of the ambient lighting would eliminate human response to those beats.

ACKNOWLEDGMENT

This work was supported by the Assistant Secretary for Conservation and Renewable Energy, Office of Building Energy Research and Development, Buildings Equipment Division of the U.S. Department of Energy under Contract No. DE-AC03-76SF00098.

CHAPTER 14

WINDOW PERFORMANCE AND BUILDING ENERGY USE:

Some Technical Options for Increasing Energy Efficiency

Stephen Selkowitz

Windows and Daylighting
Lawrence Berkeley Laboratory
University of California
Berkeley, CA 94720

ABSTRACT

Window system design and operation has a major impact on energy use in buildings as well as on occupants' thermal and visual comfort. Window performance will be a function of optical and thermal properties, window management strategies, climate and orientation, and building type and occupancy. In residences, heat loss control is a primary concern, followed by sun control in more southerly climates. In commercial buildings, the daylight provided by windows may be the major energy benefits but solar gain must be controlled so that increased cooling loads do not exceed daylighting savings. Reductions in peak electrical demand and HVAC system size may also be possible in well-designed daylighted buildings.

INTRODUCTION

Windows play many important roles in the design of buildings and strongly affect their energy use. In order to develop effective energy-conserving uses for windows, first we must carefully define the nature and magnitude of their energy problems. Our perspective on the problem, and the context in which it must be solved, will influence the solution. Personal and professional perspectives vary: an architect will bring different insights than an engineer or a scientist. There is a technical component to the problem: the relative importance of heat loss, heat gain, daylight admittance, etc. Finally, there are many non-technical or non-energy aspects such as view, comfort, appearance, health and well-being, and design aesthetics. "Saving energy" is not a problem, it is a solution. If it was the problem, an easy solution would be to close all buildings, padlock the doors, and turn off the furnaces and electrical equipment. But the real problem is *minimizing the consumption of non-renewable energy resources consistent with the functional objectives of buildings*. In houses these might include comfort and health; in offices they also include productivity.

In the United States, about 5% of total national energy consumption can be attributed to windows; this is approximately evenly split between windows in houses and windows in nonresidential buildings. By providing daylight, windows can also influence the 5% of the total national energy consumption attributable to electric lighting. Focusing on energy performance, we can define six primary factors:

1. Thermal transmission.
2. Light transmission.

0094-243X/85/1350258-12$3.00 Copyright 1985 American Institute of Physics

3. Control of solar heat gain.
4. Infiltration.
5. Ventilation.
6. Condensation.

In addition to these primary energy performance issues are a host of other critical performance issues that influence decisions regarding window design. These include sound transmission, water penetration, resistance to wind loads and operating forces, view, appearance, durability, fire-safety, security, and costs. Most decisions regarding window design must account for these latter factors as well as the energy performance issues.

From an energy perspective, window performance is distinctly different from that of insulated walls and roofs. Windows' net energy effect can balance thermal losses against useful winter solar gain and daylighting benefits. The response time to energy flows (solar gain and conductive gains or losses) is small compared to wall and roof elements. Windows are typically nonhomogeneous elements having joints and thermal bridges that influence performance. The influence of air films and air infiltrations is typically greater than for opaque building elements, and windows generally influence thermal comfort and satisfaction to a greater extent than do walls or roofs.

In northern European countries, where the winters are long and cold, and the summers rather short and mild, window performance may be equated more narrowly with control of heat loss. But in much of the United States and elsewhere, the dynamic interplays of conflicting thermal forces throughout the year are the critical factors that influence overall performance and annual energy consumption.

Our perspectives on window energy performance have changed with time. In the early 1970s, just after the initial increase in oil prices, windows were seen primarily as an energy cost. Ten years later, a different perspective is emerging. This perspective acknowledges that losses can be minimized and that the useful winter solar heat gain and benefits from daylighting can turn windows into a net benefit rather than a net cost. In fact, we make the following claim: high-performance, managed window systems carefully installed in a well-designed, energy-efficient building will provide net energy benefits for any orientation in most parts of the United States. This means that window systems will outperform the best insulated wall or roof element. The systems that meet this claim may not be cost-effective in the narrow sense of that term, but we believe that some technically promising solutions can meet those requirements.

In the residential context we look primarily at tradeoffs between gathering useful solar gain and the losses from conduction, convection, and infiltration that form the negative side of the heat balance equation. Figure 1 schematically compares the heat gain and loss characteristics of several different envelope elements using transmittance as an approximate measure of solar heat gain, and U-value as an approximate measure of heat loss. Our ideal components would lie in the upper-left quadrant of the figure, displaying low heat loss rates and relatively high potential heat gain. Not all the available heat gain is useful, of course, which is why the dynamic controls discussed earlier are important. However, Fig. 2 shows a quantitative map of net useful energy flux through a south-facing window in Madison, Wisconsin, as a function of shading coefficient and U-value. One can immediately see the specific combinations of window properties which result in either a net heat loss, a zero energy balance (the dark diagonal line), or a net energy benefit for that building module. The following sections examine existing and new options for controlling heat loss, managing solar gain, and utilizing daylight. The shading coefficient of

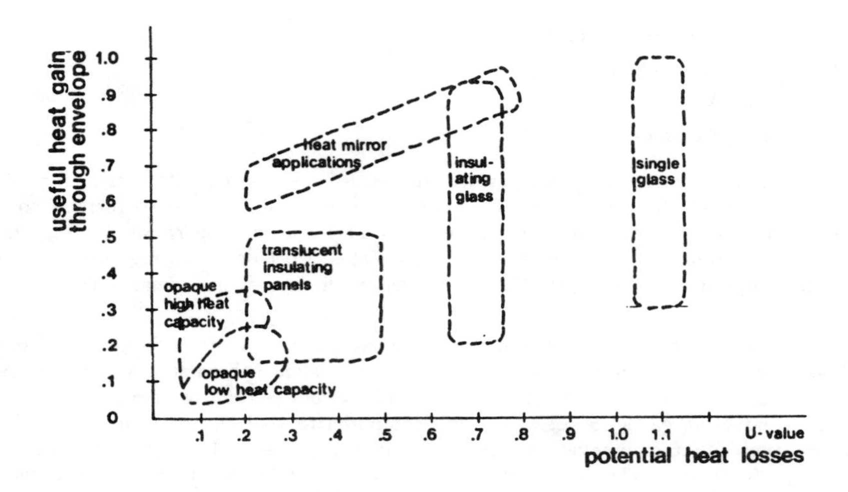

Figure 1. Solar gain and heat loss potentials of representative envelope elements. U values are in English units, BTU/ft^2-°F-hr. The maximum solar gain in winter is about 150-200 Btu/ft^2-hr at solar noon.

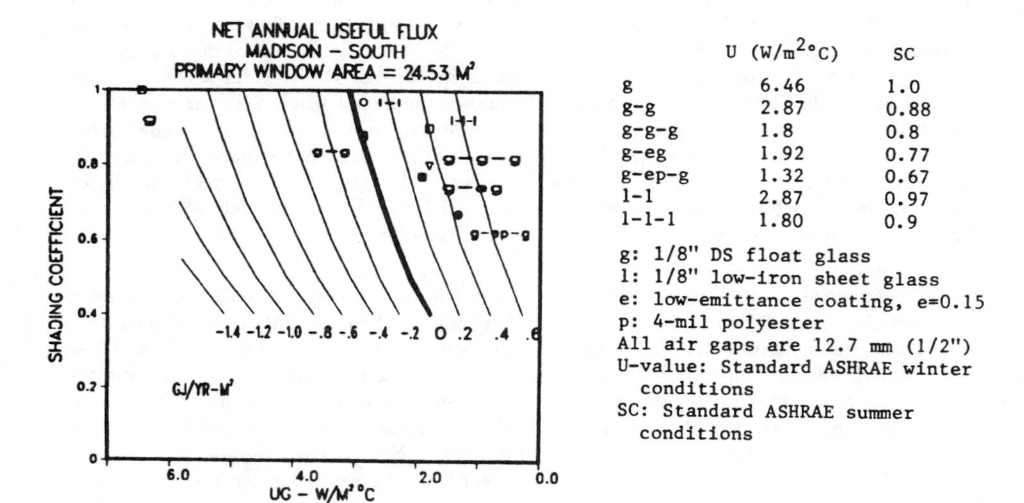

	U (W/m^2°C)	SC
g	6.46	1.0
g-g	2.87	0.88
g-g-g	1.8	0.8
g-eg	1.92	0.77
g-ep-g	1.32	0.67
l-l	2.87	0.97
l-l-l	1.80	0.9

g: 1/8" DS float glass
l: 1/8" low-iron sheet glass
e: low-emittance coating, e=0.15
p: 4-mil polyester
All air gaps are 12.7 mm (1/2")
U-value: Standard ASHRAE winter conditions
SC: Standard ASHRAE summer conditions

Figure 2. Net annual useful flux in Madison, Wisconsin, for a primary window area of 24.53 m^2 for an orientation due south. The performance of typical glazing systems is indicated for glazing properties shown above. ($U_{English} = 5.69 \times U_{metric}$).

a particular glazing system is the ratio of the solar heat gain through it compared to the solar heat gain through a single sheet of glass.

CONTROL OF HEAT LOSS

To better understand the performance of window systems that limit heat loss, we first must identify the heat loss mechanisms. Figure 3 illustrates the primary heat flows associated with double glazing and suggests the heat loss mechanisms that must be controlled to improve window performance. The primary mechanisms are: 1) radiation suppression, 2) convection suppression, 3) reduction of infiltration, and 4) movable insulating systems that completely cover the window.

The largest heat loss mechanism in a typical double-glazed window is radiative transfer. High-performance window systems can be made by introducing one or more low-emittance layers into multi-glazed systems. An ideal low-emittance coating will be transparent in the solar spectrum and highly reflective to the long wavelength infrared energy that is a component of thermal losses. After years of experimental development, these coatings have appeared as commercial products applied to both glass and plastic substrates. Many new window products incorporating these coatings are emerging on the market. The U-value of conventional single-, double-, and triple-glazed systems as a function of emissivity and coating placement is shown in Fig. 4. At present, since most low-emittance coatings are not durable, they are enclosed in air spaces of double- and triple-glazed systems. Figure 4 illustrates that a double-glazed system with a coating of emittance 0.1 to 0.2, placed on the number 2 or number 3 surface (counting from the outside), will perform as well as or better than a conventional triple-glazed window. The first primary market penetration of these systems has been in this application where equivalent or improved performance is obtained with a window system that is simpler, lighter, and has room for additional thermal improvement, such as by adding low-conductance gases to the windows.

Additional research is in progress on low-emittance coatings to further raise their solar transmittance, lower their emittance, and improve their overall durability, particularly for exposed applications. Modern vacuum deposition processes have produced relatively inexpensive coatings on glass and plastic substrates, but work continues to produce even cheaper coatings using both vacuum and non-vacuum processes. Combinations of low-emittance coated substrates and low-conductance gases in a quadruple-glazed window make it possible to obtain U-values between 0.5 and 1 W/m^2K, with a solar transmittance of 0.4 to 0.5.

Once a low-emittance coating is added to an air space, the use of a low-conductance gas such as argon, sulfur hexafluoride, or mixtures thereof will produce moderate additional reductions in heat loss rates. Many manufacturers have used such systems, although there appears to be some disagreement in the industry about how well these gases are retained in sealed-glass systems. One would expect future improvements in glass-sealing technology to further increase the opportunities for using low-conductance gases. Low-conductance spacer materials may also be required to reduce the edge losses that characterize sealed glass units with conventional aluminum spacers. In principle it is also possible to evacuate the air space completely and provide spacers to maintain glass-to-glass separation. This should result in U-values in the range of 0.5 W/m^2K as long as a low-emittance coating is provided. Providing a long-term hermetic seal that is cost-effective is still the primary stumbling block for such a system. To obtain the appropriate reduction in conductance, the air gap must be evacuated to a very hard vacuum, which places severe requirements on the seal's integrity and durability.

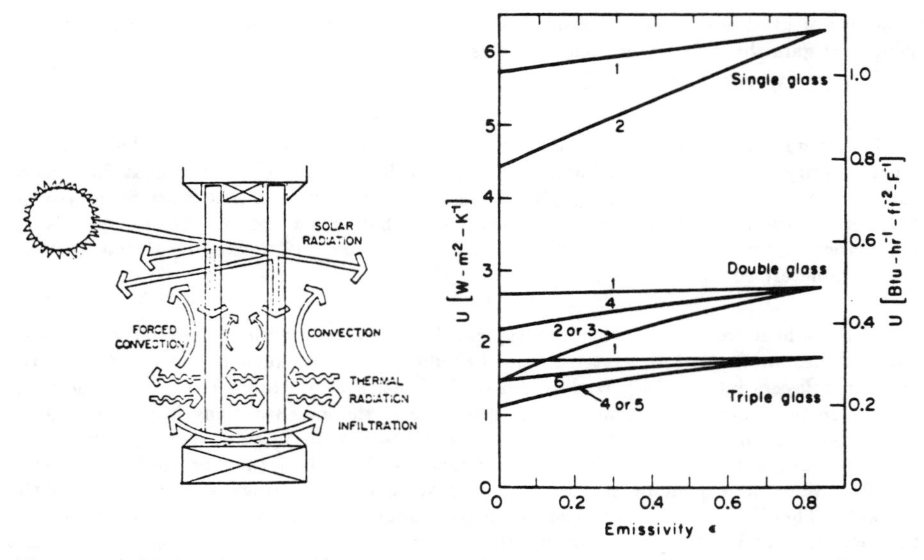

Figure 3. Major heat loss/heat gain mechanisms in windows.

Figure 4. U-value of simple, double and triple-glazing with a low emittance coating vs. coating emittance and location. Surfaces on which the low-E coating appears are given on the curves, beginning with outermost glass surface (#1), inner surface of outer glazing (#2), etc. Calculated values for standard ASHRAE winter conditions.

A variety of movable insulating devices has been developed in the last decade for use with windows. Some are designed to be deployed on the exterior of the window, others on the inside, and still others between glass. The insulating properties claimed for these systems range from negligible improvements to U-values below 0.5 W/m^2K. The performance of many of these systems has been the subject of some controversy since most involve moving elements and edge seals that may deteriorate with time and use. In addition, their thermal advantages are realized only when occupants choose to close the device, an operation that can prove unreliable. The best results occur when the device provides privacy or comfort as well as thermal control since consistent operation is more likely to occur. Operable systems can be motorized but normally at substantial additional cost. Traditional systems in the European market, such as rolling shutters, have been redesigned to improve thermal performance, and are beginning to appear in the American market as both insulating and shading devices. The most successful movable insulating devices in the American market have been the simpler devices mounted on the interior of the window, which are sold primarily for their aesthetic value but which now have improved insulating performance. Interior devices that are highly insulating and fit tightly to the window are more expensive, are less widely used, and can create problems. In winter they increase the risk of condensation and glass breakage due to thermal shock when the devices are first opened and the glass panes are very cold relative to the indoor air temperatures. During other times of the year, the heat buildup between the window and a closed insulating device can reach temperatures high enough to damage the window and/or the insulating system.

Due to the difficulties of producing highly insulating movable devices, there is increased effort to produce a highly insulating glazing material. The multilayer windows described previously achieved the desired insulating values but have themselves become relatively complex because of the additional layers involved. Another approach is to use a glazing material that is intrinsically insulating. Silica aerogel is a microporous material that has excellent insulating properties, good optical clarity, and relatively high solar transmittance. The material consists of a network of small silica particles whose size of 0.01 microns is much less than a wavelength of light, thereby reducing scattering effects. The fine pore structure of the material results in a U-value lower than that of air. Figure 5 shows the U-value of an aerogel-filled window as a function of thickness. Since at present the aerogel material is relatively fragile, it too must be protected in a hermetically sealed double-glazed unit. With further research it may be possible to produce hard surfaces on each face of the aerogel, thus simplifying this packaging requirement. Initial experiments also suggest that the aerogel window could be evacuated, resulting either in further improvements to conductivity or equivalent low conductivity with a much thinner window, and therefore a higher solar transmittance. A significant feature of an evacuated aerogel window is that the improvement in thermal properties is reached with only a modest vacuum, requiring a much simpler sealing technology than the evacuated window described earlier. Furthermore, the aerogel acts as its own transparent spacer and has sufficient structural strength to withstand the applied pressure.

CONTROL OF SOLAR GAIN

While heat loss has historically been the most important aspect of window energy use, the cooling loads resulting from uncontrolled solar gain are increasingly important. In the United States most commercial buildings, even in the northern part of the country, have central air-conditioning systems. During much of the year, heat gains from windows must be removed by these systems. Much of the new housing construction in the United States

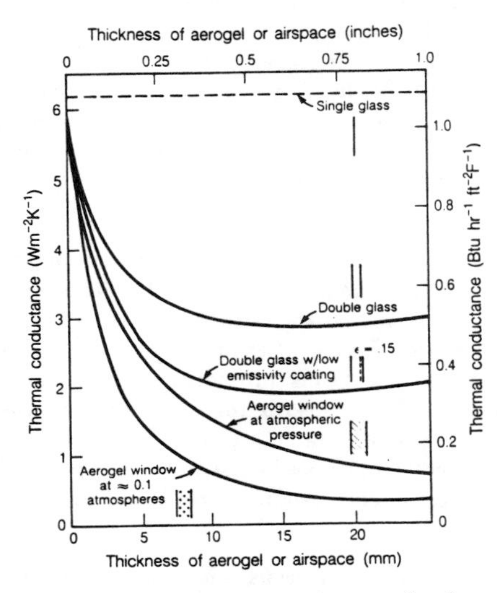

Figure 5. Conductance of aerogel window and conventional double glazing vs. airspace or aerogel thickness.

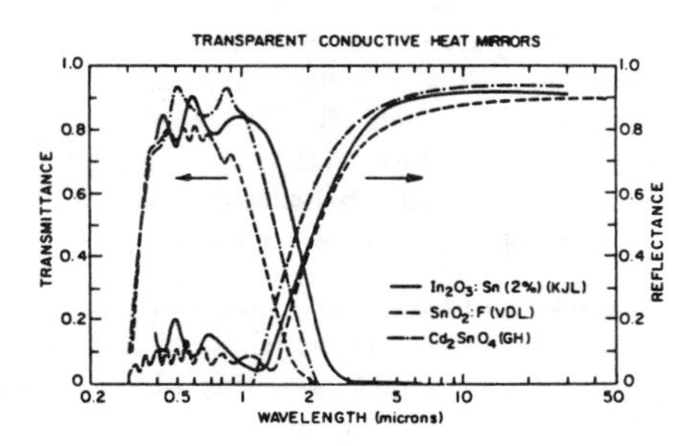

Figure 6. Spectral normal transmittance and reflectance of low-emittance coatings based on SnO_2:F, In_2O_3:Sn and Cd_27O_4, all on substrates. The solar transmittance (T_s) and infrared emittance ($E7_{IR}$) of these films are: SnO_2:F (T_s = 0.75, E_{IR} = 0.15); In_2O_3:Sn (T_s = 0.9, E_{IR} = 0.15); Cd_2SnO_4 (T_s = 0.86, E_{IR} = 0.12). Source: C. Lampert, LBL-13753.

is located in the "Sun Belt", the southern one-third of the country where cooling loads are much higher than heating loads. Almost all new construction in this region is air-conditioned. Even in the northern half of the United States, a surprisingly large percentage of new housing is built with central air-conditioning. Cooling loads from windows not only add to annual energy costs, but also add first cost to the building due to the cost of the cooling system. Furthermore, although cooling loads may be smaller than heating loads, because the cost of electricity is normally much higher than the cost of furnace fuel, the annual cost for cooling is frequently larger than for heating. In nonresidential buildings the cooling issues are even more important because of the relatively high internal heat loads from office machines, lighting, and people. A traditional solution to reduce cooling loads in office buildings is to use low-transmittance glass. This results in the sleek reflective building skins that have characterized design in the last 20 years. However, these solutions minimize not only cooling loads but also available daylight, requiring that electric lights be on whenever the building is occupied. It is possible to provide sun control and still admit daylight using a number of design approaches, which are discussed below.

Fixed exterior shading devices such as overhangs, fins, or various types of shade screen materials are often employed. These typically shade the window from sun penetration but allow some view of the sky so that daylight can be admitted. In addition they often break up and diffuse the incident solar beam so that diffused and attenuated direct sunlight is also introduced. However, because they are fixed, these solutions invariably represent a compromise between the requirements of sun control, daylight admittance, and glare control. In principle, operable sun control systems should provide better performance than fixed systems. Operable systems include a variety of interior window treatments such as shades, blinds, and drapes. Exterior systems include movable awnings, operable fins and louvers, shade systems, and exterior venetian blinds. Either type of system can be manually or automatically controlled. Automatic controls with manual overrides would appear preferable, ensuring that the systems function properly at all times. Exterior operable sun control systems have been used successfully in Europe for some time, but are only recently attracting attention in the United States. They are relatively costly compared to interior treatments or reflective glass, but since they may allow reductions in cooling system sizing as well as permitting daylight utilization, they may be economically beneficial. Operable shading systems should also provide improved thermal and visual comfort relative to most fixed shading solutions. While it is difficult to estimate the economic benefits of comfort directly, the cost of unhappy and uncomfortable office occupants is clearly large.

Window shading controls can also be located between glass. In the case of exhaust air or air-flow windows, the ventilation air from the room is exhausted between the panes of a glazing system over a venetian blind and either exhausted to the outdoors or returned to a heating and cooling system. In the winter, this provides an interior glass surface temperature that closely matches the room air temperature, thus providing good thermal comfort. In the summer, the blinds, if adjusted properly, absorb the sun's energy; the resultant heat is then carried off in the moving air stream. On sunny days in the winter, the blind acts as a solar air collector and the heat collected may be used in other parts of the building. These systems have been rather extensively used in Europe and are only now beginning to be introduced into the United States.

Window systems of the future may use optical switching materials and coatings to provide much of the solar control that now requires mechanical devices. Ideally, one would like to control the intensity of the transmitted radiation, its spectral content, and perhaps its spatial distribution in the room. Since the sun's spectrum is approximately 50% visible energy and 50% near infrared, one could reject more than half the total energy content while having only a minor effect on light transmission. Blue-green glass and some metallic coatings have spectral sensitivity, allowing higher transmittance in the visible portion of the spectrum. Future improvements in coatings should further increase spectral control. It should also be possible to use the same coating technology to apply interference coatings to produce any desirable color or tint. An ideal reflectance curve for such a coating is shown in Fig. 6. Since the visible properties and the overall solar properties of such coatings may be different, it is important to specify each separately.

Most of the shading systems described previously control the intensity of transmitted solar energy. It is possible to produce optical switching materials having transmittance properties that change from clear to reflective or absorptive as a function of exterior climate conditions such as sunlight intensity or temperature, or to use an electrical system that controls transmittance as a function of climate and building conditions. Examples of such materials are known to all of us: photochromic sunglasses switch with respect to light intensity; liquid crystal temperature indicators change optical properties in response to temperature changes, and many watch displays switch from transparent to reflective as each digit changes. However, it is difficult to scale these coating technologies up to window size and produce them at low cost in a form that will survive temperature extremes and solar exposure over many years. Research is in progress in a number of locations to produce such coatings. The most promising approach to providing active control of transmittance is based on electrochromic coatings. These multilayer coatings would be switched with a small applied current and could be continuously varied between high and low transmittance. Initial results in basic materials research look promising, but it will be some time before it is known whether this is a successful solution for building applications.

Another advanced coating application would be to produce window materials whose transmittance is a function of solar incidence angle. The coating might then perform like a series of fins or overhangs, rejecting light that arrives at greater than critical incident angles and admitting light otherwise. It may be possible to produce such effects using materials embedded within glazing substrates or with sputtered coatings or holographic films.

DAYLIGHT UTILIZATION

In the United States 30 to 50% of all energy use in nonresidential buildings is attributable to lighting. There are many ways of saving lighting energy, but once again these solutions must be consistent with maintaining or improving productivity. Despite the high cost of energy, the value of human productivity is many times greater. In the United States, with our energy costs and wages, the annual cost of providing lighting energy for a small office occupied by a single occupant is approximately equal to that worker's salary for a single hour. Thus, if a poor lighting design costs the employer even one hour's worth of productivity in return for large annual lighting savings, the employer has lost money overall.

We can say that all buildings with windows or skylights are daylighted, but no electrical energy is saved unless the lights are dimmed or turned off. A proper discussion of daylighting in buildings would consider its impacts on the following issues: electric lighting integration, energy savings, peak load impacts, HVAC systems impacts, lighting quality, view, and glare. As an architectural design element, daylighting influences the built environment at many different scales: urban planning, building form, envelope design, fenestration design, and interior design.

There are several dimensions to estimating the lighting energy savings in a daylighted building. The most important problems are to estimate the available daylight, to understand how the lighting control system responds to this available daylight, and then to estimate the overall energy impact including thermal effects. The cost implications of daylighting also should account for utility rate structures that may have additional costs for peak demand or may have time-of-day rates. We describe some insights on daylighting energy savings based on extensive use of a building simulation program, DOE-2.1B, which includes the effects of daylight in a typical office building. In these studies we determined that we can specify many daylighting effects using a new term, *effective aperture,* that includes the combined effects of window size and transmittance. The numerical value for effective aperture is simply the fraction of glass area in the wall (as a percent) times the visible transmittance of the glazing. Thus the value of effective aperture ranges from 0 to 1, although most practical values for typical wall facades and glazing types range between 0 and 0.4.

Figure 7 shows lighting energy consumption as a function of effective aperture for three cases: a nondaylighted office and two cases with daylighting, one with a dimming control, the second with on/off control. In both cases, the controls are set to provide 50 footcandles. The dimming system shows the best performance for small apertures, but eventually the two curves cross, indicating that the on/off system performs better than the dimming system. This is because the dimming feature requires a minimum power of 10% even with no light output, whereas the on/off system is turned off and consumes no power above its setpoint. Different selections of these specific control algorithms would, of course, change these results slightly.

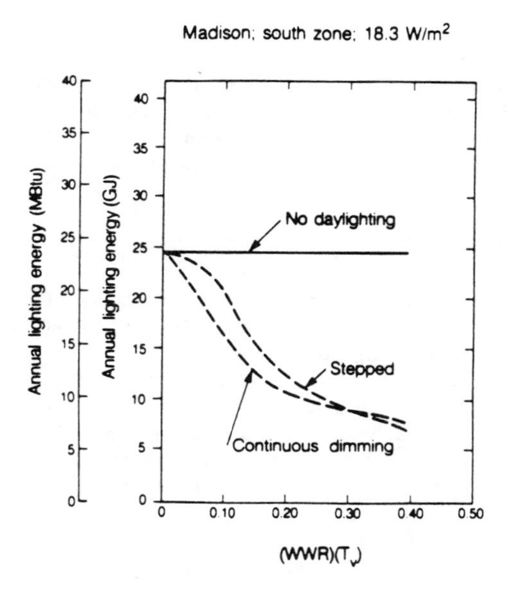

Figure 7. Annual lighting energy vs. effective aperture (window/wall ratio x visible transmittance) for a stepped switching and continuous dimming control in a south-facing office module in Madison.

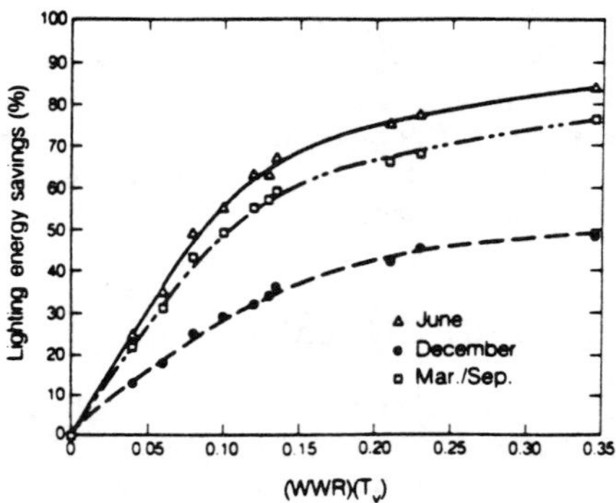

Figure 8. Seasonal variation in daylighting savings vs. effective aperture.

This annual perspective hides seasonal differences. Figure 8 shows lighting energy savings as a function of effective aperture for different seasons of the year. We see that the spring, summer, and fall curves are close together, but the winter curve shows significantly lower savings. The selection of interior illuminance level (and interior surface reflectance) as well as control type will also influence energy savings. The selection of illuminance level has a major impact at small aperture areas, decreasing in importance as window size increases.

Figure 9 examines the effect of daylighting and window management on annual electric consumption, peak electrical demand and chiller size for an office building in Madison. The demand curves show the effects of both daylighting and "window management" (i.e., use of shades or blinds) on peak demand and suggest that using daylighting combined with window management to control solar gain is the best overall strategy for controlling peak electrical demand. In the bottom curves, which show chiller size as a function of effective aperture, the daylighted case without window management (c) requires a larger chiller than the nondaylighted case with window management (b) for apertures larger than 0.2. Chiller size affects the sizing of the whole heating and cooling systems and can represent a major cost in commercial buildings. This again suggests the importance of window management to control solar gains and cooling loads.

These and other simulation studies suggest that there are large potential savings in daylighted buildings but that fenestration systems must be carefully designed and controlled to produce optimum performance. Fenestration that is too large or poorly controlled may increase energy use due to increased cooling loads that exceed the daylighting savings. This will be increasingly true as electric lighting systems become more efficient. Daylighting and window management should be of interest to building owners because they represent a potential to reduce HVAC equipment size in the building and save first costs for the owners. They should also be of interest to utilities because they may reduce peak electrical demand and thus reduce requirements for new generating capacity. In the United States a number of utilities have recognized these potential benefits and have instituted programs to accelerate the use of daylight in nonresidential buildings. Despite the potentially large savings, most estimates are based on simulation results and there are few measured data in buildings to validate these conclusions.

Successful use of daylight in buildings requires additional effort in the design of the building envelope and in the integration of the daylighting system with the electric lighting system. The lack of simple-to-use design methods that help one make critical design decisions accurately and cost-effectively throughout the design process limits current daylighting design. One approach for designing daylighting spaces is to use architectural scale models. This approach is based on the fact that a scale model will have the same illuminance levels and distribution as the real space if all the critical architectural details are faithfully reproduced. It is thus possible to use photocells to measure the illuminance distribution directly for even the most complex building design if it can be reduced to a scale model. One of the difficulties with this approach is the ever-changing sky conditions that make it difficult to compare results on succeeding days. A solution to this is to bring the sky indoors. Several types of sky simulators have been built to reproduce one or more sky conditions in a facility in which models can be tested under repeatable sky conditions. A large hemispherical sky simulator is shown in Fig. 10. This simulator, recently constructed at Lawrence Berkeley Laboratory, can produce most of the standard CIE overcast, clear, and uniform sky distributions as well as direct sunlight. A computerized data-acquisition system collects photometric information from the model in seconds and can assist with the data analysis. This simulator has been used by several design firms

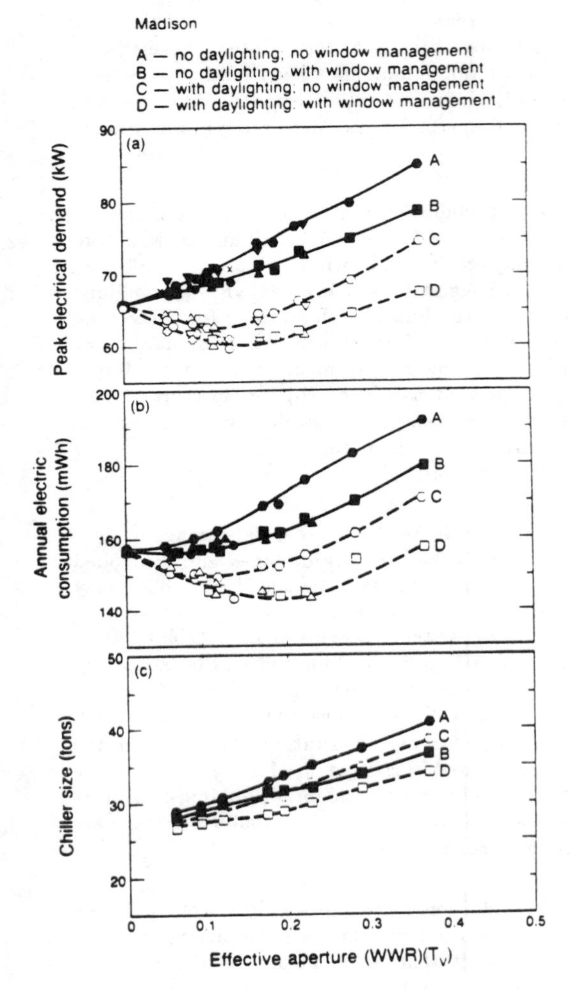

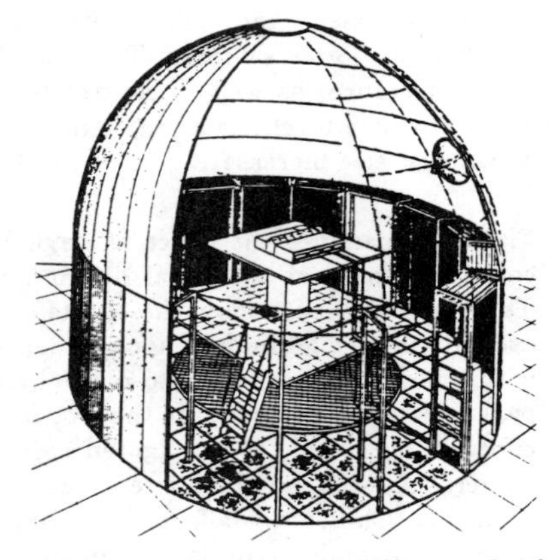

Figure 10. Schematic and photograph of LBL's hemispherical sky simulator.

Figure 9. Peak electric demand and chiller size vs. effective aperture for an office module in Madison, showing effects of window management and daylighting strategies.

for major new buildings in addition to being used as a research tool. Facilities of this type cannot readily be produced in many locations, but useful data can be collected from careful measurements made in somewhat simpler artificial skies. Analytical models for calculating daylight factors or illuminance levels have long been used in daylight predictions. A major change during the past 10 years has been the shift from simple graphic and computational techniques to computerized techniques.

Although computer programs allow more detailed and accurate calculations, their primary advantage will probably turn out to be their ability to present the resultant illuminance distribution data in a more graphic and understandable form than conventional numerical results. Computer users often get carried away with the apparent power and versatility of their models. In the future it will be important to be able to analyze luminance distribution and other aspects of lighting quality; these advanced computer models coupled with improved graphic output will provide new and more useful data to building designers. The computer also allows design solutions to be evaluated under a wide variety of conditions, which is important given that daylight changes by hour of the day and season.

In order to make detailed and accurate measurements of window system performance in a operating building, under realistic outdoor conditions, we have designed an outdoor test facility. Current performance estimates of annual energy consequences are based on laboratory tests of window characteristics, such as K-value and shading coefficient, coupled with computer models. To measure these effects directly under outdoor conditions would require the capabilities of the Mobile Window Thermal Test Facility, recently constructed at Lawrence Berkeley Laboratory, which is designed to meet most of these operating requirements. It consists of two side-by-side guarded calorimeter chambers, each of which will accept a window up to a size of about 4-m^2. The chambers will also accept skylights. In addition to the guard system to reduce losses through the non-glazed walls, we utilize novel large heat-flow sensors to determine the fate of solar radiation that enters each cell.

CONCLUSIONS

Window systems play a variety of roles in buildings, of which energy is only one. Since the primary purpose of most buildings is to provide habitation, or comfortable and productive workplaces, energy will never be more than one of the important factors that influence window design. It follows that no single simple window system will provide ideal performance under all conditions. The dynamic control of window properties, either through the fenestration materials themselves or with the addition of interior and exterior control devices, will generally provide the most versatile and effective control. In residences where heating is the primary energy factor, it should be possible to develop window systems that provide net energy benefits on even the north side of a building. In residences where cooling is the primary window design factor, it should be possible to combine static or operable shading systems to provide view and daylight while minimizing cooling load. In nonresidential buildings characterized by a large lighting load, window and skylight systems can dramatically lower lighting energy consumption. However, care must be taken to control solar gains so that increased cooling loads do not negate daylighting energy savings. The full benefits of a daylighted building would include not only energy savings but also the value of reducing peak electrical demand and the possible value of reducing the size and costs of cooling systems. New optical techniques are under development that could extend daylight utilization from the perimeter of the building to areas that are now out of reach. In most cases once a designer has used a series of tools to make design decisions, there is no feedback loop to assess how well these solutions worked. Performance data from monitored buildings and field test facilities are important to provide this quantitative and qualitative feedback. A list of publications on this subject can be obtained by writing the author.

ACKNOWLEDGEMENT

This work was supported by the Assistant Secretary for Conservation and Renewable Energy, Office of Building Energy Research and Development, Building Systems Division of the U.S. Department of Energy under Contract No. DE-ACO3-76SF00098.

CHAPTER 15

PROGRESS IN THE ENERGY EFFICIENCY OF RESIDENTIAL APPLIANCES AND SPACE CONDITIONING EQUIPMENT

Howard S. Geller
American Council for an Energy-Efficient Economy
Washington, DC 20036

ABSTRACT

This paper addresses a number of questions regarding the energy performance of residential appliances, water heaters and space conditioning equipment including the changes in the efficiency and energy consumption of new models during the past decade, the technologies used to achieve higher levels of energy efficiency, and the potential for further technical improvements. It is shown that there have been major improvements in the energy efficiency of all product types. For some products (eg., refrigerators), the average efficiency of new models has significantly risen during the past decade. For other products (eg., water heaters), major savings are possible through the purchase and use of the more-efficient models now available. A large assortment of technologies have been utilized to achieve greater efficiencies and further improvements are under development in many areas.

INTRODUCTION

Appliances, air conditioners and water heaters account for about 50% of residential energy consumption in the U.S. on a primary basis. The other 50% of residential energy consumption is used for space heating. In housing with a high level of thermal integrity, appliances present a much larger fraction of the total energy demand.

Table I shows the available data on the average efficiency of new appliances, water heaters and space conditioning equipment (hereafter referred to as appliances) produced since 1972. The efficiencies in Table I are based on the standardized test procedures developed by the U.S. Department of Energy (DOE) and the National Bureau of Standards during the 1970s. The averages are weighted according to manufacturers' shipments.

It is seen that there has been a mixed record of efficiency improvement; for some products there have been substantial gains during certain periods while for other products the documented progress has been limited. This is a consequence of a complicated set of factors including technological advances, the availability, promotion and acceptance of more efficient models, the nature of purchase decisions, and the regulatory and incentive

0094-243X/85/1350270-29$3.00 Copyright 1985 American Institute of Physics

Table I. Trends in the efficiency of new products

Product	Efficiency Parameter	1972	Efficiency (a) 1978	1980	1982	1983
Gas furnace	% seasonal efficiency (b)	63.2(e)	63.6	63.3(f)	--	69.6
Gas water heater	% overall efficiency (b)	47.4	48.2	47.9(f)	--	--
Electric water heater	% overall efficiency (b)	79.8	80.7	78.3(f)	--	--
Central air conditioner	SEER (c)	6.66	6.99	7.60	8.31	8.43
Room air conditioner	EER (c)	6.22	6.75	7.02	7.14	7.29
Refrigerator/ freezer	energy factor (d)	3.84	4.96	5.59	6.12	6.39
Freezer	energy factor (d)	7.29	9.92	10.85	11.28	11.36

(a) Average efficiencies are weighted by manufacturers' shipments.

(b) The seasonal efficiency for gas furnaces is the AFUE value and the overall efficiency for water heaters is the service efficiency as specified by the U.S. DOE test procedure.

(c) EER is the energy efficiency ratio in terms of BTU/hr of cooling output divided by watts of electrical power input. The SEER for central air conditioners is a seasonal energy efficiency ratio.

(d) Energy factor is the corrected volume divided by daily electricity consumption where corrected volume is the refrigerated space plus 1.63 times the freezer space for refrigerator/freezers and 1.73 times the freezer space for freezers.

(e) 1975 rather than 1972.

(f) These values are estimates made by manufacturers in 1979.

Source: American Council for an Energy-Efficient Economy based on information from the appliance manufacturers' associations, 1985.

programs used to stimulate the adoption of efficient appliance models.

New gas furnaces showed no gain in average efficiency during the 1970s. The average seasonal efficiency rating increased from about 63% in 1980 to nearly 70% in 1983. This improvement was due to the introduction and sales of furnaces with induced or forced draft and the condensation of flue gases beginning in 1981-82. The rating procedures and the technologies used to increase energy efficiency are discussed further in the next section.

New residential water heaters showed very little change in average efficiency throughout the 1970s. Unfortunately, more recent data on new product efficiency is not available. Although some highly efficient water heaters were developed and commercialized during the past decade, it is expected that the average efficiency of new water heaters has remained relatively constant.

For residential air conditioners, there were gradual improvements in the average efficiency of new models over the past 12 years. By and large, these improvements were achieved through the use of larger condenser and evaporator coils, more efficient motors and compressors, and improved controls.

The refrigerator and freezer efficiency ratings shown in Table I are expressed in terms of refrigerated volume per unit of electricity consumption. In these terms, the average efficiency of new refrigerators increased 66% and that of freezers 56% during this twelve-year period. In terms of electricity consumption, the typical domestic refrigerator produced in 1983 consumed 1160 KWh/yr compared to 1725 KWh/yr in 1972. Freezers on the average were down from 1460 KWh/yr in 1972 to 815 KWh/yr in 1983. These improvements were a consequence of a variety of design changes including use of better insulation and more efficient motors and compressors.

Table II lists the efficiencies of the top-rated models available as of 1984-85 and the typical models sold in 1978 and 1983. It is seen that there are major energy savings opportunities in all areas if commercially available, highly efficient appliances are purchased and used.

FUEL-FIRED SPACE HEATING

The 1982 residential energy survey showed that natural gas is the main heating fuel in about 48% of all households in the U.S. and fuel oil dominates in another 11% of households [1]. Gas-fired warm air furnaces are the most common type of heating system. In 1983, about 1.7 million gas furnaces were manufactured in the U.S.

Furnaces and boilers are rated according to their seasonal "first-law" efficiency. This rating is known as

Table II. Comparison of Typical and Highly Efficient Products

Product	Efficiency Parameter	Efficiency of typical model sold in 1978	Efficiency of typical model sold in 1983	Efficiency of top-rated model sold in 1984/85
Gas furnace	seasonal efficiency	0.64	0.70	0.94-0.97
Gas water heater	overall efficiency	0.48	0.50 (a)	0.64
Electric heat pump	heating COP	1.7 (a)	1.9 (a)	2.6
Electric water heater	overall COP	0.81	0.83 (a)	2.2
Central air conditioner	SEER	7.0	8.4	15.0
Room air conditioner	EER	6.8	7.3	12.0
Top mount R/F with auto. def.	Energy factor	4.8	6.5	9.8
Side-by-side R/F with auto. def.	Energy factor	5.0	6.1	7.6
Chest freezer with manual def.	Energy factor	11.7	11.9	22.1
Upright freezer with manual def.	Energy factor	9.3	11.4	16.5

(a) Estimates.

Source: American Council for an Energy-Efficient Economy, 1985.

the annual fuel utilization efficiency (AFUE). The procedure for determining the AFUE rating includes laboratory measurements during "warm-up", steady-state operation, "cool-down" and the off-period [2,3]. Equation (1) is the general formula for determining AFUE.

$$AFUE = 1.0 - [L(l)+L(s,on)+L(s,off)+L(i,on)+L(i,off)] \quad (1)$$

where L(l) is the latent heat loss, L(s,on) is the sensible heat loss during the on-cycle, L(s,off) is the sensible heat loss during off-periods, L(i,on) is the infiltration loss during the on-cycle and L(i,off) is the infiltration loss during off-periods. Latent losses are due to any uncondensed water vapor in the flue gas. Sensible heat losses are proportional to the temperature difference between the flue gases and indoor air. Infiltration losses represent the heating of combustion air from the outdoor to indoor temperature. Other sources provide the equations that are used to calculate the different loss factors [2,3]. It should be noted that the determination of AFUE does not account for any heat loss in the hot air or hot water distribution system.

Conventional indoor gas furnaces of the early 1970s featured a standing pilot rated at about 840 kJ/hr (800 Btu/hr), natural draft and atmospheric combustion, no stack damper, flue gas temperatures of around 315 $^{\circ}$C (600 $^{\circ}$F), a steady-state efficiency of around 76%, and AFUE ratings of 60-65% [4,5]. The major energy losses for this type of furnace include latent heat loss, sensible flue losses during the on-cycle, and off-cycle flue losses. In addition, older gas furnaces were often oversized, leading to further reductions in efficiency during actual use [4].

The common techniques used to increase the efficiency of gas furnaces and their estimated impacts on seasonal efficiency are shown in Table III. Use of an electrical intermittent ignition device (IID) eliminates the pilot loss during off-periods. Lower flue gas temperatures are achieved by reducing the amount of excess combustion air and increasing the heat exchanger area. A stack damper, vent damper or a "power burner" reduces losses during off-periods. Stack dampers are either thermally or electrically activated. A power burner consists of a blower to force the combustion gases through the furnace rather than relying on natural draft. Use of a blower can reduce the amount of excess air and lower sensible heat losses. In some cases, furnaces with induced draft are also equipped with an additional recuperative heat exchanger to provide further heat recovery from the flue gases [6].

Beginning in 1982, highly efficient condensing-type furnaces became commercially available. These furnaces

Table III. Major Options for Increasing Gas Furnace Efficiency

Design Option	Seasonal efficiency (% AFUE)
Conventional	60-65
IID	66-69
IID, lower flue gas temp.	70-72
IID, lower flue gas temp., stack damper	75-79
IID, lower flue gas temp., power burner	81-87
IID, condensing flue gases,	91-97

Sources: Refs. 4, 5 and 6.

use a power or pulse combustion burner and additional heat exchange area in order to condense and cool the flue gases to as low as 38 °C (100°F). A secondary or tertiary heat exchanger made of stainless steel is commonly used due to the corrosive condensate. Since the flue gases are relatively cool, they can be vented through a wall using plastic pipe. The design of the first condensing furnace that was marketed, the Lennox "Pulse", is illustrated in Fig. 1.

A field test was carried out in 1980-81 with pulsed combustion, condensing furnaces installed in parallel with conventional gas furnaces in 10 homes [7]. The different furnaces were alternated on a weekly basis over the course of one heating season. The condensing furnaces provided a 28% fuel savings on the average, approximately the savings expected based on relative AFUE ratings. The American Gas Association estimates that a typical pulse combustion furnace consumed about 57 GJ/yr (54 MBtu/yr) in 1982, 29% less than the average energy consumption of gas furnaces with a pilot [8].

Gas-fired boilers have characteristics that are similar to gas furnaces. Condensing-type gas boilers are now commercially available with AFUE ratings as high as

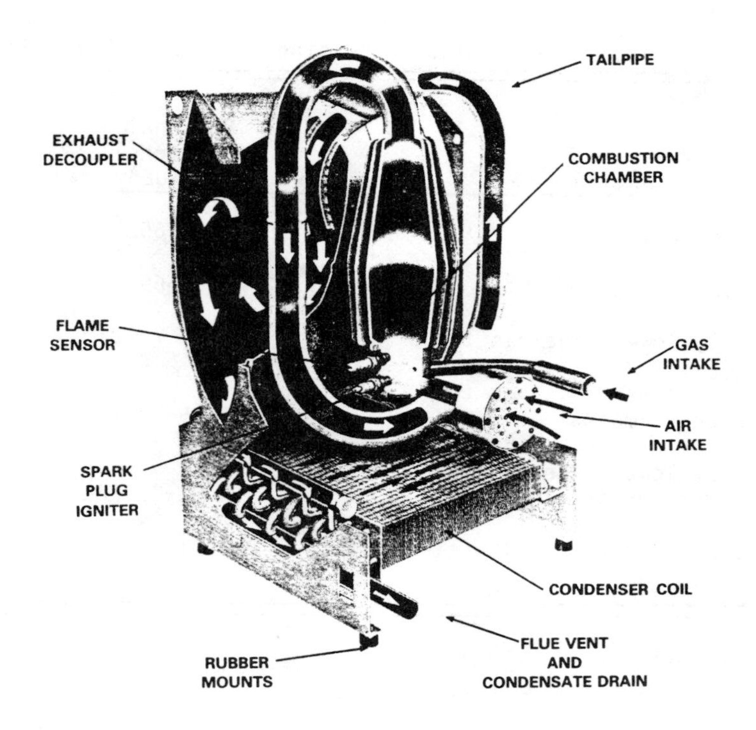

Figure 1. Design of the pulsed combustion condensing gas furnace

92%. The efficiency of condensing boilers is slightly less than that of condensing furnaces because the return water temperature in hydronic heating systems is greater than the return air temperature in warm air systems.

Gas-fired space heaters (also known as wall furnaces) are becoming increasingly popular especially in well-insulated, tight homes. Efforts to improve the efficiency of gas space heaters include the use of an IID, power burner and sealed combustion (ie., drawing combustion air from outdoors). Also, a prototype pulse combustion, condensing space heater has been developed and successfully field tested. It has a steady-state efficiency rating of 92% compared to 70-80% for conventional space heaters [9].

Oil-fired heating sytems are typically more efficient than gas-fired systems because they do not use a standing pilot and because latent heat losses are inherently lower with fuel oil than with natural gas. (Fuel oil has a hydrogen-to-carbon ratio of 1.8 compared to a ratio of 4 for natural gas.) As of 1984-85, the most efficient oil furnaces and boilers manufactured in the U.S. have AFUE ratings of 85-88%. These units include flame retention-head burners, vent dampers and recuperative heat exchangers [6].

Condensing oil-fired furnaces and boilers have not been introduced in the U.S. as of early 1985 in part because the sulfur in fuel oil causes more severe corrosion problems [6]. Also, there is less savings potential from flue gas condensation with oil-fired equipment due to the lower hydrogen content. Nevertheless, condensing oil-fired space heating systems are produced in Europe and Canada [10] and should eventually become available in the U.S.

Further improvements in the efficiency of fuel-fired heating systems may be possible through the development of heat-activated heat pumps. Both absorption cycle and heat engine-driven heat pumps are under development for residential applications. Absorption cycle heat pumps have a targeted steady-state heating COP of 1.25 and are at a relatively advanced stage of development [11]. Stirling engine-driven heat pumps have a targeted heating COP of about 1.6, but they are not likely to be commercialized before 1990 [12].

WATER HEATING

Domestic water heating consumes about 2.4 EJ (2.3 Quads) of energy on a primary basis in the U.S. This end-use accounts for nearly 15% of energy consumed in residences. According to the 1982 national residential energy survey, natural gas is the main water heating fuel in 47% of all households and electricity is the main fuel in 27% of all households. Analysis of 1978-79 survey data shows that on the average, a typical gas water heater consumes 26 GJ/yr (25 MBtu/yr) and a typical electric water heater 3900 KWh/yr [13].

The amount of energy required for heating water in a particular household depends on the amount of hot water consumed. Survey data shows that the average household in the U.S. (2.7 persons) consumes about 150 l/day (40 gal/day) of hot water [10,14]. For a four-person family, surveys during the 1970s showed a typical hot water demand of 243 l/day (64 gal/day).

Standard storage-type water heaters are rated according to their overall service efficiency, which is also known as the "energy factor" rating. The rating is based on measurements of the efficiency for water heating during heater operation (known as the recovery efficiency) and heat losses while hot water is stored (known as standby losses). The energy factor rating is also based on a hot water demand of 243 l/day.

Equation (2) is the general formula for calculating the energy factor or service efficiency of a water heater.

$$EF = H_W E_r / (H_W + S + E_r P) \qquad (2)$$

where H_W is the useful output in J/day (ie., the energy required to raise the incoming water to the desired temperature), E_r is the recovery efficiency, S is the standby loss from the tank in J/day, and P is pilot flue during off-periods also in J/day for fuel-fired water heaters with a pilot.

The recovery efficiency, given by Equation (3), is the fraction of the heat input that is used for sensible heating when a tank of cold water is heated to approximately 60°C (140°F).

$$E_r = V \rho C_p (T_f - T_i) / Q_r \qquad (3)$$

where V is the tank volume (l), ρ is the density of water (1.0 kg/l), C_p is the specific heat of water (4.18 J/kg C), T_i and T_f are the stored water temperatures at the start and end of the test respectively, and Q_r is the total energy input during the recovery test (J). The temperature difference during the test is 50°C (90°F).

Standby losses can be expressed as a combination of jacket losses and losses from the distribution pipe (convective heat transfer from the tank to the distribution pipe can occur during off-periods). Equation (4) gives the standby losses on this basis.

$$S = UA(T_W - T_a) 24 + DPL \qquad (4)$$

where U is the overall heat transfer coefficient for the tank in J/m2hr°C, A is the surface area of the tank in m^2, $T_W - T_a$ is the temperature difference between hot water and ambient air around the tank, and DPL is the distribution pipe loss in J/day.

<u>Gas water heaters</u>

Gas water heaters sold during the 1970s typically featured 1.9 cm (0.75 in) of low-density fiberglass insulation, a standing pilot consuming about 790 kJ/hr (750 Btu/hr), and a recovery efficiency of 0.75 [4, 12]. Assuming a 150 l (40 gal) tank with a surface area of 1.62 m^2 (17.4 ft^2), a 39°C (70°F) temperature difference between the tank and ambient, and a distribution pipe loss typical for bare copper pipe of 3.0 MJ/day (2.86 kBtu/day), then the standby loss as given by Eq. (4)

equals 18.9 MJ/day (18.0 kBtu/day). Sources claim that about 80% of the heat from the pilot is lost up the flue during off-periods, giving a pilot loss of 14.0 MJ/day (13.3 kBtu/day) [5].

Assuming a hot water demand of 243 l/day with a 50 °C rise in water temperature, the useful heat output equals 50.8 MJ/day (48.4 kBtu/day). Thus, from Eq. (2), the overall service efficiency in this case is 48%. Fig. 2 shows the energy balance for the gas water heater based on this example. It is seen that flue, jacket and pilot losses all account for substantial fractions of the energy input.

A number of steps have been taken to increase the efficiency of gas-fired water heaters. The best stand-alone models now commercially available, with energy factor ratings of 61-64%, include about 5.0 cm (2.0 in) of polyurethane foam insulation. This provides about six times more thermal resistance than the 1.9 cm (0.75 in) of fiberglass typically used in the past. Upgrading the tank insulation in this manner leads to about a 16% savings in gas consumption with other features held constant. (The savings are greater than the total jacket loss fraction shown in Fig. 2 because there are indirect savings through reduced burner flue losses as well as direct savings).

Standing pilots are still routinely used in gas-fired water heaters. Reducing the pilot rate from 790 kJ/hr (750 Btu/hr) to 400 kJ/hr (380 Btu/hr) would lead to about a 6% reduction in overall gas consumption. The pilot rate cannot be reduced further due to the instability of the small pilot flame [15].

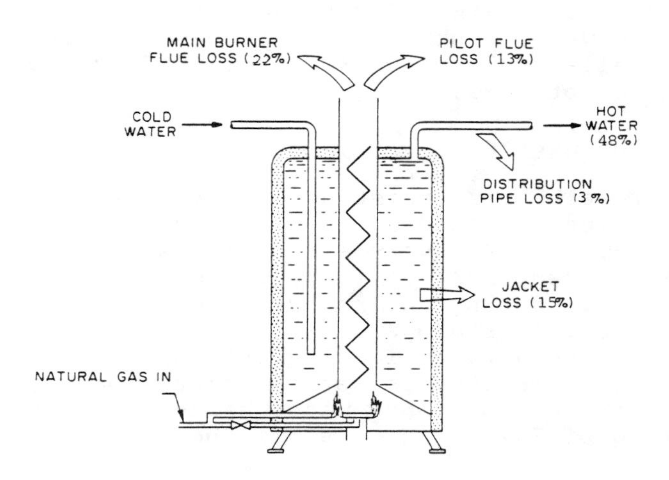

Figure 2. Energy flows in a standard gas water heater

Some prototype gas-fired water heaters include electrical ignition thereby eliminating pilot losses entirely. If electrical ignition is added to a fuel-fired water heater, it is essential to install a flue or stack damper at the same time. Othewise, there will be substantial heat loss through the internal flue during off-periods [16]. Gas water heaters with electrical ignition are not commercially available due to institutional barriers inhibiting the use of electrical connections and dampers.

Other simple techniques for increasing the overall efficiency of water heaters include installing "heat traps" and pipe insulation to reduce distribution pipe losses and lowering the thermostat setting. Heat traps prevent convective heat flow into the distribution line when hot water is not being used. Studies have shown that a 10 °C (18 °F) reduction in the thermostat setting of a standard water heater leads to 8-13% less energy consumption [15,16]. In households with dishwashers, the minimum hot water temperature is usually limited by the requirements of this appliance. Some dishwasher models include a booster heater so that the temperature of the water tank can be further reduced.

A highly efficient prototype gas-fired water heater with pulse combustion and flue gas condensation has been developed. This unit has electric ignition and a relatively high level of tank insulation. Tests on the prototype unit show a recovery efficiency of 90.3% and an energy factor rating of 83% [9].

Another means for economizing on energy use for fuel-fired water heating is to indirectly fire the water heater via a high-efficiency condensing furnace. Some condensing furnace models are marketed with an optional heat exchanger and water tank for this purpose. Using a relatively well-insulated tank, service efficiencies of 80-85% can be obtained.

Electric water heaters

Standard electric resistance water heaters produced during the 1970s included 5.0 cm (2.0 in) of fiberglass insulation [5,15]. Recovery efficiencies are typically slightly less than 100% due to to the small amount of heat loss that occurs during the recovery period [16]. Based on a 189 l (50 gal) storage tank, a 243 l/day hot water demand, a 99% recovery efficiency and the other assumptions used in the gas water heater example, a standard electric resistance water heater has an energy factor rating of 79%. Fig. 3 shows the energy balance in this case.

Electric resistance water heaters are currently available with energy factor ratings of 95% or greater.

More efficient models include thermal traps and at least 5 cm of foam insulation. Use of a highly efficient resistance water heater can reduce electricity consumption by 15-20% compared to consumption with a standard model.

Much greater reductions in electricity consumption are possible by shifting to a heat pump water heater (HPWH). HPWHs remove heat from the ambient air surrounding the heat pump and deliver it to the water in the tank. They are available either incorporated into a water tank or as a separate unit for retrofit to an ordinary water heater. Based on monitoring in homes in a number of locations, HPWHs typically provide a 50% reduction in electricity consumption compared to resistance water heaters [17].

HPWHs are tested and rated on the basis of their overall service COP. The test procedure is similar to that for ordinary water heaters, including measurements of the recovery efficiency starting with a tank of cold water and standby losses (for retrofit HPWHs, a specific standby loss rate is assumed). Commercially available models have service COP ratings of 2.0-3.4 [18]. Tests with two models show that their recovery efficiencies are typically 30% greater than the overall COP ratings as a result of including standby losses in determining the latter [19]. Thus, the estimated recovery COPs for current HPWH models are 2.6-4.4.

It is instructive to compare the actual performance to the maximum efficiency theoretically possible. Equation (5) gives the ideal (Carnot) efficiency of a heat pump in the heating mode.

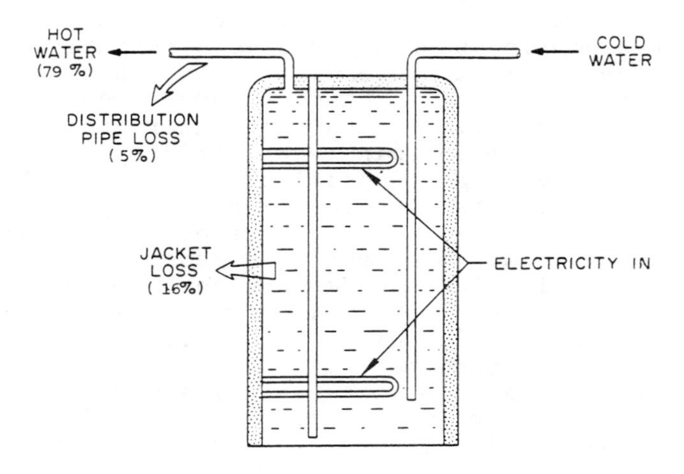

Figure 3. Energy flows in a standard electric resistance water heater

$$COP = \frac{1}{(1 - T_a/T_o)} \quad (5)$$

where T_a is the temperature of the medium from which heat is extracted and T_o is the temperature at which heat is provided expressed in degrees Kelvin. As the temperature difference between the heat "source" and sink" declines, the maximum COP increases.

In the standardized efficiency tests, the ambient air temperature is typically 19 °C (292 °K) and the mean water temperature is about 35 °C (308 °K). Based on these average temperatures, the Carnot COP is about 19. Thus, the estimated recovery COPs of 2.6-4.4 for actual models are just 14-23% of the theoretical potential. Maintaining a tank of water at 55 °C (131 °F) has a Carnot COP of 8-10, which is still much greater than actual COPs. The factors contributing to lower equipment efficiencies include motor and pump losses and the size limitations on heat exchangers.

One concern with HPWHs is that they remove heat from the surrounding air thereby increasing space conditioning requirements during the heating season. Assuming 150 l/day of hot water demand and an overall COP of 2.0, nearly 18 MJ/day (17 kBtu/day) of air cooling results (excluding the standby heat loss since it also occurs with a resistance water heater). The net impact this has on space conditioning depends on the climate and the location of the HPWH in the house [20]. If, for example, the space heating season is 150 days long and two-thirds of the air cooling has to be made up by the space heating system, then the added space heating requirement would be 1.8 GJ/yr (1.7 MBtu/yr).

In part to avoid "stealing" heat from interior space during the heating season, HPWHs can operate off of the ventilation air in a house or building where mechanical ventilation is used. It is most advantageous to remove heat from exhaust air during the winter (to provide heat recovery) and from incoming ventilation air during the summer (to provide air conditioning). Since the ventilation air stream is relatively warm, the efficiency of the HPWH is also improved compared to when it is placed in an unconditioned basement or garage.

HPWH-ventilation systems are now commonly used in new homes in Scandanavia. Fig. 4 shows a Swedish HPWH-ventilation system in which heat is recovered from exhaust air and the resulting hot water is provided for space heating as well as domestic use. The Swedes claim that electricity consumption for hot water heating is reduced by about 60% using this heat pump system rather than electric resistance heating [21]. HPWH-ventilation systems were under development in the U.S. as of early 1985.

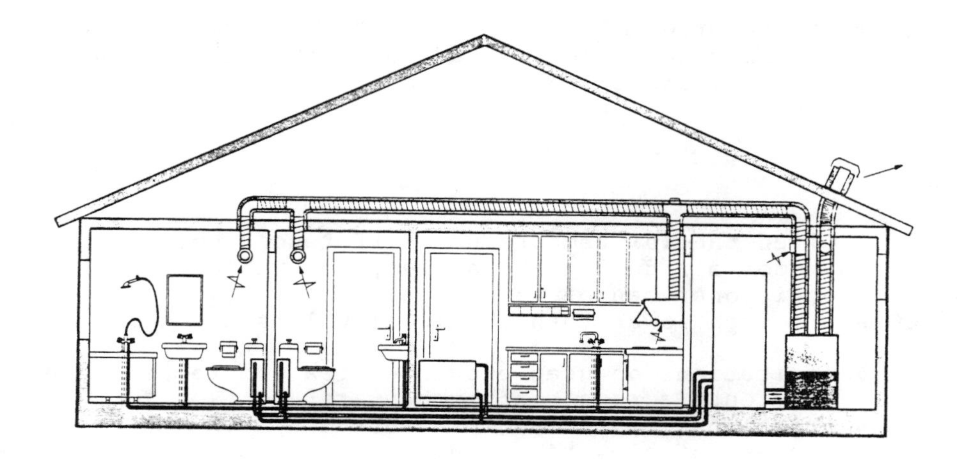

Figure 4. Design of a controlled ventilation - heat pump heat recovery system from Sweden

AIR CONDITIONERS

According to the 1982 national survey, 28% of households in the U.S. have central air conditioning (CAC) and another 30% use at least one room air conditioner (RAC) [1]. CAC systems are now being installed in about 70% of new single family homes [22]. Analysis of energy consumption data for 1978-79 shows that residential air conditioning (both CAC and RAC systems) typically consumed 2500 KWh/yr per household in households with air conditioning [13].

Conventional residential air conditioners operate based on the vapor compression cycle. A refrigerant such as R22 absorbs heat at one heat exchanger (the evaporator) and rejects heat at another (the condenser). Work is required to raise the pressure of the refrigerant vapor so that it will change phase at the condenser. An expansion device is located after the condenser to lower the pressure of the liquid refrigerant and enable it to reconvert to a liquid in the evaporator. Electric motors drive the compressor and blowers or pumps used in the system.

Equation (6) provides the idealized Carnot COP for an air conditioner or heat pump in the cooling mode, where T_a is the ambient and T_i the indoor temperature. Note that the equations for the ideal COP of vapor compression devices is slightly different for heating and cooling (compare Equations 5 and 6). In the cooling mode, the ambient environment acts as a heat sink while in the heating mode, it is a heat source.

$$COP = \frac{1}{(T_a/T_i - 1)} \tag{6}$$

Given the low temperature differences between indoor and outdoor air during the summer, the Carnot COP of an air conditioner can be very high. In practice, actual COPs are much lower than the ideal values due to:

- limitations on heat exchange area and temperature gradients across heat exchangers;
- frictional losses and pressure drops in the cycle;
- the use of blowers or pumps to enhance heat transfer and drive the air distribution system;
- motor inefficiencies.

Standardized tests for rating the efficiency of central and room air conditioners (CACs and RACs) were adopted during the 1970s [22]. For CACs, the procedure includes tests at two different ambient temperature and humidity levels and a test to determine cycling losses. Based on these tests, an overall efficiency rating, known as the seasonal energy efficiency ratio (SEER), is determined. SEER is the useful cooling output per unit of power consumption expressed in terms of Btu/Whr (ie., the SEER rating is equal to 3.414 times the COP.) For RACs, a single test is carried out to determine an energy efficiency ratio (EER), which is expressed in the same units as SEER.

In recent years, the average efficiency of new air conditioners rose at the rate of about 3%/yr for CACs and 1%/yr for RACs. However, Table II shows that the top-rated models available as of early 1985 were nearly twice as efficient as typical units produced during the past decade.

A variety of modifications were made to improve the efficiency of air conditioners. The straightforward changes include the use of larger heat exchangers and lower temperature gradients across them, use of more efficient motors, and use of improved compressors [5]. Heat exchange areas have been increased by as much as about 160% for CACs and about 75% for RACs, while fan motor efficiencies have been increased from about 60% to 70% [5].

A more innovative development has been the introduction of CAC systems with two-speed compressors. Two-speed operation provides higher efficiencies through better matching of cooling output and load. This feature is included in some of the top-rated CAC systems produced in the U.S. as of 1984-85.

A field study of energy-efficient two-speed CACs was conducted in twelve homes in Florida. This study involved alternating systems of low efficiency (SEER=6-7) and high efficiency (SEER=11.3) on a weekly basis [24]. On the average, the energy-efficient units consumed 31% less power than the conventional units (about 75% of the savings expected based solely on test ratings).

It is possible to obtain further energy savings through continous speed variation using A.C. inverters, D.C. brushless motors or advanced compressors. Some home air conditioners and heat pumps produced in Japan already include these features, and substantial R&D on variable speed units is taking place in the U.S. [25].

Dehumidification capacity is one area of concern with energy-efficient air conditioners. Because the evaporator coil temperature is increased along with the size of the heat exchanger, the latent cooling capability is reduced in many highly efficient units. At the same time, the ratio of latent-to-sensible cooling load increases as the thermal integrity of housing is improved due to the reduction of conductive, convective and solar heat gains while internal moisture sources remain approximately constant [26].

Some technical developments have been made to provide high efficiency without sacrificing dehumidification capacity. First, two-speed systems can provide high latent cooling when sensible cooling requirements are moderate. Second, heat pipe-assisted AC systems have been developed to increase the latent-to-sensible cooling ratio in an efficient manner. The heat pipe extracts some heat from incoming air and delivers it to overcooled, dry air. A small company in Florida was producing and marketing heat pipe-assisted air conditioners in early 1985 [27].

Dessicant dehumidifiers and dessicant-assisted AC systems are also available for dehumidification. However, they heat the air while drying it out and are generally undesirable for summer air conditioning. More efficient dessicant-based systems are under development. For example, a heat-activated dessicant dehumidifier has been designed with a COP of about 2.0 compared to about 0.65 for conventional dessicant systems [28].

New refrigerant mixtures are another promising line of research for increasing the efficiency of vapor compression systems. Various studies have shown that the use of nonazeotropic refrigerant mixtures can provide higher COPs than conventional single-component refrigerants and lead to energy savings of up to 25% [29].

REFRIGERATOR-FREEZERS AND FREEZERS

Refrigerator-freezers (R/Fs) and freezers (FRs) are rated for energy consumption in the U.S. based on a test in a chamber at 32°C (90°F) without any door openings or food loading [30]. The elevated ambient temperature compensates for the lack of door openings or food. The test is supposed to yield realistic estimates of electricity usage for models produced in the mid-1970s.

Some studies have compared the in-house electricity consumption and the test ratings of R/Fs [24,31,32]. All of these studies include energy-efficient R/F models as well as conventional models. The data does not show a consistent relationship between laboratory and field performance. In some cases, actual energy consumption is greater than the test ratings while in other cases, the relationship reverses. In all cases, field performance is within 30% of the test ratings.

Conventional R/Fs and FRs produced in 1970-75 were very inefficient by today's standards. Fig. 5 shows that a typical early-1970s vintage top mount R/F with automatic defrost, 450-510 l (16-18 cu ft), consumed nearly 2000 KWh/yr [33,34]. Side-by-side R/Fs from this period consumed about 2500 KWh/yr on the average. Typical chest or upright freezers with manual defrost consumed 1200-1300 KWh/yr while typical upright freezers with automatic defrost consumed over 2000 KWh/yr [34].

Figure 6 shows the estimated heat gains for the typical top mount R/F of the mid-1970s [33]. It is seen that over half of the heat gain was due to conduction, 23% of the heat gain is due to heaters and fans, and 12% is attributed to infiltration around the gasket. Heat introduced through food introduction represents only 6% of the heat gain and while door openings account for only 2% of the heat gain. (Thus it really doesn't matter if you stand for awhile with the refrigerator door open!) About 70% of the electricity consumed powers the compressor, 20% passes through various heaters, and 10% powers fans [33].

The characteristics of a circa 1975 R/F with automatic defrost include [33]:

- 6.4 cm (2.5 in) of fiberglass insulation in the sides and back;
- 3.8 cm (1.5 in) of fiberglass insulation in the doors;
- a defrost heater that consumes about 0.3 KWh/day;
- case heaters that consume about 0.17 KWh/day (for preventing condensation on the outside walls);
- a motor-compressor with an EER of about 3.1 (COP = 0.9).

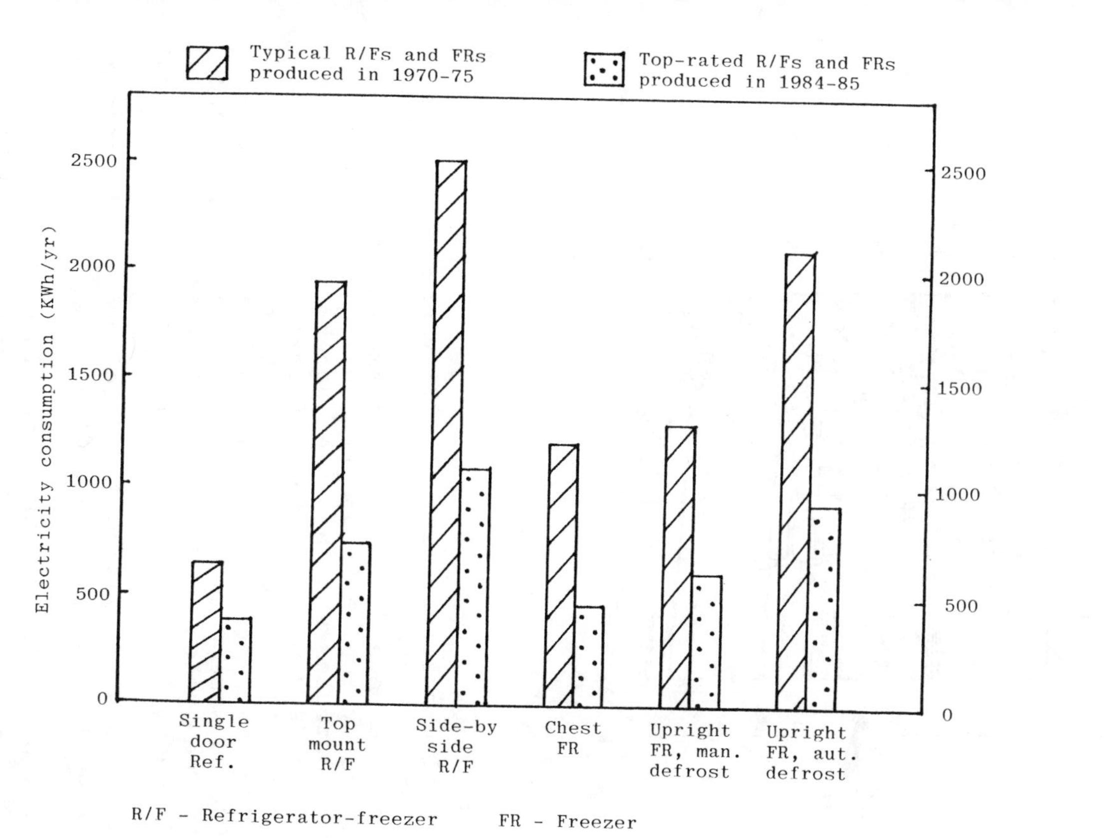

Figure 5. Comparison of the electricity consumption of typical refrigerators and freezers produced in 1970-75 and the top-rated models readily produced in 1984-85

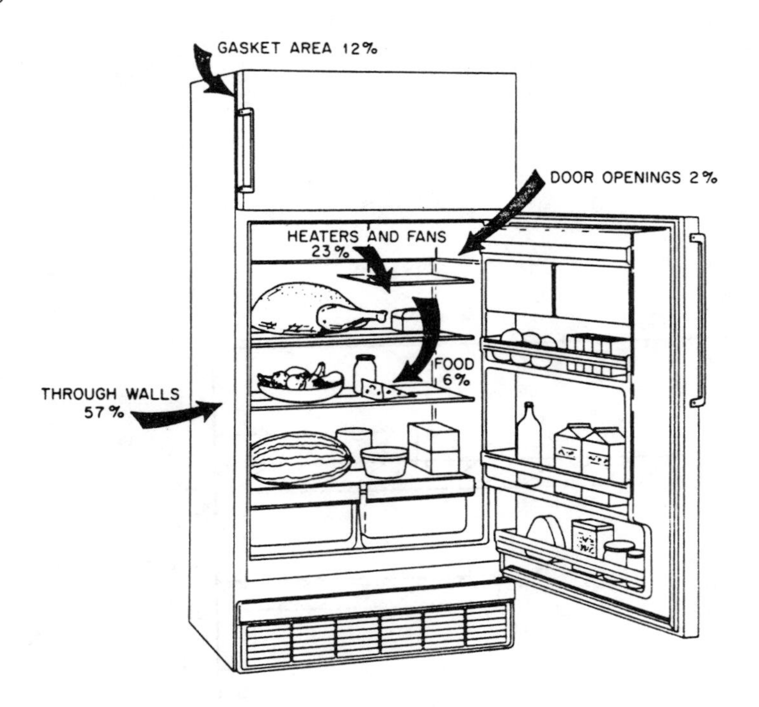

Figure 6. Refrigerator heat gains

The principal characteristics of 1975 vintage FRs were very similar to those of R/Fs.

There have been dramatic improvements in the efficiency of typical R/Fs and Frs during the past decade. Figure 7 illustrates what has been accomplished in the category of R/Fs with a top-mount freezer and automatic defrost (this category accounts for about 70% of R/F sales). The best mass-produced model in the U.S. as of April, 1985 consumed 750 KWh/yr compared to nearly 2000 KWh/yr for the typical model produced in 1972. (The top-rated 1985 model is also slightly larger than the typical 1972 model.)

Improvements in R/F and FR efficiency result from a variety of design changes, including:

- shifting to polyurethane foam insulation;
- use of more efficient motors and compressors;
- increased heat exchanger surface area;
- better refrigeration system design;
- elimination of case heaters or provision of a switch;
- use of a double gasket.

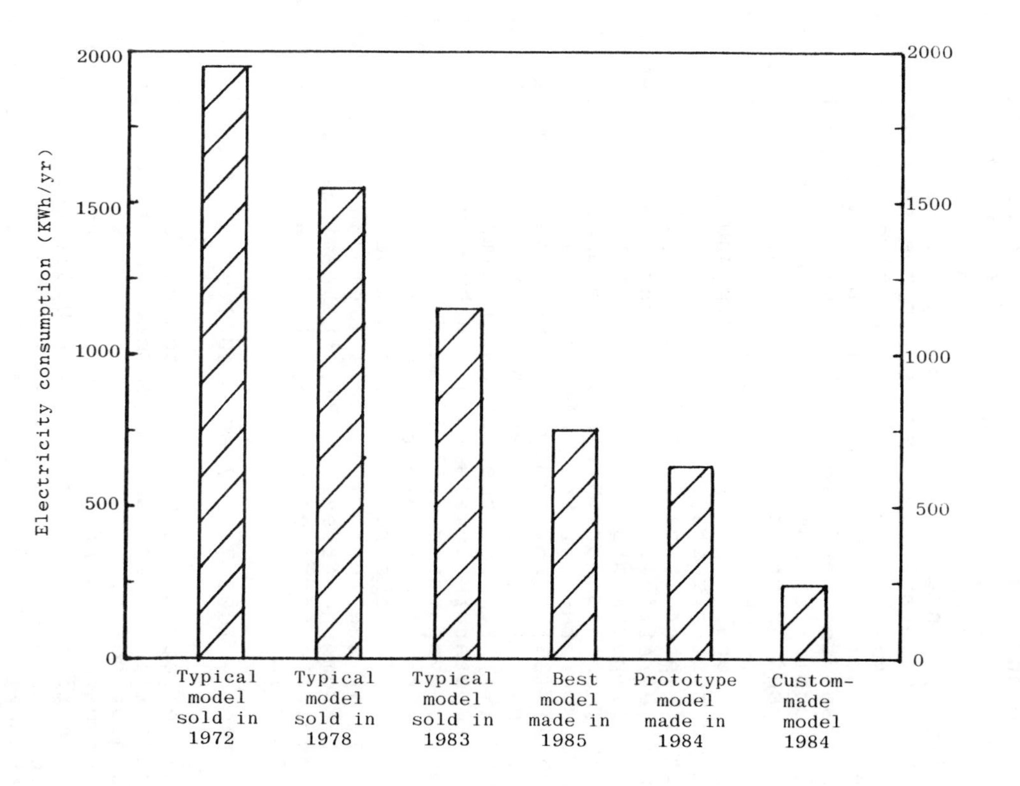

Figure 7. Progress in the electricity consumption of refrigerator-freezers (top-mount style, automatic defrost) built in the U.S.

One notable achievement was the development of a motor-compressor with an EER rating of 5.0, about 45% higher than the efficiency of motor-compressors built during the late-1970s [32]. Also, a model with separate evaporator coils in the refrigerator and freezer boxes was briefly produced during the early-1980s. The latter feature increases the overall COP and greatly cuts down on frost buildup and the operation of defrost heaters.

Well-designed, extremely energy-efficient R/Fs for use primarily with photovoltaic power systems are being custom-built by a small company in California [35]. These units feature separate compressors for the refrigerator and freezer boxes, 7.5-15 cm (3-6 in) of polyurethane foam insulation, natural convection, and top mounting of the compressors and the condensor coil to minimize heat gains. A 450 l (16 cubic ft) model with top freezer consumes only 240 KWh/yr based on the standard test procedure [35]. Although this custom-made R/F costs $2500, using it rather than a conventional R/F saves about $10,000 in photovoltaic system costs.

It is of interest to consider the implications of using separate refrigeration systems for the refrigerator and freezer compartments of a two-door R/F. At the present time, an ordinary R/F has one evaporator coil in which the temperature of the refrigerant is typically -29 $^{\circ}$C (-20 $^{\circ}$F) [33]. For analytical purposes, let us assume that heat is released at the condenser to ambient air at 35°C (95°F). The theoretical maximum COP between these two temperature levels as given by Eqn. (6) is 3.81. If a separate evaporator coil is used for the refrigerator compartment with a refrigerant temperature of 0 $^{\circ}$C (32°F), the Carnot COP for the refrigerator box would be 7.80. Assuming 60% of the thermal load is from the refrigerator box and 40% from the freezer [33], the savings from adopting separate refrigeration systems based on the ratio of ideal COPs is:

$$100\% - [60\% \times 3.81/7.80 + 40\%] = 30.7\%.$$

This estimate does not take into account additional savings that will occur due to reduced frost buildup on the evaporator coils.

Unfortunately, different test procedures are used for rating the efficiency of R/Fs in the U.S., Europe and Japan. The Japanese test is done at more realistic ambient temperatures (both 15 $^{\circ}$C and 30 $^{\circ}$C) with loading and door openings included. The use of different test procedures makes it difficult to compare models from various parts of the world.

Nevertheless, some energy-efficient R/Fs and FRs have been developed in Europe and Japan. Japanese manufacturers produce top-mount R/Fs of about 425 l (15 cubic ft)

Japanese model using the U.S. procedure yielded an energy consumption rating of 603 KWh/yr [36].

Researchers in Denmark have developed a small refrigerator (200 l in capacity without a freezer compartment) which consumes only 100 KWh/yr based on a test in an environment at 25°C (77°F) [37]. Once again, the high efficiency was achieved primarily through use of more insulation and an improved motor-compressor.

The energy performance of R/Fs and FRs in the U.S. should continue to improve as the advances described above are more widely implemented. In addition, there are good prospects for further technical advances in areas where R&D is now occuring. One such area is the use of advanced refrigerant mixtures. Another is the use of evacuated panels for insulation. Evacuated panels with an insulating value of 0.5 $m^2hr°C/kJ$ (10 $ft^2hr°F/Btu$) per inch have been developed on an experimental basis and researchers are attempting to develop panels with twice this insulating value [38].

COOKING EQUIPMENT

Cooking accounts for 5% of the end-use energy consumption in residences in the U.S., or about 2.54 GJ (2.4 MBtu) per capita as of 1978-79 [13]. About 54% of all households in the U.S. use electricity for cooking and 46% use gas (both natural gas and LPG) [1].

Analysis of the national residential energy consumption survey shows that as of 1978-79, energy consumption for cooking averaged 10 GJ/yr (9.5 MBtu/yr) in households where gas is the main cooking fuel and 1200 KWh/yr in households where electricity is the main cooking fuel [13]. The gas industry reports that the average energy consumption of gas ranges was about 7.6 GJ (7.2 MBtu) in 1982, down from 10.3 GJ (9.8 MBtu) in 1975 [8]. This 27% reduction is attributed to both technological improvements and changing cooking habits.

Cooking equipment generally consists of cooking tops and ovens either as separate devices or combined in a cooking range. The efficiency of conventional cooking tops and ovens is measured by heating an aluminum block according to standardized procedures [23]. For ovens, the test involves heating the block 130°C (234°F) above its initial temperature. For cooking tops, burners are first operated at their highest energy input while heating the block 80°C (144°F) above the initial temperature. Burners are then turned down to 25% of the maximum rate for an additional 15 minutes.

A typical electric oven produced during the late-1970s has an efficiency rating of 12% while gas ovens have efficiencies of 3.7-5.3% depending on whether pilots are included [5,39]. A typical electric cooking top has an efficiency rating of 77%, gas top with pilots 22%, and gas top without pilots 56%.

Gas pilots on a typical gas range of the late-1970s (single oven plus four burners) consume 420-525 kJ/hr (400-500 Btu/hr) [39]. This corresponds to 3.7-4.6 GJ/yr (3.5-4.4 MBtu/yr), or about 35-45% of the total energy consumption of a typical gas range with pilot lights. Most gas ranges now produced include electric ignition although ranges with standing pilots are still manufactured.

Standard ovens of the late-1970s typically featured 5 cm (2 in) of low-density fiberglass insulation [5]. Self-cleaning ovens typically have 10 cm (4 in) of insulation to provide the high surface temperatures needed during the cleaning cycle. However, the lower conductive losses are roughly cancelled out by the energy consumed during cleaning. By using higher density insulation, improved door seals, less thermal mass and a lower venting rate, it is possible to cut oven energy consumption by about 20% [5,39].

A prototype energy-efficient electric oven has been developed and demonstrated. This oven, known as the bi-radiant oven, features highly reflective walls and two heating elements which operate at relatively low temperatures [40]. Also, the bi-radiant oven has been used with cooking utensils that have high surface absorptivities. Tests show that the bi-radiant oven consumes about two-thirds of the energy used by a standard electric oven [40].

Microwave ovens are typically about 50% efficient in standardized tests involving heating a pan of water [41]. This is much greater than the efficiency of a conventional electric oven. The major energy losses in microwave ovens result from inefficiencies in the electronic components (transformer and magnetron). Studies involving the cooking of actual food items show that the availability and use of a microwave oven can reduce overall electricity consumption for cooking by 25-50% [42].

In the area of gas cooking, an advanced burner known as the infrared jet-impingement burner has been developed. The burner features a high degree of radiative heat transfer from a ceramic flame holder and the formation of gas jets as the combustion products pass through holes in a glass plate. At relatively high heat output rates, this burner is about 50% more efficient than conventional burners [43]. Besides providing about a one third fuel savings, the infrared jet-impingement burner emits much less nitrogen oxide then a conventional gas burner due to reduced combustion temperatures [43]. Field tests of stoves equipped with this innovative burner began in 1984.

CLOTHES WASHERS AND DISHWASHERS

Both clothes washers (CWs) and dishwashers (DWs) are tested and rated for direct and indirect (hot water) energy consumption. The Association of Home Appliance Manufacturers reports that the total energy consumption for the average CW produced in 1983 was 2.6 KWh/use, down from 3.8 KWh/use for the average CW made in 1972 [44]. For DWs, the reduction was from 4.2 to 2.7 KWh/use over the same 12 year period.

CWs and DWs each use about 75-150 KWh/yr directly to power motors, pumps, and heaters. The typical efficiency of motors used in CWs and DWs of the mid-1970s was only 50-55% [45]. Use of more efficient motors can reduce the direct electricity consumption in CWs and DWs by up to 25%.

The majority of the energy consumed by CWs and DWs is through hot water. CWs and DWs typically account for about one half of total hot water use in households where they are both present. For CWs, front-load models use about 40% less water than more common top-load models [46]. Also, using warm water for wash cycles and cold water for rinse cycles can significantly reduce hot water consumption. Tests have shown that these settings provide adequate cleaning unless clothes are exceptionally dirty [47].

Hot water use by DWs is in the range of approximately 38-60 l (10-16 gal) per use, and it can be reduced by 16-33% using shorter and/or fewer subcycles [46]. Many models have a "short cycle" option which eliminates one rinse cycle thereby saving 7-19 l (2-5 gal) of hot water. In addition, the energy consumed by DWs can be reduced by about 15% using natural drying rather than the resistance heaters in the DW [47].

A prototype DW has been developed which features a jet spray created by water pressure (analagous to a garden hose). Besides not having a motor, this model uses less water than conventional DWs and warm rather than hot water [48]. There are plans to commercially produce the jet spray DW in the U.S. in the near future [49].

CLOTHES DRYERS

The 1982 residential energy survey showed that 38% of all households have an electric clothes dryer (ECD) and 12% have a gas clothes dryer (GCD) [1]. A typical ECD in the appliance stock uses about 1000 KWh/yr and a typical GCD about 5 GJ/yr [10]. Thus, CDs are a major energy consumer in households where they are used. (Of course, commercial CDs are commonly used by households without their own CD).

Pilots in GCDs were replaced by electric ignition during the 1970s. The main energy savings option currently available in CDs is an automatic termination control. This device senses either the temperature or moisture level in the exhaust air. Tests with standardized loads of clothes indicate that the use of a termination control reduces overall energy consumption by up to 10-15% [45].

One obvious technique for reducing net energy use for clothes drying is to vent filtered exhaust air indoors during the heating season. However, this practice is not recommended with GCDs due to the presence of carbon monoxide and nitrogen oxides in the exhaust. Also, the high level of moisture in the exhaust with either type of dryer can cause local condensation problems.

An alternative to direct indoor venting of exhaust gases is the use of a heat exchanger to recover some of the sensible and latent energy in the exhaust. Experiments show preheating inlet air with the exhaust air can reduce overall dryer energy consumption by 26% [50]. Much greater heat recovery and savings can result during the winter if indoor air is used as the heat sink. In this case, water vapor can be condensed in the heat exchanger and the dryer exhaust air then recirculated.

It is possible to use a heat pump to remove the moisture from the exhaust air in a closed cycle. A prototype dryer of this type has been developed by a company that produces commercial-scale dehumidification dryers [51]. Tests show that this dryer uses about 60% less electricity than a conventional CD with approximately the same drying time [52]. In addition, the heat pump CD operates on 110 Volts and does not require a vent pipe.

The use of microwave radiation is another approach to reducing electricity consumption for clothes drying. A prototype microwave dryer has been developed that is reported to use only 30-50% as much electricity as a conventional CD with less drying time [53]. A major manufacturer was considering commercially producing a microwave dryer in mid-1985 [53].

TELEVISIONS

It is estimated that older color TVs with vacuum tubes draw about 310 W and that older black-and-white sets use 200 W [54]. Assuming an average on-time of 1700 hrs/yr [54], vacuum tube color TV sets consume 530 KWh/yr and black-and-white sets 340 KWh/yr.

The current generation of TVs with solid-state electronics draw much less power than vacuum tube models. It is estimated that the typical power consumption of TVs from the early-1980s is 130 W for color sets and 42 W for black-and-white sets [54].

REFERENCES

1. "Residential Energy Consumption Survey: Housing Characteristics 1982", DOE/EIA-0314(82), Energy Information Administration, U.S. Dept. of Energy, Aug. 1984.
2. G.E. Kelly, J. Chi and M.E. Kuklewicz, "Recommended Testing and Calculation Procedures for Determining the Seasonal Performance of Residential Central Furnaces and Boilers", NBSIR 78-1543, Nat. Bureau of Standards, Washington, DC, Oct. 1978.
3. J.E. Hill, "Testing to Determine the Seasonal Performance of Central Heating and Cooling Equipment", Nat. Bureau of Standards, Washington, DC, paper presented at the 1982 ACEEE Summer Study on Energy Efficiency in Buildings, Santa Cruz, Aug. 1982.
4. D.L. O'Neal, "Energy and Cost Analysis of Residential Heating Systems", ORNL/CON-25, Oak Ridge National Laboratory, July 1978.
5. "Consumer Products Efficiency Standards Engineering Analysis Document", DOE/CE-0030, U.S. Dept. of Energy, March 1982.
6. R.A. Macriss, "Efficiency Improvements in Space Heating by Gas and Oil", Ann. Rev. Energy 8:247-267, 1983.
7. G.T. Linteris, "Performance of Retrofitted and New High Efficiency Gas Equipment: Some Recent GRI Projects", What Works: Documenting Energy Conservation in Buildings, Am. Council for an Energy-Efficient Economy, 1984.
8. "Gas Consumption by Residential Appliances", American Gas Association, Arlington, VA, March 2, 1984.
9. W.H. Thrasher, R.J. Kolodgy and J.J. Fuller, "Development of a Space Heater and a Residential Water Heater Based on the Pulse Combustion Principle", Proceedings of the ACEEE 1984 Summer Study on Energy Efficiency in Buildings, Am. Council for an Energy-Efficient Economy, Aug. 1984.
10. H.S. Geller, "Energy Efficient Appliances", Am. Council for an Energy-Efficient Economy and Energy Conservation Coalition, Washington, DC, June 1983.
11. P.D. Fairchild, "A Survey of Advanced Heat Pump Developments for Space Conditioning", Oak Ridge National Laboratory, paper presented at the Eighth Energy Technology Conference, Washington D.C., March 1981.
12. T.J. Marusak, "Free-Piston Sirling Engines - an Advanced Power Conversion System for Residential and Commercial Buildings", Proceedings of the ACEEE 1984 Summer Study on Energy Efficiency in Buildings, Am. Council for an Energy-Efficient Economy, Aug. 1984.
13. S. Meyers, "Residential Energy Use and Conservation in the United States", LBL-14932, Lawrence Berkeley Laboratory, Berkeley, CA, March 1982.

14. R.D. Clear and D.B. Goldstein, "A Model for Water Heater Energy Consumption and Hot Water Use: Analysis of Survey and Test Data on Residential Hot Water Heating", LBL-10797, Lawrence Berkeley Laboratory, Berkeley, CA, May 1980.
15. R.A. Hoskins and E. Hirst, "Energy and Cost Analysis of Residential Water Heaters", ORNL/CON-10, Oak Ridge National Laboratory, Oak Ridge, Tenn., June 1977.
16. R.L. Palla, "Evaluation of Energy-Conserving Modifications for Water Heaters", NBSIR 79-1783, Nat. Bureau of Standards, Washington, DC, July 1979.
17. A. Usibelli, "Monitored Energy Use of Residential Water Heaters", Proceedings of the 1984 ACEEE Summer Study on Energy Efficiency in Buildings, Am. Council for an Energy-Efficient Economy, Aug. 1984.
18. "Consumers' Directory of Certified Water Heater Efficiency Ratings", Gas Appliance Manufacturers Association, Arlington, VA, Jan. 1985.
19. J.E. Harris, "Performance of Add-on Type Heat Pump Water Heaters Using Two Different Test Methods", NBSIR 83-2723, Nat. Bureau of Standards, Washington, DC, March 1983.
20. W.P. Levins, "Estimated Seasonal Performance of a Heat Pump Water Heater Including Effects of Climate and In-House Location", ORNL/CON-81, Oak Ridge National Laboratory, Oak Ridge, Tenn., Jan. 1982.
21. Information provided by Flakt Evaproator AB, Jonkoping, Sweden, 1984.
22. Characteristics of New Housing, U.S. Bureau of Census and Dept. of Housing and Urban Development, Washington, DC, 1984.
23. Code of Federal Regulations, Title 10, Chapter II, Part 430, U.S. Gov. Printing Office, Washington, DC, 1981.
24. W.T. Lawrence, "Field Test Measurements of Energy Savings from High Efficiency Residential Appliances", Florida Public Service Commission, Tallahassee, 1982.
25. R.V. Steele, ed., Proceedings: Seminar on Heat Pump Research and Applications, EM-3797, Electric Power Research Institute, Palo Alto, CA, Nov. 1984.
26. M.K. Khattar, "Residential Air-Conditioning Energy Calculations", FSEC-PF-55-84, Florida Solar Energy Center, Cape Canaveral, June 1984.
27. Personal communication with K. Dinh, Dinh Company, Alachua, FL, March 1985.
28. M.K. Khattar, "Dehumidification and Mechanical HVAC Systems", in Principles of Low Energy Building Design in Warm, Humid Climates, Florida Solar Energy Center, Cape Canaveral, Sept. 1983.

29. W.P. Levins, "An Assessment of the Energy Saving Potential of Nonazeotropic Refrigerant Mixtures", Proceedings of the ACEEE 1984 Summer Study on Energy Efficiency in Buildings, Am. Council for an Energy-Efficient Economy, Aug. 1984.
30. ANSI/AHAM HRF-1-1979, "American national standard for household refrigerators, combination refrigerator-freezers, and household freezers", Association of Home Appliance Manufacturers, Chicago, May 17, 1979.
31. R.F. Topping, "Development of a High-Efficiency, Automatic-Defrosting Refrigerator-Freezer. Phase II Field Test", ORNL/Sub/77-7255/3, Oak Ridge National Laboratory, Oak Ridge, TN, Dec. 1982.
32. M.G. Middleton and R.S. Sauber, "Research and Development of Energy-Efficient Appliance Motor-Compressors, Vol. IV Production Demonstration and Field Test", ORNL/Sub/78-7229/4, Oak Ridge National Laboratory, Oak Ridge, TN, Sept. 1983.
33. R.A. Hoskins and E. Hirst, "Energy and Cost Analysis of Residential Refrigerators", ORNL/CON-6, Oak Ridge National Laboratory, Oak Ridge, TN, Jan. 1977.
34. "1983 Energy Consumption and Efficiency Data for Refrigerators, Refrigerator-Freezers and Freezers", Association of Home Appliance Manufacturers, Chicago, July 1984.
35. Information provided by Mr. Larry Schlussler, Sun Frost, Arcata, CA, April 1985.
36. D.B. Goldstein, "Efficient Refrigerators in Japan: A Comparative Survey of American and Japanese Trends Towards Energy Conserving Refrigerators", Proceedings of the ACEEE 1984 Summer Study on Energy Efficiency in Buildings, Am. Council for an Energy-Efficient Economy, Aug. 1984.
37. J.S. Norgard, "Same Comfort in Buildings with One Third of Present Electricity Consumption", Proceedings of the ACEEE 1984 Summer Study on Energy Efficiency in Buildings, Am. Council for an Energy-Efficient Economy, Aug. 1984.
38. Personal communication with Mr. Ed Vineyard, Oak Ridge National Laboratory, Oak Ridge, TN, April 1985.
39. R.C. Erickson, "Household Range Energy Efficiency Improvements", Proceedings of the Conference on Major Home Appliance Technology for Energy Conservation, CONF-780238, U.S. Dept. of Energy, 1978.
40. D.P. DeWitt and M.W. Peart, "Bi-Radiant Oven - A Low-Energy Oven System", Vol. II - Executive Summary, ORNL/Sub/80-0082/2, Oak Ridge National Laboratory, Oak Ridge, TN, Oct. 1981.
41. C.R. Buffler, "Efficiency Improvement Techniques for Microwave Ovens", See Ref. 39.
42. V. Ludvigson and T. VanValkenburg, "Microwave Energy Consumption Tests", See Ref. 39.

43. K.C. Shukla and J.R. Hurley, "Development of an Efficient, Low NOx Domestic Gas Range Cooktop", GRI-81/0201, Gas Research Institute, Chicago, IL, July 1983. Also, see "Technology Profile - Advanced Four-Burner Cooktop", Gas Research Institute, Oct. 1984.
44. Data published by the Association of Home Appliance Manufacturers, Chicago, IL, July 1984.
45. W.P. Levins, "Energy and the Laundry Process", ORNL/CON-41, Oak Ridge National Laboratory, Oak Ridge, TN, April 1980.
46. R.L. Palla, "Water Usage Characteristics of Household Appliances and the Potential for Water Savings", NBSIR 80-2173, Nat. Bur. of Standards, Wash., DC, Dec. 1980.
47. A.J. Fuchs, "Energy Conservation Water Temperature Effects in Laundering and Automatic Dishwashing", See Ref. 39.
48. "Water-powered dishwasher inspired by a garden hose", Appliance Manufacturer, April 1984.
49. Personal communication with Hart Industries, Inc., Laguna Hills, CA, June 1985.
50. D. Hekmat and W.J. Fisk, "Improving the Energy Performance of Residential Clothes Dryers", LBL-17501, Lawrence Berkeley Laboratory, Berkeley, CA, Feb. 1984.
51. J. Laitin, "A New Way to Dry Clothes", Venture, Nov. 1984.
52. D.C. Lewis, "Final Technical Report - Closed Cycle Clothes Dryer", DOE/CE/15100-T1, U.S. Dept. of Energy, Washington, DC, June 1983.
53. Personal communication with Mr. James Manning, Manning Research Associates, Portland, OR, May 1985.
54. A. Meier, J. Wright and A.H. Rosenfeld, Supplying Energy Through Greater Efficiency, Univ. of California Press, 1983.

CHAPTER 16

Economics of Efficiency Improvements in Residential Appliances and Space Conditioning Equipment

M.D. Levine, J. Koomey, H. Ruderman, P. Craig[†], J. McMahon, and P. Chan

Energy Analysis Program
Lawrence Berkeley Laboratory
University of California
Berkeley, California 94720

[†]Department of Applied Science
University of California
Davis, California 95616

INTRODUCTION

The large amounts of energy consumed by residential appliances and space conditioning equipment make this sector a fruitful area for efficiency improvements. We examine eight major residential appliances that in the U.S. currently consume 9.4 exajoules/year (8.9 quads/year*) representing more than 12 percent of total 1984 U.S. energy use. Expenditures on energy for these appliances totaled over $56 billion in 1984. Our results indicate that improving the efficiency of all these appliances to economically optimal levels would reduce these annual expenditures by almost thirty percent, a savings of $17 billion per year. In steady state, the annualized additional investment cost to achieve this efficiency improvement is $7 billion, so the net savings is about $10 billion per year**.

This paper describes and analyzes energy efficiency choices for residential appliances and space conditioning equipment.*** The first section briefly

*An exajoule equals 10^{18} joules. A quad equals one quadrillion (10^{15}) Btus. In this paper, both the price and the energy value of electricity are measured as resource energy at 11,500 Btus per kWh.

**Assuming that the optimum efficiencies are calculated using a real ten percent discount rate, at current fuel prices, and in 1984 dollars. The expenditure and the energy use numbers are derived from LBL Energy Demand and Forecasting Model Runs (a new model based on the original model developed by ORNL, see reference 1). The additional investment costs are annualized by dividing by the present worth factor (PWF).

***The term appliances will henceforth include space conditioning equipment.

0094-243X/85/1350299-24$3.00 Copyright 1985 American Institute of Physics

illustrates historical trends in the average efficiencies of new appliances sold in the United States during the last decade. The second section shows results of the life-cycle cost analysis of eight major residential appliances. Our results provide striking evidence that the market is not achieving economically optimal efficiency levels.

To a physicist, optimal efficiency is defined as the maximum second-law efficiency that is theoretically attainable. However, the physicist's optimal efficiency level often cannot be obtained at reasonable cost. The economist defines optimal efficiency as the efficiency that minimizes the total cost of purchasing, operating, and maintaining a device over its lifetime. The latter definition of optimality is helpful in assessing the cost effectiveness of efficiency improvements, because it balances the cost to improve energy efficiency against the benefits of reduced fuel use.

When homeowners buy appliances, they make implicit tradeoffs between current capital expenditures and future operating expenses. Using a concept known as life-cycle cost (LCC), we can characterize the capital vs. operating cost tradeoffs that will leave the purchaser most well-off in the long run. LCC is the sum total of capital, maintenance, and operating costs over the life of the appliance, properly discounted to account for the time value of money. Results from LCC analysis reveal that purchasers often do not choose the "optimal" efficiency (i.e., the efficiency of the appliance with the lowest LCC), because they prefer to minimize present outlays at the cost of increased future expenditures. In some cases, purchasers ignore investments with simple payback times of one to two years, equivalent to a return on investment of 50 to 100 percent. One would not expect a "rational" investor to turn down such high returns, but the majority of purchasers of residential appliances do (3,4).

The simple payback time is the time required for the operating cost savings of a more efficient appliance to repay its additional capital cost. The payback times that we derive are of two types. The first type, used in the initial LCC analysis, we designate the "payback time" for an investment in a large, discrete change in efficiency from a current efficiency level to the optimum one. The second represents the approximate "incremental payback time" for an infinitesimal or marginal increase in efficiency from current market levels. The second type approximates the rate of return for the next dollar invested in efficiency improvements. We determined both payback times and life-cycle costs from data on the efficiency and purchase cost of appliances purchased between 1978 and 1984.

A major finding of this study is that the payback times for moving from the current average efficiency sold on the market today to an optimum efficiency as determined by LCC analysis range from one to nine years, with most payback times less than five years. The rates of return implied by these payback times are attractive, and they reveal that the potential for efficiency improvements in the residential appliance sector is not yet close to being realized.

Another important result of our analysis is that the incremental payback periods for investment in increasing the energy efficiency of most household appliances are less than three years, except for air conditioners. We conclude from this result that the market for energy efficiency is not performing well. In the last section, several possible explanations of the underinvestment in efficiency are proposed: 1) lack of information about the costs and benefits of energy efficiency; 2) lack of access to capital markets; 3) expected savings are too small in absolute terms to be of interest to purchasers; 4) prevalence of third party purchasers; 5) unavailability of highly efficient equipment without other features for some products; 6) long manufacturing lead times; and 7) other marketing strategies.

CHANGES IN EFFICIENCY OVER THE PAST DECADE

We focus on the efficiency choices for eight major appliances: gas central space heaters, oil central space heaters, room air conditioners, central air conditioners, electric water heaters, gas water heaters, refrigerators, and freezers. Central space heaters include boilers and furnaces. We chose these products because they account for a major part of residential energy consumption and because data on their efficiency and costs are readily available. Electric resistance heaters are not included because no significant improvement in their efficiency is possible. Data on the incremental costs of efficiency improvements for heat pumps and gas or oil room heaters were unavailable at the time of publication (incremental costs for heat pumps are likely to be similar to those of central air conditioners).

The efficiency of residential appliances has increased in the past twelve years. Figure 1 shows these improvements based on the shipment weighted energy factors (SWEFs) of products sold in the United States between 1972 and 1984. The efficiency of refrigerators and freezers increased about 60 percent from 1972 levels. The improvements for other products were less dramatic, but were significant for gas water heaters, gas furnaces, room air conditioners, and central air conditioners, ranging from 16 to 30 percent. The remaining products--oil furnaces and electric water heaters--showed less than 8 percent improvement in energy efficiency over the indicated time period. Table 1 shows the average efficiencies for selected appliances between 1972 and 1984, from which Figure 1 was derived. For an analysis of the technical changes that led to these efficiency improvements, see Howard Geller's paper in this volume.

LIFE-CYCLE COST ANALYSIS

This section provides a methodology for assessing the economic costs and benefits to the consumer who purchases appliances of varying initial costs and energy efficiency. The basis for analysis is the use of life-cycle costing.

The life-cycle cost of owning and operating an appliance is equal to the first cost or purchase price plus the operating and maintenance costs over the lifetime of the appliance. The first cost may be paid when the product is purchased or the consumer may borrow money that is paid back with interest after the purchase is made. For the purpose of this analysis it is assumed that the consumer makes a cash purchase of the appliance. We also assume that the cost of maintenance over the lifetime of the appliance is independent of efficiency; thus the maintenance cost is not included in the life-cycle cost calculation. In order to consider first cost and operating costs on a time-equivalent basis, all future operating costs are discounted to present value. Life-cycle costs are compared for classes of appliances of the

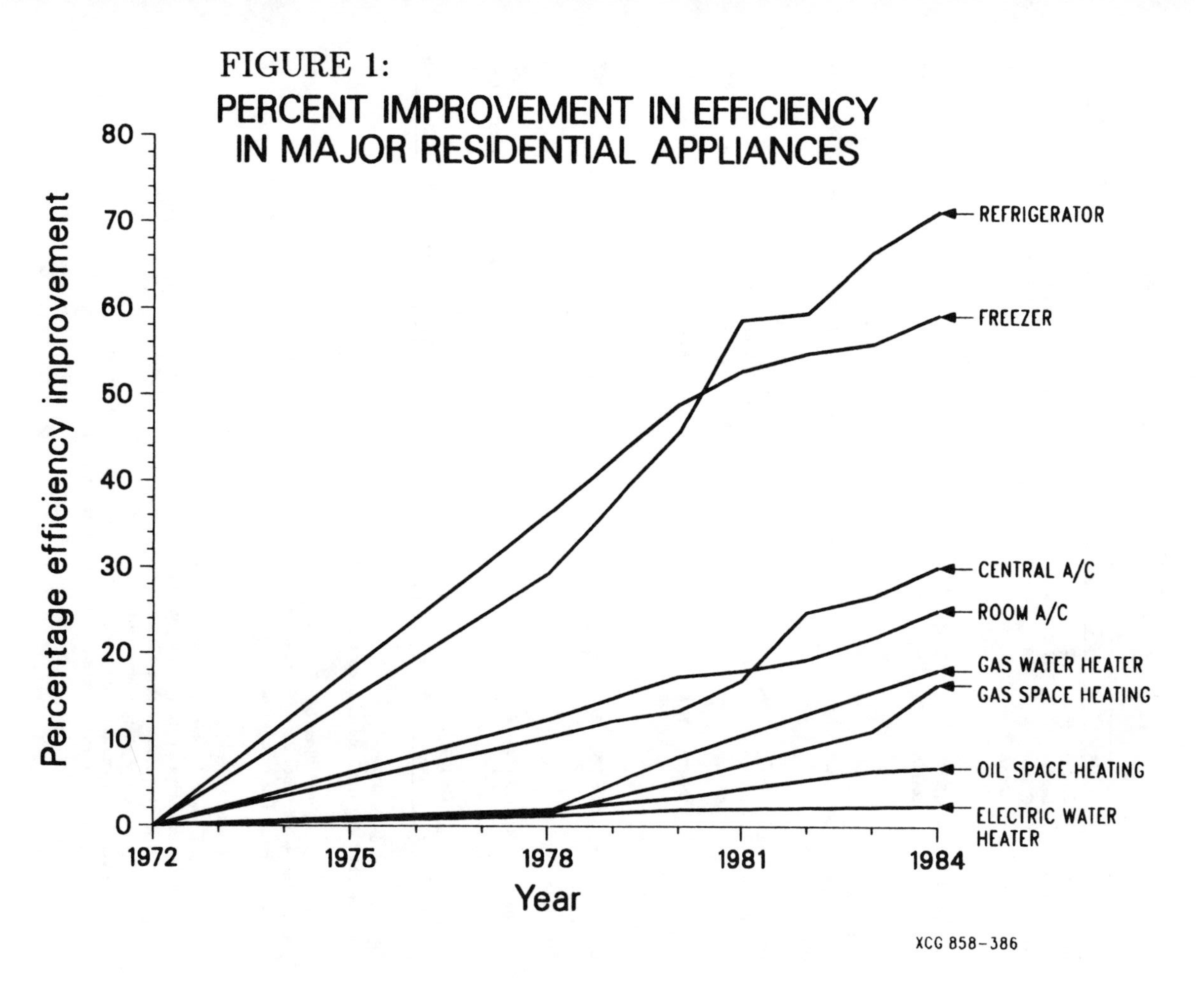

FIGURE 1:
PERCENT IMPROVEMENT IN EFFICIENCY IN MAJOR RESIDENTIAL APPLIANCES
Percentage efficiency improvement
80
70
60
50
40
30
20
10
0
1972
1975
1978
1981
1984
Year
REFRIGERATOR
FREEZER
CENTRAL A/C
ROOM A/C
GAS WATER HEATER
GAS SPACE HEATING
OIL SPACE HEATING
ELECTRIC WATER HEATER
XCG 858-386

Table 1. Shipment-Weighted Energy Factors (SWEF)

Appliance	Source	1972	1975	1976	1977	1978	1979	1980	1981	1982	1983	1984
Gas Central	CS-179	--	62.7	--	--	63.6	--	65.9*	--	--	--	--
Space Heater	Lennox	--	65.0	65.0	65.1	65.5	66.3	66.6	67.0	--	--	--
(AFUE%)	Carrier	--	63.7	--	--	65.1	66.3	66.7	66.5	--	--	--
	GAMA	--	--	--	--	--	--	--	--	--	69.6	73.0
Oil Central	CS-179	--	73.6	--	--	75.0	--	76.0*	--	--	--	--
Space Heater	GAMA	--	--	--	--	--	--	--	--	--	78.3	78.6
(AFUE%)												
Room Air	CS-179	6.22	--	--	--	6.75	--	7.03*	--	--	--	--
Conditioner	AHAM	5.98	--	--	--	6.72	--	7.02	7.06	7.14	7.29	7.48
(EER)												
Central Air	CS-179	6.66	--	--	--	6.99	--	7.76*	--	--	--	--
Conditioner	Lennox	--	6.19	6.94	7.02	7.00	7.05	7.14	7.73	8.18	--	--
(SEER)	ARI	6.66	--	7.03	7.13	7.34	7.47	7.55	7.78	8.31	8.43	8.66
Electric Water Heater (Percent)	CS-179	79.8	--	--	--	80.7	--	81.3*	--	--	--	81.7**
Gas Water Heater (Percent)	CS-179	47.4	--	--	--	48.2	--	51.2*	--	--	--	56.0**
Refrigerator	CS-179	4.22	--	--	--	5.09	--	5.72*	--	--	--	--
(cu.ft./kWh/day)	AHAM	3.84	--	--	--	4.96	--	5.59	6.09	6.12	6.39	6.57
Freezer	CS-179	8.08	--	--	--	10.07	--	10.83*	--	--	--	--
(cu.ft./kWh/day)	AHAM	7.29	--	--	--	9.92	--	10.85	11.13	11.28	11.36	11.60

*Projection made in 1979.
**Forecast by LBL/ORNL model-no data are available

AFUE - Annual Fuel Utilization Efficiency
EER - Energy Efficiency Ratio (Btus per hour/Watts)
SEER - Seasonal Energy Efficiency Ratio

Data Sources (see reference 4)
AHAM - Association of Home Appliance Manufacturers
ARI - Air-Conditioning and Refrigeration Institute
Carrier - Carrier Corporation (estimates are for manufacturer's own products)
CS-179 - Department of Energy Survey of Manufacturers
Lennox - Lennox Corporation (estimates are for manufacturer's own products)
GAMA - Gas Manufacturers Association (Derived from reference 2).

same capacity but different efficiencies. A more energy efficient product is often more expensive than one of lower efficiency, all other features of the product being equal. However, if the more energy efficient model has lower total costs to the consumer over the life of the appliance, the consumer benefits in the long run, even though the initial investment is larger.

The trade-off between the higher first costs and lower operating costs of more energy efficient products can be assessed in terms of a simple payback time. The simple payback time is the additional cost for a more energy efficient product divided by the savings in fuel costs per unit time (expressed in months or years). For example, if the payback time is one year, then the extra first cost for a more efficient product is fully recovered in reduced energy bills during the first year of operation of the product. This rate of payback is equivalent to a return on investment of 100% per year to the consumer. If, on the other hand, the payback time is 20 years, then the rate of return on the initial investment is small.

When the payback time is a few years or less, the simple payback time closely approximates the actual payback time; for our purposes, little error will be introduced by using the simple payback time. During periods of rapidly increasing energy prices, the use of simple payback time can be misleading, but it is an understandable concept that will effectively illustrate our results. A discussion of the relationship between discount rates, life-cycle costs, and payback periods is contained in Appendix 1.

Methodology and Assumptions

The life-cycle cost analysis provides a measure of the economic impact of equipment purchases on the consumer. All other things being equal, the consumer benefits in the long run from the purchase of a product with the lowest life-cycle cost. To calculate life-cycle costs, assumptions and estimates must be made about future prices of energy and the value that a consumer places on future return on investment, because a more energy efficient appliance saves money for many years(5).

The total life-cycle cost (LCC), of an appliance is given in general by:

$$LCC = PC + \sum_{t=1}^{N} ENC_t \, \frac{(FP)(1+f)^t}{(1+r)^t} \tag{1}$$

where

PC = initial purchase cost of the appliance (in dollars),

ENC_t = energy consumption in year t (in million Btus),

FP = fuel price in year 1 (in dollars per million Btus),

N = lifetime of appliance (in years),

f = annual percentage change in real fuel price*

r = discount rate in constant dollars.

The fuel price in year t is given by $FP(1+f)^t$ and the total expenditure for fuel in year t is

$$FC_t = ENC_t (FP)(1+f)^t \tag{2}$$

FC_t is the fuel cost in year t as given in Appendix 1.

In the analysis as performed, yearly energy consumption (ENC_t) and the fuel prices are assumed to be constant over the appliance lifetime. While the latter assumption may not correspond to the actual price trends of recent years, it is a conservative one that makes the results more robust. Rapidly increasing fuel prices make conservation investments even more attractive. Thus, Eq. (1) may be simplified to

$$LCC = PC + (ENC)(FP)\,PWF \tag{3}$$

where

$$PWF = \sum_{t=1}^{N} \left[\frac{(1+f)^t}{(1+r)^t} \right] \tag{4}$$

Table 2 shows the national average fuel prices used in the life-cycle cost calculations. The first cost, fuel prices, and life-cycle costs are expressed in 1984 dollars. The fuel price escalation rates and the discount rate are expressed in real dollars. Electricity prices have been adjusted to account for the fact that electric space and water heating customers usually pay lower than average prices for their electricity due to promotional rate structures, and air conditioning customers typically pay an on-peak electricity price that

*A "real" rate of change is the annual percentage change after adjusting for inflation.

is higher than the national average. This adjustment also accounts for regional differences in the distribution of these appliances across the U.S. The discount rates chosen for the analysis were 3% and 10% real. Table 3 presents the appliance lifetimes used in the LCC calculations.

Table 2: 1984 National Average Fuel Prices
(1984 Dollars per Million Btu)

	Resource Energy*	Site Energy**
Electricity (Avg):	5.89	19.9
By End Use:		
Air Conditioning	6.66	22.4
Water Heat	5.12	17.3
Space Heat	4.54	15.3
Other electrical	6.48	21.8
Natural Gas	5.89	5.89
Oil	7.71	7.71

Table 3: Appliance Lifetimes (years)

Appliance	Lifetime
Central heating	23
Water heating	13
Central air conditioners	12
Room air conditioners	15
Refrigerator	19
Freezer	21

To summarize, the analysis assumes:

(1) that national average energy prices apply (electric rates are end-use specific; see Table 2);

---

*Price per unit of resource energy consumed; 11,500 Btu/kWh includes heat rate of electricity generation plus transmission losses.

**Price per unit of electricity consumed on-site (at 3412 Btu/kWh).

(2) no escalation of energy prices above inflation

(3) real discount rates of 3% and 10%

(4) the appliance lifetimes in Table 3; and

(5) no increase in purchase price above inflation for an appliance of given efficiency.

The other input to the LCC computations is a set of exponential curves relating the initial cost and the energy use for each class of appliance, based on an update of data developed by Arthur D. Little and reported in the Engineering Analysis Technical Support Document for DOE's Consumer Products Efficiency Standards (6).

Table 4: Parameters of Purchase Cost/Unit Energy Consumption Curves

Appliance	A	$E\infty/E_o$	PC_o	E_o	Baseline Efficiency
Gas furnaces	5.76	0.65	2480	81.7	63
Oil furnaces	8.47	0.79	3750	125	76
Room air conditioners	3.77	0.51	593	12.3	6.7
Central air conditioners	2.00	0.44	1640	33.8	7.1
Electric water heaters	9.22	0.82	207	52.1	78
Gas water heaters	5.67	0.56	256	21.6	48
Refrigerator-freezers	21.6	0.39	674	14.0	4.9
Freezers	10.6	0.38	444	13.3	9.7

These exponential curves are of the form

$$E = E_\infty + (E_0 - E_\infty)\exp[-A\,(C - 1)] \quad (5)$$

where

E = unit energy consumption (UEC) million Btu/yr of resource energy

E_0 = base year UEC

E_∞ = minimum UEC attainable at infinite purchase cost

C = PC/PC_o

PC = purchase cost corresponding to E, 1984 dollars/unit

PC_o = purchase cost corresponding to E_o, 1984 dollars/unit and A is a parameter.

Table 4 shows the values for the parameters in the equation for the eight appliances. The engineering data relating cost and energy use were fitted to the above equations using standard regression techniques. Energy use is inversely proportional to efficiency, so the cost/efficiency relationship can be obtained by inserting the value of E for a given purchase cost from equation (5) into equation (6).

$$Efficiency = (E_o / E) \ (Baseline\ Efficiency) \tag{6}$$

We first look at two appliances to illustrate the return from an investment in efficiency improvements. Because these are various classes (defining different capacities and combinations of features) within an appliance type, we then use the aggregated parameters in Table 4 to describe the life-cycle costs of eight major appliances.

Results

Figures 2 and 3 illustrate cost/efficiency curves for electric water heaters and room air conditioners with less than 8000 Btu/hr capacity, respectively. Also shown are the design options chosen for analysis. Points on the curves are generated by implementing combinations of design options to varying degrees.*

Figure 2 shows that improving the insulation and installing heat traps on electric water heaters can yield an increase in efficiency of eight percentage points (from 84 to 92 percent) at a cost of about $30. Considering that an average electric water heater sold in 1984 consumes about 4300 kWh per year and costs $256 to operate annually, the $30 investment yields an *annual* return of $22.

Figure 3 shows similarly large returns on an investment in the energy efficiency of room air conditioners up to an energy efficiency ratio (EER) of 8.8. This efficiency level is achieved by a combination of measures, as described in the figure. Using current technology and manufacturing

*These design options are described in detail for each product type in Reference 6.

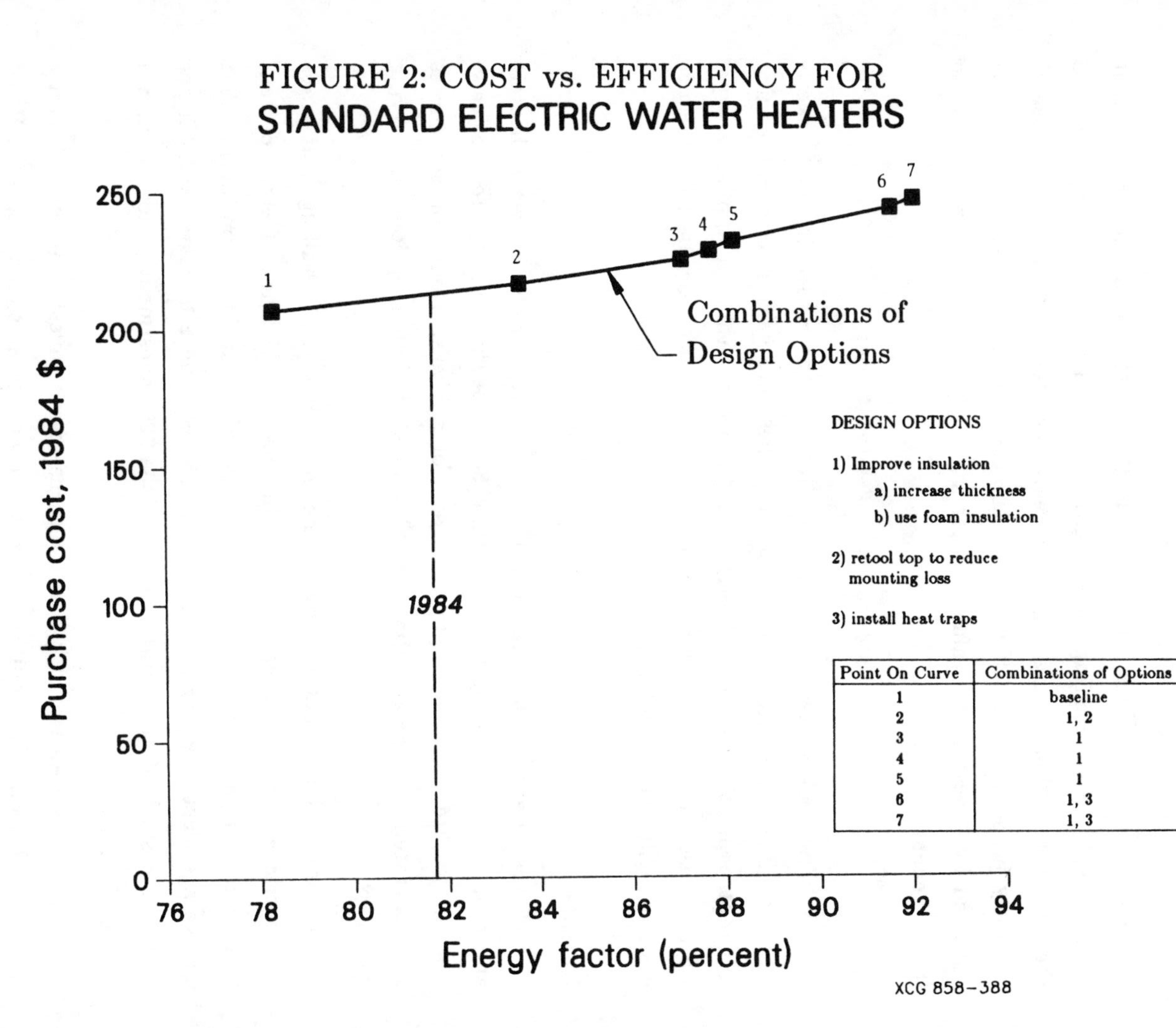
FIGURE 2: COST vs. EFFICIENCY FOR
STANDARD ELECTRIC WATER HEATERS
Purchase cost, 1984 $
250
200
150
100
50
0
76
78
80
82
84
86
88
90
92
94
Energy factor (percent)
1
2
3
4
5
6
7
1984
Combinations of
Design Options
DESIGN OPTIONS
1) Improve insulation
a) increase thickness
b) use foam insulation
2) retool top to reduce
mounting loss
3) install heat traps
Point On Curve
Combinations of Options
1 baseline
2 1, 2
3 1
4 1
5 1
6 1, 3
7 1, 3
XCG 858-388

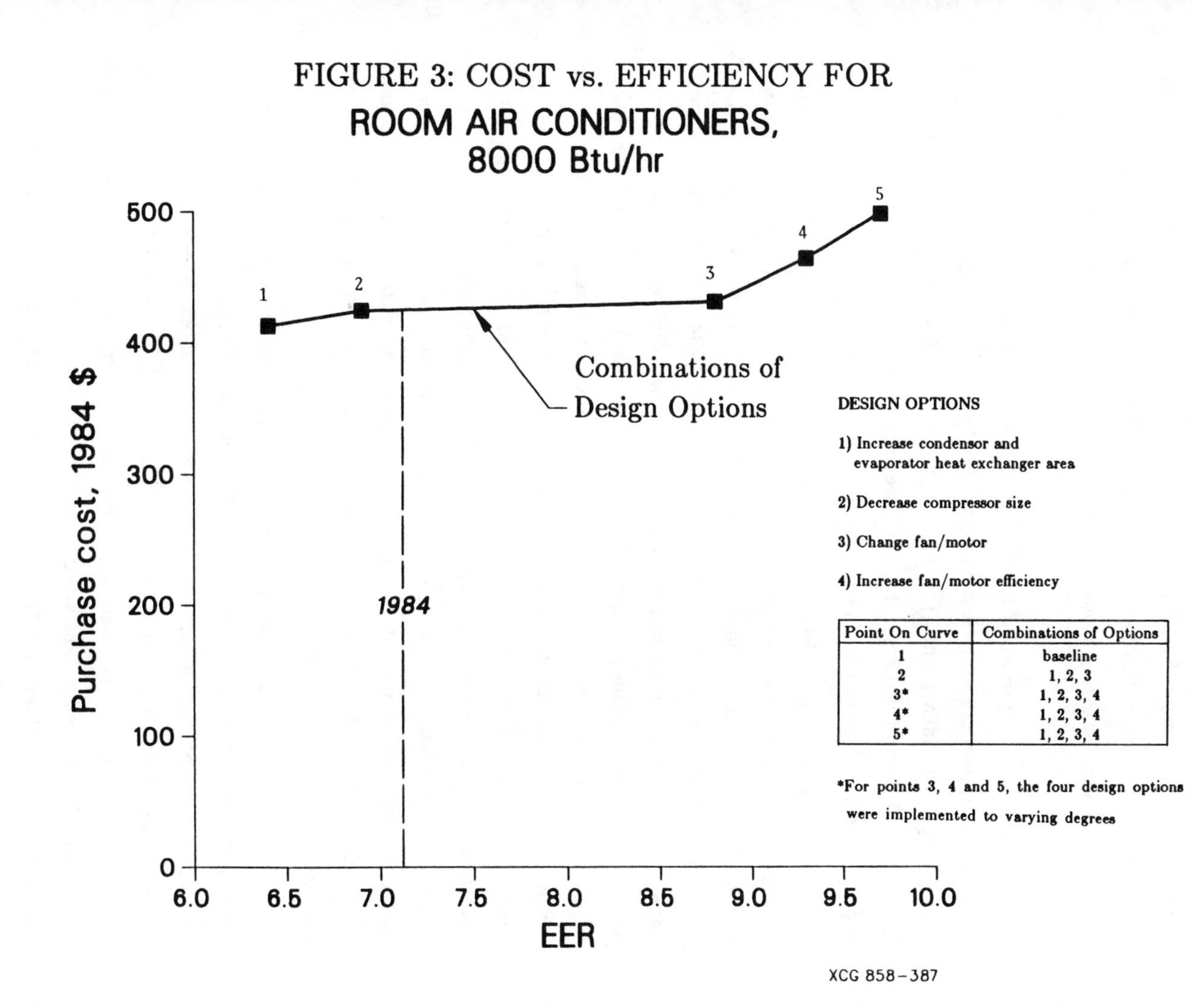

Point On Curve	Combinations of Options
1	baseline
2	1, 2, 3
3*	1, 2, 3, 4
4*	1, 2, 3, 4
5*	1, 2, 3, 4

FIGURE 3: COST vs. EFFICIENCY FOR ROOM AIR CONDITIONERS, 8000 Btu/hr

processes in the United States, the cost to increase the EER of room air conditioners in this class beyond 9 is relatively high. However, the EER of the typical new unit can be raised from 7.2 to 8.8 (an efficiency gain of 22%) at low cost.

These curves represent the cost/energy use relationships for subgroups of room air conditioners and electric water heaters. The analysis aggregates the cost/energy use data points for all subgroups within each product type (i.e., all room air conditioners), to get the aggregate cost/energy use curves represented by the parameters in Table 4.

Tables 5 through 7 illustrate the results of LCC analysis for the eight appliances. Table 5 shows the efficiency levels for each appliance corresponding to the 1984 stock efficiency, the 1984 SWEF, and the minimum LCC appliance. Table 6 compares the LCCs at the 1984 SWEF efficiency with those for the LCC minimum efficiency, using two different discount rates (3 and 10 percent). Table 7 shows the simple payback times for investing in efficiency improvements from the 1984 national average efficiency to the minimum LCC and from the 1984 SWEF to the minimum LCC, at the two discount rates. Note that these payback periods are the paybacks for an investment that improves the efficiency a discrete amount. Later we will calculate the payback for the next dollar spent on infinitesimal efficiency improvements. Because of diminishing returns, we expect the incremental payback times to be shorter than the discrete payback times, and we find this to be the case.

Table 6 reveals that all appliances examined except for central A/Cs are currently operating at life-cycle costs that are substantially higher than economically optimal levels. Freezers show the most impressive potential decrease in costs, with LCC reduced between 23% and 32% by adopting optimal efficiency levels from the 1984 national average efficiency. Freezers also show the largest potential efficiency improvement, offering a factor of 2.5 improvement in efficiency from 1984 national average levels to the LCC minimum efficiency. Note, however, that these efficiency improvements for freezers are based on prototype designs. The details of the manufacturing process are not worked out; unanticipated difficulties and costs could lower

Table 5: Efficiencies of Selected Appliances

Appliance	1984 Stock	1984 SWEF (new units)	Efficiency* of LCC min at 10%	3%
Gas Furnaces (percent)	64	73	85	90
Oil Furnaces (percent)	74	79	90	93
Room A/C (EER)	6.6	7.5	9.3	10
Central A/C (SEER)	6.9	8.7	8.0	9.5
Electric Water Heater (percent)	81	82	94	94
Gas Water Heater (percent)	50	56	80	82
Refrigerator-Freezer (ft^3/kWh/day)	4.7	6.6	11	12
Freezer(ft^3/kWh/day)	8.9	12	23	24

*For all products except gas furnaces, central air conditioners, and refrigerator-freezers, the efficiency at the LCC minimum is based on prototype designs that do not presently exist in the market.

the efficiency at the LCC minimum. The payback periods are all under nine years, and most of them are less than five. Refrigerator-freezers, freezers, and water heaters offer 1.3 to 2.6 year paybacks, excellent rates of return by any measure.

Figures 4 and 5 show the LCC curves for two representative appliances at three and ten percent discount rates: electric water heaters and room air conditioners. The solid portion of the curves corresponds to commercially available design options, while the dashed portion is an extrapolation to an estimated maximum technologically feasible efficiency point. As previously noted, for oil space heaters, gas and electric water heaters, and freezers, the appliances with LCC minimum efficiency are not commercially available. The LCC increases at low efficiencies, because of greater expenditures for fuel. It also increases at high efficiencies because of the higher cost of efficiency improvements.

FIGURE 4:
LIFE-CYCLE COST OF APPLIANCES, 1984
ELECTRIC WATER HEATERS

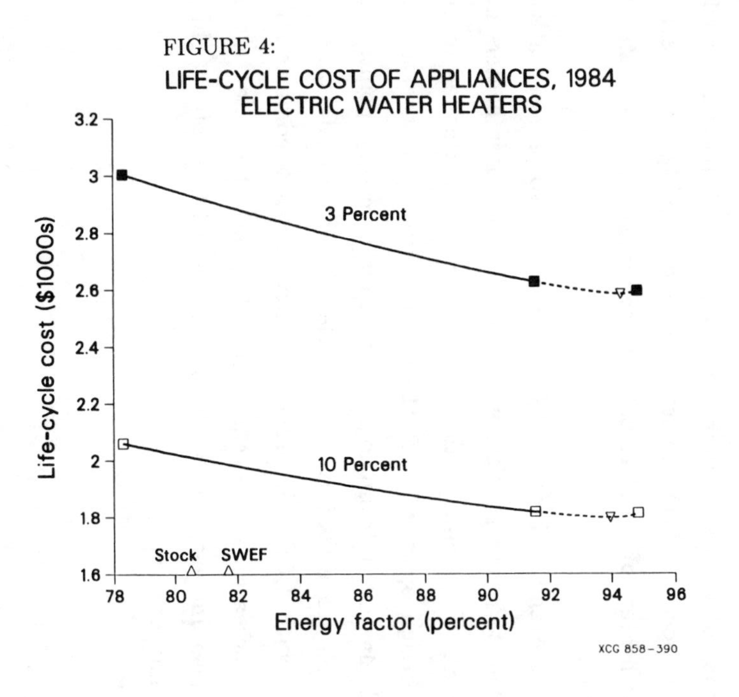

FIGURE 5:
LIFE-CYCLE COST OF APPLIANCES, 1984
ROOM AIR CONDITIONERS

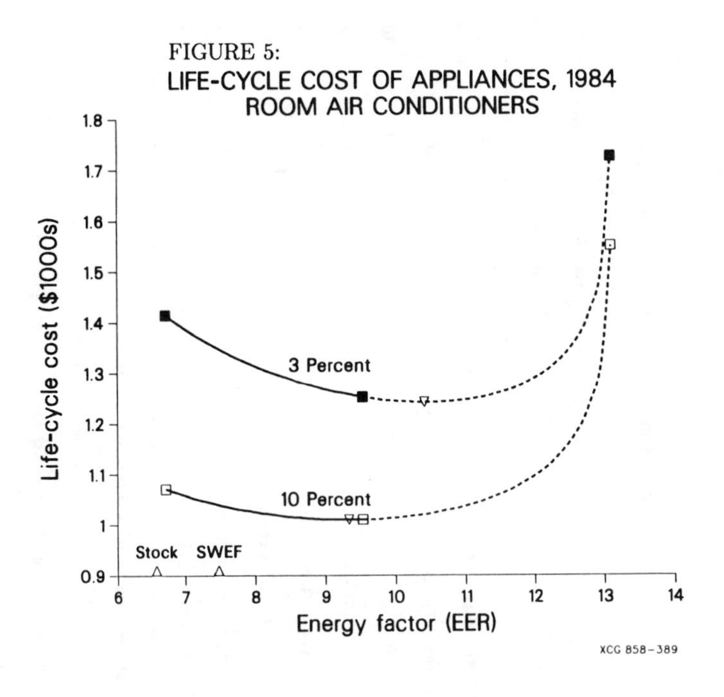

Table 6: Comparison of the LCC of Appliances with 1984 SWEF Efficiency to Those with LCC Minimum Efficiency

	3%		10%	
	LCC 1984$SWEF$ $-LCC_{min}$ (1984 dollars)	$\frac{LCC_{min}}{LCC\ SWEF}$	LCC 1984 SWEF - LCC_{min} (1984 dollars)	$\frac{LCC_{min}}{LCC\ SWEF}$
Gas Furnace	$614	0.92	158	0.97
Oil Furnace	1,240	0.92	424	0.96
Room A/C	106	0.92	30	0.97
Central A/C	16	0.99	8	1.00
Elect. Water Heater	309	0.89	190	0.90
Gas Water Heater	281	0.80	161	0.84
Refrigerator	353	0.77	179	0.84
Freezer	475	0.68	228	0.77

The location of the LCC minimum is marked on the curves. Life-cycle costs and consequently efficiency choice depend on the consumer's perception of the time value of money. Consumers with low discount rates who minimize their life-cycle costs would choose more efficient appliances than those with high discount rates. The difference is small for electric water heaters, as shown in Figure 4. For room air conditioners, on the other hand, the use of a 3 percent rather than a 10 percent discount rate will increases the efficiency of the LCC minimum appliance by almost 10 percent, as shown in Figure 5.

The economically optimum choice of electric water heater efficiency is close to the practical limit, because efficiency improvements are relatively inexpensive for this appliance. Room air conditioners show a broad minimum in their LCC curve. Consumers pay about the same LCCs when choosing air conditioners with SEERs in the range 9 to 11.5. Society as a whole, however, would benefit from the choice of higher efficiency air conditioners, because these devices reduce the need for additional peak electric generating capacity

Table 7: Payback Times for Investments in Energy Efficiency That Increase Efficiency From Existing Levels to Optimum Levels				
	Payback to Increase 1984 Existing Stock Eff. to LCCmin (in years)		Payback to Increase 1984 SWEF (new sales) to LCCmin (in years)	
Discount Rate	3%	10%	3%	10%
Gas Furnace	5.4	4.3	7.3	5.7
Oil Furnace	4.9	4.0	5.7	4.5
Room A/C	5.8	4.9	6.7	5.5
Central A/C	7.1	6.0	8.9	7.4
Electric Water Heater	1.5	1.3	1.6	1.4
Gas Water Heater	2.0	1.8	2.5	2.2
Refrigerator-Freezer	1.8	1.6	2.4	2.1
Freezer	2.3	2.0	2.6	2.2

and for the oil and gas needed to supply the peak power. Air conditioners are one example of an efficiency investment for which the economic costs and benefits to the purchaser differ markedly from the costs and benefits to society.

The numbers in Table 8 illustrate the approximate "marginal" efficiency investments for the dominant subgroups of each appliance. This table shows the simple payback times for investments that increase efficiency from the efficiency point nearest the 1984 SWEF to the next point on the cost/efficiency curve For both electric water heaters and air conditioners, the payback times calculated are for an efficiency improvement from point 2 to point 3 on the curves in Figures 2 and 3 respectively. These payback times

Table 8: Approximate Incremental Payback Times (years)	
Appliance (Dominant Class)	Incremental Payback Time
Gas Forced-Air Furnace	2.8
Oil Forced-Air Furnace	0.85
Room A/C (cap.<8000Btuh)	0.38
Split System Central A/C (cap.<39000 Btuh)	4.5
Electric Water Heaters	0.58
Gas Water Heaters	1.9
Refrigerator-Freezer (auto. def., top mount)	1.5
Chest Freezer, Manual defrost	1.1

indicate that purchasers are rejecting excellent returns on the next dollar invested in efficiency improvements.

The average appliance purchased does not include some energy efficiency measures that yield very high returns on investment. For example, an investment of $26 to include increased door insulation, a higher compressor efficiency, a double door gasket, and an anti-sweat heater switch in a refrigerator would save $28/year at 1984 fuel prices, and yield an annualized rate of return of 107 percent on the investment. Yet the average refrigerator purchased in 1984 did not have these features. Previous analysis confirms that the payback times calculated here for the dominant subgroups of each appliance type are representative of the entire market for these residential appliances* (3,4).

INTERPRETATION OF PAYBACK TIMES

The short payback times observed in this analysis suggests that the market for energy efficiency is far from rational. If a consumer demands a rate of return higher than the current loan interest rate, he or she would borrow money to purchase a more efficient appliance. The return on his/her

*The extremely short payback time calculated for room air conditioners with less than 8000 Btu/hour capacity is caused by a peculiarity in the cost/energy use curve for this subgroup. It is not representative of all room air conditioners, for which the marginal payback is closer to that of central air conditioners.

investment, namely lower fuel bills, would more than pay for the interest due on the borrowed amount. The data and previous analysis (3,4), however, indicate that this does not occur. Except for air conditioners, higher efficiency in appliances is purchased only if it pays for itself in less than three years (and much shorter time periods for several products). These results suggest that imperfections in the market inhibit economically optimal decisions. The payback times calculated in this paper for "marginal" efficiency improvements correspond to a rate of return on the investment from twenty to several hundred percent per year, far greater than real interest rates or the discount rates commonly used in LCC analysis (3,4).

Several explanations of underinvestment in energy efficiency in the residential sector can be found in the literature. (For a full discussion of this subject, see Reference 7.) These explanations include:

(1) Purchasers lack information about costs and benefits of energy efficiency improvements or may not understand how to use this information if it is available;

(2) Purchasers may not have sufficient capital to acquire funds to purchase more energy-efficient products;

(3) Purchasers may have a threshold below which savings may not be significant or worth the additional effort to obtain.

(4) The prevalence of indirect or forced purchase decisions (e.g., builder and landlord purchase of equipment for rental property; need for immediate replacement of malfunctioning equipment);

(5) The most efficient appliances may not be available in retail stores or may be available only with other features that may not be desired by most purchasers;

(6) Manufacturer's decisions to improve product efficiency are often secondary to other design changes and take several years to implement;

(7) Marketing strategies by manufacturer or retailer may intentionally lead to sales of less efficient equipment.

CONCLUSIONS

This paper demonstrates that consumers significantly underinvest in the efficiency of major household appliances. For many products, efficiency options are available that pay back in months or one to two years. These options are typically not included in new appliances.

The problem is a significant one for the nation. The appliances treated in this paper constitute more than 12 percent of U.S. energy demand, at a cost of more than $50 billion per year. Currently available, cost-effective efficiency improvements could reduce these fuel costs by $5 to $8 billion per year. Over the longer term, the national fuel bill could be reduced by $17 billion per year (net savings of $10 billion per year) through the purchase of efficiency measures at the life-cycle cost minimum. Because the market lags behind the economic "optimum," these savings are not likely to be achieved quickly without significant improvements in consumer awareness, leadership by manufacturers to produce and market cost-effective, efficient appliances, and public and private programs to strongly promote increased investment in efficient household products.

Appendix 1: Discount Rates and Life-Cycle Costs

A discount rate is a measure of the present value of money received or spent in the future. For example, if someone values an income of $110 received a year from today the same as an income of $100 received today, that person has a discount rate of 10 percent per year. Given the discount rate r, one can calculate the present value of a stream of income (or expenditures) using the formula

$$PV = \sum_{t=1}^{N} \frac{X_t}{(1+r)^t}, \tag{1}$$

where
X_t = Income in time period t
and
N = Duration of income stream.
For a constant stream of income, this formula becomes

$$PV = PWF \cdot X_t,$$

where we have defined the present worth factor PWF by

$$PWF = \sum_{t=1}^{N} \frac{1}{(1+r)^t} = \frac{1}{r}\left(1 - \frac{1}{(1+r)^N}\right). \tag{2}$$

The life-cycle cost for owning and operating an appliance is the sum of the purchase cost and the discounted operating cost. Assuming that the only operating cost is for energy, the life-cycle cost is given by

$$LCC = PC + \sum_{t=1}^{N} \frac{FC_t}{(1+r)^t}. \tag{3}$$

In this equation, PC is the purchase cost, FC_t is the fuel cost in period t, and N is the lifetime of the appliance. Maintenance costs are assumed to be independent of efficiency choice, hence they can be ignored in calculating market discount rates. For constant fuel costs, Equation 3 becomes

$$LCC = PC + PWF \cdot FP \cdot E, \tag{4}$$

where PWF is the present worth factor defined above, FP is the average fuel price, and E is the average energy consumption by the appliance. Under conditions of perfect competition, the market selects an energy use (or efficiency) that minimizes the average life-cycle cost of the appliance. Mathematically, this is equivalent to finding the energy use E_s such that

$$\left.\frac{dLCC}{dE}\right|_{E_s} = \left.\frac{dPC}{dE}\right|_{E_s} + PWF \cdot FP = 0. \tag{5}$$

Solving this for the present worth factor gives

$$PWF = \frac{-1}{FP} \left. \frac{dPC}{dE} \right|_{E_s}. \tag{6}$$

Hence, given the analytic form of the cost-efficiency curve, we can evaluate the derivative $\frac{dPC}{dE}$ at the average efficiency purchased and, using Equations 6 and 2, calculate the discount rate.

The simple payback period is defined as the time needed to recoup an initial investment in energy efficiency. Numerically, the payback period is equal to the increase in purchase cost divided by the decrease in annual operating cost. Assuming the operating costs change only because fuel use decreases, we have

$$Payback = \frac{\Delta PC}{FP \cdot \Delta E} = \frac{-1}{FP} \frac{dPC}{dE} = PWF. \tag{7}$$

Thus for a continuous cost-efficiency curve the payback period is just the present worth factor. From Equation 2, we can see that for large discount rates and long lifetimes, the payback period and discount rate are approximately reciprocal to each other.

REFERENCES

(1) J.E. McMahon, "The LBL Residential Energy Model: An Improved Policy Analysis Tool," September 1985, Unpublished Lawrence Berkeley Lab (LBL) Report: LBL-18622.

(2) "Consumer Interest in High Efficiency Appliances Grows," *Appliance Manufacturer*, May 1985, p. 10.

(3) H. Ruderman, M.D. Levine, and J.E. McMahon, "Energy Efficiency Choice in the Purchase of Residential Appliances," LBL Report LBL-17889, July 1984.

(4) H. Ruderman, M.D. Levine, and J.E. McMahon, "The Behavior of the Market for Energy Efficiency in Residential Appliances Including Heating and Cooling Equipment," LBL Report LBL-15304, Sept. 1984.

(5) I. Turiel, H. Estrada, and M.D. Levine, "Life-cycle Cost Analysis of Major Appliances," *Energy*, v. 6, no. 9, pp. 945-970, 1981.

(6) U.S. Department of Energy, Consumer Products Efficiency Standards Engineering Analysis Document, March 1982, DOE/CE-00030.

(7) M. Levine and P. Craig, "Energy Conservation and Energy Decentralization: Issues and Prospects," in *Decentralized Energy*, ed. by P. Craig and M. Levine, AAAS Selected Symposium 72, Westview Press, Inc. (1982).

CHAPTER 17

VAPOR COMPRESSION HEAT PUMP SYSTEM FIELD TESTS AT THE TECH COMPLEX

Van D. Baxter, Energy Division,
Oak Ridge National Laboratory,
Oak Ridge, Tennessee 37831

ABSTRACT

The Tennessee Energy Conservation In Housing (TECH) complex has been utilized since 1977 as a field test site for several novel and conventional heat pump systems for space conditioning and water heating. Systems tested include the Annual Cycle Energy System (ACES), solar assisted heat pumps (SAHP) both parallel and series, two conventional air-to-air heat pumps, an air-to-air heat pump with desuperheater water heater, and horizontal coil and multiple shallow vertical coil ground-coupled heat pumps (GCHP). A direct comparison of the measured annual performance of the test systems was not possible. However, a cursory examination revealed that the ACES had the best performance, however, its high cost makes it unlikely that it will achieve wide-spread use. Costs for the SAHP systems are similar to those of the ACES but their performance is not as good. Integration of water heating and space conditioning functions with a desuperheater yielded significant efficiency improvement at modest cost. The GCHP systems performed much better for heating than for cooling and may well be the most efficient alternative for residences in cold climates.

INTRODUCTION

A heat pump can be generally defined as a machine used to transfer heat from a source at one temperature to a sink at a higher temperature. The purpose of this heat transfer can be either to cool the source (e.g., the interior of a house in summer, food in a refrigerator, etc.), to heat the sink (e.g., a house interior in winter, hot water, etc.), or both (e.g. heat water while storing ice for future cooling use.) Most heat pumps fit into two basic categories: vapor compression or absorption. Since the field tests discussed in this paper involved electrically driven, vapor compression systems, only the vapor compression cycle will be considered herein.

0094-243X/85/1350323-24$3.00 Copyright 1985 American Institute of Physics

The field tests were conducted as part of the Tennessee Energy Conservation in Housing (TECH) program, initiated in 1975 as a joint venture of the Energy Research and Development Administration (forerunner of the Department of Energy (DOE)), the Oak Ridge National Laboratory (ORNL), the University of Tennessee (UT), and the Tennessee Valley Authority (TVA) with the objective of demonstrating methods of conserving energy in residences. Construction of the TECH complex was begun in 1976 when the first three houses and a garage/storage building were built. In 1979 the garage was converted into a fourth test structure and a fifth house and new storage building were added in 1981. Figure 1 is an aerial view of the site as it now exists on the UT Institute of Agriculture Experiment Station grounds near Knoxville, Tennessee.

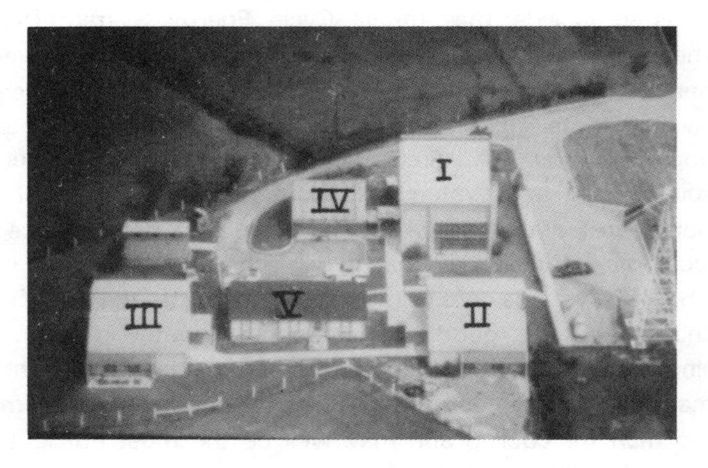

Figure 1. The Tennessee Energy Conservation in Housing Complex.

The three original houses are of identical floor plan (house III is a mirror image of houses I and II) with 170 m^2 of conditioned floor space and, insofar as possible, of identical construction. Several non-standard construction features are incorporated into the houses including exterior walls of 50.8 mm by 152.4 mm (2 in x 6 in) studs on 0.61 m (2 ft) centers to accommodate thicker wall insulation and air lock entries and magnetic weather stripping to minimize air infiltration. Complete construction details and floor plans may be found in Ref. [l] for house I and Refs. [2] and [3] for houses II and III.

THE VAPOR-COMPRESSION REFRIGERATION CYCLE

Mechanical compression systems (either electrically or engine-driven) are predominant among refrigeration cycles used by heat pumps. The coefficient of performance (COP) expresses the effectiveness of a refrigeration or heat pump system. It is a dimensionless ratio defined by the expression:

$$COP = \frac{\textit{Useful heating or cooling effect}}{\textit{Purchased energy required}}$$

A schematic of an ideal vapor-compression cycle along with temperature-entropy (T-s) and pressure-enthalpy (P-h) diagrams is shown in Fig. 2. Saturated, low-temperature, low-pressure vapor (state 3) is compressed to a higher temperature and pressure (state 4) by work input through the

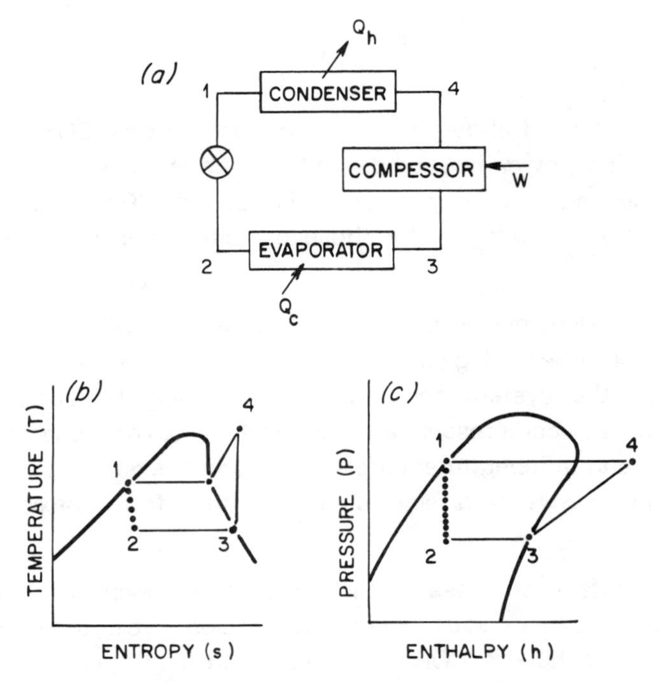

Figure 2. The ideal vapor-compression cycle.

compressor (W). Heat (Q_h) is released by the hot refrigerant through the condenser, heating its surroundings and leaving the refrigerant in a saturated liquid state (state 1). The liquid is throttled through the expansion device to state 2. Heat (Q_c) is absorbed by the cold refrigerant through the evaporator, cooling its surroundings. The refrigerant leaves the evaporator as a saturated vapor at state 3 completing the cycle. Applying the steady-state steady-flow energy equation from the first law of thermodynamics on a unit mass flow basis we have:

$$W = h_4 - h_3,$$

$$Q_h = h_4 - h_1,$$

$$Q_c = h_2 - h_3, \text{ and}$$

$$h_1 = h_2.$$

COP's for heating (COP_h) and cooling (COP_c) are:

$$COP_h = \frac{Q_h}{W} = \frac{h_4 - h_1}{h_4 - h_3}$$

and

$$COP_c = \frac{Q_c}{W} = \frac{h_3 - h_2}{h_4 - h_3}$$

The COPs as defined above are ideal compressor-only COPs. In an actual system parasitic power for pumps and/or fans is required to transport the deliverable heating or cooling where it is needed. Thus, for a real system W would include power use for those parasitic devices as well as for the compressor.

The thermodynamic cycle of an actual system will depart significantly from the ideal cycle of Fig. 2. This departure occurs because of pressure drops within the system components (principally the compressor), heat losses from the compressor and connecting piping, compressor motor inefficiency, and temperature differences across system heat exchangers. These factors result in a significant departure from maximum potential performance.

The net effect of these departure or loss mechanisms is illustrated clearly in Fig. 3. An ideal vapor-compression cycle using R-22 as the refrigerant would have a COP of 21.45 when it is operating between the temperature limits of 21 and 8°C. Temperature differences across the heat exchangers, a necessary condition for operation of a real system, reduce the computed COP substantially. Heat exchanger refrigerant temperatures observed during steady-state tests of an air-to-air heat

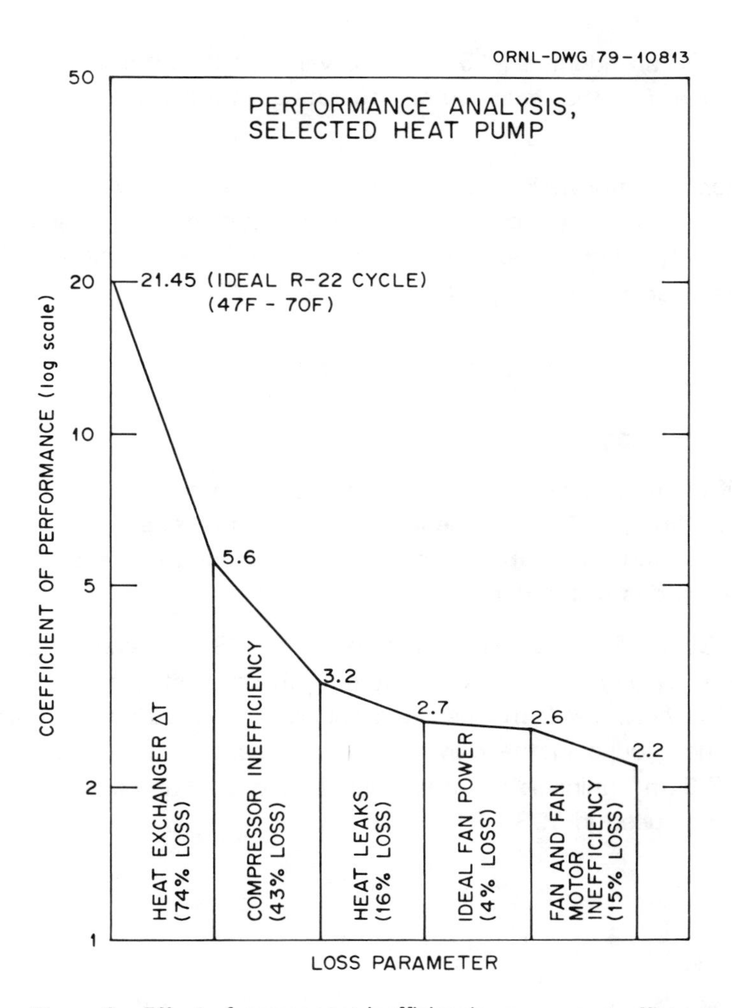

Figure 3. Effect of component inefficiencies on system efficiency.

pump, typical of those sold before 1978, at 8°C outdoor air temperature and 21°C indoor air temperature were 50°C condensing and -1.2°C evaporating. The computed COP using these temperature differences in an otherwise ideal system is 5.6, a 74% reduction from the maximum potential COP.

Departure from ideal operation within the compressor reduces the COP to 3.2, an additional 43% loss. Heat losses observed at the compressor shell reduce the COP an additional 16% to 2.7. Ideal fan power consumption (necessary to transport the heat throughout a house) would reduce the COP to 2.6; the observed fan and fan-motor inefficiencies reduce the COP an additional 15% to the value of 2.2, which is about the observed value for the heat pump operating in steady state at these conditions [4].

More general information about vapor-compression and other heating, refrigeration, and air-conditioning systems may be obtained from references [5-7]. For an excellent overview of heat pumps and factors affecting their performance, see [8].

THE FIELD TESTS

Tech House I Tests

Two different heat pump systems have been tested in TECH house I. From 1979 through 1981 a series solar assisted heat pump (SSAHP) system was tested. In 1981 a horizontal coil ground coupled heat pump was installed and is still under test.

Series Solar Assisted Heat Pump. The SSAHP system is shown schematically in Fig. 4. Two different thermal storage tanks were available. The first, a 4.2 m^3 insulated concrete tank buried just below the surface of the ground in the crawlspace, was used in 1979-80 while the second, a 7.6 m^3 uninsulated steel tank buried 2.1 m deep just outside the house, was used in 1980-81.

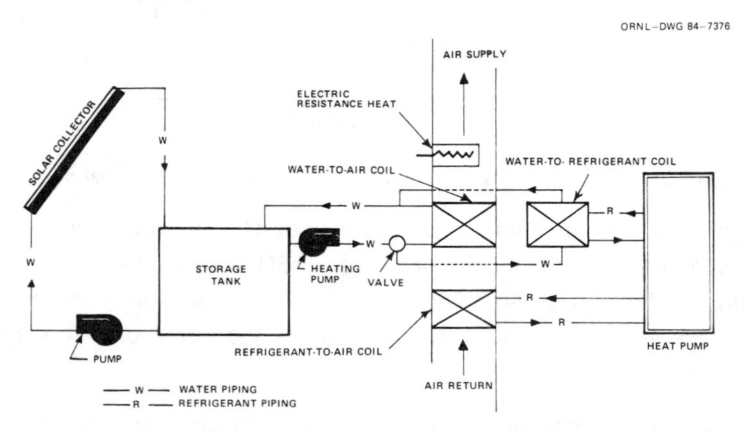

Figure 4. Series Solar-assisted heat pump (SSAHP) system schematic.

The system had a modified air-to-air heat pump added to a direct solar system to supply auxiliary heat in lieu of electric resistance heat. A refrigerant-to-water heat exchanger was added to the heat pump to serve as the evaporator in the space heating mode. Water was circulated to this heat exchanger from the solar storage tank. An 8 kW electric resistance heater was used for third stage heating with the SSAHP system and the outdoor coil of the heat pump was utilized as the heat pump condenser in the cooling mode. Thus, the SSAHP could heat the space by extracting heat from the solar storage tank through the water-to-refrigerant coil and cool the space while rejecting heat through the outdoor coil. No domestic water heating was done by the SSAHP system.

Table I gives a summary of the SSAHP monthly and seasonal performance factors (SPF) for space heating-only. These SPFs are calculated by:

$$\text{SPF} = \frac{Q_t}{EC_t}$$

where

Q_t = total heating/cooling energy delivered (monthly or seasonal)

and

EC_t = total energy consumed to provide Q_t (monthly or seasonal)

Table I. Seasonal Performance Factors (SPF) for SSAHP system.

	Nov.	Dec.	Jan.	Feb.	Mar.	Season
1979-80 Season	4.55	4.43	1.70	1.80	3.27	2.42
1980-81 Season	--	2.74	2.67	3.06	5.43	3.16

Cooling load was not measured during the test period; therefore, cooling performance factors are not presented. When the system was switched from the insulated tank to the uninsulated tank a significant reduction in peak power consumption (11.2 kW to 4.3 kW) as well as an increase in SPF was realized because no electric resistance heat was required during the second test season. Heat transfer from the ground through the uninsulated steel tank and the increased storage volume appeared to contribute to this peak reduction. Other unquantified variables (weather severity, etc.) also played a role [12]. SSAHP heating SPFs for both years were higher than those of the air-source heat pumps tested in house III.

However, installation cost of an SSAHP is likely to be much higher than that of the conventional system. A more complete description of the SSAHP system and test results can be found in Refs. [12-14].

Horizontal Coil Ground-Coupled Heat Pump System. In 1981 and 1982 a horizontal coil ground-coupled heat pump (GCHP) system was installed. The system consisted of a unidirectional, water-to-water heat pump, an air handler with a water-to-air heat exchanger for distributing conditioned air to the space, and a ground-coupled heat exchanger. A 20% by weight methanol-water brine solution was used in the ground and fancoil loops to guard against freezing of the heat exchangers. Manually controlled three-way valves were used to direct the hot and cold brine streams to the proper heat exchangers for space cooling and heating. Fig. 5 is a schematic of the heat pump package.

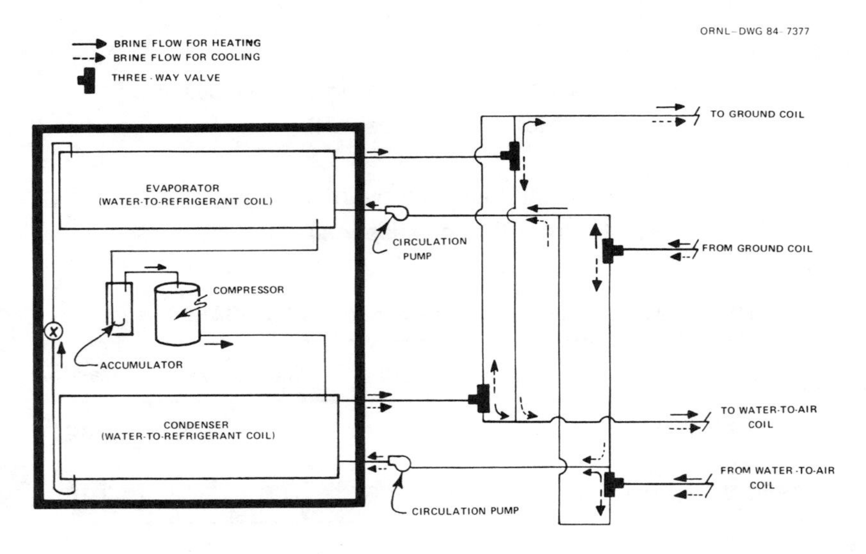

Figure 5. Ground-coupled heat pump (GCHP) mechanical package schematic.

The ground coil is 206 m long and consists of 32 mm nominal diameter polybutylene pipe buried approximately 1.2 m deep. A schematic of the coil layout is shown in Fig. 6.

Table II presents monthly and seasonal loads and performance factors for the 1983-1984 heating season and the 1984 cooling season. Since the heat pump mechanical package could be located either inside or outside the conditioned space, two seasonal performance factors (for space conditioning only) are presented [15].

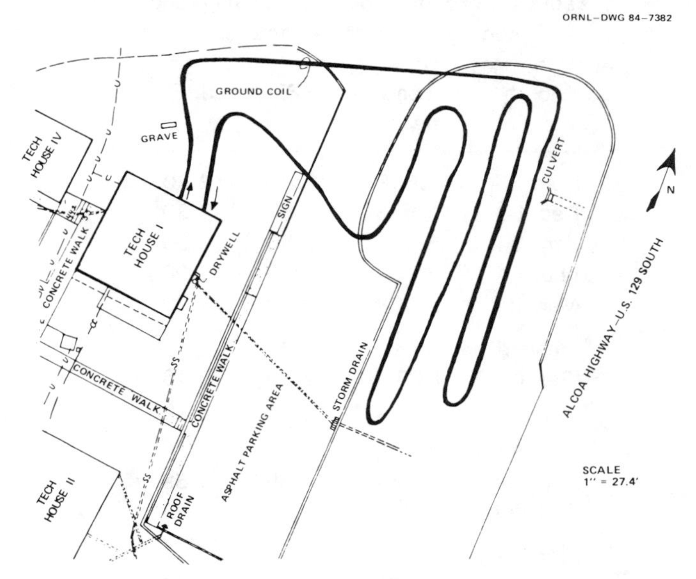

Figure 6. Layout of TECH house I ground coil.

SPFi is the performance factor for the case in which the heat pump package is located within the conditioned space as is the case for the test installation and is calculated by:

$$SPF_i = \frac{Q_{t_n}}{EC_t}$$

where Q_{t_n} = net heating/cooling delivered.

For the heating season

$$Q_{t_n} = Q_g + EC_t$$

where Q_g = heat extracted from the ground,

and for the cooling season

$$Q_{t_n} = Ql_c - Q_l$$

where Ql_c = cooling energy delivered through air handling unit

and

Q_l = heat losses to internal space from circulation pump(s) and compressor.

Table II. Horizontal coil GCHP heating and cooling performance factors.

Month	Load (kWh)	SPFi	SPFo
Nov. 83	1,609	2.80	2.30
Dec. 83	3,633	2.80	2.30
Jan. 84	3,401	2.50	2.00
Feb. 84	2,367	2.50	2.00
Mar. 84	2,308	2.40	2.20
Heating Season	13,318	2.59	2.16
June 84	1,242	1.34	1.72
July 84	1,296	1.43	1.70
Aug. 84	1,661	1.35	1.69
Sep. 83	416	1.20	1.79
Cooling Season	4,604	1.35	1.71
Annual Total	17,922	2.10	1.98

SPFo is the performance factor for the case where the heat pump package is located in an unconditioned space (i.e. a garage). It is calculated by

$$SPFo = \frac{Ql_{h/c}}{EC_t}$$

where

$Ql_{h/c}$ = heating/cooling delivered through air handling unit

As can be seen from Table II, locating the heat pump package within the conditioned space is clearly beneficial during the heating season but is a disadvantage for cooling operation. It is also clear that the GCHP's heating performance is very good but its cooling performance (even had the heat pump been outside the space) was poor.

This poor cooling performance was due to inadequate ground heat transfer. The primary causes were (1) the ground tended to dry out over the course of the summer resulting in reduced soil thermal conductivity, and (2) the ground coil was undersized for cooling [16]. Cooling performance was also adversely affected by the fact that the heat pump used is not designed for space cooling applications.

Tech House II Tests

Systems tested in house II were the Annual Cycle Energy System (ACES) from 1976 through 1982 and a multiple vertical coil GCHP from 1983 to the present.

ACES System Tests. From November, 1977, through September, 1982, TECH house II was used for testing of a residential ACES [9,10,11]. ACES is an integrated space conditioning and water heating system consisting of a unidirectional heat pump with low-side thermal storage and, if needed, an auxiliary solar collector or outdoor fancoil. Fig. 7 is a schematic showing the basic ACES components. During winter the heat pump extracts energy from an insulated storage tank filled with water

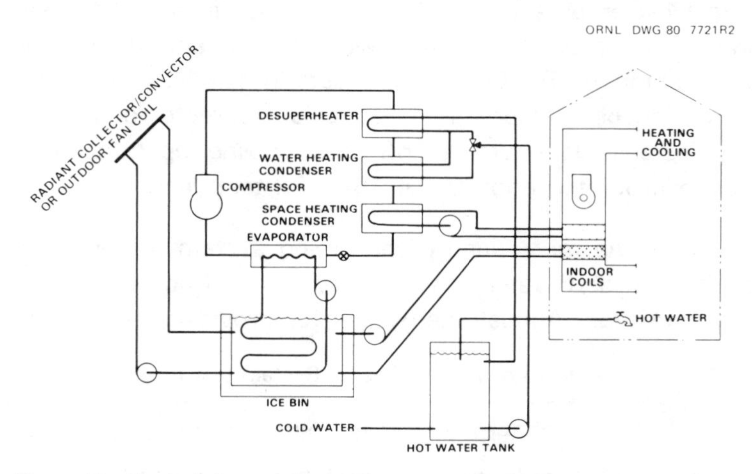

Figure 7. Typical Annual Cycle Energy System (ACES) schematic.

and boosts it to provide space and water heating. As the heating season progresses, the water is converted into ice to be used for space cooling the following summer. The solar collector or fancoil would be used to supply extra heat to the storage tank (in areas where more ice could be made than was needed) or to provide a heat sink for late-summer, off-peak air-conditioning operation (for areas where the winter's ice production is insufficient to meet the building cooling needs). Use of the thermal storage gives the ACES its two primary energy conserving features: (1) a constant-capacity heat pump (eliminating the need for electric resistance backup heating) and (2) interseasonal transfer of cooling energy from winter to summer (greatly reducing energy consumption for cooling). A review of the theory and design requirements of residential ACES is given in [17].

Table III gives the results of four annual test cycles performed at house II. The annual performance factors (APFs) given in Table III are for space conditioning and water heating and are calculated by

$$APF = \frac{\text{Space heating} + \text{space cooling} + \text{water heating}}{\text{annual purchased energy consumption}}$$

During the 1977-78 test it became apparent that heat transfer from the earth to the thermal storage bin was much higher than anticipated causing early summer ice exhaustion [9]. Insulation levels on the bin walls, floor, and ceiling were increased in October, 1978, to reduce this heat transfer level. Stored ice was maintained until late August of the 1978-79 and 1979-80 test cycles as a result [10]. Indoor air flow and fan energy use for cooling were reduced in 1980 with a resultant increase in annual performance factor (APF). Component tests in 1981 indicated that use of an outdoor fancoil rather than a solar collector/convector would significantly increase late-summer off-peak cooling operation [11]. This feature was incorporated prior to the 1981-82 test year.

The ACES is the most energy conserving system for providing annual space conditioning and water heating needs for a Knoxville area residence that we have tested. Annual energy use for providing these services was

Table III. Annual Cycle Energy System test results.

Test Year	1977-1978	1978-1979	1979-1980	1981-1982
Test Period	11/1/77- 9/18/78	12/1/78- 9/30/79	12/1/79- 9/1/80	9/28/81- 9/28/82
Space Heating (GJ)	43.07	32.85	34.40	37.31
Space Cooling (GJ)	26.21	19.17	22.65	24.40
Water Heating (GJ)	20.84	15.81	14.34	14.16
Annual Energy Use (kWh)	9012	6719	6447	7360
APF	2.78	2.80	3.08	3.05

about 50% less than that required by a high efficiency air-source heat pump/electric water heater system in TECH house III [10]. A survey of field test results of numerous solar systems indicated that ACES has a better APF than solar systems as well [18]. Economic analyses have shown that energy savings for ACES will take about 25 years to pay back its incremental first cost (about $7,000-$9,000 for the Knoxville system in a mature industry) relative to air-source heat pump systems under present economic conditions [19-21]. The estimated performance of ACES across the United States is shown in Fig. 8. This map shows that ACES is most efficient in the central U.S. where there is a good balance between heating and cooling needs.

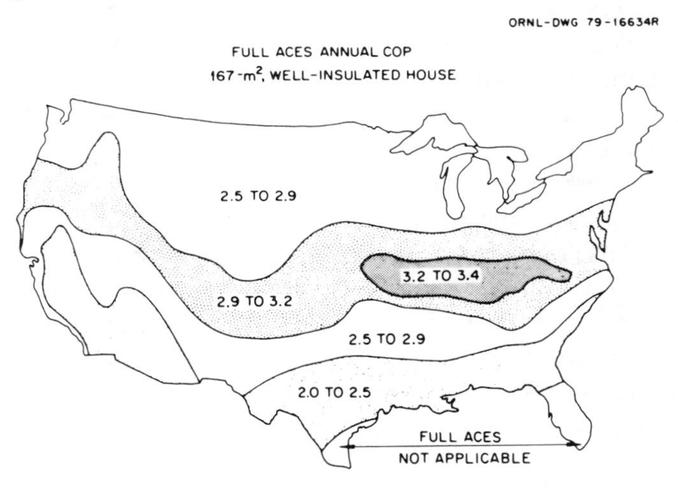

Figure 8. Full ACES APFs for the United States for a 167-m^2, well-insulated house.

Multiple Shallow Vertical Coil Ground-Coupled Heat Pump. Between October, 1982, and June, 1983, the GCHP system was installed in TECH house II. As illustrated schematically in Fig. 9, this system consisted of a water-to-water heat pump with a water-to-air fancoil, identical to those used for the TECH house I GCHP, and a ground coil composed of six vertical heat exchangers. These heat exchangers ranged in depth from 18.3 to 30.5 m and were composed of 38.1 mm diameter high density polyethylene tubing arranged in a U-tube configuration. As for the house I system, a 20% by weight methanol-water brine was used in the ground and fancoil loops. Testing has just been completed and reports will be available in the future.

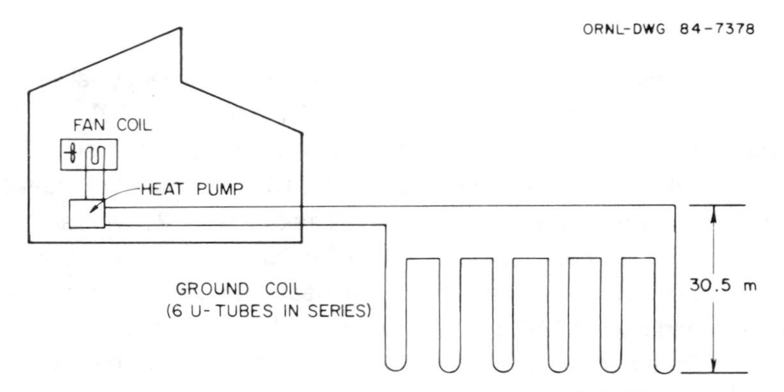

Figure 9. TECH house II ground coupled heat pump (GCHP) schematic.

Table IV gives the results of seasonal performance factor tests (for space conditioning only) from July, 1983, through April, 1984. The SPFi's include the effect of the intermediate brine loop between the heat pump and the air handler and the crankcase heater energy consumption. Actual installations would probably use water-to-air heat pumps so SPFi would be somewhat better than that of this experimental system. The same, of course, holds true for the horizontal coil GCHP. Both the heating and cooling SPFi's were better than those of the TECH house I system primarily because most of the house II ground coil was in saturated soil or rock and, thus, was not subject to wide temperature swings or summertime soil dryout associated with relatively shallow pipe burial depths. Heating performance was better than that of any of the air-source heat pumps tested in house III as well, but cooling performance was not as good as that of the high efficiency air-source heat pump. Unlike that of the house I system, however, the cooling performance of the vertical coil GCHP was reasonable. In addition, the overall APF (for space conditioning only) of the vertical coil GCHP was superior to that of the air-source systems.

Tech House III Experiments

Air-Source Heat Pump System Tests. During the 1978-79 and 1979-80 test years, two air-source heat pump/electric water heater systems were tested in house III [22]. A standard efficiency heat pump (COP = 2.46 at 8.3°C) was used for the 1978-79 season (heat pump I) while a high efficiency (COP = 3.11 at 8.3°C) unit from the same manufacturer (heat pump II) was used during subsequent years. Table V summarizes the results of the annual performance of each system.

Table IV. Multiple shallow vertical coil GCHP heating and cooling performance.

Period	Load (kWh)	Energy Consumption (kWh)	SPFi
6/27-7/29/83	2,094	1,174	1.78
7/29-8/24/83	2,200	1,284	1.71
8/24-9/30/83	1,233	701	1.76
9/30-10/10/83	122	76	1.73
Cooling Season	5,649	3,235	1.75
10/31-11/28/83	1,206	390	3.09
11/28/83-1/3/84 [a]	2,383	784	3.04
1/3-1/30/84	3,475	1,243	2.80
1/30-3/2/84	3,214	1,157	2.78
3/2-4/2/84	2,133	771	2.77
4/2-4/26/84	383	150	2.55
Heating Season	12,994	4,495	2.85
Annual Totals	18,443	7,730	2.39

[a] Does not include 12/22-12/30.

Comparing Tables III and V shows that the APFs (including water heating) of the air-source systems are lower than those of the ACES.

In addition to the system comparisons with ACES, it was desired to measure the actual seasonal performance of air-source heat pumps and compare it to their rated performance. Fig. 10 shows measured monthly heating COPs and rated steady-state COPs plotted against outdoor temperature for both heat pumps tested. It is interesting to note that, while the steady-state COP values increased with increasing temperature the field measured monthly values tended to decrease. Cycling losses at high ambient temperatures degraded performance by as much as 45% compared to rated steady-state performance.

After the 1980 test year, instrumentation was added to heat pump II in an attempt to measure directly the magnitude of the different dynamic losses that affect a heat pump's field performance (frosting, defrosting, and cycling) [23]. Table VI illustrates the fraction of measured annual energy consumption due to the various losses.

Table V. Air-source heat pump/electric water heater system test results.

	1978-1979	1979-1980
Test Period	12/1/78-9/30/79	12/1/79-9/15/80
Space Heating (GJ)	32.85	34.29
Space Cooling (GJ)	19.18	22.70
Water Heating (GJ)	13.74	13.57
Energy Cons. (kWh)	12853	11358
System APF	1.42	1.73
Heat Pump Only[a]		
SPF, heating	1.58	1.99
SPF, cooling	1.64	2.27
APF	1.60	2.10

[a] not including water heating.

Table VI. Heat pump II energy use due to dynamic losses 1982-83 heating and cooling seasons

Component	% Total Energy Consumed
Steady-state Energy Use to Meet Load	75.8
Frosting Losses	2.3
Defrosting Losses	5.9
Start-up Transient Losses	9.8
Off-cycle Parastics	6.2
Total	100.0

Air-Source Heat Pump/Desuperheater Water Heater System. During the 1981-82 test year, in alternate months with the dynamic loss measurement test, heat pump II was operated with a desuperheater water heater to determine its effect on system performance compared to that of the same heat pump with an electric resistance water heater [24]. The desuperheater was a small heat exchanger designed to recover excess or wasted heat from the refrigeration cycle and use it to heat domestic hot water. A schematic of the system is shown in Fig. 11, with test results given in Table VII.

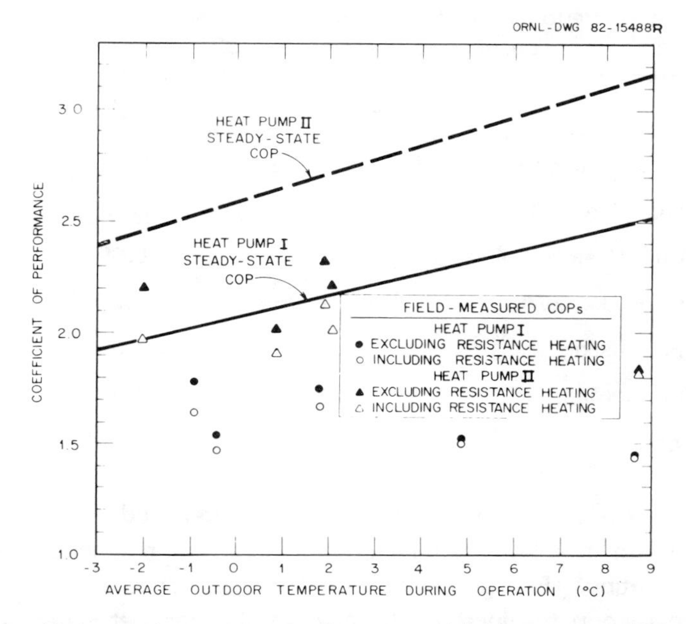

Figure 10. Air-source heat pump steady-state and field-measured COP vs. temperature.

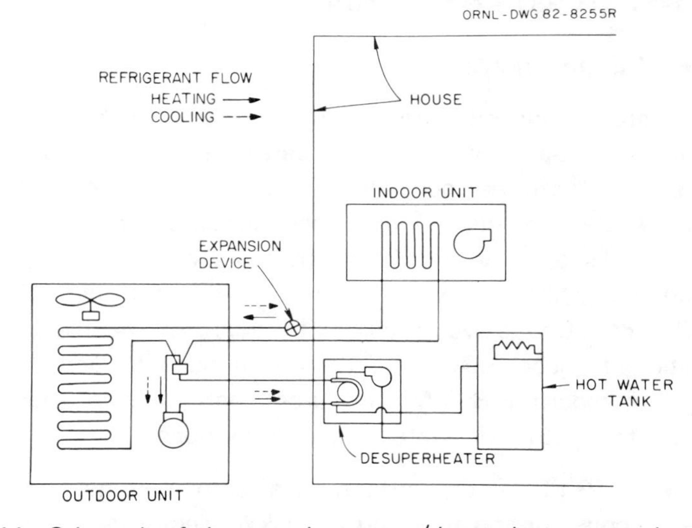

Figure 11. Schematic of air-source heat pump/desuperheater water heater system.

Test results showed a 14% improvement in APF for the desuperheater system over the electric water heater system due almost entirely to summertime heat recovery water heating by the desuperheater. Heating seasonal performance of the systems proved to be nearly identical, a result

Table VII. Performance of heat pump II with desuperheater vs. electric water heater, 1981-82

	Desuperheater	Electric
Space Heating (GJ)	18.58	19.65
Space Cooling (GJ)	13.85	11.96
Water Heating (GJ)	8.26	8.69
Energy Use (kWh)	5782	6525
SPF, Heating	1.68	1.63
SPF, Cooling	2.50	1.90
APF	1.96	1.72

not unexpected since in winter the desuperheater reduces the space heating capacity of the heat pump forcing increased reliance on electric backup space heating. Energy savings of the desuperheater system could be expected to recoup the incremental cost of the desuperheater unit in 7 to 9 years based on 1981 electricity costs.

Tech House IV Experiments

Parallel Solar Assisted Heat Pump System With and Without Off-peak Electric Heat Storage. The parallel solar assisted heat pump (PSAHP) system, illustrated schematically in Fig. 12, has three main components. A 29.4 m^2 air heating solar collector array collected solar energy when available. The individual collectors had a double glazing of fiberglass and teflon film and were inclined at 45° and oriented 10° west of south. Thermal storage was provided by a 7.4 m^3 pebble bed (located in the conditioned space) containing 25.4 mm nominal diameter crushed limestone. A standard 8.8 kW air-source heat pump provided backup heat for the system [25]. No water heating was done.

During the 1979-80 heating season an attempt was made to minimize on-peak energy consumption by using a 15 kW electric heater to heat the pebble bed during utility off-peak hours. In on-peak periods, the system operated like a conventional PSAHP. That is, solar heat or heat from storage, if available, was used to heat the space. If neither was available the heat pump provided space heating. When no space heating was required and solar energy was available, it was used to charge the pebble

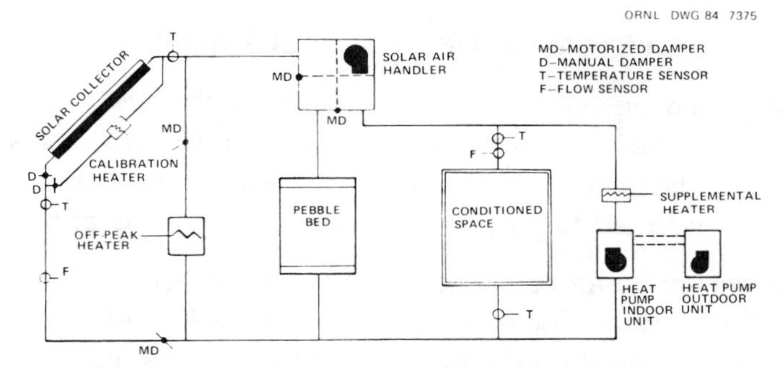

Figure 12. Parallel solar-assisted heat pump (PSAHP) system schematic.

bed storage. During the 1980-81 and 1981-82 heating seasons, the system was operated as a straight PSAHP without off-peak storage heating.

Table VIII is a summary of the system performance factors for all three heating seasons. The 1979-80 overall SPF of 1.61 was lower than that of subsequent years due to the off-peak electric heat storage feature. Eliminating that option resulted in much improved overall efficiency. From these results it is obvious that an overall SPF penalty must be paid in order to incorporate off-peak electric heating into the system. An on-peak to off-peak electric rate differential of greater than 7 to 1 would be required in order to offset the added energy consumption imposed by using electric heat for off-peak storage charging [26].

Table VIII. PSAHP system seasonal performance factors (space heating only).

Month	1979 - 1980 Season	1980 - 1981 Season	1981 - 1982 Season
November	2.59	2.65	2.95
December	1.77	2.52	2.16
January	1.32	2.01	1.39
February	1.46	2.39	2.55
March	1.68	3.59	3.87
Season	1.61	2.56	2.11

SYSTEM COMPARISON SUMMARY

A cursory comparison of the performance of the field tested systems is included in this section. It must be clearly understood, however, that this comparison is based on test results from a single location. They should not be construed to be generally applicable to all locations.

A direct annual performance comparison of the various systems tested at the TECH site is not possible since not all delivered the same loads under the same weather conditions. However, a rough comparison, made by assuming certain load deliveries, is offered in Table IX. For this comparison, all systems that delivered no water heating were assumed to have an electric water heater delivering 14.2 GJ/yr (same as ACES during 1981-82). Similarly, for the solar systems, a cooling load of 24.4 GJ/yr was assumed at a cooling SPF of 2.27 (same as heat pump II tested in house III). In addition, the two solar systems were assumed to deliver a 37.3 GJ/yr heating load at their demonstrated heating SPFs. Measured heating and cooling SPFs (sans water heating) and estimated system costs are also given in Table IX for comparison.

Several observations can be drawn from the comparison in Table IX and the test results presented in this paper. One is that integration of the water heating and space conditioning functions (via use of a desuperheater) yields significant performance benefits for a modest investment. Use of a desuperheater could have improved the performance of the other heat pump systems by at least as much as, if not more than, it did that of the air-source heat pump.

The GCHP systems performed only marginally better than the air-source heat pump in Knoxville. Had this comparison been done at a more northern location, however, the GCHPs would have compared more favorably due to their much better heating performance. Indeed, field tests of GCHP systems in the Syracuse, NY, area yielded heating and cooling SPFs in excess of 3.0.[27] Similarly, test results for the solar assisted systems show that they perform better during warmer months. Thus, their heating performance should be better in warmer locations and in areas that receive more sunshine. The SAHP systems tested offer no advantages over air source heat pumps for cooling.

The ACES is by far the most energy-efficient system tested at the TECH site. This may not be the case in the extreme southern and

Table IX. System performance and cost comparison for Knoxville, TN.

System	APF	Heating SPF[a]	Cooling SPF[a]	Estimated Costs (1985 $)
SSAHP	2.07[b,c,d]	3.16	-	15000+
PSAHP with no off peak electric heating	1.92[b,c,d]	2.56	-	15000+
Air Source Heat Pump[e]	1.73	1.99	2.27	4000
Air Source Heat Pump with desuperheater	1.96	1.99	2.27	4500
Ground coupled heat pumps[f]				5000-6000
Horizontal coil	1.75[d]	2.59	1.35	
Vertical coil	1.92[d]	2.85	1.75	
ACES	3.05	-	-	15000+

[a] assumed 37.3 GJ heating load delivered at demonstrated system SPFH.

[b] assumed 14.2 GJ hot water load delivered at COP of 1.0.

[c] high efficiency heat pump system with electric water heater.

[d] Water-to-water heat pump systems (energy consumption of two circulation pumps included), located inside conditioned space.

[e] Space conditioning service only (no water heating).

[f] assumed 24.4 GJ cooling load delivered at SPFC of 2.27

northern areas of the U.S. It is unlikely that either ACES or the solar assisted heat pumps will achieve wide-spread use due to their much greater complexity and cost of installation [14]. It is probable that the air source heat pump/desuperheater combination will be difficult to match economically for climates similar to and warmer than Knoxville. For colder climates, GCHP systems (with desuperheaters) may well offer the best alternative of the systems discussed in this paper.

ACKNOWLEDGEMENT

The author would like to thank Mr. Bernie A. McGraw of the University of Tennessee for his assistance in compiling the performance information on the solar assisted heat pumps for this paper.

Research sponsored by the Building Equipment Research Division, U. S. Department of Energy under contract DE-AC05-84OR-21400 with Martin Marietta Energy Systems, Inc.

REFERENCES

1. *Solar House Design Report,* Tennessee Valley Authority, Knoxville, Tennessee.
2. J. C. Moyers et al., *Design Report for the ACES Demonstration House,* ORNL/CON-1, Oak Ridge National Laboratory, Oak Ridge, Tennessee (October 1976).
3. A. S. Holman and V. R. Brantley, *ACES Demonstration: Construction, Startup, and Performance Report,* ORNL/CON-26, Oak Ridge National Laboratory, Oak Ridge, Tennessee (October 1978).
4. A. A. Domingorena and S. J. Ball, *Performance Evaluation of a Selected Three-Ton Air-to-Air Heat Pump in the Heating Mode,* ORNL/CON-34, Oak Ridge National Laboratory, Oak Ridge, Tennessee (January 1980).
5. ASHRAE Handbook, 1981 Fundamentals, American Society of Heating, Refrigerating and Air-Conditioning Engineers, Atlanta, Georgia.
6. ASHRAE Handbook, 1983 Equipment, American Society of Heating, Refrigerating and Air-Conditioning Engineers, Atlanta, Georgia.
7. J. L. Threlkeld, *Thermal Environmental Engineering,* Second Edition, Prentice-Hall, Inc., Englewood Cliffs, New Jersey, 1970.
8. V. W. Goldschmidt, *Heat Pumps: Basics, Types, and Performance Characteristics,* Annual Review Energy 1984.9, pp. 447-472. Annual Reviews, Inc. 1984.

9. A. S. Holman et al., *Annual Cycle Energy System (ACES) Performance Report November 1977 through September 1978*, ORNL/CON-42, Oak Ridge National Laboratory, Oak Ridge, Tennessee (May 1980).

10. V. D. Baxter, *ACES: Final Performance Report December 1, 1978 through September 15, 1980*, ORNL/CON-64, Oak Ridge National Laboratory, Oak Ridge, Tennessee (April 1981).

11. V. D. Baxter, *ACES Tests at the TECH Site: 1981*, ORNL/CON-96, Oak Ridge National Laboratory, Oak Ridge, Tennessee (August 1983).

12. D. J. Roeder, *A Computer Simulation of a Residential Solar Assisted Heat Pump System with Ground Coupled Thermal Storage.* M.S. Thesis, University of Tennessee, 1982.

13. R. L. Reid, et al., *Final Report - Evaluation of Solar Assisted Heat Pumps and Solar House*, Report to TVA on Contract No. TV-42885A Supplements 2, 3, 4, 1981.

14. B. A. McGraw, et al., "Experimental Evaluation of a Series Solar Assisted Heat Pump System," *Proceedings of the 1981 Annual AS/ISES Meeting in Philadelphia, Pennsylvania*, pp. 562-566.

15. S. Johnson, B. A. McGraw, et al., *TECH House I Horizontal Ground Coupled Heat Pump: 1982-83 Heating Season Performance*, ORNL/Sub/82-7685/1&92.

16. S. Johnson, et al., *TECH House I Horizontal Ground Coupled Heat Pump: 1983 Cooling Season Performance*, ORNL/Sub/82-7685/2&92.

17. E. A. Nephew, L. A. Abbatiello, and M. L. Ballou, *Theory and Design of an Annual Cycle Energy System (ACES) for Residences*, ORNL/CON-43, Oak Ridge National Laboratory, Oak Ridge, Tennessee (May 1980).

18. R. L. Reid and L. A. Abbatiello, "Seasonal Performance Factors of Active Solar Systems and Heat Pump Systems," *Proceedings of the 16th Intersociety Energy Conversion Engineering Conference.* Atlanta, Georgia, August 9-14, 1981, pp. 1684-1690.

19. L. A. Abbatiello, et al., *Performance and Economics of the ACES and Alternative Residential Heating and Air-Conditioning Systems in 115 U.S. Cities,* ORNL/CON-52, Oak Ridge National Laboratory, Oak Ridge, Tennessee (March 1981).

20. E. A. Nephew and L. A. Abbatiello, *Performance and Economics of Eight Alternative Systems for Residential Heating, Cooling, and Water Heating in 115 U. S. Cities,* ORNL/CON-89, Oak Ridge National Laboratory, Oak Ridge, Tennessee (November 1982).

21. V. D. Baxter, et al., *Annual Cycle Energy System Performance and National Economic Comparisons with Competitive Residential HVAC Systems,* ASHRAE Transactions 1982, Vol. 88, Pt. 1, pp. 1279-1294.

22. V. D. Baxter, L. A. Abbatiello, and R. E. Minturn, *Comparison of Field Performance to Steady State Performance for Two Dealer-Installed Air-to-Air Heat Pumps,* ASHRAE Transactions 1982, Vol. 88, Pt. 2, pp. 941-953.

23. V. D. Baxter and J. C. Moyers, *Air Source Heat Pump: Field Measured Frosting, Defrosting, and Cycling Loses, 1981-83,* ORNL/CON-150, Oak Ridge National Laboratory, Oak Ridge, Tennessee (November 1984).

24. V. D. Baxter, *Comparison of Field Performance of a High Efficiency Air-to-Air Heat Pump With and Without a Desuperheater Water Heater,* ASHRAE Transactions 1984, Vol. 90, Pt. 1, pp. 180-190.

25. R. L. Reid, J. J. Tomlinson, et al., *Performance Evaluation of Parallel Solar Augmented Heat Pump with Off-Peak Heat Storage,* Proceedings AS/ISES 1980 Annual Meeting, Phoenix, Arizona, June 2-6, 1980, pp.44-47. CONF-08155, NTIS.

26. D. J. Chaffin and R. L. Reid, "Experimentally Verified Simulation of a Parallel Solar/Off-Peak/Heat Pump Air Heating System," *Proceedings of the 1980 Annual Meeting of the American Section of IES,* Phoenix, Arizona, June 2-6, l980, pp. 339-343.

27. P. J. Hughes, R. A. O'Neil, J. Rizzuto, and L. Loomis, *Results of the Residential Earth Coupled Heat Pump Demonstration in Upstate New York,* ASHRAE Transactions 1985, Vol. 91, Pt. 2B.

CHAPTER 18

INDUSTRIAL ENERGY CONSERVATION*

Marc Ross
Physics Department
University of Michigan
Ann Arbor, Michigan 48109

ABSTRACT

Industrial processes are extraordinarily diverse, so the physical conditions for energy conservation are diverse. The opportunities for conservation also depend on the economic and technological outlook of each industry. This chapter briefly examines the subject as a whole and explores a few examples to a little depth. These issues are touched upon: the structure and trends in use of energy, the thermodynamic factors that influence energy intensity,and examples of technical change enabling the reduction of energy intensity--from operations to conservation equipment to revolutionary changes in manufacturing process. The future of the energy-intensive industries and energy conservation is also discussed.

I. INTRODUCTION

During the first seven decades of the twentieth century energy prices fell dramatically and the energy forms purchased by final consumers became much easier to use. Energy use rose six fold in the United States during this period, an average growth of over 2 percent per annum. The energy supply industry created an enormous capital base: oil and gas fields, refineries, power plants and energy transportation systems. A huge construction industry grew up to create these kinds of facilities. Moreover, energy users adapted their capital to the cheap easy-to-use energy: heavy manufacturers came to rely primarily on relatively simple natural gas-fueled equipment; commercial buildings were designed to overwhelm user and climatic variations with energy rather than through efficient design; and during the 1950s and 1960s automobile and appliance energy efficiencies fell as the real prices of gasoline and electricity fell.

---

*Notation: When customary U.S. units are used, to avoid confusion I use $\overline{M}$ to represent one million, and K one thousand, tons are short tons. I denote the metric ton by tonne. Thus 1 $\overline{M}$Btu/ton = 278 Mcal/tonne = 1.163 GJ/tonne.

0094-243X/85/1350347-21$3.00 Copyright 1985 American Institute of Physics

This pattern dramatically changed in the 1970s: Fuel costs increased, energy conversion costs significantly increased, and certain energy supplies were temporarily interrupted.

The nation's response has been equally dramatic if not as well publicized: U.S. energy use doubled in the 17 years preceding 1973; but in 1984 it was about 1% less than in 1973.[1] This halt in energy growth is not primarily associated with a slowdown in the growth of economic activity. Energy consumption per unit of economic activity has been falling rapidly since 1970.

In Figure 1.1 the recent history of energy consumption by industry is shown. Absolute energy use declined 12% from 1973 to 1984. The ratio of energy consumed by industry to the constant-dollar Gross National Product (GNP) declined a startling 32% during the same period. In Western Europe and Japan, however, energy prices rose even more sharply and even more energy conservation has been carried out. Most of the cost-effective conservation investments remain to be made in U.S. industries.

Declining energy use per unit of economic activity is and will remain in the forseeable future a much more important factor than increased energy production. For this reason I believe that society should give at least as high a priority, in education and research and in capital spending, to improving the efficiency of energy use, as it gives to new sources and supplies of energy. In this article I will discuss the factors that affect the use of energy by industry and the potential for continued reductions in energy use per unit of production.

THE STRUCTURE OF INDUSTRIAL ENERGY USE

The US Department of Energy includes manufacturing, mining, agriculture, and construction in the industrial sector.[1] Industrial energy use divides roughly: manufacturing 78 percent, mining 10 percent, agriculture 6 percent, construction 6 percent.[2] Overall, industrial energy use is 38% of total energy use in the US (1984).[1]

Industry mainly uses gas, electricity and oil. Gas and oil consumption have been dropping rapidly while electricity consumption climbed gradually from 1973 to 1979 and has been stagnant in the five years since. (Figure 1.2). Oil use is somewhat specialized; it is primarily used as a fuel in petroleum refining, as a feedstock in making organic chemicals and as a motor fuel in mining, construction, and agriculture. Coal is used primarily in steelmaking. Coal is also, of course, a major fuel in generating electricity. Wood is an important source of energy for the forest products, especially the paper, industry.

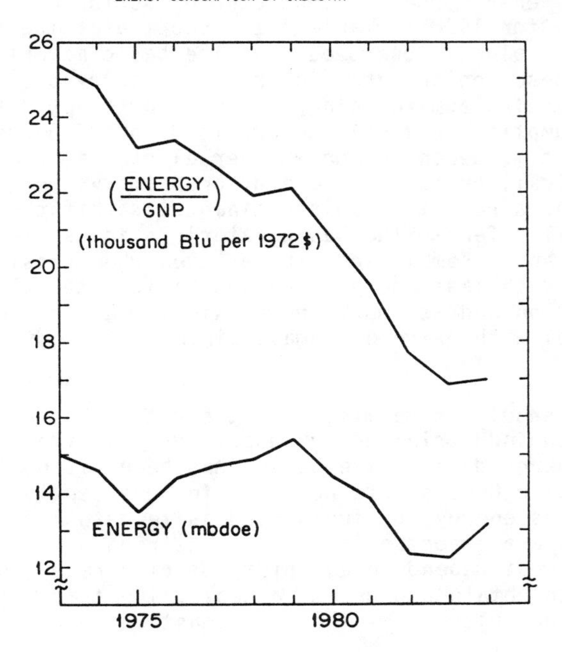

Fig. 1.1. Energy Consumption by Industry.
1 Mbdoe = 2.12 quads/yr = 70.8 GW
1 thousand Btu = 1.055 MJ
Source: Monthly Energy Review, Ref. 1.

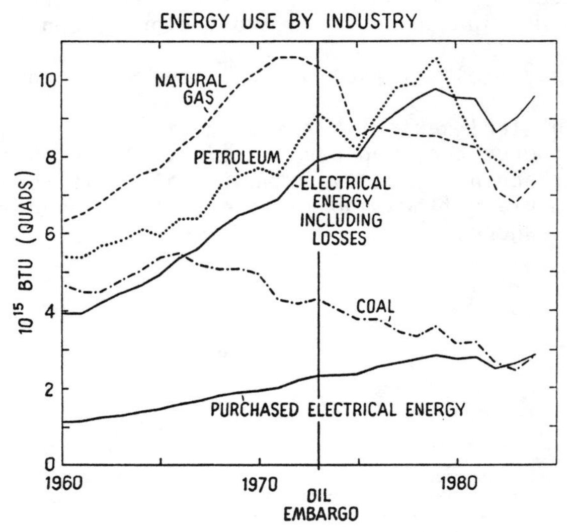

Fig. 1.2. Source: Ref. 4

A matrix of energy use by manufacturing sector and by energy carrier is shown for 1980 in Table 1.1. Looking at the second from last or the last columns, one sees that the basic materials sectors predominate. These sectors are (in order of energy use): chemicals, primary metals, petroleum refining, paper, and stone, clay and glass. The energy consumption in these sectors is 81 or 77 percent of the manufacturing total, depending on whether electrical use is counted in terms of electrical, or carrier, energy or primary energy, respectively. As a result of this dominance, specific discussion in this article will refer to the basic materials industries. The reason why fabrication and assembly industries, even when considered heavy industries, use much less energy than the basic materials industries is that fabrication and assembly are physical rearrangements at the macroscopic level with extremely small minimum thermodynamic requirements.

The reader shouldn't be misled by the detail in Table 1.1. The data available on industrial energy use is very limited. Annual collection of energy data by the Census has been discontinued since 1981. In any case Census data does not include captive energy (including biomass energy) or fuels used as feedstock for organic chemicals, which are important in some industries as seen in Table 1.1. One must depend on a variety of disparate sources and on extrapolations to obtain and update a table like this. The information on agriculture, mining, and construction is much poorer.

In this brief report most of the discussion will refer to aggregate energy use. In practice the different forms are substitutable one for the other only to a limited extent (e.g. using boilers which can burn more than one fuel), unless major new investments are made.

It will be useful to organize our discussion of changes in industrial energy use in terms of the two factors shown by the equation:

$$\text{(energy use)} = \text{(level of activity)} \times \text{(energy intensity)}$$

Thus energy use in steel making is the product of the tons of steel produced and the energy consumed per ton of steel. Both factors have been and will be changing. Energy intensity alone is addressed in this article. Materials flows and their effects on energy use have been addressed elsewhere.[5,6]

TABLE 1.1

ENERGY USE IN MANUFACTURING IN 1980 (Quadrillion Btu)

SIC	Manufacturing Sector	Coal & Coke	Petroleum Products	Natural Gas	Other	Total Fuels	Purchased Electricity	Total Carriers	Generation & Transmission Losses	Total Primary
20	Food	0.13	0.12	0.52	0.03	0.81	0.140	0.95	0.33	1.28
21	Tobacco	0.01	0.00	0.00	0.00	0.02	0.005	0.02	0.01	0.03
22	Textiles	0.04	0.05	0.11	0.01	0.21	0.088	0.29	0.21	0.50
23	Apparel	0.00	0.01	0.02	----	0.04	0.021	0.06	0.05	0.11
24	Lumber	0.00	0.05	0.07	0.03	0.15	0.050	0.20	0.12	0.32
25	Furniture	0.00	0.01	0.02	0.00	0.03	0.013	0.05	0.03	0.08
26	Paper - purchased f.f.&e.[a]	0.24	0.38	0.43	0.06	1.11	0.170	1.28	0.40	1.68
	- wood derived[b]				[1.05]	[1.05]		[1.05]		[1.05]
27	Publishing	----	0.01	0.04	0.00	0.06	0.033	0.09	0.08	0.17
28	Chemicals - excl. feedstock	[0.35]	[0.20]	[1.59]	[0.12]	[2.26]	[0.454]	[2.72]	[1.08]	[3.80]
	- feedstock		[2.11]	[0.59]		[2.70]		[2.70]		[2.70]
	- total	0.35	2.31	2.18	0.12	4.96	0.454	5.42	1.08	6.50
29	Petroleum - purchased	[0.01]	[0.06]	[0.97]	[0.03]	[1.07]	[0.110]	[1.18]	[0.26]	[1.44]
	- captive[c]		[1.68]			[1.68]		[1.68]		[1.68]
	- total	0.01	1.74	0.97	0.03	2.75	0.110	2.86	0.26	3.12
30	Rubber	0.02	0.03	0.09	0.00	0.15	0.074	0.22	0.18	0.40
31	Leather	----	0.00	0.01	0.00	0.01	0.005	0.02	0.01	0.03
32	Stone, clay & glass	0.37	0.07	0.57	0.01	1.02	0.104	1.12	0.25	1.37
331	Steel - purchased	[0.45]	[0.11]	[0.60]	[0.04]	[1.19]	[0.193]	[1.38]	[0.46]	[1.84]
	- captive[c]	[1.00]				[1.00]		[1.00]		[1.00]
	- total	[1.45]	[0.11]	[0.60]	[0.04]	[2.19]	[0.193]	[2.38]	[0.46]	[2.84]
3334	Aluminum	----	[0.08]	[0.17]	[0.01]	[0.27]	[0.265]	[0.54]	[0.63]	[1.17]
33	Primary metals total	1.53	0.23	0.99	0.05	2.80	0.560	3.36	1.33	4.69
34	Fabricated metal prod.	0.01	0.04	0.22	0.01	0.27	0.086	0.36	0.20	0.56
35	Non-electrical machinery	0.03	0.03	0.16	0.01	0.23	0.104	0.33	0.25	0.58
36	Electrical equip.	0.01	0.02	0.11	0.01	0.15	0.093	0.24	0.22	0.46
37	Transportation equip.	0.05	0.04	0.14	0.02	0.24	0.102	0.34	0.24	0.59
38	Instruments	----	0.01	0.03	0.00	0.06	0.020	0.08	0.05	0.13
39	Misc. Mfr.	0.00	0.01	0.01	0.00	0.03	0.012	0.05	0.03	0.07
	TOTAL MANUFACTURING	2.83	5.16	6.70	0.39	15.09	2.245	17.34	5.34	22.68

Notes:

[a]fossil fuel and electricity.

[b]wood-derived fuels are not included in industry totals.

[c]captive fuels are fuel materials extracted by the firm that burns them.
(The Census omits coal used for coking and petroleum at refineries diverted to fuel use, the two major categories of captive fuels.)
Source: Marc Ross, Natural Resources Journal (Ref. 3).

II. THE ENERGY INTENSITY OF MANUFACTURING

A. DETERMINANTS OF ENERGY INTENSITY

The energy intensities for certain basic materials are shown in Table 2.1. This is the energy used within each manufacturing sector to produce an average ton of product. (Note that the particular numbers depend on accounting conventions.) Ideal thermodynamic minimum energies to manufacture these materials are also shown in Table 2.1. Only in the case of reduction of metal ore are the availability requirements really large.[7] Some chemical rearrangements in petroleum refining and petrochemical production also have significant availability requirements. Physical rearrangement, such as the separation of components in wood to make paper and the shaping of metals, requires very little energy ideally. Typically those process stages with large availability requirements like reducing iron ore in a blast furnace or reducing alumina in an electrolytic cell not only require a great deal of energy use, but are carried out fairly efficiently. (Carrier energy use is roughly 50% efficient or higher in these cases.) Because so much energy is involved, it nevertheless pays to continue striving to make the process more energy efficient. The great majority of process stages in industry do not, however, have substantial availability requirements and are in this ideal sense astonishingly energy inefficient. In these cases huge relative energy savings are, in principle, possible and are in some cases are being achieved.

Not only is the energy efficiency poor for almost all processes which are exothermic or weakly endothermic, but it often costs very little to substantially reduce energy use in accomplishing the same purpose. That is, the cost as a function of energy efficiency tends to have a very broad minimum. This is illustrated by Figure 2.1, which presents a simplified picture of a rather complex situation. If the availability required per unit of service is low or negative (part a of the figure), very large decreases in the energy intensity of the process can be achieved economically (especially through technological change as suggested by shifting from t_1 to t_2). Since the cost curve has a broad minimum, the optimum, and thus the desirable level of conservation is not well determined. In other words higher energy prices motivate a decrease in the energy intensity as suggested by the steep curve at upper left in Figure 2.1a, but they do not determine how far one should strive to decrease the energy intensity. For this reason corporate and public policies often play a strong role in controlling the level of energy conservation. They, rather than costs, often determine the energy intensity which will be achieved.

The situation is rather different when the laws of thermodynamics require the use of a lot of availability (part b of the figure). Here the scope for percentage reduction of energy intensity is relatively

TABLE 2.1

ENERGY INTENSITIES FOR SELECTED BASIC-MATERIALS

	Energy Intensity 1980 (Gcal/tonne)		Thermodynamic Minimum[c] (Gcal/tonne)
	primary energy[a]	carrier energy[b]	
Paper	7.3[d]	5.6[d]	--[e]
Steel	9.4	7.9	1.7
Chemicals	---	17 (for polyethylene)	--
Aluminum	46[f]	21[f]	7.0[h]
Petroleum Refining	1.17	1.08	0.1
Cement	1.7	1.4	0.2

[a]with purchased electricity evaluated at about 2.9 Mcal/kwh (11,500 Btu/kwh) as per Table 1.1.
[b]with electricity evaluated at 0.86 Mcal/kwh (3413 Btu/kwh).
[c]Gyftopoulos et al, Ref. 6.
[d]wood derived fuels not included.
[e]For paper the absolute value of the minimum is small and its sign depends on accounting conventions and product.
[f]The energy intensities are per tonne of shipped product. If the base is taken to be tonnes of primary plus secondary metal the energy intensities are 17% higher and if the base is tonnes of primary metal it is 37% higher.
[h]per tonne of primary metal.

FIGURE 2.1

COST OF SERVICE VS. ENERGY INTENSITY (SCHEMATIC)

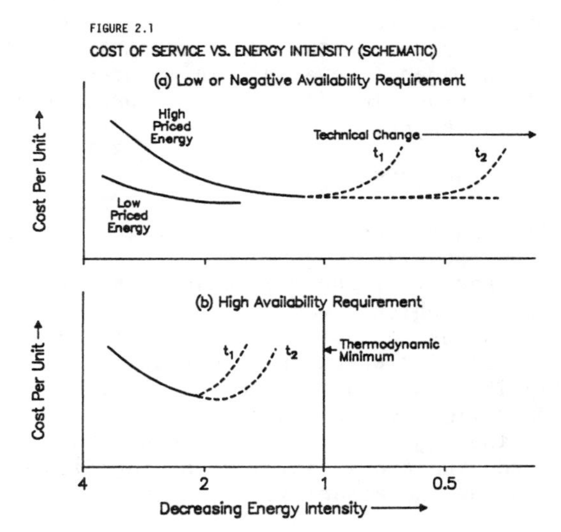

small because most processes are already fairly efficient. Concomitantly, cost considerations rather powerfully determine the optimum level of energy intensity.

Because of conservation efforts, the energy used to produce a unit of a given material has been declining. This decline began in some areas before the oil embargo of 1973. It has accelerated since then (Table 2.2). The rate of decline in the average energy intensity since 1972 has been about 2% per year. This decreasing energy use per ton has been largely driven by increasing prices for energy (Figure 2.2.). Energy prices paid by industrial customers roughly tripled relative to the average price of other purchases 1973-1982. Compared to industrial value added--the cost of labor, management and capital--the average cost of energy to industry has risen from 5% in the late '60s and early '70s to over 10%.

The cost of energy in 1980 compared to value added in particular basic materials manufacturing sectors is shown in Table 2.3. While manufacturing exclusive of basic materials has a cost of energy to value added ratio of only three percent, several major basic materials sectors have ratios of 1/4, 1/3, or more. It is seen that several basic materials sectors (as defined by two-digit Standard Industrial Classifications) have high energy-cost subsectors. The pattern tends to be that upstream activities are energy- intensive (in this context, high energy use per dollar of value added) and that downstream activities are labor-intensive. Energy analyses based on all-industry averages or even on 2-digit SIC averages must be examined critically because of these order-of-magnitude differences in the energy-cost ratio among various subsectors of industry.

I repeat, however, that direct cost considerations are not the only important motivation for industrialists to increase energy efficiency. The threat of energy shortages is another important motivation as is the societal goal to reduce the dependence on imported oil. Some manufacturers have a technological orientation; they like to do things right, within cost constraints. In addition, there is a pattern to major innovations in manufacturing processes: they tend to create savings in all factors of production: labor, capital, materials and energy.

B. TECHNICAL CHANGE AND ENERGY INTENSITY

The kinds of technical change which lead to improved efficiency of energy use can be roughly categorized:

1) Changes in operations and maintenance, and retrofits with low cost equipment, which lower energy use.
2) Changes in energy-intensive equipment or energy conservation add-on technologies which involve significant investment (typically $50,000 to a few tens of million dollars) and are largely justified by reduced energy costs.
3) Changes in the major processes of production. Often major processes require a new facility costing $100 million or more, but not necessarily.

TABLE 2.2

REDUCTION OF ENERGY INTENSITY[a]
IN THE BASIC MATERIALS INDUSTRIES (1972-1983)

	Percent
Chemicals[b]	31
Steel	18
Aluminum	17
Paper[c]	26
Petroleum refining[d]	10
Energy Weighted Reduction	21

[a]Generally energy per pound of product, unadjusted for for environmental and other changes. Purchased electricity accounted for at 10,000 Btu/kwh (2.5 Mcal/kwh).
[b]Not including fuels used as feedstock.
[c]Not including wood-based fuels.
[d]Changes in inputs and outputs and environmental regulations have had a particularly strong impact on petroleum-refining energy. Adjusted for such changes, energy intensity was reduced 26%.

SOURCE: Trade Association Reports (Ref. 9).

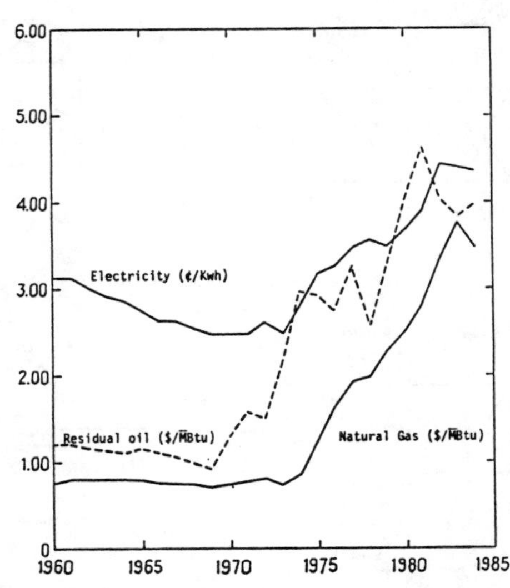

Fig. 2.2. Price of Energy to Industrial Users. National Average, 1980$. Source: Ref. 4

TABLE 2.3

THE RATIO OF THE COST OF ENERGY TO VALUE ADDED, AND TO VALUE OF SHIPMENTS, PERCENT (1980)

SIC	Industrial Sector	Compared to Value Added	Compared to Value of Shipments
26	Paper & Allied Products[a]	16	6
261-3	Pulp & Paper Mills[a]	31	11
28	Chemicals & Allied Products[b]	12,23	
281	Industrial Inorganics	27	
286	Industrial Organics[b]	21,76	
29	Petroleum Refining	35	4
32	Stone, Clay, & Glass	15	8
3241	Hydraulic Cement	45	
33	Primary Metals	23	8
331	Basic Steel	32	
3334	Primary Aluminum	46	
20-39	Manufacturing except sectors 26, 28, 29, 32 and 33	3	2

[a]Cost of wood-derived fuels not included.
[b]Energy cost without cost of organic feedstocks and with cost of organic feedstock are shown, respectively.

Source: Marc Ross, Natural Resources Journal (Ref. 3).

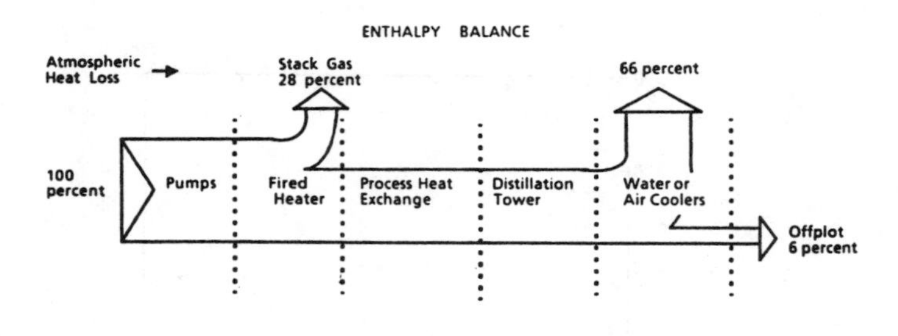

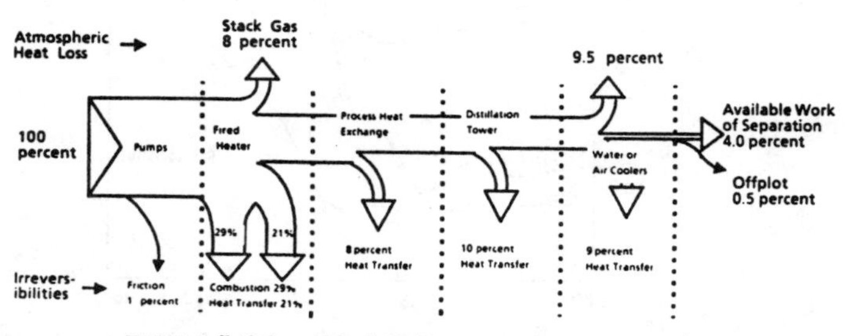

FIGURE 2.3 **Enthalpy and Available Work Balance for a Crude Separation Unit**

The available data from the first decade after the oil embargo suggest that comparable reductions in energy intensity have occurred through operational improvements, and through a combination of investments in major process change and energy-conservation equipment.

1. IMPROVED OPERATIONS

This is, in part, what is called housekeeping. In order to make good progress a well-qualified staff is needed to carry out energy conservation activities, with top management leadership and support. Among general practices and technical changes are management practices such as:

- inspections to encourage conservation activity,
- training programs for operation of energy-intensive equipment,
- scheduling of energy-intensive activities, such as turning off motors when not in use, and turning down heaters as appropriate,
- systematic maintenance programs,
- accounting procedures to charge energy costs to production departments, not to general overhead, and

low-level investment programs such as:

- direct metering of major energy-using facilities, and
- sophisticated inspection and maintenance equipment such as infrared scanners.

One way to improve operations which has proved successful at some plants is employee participation in energy conservation, including systematic solicitation of employee suggestions for technical changes (e.g. using quality circles).

2. ADD-ON EQUIPMENT
HEAT AND POWER RECOVERY

Heat recovery is one of the most important conservation technologies, but its importance can be exaggerated as shown by second-law analysis.[10] In Fig. 2.3 energy use in a crude distillation unit at a petroleum refinery is shown from first-law (enthalpy) and second-law (availability) perspectives. The second-law analysis shows that about 30% of the availability of a fuel is lost in the (irreversible) process of combustion. Most of the rest of the availability is lost in the thermal degredation of heat in distillation. That is the essence of the distillation process: the entire mass of material is raised to the maximum temperature by direct heating and then various components decline in temperature as they rise through the tower. A moderate amount of availability (8%) is lost with the hot gases up the heater stack. Most of the enthalpy (66%) but relatively little availability (9%) is lost in cooling the product streams. This discrepancy results from the relation of the availability, B, the work available in principle from the heat Q, and Q. For an infinite reservoir, at temperature T:

$$B = \frac{T - T_0}{T} Q,$$

where T_0 is the temperature to which materials can be cooled.

In the case at issue where the reservoir is finite (assuming constant heat capacity)

$$B = C\int_{T_0}^{T} dT' \ (T'-T_0)/T' = Q \ (1- \ \frac{T_0}{T-T_0} \ \ell n(T/T_0))$$

If for example the "dead state" is at 38°C (100°F), and the typical temperature of products is 150°C (300°F), then B/Q = .145, corresponding to the result shown for heat rejected by air and water coolers in Fig. 2.3.

The implication of this is that heat recovery from stack gas tends to be economically justified because the temperature is high. Heat recovered at high temperature can, for example, be transformed into steam and used elsewhere. On the other hand, if there is no use for low-temperature heat very nearby, it probably doesn't pay to recover it. Generation of electricity with organic Rankine cycle equipment is a possibile way to use excess low-temperature waste heat for which there is no nearby use; but it is marginal economically even if there is a large concentrated source of heat.

Power recovery from pressurized gas streams is also important. In many cases in present practice, steam or product gases are throttled, reduced in pressure through pressure reduction valves, but can instead undergo pressure reduction through turbines, generating electricity (or shaft power to be used nearby).

UTILITY SYSTEM IMPROVEMENTS

The energy utilities, steam and electricity, are often the first target for overall automatic control. Such a control system can keep instantaneous purchased power use below a pre-set goal, i.e. by controlling loads which have been identified as interruptible or temporarily reducible to a predetermined set point. (In the U.S., roughly one half of the cost of electric power at industrial plants is typically based on the peak power or kilowatt use, as contrasted with total energy or kilowatt-hour use.) It can also select rates of steam production by different boilers. Such general energy management systems are also sometimes designed to be centers for on-line information on the general status of each plant. Such an information center can be effective for dispatching maintenance personnel.

Cogeneration of work and heat, usually but not always electricity and steam, is frequently found at paper mills, petroleum refineries, chemical and other plants. Cogeneration now in place often involves the production of moderate pressure steam, perhaps 40 atmospheres (4 MPa or about 600 psig),in boilers, which is let down through a back pressure turbine (as contrasted with a turbine leading to a condenser) to produce work and lower-pressure steam. In the paper industry 100 atmosphere (10 MPa or 1500 psig) steam is often provided, enabling higher efficiency. Gas turbine systems also provide higher fuel savings per unit of process steam provided (i.e. assuming the additional electricity or work is needed). Although gas turbine technology is widely available it is only beginning to be very widely adopted by industry.

All the changes that might effect the utility system in an energy conservation program may call for substantial redesign of the system. Much less low temperature process heat may be needed because of applications of heat recovery and other forms of conservation. Substantial reductions in use of boilers and in opportunities for cogeneration may result.

COMBUSTION CONTROL

Accurate on-line sensing of CO and O_2 in the stack, digital analysis of the information and modification of fuel flow and air dampers enable combustion to be accurately stabilized much nearer to the ideal than manual or semi-automatic controls. Near stoichiometric conditions,the CO concentration in stack gas is a very sensitive indicator of the oxygen concentration in the combustion area. With advanced automatic controls the oxygen concentration can be reduced below 1% (5% excess air) in the combustion chamber. (It's higher in the stack due to leakage.) With coarser sensors, but with systematic attention to operations, the average oxygen concentration is typically higher, corresponding, perhaps, to 15%-25% excess air. (The average concentration is kept high to avoid major excursions to oxygen levels below stoichiometric which cause smoke, which fouls equipment and which is an air pollutant.) If excess air is reduced through the use of controllers from, say, 25% to 5% then the efficiency is improved roughly 2% for a stack gas temperature of about 260°C (500°F).

MOTOR-RELATED IMPROVEMENTS

The energy used for mechanical drive can be reduced: (1) at the motor by using high-efficiency motors properly sized to the load, and by using power factor and variable speed controllers, and (2) away from the motor by redesigning the load and the powered equipment (such as pumps and fans). I will briefly discuss variable speed control (VSC) applications.

The flow from many pumps in industry is controlled by throttling valves. The motor-impellor system is designed for higher flow than required; the required rate being achieved by throttling. Friction is also commonly used to control speed of flow at fans and compressors. In cases where required flow rates vary substantially with time and induction motors are used, replacement of variable throttling by VSC is is often cost effective. The energy savings increase with decreasing ratio of average actual flow to design flow.

The newest VSCs create an alternating wave form (of adjustable voltage and frequency) using digital synthesis of the wave form, solid state switching and rectifiers. Not only do such devices eliminate the energy waste inherent in throttling; they enable sensitive control of flow and reduce pump wear. Pumping capacity may also be increased and pump cavitation avoided at very low flows.

ADVANCED CONTROLS

Advanced automatic controls encompass: (1) sensing critical physical characteristics of production, (2) rapid analysis of those characteristics and determination of desired actions to modify the process (upstream or downstream), and (3) automatic implementation of some of these actions. At the same time, information is made conveniently available to operators so they can make an informed judgement on the state of the process and intervene as appropriate.

The critical element in developing these controls is typically the sensors. These devices must be accurate and respond rapidly. Often they have to operate in harsh environments (e.g. in corrosive atmospheres at high temperatures). Computational capabilities enable one to rapidly interpret signals thereby greatly expanding the effects which can serve practical sensing needs.

Two general approaches to system design have been made: Programmable controllers have evolved from the rack of relays or pneumatic controls in older plants. They have the advantage that the structure of control is familiar to operators. (It is essential that operators be able to learn and use the new techniques.) The other approach uses microprocessors which convert analog to digital signals and mathematically process the information, a technique of great power and flexibility. This approach has evolved from laboratory applications. The two approaches are growing together as programmable controllers acquire more mathematical capabilities, as microcomputer software becomes easier to work with, and as operating personnel become more sophisticated.

An important outgrowth of advanced controls is that through them one can learn in detail about the performance of the production process at the plant. By this means, all aspects of production can be scientifically examined and improved, or replaced by a better process.

3. MAJOR PROCESS CHANGE

Typically, process change is not primarily motivated by energy conservation but in many cases the conservation benefits are very large. Let us briefly consider two potentially revolutionary process changes. About 40% of the energy used in iron and steel mills (Table 2.1) is involved in shaping and treating starting from liquid steel.[12] No energy is required in principle because the thermal energy of the melt is much greater than any energy of rearrangement (which is small because essentially physical not chemical). As shown by Eketorp,[13] the series of reheatings and rollings which are carried out at present are required both to obtain the desired shape and to obtain the desired internal structure. (The uncontrolled solidification of thick shapes does not enable one to obtain a desired internal structure directly.) Controlled solidification, perhaps very rapid, of thin castings near their final shape offers revolutionary opportunities to directly determine internal structure in mass production. When the technology is fully developed it will eliminate almost all the energy use which now characterizes shaping and treating. The very large energy savings would be only one of the benefits. Some others would be increased yield, reduction of inventories and immediate feedback to steelmakers on the quality of steelmaking. This technology is now under development, primarily in Sweden, Germany and Japan; the opportunities are sill wide open.

Petroleum refining consists of two broad categories of process: (1) physical separation of molecules, broadly according to their molecular weight, and (2) chemical rearrangements such as breaking up heavy molecules and fusing light molecules. Let us consider a separation process. The physical mixing of n different kinds of molecules (without intermolecular interaction) involves an entropy increase, per mole of material, of

$$S_1 = -\sum_{i=1}^{n} Rx_i \ln x_i$$

Where the x_i are the mole fractions of each species. If each kind of molecule is present in equal amount, $x_i = 1/n$ and $S_1 = R \ln n$. Now suppose that a refinery separation process for crude oil involves the separation of a mixture of n kinds of molecules, present in equal number, into m mixtures such that each mixture has n/m kinds of molecules. The entropy of the m separate mixtures is

$$S_2 = -m\,[R(1/m)\ln\,(m/n)] = R\ln(n/m)$$

The entropy change going from the single mixture to the m separate mixtures is

$$\Delta S = S_2 - S_1 = R\ln\,(1/m)$$

The minimum availability, or energy, needed to achieve such a separation is

$$\Delta B = - T\Delta S = RT \ln m$$

The separation of crude oil achieved by a crude distillation unit is roughly described by this analysis. With m ≃ 10 and T near ambient, say 300° K, ΔB = 1.4 kcal/mole. The averabe molecular weight in crude oil is near 200 fo the absolute minimum energy to separate the crude is about

$$\Delta B = 4 \text{ kcal/kg}$$

A typical crude distillation unit consumes about 25 times as much energy so its second-law efficiency is 4% The losses responsible for this low efficiency were illustrated in Figure 2.3 and discussed at that point. Although the losses can be reduced, the larger part of them are inherent in the design of distillation. There is the challenge: Can a new process be invented, which would save energy and also be flexible in its handling of materials, offer good control of product qualities, be easy to maintain, etc.[14] No obvious candidate is in view at this time but I believe that the technological opportunity is very good.

Brief descriptions of many of the revolutionary process changes (for basic-materials manufacture) which are the focus of research and development have been provided by Hane et al.[15] Since process changes often dramatically change the thermodynamics of production, the greatest energy-conservation opportunities may be realized through them. R&D on production processes should thus be a key part of any comprehensive long-term conservation program.

III. CONCLUSIONS

A. THE MEDIUM-TERM PERSPECTIVE

THE VALUE OF SMALL PROJECTS

Through a wide variety of technical efforts the energy intensity in each of the energy-intensive industries has been reduced an average of about 20% from 1972 to 1983, and can be further reduced very substantially. The largest part of the energy-intensity reduction from 1972 to date has been due to improvements in production operations not requiring substantial investment. Two kinds of investment will play a larger role in the future: conservation equipment investments during the 1980s and '90s,and,more gradually, investments in radically new production processes (including R&D and innovation).

Engineers at large process plants have learned that comprehensive programs consisting primarily of smaller conservation projects (roughly $20 million and less) can enable existing plants to begin to approach the energy-intensity performance of state-of-the art plants. Let me digress to discuss how a good plan is developed. The first challenge is to identify as many opportunities for applying the diverse approaches to conservation as is practical. The second is to design and cost each promising project. The third is to sell the good projects to influential operators and managers. I comment only on the first. Typically it is detective work because at a factory one usually begins without adequately detailed information on the energy use and other physical physical parameters of a process step. While some conservation opportunities are evident to an experienced investigator, one generally also needs to measure energy use and a few other key parameters and their time dependence. The dependence of energy use on production rate, for example, will often reveal important opportunities for savings through management of energy use at reduced levels of production. One can also carry out a thorough parametric study of the variation in performance of a process unit. Although the cost of such an investigation may be high, major savings have often been realized through the resulting ability to identify conservation projects.[16]

The capital cost of a major program of small projects is of course far less than that of a new plant. The cost reduction which can be achieveed with such a program of small projects is substantial: In two sample programs energy use in a petroleum refinery would be reduced 28% and that in a steel mill 20% (Table 3.1). The overall cost of petroleum products at this refinery (including capital charges for the program) would be reduced 60¢/barrel, about 2% of sales price, and that of steel products $12/ton, about 2 1/2% of sales price. While not enough to redress the cost advantages held by some foreign producers, these costs reductions would be very significant to the earnings of the manufacturers. In other words, these investments in the firms' own facilities typically offer excellent returns.

TABLE 3.1. TWO SAMPLE ENERGY-CONSERVATION PROGRAMS*

	Steel mill	Petroleum Refinery
Reduction in energy use	20%	28%
Energy intensity with program[a]	22.6 MBtu/ton[b]	422 KBtu/bbl[c]
Capital cost per unit of production capacity	$48/annual ton[d]	$650/bbl per day[e]
Simple payback overall	1.7 years	2.6 years
Simple payback of marginal projects	3.5 years	4.5 years
Net reduction in cost of production	$12.00/ton	60¢/bbl
Cost reduction compared to sales revenue	2 1/2 %	2%

*Source: References 12 and 17

[a]energy intensity with production at design rates. Electricity is evaluated at 10,000 Btu/kwh (2.5 Mcal/kwh).

[b]Purchased coke is evaluated at 1.33 times its heating value.

[c]includes coke combustion, but not hydrogen feedstock.

[d]annual ton of mill products.

[e]barrel per stream day of crude capacity.

DIFFICULTIES OF IMPLEMENTATION

While some firms in energy-intensive industries have made large investments to reduce their energy-related costs, most are proceeding very slowly. This can be frustrating for engineers who develop good energy-conservation projects. Why are the investments slow in coming?

One can view the underlying cause to be the slow growth or even decline of basic materials production in the U.S. This means that few new production facilities are being built. Suppliers to these industries lack the stimulus of new plant construction. Industrial R&D labs have been redirected, and technical staff has been reduced. Top management has become preoccupied with financial manipulation. The strategy adopted by many firms in these industries gives high priority to diversification into new businesses. While major efforts have been made to reduce costs, this has been accomplished by closing less efficient facilities and by operational changes. Most of these firms do not pay much attention to the opportunity to cut costs through investments to modernize existing plants.

Two specific characteristics of many of these firms which may help us understand the relative lack of investment in smaller modernization projects are (1) their financial perspective and (2) their centralized management. Most businesses based on energy-intensive manufacturing are no longer growing rapidly and many face strong foreign competition. Moreover businesses in the U.S. are being pressed to focus on short-term goals. (For example institutional investors typically hold common stocks only about half a year.) It is not surprising, then, that most firms in energy-intensive industries have assigned a low priority to technology while emphasizing financial measures such as refinancing. restructuring, and diversification. (A technical orientation is more common however, in the chemical industry.)

Most of the firms in question concentrate investment decision making at the top. The effect of this is not that top management pours over a huge number of small-project proposals. Instead the typical managerial procedure is to severely ration capital to divisions and plants while giving them responsibility for effective decisions on smaller projects, with the result that smaller discretionary projects (i.e. for cost cutting) face high *de facto* hurdle rates.*

Not all firms in these industries have these characteristics. Some are well staffed with engineers at their plants and they give these engineers considerable scope. I believe we may see the more

*The prevalence of capital rationing is well known. The relationship between size of project (and locus of primary decision making) and the effective hurdle rate was observed in an Alliance to Save Energy field study.[18] The high effective hurdle rates for smaller projects observed at these firms is a phenomenon quite separate from discrimination against projects in less favored plants and on less favored product lines. The small project-high hurdle rate correlation was observed for the best plants and product lines.

technologically oriented and decentralized firms achieve some success, even in the difficult business conditions which exist. My reason is that, although these industries are largely mature in terms of overall sales, revolutionary process innovations are being developed. Those firms with strong technical capabilities which are open to technical opportunities may do very well.

B. THE LONG-TERM PERSPECTIVE

Some of the most energy intensive industries in the Unites States face a grim future because there are isolated sites in other countries with cheap and hard-to-transport energy resources. The most important examples are hydropower and natural gas in Canada, low quality coal in Australia, hydropower in remote parts of Brazil, and especially, natural gas in the Middle East, Indonesia, North Africa and other sites remote from present concentrations of industry. Industries like primary aluminum and certain base organic chemicals will move to those sites.[6]

The competitive position of related downstream producers and of other energy-intensive industries based on more easily transported energy forms will depend to a large extent on their manufacturing technology. Plants located in the United States will continue to enjoy good access to many materials, especially coal, recycled materials and biomass. They also are close to a very big market and so have low transportation costs and close contacts with customers. If process technology is developed which sharply cuts capital, energy, and labor costs and if this technology is effectively adopted in the U.S. I believe the cost advantage now enjoyed, for example, by foreign producers of steel would be overcome (even though the foreign producers would also adopt new technology). Domestic manufacturers would be the primary suppliers for this country for all processes where labor requirements are not very high or where close contacts with customers are especially important. Of the uncertainties mentioned the most important is whether U.S. manufactureres will help develop and will adopt the best process technologies. On this point, the trends of the last two decades are not encouraging. Federal research and development policies are tending to drain talent away from research relevant to industry. While private firms do a lot of specific product R&D, few do research an basic technologies. Most of the research on the basic technologies which will become the industrial processes of the next century is being done in Europe and Japan.

REFERENCES

1. Energy Information Administration, Monthly Energy Review, U.S. Dept. of Energy, Washington, D.C.

2. Energy and Environmental Analysis, Inc., "The Industrial Sector, Energy Consumption Data Base for 1975 and 1976," a report to the Energy Information Administration, USDOE (1980). See also Solar Energy Research Institute, A New Prosperity: Building a Sustainable Energy Future, Brick House, Andover, MA (1981), p.385ff.

3. Marc Ross, "Industrial Energy Conservation," Natural Resources Journal 24, 369 (April 1984).

4. Reference 1 and The Energy Factbook, Committee on Interstate and Foreign Commerce, U.S. House of Representatives, Committee Print 98-IFC-60 (1980).

5. Robert E. Marlay, "Trends in Industrial Use of Energy," Science 226, 1277 (Dec. 14, 1984).

6. Marc Ross, Eric Larson, and Robert H. Williams, "Energy Demand and Materials Flows in the Economy" to be published in Energy-The International Journal.

7. For a discussion of availability requirements, or minimum available work, see reference 10.

8. Elias P. Gyftopoulos, Lazaros J. Lazarides, and Thomas F. Widmer, Potential Fuel Effectiveness in Industry, a report to the Energy Policy Project of the Ford Foundation, Ballinger (1974).

9. Reports issued by: American Paper Institute, New York, NY, and American Petroleum Institute, Chemical Manufacturers Association, American Iron and Steel Institute and Aluminum Association, all of Washington, D.C.

10. W.H. Carnahan et al., "Efficient Use of Energy, A Physics Perspective," AIP Conference Proceedings, Vol. 25 (1975).

11. Figure provided by Willilam G. Larsen

12. Marc Ross "Industrial Energy Conservation and the Steel Industry," to be published in Energy The International Journal.

13. Sven Eketorp, "Energy Considerations of Classical and New Iron and Steelmaking Technology," to be published in Energy-The International Journal.

14. Thomas W. Mix, "Low Energy Separations," in Kirk-Othmer Encyclopedia of Chemical Technology, Supplement Volume, Third Edition, John Wiley (1984)

15. G.J. Hane et al., "A Preliminary Overview of Innovative Industrial Materials Processing," a report by the Pacific Northwest Laboratory for the U.S. Department of Energy, PNL-4505/UC-95f (1983).

16. W.F. Kenney, Energy Conservation in the Process Industries, Academic Press (1984)

17. William G. Larsen and Marc Ross , "Energy Conservation in Petroleum Refining", to be published in Energy-The International Journal

18. Alliance to Save Energy, "Industrial Investment in Energy Efficiency: Opportunities, Management Practices and Tax Incentives," Washington, DC (1983); Marc Ross "Energy Conservation Investment Practices of Large Manufactureers" in The Energy Industries in Transition, 1985-2000, J.P. Weyant and D.B. Sheffield, Eds., International Association of Energy Economists, Washington DC (1984).

CHAPTER 19

POTENTIAL FOR ENERGY SAVINGS IN OLD AND NEW AUTO ENGINES

John R. Reitz
Research Staff, Ford Motor Company
Dearborn, Michigan 48121

ABSTRACT

This paper discusses the potential for energy savings in the transportation sector through the use of both improved and entirely new automotive engines. Although spark-ignition and diesel internal combustion engines will remain the dominant choices for passenger-car use throughout the rest of this century, improved versions of these engines (lean-burn, low-friction spark-ignition and adiabatic, low-friction diesel engines) could, in the long term, provide a 20-30 percent improvement in fuel economy over what is currently available. The use of new materials, and modifications to both vehicle structure and vehicle transmissions may yield further improvements. Over a longer time frame, the introduction of the high-temperature gas-turbine engine and the use of new synfuels may provide further opportunities for energy conservation.

The transportation sector accounts for roughly 25 percent of total energy use in the United States, and this sector is responsible for somewhat over half of U.S. total petroleum demand. (1) Thus, even modest improvements in vehicle powerplants and overall vehicle design offer a potential for a substantial energy savings. This paper will address both near- and long-term innovations in automotive engines/vehicle design which could improve the fuel economy of this part of the transportation sector. Although the focus of the paper will be on automotive engines, we shall also address structure and weight of automotive vehicles, transmissions, exterior vehicle design and alternate fuels, since these all influence the bottom line in improved energy usage.

The past decade has seen a substantial improvement in the fuel economy of automobiles produced in the United States. Between 1974 and 1984 the fuel economy of U.S.-built passenger cars, as measured by prescribed federal test procedure, improved 92 percent from a value of 13.2 miles per gallon (mpg) in 1974. (2,3) This year,if the industry achieves the federally-mandated Corporate Average Fuel Economy (CAFE) level of 27.5 mpg, the

0094-243X/85/1350368-14$3.00 Copyright 1985 American Institute of Physics

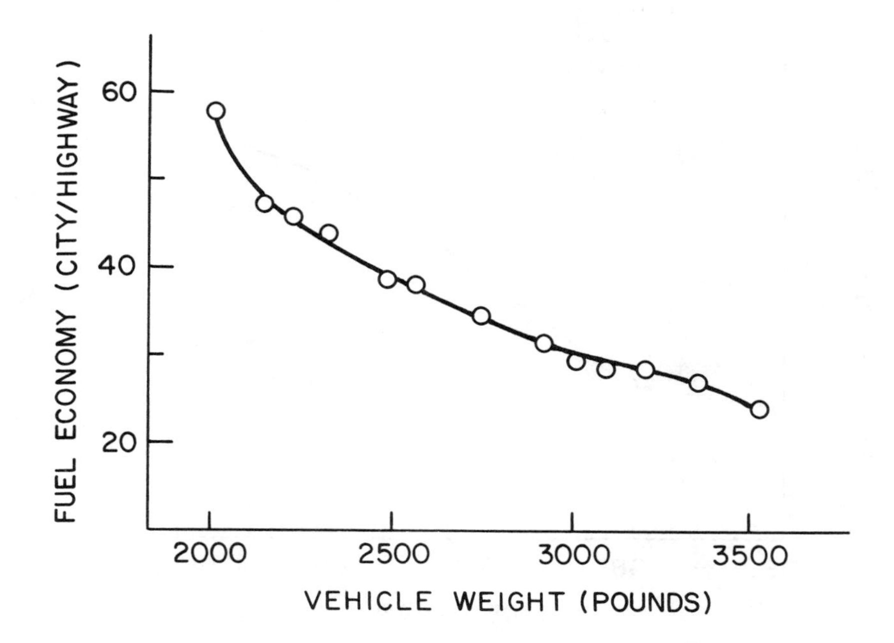

Fig. 1. Fuel economy (mpg) of best gasoline-powered vehicle in each inertia weight class (1984). (4)

improvement will be over 100 percent. The achievement of this rather dramatic improvement in fuel economy over a period of 10 - 12 years has come about through the introduction of a number of technological advances (4) into automotive design and manufacture: (i) vehicle "downsizing" and material substitution to achieve weight reduction, (ii) electronic engine control for better control of the combustion process, (iii) use of more efficient transmissions such as the front-wheel drive transaxle, and (iv) improved aerodynamic styling and rolling resistance. The separate effects of weight reduction and other technologies [(ii)-(iv)] on fuel economy are shown in Figs. 1 and 2. In the years ahead we can expect continued improvements in the above four areas, but in addition there is the potential for larger fuel-economy gains through the introduction of new automotive powerplants, such as the lean-burn (lean air/fuel) gasoline engine and the adiabatic diesel which uses ceramic components to minimize heat losses.

Although fuel economy has become a more important consideration in recent years, it is not the most important attribute of a personal vehicle in the eyes of the average American customer. Table I ranks the requirements in the order of perceived importance. (5) A soci-

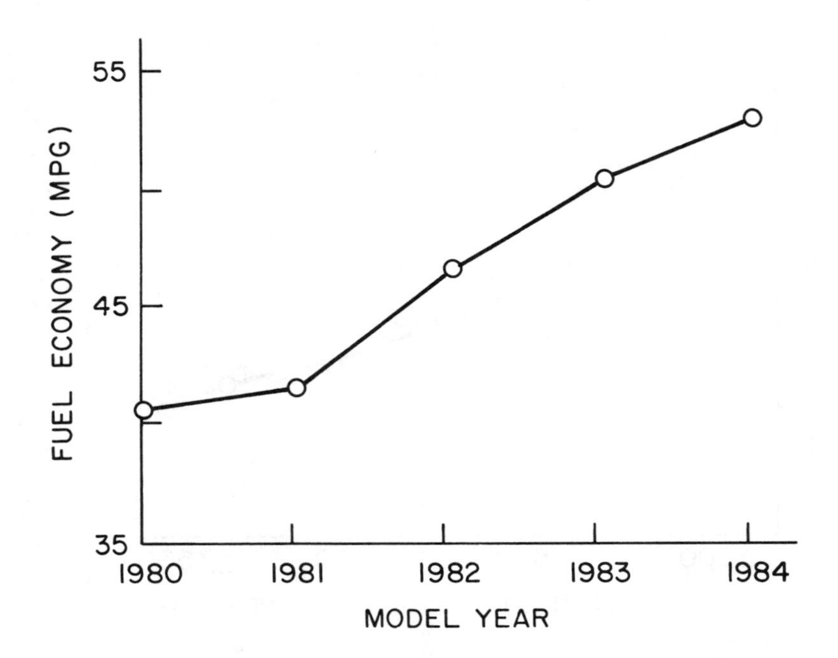

Fig. 2. Fuel economy of best vehicle in 2125 lb. inertia weight class. (4)

etal ranking of these requirements would, of course, place low exhaust emissions and fuel economy much higher in the list; nevertheless, the vehicle manufacturer must consider all of these requirements and cannot focus on one (such as fuel economy) to the detriment of the others. The impressive improvement in fuel economy of the past decade has been accomplished without sacrificing vehicle performance and reliability, without a major change in passenger comfort, and under the simultaneous imposition of stringent emission controls.

Let us now consider some of the areas which have had a major impact on improved fuel economy and those which will have an impact in the future.

VEHICLE STRUCTURE: THE USE OF NEW MATERIALS

Vehicle weight is a primary factor affecting fuel economy, as is evident from Fig. 1. The figure shows the fuel economy (city/highway mpg) of the best gasoline-powered vehicle in each inertia class over the range of 2000 to 3500 lbs. Vehicle downsizing and weight reduction has been an important factor in the improvement in CAFE mpg over the past ten years; in 1975 the average car weighed about 4050 lbs., and today weighs approximately

Table I. Market/Societal Requirements of a Motor Vehicle (in order of Customer-perceived Importance). (5)

1. Convenience, Reliability
2. Overall Transportation Cost (including maintenance and depreciation)
3. Safety
4. Fuel Economy
5. Low Emissions

3080 lbs. (2)

Weight reduction has been accomplished through vehicle redesign -- substituting, when possible, new materials such as high-strength low-alloy (HSLA) steels, aluminum and plastics. Complete design analysis using finite-element theoretical modelling is being carried out for some structural members to ensure that these members have adequate strength but are not contributing unneeded weight. The use of the computer for carrying out finite-element analyses can greatly reduce the need for building prototype vehicles, which is a slow, time-consuming process. In addition, plastics have replaced metal components in most decorative applications; they also find use in other areas such as body panels and fuel tanks.

Materials sustitution and vehicle redesign will continue to impact vehicle weight, and hence fuel economy, in the future. Structural composites using fiberglass are expected to replace metal for many applications. The use of graphite composites for components requiring high stiffness or subjected to high fatigue could have an important impact but, unfortunately, these materials are quite expensive at the present time. Only if their production costs can be brought down substantially, will they be used in large enough quantities to cause an appreciable increase in fuel economy.

AUTOMOTIVE ENGINES

Contemporary automotive engines use the spark-ignition internal-combustion engine (based on the thermodynamic Otto cycle, see Appendix) or the diesel engine. These engines, or variants thereof, are quite likely to

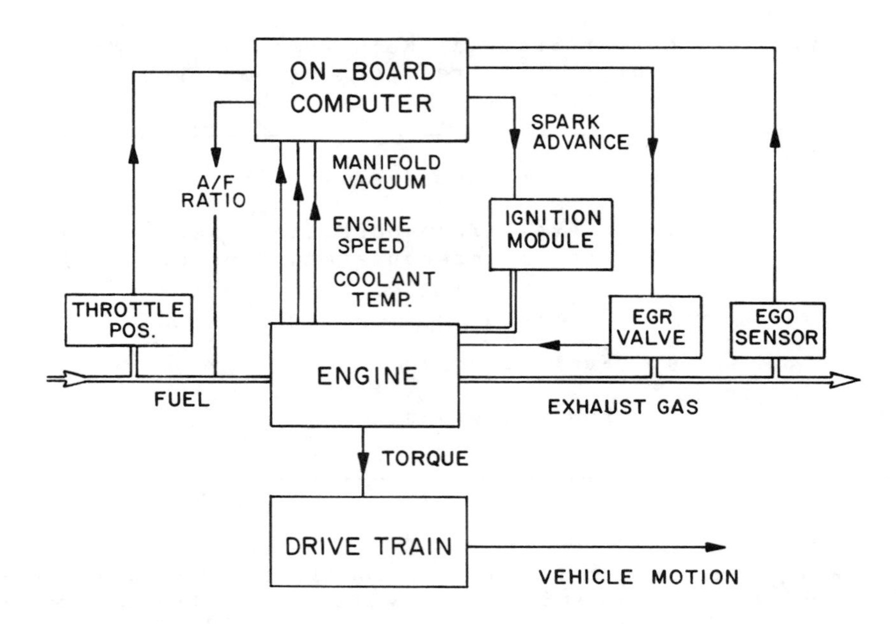

Fig. 3. Electronic engine control

remain the dominant choices for passenger car use through the rest of this century.(5,6)

Since the Otto cycle engine predominates in the United States, most federal regulations (exhaust emissions, CAFE requirements) have been written with this engine in mind. Federal fuel economy legislation, which was introduced in the mid-1970's, made it necessary to improve automobile mileage, concurrent with a tightening of exhaust emission control. This was accomplished through a variety of techniques. The use of the 3-way catalyst and implementation of stoichiometric engine control was a big factor in allowing improved fuel economy at today's low emissions. However, this dictated electronic engine control (EEC). EEC with an on-board electronic computer allows a more precise control of engine parameters under a wide range of operating conditions. A schematic highlighting the central function of the on-board computer in electronic engine control is shown in Fig. 3. The use of EEC has allowed improved driveability in addition to better fuel economy. With the keep-alive memory now available in microprocessor chips, it is possible to include adaptive control strategies to allow the computer to adapt to variations in manufacturing tolerances, in-service wear and varying enviromental conditions, and thus cause the engines to function in a more efficient manner.

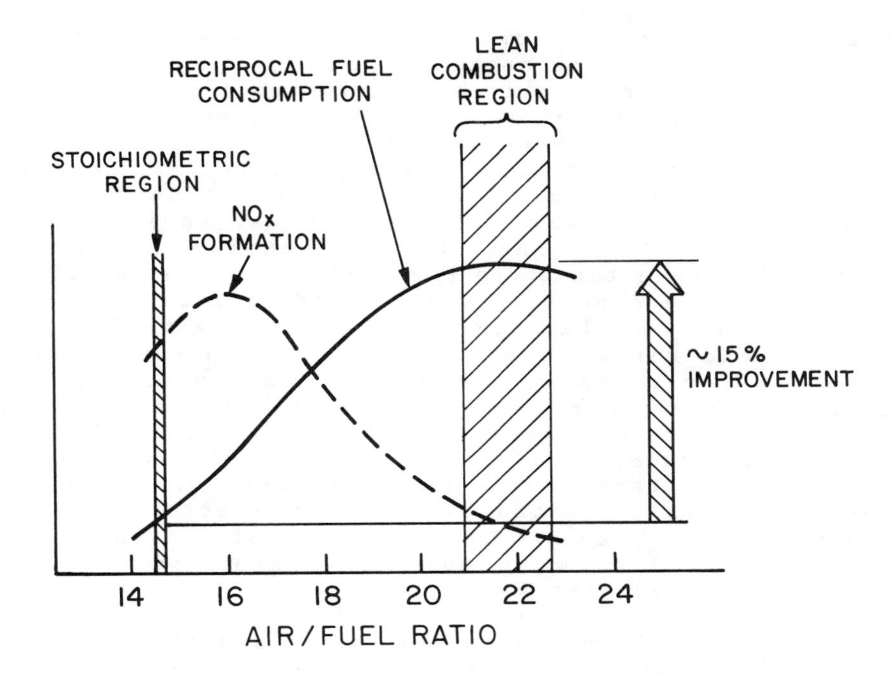

Fig. 4. Increase in fuel economy of a lean-burn engine. (7)

The continued search for more fuel-efficient, spark-ignition internal combustion engines has focused on a variant of the Otto cycle, namely the lean-burn engine. The term lean-burn refers to combustion taking place with more air than is required to combine with the fuel stoichiometrically. This excess of air reduces the combustion temperature and improves the thermodynamic efficiency. At an air/fuel ratio of 22, an increase in fuel economy over stoichiometric operation of about 15 percent is theoretically attainable (see Fig. 4). Another advantage of lean combustion is that the oxides of nitrogen (NO_x) are reduced substantially; and on small cars it may be possible to remove exhaust-gas recirculation (EGR) which is the usual NO_x-control strategy. On the other hand the hydrocarbon emissions are somewhat higher, and, to maintain driveability, sophisticated control systems might be necessary.

Lean mixtures normally burn more slowly than stoichiometric ones, thus introduce the potential for both unstable combustion (with possible misfire) and extraneous heat loss. It is therefore important to restore, or if possible increase, the burning speed. This is being done by (i) introducing turbulence and swirl into the air/fuel flow in the engine cylinder, (ii) increasing the electrical energy in the spark ignition system, and (iii) a combination of more precise fuel metering and the use of a richer A/F near the spark plug.

To further improve engine performance, most automobile manufacturers have introduced turbochargers on at least one model. This device provides additional air flow to meet the engine's greater requirements at high speed. The advantage from the fuel-economy point-of-view is that one can operate overall with a smaller (lower-powered) engine, which is augmented by the turbocharger when needed for acceleration or high-speed operation. Although turbochargers use waste energy from the hot exhaust gases to provide the increase in engine power, this energy is not completely without cost since the device introduces increased back pressure in the exhaust. Ceramic turbochargers and turbochargers with variable-geometry vanes, controlled by the electronic engine control system, can provide higher boost pressures and a faster response, with the potential for a further increase in fuel economy.

The diesel engine is a very efficient powerplant because of its high compression ratio and the capability of running without throttling. The diesel is about 15 percent more efficient than a comparable-sized spark-ignited engine (on an mpg basis), but this efficiency advantage is substantially reduced if we normalize to the same weight of fuel. Nevertheless, the diesel has further advantages in that it uses less-refined petroleum products and can use a somewhat wider range of fuels than the spark-ignition engine. The diesel is inherently quite efficient, and a number of developments (described below) are in progress which will increase the efficiency of the present diesel engine.

The disadvantages of the diesel compared to the spark-ignition engine are that it is heavier, noisy, has lower performance than a gasoline engine per unit size, has cold-start problems in cold climates, and because of its expensive fuel-injection equipment, more costly.(6) Also, the friction losses are greater. And finally, the emission of exhaust particulates (or smoke) is of concern and is under federal (and in some cases, state) regulation. Meeting the 1986 California and 1987 Federal 0.2 gm/mile particulate emission standard for passenger cars poses a major problem; only the light-weight vehicles can achieve this objective now.(8) In addition, vehicle manufacturers are very concerned about being able to meet the 1989 California 0.08 gm/mile standard.

Most diesel engines in passenger cars today use an indirect injection process in which the fuel and air mixture is ignited in a prechamber from which the flame travels to the main combustion chamber where it receives the main charge of air. This type of engine provides for lower-pressure fuel injection equipment, and is compatible with gasoline-engine peripherals. The next generation of diesel engine will be a direct-injection engine which eliminates the prechamber; this is considerably more efficient and has been found to give a 10-15 % increase in fuel economy. (9)

Of even more interest over the longer term is the so-called adiabatic diesel. This engine will use ceramic components to greatly reduce heat losses. This concept, while not practical for spark-ignition engines, is particularly suited to the diesel engine which uses compression ignition, because the combustion chamber can be kept much hotter. A 10 to 15 percent improvement in fuel economy may be possible for the adiabatic diesel over a conventional water-cooled direct-injection diesel (10); with the incorporation of various low-friction features, such and ringless pistons and ball-bearing connecting joints, this improvement could possibly be as large as 15 - 25 %.

What about other candidate engines for automobiles? Frequently mentioned are the high-temperature gas-turbine engine (which operates on the thermodynamic Brayton cycle) and the Stirling engine. Both of these engines use continuous combustion and thus offer the advantage that they can operate with a wide range of fuels. Although both have potential for improved fuel economy over conventional automotive engines, they each have technical problems which require substantial research and development work before a decision about their technical feasibility can be made. For the gas turbine, the problem is high strength/high temperature, reliable ceramic materials which will allow turbine operation at temperatures above those suited to metallic superalloys; in the case of the Stirling engine, the problems are reliable seals for the working fluid (which is hydrogen), and inefficiencies related to changing the output power level. The gas turbine's problems are closer to solution than those of the Stirling engine, and ceramic materials/gas turbine research is the focus of several Department of Energy sponsored industrial contracts. For a technical review of both the turbine and Stirling engine, see Nunz and Moore.(11)

POWERTRAIN MODIFICATIONS

The automobile operates under a variety of driving conditions: acceleration, deceleration/braking, steady speed, and idle; and the engine cannot perform efficiently in all circumstances. In fact, in some urban driving cycles, the fuel used during idling and deceleration may exceed the fuel used to propel the vehicle forward. This has led to a number of novel proposals for engine or powertrain modifications, such as (i) cylinder cut-off, (ii) engine shut-down at idle, and (iii) regenerative braking -- as well as to continued improvement in transmissions.

Let me address the engine transmission first. Front-wheel drive transaxles, with their more efficient helical and spur gears replacing the hypoid gears in conventional rear axles, have contributed to the improved fuel economy of front-wheel drive cars. (2) Five-speed manual transmissions with appropriately chosen gear ratios have been particularly effective. But automatic transmissions are very popular in the United States; here new developments include locking features to bypass the less efficient (but smooth) torque converter after start-up, and fourth gear overdrive. In the near future, continuously-variable transmissions (or CVT's) are expected to be available. A properly designed CVT should provide the same fuel economy as a five-speed manual transmission.

Cylinder Cut-off. The maximum power output of a passenger car's engine is important for its performance rating, e.g. its acceleration and how well it climbs hills. However, under most driving conditions, urban and freeway, the power requirement is much lower and would be more efficiently handled by a smaller engine. Ideally, what one wants is a variable displacement engine which can be adjusted to the load requirement. One technique for accomplishing this is by shutting off some of the cylinders under reduced load conditions, either by interrupting the fuel to the suspended cylinders or by closing their valves. Thus, e.g., in a six-cylinder engine, three cylinders might be shut off during steady freeway driving but would be reactivated when one wanted to accelerate for lane changing or passing.

Experimental studies have shown that cylinder cut-off is an effective method for improving fuel economy, (12) although it entails additional cost and adds complexity to the engine control system. And in addition there are driveability problems which have not been completely solved, an early attempt by Cadillac to use this technique was discontinued after one year. Furthermore, the results obtained are to some extent similar to what one gets through turbocharging or by using continuously-varying transmissions, ways which seem to be more cost-effective solutions to energy/load matching.

Engine Shut-down at Idle. In urban driving, a substantial fraction of engine time is spent idling, and this of course contributes to the poorer fuel economy of the urban cycle compared with freeway driving. A complete shut-down of the engine during idle could produce higher fuel economy (up to 12 percent increase in some urban cycles). However, there are problems associated with this procedure: first, there are peripheral power requirements during idle (such as air conditioning, headlights, etc) and idle speed is normally set to maintain battery charge under a variety of urban conditions. Second, engine shut-down is not advisable during the warm-up period because it tends to maintain the engine in the less-efficient warm-up phase (which is also a period of excessive hydrocarbon emissions). And thirdly, there are concerns about customer acceptance, in particular in regard to the reliability of re-start under a wide range of operating conditions.

Regenerative Braking. Again, in the urban driving cycle, it is frequently necessary to decelerate using the brakes. This procedure wastes energy since the car's energy of forward motion is transformed into heat in the brake linings. What one would like to have is a regenerative braking system in which this energy is stored in another form which can later be extracted and fed back into forward motion. A number of different systems have been considered for this application, including motor/generator/battery combinations, but the one which appears to have the greatest potential for regenerative braking is the flywheel. Flywheels made of strong, light-weight fiber-composites, which are capable of spinning at more than 30,000 rpm, are required in order to store the energy in a reasonable weight and volume (energy density of 50 W hr/kg). The potential for flywheel energy storage in automobiles and other vehicles has been discussed by Post and Post. (13)

AERODYNAMIC DRAG AND ROLLING LOSSES

The total road load on a vehicle is the sum of aerodynamic drag losses and rolling losses. The latter include brake and bearing drag, driveline losses and tire friction. Aerodynamic drag losses, which predominate at high speed, are given by

$$\text{Drag Losses} = (1/2)\, \rho\, C_d A v^3, \qquad (1)$$

where ρ is the density of air, A is the projected frontal area of the vehicle and v is the vehicle velocity. C_d, the drag coefficient, is a dimensionless quantity and is a measure of the "shape efficiency" of the vehicle.

Reduction in the drag coefficient will lead directly to improved fuel economy in high-speed operation. For example, at 55 miles per hour (89 km/hr), the aerodynamic drag losses on a 1985 Ford Escort with a $C_d = 0.40$ are 8.7 horsepower (6.5 kW). A reduction of 10 % in C_d to 0.36 would lead to a 0.87 hp reduction in aerodynamic losses; total road loss would be reduced about 6% and this would also produce a 6% increase in fuel economy for 55-mph, steady operation, or a 2% increase for overall (city/highway) driving. A variety of improvements in aerodynamic styling could lead to a practical C_d coefficient as low as 0.30. This would require a low, sloping hood, spoilers, elimination of drip molding over the doors, flush-glass windows, and "aerodynamic" head lamps. (4)

Improvements in the rolling losses of vehicles can also be expected, and these will have impact at low and intermediate speeds. Even tire friction is something that has been and may continue to be improved; higher inflation pressure, better tire materials, and the switch to p-metric radial tires increased average fuel economy by about 1.5 mpg between 1978 and 1983. (2) On the negative side, however, very low tire friction reduces the vehicle′s road-holding and braking ability.

ALTERNATE FUELS

The use of alternate fuels is a conservation measure only in the sense that it allows a plentiful fuel (or in some cases a fuel from renewable resources) to substitute for a scarce one (e.g., petroleum).(14) Of particular interest for spark-ignition engines are the alcohols (ethanol and methanol). These are clean-burning fuels (i.e. fuels which do not produce many particulates) which can be used in the near-neat state or as blends (with gasoline). In the latter case they are frequently used to boost the octane rating of the gasoline. Developments in lean-burn, high-compression engines seem particularly suited to the alcohol fuels; based on Ford experience with alcohol fuels, it is believed that fully-optimized alcohol engines can have efficiencies as good as, or better than, the IDI diesel, and approaching that of the direct-injection diesel in system performance. (15)

The energy density of the alcohols is less than that of gasoline. Methanol, e.g., has only one-half the energy density of gasoline. This means that a full fuel tank has only half as much energy when the fuel is methanol compared to gasoline. In addition, the exhaust emissions from engines using alcohol fuels have not been adequately studied, nor has the long-term compatibility of engine materials and alcohol. On the plus side, both of these alcohols can be made from non-petroleum resources. In Brazil, ethanol derived from sugarcane has already become a major energy source for the transportation sector; and in the United States, methanol made from our large resources of coal could be the dominant fuel of the future.

The diesel, turbine and Stirling engines can each operate with a range of fuels, including vegetable oils. At the present time such substitution has not proved economic, but this does point the way toward possible future opportunities.

CONCLUSIONS

The spark-ignition and, to a lesser degree, the diesel internal-combustion engines are likely to remain the dominant choices for passenger-car use throughout the rest of this century. But even here there are many opportunities for energy savings. Both the lean-burn, low-friction spark-ignition engine and the adiabatic diesel engine have potential for improvement in fuel economy over current engines. In the spark-ignition case, the lean-burn low-friction engine offers potential for a 12-15% improvement; and in the diesel area, the adiabatic, low-friction diesel engine offers the potential for a 20-30% improvement over current water-cooled, indirect-injection diesels. (4,9) Improvements in the aerodynamics and rolling friction of vehicles, improved transmissions, and further materials substitution and downsizing are anticipated; these should produce additional increases in fuel economy. Improved lubricants could also have an impact. (2)

Over the long term, if both the high-temperature gas turbine engine and synfuels are introduced into the marketplace, there could be further opportunities for energy conservationwithin the transportation sector.

The author would like to thank his colleagues at Ford Motor Company for providing assistance in preparing this paper, and in particular he would like to thank Mssrs. W. R. Wade and A. O. Simko for a thorough reading of the manuscript.

APPENDIX

The ideal Otto cycle is a thermodynamic cycle which absorbs and discharges its heat at constant volume (V_1 and V_2); the other two parts of the cycle are adiabatic expansion and compression. The efficiency of the ideal Otto cycle (16) is

$$\text{Eff} = 1 - r^{1-\gamma} , \qquad (2)$$

where $r = V_2/V_1$ is the expansion (compression) ratio and

γ is the ratio of specific heats for the working fluid. With $r = 10$, $\gamma = 1.4$, Eq. (2) yields an efficiency of 60%; gas pumping losses, heat and pressure losses in the cylinder, intake and other pumping losses, and engine friction will reduce this to between 35 and 40%. This, then, is a typical automotive (gasoline) engine efficiency to the beginning of the drivetrain.

REFERENCES

1. S. D. Berwager in "New Energy Conservation Technologies and their Commercialization," J. P. Millhone and E. H. Willis, Eds. (Springer-Verlag, New York 1981), pp 242-250.
2. B. H. Simpson, "Trends and New Developments in Automotive Fuel Economy," 49th American Petroleum Inst. Meeting, Paper 820-00025, New Orleans (1984).
3. R. M. Heavenrich et al, "Light Duty Automotive Fuel Economy," Soc. Automot. Eng. Paper 850550 (Feb.1985)
4. E. J. Horton and W. D. Compton, "Technological Trends in Automobiles," Science **225**, 587-593 (1984). See also C. L. Gray, Jr. and F. von Hippel, "The Fuel Economy of Light Vehicles," Scientific American **224**, 48-59 (May 1981).
5. E. E. Ecklund in "New Energy Conservation Technologies," op cit, pp 2269-2277.
6. J. H. Weaving in "New Energy Conservation Technologies," op cit, pp 2293-2304.
7. N. Kobayashi et al, JSAE Review (Japan, July 1984), p 106; D. R. Hamburg and J. E. Hyland, Soc. Automot. Eng. Paper 760288 (Feb. 1976).
8. V. D. Rao et al, "Advanced Techniques for Thermal and Catalytic Diesel Particulate Trap Regeneration," Soc. Automot. Eng. Paper 850014 (Feb. 1985).
9. W. R. Wade, T. Idzikowski, C. A. Kukkonen and L. A. Reames, "Direct Injection Diesel Capabilities for Passenger Cars." Soc. Automot. Eng. Paper 850552 (Feb. 1985).
10. W. R. Wade and C. M. Jones, "Current and Future Light Duty Diesel Engines and their Fuels," Soc. Automot. Eng. Paper 840105 (Feb. 1984).
11. G. J. Nunz and N. R. Moore, "Should We Have a New Engine? An Automobile Power Systems Evaluation." Vol II. JET Propulsion Laboratory, California Institute of Technology (1975). Chaps. 5 & 6.
12. G. L. Berta in "New Energy Conservation Technologies," op cit, pp 2402-2410.
13. R. E. Post and S. F. Post, "Flywheels" in Scientific American **229**, 17-23 (December 1973).
14. See Chapter by Tom Bull in this book.

15. R. J. Nichols and D. R. Buist, "Ford's Development in Alcohol-Fueled Vehicles," Proc. 48th Amer. Petrol. Inst. Meeting, Paper 820-00019, Los Angeles 1983).
16. M. W. Zemansky, "Heat and Thermodynamics", Fourth Ed., (McGraw-Hill, New York, 1957).

CHAPTER 20

MANAGING ELECTRICITY DEMAND THROUGH DYNAMIC PRICING

R. A. Peddie and D. A. Bulleit

Introduction by A. H. Rosenfeld

Editor's Note: Neither Peddie nor Bulleit was able to be present at the APS Short Course on which this book is based, so Professor Rosenfeld gave their paper, with the following Introduction to put Peddie's CALMU (Credit and Load Management Unit) into a U.S. context.

ABSTRACT

Responsive microprocessor meters are now cheaper than electro-mechanical time-of-use meters. The English residential Credit and Load Management Unit (CALMU) responds to dynamic prices broadcast by the British Broadcasting Corporation and turns off individual equipment at prices preselected by the homeowner. In the U.S. this meter could (with the thermal storage it will incite) displace one-third of the U.S. residential peak demand of 80 gigawatts (GW) and could displace the demand of new homes by 1 GW/year.

INTRODUCTION

It is a pleasure to describe the pioneering developments at SEEBoard (South Eastern Electricity Board) and at ICS (Integrated Communications Systems, Atlanta) and to relate their hardware to the potential for U.S. load management which I have discussed in Chap. 8. In particular I showed in Table II of Chap. 8 that the U.S. peak residential air conditioning demand is growing 2 GW each year, and is out of control because the residential (and small commercial) sector has no time-of-use rates or other surcharges on scare peak power. The CALMU technology discussed here has the potential to halve this growth of new a/c demand, and probably to shave one-third of the 80 GW peak demand of existing residences.

When Peddie was chairman of SEEBoard, he was most interested in managing the British winter peak, but of course in America we will first address the summer cooling peak, which is illustrated, for U.S. North Central utilities, in *Fig. 1* from the 1983 OTA report, INDUSTRIAL AND COMMERCIAL COGENERATION.

The English Credit and Load Management System (CALMS)

CALMS, summarized in *Fig. 2*, is the brainchild of Robert Peddie, a nuclear engineer, who served on the British Central Electricity Generating Board, and later as Chairman of the South Eastern Electricity Board (SEEBOARD). While the whole system is called CALMS, its electronic heart is called the Credit and Load Management Unit (CALMU). It has gone through several generations of development, and is now under test with 300 homes. The BBC already broadcasts a (fake) electricity price every five minutes. Depending on the interests of an

0094-243X/85/1350382-4$3.00 Copyright 1985 American Institute of Physics

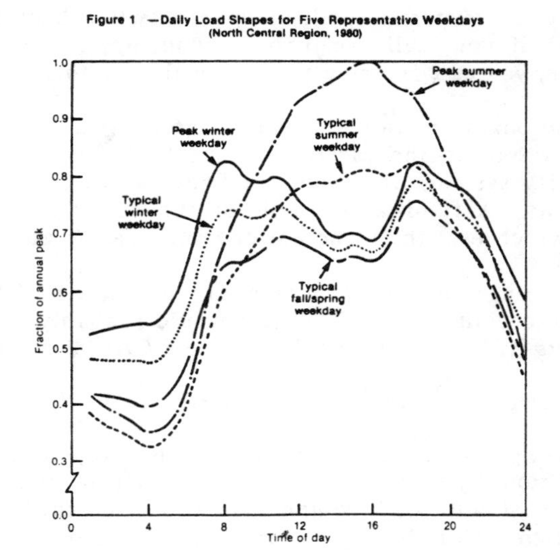

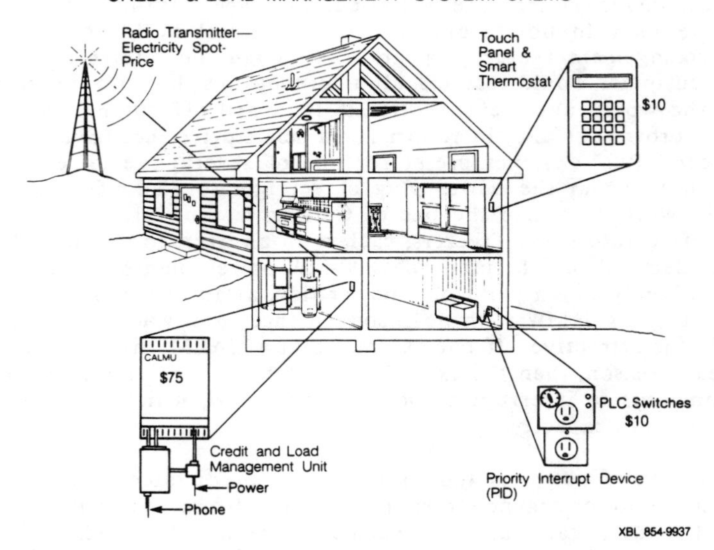

Fig. 2. English CALMS price-responsive meter system.

American utility, this price could be a time-of-use price, a "spot" price, or any other experimental price. CALMU immediately "transponds" this price to a 30-kHz power-line carrier (PLC) signal which is distributed over the home power lines. Each major appliance is then plugged into a little PLC switch box (about $10), which is in turn plugged into the wall. (If the appliance is hard-wired, like a central air conditioner or a water heater, the PLC switch is associated with the circuit breakers.) The PLC box has a nine-position selector switch, labelled in pennies/kWh, and will switch off when the price of electricity exceeds the threshold selected by the customer (and, of course, switch back to "on" when the price is right). (And, if you really need to run your appliance and are willing to pay the current price, you can override your normal threshold.)

In England, these boxes are called "priority interrupt devices" (PID's). In the U.S., they are called "power-line carrier switches" (PLC's). Familiar examples include homes security systems, in which the homeowner, or a clock, can control lights and coffee pots. A familiar example of the whole technology is the $99 cordless telephone, which uses the home power lines as a big antenna.

A note on the prices in Fig. 2, which are estimated 1984 English costs, translated directly into dollars. A U.S. system would probably cost twice as much. For example, ICS lists its "grey box" (equivalent to a CALMU) for $200.

Emergencies and Spinning Reserve

The first major appliances to be controlled by responsive meters will, of course, be air conditioners, electric water heaters, dryers, pool pumps, etc., and their response to price should work wonders for to flatten the summer peak. But there is also the question of reducing the utility's margin of spinning reserve, i.e., the capacity standing by for an emergency.

Once responsive meters are installed, and are paying for themselves by managing the load in homes and small commercial buildings, an economical emergency management feature appears. Although the refrigerator is about the last thing a customer would turn off all afternoon, just because the price of power is high, he/she would never even notice if it turned off for half an hour just to help avoid a brown-out. So, in return for a small price incentive, the utility can give out special PLC switches preset for one of several special emergency "high prices" for insertion at the power plug of refrigerators and freezers. Note how well this will work on a grid the size of California. In California there are over 10 million refrigerators and freezers, each with an average demand of 180 W, i.e., with a total demand of 1.8 standard 1000-MW power plants. Thus, a California utility can suddenly lose a plant or a power line, at any time of year, and reliably respond by dumping 1 MW with no inconvenience to any user. The economics of this switching is attractive. If one assumes that a CALMS-like system is installed for its primary reason, then the extra PLC switch for a refrigerator will cost only $10. This amounts to interrupting power for $60/kW, which is far cheaper than investing in reserve capacity.

This dumping of power is 100% reliable and predictable. The utility can test price response and emergency response several times every day, so it knows exactly what loads it can dump for various price signals. And, since CALMS is telephone connected, it can check on this response down to the individual home.

Even without the emergency argument, responsive meters will pay for themselves because they are connected to the telephone, and thus displace meter readers. SEEBOARD estimates it will save a dollar a month on meter reading--$12/year, which it capitalizes at $120 first cost. Since a complete CALMS system, retrofit to an existing home, should cost less than $120, we can hardly afford not to start using them. On a new house, where we avoid the cost of the conventional meter, the economics are even more favorable.

Applications to America: The ICS Transtext System

The U.S. (second generation) version of CALMS is called Transtext and was conceived by Douglas Bulleit, who founded an Atlanta company called ICS (Integrated Communication Systems) to develop it. Transtext is more ambitious than CALMS both in hardware and services offered.

1. The original concept targeted only homes with cable TV, so it had a TV screen and audio messages available. For its touch panel, it uses the customer's phone. Like CALMS, it uses the phone for meter reading, bill paying, and remote control by the customer. Later versions no longer depend on cable TV, but rely on telephone communications and are becoming more like CALMS.

2. In addition to cable TV and load management, Transtext will pay for itself by offering many other services: financial, security, banking, etc.

Future Smart Homes and Smart Appliances

The discussion so far concerns hardware we can already build and innovations we can already implement. Even more exciting is the resulting improvements we can make in appliances, homes, and small commercial buildings before the turn of the century.

All the California utilities are planning experiments with responsive meters. My ambition is the have the State (or any other state that can beat us to it) quickly announce a protocol for price/time/weather information to be put on the power lines of *all* buildings. The list of data, updated every five minutes (or immediately, in the case of an emergency) would include prices of electricity and gas, date and time, three temperatures (dry- and wet- bulb and thermostat), and perhaps the utility's weather forecast to help the logic of thermal storage systems. Then major changes would occur in two industries.

Smart Appliances. Manufacturers would offer room air conditioners with built-in PLC switches, and central a/c with several switches to control different zones of the house. The zoning would be introduced because at one price P1 the resident might want to knock off cooling of an unoccupied room, but wait for P2 before giving up the cooling of his study, ...

Refrigerators would respond to two prices. At a low price P1 the defrost cycle would wait. At an emergency price P2 (guaranteed not to last more than an hour) the whole refrigerator/freezer would wait.

Thermostats would become portable and would be moved to whichever room the resident wanted most comfortable.

Even clocks would reset themselves after a power failure.

Smart Homes

Home designers could now add whatever becomes cost effective under dynamic pricing: thermal storage, whole- house fans, evaporative coolers, exhaust-air recuperators, all of which could be efficiently linked with the conventional "dumb" air conditioner and controlled by the protocol PLC data. There would be similar revolution in the design of small commercial buildings. It's an interesting challenge to the imagination. And now on the to main paper.

CHAPTER 20

MANAGING ELECTRICITY DEMAND THROUGH DYNAMIC PRICING

Robert A. Peddie
South Eastern Electricity Board
5 The Mount Drive, Reigate
Surrey, UK RH2 ØEZ

Douglas A. Bulleit
President, Integrated Communication Systems, Inc.
1ØØØ Holcomb Woods Parkway, Suite 412
Atlanta, Georgia 3ØØ76

ABSTRACT

As electrical energy cannot be stored in large quantities with current technology, the energy balance of the electricity supply system has to be maintained continually by adjustments in supply to meet an unrestricted customer demand.

Electricity over the projected peak tends to be very expensive, so utilities have sought to restrict demand during such periods to improve internal economic efficiency. The techniques used can be shown to be inefficient and disruptive if widely applied. Due to the way the utility and the regulators have homogenized costs, rate structures provide the customer no useful cost message as motivation to economically control utilization.

Advances in microelectronics and communications remove these restrictions by allowing the customer to be informed continually of the cost of a kiloWatt hour (kWh) at the time of use. For the first time, the ensuing control of demand by the customer enables efficient utilization and simplifies the dynamic control of the electric system. This paper describes the method of formulating dynamic prices, the main elements and examples of the system, and how it can be introduced. The enumerated benefits show that greater customer satisfaction and improved economic management of the nation's resources and the utility's assets would result.

1. INTRODUCTION

In order to set the backdrop to dynamic pricing, the title of this paper requires closer examination. While load management is a freely used expression among utility professionals, it carries an implied assumption of being a desirable objective. This is questionable in the context often used and needs examining in more detail.

Since electrical energy is the ultimate perishable commodity and cannot be stored in large quantities with present technology, supply is continually being adjusted to match an uncontrolled customer demand. For the majority of electrical

0094-243X/85/1350386-10$3.00 Copyright 1985 American Institute of Physics

utilities, production of the marginal kWh -- particularly at the time of peak demand -- is very expensive. [See Fig. 4.] This gives rise to their desire to "manage the load" in order to improve the internal operating efficiency of the utility.

Utility managers reason that if they can reduce demand when the marginal kWh is expensive, then (a) the utility's costs are reduced, and (b) the overall efficiency of energy conversion rises, so the average cost of energy to all customers falls.

While in global terms this might prove true (trials have dealt with limited population samples), it does not follow that the individuals, whose loads have been "managed," have received the full benefit of the saving. Nor does it follow that the overall economic use of energy has improved, since only a part of the total energy cycle of production/transmission/utilization has been affected.

Generalities aside, Direct Load Control (DLC) techniques do not encourage conservation in the consumer sector with the current operating practice of **GENERATION FOLLOWING AN OPEN-ENDED CUSTOMER DEMAND IRRESPECTIVE OF COST.** (See Fig. 1 below.)

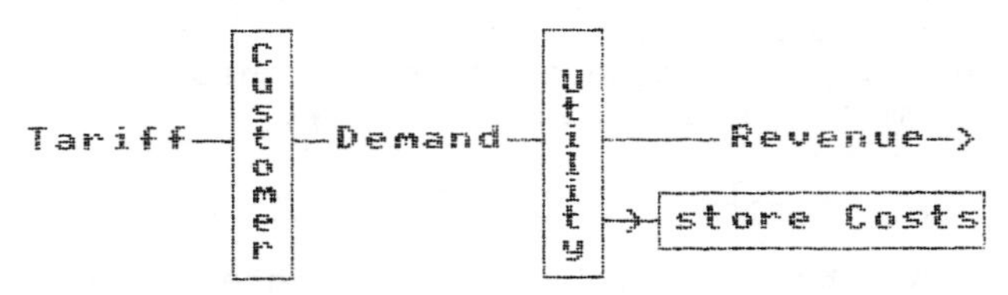

Figure 1

Current research illustrates, however, that if **DEMAND CAN BE TAILORED BY THE CUSTOMER TO PRODUCE ECONOMIC GENERATION** (See Fig. 2 below.) then conservation can be encouraged over the whole energy cycle. For this to occur, the customer must be informed of the price of a kWh (which accurately reflects the utility's costs at that time) continually and in real time. This in turn requires an open communication channel to each customer, which is now technologically and economically feasible.

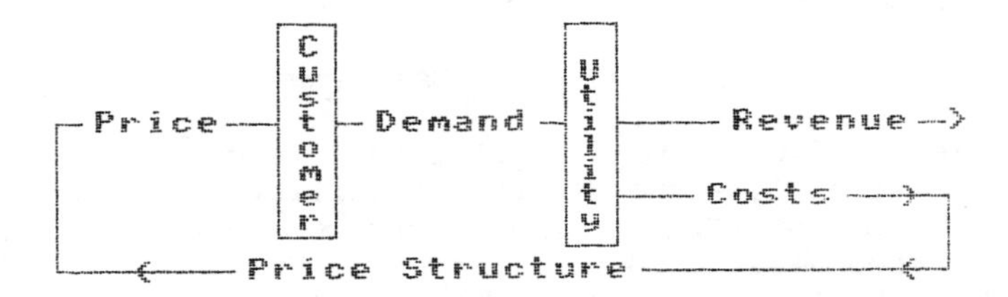

Figure 2

[The above are expected from larger networks. See Ref. 8]

It can be confidently stated, based on Great Britain's experience with its Credit and Load Management Unit (CALMU) equipment[1] now installed in customer's premises (discussed in section 4.3.3), that there are no technical barriers to the economic conservation of electrical energy. The barriers are institutional, political, and social.

2. IMPEDIMENTS TO THE ECONOMIC USE OF ELECTRICAL ENERGY

In order to identify and understand the changes needed to effect improved economic conservation of electrical energy, historical restrictions have to be identified. These restrictions have been due mainly to lack of appropriate technology, but no longer apply due to the rapid advances in microelectronics and communications.

2.1 Load Management

Current tariffs generally do not reflect the variable cost of meeting demand at different times or supply problems near system peak. Thus, customer demand patterns are not sensitive to cost or peak availability problems.

Two primary load management techniques are used at the moment to compensate for this:

2.1.1 Interruptible Loads

With this technique, the utility interrupts customers' loads either directly or, under a rate agreement, by requesting the customer to reduce load at a time of the utility's choosing. This is crude, but it's effective on a small scale. The financial inducements are frequently arbitrary and rarely related to the utility's savings at the time the load is shed. A further disadvantage to the large scale application of this technique is that, to maintain system stability, the restoration of the load has to be carefully controlled at all voltage levels. This technique is being increasingly criticized as an unnecessary infringement of customers' sovereignty, i.e., the utility crosses the "meter line". Finally, as it takes no account of utilization, it might or might not improve overall energy conservation or generating efficiency.

2.1.2 Time-of-Use Rates

This world-wide trend in utility thinking[2], supported by massive studies like the Electrical Utility Rate Design Study undertaken in the U.S.[3], all point to the need to convey a more cost-related message to the customer. Utility rate experts have therefore focused their attention on Time-of-Use rates, which call for a small technological changeover to a multi-rate meter.

While this is a move in the right direction, insomuch as it begins to influence the economic utilization of electricity, it has two drawbacks:

Firstly, applied on a large scale, it introduces synchronized step changes in total demand, which can present significant operating and system stability problems. In the U.K., there is considerable field experience of the inefficiencies introduced by step changes in demand, arising from the fact that the demand of 20 million customers can be synchronized by popular national television programs, e.g., Princess Diana's wedding.

In order to avoid step changes in demand, current U.K. day/night rates quote no specific changeover time and simply offer the night rate for a stated number of hours. This allows the Distribution Boards* to stagger metering periods and disconnections/reconnections.

CENTRAL ELECTRICITY GENERATING BOARD@
HALF HOURLY DYNAMIC PRICES (24 HOURS)

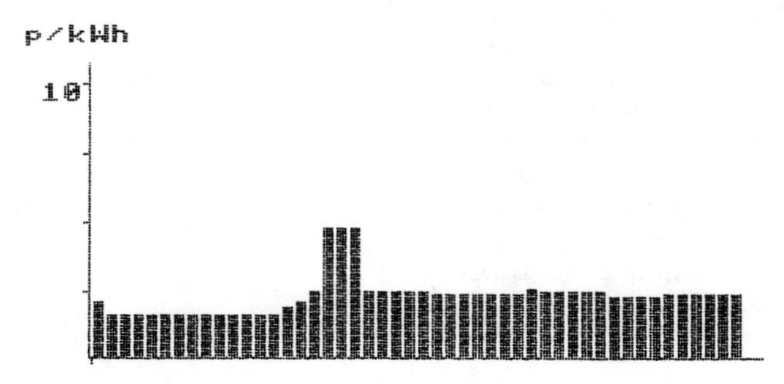

Figure 3
Typical Summer Load Curve
June 1, 1983

* U.K. jurisdictions/service areas

@ C.E.G.B. generates and transmits electricity to the 12 Area Distribution Boards in England and Wales. Prices quoted in Figures 3 & 4 are in British pennies per kWh of electricity.

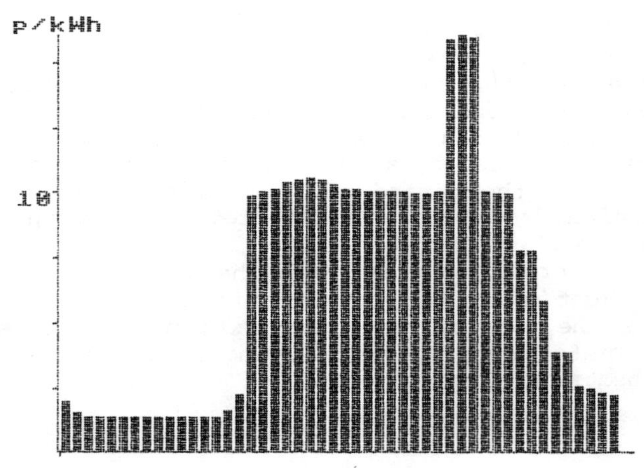

Figure 4
Typical Winter Load curve
January 19, 1984

Such a time duration rate, implemented at the discretion of the utility and utilizing a multi-rate meter, would not be acceptable where more than two time periods per day were involved. The alternative of staggering rate times and prices to the majority of customers would involve difficult managerial and regulatory problems.

Secondly, rates per se do not inform the customer of the large diversity of utility cost which occurs throughout the day and the seasons (Figuers 3 & 4 display the C.E.G.B.'s half-hour dynamic prices for transmitting voltage supplies and reflect costs in that time interval.)

2.2 Rates

While rates are the main technique of motivating customers to conserve energy, their main objectives from a utility's viewpoint are:

- revenue acquisition to maintain a viable business; and
- load management to improve their economic operation.

For the majority of utilities, both objectives are subject to some limitations, examples of which are:

- no price discrimination between customer groups;
- prices to reflect costs; and
- continuity and quality of electrical supply.

@ C.E.G.B. generates and transmits electricity to the 12 Area Distribution Boards in England and Wales. Prices quoted in Figures 3 & 4 are in Britich pennies per kWh of electricity.

2.2.1 Shortcomings of the Present Rate Systems

In periods of economic instability, the effect of inflation, varying interest rates, salaries, wages and other operating costs, government policies on taxation and lending, and many other variables all make revenue requirements difficult to estimate.

Due to the large number of uncertainties, setting rates to ensure adequate revenue frequently takes the form of an educated guessing game, e.g., customer demand (kWh's) and incidence (kW), plant availability, external economic events, sectional customer pressures, and government/regulatory policy on price control.

The resulting rate based on annual costs, homogenized by using many arbitrary assumptions, produces no real cost message which would assist in conservation. By averaging costs, the economic integration of renewable energy sources into the electricity supply system becomes more difficult.

2.2.2 Regulation

Since regulators have different objectives than utility management and make equally valid but different estimates of all the variables which are fed into rate formulations, it is not surprising that differences of opinion arise. It will be seen later that dynamic pricing can eliminate many areas of contention, allowing a mutual concentration on the efficient use of energy.

2.2.3 Telestrategies for AEM and Dynamic Pricing

The advances in microelectronics and communication technology now make it possible to transmit prices more related to costs at any chosen time interval. The literature on dynamic pricing effectively starts with Vickery's article in 1971[4] and has been advocated a number of times since under different labels, e.g., Spot, Real-time, Flexible, or Load Adapted Pricing. F.C. Schweppe and colleagues at M.I.T. have produced numerous related papers[5,6], and world-wide interest in the subject is growing, particularly in the U.K.[7]. The rest of this paper examines the implication of this changeover and how it leads to improvements in almost all aspects of electrical utility operations and customer utilization encouraging the conservation of energy.

3. FORMULATION OF DYNAMIC PRICES

There is universal agreement that the economical use of energy in all its forms is of importance to everyone, not only for the present, but for future generations as well. In developed countries where a safe and reliable supply of electrical energy is considered basic to the functioning of society, it is generally accepted that the following criteria are applied:

- prices to be based on marginal costs;
- no price discrimination between user groups;
- best financial practices; and
- maintenance of continuity and quality of supply.

While the principles of marginal cost theory are clear, there are differences among economists as to how the various costs are to be treated in order to obtain short and long-run marginal costs. The final crunch comes when a utility manager has to reconcile the financial outcome that results from the application of economic theory with that called for by the best accounting practice and political/social opinion.

Fortunately, many utilities already have considerable experience in the application of marginal cost principles which are used in evaluating generator loading programs to operate the electrical system. They underlie many inter-utility trading agreements.

A small extension in the use of this data and associated computer programs enable the Short Run Marginal Cost (SRMC) of a demand increment to be determined at any time. If units priced at this value do not result in the required revenue, the SRMC price may be modified to the required value by the inverse of the price elasticity of demand. This technique is used by economists to reconcile economic theory with accountancy, in a manner which least perturbs the customer's economic decision-making.[8]

Data on price elasticity of demand is currently very poor, so assumptions would have to be made in the initial implementation. However, the availability of communications between the customer and the utility would enable measurement of customer response to a price increment, thus rapidly eliminating this deficiency. Such measured data would naturally be exchanged through professional institutions and societies, further improving the price elasticity data base. Since it would require at least a decade to change customers over to dynamic pricing in a large utility, adequate price elasticity data would be available before large numbers of customers were involved.

4. SYSTEM TO SUPPORT DYNAMIC PRICING

Dynamic pricing cannot be considered in semi-isolation when formulating rate prices, as it introduces control of utilization by the customer into the operation of the electrical system. For this reason, an integrated electronic management system must be designed for compatibility with safe and reliable operation of the electrical supply system and provide information on which customers can act[9].

The main elements of such a system are Management, Communications, and Customer Premise Equipment.

4.1 Management

Such a system should preferably improve the safe and reliable operation of the electrical system, not reduce it. Any system must efficiently meet the requirements stated in paragraph 2.2. The CALMS system (See section 4.3.3.) under trial in the U.K. meets both requirements.

4.1.1 System Control

Because of the economic and social repercussions engendered by failure of electricity supplies (New York, France, etc.), operators controlling the electrical system are apprehensive of "delegating" control of demand to customers, hence their attachment to utility interruptible and similar tariffs.

Operators' fear of the disruptive effects of customers controlling demand are misplaced. People en masse tend to demand energy in a predictable way when responding to their environment (weather, financial etc.) and only slowly change their habits. After all, the whole existence of load forecasting, system operation, and planning is based on this fact.

For the customer to derive the maximum benefit from dynamic pricing, as well as for system stability reasons, automated control of utilization by the customer and supplies by the utility will be desirable and can now be accomplished cheaply. With these facilities, the price elasticity data for groups of customers is unlikely to change in a random manner, so the smooth slow change in the characteristics of consumer demand will allow ample time to adapt control techniques.

Currently, in a large integrated electrical system like the one in the U.K., with generation following normal demand, the response to an increase in generation following a fall in frequency can vary from minutes to tens of minutes due to time lags in the chain, controllers' observations (telemeter generating station), and boiler house response (human and thermal).

With industrial consumers using energy management systems and simple price-responsive plugs attached by domestic customers to their appliances, the customer-chosen response is effected immediately. As the price message can be transmitted to such automatic equipment in seconds, an order-of-magnitude improvement in response time is achieved. Since the individual customer's price elasticity of demand will be different and measurable, the system controller will be able to select the appropriate price interval to evoke the required response. Such a management change would allow both supply and demand to be used to control the energy balance of the electrical system.

The installation of electronic Customer Premise Equipment (CPE) also permits a greater degree of control under abnormal system conditions, i.e., energy input shortage, abnormal instantaneous changes in energy flows, or catastrophic failure. In so doing, the economic and socially damaging brown-outs and black-outs will be eliminated.[10]

4.1.2 Price Levels and Time Intervals

The number of price levels used would be a matter of judgement related to the shape of the cost curve and the degree of control sought. These price levels would be expected to be retained for a minimum of a year, but in practice they would be relevant for much longer. The chosen discrete prices could be used as one of the many checks on the validity of the received price. The other characteristic of the price levels is that they would have a non-linear structure in order to facilitate electrical system control.

The selection of the time interval, number, and level of prices would vary among utilities. It would be achieved by a series of rapidly converging estimates, bounded by customer price elasticity, communication costs, and utility economics. Prices would need to be dispatched at intervals of less than six minutes, even with a highly variable demand.

4.2 Customer Premise Equipment (CPE)

Since the Customer Premise Equipment terminal is built around an inexpensive, programmable microprocessor, many facilities are considered desirable and can be made available.[II] This section confines itself to the way the CPE handles the operation of dynamic pricing.

The following information could be displayed on demand when the customer enters simple commands on the CPE touch panel:

- kWh used at each price level
- Total account
- Flexible 24-hour sub-accounts, covering the immediate 24-hour period, plus the previous or projected 24 hours (for example, detailing price, kWh's used and account increment, each group with a time tag).
- Price for each time interval for immediate, previous and next 24-hour periods.

A CPE with a standard communication Input/Output (I/O) port for customer use would provide a convenient method for this information to be loaded on demand into a peripheral, e.g., a home computer, simple printer, energy management system, or displayed on the TV set.

The kWh used in each time interval would be (1) incremented into the "energy register" for that price, (2) multiplied by the price, and (3) the result posted to the total account register. The price, associated kWh's used, and the resulting increment to the account, with a time tag, would be added into one of the 24-hour sub-account registers. The utility could poll the CPE via whatever communications link is prescribed.

The price with a time tag could likewise be stored for the previous 48 hours. Initially, this price stream would be compared each day with the centrally transmitted price. As experience was gained on the system's reliability, this could change to automatic random checks.

With the CPE capable of displaying the data or loading it into a customer-owned peripheral, customers would have all the data necessary to develop an energy utilization strategy to suit their production process or lifestyle. In the domestic sector, since the level at which the "price plug" operates is set by the user -- like a room thermostat -- it could readily be adjusted as experience is gained and/or circumstances change.

By reflecting the wide variations in true costs over a range of ten to one or more in some areas, dynamic pricing will have a significant and beneficial long-term effect on the design and procurement of new equipment and processes. The incremental cost of energy efficient equipment and processes is usually small when specified at the design stage. This price increment can be recouped in a short time by avoiding the high-priced energy. In many cases, there is secondary conservation. For example, increased thermal insulation to maintain temperature with no energy input is equally effective at a time of lower energy costs.

4.3 Communications

Needless to say, we are witnessing explosive growth in communication technology. Service industries which previously had only person-to-person or postal contact with their customers now face a veritable cornucopia of techniques and systems to "communicate at a distance." Decision-making thus becomes difficult. How does one invest in communication facilities while retaining the freedom to take advantage of new and more economic systems as they become available? Fortunately, it is possible to design systems which are virtually communications-independent. The appreciable efforts to produce communication standards throughout the world reduce this problem somewhat.

4.3.1 Communication Economics

In considering the application of dynamic pricing in the supply of electrical energy, real and potential advances in communication offer a variety of means for communicating with the customer, i.e., via power line carrier, radio, telephone, and broadband cable. Nor can the potential for fiber optic and satellite links be ignored.

The simplest economic appraisals show that using any system in a dedicated mode solely to transmit dynamic prices is not viable. With a total utility data flow of only 1.5 to 2 kilobytes per day/customer, the only cost-effective way is to piggy-back on other types of investment already made, in a manner which does not interfere with the system's primary use, e.g., signalling over the power mains or phase-modulating a commercial radio broadcast signal. In contrast to the insignificant data traffic, the quality and integrity of the communication system is very important if the customer's utilization is to be incorporated into the economic operation of the electrical system.

The next logical step is to seek lower costs by sharing communication channels with others who also wish to provide a service to the customer. In order to minimize investment and maximize reliable service, this secondary use of existing systems encourages examination of the efficacy of each technique for the various types of communications required. For example, radio could be employed for broadcasting the same message to all, telephone for individual communication, and the power line carrier for communicating with individual appliances within the home.

Hybrid communication systems are increasingly being constructed so that service providers have a cheaper and easier means of increasing the value of their customer service.

4.3.2 Types of Messages

Data traffic falls into the following four broad message categories:

1. **BROADCAST MESSAGES** of about five bytes could be used to transmit a price at defined time intervals and load-limiting messages in times of emergency. Both types of messages call for the total time between initiation and action to be short, preferably under 10 seconds.

2. **STATUS MONITORING** of about 15 bytes consists of a routine electronic check on the system and its management functioning (interruptions of main supply or communications, ground faults on equipment/appliances, payments made, tampering, etc.). The frequency of monitoring the fault flags in the early stages for the U.K. would be once per hour. A response time of under 10 seconds is looked for.

3. **DATA ACQUISITIONS** tend to be carried out on a batch basis overnight, when the communication systems are lightly loaded. This activity is not time-critical. The data flow will be dependent on the number of price intervals chosen (80 to 500 bytes per message) and the billing frequency (weekly/monthly/quarterly).

4. **CONSUMER-INITIATED SIGNALS**, in the context of electrical utility operations, are confined to electronic payment of accounts. Since rapid response is required -- 2/3 seconds in the U.K. system -- this is first activated within the CPE, which subsequently completes the transaction on the hourly status pollings. This is only one of many ways to handle this infrequent operation.

It should be noted that with low total data flow spread out over 24 hours by the different types of messaging, the costs of using a shared communication system should not be a deterrent to the implementation of dynamic pricing.

4.3.3 Communication System Examples

In the introduction, it was stated that the main barriers to the conservation of energy through dynamic pricing were institutional, political, and social. Nowhere is this more evident than in the service industries. Each has a clear historical concept of their areas of jurisdiction and desires to purvey a specialized service to the same customers. Despite this, solutions are being found to these difficult organizational and commercial problems.

As reams could be written on the ownership, use, cost allocation, signalling priorities, etc., of a shared communication resource, the rest of this section confines itself to describing the evolution of two systems, one in the U.K., the other in the U.S., each of which has the potential to implement dynamic pricing.

It is interesting to observe the convergence of thinking of these two groups, each desiring to offer the customer an improved service, starting with different objectives and working independently in two countries.

In the U.K., the South Eastern Electricity Board (SEEB) started from the concept of designing an electronic Credit and Load Management System (CALMS). Early in the design phase, management congruence of water, gas, and electrical utility customer activities -- in the areas of measurement, rate structures, data storage, and communication -- became obvious. In light of this, the common customer Credit and Load Management Unit (CALMU) was designed.

The wide range of communication systems available, coupled with the rapid changes in this field, resulted in the design of a software-based communication interface capable of utilizing any of the current or projected systems referred to in 4.3.1. In order to minimize costs, the CPE was designed to operate in a passive mode, i.e., it did not initiate signals but replied to them, by utilizing idle time on the telephone and receiving phase modulated signals transmitted within the commercial B.B.C. 200kHz radio broadcasts. The CALMU communicates with individual appliances by signalling over the house power wiring.

Rapid advances in telecommunications have made voice-over-data transmission practical. Accordingly, the CPE can readily be changed to initiate communications, thus extending its functionality to cover other activities like alarm systems, electronic mail, etc.

Integrated Communication Systems, Inc., of Atlanta, Georgia, started from the concept that there were many organizations wishing to offer their customers interactive services which provided more control in their home environment. ICS considered the logical communication medium for such "value-added" services to be the telephone. Starting with this premise, ICS adopted the technique of transmitting voice-over-data using BellSouth's PulseLink^{SM} public packet-switched telephone network.

The individual organizations participating in ICS' interactive home management and communication system - TranstexT® - ensure that their end-use equipment (i.e., electric meters and wiring, cable plant, television and telephone equipment) are compatible with the microelectronic Com'Set (formerly known as the Grey Box) communications interface. Such a system can readily convey dynamic price data to individual customers and carry out all the advanced energy functions previously referred to.

5. OPERATION OF A DYNAMIC PRICING SYSTEM

Dynamic pricing requires a continually updated flow of information upon which decisions are made to achieve economic utilization. Customers continually make two types of decisions which affect energy conservation (1) through their use of existing electrical equipment, and (2) through the purchase of new equipment. Both require a knowledge not only of past prices but current and projected price curves. Fortunately, as indicated earlier, the price profiles for weekdays/weekends or seasons of the year will only change slowly, so that projections for the month/year ahead could be published regularly.

Customer familiarity could be enhanced by making available estimates of the price profile for the ensuing day as a subset of the evening TV weather forecast. This data could also be down-loaded for display locally on the CPE. This latter facility would be of particular use to the industrialist in scheduling energy management for the next day's production.

The significance of the revolution in utility thinking and operation -- implied by the expression "Let the Customer Decide" -- must not be underestimated, as the ingenuity of individuals is almost infinite. Currently, utilities concentrate their load management efforts on the obvious - like thermal lag loads and water/house heating and cooling - together with loads like water usage, whose time of use is not critical.

Once the customer has to decide how to use a product whose price varies in a predictable manner with time, innumerable new ways of conserving energy will emerge. Two which emerged from the CALMS U.K. field trials were 1) the way the batteries were used in the telephone exchanges (an M.W. load for the whole of the U.K.), and 2) in the home, customers have responded to multi-time-of-day prices by buying "jug-shaped" domestic kettles to minimize the amount of water boiled.

6. INTRODUCING DYNAMIC PRICING

Dynamic pricing would initially be offered to industrial and commercial customers, since they have the following characteristics:

- Small in number, but consume large percent of kWh's sold
- Interested in reducing costs
- Own a wide spectrum of controllable energy use
- Are on the telephone system
- Located in urban areas and receive commercial radio broadcasts
- Have metering systems of varying complexity and cost, all more expensive than projected cost of a three-phase electronic management unit

In many utilities serving urban/metropolitan areas, 40-50% of the total electrical demand is taken by the industrial customers who comprise less than 1% of all customers. The changeover of this group could therefore be accomplished in three years or less.

As the architecture of the majority of communication systems would require a data concentrator at the first nodal point beyond the customer, the conversion of a single industrial user would result in the installation of the data concentrator equipment necessary to serve all the customers in that vicinity, thus enabling dynamic pricing to be rapidly taken up by other nearby groups of customers.

The CPE must be able to operate in a stand-alone mode. Without communications capabilities, the customer is bereft of many of the benefits stemming from an interactive system. Nevertheless, the CPE can operate on a predetermined schedule of prices derived from a simulated demand.

Although the existing procedure of a periodic "meter reading" by the utility would continue, a hand-held electronic unit would be used to "read" the kWh and price registers, together with the flag data. The revised price/time schedule would then be installed. As far as the customer was concerned, the CPE would then display the current dynamic price, motivating them to utilize energy economically.

The half-way house between the above and the complete system would be to utilize one-way communication over the radio, which would have the advantage of conveying the actual dynamic price but the disadvantage of not being able to check on missing or corrupted signals. Thus, it would not be "lawyer proof". The expensive personal visits by a utility would still be required.

With this range of approaches to the conversion, an orderly introduction of dynamic prices could be achieved.

7. BENEFITS

With a flexible integrated electronic management system involving the customer's automatic demand control and responding to a communicated dynamic price stream, many benefits flow:

1. Prices truly reflecting the rapidly changing marginal costs of producton can be simultaneously transmitted to all customers.

2. Non-discrimination is clear with allowances in the individual dynamic price for differing costs of supply H.V., etc. (In the U.K., large industrial customers could receive their prices via the telephone. Domestic and other large groupings could receive theirs via a 200kHz radio broadcast).

3. All changes in costs are immediately incorporated and reflected in prices, i.e., fuel, interest rates, taxation, borrowing, wage rates, government policies, etc.

4. All changes in operating conditions are likewise immediately incorporated into prices, i.e., weather conditions, plant availability, etc.

5. As customer demand control plays a greater part in the system's balance, and as experience of system response dynamics and customer price elasticity is acquired, the utility's planning, operational plant margins, and spinning reserve could be reduced.

6. Price elasticities of all customers under all conditions can be determined to any degree of detail.

7. The financial books of account can be balanced with any chosen margin, as the majority of the present unknowns in the revenue balance sheet are avoided using dynamic pricing.

8. With each dynamic price reflecting the utility's total costs at that instant, the customer is better able to make economic decisions - not only on usage but on the energy characteristics of the equipment they purchase, resulting in conservation and a rational use of energy.

9. Since dynamic prices would be based on marginal costs, this would simplify the contractual relationship with autoproducers and alternative energy sources. The utility would buy and sell at the margin.

10. Rate harmonization is automatic if all are on dynamic pricing.

In addition to the above, a less obvious benefit could be improved relations between regulator and regulated. Because of their key role in the economy, most electrical utilities are regulated in some way, with particular emphasis on rates. Traditional rate formulations and costing methods, coupled with the different objectives between regulator and regulated, make differences of opinion on rates inevitable. Whereas once the formula for derivation of the dynamic price was agreed upon, the focus of both parties would change to a common interest - the internal efficiency of the utility - to the benefit of customers, shareholders, regulators, and regulated.

In short, the benefits of dynamic pricing could lead to greater customer satisfaction as well as improved economic management of the nation's resources and the utilities' assets.

REFERENCES

1. R. A. Peddie, I.E.E.E. Trans. on Power Sys. & App.. Vol. 1Ø2, No. 8 (1983).

2. R. A. Peddie, Elec. Rev. Vol. 215, No. 6, p. 7 (1984).

3. R. A. Peddie, Elec. Util. Rate Design Study (1982).

4. W. Vickery, Bell Journ. of Econ. & Mgmt. Sciences Vol. 2, No. 1 (1971).

5. R. E. Bohn, M.I.T. Energy Lab. Rep., MIT-EL 82-Ø31 (1982).

6. F. C. Schweppe, I.E.E.E., Power Eng. Soc., Winter Meeting (1982).

7. R. A. Peddie, Homeostatic Control & SEEB Credit & Load Mgt. Sys. (1983).

8. R. A. Peddie, G. Frewer, A. Goulcher, The App. of Econ. Theory Utilizing New Tech for the Benefit of the Consumer (1983).

9. P. Ellis, G. Gaskell, A Rev. of Soc. Research on the Ind. Energy Consumer (1978).

10. R. A. Peddie, Sys. Operation a New Dimension (1983).

11. R. A. Peddie, Energy Mgt. & Load Control Rep. of the Energy Mgt. Task Force (1981).

CHAPTER 21

TECHNICAL AND ECONOMIC ANALYSIS OF STEAM-INJECTED GAS-TURBINE COGENERATION

Eric D. Larson
Robert H. Williams
Center for Energy and Environmental Studies, Princeton University,
Engineering Quadrangle, Princeton, NJ 08544

ABSTRACT

Industrial cogeneration is gaining popularity as an energy and money saving alternative to separate steam and electricity generation. Among cogeneration technologies, gas-turbine systems are attractive largely because of their lower capital cost and high thermodynamic efficiency. However, at industrial plants where steam and electricity loads vary daily, seasonally, or unpredictably, the economics of conventional gas turbines are often unfavorable due to low capacity utilization.

Steam-injected gas-turbine cogeneration overcomes the part-load problem by providing for excess steam to be injected back into the turbine to raise electrical output and generating efficiency. Under provisions of the Public Utilities Regulatory Policies Act, any excess electricity can be sold to the local grid at the prevailing avoided cost of electricity. Steam-injected gas-turbine cogeneration can result in a consistently high rate of return on investment over a wide range of variation in process steam loads. Moreover, this technology can also give rise to greater annual electricity production and fuel savings per unit of process steam generated, compared to simple-cycle cogeneration, making the technology attractive from the perspective of society, as well as that of the user.

Steam-injected gas-turbines may soon find applications in electric utility base-load generation, as well, since it appears that electrical generating efficiencies in excess of 50% can be obtained from turbines producing of the order of 100 MW of electricity at a fully-installed capital cost as low as $500/kW.

INTRODUCTION

Industrial cogeneration was made economically more attractive by the 1978 passage of the Public Utilities Regulatory Policies Act (PURPA). The Act, and the rules implementing it, which have been upheld in Supreme Court decisions of 1982 and 1983, require electric utilities to (i) purchase power from qualifying cogenerators at a price that reflects the costs the utilities would avoid by not having to provide the electricity themselves and (ii) provide back-up power at rates that do not discriminate against cogenerators.

Commercially available cogeneration technologies include gas turbines, steam turbines, gas turbine/steam turbine combined cycles,

0094-243X/85/1350402-24$3.00 Copyright 1985 American Institute of Physics

and diesel engines. One technology is chosen over another primarily based on economics, as influenced by factors such as relative fuel prices, security of fuel supply, the size and variability of particular steam and electricity loads, the reliability of the system, and environmental constraints.

Gas-turbine cogeneration is attractive largely because of its low installed capital costs compared to competing technologies (see Fig. 1), and its high thermodynamic efficiency in baseload applications (e.g., for meeting process heat loads in capital-intensive, energy-intensive industries characterized by relatively constant steam loads).

The recently introduced steam-injected gas-turbine extends the range of applicability of gas turbines to cogeneration applications characterized by variable heating loads -- applications for which conventional gas turbines are generally not well suited.

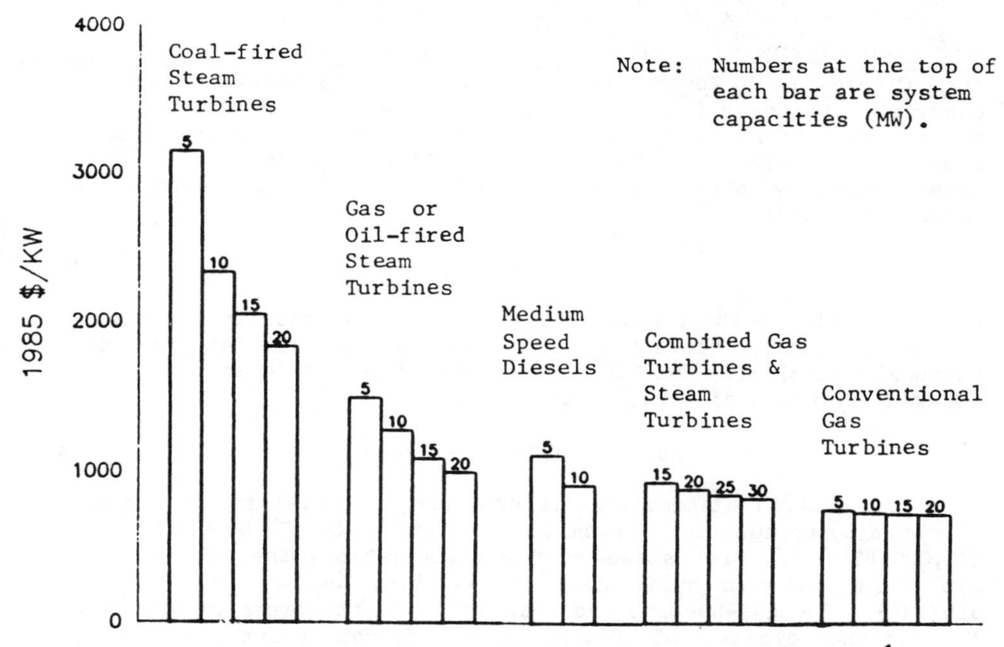

Fig. 1. Installed capital costs of small cogeneration systems.[1]

TECHNICAL PERFORMANCE MEASURES OF COGENERATION

Several technical performance indicators provide a basis for analyzing all cogeneration technologies.

First-Law Efficiency: The most universally understood measure is probably the overall first-law efficiency: the ratio of the energy in the useful heat plus the electricity to the heating value of the fuel consumed. For the separate production of electricity and steam (e.g., in a central station power plant and a stand-alone boiler), first-law efficiencies range from 60 to 70%. Steam-turbine cogeneration efficiencies are typically 80 to 90%; those for gas turbines and combined cycles, 70-80%; and for diesel cycles, 60-70% (see Table 1).

TABLE 1. TYPICAL ENERGY PERFORMANCE INDICATORS FOR ALTERNATIVE COGENERATION SYSTEMS.[2]

	FUEL FRACTION CONVERTED TO		OVERALL 1st LAW EFF.	FUEL CHARGE-ABLE TO POWER IN kJ/kJ AND (BTU/kWh)(b)	ELECTRICITY-STEAM RATIO kJ/kJ AND (kWh/MBTU)	SECOND LAW EFF.(c)	FUEL SAVINGS RATE (kJ FUEL PER kJ TO PROCESS STEAM)
	ELEC	STEAM(a)					
STEAM TURBINE(d)	0.12	0.73	0.85	1.38 (4700)	0.17 (50)	0.40 [0.34]	0.27
GAS TURBINE(e)	0.30	0.46	0.76	1.60 (5450)	0.64 (188)	0.48 [0.35]	0.86
COMBINED CYCLE(f)	0.35	0.42	0.77	1.49 (5080)	0.84 (245)	0.52 [0.35]	1.21
DIESEL(g)	0.35	0.25	0.60	2.02 (6900)	1.38 (405)	0.45 [0.35]	1.01

(a) SATURATED STEAM AT 1 MPa (150 PSIG).
(b) ASSUMING A STAND-ALONE BOILER EFFICIENCY OF 88%.
(c) ASSUMING FEEDWATER AT 100°C (212°F) AND AN AMBIENT TEMPERATURE OF 15°C (59°F). BRACKETED NUMBERS ARE SECOND-LAW EFFICIENCIES FOR SEPARATE STEAM PRODUCTION AND ELECTRICITY GENERATION AT A CENTRAL STATION PLANT.
(d) FOR SYSTEMS PRODUCING OF THE ORDER OF 50-100 MW OF ELECTRICITY.
(e) FOR A GAS TURBINE/STEAM TURBINE SYSTEM PRODUCING 73.8 MW OF ELECTRICITY.
(f) FOR A SYSTEM PRODUCING 87 MW OF ELECTRICITY.
(g) FOR UNJACKETED DIESELS PRODUCING OF THE ORDER OF 20 MW OF ELECTRICITY.

Fuel Chargeable To Power: Another measure is the amount of fuel chargeable to power (FCP), defined as the fuel required in a cogeneration facility in excess of that required to produce steam in a separate (stand-alone) boiler. The FCP is expressed dimensionlessly as the fuel energy charged per unit of electrical energy produced:

$$FCP = (1 - S/\eta_b)/E, \quad (1)$$

where S and E are the fractions of the fuel converted to steam and electricity, respectively, and η_b is an assumed efficiency for the stand-alone boiler. The FCP is commonly expressed in units of BTU/kWh, in which case Eqn. 1 becomes

$$FCP = 3413 * (1 - S/\eta_b)/E. \quad (1a)$$

At a central station generating plant, all fuel is charged to electricity production, typically yielding a FCP of about 2.93 kJ/kJ (10,000 BTU/kWh). For steam-turbine cogeneration, the FCP is typically less than half this value -- about 1.4 kJ/kJ (4700 BTU/kWh). For gas-turbine systems, it is of the order of 1.6 (5500), for combined cycles, 1.5 (5000), and for diesel engines, 2.0 (7000) (see Table 1).

The first-law efficiency and the FCP can be somewhat misleading indicators of overall thermodynamic performance because they account primarily for quantity rather than quality of energy. Thus, they tend to favor cogeneration schemes that produce a relatively small amount of electricity per unit of process steam, such as steam-turbine systems, which typically convert only 10-15% of the fuel into electricity, but 65-75% into process steam (see Table 1). The higher electricity-to-steam ratios of gas-turbines, combined cycles, and diesels mean they typically produce 4-8 times more electricity per unit of process steam than steam-turbines (see Table 1), and thus have the largest potential for generating electricity in excess of onsite needs for sale back to the grid.

Second-Law Efficiency: A better measure of overall thermodynamic performance, which takes into account the high thermodynamic quality of electricity, is the second law efficiency, defined as the ratio of the minimum available work required to produce steam plus electricity, B_{min}, to the actual available work used up in the process, B_{act}.[3] For a cogeneration system producing saturated process steam at a temperature, T, and pressure, P,[4]

$$B_{min} = mv(p-p_o) + mc_p[(T-T_1)-T_o\ln(T/T_1)] + mh_v[1-(T_o/T)] + E, \quad (2)$$

where m is the mass of saturated process steam at p, p_o and T_1 are the inlet pressure and temperature of the feedwater, v is the specific volume of the feedwater, c_p is the specific heat (at constant pressure) of the feedwater, and h_v is the enthalpy of vaporization of water at pressure p. The available work in fossil fuels roughly equals the lower heating value (LHV)[3] -- the heat released by complete stoichiometric combustion of the fuel, with water in the products remaining in vapor form (The higher heating value (HHV) is equal to the lower heating value plus the heat released by condensing the product water.)

As shown in Table 1, second law efficiencies for separate steam and central station electricity generation are all about 0.35, while those for cogeneration range from 0.40-0.52, with the highest of these values corresponding to systems characterized by the highest electricity-to-steam ratios -- gas-turbines, combined cycles, and diesels.

The Fuel Savings Rate: One additional measure, the fuel savings rate (FSR), also takes account of energy quality, but is somewhat less abstract than the second law efficiency. Defined as the amount of fuel saved in producing electricity (by cogeneration rather than central station generation) for each unit of process steam generated,[4] the FSR is commonly expressed in dimensionless terms (e.g., kJ of fuel per kJ to steam) and is calculated from

$$FSR = (2.93 - FCP) * ESR, \quad (3)$$

in which ESR is the electricity-to-steam ratio.

As Table 1 indicates, the greatest fuel savings come from the cogeneration technologies that produce the largest amount of electricity per unit of process steam -- gas-turbines, combined cycles, and diesels.

PART-LOAD PERFORMANCE OF GAS-TURBINE COGENERATION SYSTEMS

The attractive thermodynamics of gas-turbine cogeneration systems operating at full-load, as discussed in the preceding section, tends to degrade when they are operated at part-load for a sizeable fraction of the time, as is the case in many smaller scale industrial and commercial applications.

The second law efficiency and the FSR drop precipitously in conventional gas-turbine systems when the process steam demand drops

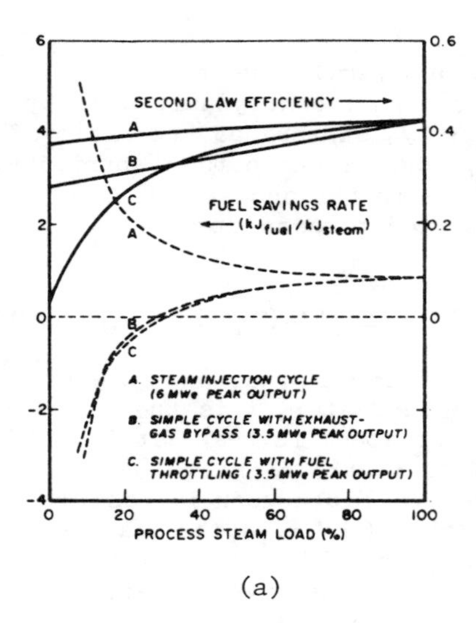

(a)

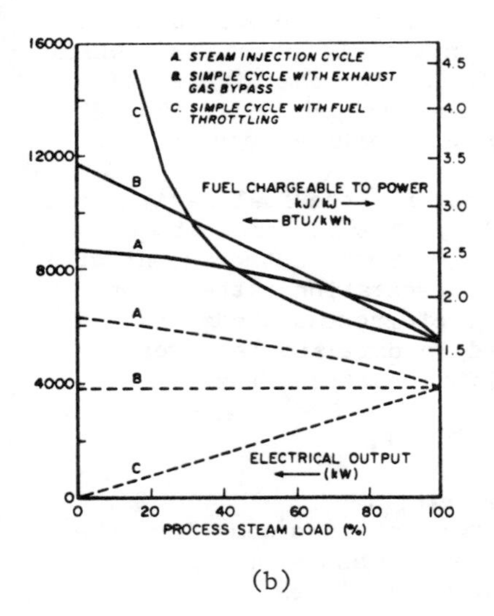

(b)

Fig. 2. Part-load gas-turbine cogeneration performance, based on the Detroit Diesel Allison 501-KB turbine.[5] Case A is with steam injection (6 MW peak output); Case B is a conventional system with exhaust-gas bypass (3.5 MW peak); Case C is a conventional system with the fuel flow throttled (3.5 MW peak).

and the fuel input to the turbine is throttled, as shown in Fig. 2a (case C)[6]. In addition, the FCP rises sharply and the electrical output drops off (see Fig. 2b). Because of these poor part-load performance characteristics, the more commmon mode of operation is to bypass the steam generating system with the hot exhaust gases, allowing full electrical production to continue at the expense of an increasing FCP (see Fig. 2b, case B). The second law efficiency falls off somewhat more slowly (see Fig. 2a, case B), but the FSR drop is equally sharp. The unfavorable economics resulting from poor part-load performance has restricted the use of conventional gas turbines largely to applications where steam loads are relatively constant.

The recently-commercialized steam-injected gas-turbine, however, overcomes problems associated with part-load operation by providing (through only minor hardware modifications) for excess steam to be injected into the combustor of the turbine system. As a result of the greater mass and energy flows through the turbine, obtained without additional compressor work, net electricity production increases significantly, but the FCP increases only moderately (see Fig. 2b, case A). The second law efficiency remains essentially constant for all process steam loads, while the FSR actually rises as the process load drops (Fig. 2a).

One of the keys to the viability of steam-injected gas turbines in cogeneration applications is the availability of markets for excess electricity, which now exist as a result of the PURPA legislation. In addition, general apprehensions several years ago to investing in natural gas-fired equipment, arising from fears of significant increases in natural gas prices, have proved to be unfounded. The average price of natural gas paid by industry in the U.S. (in inflation-corrected dollars) is currently projected to fall by about 12%, 1983-1995.[7]

There is also interest in the use of steam injection by utilities for the generation of power only, due to the fact that the FCP when no process steam is required is significantly lower than that for most large central station power plants, even for the very small gas turbines focussed on in this article (see Fig. 2b)! For moderately-sized systems (~100MW), it appears that FCPs of the order of 1.8 kJ/kJ (6200 BTU/kWh) are attainable.[8]

A REVIEW OF CONVENTIONAL GAS-TURBINE COGENERATION THERMODYNAMICS

Conventional gas-turbine cogeneration systems embody an open Brayton cycle, with heat in the turbine exhaust used to produce steam in a heat-recovery steam generator (HRSG).

Brayton Cycle Calculations:[9] As indicated in Fig. 3, the standard open-cycle gas turbine system incorporates: a compression process (state 1 to 2), requiring an input of work; a heat addition process (2 to 3); and an expansion process (3 - 4), by which work is produced. Hardware constraints will dictate: compressor pressure ratio (P_{rC} = P_2/P_1); turbine inlet temperature (T_3); the mass flow rate through the compressor (M_C); compressor adiabatic efficiency (η_C), defined as the work input required to isentropically compress a unit mass (w_{Cs}) divided by the actual work (w_C); turbine adiabatic efficiency (η_T), defined as the actual work output per unit mass (w_T) divided by the work output for isentropic expansion (w_{Ts}); and gearbox-generator efficiency (η_g), defined as the ratio of the electrical work output of the system divided by the net shaft work output.

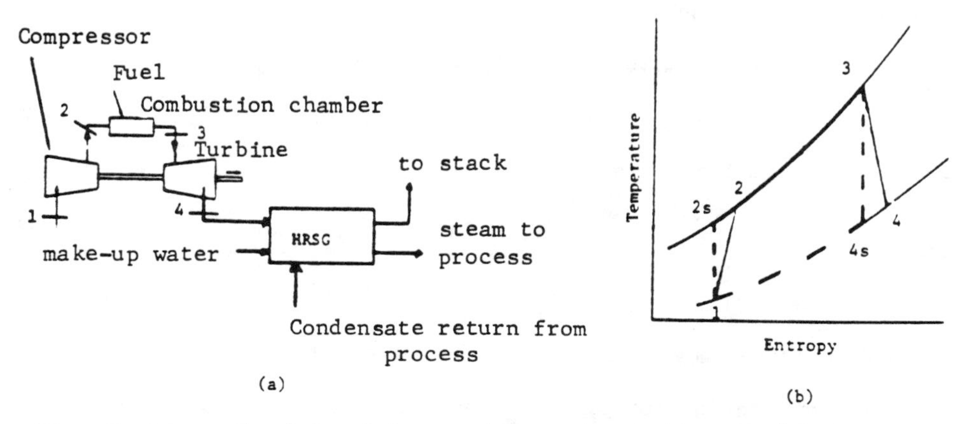

Fig. 3. Schematic (a) and temperature-entropy diagram (b) for a gas-turbine simple-cycle cogeneration system.

To calculate the net power output and generating efficiency of this cycle, the temperature of the compressor discharge is first determined. Assuming that the working fluid (usually air) behaves as an ideal gas (an assumption used throughout this article), so that its enthalpy is a function of temperature only [h = h(T)], and that its specific heat at constant pressure, c_{pC} is constant (so that $\Delta h = c_{pC} * \Delta T$), the compressor outlet temperature is obtained from the isentropic compressor efficiency:

$$\eta_C = w_{Cs}/w_C = c_{pC}*(T_{2s}-T_1)/c_{pC}*(T_2-T_1) \quad (4)$$

or

$$T_2 = T_1 + (T_{2s} - T_1)/\eta_C. \quad (5)$$

In Eqn. 4, the specific heats are evaluated at the mean of the temperature range with which they are associated. In this case, since the difference between T_2 and T_{2s} is small, the specific heats cancel, as indicated by Eqn. 5. For isentropic compression,

$$T_{2s} = T_1 * (P_2/P_1)^{[(k-1)/k]} = T_1 * P_{rC}^{[(k-1)/k]}, \quad (6)$$

where k is the ratio of the specific heat at constant pressure to that at constant volume.

The heat input required to the combustor per unit mass of working fluid (q_{in}), assuming constant pressure combustion and a constant specific heat, c_{pcb}, is

$$q_{in} = c_{pcb} * (T_3 - T_2). \quad (7)$$

In practice, heat addition adds mass -- the products of combustion -- to the working fluid and changes its specific heat. However, since typical turbine air-to-fuel ratios are in the neighborhood of 40:1, for a first-order analysis, the changes in mass flow and specific heat can be neglected.

The turbine exhaust temperature is calculated from the turbine isentropic efficiency,

$$\eta_T = w_T/w_{Ts} = c_{pT}*(T_4-T_3)/c_{pT}*(T_{4s}-T_3) \quad (8)$$

or

$$T_4 = T_3 - \eta_T *(T_3 - T_{4s}). \quad (9)$$

Defining the turbine pressure ratio as

$$P_{rT} = P_4/P_3, \quad (10)$$

then

$$T_{4s} = T_3 * (P_{rT})^{[(k-1)/k]}. \quad (11)$$

For a first order analysis, P_{rT} can be assumed to equal the inverse of P_{rC}.

The net work output per unit mass is simply the sum of the actual compressor and turbine work, expressions for which are given in Eqns. 4 and 8:

$$w_{net} = w_T + w_C \quad (12)$$

(Note that work and heat input to the system are defined to be positive.) The net shaft and electrical power outputs are

$$P_s = M_T * w_T + M_C * w_C \quad (13)$$

$$P_e = P_s * \eta_g \quad (14)$$

and the cycle shaft and electrical power production efficiencies are

$$\eta_s = |w_{net}/q_{in}| \quad (15)$$

$$\eta_e = \eta_s * \eta_g \quad (16)$$

Equations 4-16 constitute a useful set of equations for doing gas-turbine analysis, as illustrated in Fig. 4, which indicates that at a fixed TIT, one value of P_{rC} maximizes work output, and another maximizes efficiency (see Fig. 4).

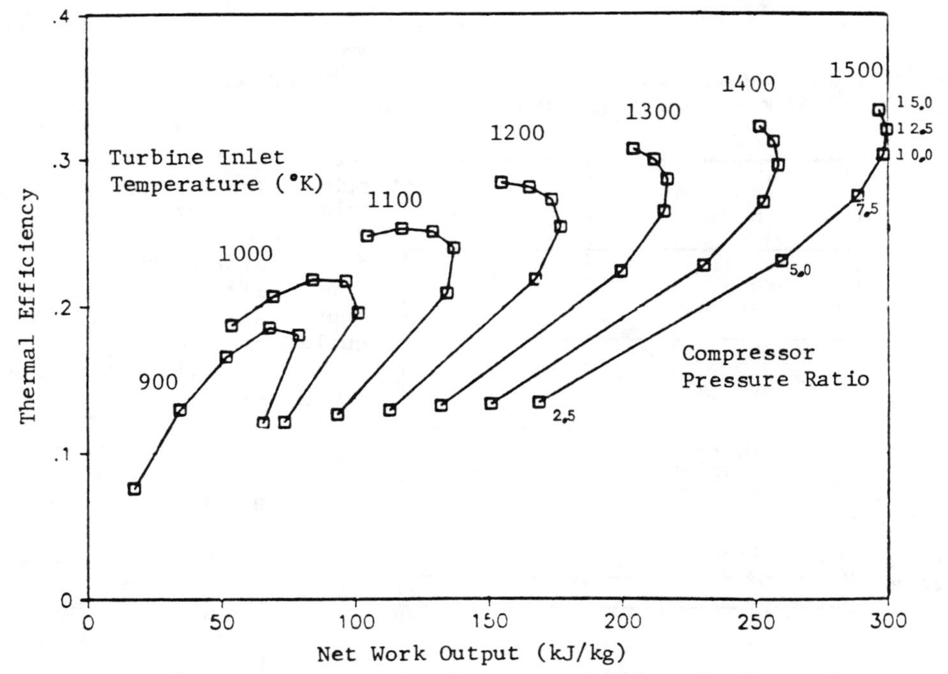

Fig. 4. Performance map for a gas-turbine simple-cycle.

<u>The Heat-Recovery Steam Generator</u>:[10] The heat-recovery steam generator (HRSG) usually consists of an economizer, in which the feedwater (part cold make-up and part condensate return) is heated to near its saturation temperature, and a boiler in which the water is converted to saturated steam. Most industrial processes use

saturated steam, but if superheated steam is required, a superheater follows the boiler.

In some applications, a supplemental firing system -- a duct burner -- is inserted between the turbine exit and the HRSG to increase the steam-generating capacity of the HRSG. Ample oxygen exists in the turbine exhaust to permit supplementary fuel to be completely combusted in the duct burner. The high temperature of the turbine exhaust gases results in very high combustion efficiencies and correspondingly high steam-generating efficiencies[11,12].

Energy Balances: For pedagogical purposes, the three sections of an HRSG can be considered as a single counterflow shell-and-tube heat exchanger, with the hot exhaust gases (from the turbine or the duct burner) passing in one direction on the shell side and water and/or steam traveling through tubes in the opposite direction. The temperature profiles in the HRSG as functions of the percent of heat transferred are shown in Fig. 5.

The gas-side temperature profile is essentially linear in this figure, since the heat tranferred per unit time from the gas, Δq, along any section from one point to another can be expressed as

$$\Delta q = M_g * c_{pg} * \Delta T \tag{17}$$

where M_g is the gas mass flow rate, c_{pg} is the specific heat of the gas per unit mass (which is assumed constant at its average value in the HRSG), and ΔT is the change in gas temperature between the two points. The water-side profiles will also be essentially linear, but with a different slope for each section of the HRSG.

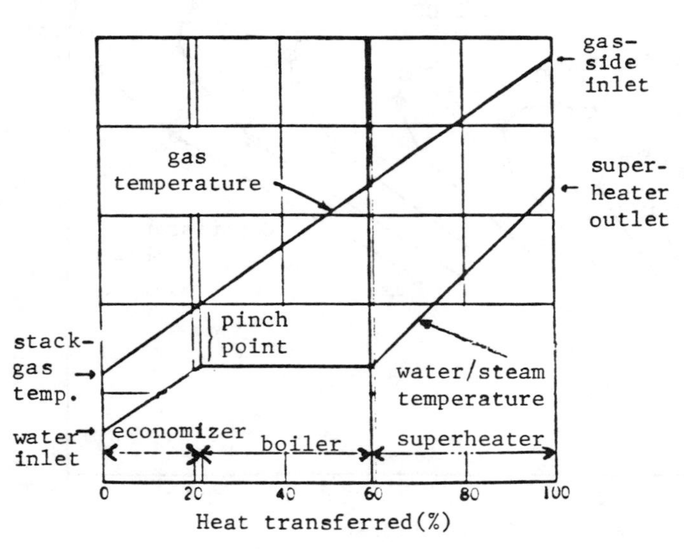

Fig. 5. Typical temperature profiles in a heat recovery steam generator.

In the HRSG of most cogeneration systems using air and water as the working fluids, the closest point of approach between the gas and liquid temperatures at any point will be at the inlet to the boiler, the so-called pinch point (see Fig. 5), where the temperature difference between the gas side and the water side is defined as ΔT_{pp}. From the gas-side inlet, where $T=T_{gi}$, to the pinch point, the total heat released per unit time by the gas is

$$Q_g = M_g * c_{pg} * [T_{gi} - (T_{sat} + \Delta T_{pp})] \tag{18}$$

Typically 1-2% of this energy is lost by radiation or other means,[13] so that the heat absorbed by the water is

$$Q_{ws} = 0.98 * Q_g \tag{19}$$

The enthalpy change per unit mass of water between the pinch point and the point it leaves the HRSG is

$$\Delta h = h_s - h_f \tag{20}$$

where h_s is the enthalpy of the superheated steam exiting the HRSG and h_f is the enthalpy of saturated liquid at the HRSG pressure. Note that for any given system, the upper limit on the steam temperature at the superheater outlet (and hence on h_s) is determined by the temperature of the turbine exhaust gas at that point. Based on practical considerations,[2] the minimum difference between these two temperatures, $(\Delta T_{so})_{min}$, is commonly in the range 15-30°C (30-50°F).

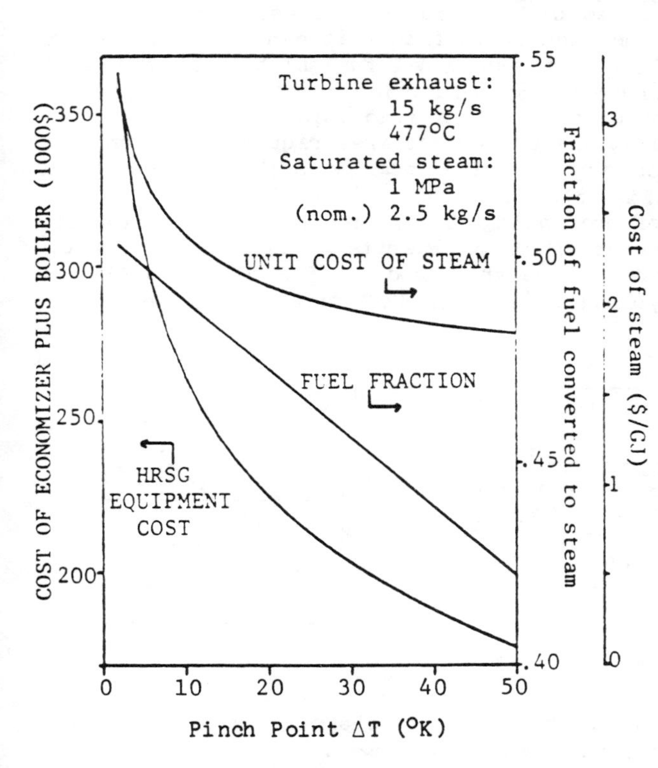

Fig. 6. The influence of the pinch point on a gas-turbine simple cycle cogeneration system.[2] The unit cost of steam represents the capital equipment fraction of the total cost of steam.

The mass flow rate of steam is calculated from Eqns. 19 and 20:

$$M_{st} = Q_{ws}/\Delta h \tag{21}$$

Decreasing $(\Delta T_{so})_{min}$ (i.e., increasing the superheat of the steam) has only a relatively minor effect on the mass of steam that is generated.[2] The stack-gas temperature can be obtained from an energy balance on the economizer:

$$T_{stack} = T_{sat} + \Delta T_{pp} - \{M_{st}*(h_f - h_{fw})*(1 + B)/[(M_g*c_{pg})*0.98]\}. \quad (22)$$

Here h_{fw} is the enthalpy of saturated water at T_{fw} (a weighted average of the known make-up and condensate return temperatures) and B is the continuous blow-down fraction -- an amount of saturated liquid at T_{sat} (expressed as a fraction of the steam flow) that is continuously added to and subsequently flushed from the system to prevent mineral buildup in the boiler. Typically, B is about 0.05.[13] To ensure that condensation does not occur, T_{stack} must be above the dew point of the exhaust gas, which is typically in the neighborhood of 60-70°C (140-160°F). Stack gas temperatures are generally maintained at 150°C (300°F). (If there is a use for low temperature heat beyond that needed in the economizer, additional energy equivalent to up to 20% of that already extracted in the HRSG can be extracted from the turbine exhaust before condensation occurs.)

<u>The Influence of the Pinch Point</u>: As is evident from Eqns. 18-22, for a fixed process steam quality and HRSG gas-side inlet temperature, ΔT_{pp} determines the steam flow that can be generated, or alternatively, the fraction of the original fuel that is converted to steam, which is a linearly decreasing function of ΔT_{pp} in a typical system (see Fig. 6). The pinch point is also important in determining the HRSG heat-transfer surface area requirements, and consequently capital cost, which varies roughly as the inverse of the logarithm of ΔT_{pp} (see Fig. 6).

The net influence of decreasing ΔT_{pp} is to increase the capital investment required per unit of steam generated. As seen in Fig. 6, this cost rises sharply in the neighborhood of 15-20°C (27-36°F) pinch points, a range typical of current HRSG design practice.[14]

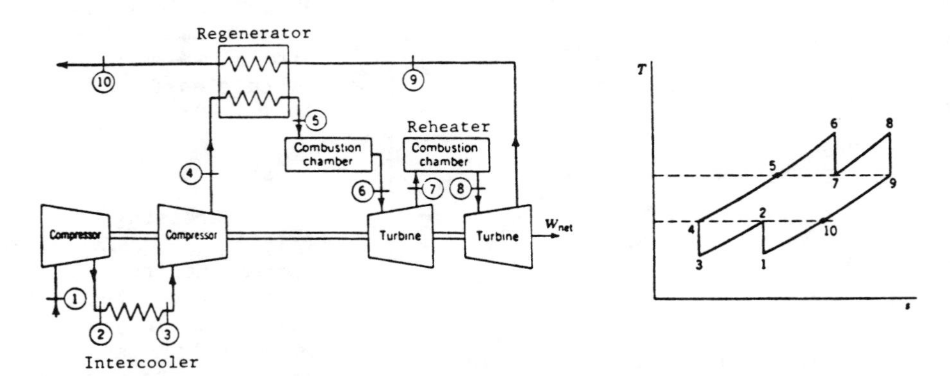

Fig. 7. Schematic and temperature-entropy diagram for alternative gas-turbine configurations for generating power.

<u>Textbook Variations of Simple-Cycle Cogeneration</u>: Simple-cycle gas turbine systems are by far the most commonly used in practice, but the basic analyses of the Brayton cycle and the HRSG presented above can be applied to alternative gas-turbine cycle configurations involving combinations of: heat exchange between the turbine exhaust

gases and the cooler compressor outlet gases through use of a regenerator to reduce overall fuel requirements; multiple-stage compression with intercooling between each stage to decrease overall compressor work; and multiple-stage expansion with reheating between turbine stages to increase overall turbine work. These alternatives are shown schematically in Fig. 7.

Results of calculations based on these alternatives (Fig. 8) indicate that gas-turbine systems can operate over a wide range of conditions. Such more-elaborate systems could gain greater popularity in the future, but it appears that the higher cost and greater complexity (and hence more difficult maintenance problems) associated with regeneration and multiple-stage compression and expansion cycles have acted thus far to limit the actual implementation of these often higher efficiency and/or output systems[15].

THE THERMODYNAMICS OF STEAM-INJECTED GAS-TURBINE COGENERATION

Injecting water or steam into a gas turbine is not a new idea, but only recently have steam-injection cogeneration systems been introduced. The steam-injected gas-turbine (STIG) cycle is described in textbooks[16,17] and in a number of other publications (see, e.g., references 18-21). The use of water (not steam) injection for short periods was commmon in the past for thrust augmentation in jet-aircraft engines, although this is now usually done with afterburners.[15] Water is injected into the combustors of many stationary gas turbines used today to keep combustion temperatures low, thereby suppressing the formation of NO_x pollutants,[22,23] to comply with pollutant emissions regulations in many states.

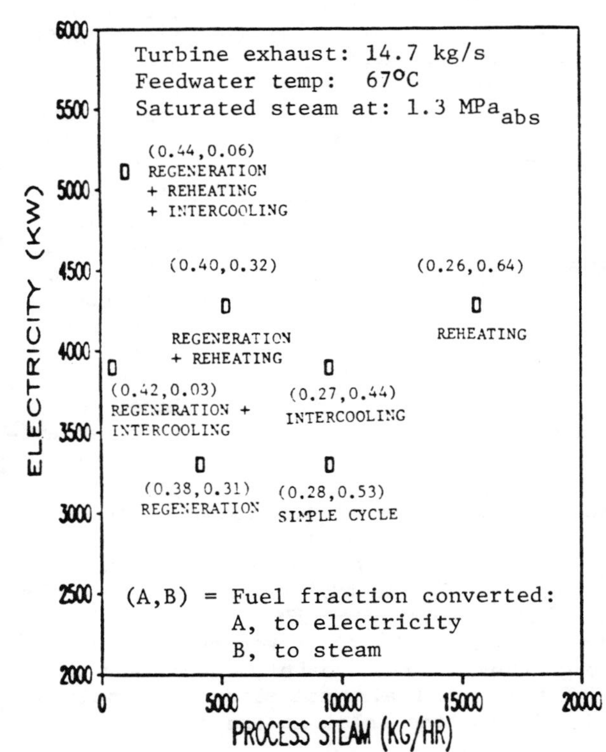

Fig. 8. Performance characteristics for alternative simple-cycle cogeneration configurations.[2]

The passage of PURPA has focussed attention on the steam-injection technology for cogeneration applications, primarily because of its flexible and highly efficient operation.[24,25] Two US companies -- International Power Technology (IPT), Palo Alto, CA[26] and Mechanical Technology, Inc. (MTI), Latham, NY[27] -- now offer packaged STIG cogeneration systems based on the Allison 501-KH turbine (the 501-KB modified for steam injection). IPT has installed a single unit at San Jose State University, San Jose, California and dual units at a Sunkist Growers, Inc. processing plant in Ontario, CA.[26] Dah Yu Cheng of IPT holds a patent[27] which claims rights to the operation of any STIG cycle in the region of its peak electrical efficiency, which as subsequent calculations will illustrate, is defined uniquely for each STIG cycle.

A "Back-of-the-Envelope" Approach to the STIG Cycle: The basic operation of the STIG cycle involves generating steam with the turbine exhaust heat and injecting some or all of it back into the combustor (see Fig. 9). Thus, unlike the simple-cycle cogeneration calculations, the HRSG and turbine analyses are coupled. As subsequent calculations will show, the peak electrical efficiency for a DDA-501 based system occurs when a mass flow of steam somewhat greater than 15% of the air flow is injected into the combustor. Based on this percentage, a 4-step "back-of-the-envelope" calculation can be made to illuminate the thermodynamics of steam-injected gas-turbine cycles.[30]

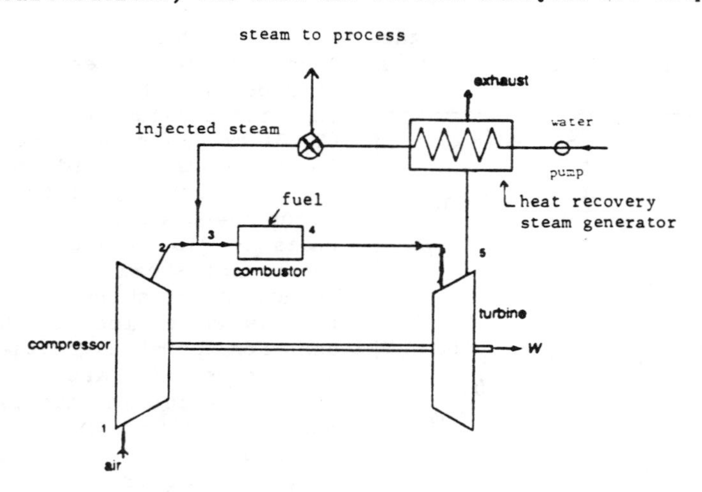

Fig. 9. Schematic of a steam-injected gas-turbine cogeneration cycle.

Step 1: Reference Case: Based on a simplified calculation using air as the working fluid, an Allison 501-KB turbine in a simple cycle will produce about 3350 kW at an efficiency of 27%.

Step 2: "Free" Extra Mass: One result of injecting steam is to increase the mass flow through the turbine. To help understand the mass effect, suppose that 15% additional air (not steam) is supplied (in an unspecified manner) to the turbine inlet at the required temperature and pressure.

In most gas turbines, the working fluid flows from the combustor to the turbine through a nozzle in which the flow is sonic, and thus choked, so that additional mass cannot pass without an increase in the pressure upstream of the nozzle.[15] Introducing additional mass there induces a pressure rise to the level which will permit the injected mass to flow through the nozzle. This pressure rise is felt at the compressor outlet and results in a rise in the compressor pressure ratio, since compressors are designed to operate with a constant mass flow.

As a result, the pressure ratio of the compressor for the Allison 501 turbine is increased according to:[31]

$$\Delta P_{rC} = (13.47 / M_C)*\Delta M_T, \tag{23}$$

where ΔM_T is the increase in the turbine mass flow with steam injection (in kg/s).

Using Eqn. 23 together with differentiated expressions for power and efficiency developed from Eqns. 4-16, changes in power output and efficiency that accompany changes in the turbine mass flow can be determined:

$$\Delta(P) = 574*\Delta M_T \tag{24}$$

$$\Delta(\eta) = 0.056*\Delta M_T \tag{25}$$

where ΔP is in kW when ΔM_T is expressed in kg/s. Adding the increments predicted by these equations to the power and efficiency calculated in Step 1 yields an output of 4625 kW and an efficiency of 38%, both of which are significantly higher than the reference case.

Step 3: Paying for the Extra Mass: The efficiency increase is large in step 2 because no account was taken of the work and heat needed to raise the extra mass to the turbine inlet conditions.

Since we are ultimately interested in steam, in this step we will continue to neglect the work required to raise the mass to the turbine inlet pressure (pumping a liquid requires negligible work compared to compressing air). However, despite the fact that we can recover enough energy from the turbine exhaust to create steam for injection, additional heat must be supplied in the combustor to heat the steam from its injection temperature, T_{inj}, to the turbine inlet temperature, TIT. The total heat added, is, therefore,

$$Q_{in} = M_C*c_{pcb}*(TIT - T_2) + (\Delta M_T)*c_{pinj}*(TIT - T_{inj}), \tag{26}$$

where T_2 is the compressor outlet temperature, and c_{pcb} and c_{pinj} are the specific heats of air and the injected fluid at their respective average temperatures in the combustor.

Equation 24 for the change in power output is applicable for this step, but Eqn. 25 is replaced by

$$\Delta(\eta) = 0.028*\Delta M_T \tag{27}$$

where ΔM_T is expressed in kg/s. Because of the additional heating that is required for the same output, the efficiency drops from 38 to 34%.

Step 4: The Specific Heat Effect: In this last step, we account for the increase in specific heat of the mass flowing through the turbine that occurs when steam is injected. The specific heat per unit mass of air flow in the turbine will be

$$c_{pT} = S/A * c_{ps} + c_{pair} \quad (28)$$

where S/A is the mass ratio of steam to air. Values of c_p for steam[32] and air[33] (in units of kJ/kg-K) are given by

$$c_{ps} = 4.6 - 103.6*T^{-0.5} + 967.2*T^{-1} \quad (29)$$

$$c_{pair} = 1.003 + 1.816x10^{-4}*T \quad (30)$$

where T (in Kelvin) is the average temperature at which the process occurs. For a S/A = 0.15, c_{pT} is about 25% higher than c_{pair}.

Since 2 variables are changing in this case, the changes in power output and efficiency are

$$\Delta(P) = 768*\Delta M_T + 2027*\Delta c_{pT} \quad (31)$$

$$\Delta(\eta) = 0.028*\Delta M_T + 0.176*\Delta c_{pT}, \quad (32)$$

where ΔP is in kW when ΔM_T is in kg/s and Δc_{pT} is in kJ/kg-K. Because the specific heat of steam is roughly double that of air, both the turbine work output and the heat input requirements increase, with a net result that 5420 kW of electricity are produced at an efficiency of 39%

Summary: Steps 2 and 3 of this "back-of-the-envelope" calculation demonstrate that since the compressor typically consumes 1/2-2/3 of the total turbine work output in a simple-cycle gas-turbine, if extra mass can be provided to the turbine without requiring additional compression work (as is the case with steam), both efficiency and net cycle output will increase. Step 4 demonstrates that the increase in specific heat of the fluid passing through the turbine has by far the most important effect on improving cycle performance. (For accurate calculations, therefore, accurate values for the specific heat should be used in the calculation.)

More Careful STIG Cycle Calculations: Accurately calculating actual STIG-cycle performance requires closer attention to detail than the simplified approach just described, but conceptually there are no differences, and indeed, the results of the "back-of-the-envelope" calculation turn out to be surprisingly accurate.

To calculate cycle efficiency and net power output as functions of S/A, the compressor discharge temperature and work requirements are first determined, as in the case of the simple Brayton cycle. Then (passing over the combustor for the moment), with the turbine inlet temperature (TIT) and S/A specified, the turbine calculation can be performed to give the turbine outlet temperature (TOT) and work output, also using the simple-cycle procedure, but with adjusted values for the specific heat parameters.

With the turbine exhaust temperature, the HRSG pressure, and the steam flow rate specified, the enthalpy of the superheated steam is determined by the pinch point temperature difference. The steam injection temperature (T_{inj}) at this enthalpy and pressure is obtained from the steam tables. If the difference between TOT and the calculated T_{inj} is less than the specified $(\Delta T_{so})_{min}$, then T_{inj} is set equal to TOT - $(\Delta T_{so})_{min}$ (which implies an increase in the pinch point temperature difference).

The total heat addition to the cycle can now be calculated using Eqn. 26, based on which cycle efficiency can be determined.

For an Allison 501-KH system, with TIT = 982°C (1800°F), ΔT_{pp} = 10°C (18°F), ΔT_{so} = 30°C (54°F), and T_{stack} > 150°C (300°F), the calculated peak cycle efficiency occurs at a S/A of about 0.17, as shown in Fig. 10. For S/A up to this value, the $(T_{so})_{min}$ constraint determines the superheat temperature of the steam, as there is enough energy and "temperature" in the turbine exhaust to heat all of the steam to the maximum specified value [=TOT-$(\Delta T_{so})_{min}$]. As S/A continues to increase beyond this critical point, there is insufficient energy in the turbine exhaust to raise all of the steam to this temperature, so additional energy input to the combustor is required. Beyond this critical point, the additional work derived from the larger mass flow through the turbine is more than offset by the increased fuel requirement, so that cycle efficiency begins to fall.

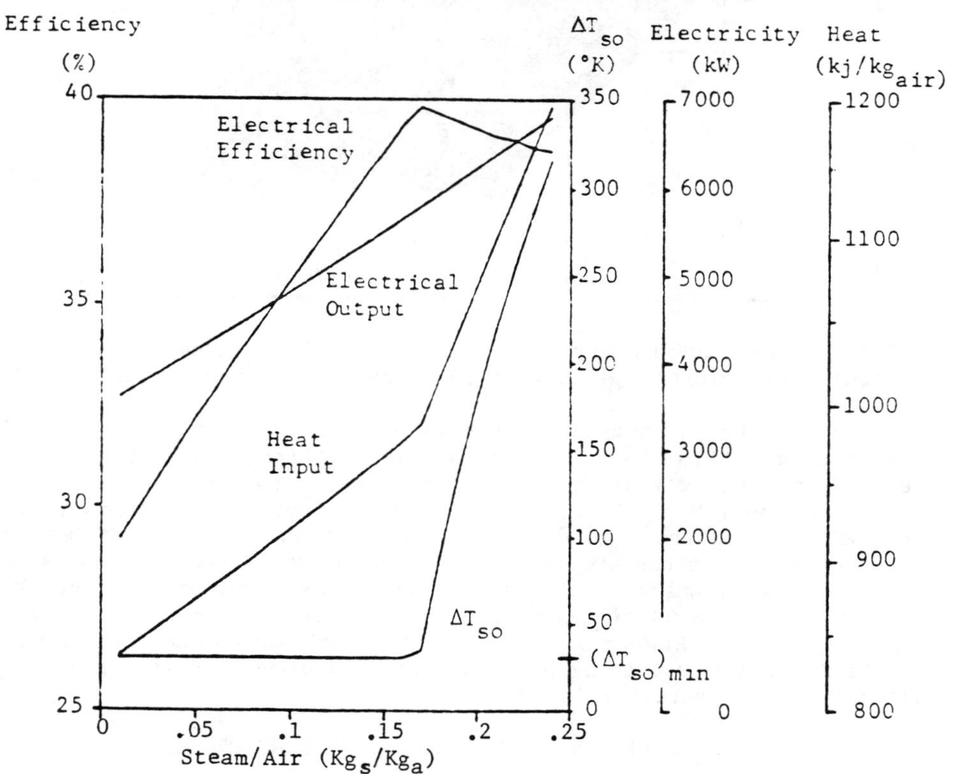

Fig. 10. Operating characteristics of a steam-injected gas-turbine cogeneration system based on the Allison 501-KH turbine.[5]

In a cogeneration application, as the process steam demand decreases, S/A increases. The corresponding increases in efficiency and output account for the attractive part-load performance seen earlier in Fig. 2.

STIG Hardware-Related Considerations: The operating regime of a STIG cogeneration system, defined in terms of electrical output and steam generation, can be considerably expanded by the use of supplemental firing in a duct burner, which is standard equipment on the systems offered by IPT and MTI. IPT claims their system will operate anywhere within "Region A" in Fig. 11, producing from 3.5 to 6 MW of electricity and from 0 to 13 MW (45 MBTU/hr) of process steam.[26]

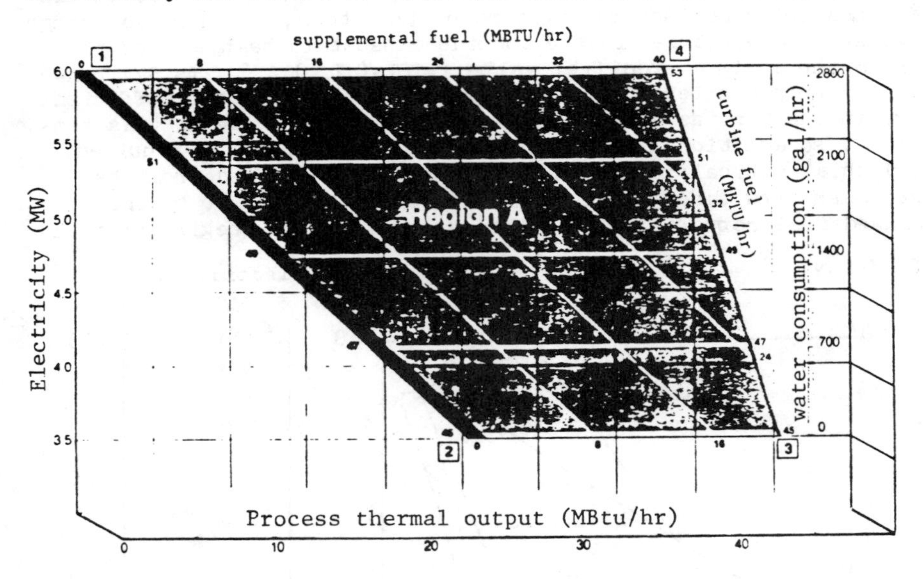

Fig. 11. Performance map for a commercial steam-injected gas-turbine cogeneration system with supplementary firing capability.[26]

STIG cycle thermodynamics are quite attractive, but a number of practical benefits also accompany steam-injection,[2] including generally lower peak operating temperatures, better cooling of turbine blades, NO_x pollution reduction, and better heat recovery in the HRSG (due to the higher specific heat of the turbine exhaust).

Not all turbines are as easily adapted for STIG operation as the Allison 501-KB, which is an aircraft-derivative turbine, and thus has a torque limit (6190 kW) considerably in excess of its design peak value (3500 kW). However, other aircraft-derivative turbines, such as the General Electric LM series, appear to be well suited for steam-injected operation.[8]

THE ECONOMICS OF STEAM-INJECTED GAS-TURBINE COGENERATION

Businesses often make investments based on an expected internal rate of return, determined by solving for the discount rate in an equation which equates the total initial investment required for the new equipment to the total discounted operating-cost savings that are expected to result from replacment of the existing facility.[34] To illustrate this method of assessing the economics of STIG cogeneration, we consider a hypothetical plant in California (where 3 STIG systems have already been installed) currently purchasing electricity from the grid and operating a stand-alone natural-gas fired boiler (83% efficient) to provide its process steam. To simplify the analysis, we assume the plant's full steam demand is 9090 kg/hr (20000 lb/hr) of 1.4 MPa (205 psig) saturated steam (at the HRSG) and full electrical demand is 3500 kW (corresponding to the maximum outputs of the IPT STIG-cycle cogeneration system with no steam injection and no supplemental firing -- point 2 in Fig. 11). We also assume an industrial plant load profile: full steam and electricity demand Monday-Friday from 7a.m. - 6p.m., and half demand at all other times. Finally, we neglect all subsidies and taxes in our analysis (investment tax credit, property tax, etc.)

TABLE 2. ESTIMATED CAPITAL AND INCREMENTAL OPERATING COSTS FOR AN ALLISON 501-KH-BASED STEAM-INJECTED GAS-TURBINE COGENERATION SYSTEM (6000 kW PEAK OUTPUT).[2]

FULLY INSTALLED CAPITAL COST	$5 MILLION
INCLUDES: GAS-TURBINE GENERATOR HRSG WITH DUCT BURNER CONTROL SYSTEM BUILDING AND MISCELLANEOUS INSTALLATION	
INCREMENTAL OPERATING COSTS OVER A STEAM BOILER PLANT	
TURBINE OVERHAUL ONCE EVERY 3 YEARS @	$ 220,000/OVERHAUL
WATER CONSUMED DURING STEAM-INJECTION OPERATION	$ 2/1000 GALLONS
NON-TURBINE MAINTENANCE	$ 60,000/YEAR
ADDITIONAL TECHNICAL SUPERVISION	$ 40,000/YEAR
INSURANCE	$ 37,500/YEAR
PLANT LIFE IS ESTIMATED TO BE 20 YEARS.	

The initial cost incurred to replace the existing stand-alone boiler is the estimated installed capital cost of $5 million for the STIG system. (Here it is assumed that the construction period is sufficiently short that interest charges accumulated during construction can be neglected.) Operating the STIG system requires some additional expenditures -- maintenance, fuel, water, technical supervision, and insurance (see Table 2). But savings accrue from no longer having to purchase electricity and from being able to sell electricity to the grid at the utility's avoided cost.

The calculation is complicated by the fact that purchased electricity prices and avoided costs vary with the time of day and season of the year. For example, in the Pacific Gas & Electric (PG&E) territory in California, the purchased-electricity price structure includes peak, mid-peak, and off-peak rates, as does the avoided cost structure. In addition, there is a capacity charge applied to the peak power level reached each month by a purchaser of electricity and a capacity payment to a cogenerator if it can guarantee continuous delivery of firm power to the utility.[35]

At PG&E's prices for gas -- $4.70/GJ ($4.96/MBTU) (LHV) -- and electricity ($.106/kWh peak) and avoided cost paid for cogenerated

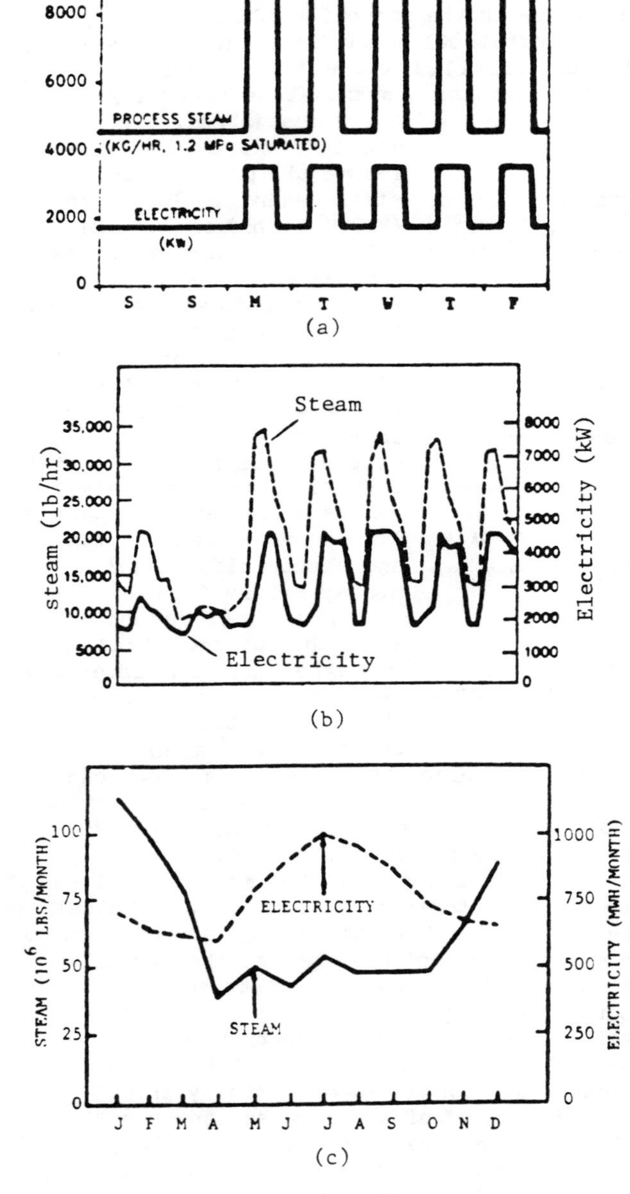

Fig. 12. Plant steam and electricity (a) idealized weekly load used in the analysis, (b) typical weekly load,[12] (c) example seasonal load profile.[36]

electricity ($0.086 peak),[35] the inflation-corrected internal rate of return on an investment in a STIG cogeneration system is about 21%/year, a respectable real rate of return. However, the rate of return for a conventional gas-turbine system with 3500 kW of electrical capacity (at a capital cost of $3.6 million) is about the same. Thus, for the rate structure and load profile we assumed, there appears to be no great incentive for a plant owner to invest in a STIG system.

However, the steam and electricity loads we assumed were very predictable (Fig. 12a), and the size of the cogeneration systems were such that they were operating near their maximum capacity all the time, a rather idealized situation. More typical load profiles are shown in Fig. 12b and 12c, but even these do not account for unexpected shut-downs or other load changes that may occur in the future, e.g., due to the installation of energy conserving equipment.

With unexpected load drops or plant shut-downs, the internal rate of return on an investment in a conventional system can drop substantially, but in a STIG system, since process steam can be redirected to produce additional electricty for sale to the grid, the rate of return is unaffected (Fig. 13). Thus, investing in a STIG system instead of a conventional one eliminates the financial risk associated with unpredicted load changes or plant shut-downs.

Even under a "no surprise" scenario, the STIG system will annually produce about 25% more electricity and save nearly twice as much fuel per unit of process steam generated as the conventional system. As down-time rises, both the electricity produced and the fuel savings rate increase still further. For a 20-week shut-down, the STIG system will produce over 40% more electricity and save over 8 times as much fuel per unit of process steam.

Thus the user's benefit from STIG -- reduced financial risk -- is complemented by the societal benefits of a greater power generating and greater fuel savings potential than what is offered by the simple-cycle gas-turbine system.

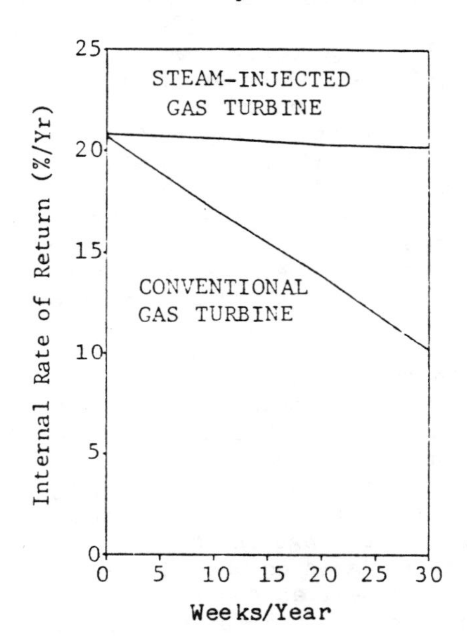

Fig. 13. Internal rates of return on investments in STIG or in conventional gas turbine cogeneration as a function of plant idle time.

FUTURE PROSPECTS FOR STEAM-INJECTED GAS TURBINES

Larger steam injected gas turbine systems, operating at higher pressure ratios and turbine inlet temperatures, are under active development. Based on modelling and some preliminary tests, the General Electric Company (GE) indicates that with only minor modifications, the electrical output of their LM-5000 turbine could be raised from 33 MW in the conventional mode to 50 MW in the steam-injection mode, while the efficiency in generating electricity alone would rise from 37% (LHV) to 44% (LHV).[8] With another 2-3 years of developmental work, GE indicates that 108 MW could be produced at an efficiency of 55% by a system using intercooling with steam injection.[37] A back-of-the-envelope calculation indicates that this level of performance is thermodynamically attainable.

The Simpson Paper Company in Anderson, California currently operates an LM-5000 turbine in a conventional cogeneration configuration. Before the end of the year, they will be making the minor modifications necessary to inject steam, and thus produce electricity at up to 44% efficiency (LHV).[38]

Steam-injected gas-turbines operating on natural gas may soon come to compete successfully with large coal-fired power plants for utility base-load generating capacity in many regions of the country. The favorable economics for central station STIG plants arise from very low capital costs [installed capital costs (in 1984$) as low as $500/kW,[5] compared to $1200-1500/kW for large coal-fired facilities[39]], and relatively low fuel costs [because of efficiencies as high as 55% and expectations that natural gas prices will remain essentially constant for the next 10-15 years in many regions of the US[7]]. Finally, much shorter construction lead times and the capability to make incremental additions (~100 MW) to the generating base with STIG systems would help utilities avoid the inherent risk-taking involved in embarking on expensive 7-10 year construction projects, which may eventually not be needed due to slower-than-expected growth in demand for electricity during the construction period.

Utilities might also utilize STIG cogeneration systems as "electrical peaking systems." In the non-supplementary-fired mode, these systems could provide steam for process needs and base-load electricity, while supplementary firing could be used to provide extra steam for injection to meet peak electrical demands.

PG&E is pursuing still another STIG application. According to a PG&E plan under active discussion,[38] aging boilers which produce steam for heating about 100 commercial buildings in San Francisco will be replaced by a single STIG system based on the LM-5000 turbine, allowing PG&E to increase its base-load generating capacity during the warmer months of the year when steam demands drop and electricity demands rise.

REFERENCES AND NOTES

1. For this figure, cost data in 1980 dollars (Dun & Bradstreet Technical Economic Services and TRW Energy Development Group, "Industrial Cogeneration Potential, 1980-2000, for Application of Four Commercially Available Prime Movers at the Plant Site," prepared for U.S. Department of Energy, Office of Industrial Programs, Washington, D.C., August 1984) have been converted to 1985 dollars using the GNP deflator (Council of Economic Advisors, "Economic Indicators," prepared for the Joint Economic Committee, U.S. Government Printing Office, Washington, D.C., April 1985).
2. Larson, E.D. and Williams, R.H., "A Primer on the Thermodynamics and Economics of Steam-Injected Gas-Turbine Cogeneration," Center for Energy and Environmental Studies Report No. 192, Princeton University, Princeton, NJ, June 1985.
3. American Institute of Physics, Efficient Use of Energy, AIP Conference Proceedings No. 25, AIP, New York, 1975.
4. Williams, R.H., "Industrial Cogeneration," The Annual Review of Energy (3), 1978.
5. The Detroit-Diesel Allison 501-KB turbine is estimated to have the following characteristics[2]: compression ratio of 9.3, air flow of 14.7 kg/s (32.3 lb/s), turbine inlet temperature of 982°C (1800°F), peak electrical output of 3500 kW, compressor efficiency of 0.833, turbine efficiency of 0.897, and a gearbox-generator efficiency of 0.93.
6. To reduce process-steam production in conventional gas-turbine cogeneration systems, the fuel flow to the turbine can be throttled, resulting in lower turbine inlet and outlet temperatures, or the hot turbine exhaust can be bypassed around the heat recovery system. These are the only two alternatives in nearly all gas-turbine systems used to generate electricity, since most incorporate only a single shaft for the compressor, turbine and generator. Since the generator speed must be maintained constant (and turbo-machinery is designed to move a constant volume of air at a fixed speed), the mass flow through the system cannot be throttled.
7. American Gas Association, "Energy Analysis -- Historical and Projected Natural Gas Prices: 1985 Update," AGA, Arlington, VA, March 15, 1985.
8. General Electric Company, "Scoping Study: LM5000 Steam-Injected Gas Turbine," work performed for Pacific Gas and Electric Co., San Francisco, California, July 1984.
9. Standard thermodynamics textbooks describe gas-turbine cycle analysis. For example, see Sonntag, R.E. and Van Wylen, G., Introduction to Thermodynamics, 2nd Edition, Wiley, NYC, 1982 or Wark, K. Thermodynamics, 3rd Edition, McGraw-Hill, NYC, 1977.
10. The analysis of heat recovery steam generators can be found in some textbooks, e.g. Fraas, A.P., Engineering Evaluation of Energy Systems, McGraw-Hill, NYC, 1982. An excellent review can be found in reference 10.

11. Wilson, W.B., "Gas Turbine Cycle Flexibility for the Process Industry," GER-2229M, General Electric Company, Schenectady, NY, 1978.
12. Kosla, L., Hamill, J., and Strothers, J., "Inject steam in a gas turbine -- but not just for NO_x control," Power, February 1983.
13. Eriksen, V.L., Froemming, J.M., and Carroll, M.R., "Design of Gas Turbine Exhaust Heat Recovery Boiler Systems," American Society of Mechanical Engineers, paper No. 84-GT-126, New York, 1984.
14. Cook, D.S., Design Engineer, Deltak Corporation, Minneapolis, Minnesota, personal communicaton, April 1985.
15. Leibowitz, H., Mechanical Technology, Inc., Latham, New York, personal communication, May 1985.
16. Haywood, R.W., Analysis of Engineering Cycles, 3rd Edition, Pergamon Press, Oxford, 1980.
17. Diamant, R.M.E., Total Energy, Pergamon Press, Oxford, 1970.
18. Maslennikov, V. and Shterenberg, V.Y., "Power Generating Steam Turbine-Gas Turbine Plant for Covering Peak Loads," Thermal Engineering, 21(4), 1974.
19. Fraize, W.E. and Kinney, C., "Effects of Steam Injection on the Performance of Gas Turbine Power Cycles," Journal of Engineering for Power, (101), April 1979.
20. Davis, F. and Fraize, W., "Steam-Injected Coal-Fired Gas Turbine Power Cycles," MTR-79W00208, Mitre Corporation, McLean, Virginia, July 1979.
21. Brown, D.H. and Cohn, A., "An Evaluation of Steam Injected Combustion Turbine Systems," Journal of Engineering for Power (103), January 1981.
22. Allen, R.P. and Kovacik, J.M., "Gas Turbine Cogneration -- Principles and Practice," Journal of Engineering for Gas Turbines and Power, (106), October 1984.
23. Boyce, M.P., Gas Turbine Engineering Handbook, Gulf Publishing Co., Houston, 1982.
24. Leibowitz, H. and Tabb, E., "The Integrated Approach to a Gas Turbine Topping Cycle Cogeneration System," Journal of Engineering for Gas Turbines and Power, (106), October 1984.
25. Jones, J.L., Flynn, B.R., and Strother, J.R., "Operating Flexibility and Economic Benefits of a Dual-Fluid Cycle 501-KB Gas Turbine Engine in Cogeneration Applications," American Society of Mechanical Engineers, paper No. 82-GT-298, NYC, 1984.
26. International Power Technology, "Cheng Cycle Series 7-Cogen," company publication, Palo Alto, CA, received March 1984.
27. Mechanical Technology, Inc. "Turbonetics Energy, Inc: Gas Turbine Cogeneration System," company publication, Latham, NY, received March 1984.
28. Koloseus, C., Director of Market Development, International Power Technology, Inc., Palo Alto, CA, personal communication, April 1985.
29. U.S. Patent No. 4,128,994, December 12, 1978.
30. Details of the "back-of-the-envelope" calculations can be found in reference 2.
31. Jones, J.L., Program Manager, International Power Technology, Inc., Palo Alto, CA, personal communication, April 1985.

32. Mark's Handbook for Mechanical Engineers, 8th edition, Baumeister and Marks (eds.), McGraw-Hill, NYC, 1978.
33. Cain, G., Mechanical Technology, Inc., Latham, New York, personal communication, April 1985.
34. Goldemberg, J. and Williams, R.H., "The Economics of Energy Conservation in Developing Countries: A Case Study for the Electrical Sector in Brazil," this volume.
35. Pacific Gas and Electric Company, company rate schedules, San Francisco, California, January 1985.
36. Pacific Northwest Laboratories, "Cogeneration Handbook for the Food Processing Industry," US Department of Energy, DOE/NBB-0061, Washington, D.C., February 1984.
37. The GE LM-5000 permits an intercooler to be used in the system. GE claims that since the turbine blades are cooled with compressor bleed air, higher gas temperatures at the turbine inlet can be obtained without equally high turbine blade-metal temperatures, if intercooling were used. An electrical output of 108 MW at 55% efficiency (LHV) is claimed to be obtainable with an overall compressor pressure ratio of 33.5, a turbine inlet temperature of 1355°C (2470°F), and a steam-to-air ratio of 0.146.[5]
38. Price, D., Senior Engineer, Pacific Gas and Electric Company, San Francisco, California, personal communication, May 1985.
39. Costs given in 1978 dollars (Bechtel Power Corporation, "Coal-Fired Power Plant Capital Cost Estimates," prepared for the Electric Power Research Institute, San Francisco, California, May 1981) were converted to 1984 dollars using the GNP deflator (see reference 1).

CHAPTER 22

PROGRESS ON PHOTOVOLTAIC TECHNOLOGIES

Paul Maycock
Photovoltaic Energy Systems
P.O. Box 290
Casanova, VA 22017

INTRODUCTION

Worldwide module shipments of photovoltaics increased by 95 percent, from 21.7 MW in 1983 to 25 MW in 1984. U.S. shipments unfortunately dropped 10 percent last year, from 13.1 MW to 11.7 MW, while Japanese production grew from 5 MW to 8.9 MW - a whopping 80 percent increase! The Japanese increase was strictly due to explosive growth in amorphous silicon consumer products. European sales remained flat at about 3.6 MW.

Table I: WORLD PV MODULE SHIPMENTS

	1983		1984	
	MW SHIPPED	PERCENT	MW SHIPPED	PERCENT
UNITED STATES	13.1	60.4	11.7	46.8
JAPAN	5.0	23.0	8.9	35.6
EUROPE	3.3	15.2	3.6	14.4
OTHER	0.3	1.4	0.8	3.3
TOTALS	21.7 MW	100%	25.0 MW	100%

0094-243X/85/1350426-22$3.00 Copyright 1985 American Institute of Physics

The 1984 world market was divided into the following end-use sectors:

Table II: 1984 WORLD MARKET END-USE SECTORS (MW)

WORLD CONSUMER PRODUCTS	5
U.S. OFF-THE-GRID RESIDENTIAL	2
WORLD OFF-THE-GRID RURAL	0.2
WORLDWIDE COMMUNICATIONS	5
WORLDWIDE PV/DIESEL	2
U.S. GRID-CONNECTED RESIDENTIAL	0.1
U.S. CENTRAL STATION & THIRD-PARTY FINANCED PROJECTS	10

During 1984, there was continued erosion of single crystal silicon world-wide market share, from 50 percent to 43 percent, while amorphous silicon leaped from 14 to 28 percent -- the latter all in the Japanese consumer products area, not in terrestrial power products. With ARCO Solar, Solarex, Chronar and ECD/SOHIO expected to ship amorphous silicon in 1985, we expect amorphous silicon to pass single crystal silicon in 1985. In Table III below, we have listed the 1983 and 1984 PV shipments by module type and in Table IV, the shipments by U.S. companies.

Table III: WORLD PV MODULE SHIPMENTS BY MODULE TYPE

	1983		1984	
	MW SHIPPED	PERCENT	MW SHIPPED	PERCENT
SINGLE CRYSTAL FLAT PLATE	10.9	50.2	10.8	43.2
AMORPHOUS	3.1	14.3	6.95	27.8
POLYCRYSTAL	3.1	14.3	3.95	15.8
SINGLE CRYSTAL CONCENTRATORS	4.5	20.7	3.1	12.4
RIBBON	0.1	0.5	0.2	0.8
TOTAL	21.7 MW	100%	25.0 MW	100%

Table IV: U.S. PV MODULE SHIPMENTS BY COMPANY

	1983		1984	
	MW	MODULE	MW	MODULE
ARCO SOLAR	6.0	SINGLE	5.5	SINGLE
UNITED ENERGY	4.5	CONC.	3.0	CONC.
SOLAREX	1.3	POLY	2.1	1.5 POLY 0.5 SINGLE[1] 0.1 A-Si
SOLAR POWER	0.4	SINGLE	0.0[2]	
PHOTOWATT	0.3	SINGLE	0.0[2]	
SOLEC INT'L	0.3	SINGLE	0.4	SINGLE
SOLENERGY	0.05	SINGLE	0.1	SINGLE
MOTOROLA	0.1	SINGLE	0.3	POLY
MOBIL SOLAR	0.07	RIBBON	0.08	RIBBON
OTHER (INTERSOL, ENTECH, CHRONAR, WESTINGHOUSE, TIDELAND, SILICON SENSOR)	0.05	60% SINGLE 40% CONC.	0.2	20% RIBBON 40% SINGLE 40% CONC.
TOTAL	13.07 MW		11.68 MW	

1 - SOLAR POWER 2 - TERMINATED BUSINESS

U.S. module shipments in 1984 were still attributable in no small measure to tax credit stimulation and government-supported sales. Over 50 percent of product went to projects that were underwritten at least in part by federal agencies. An encouraging note was that exports grew by 45 percent, from 2 MW to 2.9 MW. Solarex and ARCO Solar were prime movers in this growth. In fact, we estimate that the U.S. share of exports on a global basis increased despite Japanese pressure. Table V below lists shipments by type of application. Figure I displays the cumulative PV shipments (MWp and year) and their costs ($/Wp in 1975$).

Table V: U.S. PV MODULE SHIPMENTS BY APPLICATION

APPLICATION	1983 MW	1984 MW
THIRD-PARTY-FINANCED (<5 kW)	4.5	3.0
THIRD-PARTY-FINANCED CENTRAL STATION (1 MW+)	4.5	2.1
EXPORTS	2.0	2.9
OFF-THE-GRID RESIDENTIAL	0.8	1.5
U.S. GOVERNMENT PROJECTS (SMUD, DOD)	0.7	1.2
OFF-THE-GRID INDUSTRIAL AND COMMERCIAL	0.6	1.0
TOTAL	13.1 MW	11.7 MW

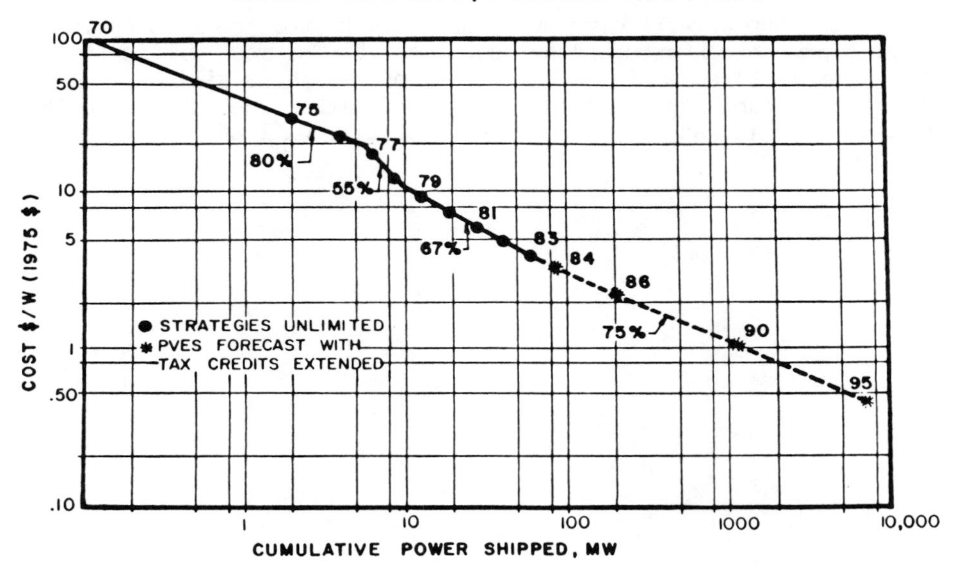

We will discuss the five silicon options that are now commercially available. For each of the options we will present a 1985 status, key cost reduction factors, a technical forecast and a summary chart forecasting our present view on the performance and cost of each item.

Table VI summarizes our present forecast to 1995. Obviously the forecast depends on many other parameters than those covered here. We have had to estimate many key variables. These include:

What size plant will be built? In most cases we assumed 10 MW, 2.5 shift plants.

What asset productivity can we expect? In most semiconductor processes an asset to annual billings ratio of 0.5 hilds. We used this ratio when in doubt.

What process yields can we expect at a given efficiency? We used 30% raw material yield (silicon) for single crystal and 80% material (silicon) yield for the casting and ribbon processes, 20% for amorphous silicon.

Silicon Cost. We used $60/kg in this analysis in 1985 and $40/kg in 1988. Purities close to semiconductor grade were assumed.

Plant Construction and Long Lead Times. We found one year to be a reasonable time from empty floor to production. This would require early purchase of long lead time equipment.

This chapter will give our best estimate for the five options without any "breakthroughs" or "surprises." The most difficult parameter to forecast is the stability of amorphous silicon. We expect the stability issue to be resolved within the next 12 to 18 months; that is, at least one major U.S. firm will offer a 5-year warranty.

Table VI: SUMMARY OF TECHNOLOGY/COST FOR KEY SILICON-BASED OPTIONS
(1984$)

	1985	1990	1995
Single Crystal Silicon			
- Module Efficiency (%)	11	15	16
- Profitable Price ($/Wp)	6.50	4-5	3.00
Concentrators			
- Module Efficiency (%)	14	17	20
- Profitable Price ($/Wp)	5-6	3.30-4.00	2.50
Ribbon/Sheet			
- Module Efficiency (%)	10	11	14
- Profitable Price ($/Wp)	7.50	3.30	2-3
Cast Ingot			
- Module Efficiency (%)	11	13	15
- Profitable Price ($/Wp)	7.00	3.50	3.00
Amorphous Silicon			
- Module Efficiency (%)	5	8	10
- Profitable Price ($/Wp)	5-6	2-3	1.66-2.50
AVERAGE MODULE PRICE ($)	6.50	3.50	2.50

THE FIVE SILICON-BASED OPTIONS

Five silicon-based technologies now serve the world PV market:

Single Crystal Silicon: Sliced into 4" or 5" wafers which are made into circular or square solar cells which are wired together into modules.

Multicrystal-Cast or Semicrystal-Cast Silicon ingots are sliced into 4" squares which are made into cells, then the cells are wired together to form modules.

Ribbon Sheets (2" wide) of Silicon are pulled directly from molten silicon and are cut into 2" x 4" rectangles which are made into cells which are wired together to make modules.

Optical Concentrators focus sunlight into small single crystal solar cells. Cells are mounted in modules on two-axis tracking pedestals.

Amorphous Silicon: Very thin layers of non-crystalline silicon. A 3-layer film is processed to make cells with areas ranging from 1 cm^2 to 1000 cm^2.

Hundreds of new materials are being investigated for lower cost, high efficiency cells. Cadmium telluride thin films and copper oxide/copper sulfide films are being produced in sample quantities. It will be several years before these new cell materials are commercially available. We will cover the non-silicon options only briefly as the date of commercial significance appears to be well past 1995. The key factors limiting market penetration of these new materials are:

- The continued improvement of silicon technology at reduced cost;
- The long time period that is required to turn a research result on a new material into high volume, high yield cells/modules with proven reliability.

THE SINGLE CRYSTAL SILICON OPTION

Sixty percent of the PV market of 25 MWp (1984) is served by single crystal silicon wafers which are usually .015" in thickness and 3" to 5" in diameter. Each cell puts out about one Watt at 0.5V. ARCO Solar introduced 4" square wafers in 1983 cut from 5 1/2" diameter round stock. The key attributes of single crystal silicon cells/modules are:

- The range of power or voltage, from 1 Watt to 100 Watts, 0.5 Volt to 24 Volts;

- Conversion efficiencies: 13% for cells, 11% for modules;
- Price*: Factory wholesale price typically $6.50/W for modules;
- Using semiconductor grade silicon (one impurity per billion parts), the 4" slice costs $2.50 to $3.00/Wp;
- Energy payback period of today's modules is 4 to 6 years.

* Volume Large Order Sales: $4.50/W (1985$)

The following table shows the present costs for the key steps in the manufacturing of a single crystal silicon module. The raw semiconductor grade silicon costs about 30% of the total. A typical one Watt 4" x 4" cell uses .040" to .045" of silicon thickness for a .015" thick slice. Virtually all other options address the silicon consumption as their key to cost reduction.

Table VII: SINGLE CRYSTAL CELLS: MANUFACTURING PROCESS AND COSTS*

Product/Process	Market Price/kg	Incremental Process Cost/W (1985$ per Watt)
Sand	$1.00 - 2.00	Not applicable
Metallurgical Grade Silicon	$3.00 - 5.00	Not applicable
Solar Grade Silicon**	$5.00 -10.00	
or		
Semiconductor Grade Silicon	$40 - 60	$1.00 - 1.50
Slices of Silicon (4" square)	$200 - 300	+1.00 - 1.50
Single Crystal Cells (13% Eff.)	$350 - 400	+1.00 - 1.25
Hermetically Sealed Package	Not applicable	+1.00 - 1.25
Total Module Cost	Not applicable	+4.00 - 5.50

* The costs presented in this chart are to be viewed as the base costs against which the costs associated with producing semicrystal and ribbon silicon will be compared.

** Solar grade silicon is a less expensive, less pure material with possible application to PV cells. Solar grade silicon is made from metallurgical grade silicon. No solar grade silicon is yet marketed. Wacker, Solarex, Pragma, Siemens are all working on lower cost polysilicon. We do not see PV grade silicon at prices lower than $40/kg in 1985$. Lower cost off-spec material is available, but 13% efficient cells with 95% yield are unlikely from scrap.

Forecast for Single Crystal Silicon: 1990, 1995

The performance of single crystal cells cannot be changed dramatically. Present production line cell efficiencies are 13%. We forecast 16% to 17% in 1990, only if new processes are developed. Modules are now only 11% efficient because the cells effectively cover only 80% to 90% of a module's surface area. Square cell modules with 12% efficiency are being made. We forecast 16% square cells, resulting in 15% efficient modules by 1990. By 1995 module efficiency will have leveled off at 16-17% as the cells reach theoretical limits (19-20%). In 1990 to 1995, single crystal processing can use some combination of the following cost reduction options:

- Continuous melt replenishment;
- Multiple crystal pulling from one crucible with recycling of scrap;
- Ion implant for junction formation;
- Less expensive silicon;
- More automation in the fabrication and assembling steps.

These process changes can lead to a cost reduction of a factor of two by 1990. A $4.00/Watt single crystal module price is very likely by 1990. We see continued cost reduction to permit a $3.00/Watt module price by 1995.

Potential Impact of Hoxan Automated Plant on Thick Crystal Module Cost

The July, 1984 announcement by Hoxan of Japan, and the April 5, 1985 dedication of a fully automated ion implant, nine megawatt PV module plant costing $8.3 million, leads one to speculate as to how low the Hoxan costs could go. We will attempt to compare the possible Hoxan costs with a 5 MW, 2 1/2 shift plant based on single crystal silicon operating in the U.S. producing 10% efficient modules. Our analysis indicates that a well managed U.S. PV plant that starts with 4" x 4" slices will have module manufacturing costs of $4.00 to $5.50 per Watt. By manufacturing cost we mean the cost of all materials, labor, overhead and payback of fixed assets. Most analysts agree that a 4" square single or polycrystal cell blank costs from $2 to $2.25 per slice. A labor cost of $5 million per year would keep one hundred, $50,000 per year (with all fringes) persons on the line. The equipment for a 5 MW line using diffusion, cell test and sort, printed metallization, automatic cell tabbing, lamination with ethyl vinyl acetate and 100% panel test would run about $10 million. Our cost for this plant would be as follows:

Table VIII: Manufactured Cost per Watt (10% Module): U.S. 5MWp

Silicon Blank	$2.00-2.25
Other material, glass, EVA, metallization	.50
Burdened labor including incoming QC, manufacture and test and shipping	1.00
Amortize Equipment - $10 million total; $3 million per year (30%); 20% cost of money and 5 year life	.60
Total Material, Labor, Overhead (MLO)	$4.10-4.35

Note: This includes no research, engineering, marketing or corporate G&A costs. A 40% of sales gross profit margin to cover these costs and profit would lead to a profitable price of $6.83 to $7.25/Wp which is near today's price.

In order to understand what Hoxan meant by "being competitive with all prices," recent Spire results indicate that 16-17% efficient single crystal silicon cells can be made by their new ion implantation equipment that leads to 15% modules. If Hoxan could produce 13% efficient modules instead of the 10% used in our base case, then the materials cost would be reduced by 30%. Hoxan claims "one operator per shift" to run the 9 MW plant. If we assume as high as 30 people for all functions (incoming inspection, production, maintenance and shipping), labor costs would be reduced by automation to one-third the present U.S. batch factory. The $8.3 million in equipment costs less to pay back than the U.S. case because the Japanese cost of money for an "assisted" industry is 5-8% versus 14-18% for a U.S. manufacturer. Taking all of these factors into account, we could envision the following cost structure for the Japanese plant.

Table IX: MANUFACTURED COST PER WATT (13% MODULES): Hoxan 9MW

Silicon Blank	$1.40-1.50
Other material	.35
Labor (1/3 of U.S.)	.33
Equipment - $8.3 million (20%); $1.6 million per year over 8 years	.18
Total MLO	$2.26-2.36
Profitable Price: 40% Gross Profit - - - - - - -	$3.76-3.93

CONCLUSION

The following table is our forecast for the single crystal moving target. The 1995, 16% modules with profitable prices of $3/Watt are 80% probable and represent a challenge to other options.

Table X: SINGLE CRYSTAL TECHNOLOGY FORECAST

	1985	1990	1995
CELL EFFICIENCY (%)	12.5	16	18
MODULE EFFICIENCY (%)	11	15	16
WARRANTY	5 year	5-10 year	10 year
MANUFACTURED COST ($/Wp)	4	2.50-3.00	1.80
PROFITABLE PRICE (40% Profit, $/Wp)	6.60	4-5	3

U.S. Single Crystal Flat Plate Status and R&D Outlook

ARCO Solar Thick Crystal Progress: ARCO Solar has completed significant in-house research on single crystal silicon. Their major recent accomplishments include:

- Raising production 95% yield cell efficiencies to 13% and module efficiencies to nearly 12%.
- Reducing costs so they could bid the recent Sacramento Municipal Utility District's 6 MW project at prices very near $4/Watt for modules in racks delivered to the site.

- Being awarded a contract to study excimer-laser annealing and processing of cells.

- Extensive research on ion implant of silicon cells.

- Increasing their reliability to permit a 10-year warranty.

- Obtaining Underwriter's Laboratory listing for modules used on roofs.

ARCO Solar is the world leader in single crystal silicon technology, cost, performance and has the largest capacity - 5 to 7 megawatts per year. In 1985 they have priced product from $5 to $6.50/W depending on volume and credit of purchaser. They do all their pulling, slicing, cell manufacturing, and module manufacturing in one semi- automated batch process plant at Camarillo, California.

Spire Corporation: Spire has grown its PV R&D and business to several million dollars in billings and nearly 20 professionals are working on single crystal technology. Spire sells PV manufacturing equipment and lines. These include:

- Cell and module testers;

- Laminators;

- Cell tabbers;

- Cell diffusion equipment.

The most recent equipment is a cartridge-loaded ion implant machine that can implant 2 million four-inch slices per year. The non-mass-ionized ion-beam uniformly implants the 4" slice. The Spire research in single crystal silicon is primarily aimed at increasing cell efficiency through ion implant, advanced device structure, improved metallization and use of carefully characterized material. In 1984 Spire reported 19% efficiency in single crystal silicon. Their goal is to develop a 16-17% efficient, 98% yield production process using ion implant. If this could be accomplished, the forecast of 16% modules with profitable $3/Watt prices could be possible.

OPTICAL CONCENTRATORS USING SILICON SOLAR CELLS

Concentrators use lenses or reflectors to focus sunlight onto small solar cells of single crystal silicon. Because optical concentrators must have direct solar radiation and are unable to focus scattered light, their use is limited to arid, sunny, non-cloudy climates. As the concentration ratio increases, concentrator cells become more efficient. Conversely, each type of concentrator cell has an optimum point of concentration ratio and temperature.

Cooling Mechanism: Passive cooling can keep cells at a concentration ratio of 150 or less at 100°C at maximum efficiency. Higher concentrations (those greater than 150) require active cooling (i.e., cooling with water) to bring the temperature down to 100°C.

Waste Heat from concentrators requiring active cooling can be captured and used to produce both electrical and low temperature thermal energy. Sanyo, for example, has developed an integrated PV evacuated tube which utilizes the waste heat. United Energy Corporation has sold over 10 MW of hybrid electricity and heat concentrators.

Key attributes of concentrators are:

- Use of very efficient small cells. Efficiencies of 12% to 13% are common compared to 10% for flat plate modules.

- Costs are 10% to 30% less than current flat plate configurations and can be reduced to one-half the cost of single crystal flat plate collectors by 1995.

- Sun tracking requires the use of moving parts which leads to potential reliability problems and higher capital costs. (The design goal is to eliminate reliability problems and to offset higher costs through increased output.)

- Will not work in overcast climatic conditions of cloud, smog, haze.

Forecast for Concentrators - 1990, 1995

Performance - Concentrators are now 12% to 15% efficient and require 20% to 50% more land per kilowatt than flat plate collec-tors. It is very likely that concentrators with 20% efficiencies will be built by 1995. This outcome will be the key factor for central station power.

Cost - Present prices are $6 to $7/W installed. In perfect climates, the concentrator generates 20% to 30% more energy per year than flat plate (fixed) collectors. We forecast concentrator prices at $2/W by 1995. Concentrators will become the main option for large (100kW) ground-mounted systems by 1990 and will become the most popular options because of very high efficiency and very low cost.

Reliability - This is a key issue that limits present sales. The third generation concentrators of 1984-85 have increased reliability, requiring less maintenance and less down time than 1982-83 concentrator systems. They have about 1/50th the part count per Watt and no seasonal adjustments will be required. Systems will be built larger to minimize maintenance. The reliability issue has largely disappeared as a concentrator issue. Three U.S. companies, United Energy Corporation, Intersol and Entech, are developing and marketing PV concentrators.

Table XI: Concentrators	**1985**	**1990**	**1995**
CELL EFFICIENCY (%)	17	19	23
MODULE EFFICIENCY	14	17	20
WARRANTY	2-5 year	5-year	10-year
MANUFACTURED COST ($/Wp) (Includes Pedestal/Tracker)	3-3.50	2-2.50	1.50
PROFITABLE PRICE (40% GPM)($/Wp)	5-6.00	3.30-4	2.50

U.S. Concentrator R&D

United Energy Corporation:

UEC has designed and installed a concentrator cell processing line which obtains over 80 Watts cell capacity from a four-inch slice. UEC uses the highest quality, low oxygen silicon, diffusion and metallizes using precision masks. UEC cell efficiency is in the 15-17% range. A new silicon cell will permit 300 sun concentration and result in cell costs less than $0.20/Wp. UEC is also performing MOCVD gallium arsenide cell research. The goal is 25% efficient cells at 300X. As we write this, UEC is considering reorganization under Chapter XI (bankruptcy) provision. It is not clear what will happen to this excellent concen-trator cell R&D.

Intersol Power Corporation:

Intersol purchases cells from Applied Solar Energy, Fresnel lenses from 3M Corporation and designs and assembles concentrator modules. R&D covers the cell, thermal reservoir structure and advanced tracker designs. Recently Intersol bid the Sacramento Municipal Utility District (SMUD) Phase III, 5 MW project at $7/Watt for an installed system. Because SMUD requested only one-axis tracking flat plate, the Intersol bid was rejected. Intersol was the first U.S. concentrator company to reach 15% operating module efficiency.

Entech:

Entech is also an assembler of concentrators. Entech uses Linear Fresnel lenses (line focus). Cells are purchased from Applied Solar. Recently Entech has developed with ARCO Solar a "black" cell that uses prisms over the metal to direct all light to the cell junction. This appears to permit 20% silicon cell efficiency from 100X concentrators at little or no cost disadvantage. This idea is available for license. Entech is working with 3M on advanced Fresnel lens R&D. Despite the small R&D effort, we believe that profitable concentrator module prices well below $3/Wp are very likely. The Electric Power Research Institute (EPRI) has chosen concentrators as the "Preferred Option" for central PV generation. EPRI will spend $10 million in the next 5 years on concentrator test and development.

SILICON RIBBON OR SHEET

A sheet of crystal silicon is pulled directly from the melted pure silicon. This pulling procedure eliminates sawing, sawdust (kerf loss), and results in square cells rather than circular cells. Silicon ribbon cells comprise about 6% of today's market. Key attributes of silicon ribbon cells/modules are:

- Throughput and cell quality improvements have resulted in ribbon material costs which are competitive with sliced single crystal silicon.

- Cells now in production have efficiencies of 12%; modules are 10% to 11% efficient.

- By increasing the "pulling" rate and by using thinner material, a 3-fold reduction in cost can be achieved by 1995.

- If .012" ribbons can be produced, with 80% overall device yield, at 12% efficiency, then .015" of silicon will be used (versus the .045" for sliced single and polysilicon).

Forecast of Ribbon/Sheet - 1990, 1995

Various companies and institutions are engaged in research and development within this area. These efforts will contribute to the development of low cost, 15% efficient cells by 1995.

- Mobil Solar: Multiple ribbon pulling; now pulling 9-sided cylinders.

- Westinghouse: Dendritic Web; pilot plant designed.

- Solar Energy Research Institute: Edge-supported pulling.

- Energy Materials Corp: High speed horizontal growth.

- Motorola/Shell: Joint Venture (Solavolt) Ribbon-to-ribbon (RTR)

- Japan Solar Energy Corp: Joint venture between Mobil and Kyocera; the ribbon project was terminated in 1984.

Performance - Present cells are 12% efficient at low production speeds and 8% to 9% efficient at high speeds and low cost pull rates. Extensive R&D is needed to break this speed/efficiency barrier.
Cost - Ribbon is now priced competitively with single crystal silicon ($6-7/Wp). We forecast prices at $3/W in 1995.
Reliability - High speed (10 cm or more/minute) production causes ribbon to warp, buckle and grow into multicrystal rather than single crystal. We see no reliability problems after buckling is controlled. Reliability equal to that of single crystal or semicrystal cells is expected.

Technology Forecast for Silicon Ribbon/Sheet

Includes: Mobil Solar - Edge-Defined Nonagon Growth
Westinghouse - Dendritic Web
Energy Material Corporation - Low Angle Sheet
Motorola - Ribbon-to-Ribbon

Table XII: Si Ribbon	**1985**	**1990**	**1995**
RIBBON THICKNESS	.012"	.010"	.008"
MATERIAL YIELD (%)	60	80	90
CELL EFFICIENCY (%)	11	12	15
MODULE EFFICIENCY (%)	10	11	14
WARRANTY	2-5 year	5-year	5-year
MANUFACTURED COST ($/Wp)	4.50	2.00	1.20-1.80
PROFITABLE PRICE ($/Wp)	7.50	3.30	2.00-3.00

If ribbon/sheets are to succeed in the market place, serious commitment and investment from current leaders must come within the next 2-3 years. Scaling to larger production plans is necessary to realize the low cost potential. Management of these corporations have been holding back up to now to see how the market develops. It is our belief that the "window" on ribbons/sheets is closing rapidly. If this technology is not transferred from laboratory to manufacturing in the near future, it may not materialize as a potent option in the 1990s.

U.S. Research and Development on Ribbon and Sheet

Mobil Solar Corporation:

We estimate Mobil Solar has invested since 1972 over $50 million on the edge-defined film process to grow ribbons of silicon directly from the melt. Recent successes include:

- Converting all pullers to 9-sided cylinders with 0.015 in-wall thickness and 2 in-wide strips;
- Cutting slices using a laser from a nonagon;
- Obtaining 10% modules from 11-12% efficient cells;
- Use of hydrogen passivation to increase efficiency and reduce stress induced breakage;
- Designed and built a module packaging line capable of 1 MW/year production.

Mobil Solar has deferred entering the PV business (beyond the 75 kW shipped in 1984) one year for the past 3 years.

Westinghouse:

Westinghouse has been attempting to develop the dendritic web ribbon process to production since 1957. Westinghouse can pull 2" wide ribbons at slow speeds of 1 cm per minute in a batch mode that produces cells with 14% efficiency. When pulling speeds are increased and melt replenishment attempted, quality of product and efficiency drops. In 1984, 1985 Westinghouse has made a critical attempt to do continuous runs producing 50,000 cm2 per run per machine which result in 12% cells. The present status is 25,000 cm2. A consortium of DOE, EPRI, Westinghouse, Southern California Edison and Pacific Gas & Electric are investing nearly $3 million per year to permit Westinghouse to enter the market with a lower cost PV module based on ribbon. Recent negotiations with the Republic of China (Taiwan), if successful, could lead to a 1 MW plant late in 1986. If the cell efficiency of 15% can be maintained while obtaining the large area pulling, then costs will permit a profitable price in the $3/Watt range.

Other Sheet and Ribbon R&D

Motorola-Shell Joint Venture (Solavolt)

This joint venture is developing the so-called ribbon-to-ribbon process. A thick film of very pure polysilicon is deposited (from silane or trichlorosilane) onto a metal or glass belt or drum. A laser is used to melt the film and a large crystal sheet is peeled off of the substrate. While developing the new source of sheet material, a 2 MW module packaging line has been built. About 500 kW of polycrystal cells using material from Wacker and Crystal Systems (Salem, MA) will be delivered in 1985. Solavolt prices for polycrystal modules are competitive with other producers.

SEMICRYSTAL-CAST SILICON

About 20% of today's modules are made from pure silicon ingots which are sliced into 4" square cells. The casting process makes silicon that is multicrystal rather than single crystal. Another casting method, using a heat exchanger to form a silicon crystal in the form of a square block, has been developed. The key attributes of semicrystal-cast silicon cells/modules are:

- Through the use of cast semiconductor-grade silicon, costs could be reduced by 20% to 40% from the cost of "pulled" single crystal silicon.

- Cell efficiency is presently 10% to 15% less than that of single crystal silicon.

- Module efficiency is as high or higher than single crystal modules because the cell squares cover 100% of a given area, whereas circles cover only 70%.

- Present costs of semicrystal silicon cells can be about 20% less than single crystal silicon cells if efficiencies are equivalent.

- If the semicrystal cells can be made from less pure silicon, costs can be a factor of two less than single crystal. This development should occur by 1990.

The following table summarizes the players, processes and cost reduction opportunities for cast silicon technology.

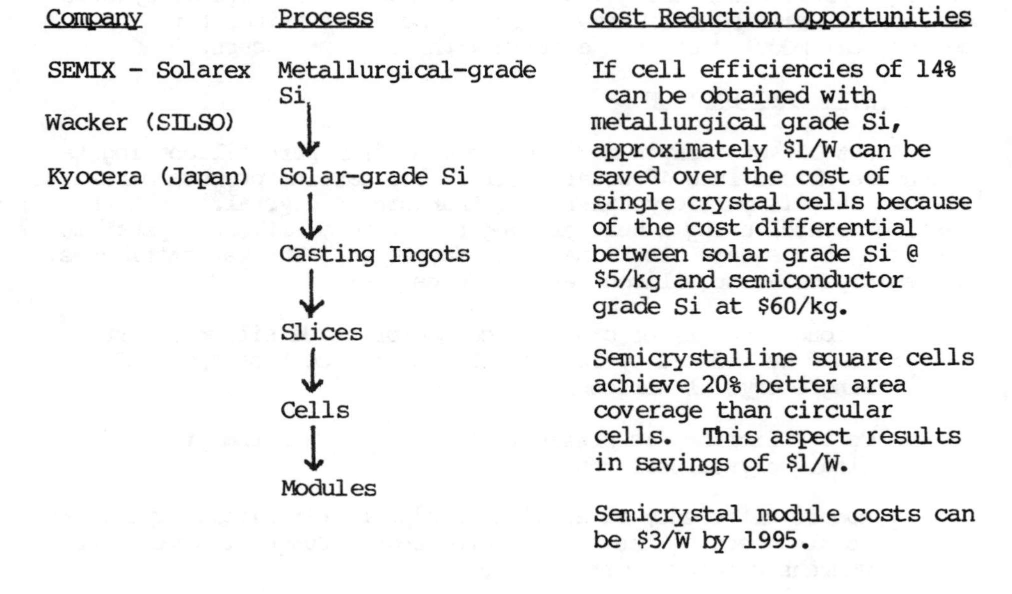

Company	Process	Cost Reduction Opportunities
SEMIX - Solarex Wacker (SILSO) Kyocera (Japan)	Metallurgical-grade Si ↓ Solar-grade Si ↓ Casting Ingots ↓ Slices ↓ Cells ↓ Modules	If cell efficiencies of 14% can be obtained with metallurgical grade Si, approximately $1/W can be saved over the cost of single crystal cells because of the cost differential between solar grade Si @ $5/kg and semiconductor grade Si at $60/kg. Semicrystalline square cells achieve 20% better area coverage than circular cells. This aspect results in savings of $1/W. Semicrystal module costs can be $3/W by 1995.

Our forecast of $2-3/W in 1995 is predicated on polysilicon cost reduction for 16% cells to the $20/kg (1984$) level. This is about 60% probable. If $40/kg is the cost, then the $3/Wp price is more likely.

TECHNOLOGY FORECAST FOR CAST POLYSILICON (SEMICRYSTAL)

INCLUDES: WACKER, SOLAREX, CRYSTAL SYSTEMS

Table XIII: Polysilicon	**1985**	**1990**	**1995**
CELL EFFICIENCY (%)	12	15	16
MODULE EFFICIENCY (%)	11	14	15
POLYSILICON COST ($/kg)	40	20	10
WARRANTY	2-5 year	5-year	<10 years
MANUFACTURED COST ($/W)	4.50	2-2.50	1.20-1.80
PROFITABLE PRICE	7.50	3.30-4.00	2-3
POLYSILICON COST ($/kg)	40	20	10

AMORPHOUS SILICON

This thin film sheet of silicon uses 1/50th to 1/100th of the silicon that is used in semicrystal or single crystal silicon and is now about 9-10% efficient on very small cells and 5-6% on 10 cm x 10 cm (4" x 4") cells.

Amorphous silicon cells are formed in a vacuum furnace by a process which basically involves the application of successive layers of vapor-deposited silicon onto a substrate (typically glass, metal or stainless steel). Each layer that is applied is only about one micron thick. A-silicon cells are cut from a large (1-2 square foot) sheet of material. Obtaining high yields from a-Si sheets is problematic.

Yield refers to the number of working devices relative to the total number manufactured. Defects in a-silicon tend to be confined to small areas on the overall sheet. Consequently, high yields are realized by creating a large number of small cells. For example, if a 2 square foot sheet has one "pin prick" defect, the sheet could be cut into two pieces for a 50% yield or it could be cut into 100 pieces for a 99% yield. In the latter case, the small defective area is simply thrown away. Maximum yield currently occurs in the range of 1 cm^2 to 3 cm^2. Key attributes are:

- Intrinsically very low cost.
- Efficiencies 1/2 of other silicon options.
- Stability of very thin films in sunlight must still be proven.
- Process technology needs extensive development.
- Consumer applications of watches, calculators, portable sound equipment, etc.
- Will be sold for larger PV system applications around late 1985.

Performance: Nearly 300 professionals are working worldwide to increase the efficiency of a-silicon. Currently, laboratory work is producing 1 cm^2 cells of 10% efficiency and 1 to 3 cm^2 cells with 90% production yield at 6% efficiency. Cells of 100 cm^2 with 5-6% efficiency can also currently be produced, but the yield is low. We forecast that 100 cm^2 cells with 8-10% efficiency and a 90% yield will be produced by 1990.

Summary on Amorphous Silicon

The a-silicon competition is very intense with $50 to $70 million worth of funds being spent to develop increased efficiency and increased stability devices. Most of the companies are using RF glow discharge techniques. We are cautiously optimistic that the intense R&D on a-silicon can result in a stable, moderately efficient product. Our 80% probability forecast for a-silicon follows:

TECHNOLOGY FORECAST FOR AMORPHOUS SILICON (1984$)

Table XIV: Amorphous Si	1985	1990	1995
4" x 4" CELL EFFICIENCY (%)	6	9	12
MODULE EFFICIENCY (%)	5	8	10
AREA YIELD (%)	50	70	90
WARRANTY	1 year	2 year	5 year
MANUFACTURED COST ($/Wp)	3-4	1.20-1.80	1-1.50
PROFITABLE PRICE (40% GPM) ($/Wp)	5-6.50	2-3	1.66-2.50

Major Conclusions for the Silicon Options

- Single Crystal Silicon will cease to dominate the world PV market to 1990. Market share will drop to 30-40%. Cell efficiency will increase to 16%, modules 15% (square cells). Module prices will be at $3/Watt in 1995. Major U.S. producers: ARCO Solar, Solenergy, Spire. Japanese manufacturers: NEC, Sharp. European: Pragma, Ansaldo.

- Optical Concentrators (with thermal energy output utilized) will increase market share to 20% by 1990. Efficiencies will exceed 20% by 1995 and $2/Watt prices will be achieved. (The market share of concentrators is included in that of single crystal.) Major producers: United Energy Corp, Intersol, Entech.

- Semicrystal Silicon can gain market share to at least 15% of the total in 1990. Efficiency will reach 15% for cells; 14% for modules. Prices as low as $2/W in 1995 are possible if low cost silicon can be used. Major producers: Solarex, Kyocera (Japan), Wacker (wafers only), AEG Telefunken, Crystal Systems (wafers).

- Amorphous Silicon or thin films of silicon will gain 40-50% of the world market in 1990. Efficiency will reach 10% and prices of $1.50/W will emerge by 1995 if stability is not a fundamental problem. A-silicon will dominate the roof-mounted market by 1995. Major producers: ARCO Solar,

ECD-SOHIO, Solarex, Chronar, Sharp, Fuji, Sanyo and Spire with technology/equipment.

- Ribbon or Sheet Silicon will gain 5% of the world market in 1990 with efficiencies of 15% in 1995 and prices of $2/W. Throughput versus quality will be the key issue. Likely producers: Mobil Solar, Westinghouse, Motorola-Shell.

- Thin Film or Materials Other than Silicon will be limited to 5% of the world market due to low efficiency, lifetime problems, and market reluctance to risk using an unproven material. Likely producers: ARCO Solar, Boeing, Spire.

U.S. PV RESEARCH AND DEVELOPMENT STATUS

This section will discuss the 1985 status of the U.S. cell research and development effort in amorphous silicon. We will attempt to forecast the efficiency cost and lifetime for the amorphous silicon product. We estimate that 500 professionals are performing research and development on advancing cell efficiency increasing film lifetimes and leading to lower costs. The U.S. industry invested nearly $80 million in 1984 in PV R&D. The DOE invested about $20 million more with the industry and the universities. The 500 professionals were working on five major areas of R&D. The amorphous silicon R&D clearly dominates the U.S. R&D strategy.

Table XV: Option	**# of Professionals**
Single crystal cells	30
Polycrystal (casting)	30
Ribbon/sheet	40
Concentrator cells	10
Amorphous silicon	300-350
II-VI materials	20
III-V Compounds for PV	15
Stacked Cells	20
	465-515 professionals

CHAPTER 23

APPLICATIONS OF MAXIMALLY CONCENTRATING OPTICS FOR SOLAR ENERGY COLLECTION

J. O'Gallagher and R. Winston

University of Chicago
Enrico Fermi Institute
Chicago, Illinois 60637

ABSTRACT

A new family of optical concentrators based on a general nonimaging design principle for maximizing the geometric concentration, C, for radiation within a given acceptance half angle $\pm\theta_a$ has been developed. The maximum limit exceeds by factors of 2 to 10 that attainable by systems using focusing optics. The wide acceptance angles permitted using these techniques have several unique advantages for solar concentrators including the elimination of the diurnal tracking requirement at intermediate concentrations (up to ∿10x), collection of circumsolar and some diffuse radiation, and relaxed tolerances. Because of these advantages, these types of concentrators have applications in solar energy wherever concentration is desired, e.g. for a wide variety of both thermal and photovoltaic uses. The basic principles of nonimaging optical design are reviewed. Selected configurations for thermal collector applications are discussed and the use of nonimaging elements as secondary concentrators is illustrated in the context of higher concentration applications.

I. INTRODUCTION

The Purpose of Concentration

The role of optical concentration in solar collector design is often misunderstood. The simplest and most widely deployed collectors employ no concentration at all and are simply flat panels. This is true for both of the conventional applications, i.e. photothermal and photovoltaic conversion systems. For both flat panels and concentrators the amount of energy collected depends on the area of intercepted radiation. Obviously the concentrator cannot "intensify the energy" received or "amplify the energy flux" as is sometimes stated in popular articles on the subject. What the concentrator does is to collect energy over some large area A_1 and deliver it to some smaller area A_2. This provides a "geometric concentration ratio"

$$C \equiv A_1/A_2 \qquad (1)$$

and in the process sacrifices some angular field of view which in turn requires, in many cases, that the collector be "tracked" or moved to follow the sun. Only relatively recently, through the use

0094-243X/85/1350448-24$3.00 Copyright 1985 American Institute of Physics

of a class of nonimaging concentrators usually referred to as Compound Parabolic Concentrators, or CPC's(1), has it become possible to relax and sometimes eliminate this tracking requirement.

The motivation for increasing concentration above unity (corresponding to the flat panel case) is two-fold:

1) *Thermal performance:* For a solar thermal collector the heat losses depend on the area of the hot absorber. By reducing the area of the thermal transducer (A_2) relative to the collecting area (A_1), one can achieve respectable efficiencies at higher temperatures than could otherwise be attained. Very high concentrations can be used to generate very high temperatures -- in principle approaching that of the surface of the sun.

2) *Economic cost effectiveness:* If the cost-per-unit area of the energy transducer (i.e. a solar cell) is very much greater than that for the concentrating optics comprised of concentrating mirrors or lenses, then the overall cost-per-unit energy collected can be dramatically reduced. However, the cost of any required tracking system must be included in the economic analysis.

These are the only two reasons for employing concentration. Clearly, for photovoltaic applications only the second applies; while for thermal applications both apply, although the economic benefit is not usually emphasized.

The Thermodynamic Limit

If one wishes to concentrate radiation according to equation (1), it is evident qualitatively that there must be some sacrifice in view angle (the smaller absorbing surface cannot "see" the full field of view comprising the hemisphere visible to the larger aperture). The quantitative relationship between this reduction in view angle and increasing concentration ratio is not intuitively obvious, but can be derived in a straightforward manner from thermodynamic arguments(2). These arguments are based on the fact that in thermodynamic equilibrium, the absorber must re-radiate back to the environment the same amount of energy as it receives. In particular, for a given acceptance half-angle ($\pm\theta_a$) at the collecting aperture, this condition defines a minimum absorber area and corresponding maximum possible geometrical concentration given by

$$C \leq \frac{1}{\sin \theta_a} \qquad (2a)$$

if the concentration is done in only two dimensions (trough-like geometries) and

$$C \leq \frac{1}{\sin^2\theta_a} \qquad (2b)$$

if the concentration takes place in three dimensions (cone-like geometries). This is referred to as the thermodynamic limit since, if one could make a concentrator which transferred all the incident radiation to an absorber smaller than that given by equations (2a) or (2b), it would not have sufficient area to re-radiate this energy in thermodynamic equilibrium and its temperature would begin to rise above its surroundings in violation of the Second Law. Any concentrator system which can attain this limit is referred to as "ideal." Concentrating systems based on imaging or focusing optics fall short of this limit by a factor of 3 to 4. The CPC and other concentrator shapes determined by the principles of nonimaging optics(3) actually achieve this limit in two dimensions and closely approach it in three dimensions.

Equations (2) can also be derived by application of the principles of etendue or phase space conservation(4). A bundle of rays propagating in the z direction can be characterized by a distribution of points in a four-dimensional (two-position and two-directional coordinates) phase space. Phase space conservation then requires that the volume representing an ensemble of rays propagating through an optical system along the z axis, must remain constant. If one then considers a distribution uniform in position across an aperture $A_1(z_1)$ and consisting of rays isotropically filling an acceptance angle θ_a with rays of $\theta < \theta_a$ and asks what is the smallest cross-sectional area $A_2(z_2)$ to which these rays can be reduced while allowing the directional distribution to expand to fill the semicircle in direction coordinate space, one is led directly to equations (2) in those geometries.

Neither the thermodynamic nor the phase-space conservation arguments tell us actually how to achieve the maximal or "ideal" concentration, but simply that these are the limits which no optical system can exceed.

Nonimaging vs. Imaging Optics

Concentrators designed according to classical imaging optical principles are optimized for paraxial rays and in general fall far short of the limits of equation (2). Nonimaging concentrators, on the other hand, are optimized for the extreme angles desired to be included in the acceptance of the optical system and can be maximally concentrating, i.e. approach and in some cases attain the thermodynamic limit. The difference in design techniques can best be explained by reference to typical image-forming optical systems, e.g. camera lenses, slide projector lenses, etc. In these systems there is an axis of symmetry on which the centers of all the lens components lie. Likewise the object, e.g. a scene to be photographed or a slide to be projected, has usually its center or most important point on or near the axis. The designer of the image-forming system then attends first to the region near the axis: the basic properties such as focal length and magnification are determined by the properties of the lens components near the axis and it follows that the image quality or sharpness must be best near the axis. By adding or changing lens components, the designer can improve the image quality away from the axis; but it inevitably deteriorates gradually until the image is so fuzzy at some distance from the axis that the optical system is unusable beyond that point. This, of course, determines the "field of view."

Now in nonimaging concentrators, there is a complete opposite emphasis. These systems can have either an axis or a plane of symmetry, or they may even have no symmetry at all; but in any case, they are designed to collect entering flux over a certain angular range and larger aperture and concentrate it so that it emerges from a smaller exit aperture or strikes an absorber with a surface smaller than the entry aperture. The design principle to be followed in order to get the maximum concentration is that all rays entering at the extremes of the angular range must emerge just grazing the edge of the exit aperture or absorber surface. This is a concise statement of the "edge-ray principle". The edge-ray principle is the basis of the design of all efficient nonimaging concentrators, together with the laws of reflection and refraction. The edge-ray principle ignores rays from the regions near the axis (if there is an axis) or near the center of the entering angular range; there is no question of sharpness of images anywhere; all that is required is that extreme entering-rays meet certain operationally-defined conditions -- for example, graze the edge of the exit aperture as they emerge -- whereas in image-forming systems, the extent of the field of view gradually evolves *post hoc* when the center of the field has been taken care of.

A more specific formulation of the ideal concentrator design principle, particularly applied to systems where the active optical surfaces are reflectors, is the "maximum slope principle." It says that each element of the reflector profile curve should be made to assume the maximum possible slope (i.e. rate of increase of the entrance aperture with increasing concentrator depth) consistent with the requirement that rays incident on the aperture at the desired extreme angle (i.e. "edge ray") illuminate the absorber. Given a particular absorber shape and a desired acceptance θ_a the application of this principle yields a differential equation which allows generation of ideal two-dimensional reflector shapes. Examples of solutions in two dimensions are shown in Figure 1.

If one considers the two-dimensional geometry defined by two parallel surfaces (A_1 and A_2 in Figure 1a) perpendicular to a common optical axis, one can show that reflectors connecting these two surfaces with parabolic segments as shown in Figure 1a actually achieve the limit defined by equation (1). The left-hand reflecting surface is a segment of a parabola with its axis tilted by an angle θ_a with respect to the optical axis and positioned so that its focus lies at the lower edge of the right reflector. All rays entering the entrance aperture at an angle θ_a will then strike one reflector and be brought to a focus at the opposite edge of the exit aperture. All rays entering between $\pm\theta_a$ must be reflected to points between the lower mirror edges. This is the simplest manifestation of an "ideal" concentrator in two dimensions and is the solution which gave rise to the name Com-

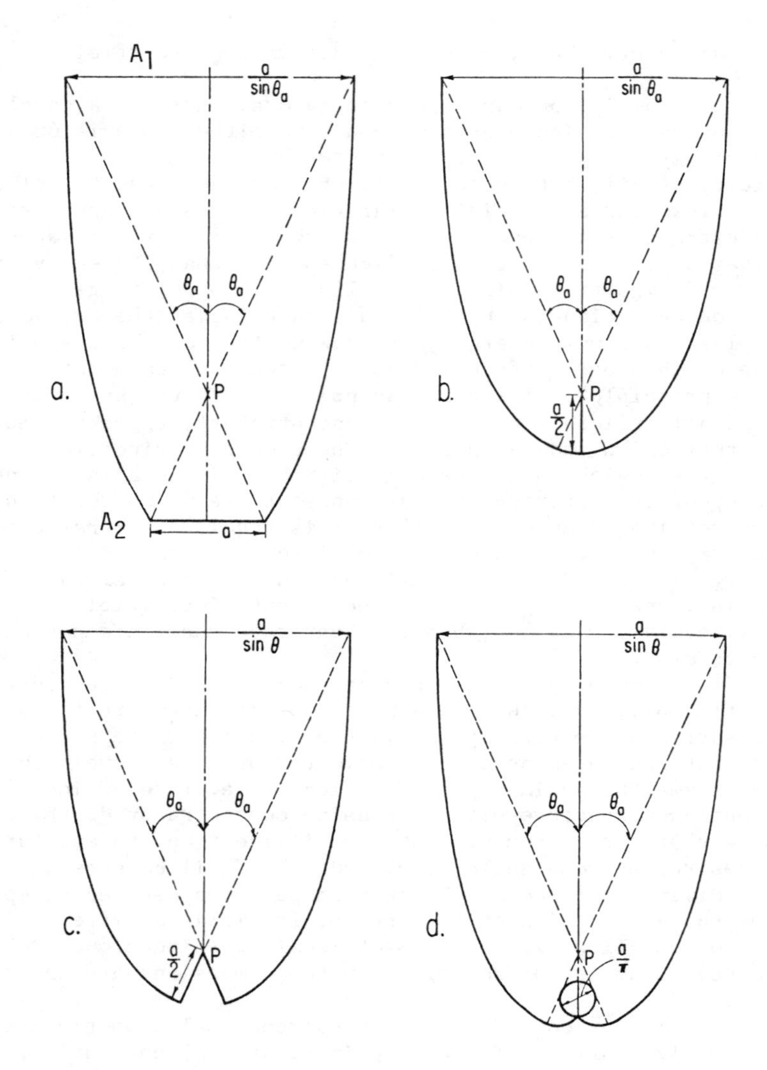

Figure 1. Cross-sectional profiles of ideal trough concentrators generalized to absorbers of different shapes. In practice the reflectors are usually truncated to about one-half their full height to save reflector material with only negligible loss of concentration.

pound Parabolic Concentrator, which is now often used to include the other solutions as well. A figure of revolution formed by rotating the shape of Figure 1a about the concentrator axis does not perform as a true ideal concentrator, essentially because in three dimensions, rays which are not in the meridian plane have an additional quantity (angular momentum) which must be conserved as well as the phase space so that the system is overdetermined. However, such three-dimensional cones come very close to ideal performance. Solutions for the other common absorber shapes are shown in Figures 1b, 1c, and 1d. Three dimensional versions of these latter profiles do *not* work as concentrators because angular momentum would not be conserved for skew rays.

The original conical CPC's were first developed in the mid 1960s to maximize the light collection of a system of Cerenkov detectors in a hyperon decay experiment at Argonne National Laboratory(5). Similar devices were proposed independently around the same time in Germany(6) and the Soviet Union(7). In 1974 it was determined that a trough with the profile of the Cerenkov Light Coupler had unique advantages for solar energy concentration(1). In particular, the wider acceptance angles made possible low and intermediate levels of concentration without continuous tracking. If the long axis of the trough is oriented in an east-west direction and tilted towards the Southern sky (in the Northern Hemisphere), a completely stationary concentrator can attain geometrical concentrations up to about 2X and with seasonal adjustments concentrations up to about 10X are achieved. Other advantages of nonimaging solar concentrators are greatly relaxed optical tolerances, collection of the circumsolar and a large fraction of the diffuse component of the insolation and reduced sensitivity to small angle scattering from dust or scratches.

The application of CPC's for solar thermal collection underwent rapid development in the late 1970s at Argonne National Laboratories and the University of Chicago. In subsequent sections of this review, we survey some of the highlights of this development period. In addition, we present some of the more recent advanced concepts in which nonimaging techniques are used to design two-stage concentrators employing an imaging primary and a CPC-like device in the focal zone. Such hybrid devices can in fact be made to approach the maximally concentrating limit and have many interesting applications.

II. SOLAR THERMAL APPLICATION OF CPC'S

The preferred configurations are those which eliminate heat losses through the back of the absorber such as shown in Figure 1b and 1d. Note that the geometric concentration is defined relative to the full surface area of the fin absorber (both sides) or tube (full circumference), and that these designs effectively have no back.

In recent years work has centered on the development of CPC's for use with spectrally selective absorbers enclosed in vacuum. Work at Argonne National Laboratory after 1975 was strongly influenced by the emergence of the dewar-type evacuated absorber. This thermally efficient device, developed by two major U.S. glass manufacturers, the General Electric Co. and Owens Illinois, could, when coupled to CPC reflectors, supply heat at higher temperatures than flat plate collectors, while retaining the latter's advantage of a fixed mount. The heat losses associated with such absorbers are so low that the moderate levels of concentration associated with CPC's provide a dramatic improvement in thermal performance and achieve excellent efficiencies up to 300°C. In fact, the gains associated with even higher concentration ratios are marginal and probably negligible when the added complication and expense of active tracking are considered. This fact is illustrated in Figure 2, which compares the calculated efficiency relative to total insolation of a collector consisting of evacuated tubes with selective absorber coating under increasing levels of concentration.

In characterizing the thermal performance of CPC's in general, we describe the thermal collection efficiency, η, as a function of operating temperature T by

$$\eta(T) = \eta_0 - \frac{U(T - T_a)}{I} - \frac{\sigma\varepsilon(T^4 - T_a^4)}{CI} \qquad (3)$$

where

- C = geometric concentration ratio
- T_a = ambient temperature
- I = total insolation
- σ = Stefan-Boltzmann constant
- ε = absorber surface emittance
- U = linear heat loss co-efficient.

The optical efficiency η_0 is the fraction of the incident solar radiation (insolation) which is actually absorbed by the receiver surface after transmission and reflection losses. The other terms represent parasitic conduction losses and radiation losses from the absorber surface, both of which increase substantially as the collector working fluid temperature is increased. The most important point to be seen from Figure 2 is that the dramatic reduction in relative thermal losses produced by adding a 1.5X reflector or again in effectively tripling the concentration from 1.5X to 5X is not continued as one increases the concentration much beyond 5X. This is because at, say, $\sim$300°C ($\Delta T/I \cong 0.3$), the thermal losses for a 5X are already so low that a further factor of 5 reduction corresponds to only a negligible fraction of the operating efficiency. Thus it is not necessary or even desirable to increase the concentration much beyond 5X when using an evacuated selective absorber at these temperatures. Thus the combination of CPC's up to $\cong$5X with these thermally efficient absorber tubes represents a nontracking strategy for practical solar thermal collection up to power generation temperatures with many unique advantages.

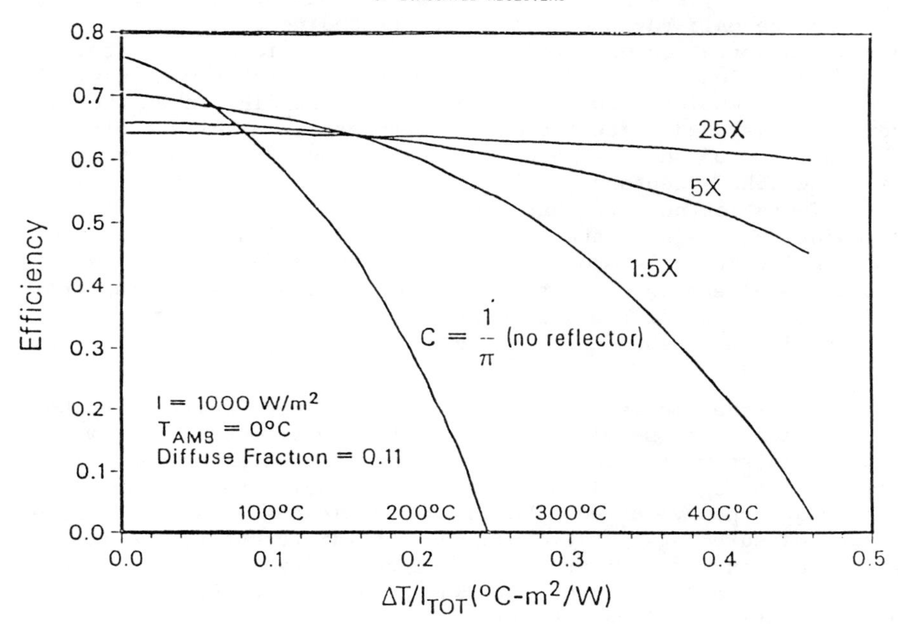

Figure 2. Calculated thermal performance curves for evacuated tubular absorbers under increasing levels of concentration. Note that the improvement in increasing the concentration above 5X is marginal.

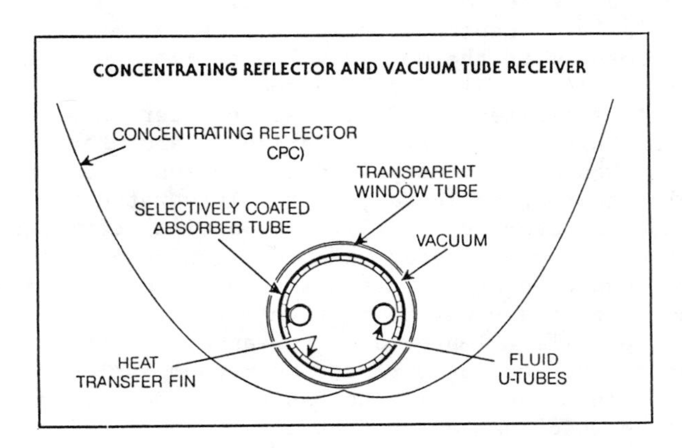

Figure 3. Cross section of a contemporary commercial evacuated tube CPC according to the basic design developed by Argonne National Laboratories(3).

The most developed evacuated CPC design is that optimized for totally stationary collection throughout the year, having an acceptance half-angle θ = ±35° corresponding to a maximum "ideal" concentration of ≃1.8X. Commercial versions, truncated to net concentrations of ∿1.1X to 1.4X are now available with typical optical efficiencies between 0.52 and 0.62 depending on whether a cover glass is used. The thermal performance is excellent with values of $\varepsilon \simeq 0.05$ and $U \simeq 0.5$ W/m^2K in equation (3). Other versions have opened the acceptance angle to ±50° with C ≃ 1.1X to allow polar orientation and are characterized by U = 1.3 W/m^2K (lumping the radiative losses in the linear term). An illustration of the basic configuration is shown in Figure 3. Several manufacturers introduced collectors of these types for applications ranging from heating to absorption cooling to driving Rankine Cycle engines. A number of installations each in excess of 1000m^2 have been successfully deployed.

Experimental prototype CPC's have been built in higher concentrations for use with evacuated absorbers. One version with C = 5.25X is shown in Figure 4a. This CPC, studied at the University of Chicago, is a large trough coupled to the same glass dewar-type evacuated tube as is used in the 1.5X above. Performance measurements for two modules with different reflecting surfaces are shown in Figure 4b. The upper curve is for a module with a silver foil reflector. It has been operated at 60% efficiency (relative to direct beam) at 220°C above ambient. This is to be compared with the measured performance of a fully-tracking parabolic trough tested by Sandia Laboratories as shown by the dashed line in Figure 4b. The performance of the CPC is comparable to that of the parabolic trough at all temperatures tested. The lower curve is for a module with aluminized mylar reflectors, and even with poorer reflectors, exhibits quite respectable performance. The angular acceptance properties of the module are in excellent agreement with the design value of ±8°, which allows collection with 12-14 annual tilt adjustments.

An experimental CPC collector under development at the University of Chicago and Argonne National Laboratory in recent years, which should ultimately lead to the most practical general purpose solar thermal collector, is the Integrated Stationary Evacuated Concentrator or ISEC(9). The optical efficiency of evacuated CPC solar collectors can be significantly improved over that of contemporary commercial versions discussed above by shaping the outer glass envelope of the evacuated tube into the concentrator profile. Improved performance results directly from integrating the reflecting surface and vacuum enclosure into a single unit. This concept is the basis of a new evacuated CPC collector tube which has a substantially higher optical efficiency and a significantly slower rate of exposure-induced degradation than external reflector versions. These performance gains are a consequence of two obvious advantages of the integrated design:

a) Placing the reflecting surface in vacuum eliminates degradation of the mirror's reflectance and thus permits high

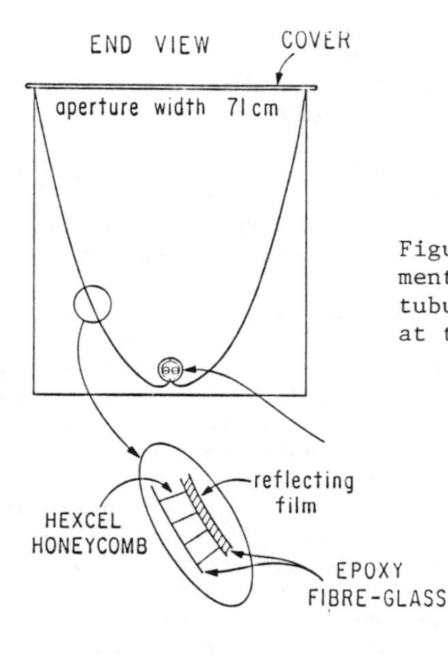

Figure 4a. Profile of an experimental 5X CPC for an evacuated tubular absorber built and tested at the University of Chicago.

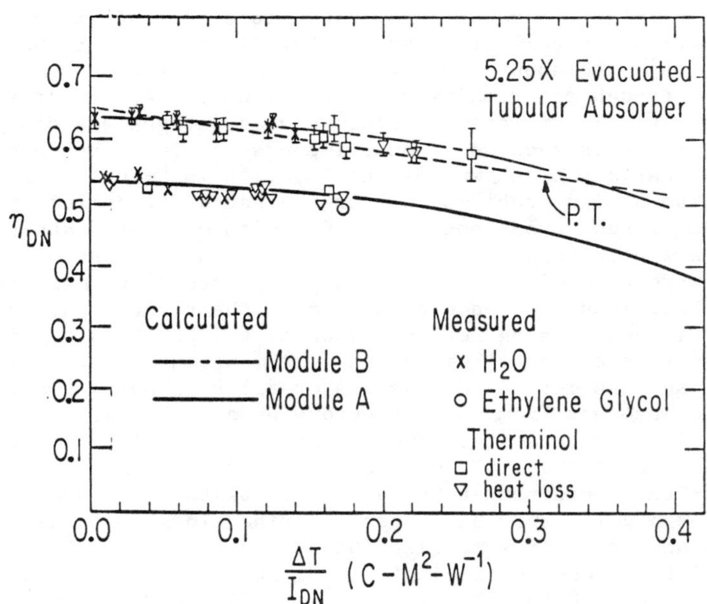

Figure 4b. Measured performance curves for two 5.25X CPC prototype modules. The performance is comparable to that of a commercial parabolic trough shown by the dashed line.

quality (silver or aluminum with reflectances ρ = 0.91-0.96) first surface mirrors to be used instead of anodized aluminum sheet metal or thin film reflectors (ρ = 0.80-0.85) typical of the external reflector designs.

b) The transparent part of the glass vacuum enclosure also functions as an entrance window and thus eliminates the need for an external cover glazing. This increases the initial optical efficiency by a factor of $\frac{1}{\tau}$, where typical transmittances τ = 0.88-0.92.

For the past several years, the solar energy group at the University of Chicago has been developing this concept in collaboration with the staff of GTE Laboratories which fabricated the tubes. Eighty prototype tubes were built and forty-five of these assembled into a panel with $\sim 2 m^2$ net collecting area.

The Integrated Stationary Evacuated Concentrator (ISEC) shown in Figure 5, is an extended cusp tube CPC(10) matched to a circular absorber of diameter 9.5 mm. The design acceptance half-angle of $\theta_a = \pm 35°$ was chosen to permit stationary operation throughout the year. After truncating the CPC the net concentration was 1.64X. This collector was tested at Chicago for three years and routinely achieved the highest high-temperature performance yet measured for a fixed stationary mount collector. Performance curves based on these tests are shown in Figure 6 along with curves for three other collector types. Note in particular that for temperatures up to about 200°C the ISEC is comparable to a fully tracking trough and remains respectable up to temperatures approaching 300°C. The relative performance advantages are similar when the comparison is made on an annual energy delivery basis at a variety of locations as shown for one location in Figure 7.

The design problems for non-evacuated CPC collectors are entirely different from those for CPC's with evacuated absorbers. These are particularly vulnerable to high heat losses if one uses an improper design. One must be very careful to minimize or eliminate heat loss via conduction through the reflectors. This can be accomplished by using reflectors whose thickness is negligible compared to the ovarall dimension (e.g., height, aperture) of the trough such as metalized plastics or films or by thermally decoupling the absorber from the reflectors by a small gap maintained by insulating standoffs. Two prototype CPC's with non-evacuated absorbers, a 3X and a 6X, were built and tested extensively as part of our early program at Chicago. The features of and performance of these collectors have been described in some detail elsewhere(11) and are only summarized here. The optical efficiencies and total heat-loss coefficients were $\eta_o = 0.68 \pm 0.01$ and $U = 1.85 \pm 0.1$ W/m^2- °C for the 6X and $\eta^0 = 0.61 \pm 0.03$ and $U = 2.7 \pm 0.02$ W/m^2 - °C for the 3X.

The efficiencies of these non-evacuated CPC's are to be compared with typical values for flat plates with η_0 from 0.70-0.78 and U from 4.5-7 W/m^2K. Despite lower optical efficiencies, the CPC's outperform typical flat-plate collectors above temperatures of as low as 10°C above ambient (for the 6X) to about 35° above ambient (for the 3X). This is particularly important when it

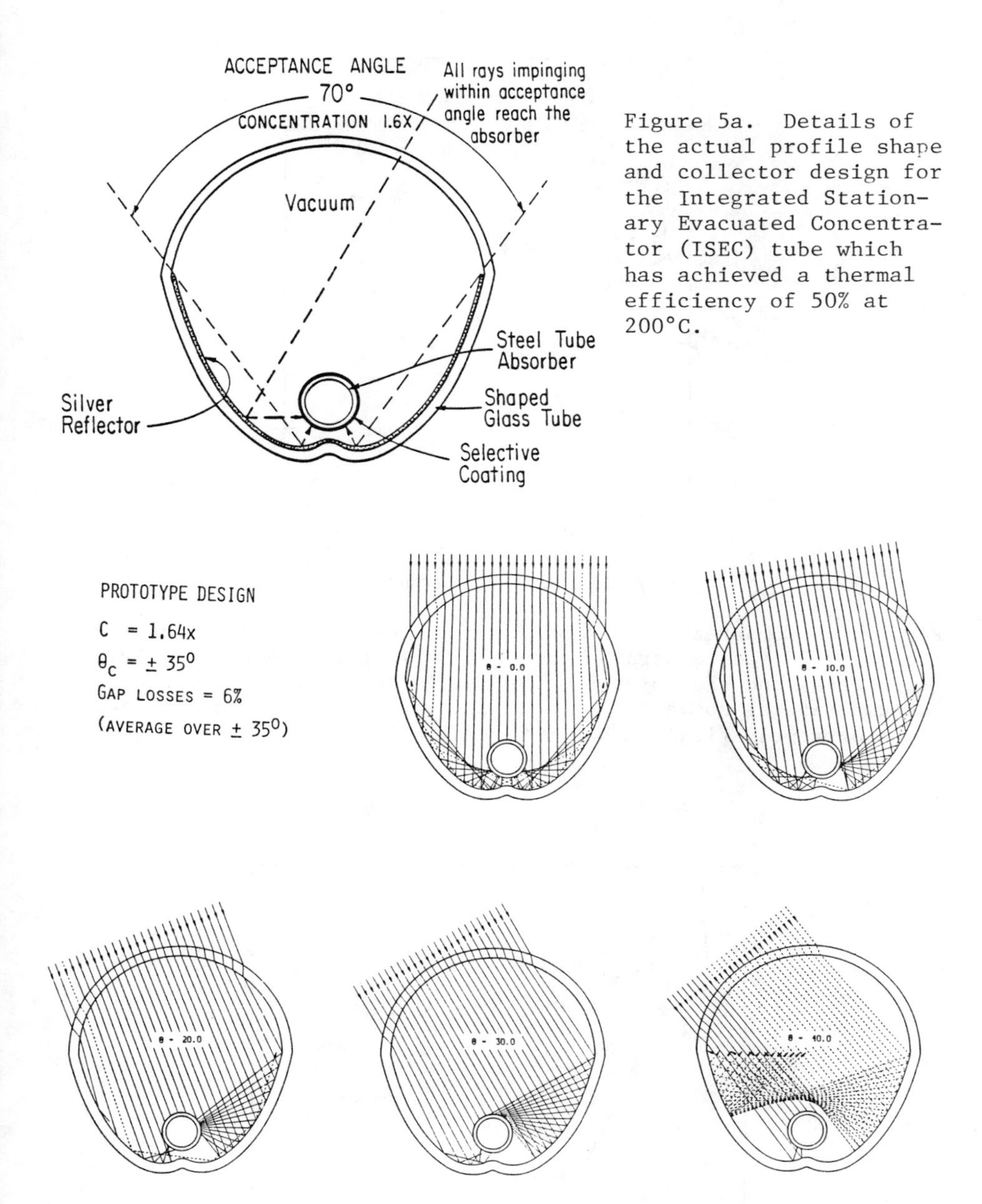

Figure 5a. Details of the actual profile shape and collector design for the Integrated Stationary Evacuated Concentrator (ISEC) tube which has achieved a thermal efficiency of 50% at 200°C.

Figure 5b. Ray trace diagram showing how essentially all the solar energy incident within ±35° is directed onto the absorber tube. Since the reflector cannot physically touch the absorber as required for an "ideal" concentrator, a small fraction is lost in the gap between the reflectors and the absorber.

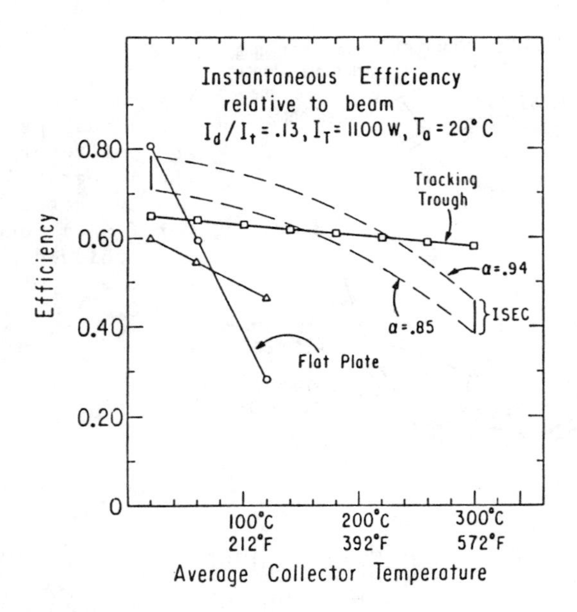

Figure 6. Comparative peak performance for a tracking parabolic trough, ISEC, contemporary (external reflector) evacuated CPC's (triangles) and flat plates. The ISEC's superior performance up to temperatures above 200°C, achieved with no miving parts, makes it an extremely flexible solar thermal collector.

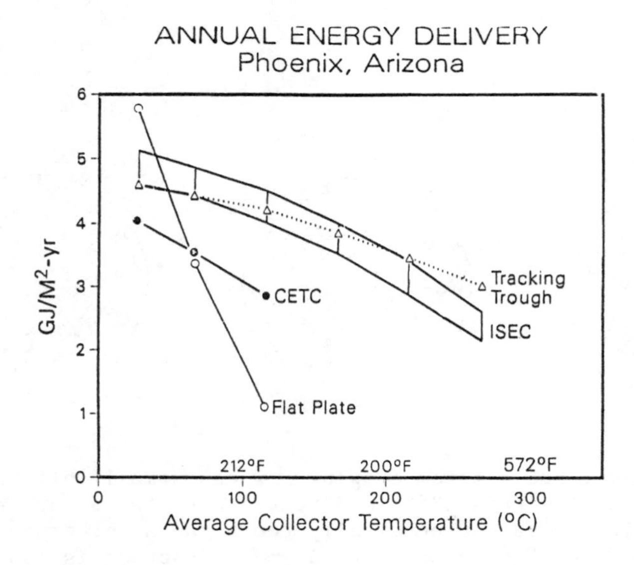

Figure 7. Comparative annual energy delivery for the same collector types as in Figure 6.

is recognized that the 3X should represent a relatively inexpensive collector design. While a detailed economic analysis cannot be based on the prototype construction methods used here, there are several unique features which contribute to its low cost potential, among them the relatively small absorber cost and very limited insulation requirements.

III. TWO STAGE MAXIMALLY CONCENTRATING SYSTEMS

The principal motivation for employing optical concentration with photovoltaic cells in solar energy is economic. By using what one hopes are relatively inexpensive lenses or mirrors to collect the sun's energy over a large area and redirect it to the expensive but much smaller energy conversion device, the net cost per unit total area of collection can be reduced substantially. Alternatively, if one desires to generate electricity through the thermodynamic conversion of solar heat to mechanical energy, one requires concentration to achieve the high temperatures necessary to drive a heat engine with reasonable efficiency. Here one directs the solar flux to an absorber (often a cavity) small enough so that the heat losses, even at high temperatures, remain relatively small. It often turns out in both the photovoltaic and the thermal conversion cases that the desired concentration is much higher than can be achieved with non-tracking CPC-type devices. The conventional means for achieving these higher concentrations is by means of some kind of focusing lens or paraboloidal mirror which are not maximally concentrating.

It is not widely recognized that the same nonimaging techniques described in the previous section can be used to design secondary elements(12), which can augment the concentrations of more conventional focussing elements used as primaries, and that such a hybrid optical system can also approach the allowable limit. Applications of such two-stage designs lie in the higher concentration, small angular acceptance regime where the geometry of a single stage CPC becomes impractical. The fundamental advantage is the same as in lower concentration applications and may be expressed in complementary ways: either significant additional system concentration can be attained (i.e., a smaller, lower cost absorber) or the angular tolerances and precision can be relaxed while maintaining the same level of concentration.

The limits to achievable levels of solar concentration are represented in Figure 8 for both line focus and point focus geometries across the whole range of possible desired angular acceptances from wide angles permitting stationary or seasonal adjusting CPC's down to the angular subtense of the sun ($\pm$4.6 milliradians). If one could achieve the thermodynamic limit in a point focus geometry (no configuration solution which could accomplish this is known) with no slope or alignment errors, one could reach the thermodynamic limit of 46,000 suns and in principle reach the sun's surface temperature of 6000°K. In practice slope and alignment tolerances, typically $\sim\pm 0.5°$, and the aberrations associated with focusing designs limit the actual values to 30-70X in line focus and 1000-5000X in point focus geometries. Use of a nonimaging secondary can increase these limits (or tolerances) as indicated.

It has been proposed by some that holographic optical elements (HOE's) could achieve concentrations in the range 10X-20X without tracking by stacking individual elements designed to be effective at different times of the day. This is clearly impossible as indicated by the dashed box in Figure 8, because it would violate the thermodynamic limit as has been discussed in some detail elsewhere.(13)

In this section we describe both photovoltaic applications where the primary is a lens and thermal electric applications where the primary is a paraboloidal mirror. In each case we discuss only the point focus configuration. Schematic drawings of the basic elements for the two cases are shown in Figure 9.

For the thermal application (Figure 9a), the primary is characterized by its focal length F and aperture diameter D which defines the rim angle

$$\tan\phi = \left[2f - \frac{1}{8f}\right]^{-1} \tag{4}$$

where $f \equiv F/D$. The secondary is a nonimaging concentrator of either the Compound Elliptical Concentrator (CEC)(12), a variant of the more familiar Compound Parabolic Concentrator or hyperbolic trumpet(14) types. It is convenient to simplify the analysis by characterizing the primary as having a conical angular field of view of half-angle $\pm\theta_I$. This is chosen to accommodate the angular tolerance budget of the primary including concentrator slope errors, specularity spread, pointing error, and incoming direct sunlight.

The thermodynamic limit in a point focus reflecting geometry is given by equation (2b) while geometrical arguments show that the geometrical concentration of the primary alone must be

$$C_1 < \frac{\sin^2\phi\cos^2\phi}{\sin^2\theta_I} \tag{5}$$

to intercept all of the energy incident within $\pm\theta_I$. The limiting concentration for the secondary is

$$C_2 < \frac{1}{\sin^2\phi} \ . \tag{6}$$

Therefore, the maximum combined concentration is

$$C_1 \cdot C_2 = \frac{\cos^2\phi}{\sin^2\theta_I} \ . \tag{7}$$

which approaches the maximum limit for small ϕ (large f).

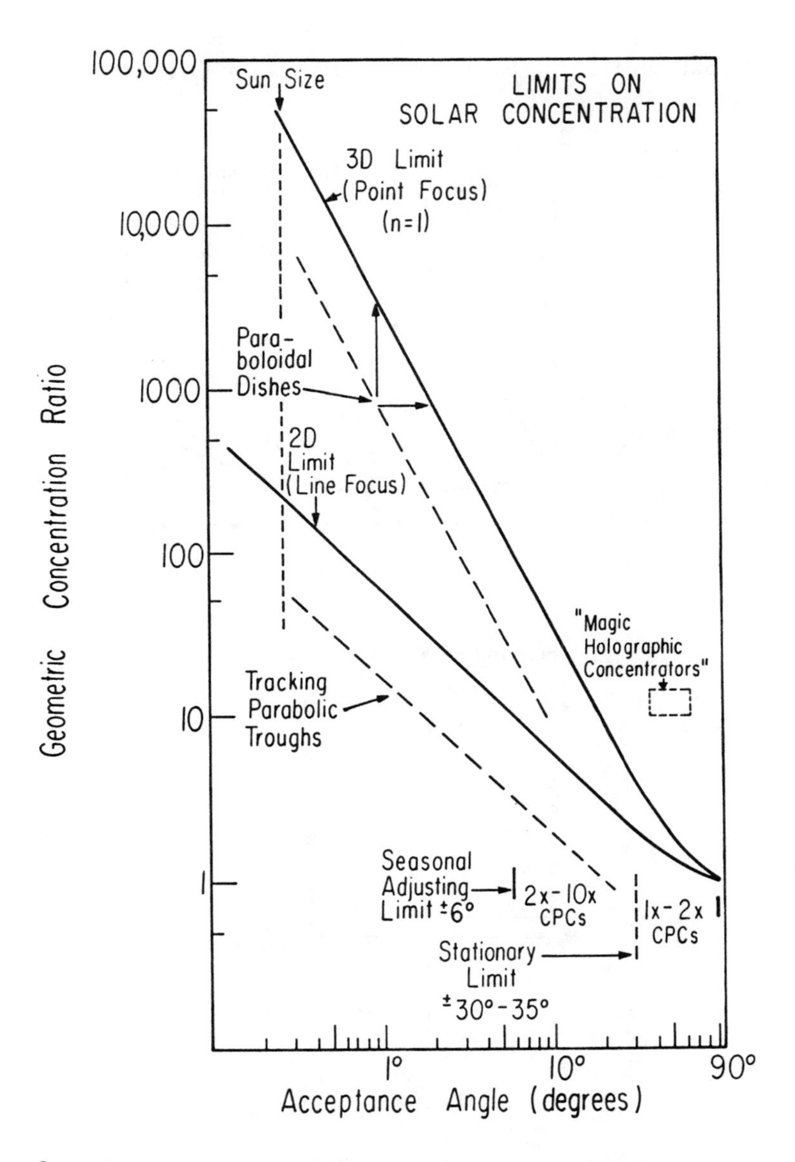

Figure 8. The maximum geometric concentration ratio corresponding to a given half angle of incidence permitted by physical conservation laws ("The Thermodynamic Limit") is shown by the solid lines. Traditional focusing concentrators fall about a factor of 4 below this limit (dashed lines). Proposed concentrator designs purporting to achieve 10:1 ratios with no tracking based on holographic techniques cannot work since they violate this limit.

a.

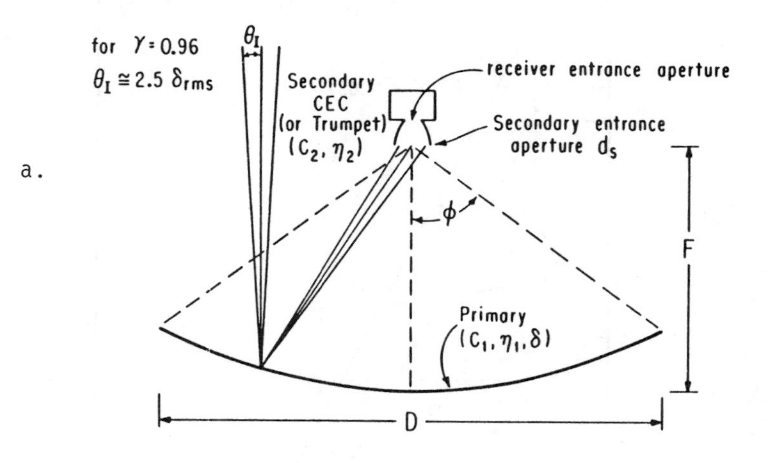

b.

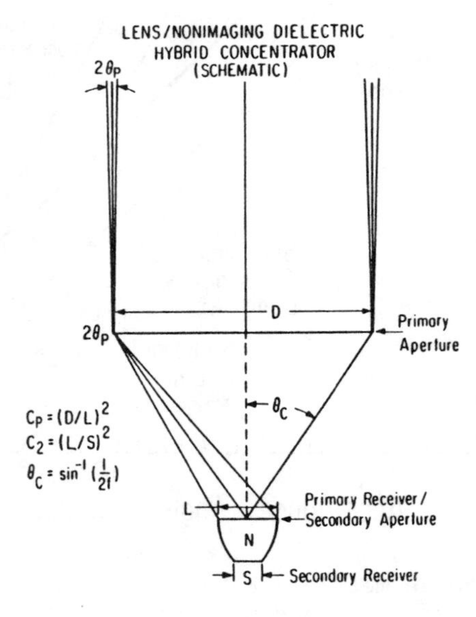

Figure 9. The geometric and optical parameters characterizing the two-stage point focus reflecting dish (a) and lens concentrators (b) discussed in the text.

For the photovoltaic concentrator (Figure 9b) the nonimaging secondary is formed from a transparent dielectric material with an index of refraction n = 1.3-1.5 in contact with the solar cell. It is usually possible to ensure that total internal reflection (TIR) occurs for all rays accepted by the secondary,(15) thus saving the cost of applying and protecting metallic reflecting surfaces. In such a totally internally reflecting dielectric CPC (DCPC), the refractive power of the dielectric operates in combination with the reflecting profile shape so that secondary concentration ratios are achieved which are a factor of n^2 larger than is possible using conventional reflecting secondaries(2). The primary is a lens (usually a fresnel lens) of aperture diameter D which focuses normally incident rays at a point which defines the center of the primary receiver. Rays from outer edges of the primary are brought together with a convergence angle $2\theta_c$. If the incident rays subtend an angle of $\pm\theta_p$ on either side of the aperture normal, one can show that the maximum possible geometric concentration C_{max} attainable is

$$C_{max} = n^2/\sin^2\theta_p. \qquad (8)$$

Here n is the relative index of refraction at the final absorber surface relative to that just outside the collecting aperture. Equation (8) for a dielectric secondary corresponds to equation (2b) for a reflecting secondary in that it defines the "ideal" limiting concentration allowed by physical conservation laws. For a focusing lens of focal ratio f', where

$$f' \equiv \frac{1}{2\sin\theta_c}, \qquad (9)$$

such that it corresponds to the generalized focal ratio used to express the Abbe sine condition for off-axis imaging,(3) one can show that the actual concentration achieved by the lens alone is

$$C_p = \frac{1}{4f'^2\sin^2\theta_p}. \qquad (10)$$

Comparing equations (10) and (8), we see that for practical lens systems systems where $f' \gtrsim 1$, C_p falls short of the limit by a factor of $(4f'^2n^2)^{-1}$.

If a DCPC type secondary with entrance aperture diameter L and exit aperture diameter S is placed at the focal spot of the imaging primary lens, it can achieve an additional geometric concentration C = L /S which is given by equation (8) with θ_c replacing θ_p or

$$C_2 = \frac{n^2}{\sin^2\theta_c} = 4f'^2\,n^2. \qquad (11)$$

Combining equations (10) and (11), one sees that in principle the two stage system can attain an overall geometric concentration equal to the thermodynamic limit. In practice, certain compromises need to be made which reduce this somewhat; but typically secondary concentrations in the range 7X-10X are readily achievable.

The limits of concentration for both reflecting and dielectric refracting systems with and without optimized nonimaging concentrators are shown in Figure 10a and 10b as a function of the focal ratio of the primary.

Applications

Figure 11 shows examples of practical nonimaging secondaries used to date for both kinds of applications. The flow line or "trumpet"-shaped secondary is a recent development(14) with particular advantages in a retrofit mode: i.e., to increase the concentration of a dish which is already designed and built. Figure 11b illustrates how introducing a small amount of curvature into the front surface of the secondary provides some of the concentration so that the overall height of the side walls can be reduced yielding substantial savings in material.

For thermal applications with a CPC type secondary, a particularly attractive option is the use of a primary of longer focal ratio (f/no $\gtrsim$1.0) with flat mirror facets in which case the number of required facets can be reduced by a factor approximately equal to the secondary concentration ratio. For example, with a secondary with C_2 = 5X the number of flat facets required to achieve 150X can be reduced from nearly 200 to less than 40. Details of such a design are being developed currently.

A two-stage photovoltaic concentrator presently being developed is shown in Figure 12. This system has no real analog in contemporary solar concentrator configurations, since conventional 100X concentrators must track quite accurately ($\lesssim\pm1°$) and crude tracking devices have much lower concentrations ($\sim$5-15X). Here a good primary is designed to provide a concentration of about 13X with an acceptance angle wide enough to accommodate the sun's movement for close to one hour. A TIR secondary provides the additional concentration (here 7.7X) required to make the economic savings associated with reduced cell area really worthwhile. No scale is shown in the figure, since a study of the trade-offs associated with size is one of the objectives of the present work. For reference, note that an 0.5" cell would correspond to a 5" diameter circular collecting lens. These individual elements could be arranged in a hexagonal close-packed geometry if a high packing density is required, or simply on a square lattice if not. Finally, note that in this geometry the function of the secondary in redistributing the concentrated sunlight more uniformly on the cell is especially useful.

IV. SUMMARY

Nonimaging optics departs from the methods of traditional optical design to develop instead techniques for maximizing the

a.

b.

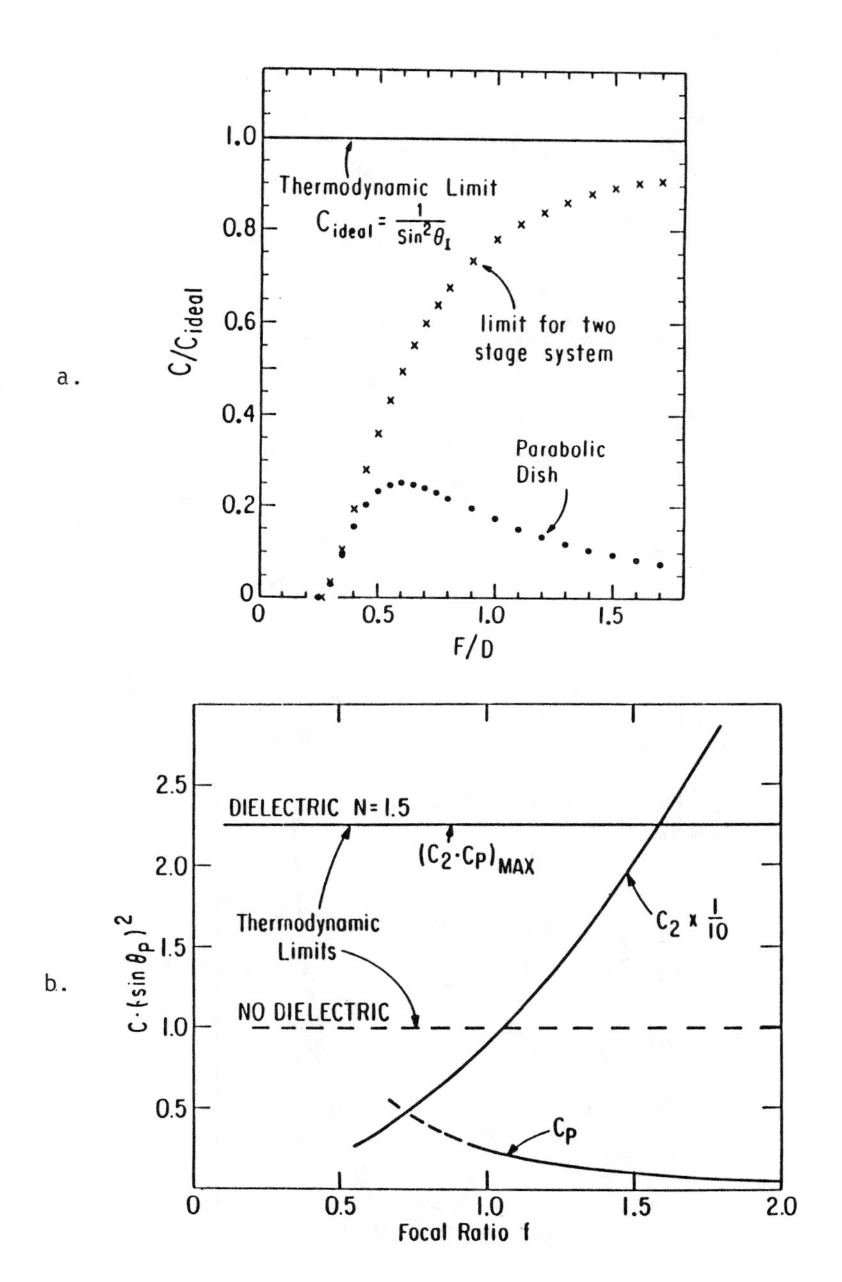

Figure 10. The maximum geometric concentration achieved by (a) reflecting dish primaries with nonimaging CPC type secondaries, and (b) lens primaries combined with refracting dielectric CPC (DCPC) type secondaries.

a.

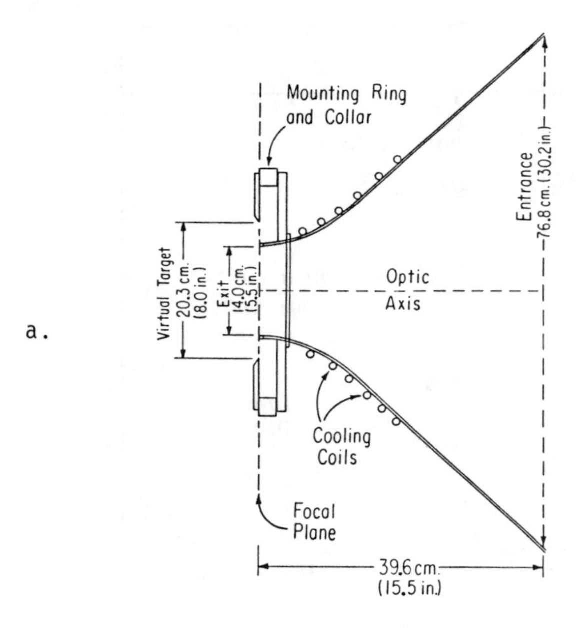

b.

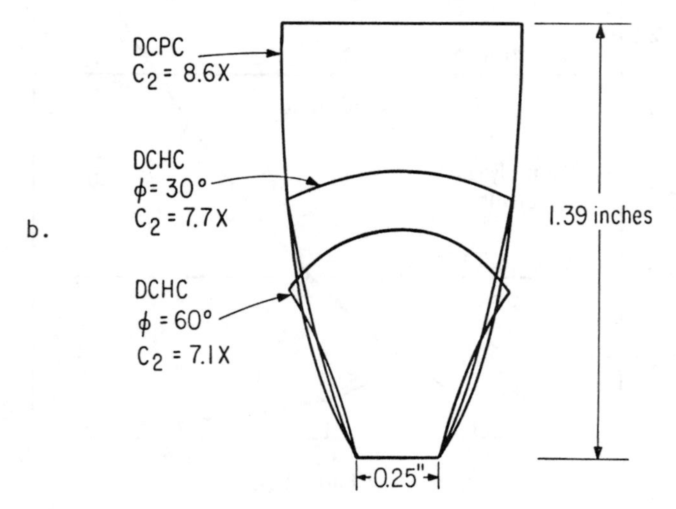

Figure 11. (a) A practical thermal secondary concentrator referred to as the "trumpet" which has been built and tested by our laboratory, and (b) profiles for a DCPC and two dielectric Compound Hyperbolic Concentrators, or DCHS's, used for photovoltaic secondaries.

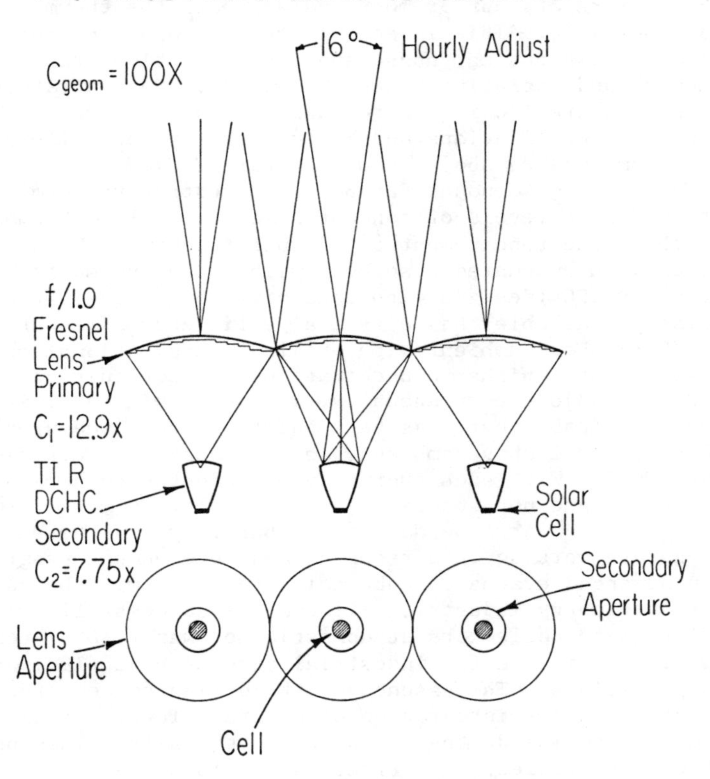

Figure 12. Illustration of a lens/DCHC combination which could provide 100:1 concentration while requiring only approximately hourly adjustment. Prototypes of actual devices based on this concept are presently under development.

collecting power of concentrating elements and systems. Designs which exceed the concentration attainable with focusing techniques by factors of four or more and approach the thermodynamic limit, are possible. This is accomplished by applying the concepts of Hamiltonian optics, phase space conservation, thermodynamic arguments, and radiative transfer methods. Concentrators based on this approach are finding increasing use in solar energy and a variety of other applications and include the by now well-known nontracking Compound Parabolic Concentrator (CPC).

Compound Parabolic Concentrators permit the use of low to moderate levels of concentration for solar thermal collectors without the requirement of diurnal tracking. When used in conjunction with evacuated absorbers with selective surfaces, a fully stationary CPC (designs with concentrations 1.1X-1.4X are now commercially available) has a typical efficiency of about 40% at 150°C (270°F) above ambient with available conventional materials. An experimental higher concentration 5X CPC requiring approximately 12 tilt adjustments annually, when used with a similar available evacuated absorber, has been built having a measured efficiency of 60% at 220°C above ambient, and is capable of efficiencies near 50% at 300°C. With such thermally efficient absorbers, higher concentrations are not necessary or desirable.

Argonne National Laboratory and the University of Chicago are working on a research and development program to develop an advanced evacuated tube collector that will be suitable for mass production by industry, will compete successfully with conventional flat-plate collectors at domestic hot water (DHW) temperatures, and will be suitable for industrial process heat (IPH) and/or cooling applications. The essence of the design concept for these new collectors is the integration of moderate levels of nonimaging concentration inside the evacuated tube itself. This permanently protects the reflecting surfaces and allows the use of highly reflecting front-surface mirrors with reflectances greater than 95%. Previous fabrication and long-term testing of a proof-of-concept prototype has established the technical success of the concept. Present work is directed toward the development of a manufacturable unit that will be suitable for the widest possible range of applications.

The temperature capabilities for CPC's with non-evacuated absorbers are somewhat more limited. However, with proper design, taking care to reduce parasitic thermal losses through the reflectors, non-evacuated CPC's will outperform the best available flat-plate collectors at temperatures above about 50°C-70°C.

Nonimaging near "ideal" type secondary or terminal concentrators have advantages for any solar concentrating application. In the near term, their use is being applied to increase the angular tracking, alignment, and slope definition requirements for the primaries in an effort to reduce system cost and complexity. In future applications, very high geometric concentrations which cannot otherwise be attained can be achieved through the use of these devices.

ACKNOWLEDGEMENT

Work described in this review was supported in large part by the United States Department of Energy under a number of grants and contracts including most recently Contracts DE-AC02-80-ER10575, DE-AC02-ER10558, DE-AC03-82SF11655, and DE-FG02-84CH10201.

REFERENCES

1. R. Winston, Principles of solar concentrators of a novel design. Solar Energy 16, 89 (1974).
2. Ari Rabl, Comparison of solar concentrators, Solar Energy 18, 93 (1976).
3. W. T. Welford and R. Winston, The Optics of Nonimaging Concentrators. Academic Press, New York (1978).
4. R. Winston, Light collection within the framework of geometrical optics, J. Optic. Soc. Am. 60, 245 (1970).
5. H. Hinterberger and R. Winston, Efficient light coupler for threshold Cerenkov counters, Rev. Sci. Inst. 37, 1094 (1966).
6. J. Ploke, Light collectors with high concentration efficiency, Optik 25, 31 (1967).
7. V. K. Baranov and G. K. Melnikov, Soviet Journal of Optical Technology 33, 408 (1966).
8. Configuration design illustrated is that for the commercial collector Model XE-300 manufactured by the Energy Design Corporation, Memphis, Tenn. 38134.
9. K. A. Snail, J. J. O'Gallagher and R. Winston, A stationary evacuated collector with integrated concentrator, Solar Energy 33, 441 (1984).
10. R. Winston, Ideal flux concentrators with reflector gaps, Appl. Opt. 17, 1668 (1978).
11. A. Rabl, J. O'Gallagher, and R. Winston, Design and test of non-evacuated solar collectors with Compound Parabolic Concentrators, Solar Energy 25, 335 (1980).
12. A. Rabl and R. Winston, Ideal concentrators for finite sources and restricted exit angles, Appl. Opt. 15, 2880 (1976).
13. W. T. Welford and R. Winston, Non-conventional optical systems and the brightness theorem, Appl. Opt. 21, 1531 (1982).
14. J. O'Gallagher and R. Winston, Test of a "Trumpet" secondary concentrator with a paraboloidal dish primary. Solar Energy (in press).
15. R. Winston, Dielectric Compound Parabolic Concentrators, Appl. Opt. 15, 291 (1976).

CHAPTER 24

CURRENT METHODS FOR THE DYNAMIC ANALYSIS OF HORIZONTAL AXIS WIND TURBINES

Robert W. Thresher, Principal Scientist
Solar Energy Research Institute
Wind Energy Research Center
1617 Cole Boulevard
Golden, CO 80401

ABSTRACT

Current methods for the dynamic analysis of horizontal axis wind turbines are explored. The paper first introduces the reader to typical wind turbine configurations and provides background on the current status of the wind industry. Blade element aerodynamic analysis is covered first, with insight into the current status of aerodynamic analysis methods. Next, the importance of resonant vibration analysis is discussed. Forced dynamic analysis is the final analysis area to be addressed. The dynamic analysis computer code FLAP (Force and Loads Analysis Program) is discussed in some detail to give the reader a clear idea of the level and complexity of blade dynamics. Current areas of research and advanced wind turbine designs are the final subjects considered. These are intended to give the reader some idea of how the technology will progress during the next five to ten years. In closing, the reason that lead researchers to believe that wind technology can be signficantly improved through a better knowledge of their dynamic behavior are discussed.

0094-243X/85/1350472-25$3.00 Copyright 1985 American Institute of Physics

AN INTRODUCTION TO WIND TECHNOLOGY (1)

The two basic approaches for converting wind to useful energy are vertical-axis wind turbines, in which the axis of the rotor's rotation is perpendicular to the wind stream and the ground; and horizontal-axis machines, whose axis of rotation is parallel to the wind stream and the ground. Both types are illustrated in Figure 1. They contain five basic subsystems: (1) a blade or rotor, which is the energy conversion device; (2) a drive train, usually including a gearbox and generator; (3) a tower that supports the rotor; (4) various turbine supporting systems including controls, electrical cables, etc.; and (5) "balance-of-station" subsystems, which, depending on the application, might include ground support equipment, interconnection equipment, etc.

Many variations in configuration are possible for both vertical- and horizontal-axis concepts. Several vertical-axis machines are available from U.S. manufacturers. They are similar in design and size, ranging from 100 to 300 kilowatts (kW). In contrast, there are over 50 horizontal-axis machines which vary widely in size, from several watts to 4 megawatts (MW), and in design configuration. For example, yawing mechanisms on horizontal-axis machines are designed to keep the rotor oriented properly in the wind stream. Some machines simply have a tail vane or rudder to control yawing motion;

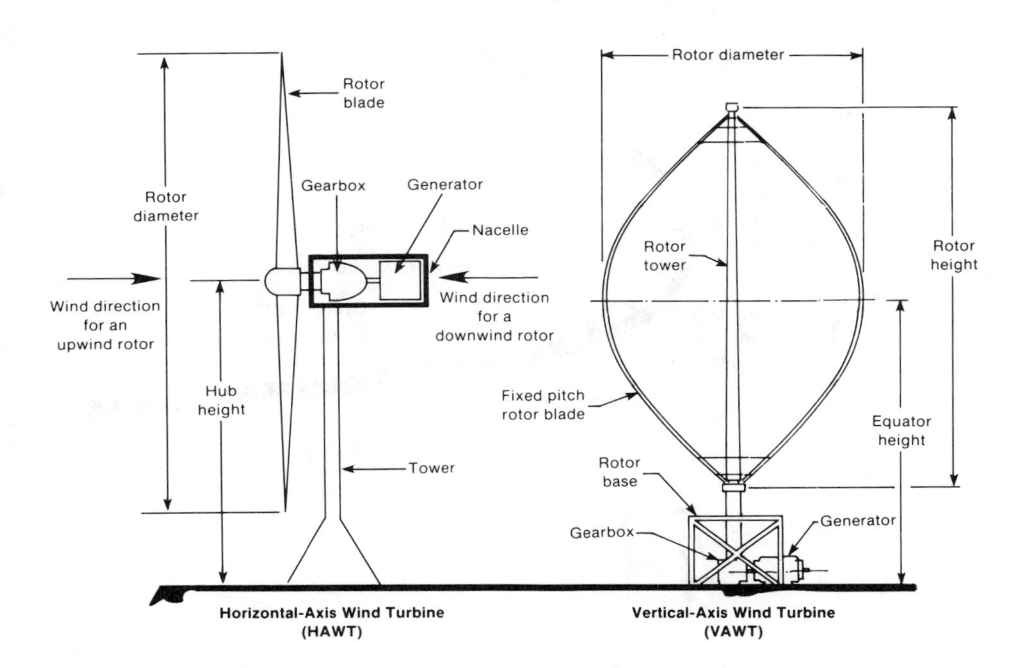

Figure 1. Wind Turbine Configurations

others (typically larger machines) have active (driven) yaw systems controlled by microprocessors. On some machines the blades are located upwind of the tower; others are downwind. Finally, some machines have fixed-pitch blades that reduce design complexity; others have variable-pitch blades that aid in starting and stopping and regulate power output by changing the angle at which the blades cut through the air.

Figure 2 illustrates the historical development of wind technology, beginning with rural applications like water pumping in the last century. In the first half of this century, wind turbines were also widely used for electricity generation in rural areas of the United States. Between 1850 and 1970, over six million small (less than one 1 kW) wind machines were installed in the United States. The largest and most advanced machine built during this period was the 1.25-MW

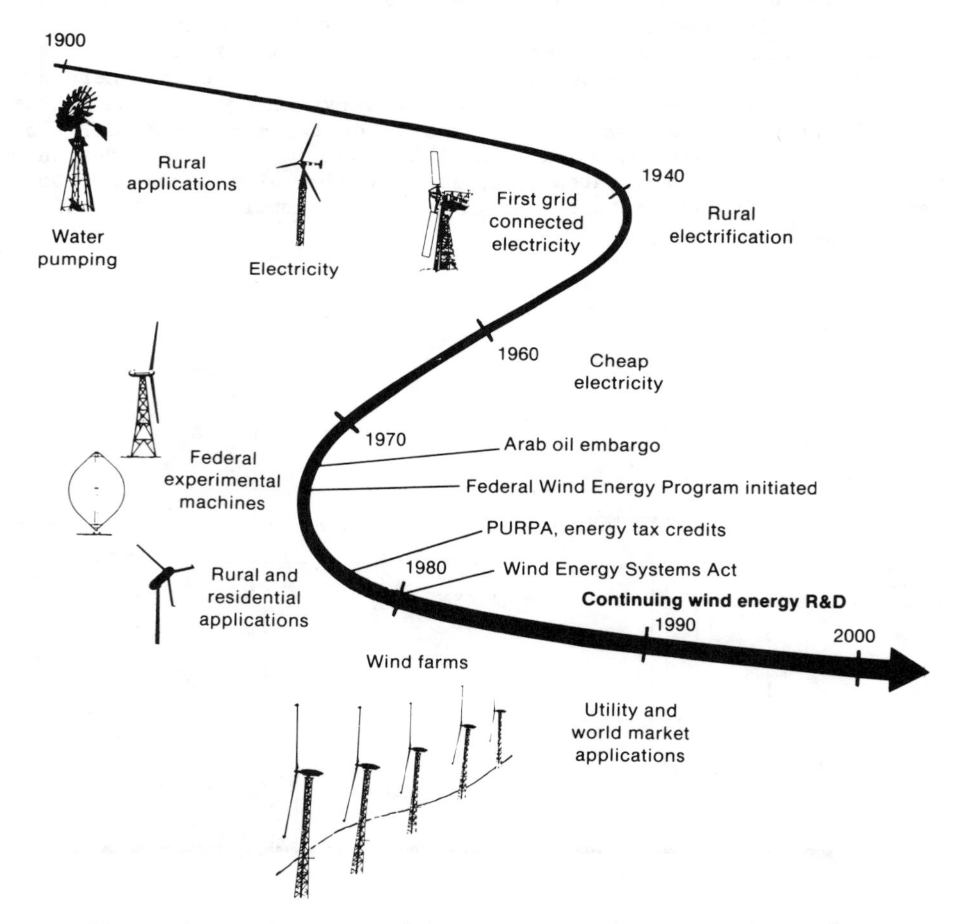

Figure 2. Historical Perspective on Wind Energy Technology in the United States

Smith-Putnam machine, which was completed in 1941. However, when the Rural Electrification Administration brought inexpensive utility service to farms and small villages in the 1930s and '40s, the U.S. market for wind turbines virtually disappeared. By 1973, there were only a few firms selling wind turbines in the United States. In 1974, the costs of conventional fuels began rising rapidly, and developing wind and other renewable energy technologies became a national priority. The Federal Wind Energy Program was initiated at that time.

The Federal Wind Energy Program stimulated research and activities that matured the technology, but in the late 1970s the federal government took two additional actions that accelerated the use of wind systems in electric utility systems. First, the federal government made tax credits available for renewable energy technologies, including wind turbines. Second, through the Public Utility Regulatory Policies Act of 1978 (PURPA) it required electrical utilities to buy power from owners of renewable energy projects at rates based on the utility's avoided (marginal) cost of energy.

The improvements in the technology, combined with PURPA and the tax credits, have resulted in a large market for grid-connected wind turbines. Most of the wind machines installed in the last three years have been in wind farms, where many machines are installed at a single site to generate power for sale to electric utilities. A 1984 survey estimated that more than 8000 units had been sold by the end of 1983, for a total capacity of more than 300 MW (2). Most of this capacity is in wind farms in California, as illustrated in Figure 3, where high avoided costs, excellent wind resources, and state and federal tax credits have stimulated an early market for wind machines.

Although the federal tax credits are scheduled to expire at the end of 1985, the current technology with PURPA is sufficient to make wind systems viable in regions with good wind resources and high avoided costs. In addition, there is the general feeling within the private sector that the technology can be significantly improved, making it competitive in a much wider market.

This paper examines the technical status of wind technology by reviewing the methods used for dynamic analysis, and by looking at the current focus of research to see where the technology can be improved further. The paper considers only the horizontal-axis wind turbines. There are many similarities between the horizontal- and vertical-axis technologies and they face common problems, but the detailed considerations are quite different.

For convenience, we will separate the fundamentals of wind turbine dynamic analysis into three elements:

1. Aerodynamic Analysis

Figure 3. Horizontal- and Vertical-Axis Wind Turbines in Wind Farms in California

2. Resonant Vibration Analysis

3. Forced Dynamic Analysis

These elements are considered and discussed in succession, including current areas of research activity. The paper concludes with a short discussion of possible configurations for advanced turbines. The above list of wind turbine dynamic analysis elements is clearly incomplete, since it ignores the control system and the electrical system dynamics. At this time, the control and electrical system dynamics are considered as separate problems and are not yet dealt with as an integrated design task.

AERODYNAMIC ANALYSIS

The primary method used for computing the aerodynamic forces that act on a wind turbine rotor is called "blade element theory" or "strip theory." The flow through the rotor disk plane is assumed to occur in noninteracting annular stream tubes, as shown in Figure 4. The flow enters the stream tube far upwind at velocity, V_w, and slows to velocity, u, at the rotor disk. It exits far downstream of the rotor disk at a further reduced velocity u_1. The rotor forces that act on the air moving through the rotor plane both slow the flow and impart a rotational motion.

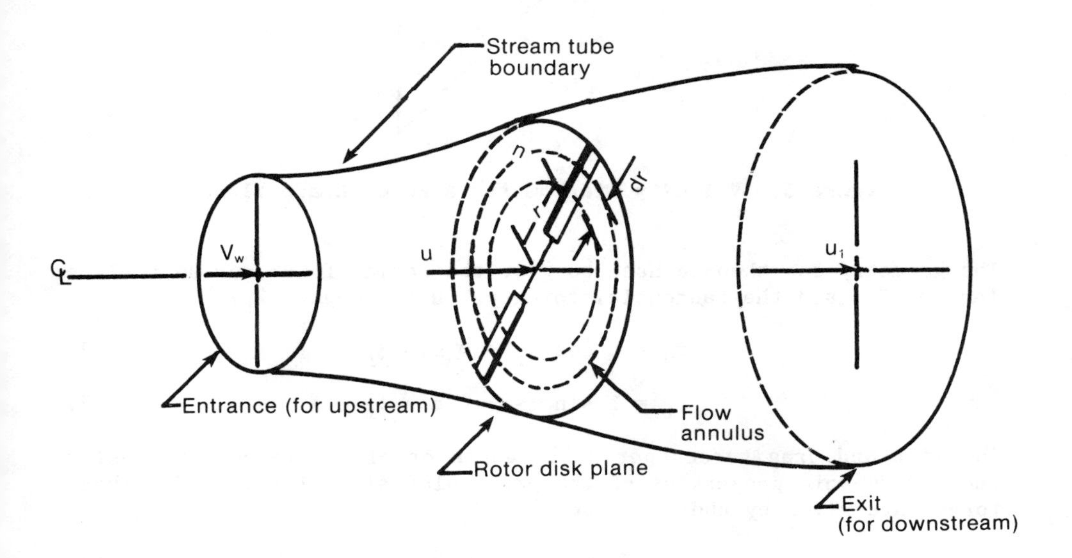

Figure 4. Stream Tube for Aerodynamic Analysis

Figure 5 is the velocity diagram for a blade element at some radial station r. The velocity $V_w(1-a)$ is the retarded velocity u of the flow through the rotor, while the quantity $r\Omega(1 + a')$ is the rigid body motion, $r\Omega$, combined with the swirl velocity of the fluid. The factors a and a' are aptly named the aerodynamic interference factors. The figure also shows the aerodynamic lift L, and the drag, D, per unit of blade length. The lift and drag per unit length of blade are also resolved into a force normal to the rotor disk plane F_n and the torque producing force F_t. The blade has a pitch setting θ, and the angle of the resultant velocity with respect to the blade element is ϕ_f. The angle of attack is then given by α.

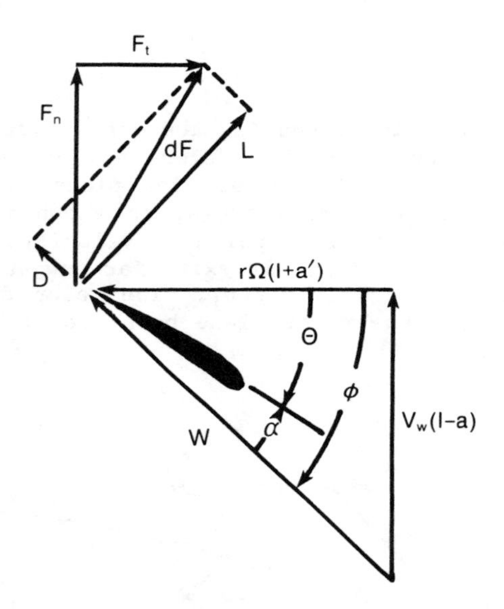

Figure 5. Velocity Diagram for a Rotor Blade Element

The geometry can then be used to give the normal force per unit blade length, F_n, and the tangential force per unit length, F_t:

$$F_n = L \cos \phi_f + D \sin \phi_f \tag{1}$$

$$F_t = L \sin \phi_f - D \cos \phi_f \tag{2}$$

The lift and drag forces per unit length of blade are computed using the aerodynamic properties of the particular airfoil involved. These forces are given by the relationships

$$L = C_L \frac{1}{2} \rho c W^2 \tag{3}$$

$$D = C_D \frac{1}{2} \rho c W^2 \tag{4}$$

where C_L and C_D are the dimensionless lift and drag coefficients, ρ is the air density, c is the blade chord, and W is the magnitude of the relative velocity vector from Figure 5.

The interference factors a and a' are computed from momentum considerations. The fluid momentum in the flow direction, for an annulus of width dr, is balanced by the combined force of n blades to give

$$dF_{fluid} = \rho(2\pi r dr)\, u\, (V_w - u_1) = nc\, \frac{1}{2}\, \rho W^2\, F_n\, dr = dF_{blades}\,. \quad (5)$$

Using the assumption that the overall retardation of the flow between the entrance and exit in the stream tube is twice the value between the entrance and the rotor, i.e., $u_1 = V_w(1-2a)$, the geometric relationships of Figure 5 give the equation

$$\frac{a}{1-a} = \frac{nc}{8\pi r \sin^2\phi_f}\left(\frac{F_n}{\frac{1}{2}\rho c W^2}\right). \quad (6)$$

In a similar manner, angular momentum can be conserved to develop a relationship for the swirl velocity. This is written as

$$\frac{a'}{1+a'} = \frac{nc}{4\pi r \sin 2\phi_f}\left(\frac{F_t}{\frac{1}{2}\rho c W^2}\right). \quad (7)$$

The interested reader is referred to Wilson and Lissaman (3) for the details of the development. To determine the flow state at the rotor and to compute the forces acting on the rotor, these two nonlinear equations must be solved iteratively for all radial stations of interest along the rotor blade. Almost all computational schemes to do this are modifications of a computer code called PROP originally developed by Wilson and Lissaman (3) in 1974.

Modifications to this computation scheme accounting for the change in wind velocity as a function of height have been made to account for the influence of wind shear in the atmospheric boundary layer. For rotors positioned downwind of the tower it was quickly recognized that the "tower shadow" region had a significant effect on the blade aerodynamic forces. More recently, because of their importance, dynamic effects have been the most active research area, as fatigue problems have been discovered in wind farm turbines and on some test machines. In the foregoing analysis it was assumed that the inflow was steady, but field experience indicates that the turbulence inputs are significant and that the flow field is quite variable.

The aerodynamic properties for a particular airfoil are usually given as curves of lift coefficient, C_L, versus angle of attack, and drag coefficient, C_D, as a function of lift coefficient. Figure 6 shows this type of data for a popular airfoil used in wind turbines. Notice that the lift coefficient for smooth airfoils varies linearly with angle of attack over the range from -8° to 10°, but the lift coefficient drops abruptly for angles of attack above 16°. This abrupt drop is called stall, and occurs because of separation of the flow from the airfoil. Most of the airfoils in current use on wind turbines were designed for aircraft, where stall can be avoided. For wind turbines, it is impossible to avoid stall at inbound locations near the hub; at high wind speeds the region of stall will spread outward so that a large portion of the blade may actually be stalled. Some turbine designs depend on stall to regulate power output at high wind speeds. For this reason, the aerodynamic properties of stall are quite important. Recently there has been considerable interest in designing special purpose airfoils for wind turbines, where the stall behavior has been tailored so that power output is regulated while maintaining a high overall energy capture. In addition, there is interest in dynamic stall. During dynamic stall the angle-of-attack varies rapidly and the sectional lift and drag coefficients can be quite different from the static values of Figure 6. Dynamic stall will be discussed again later.

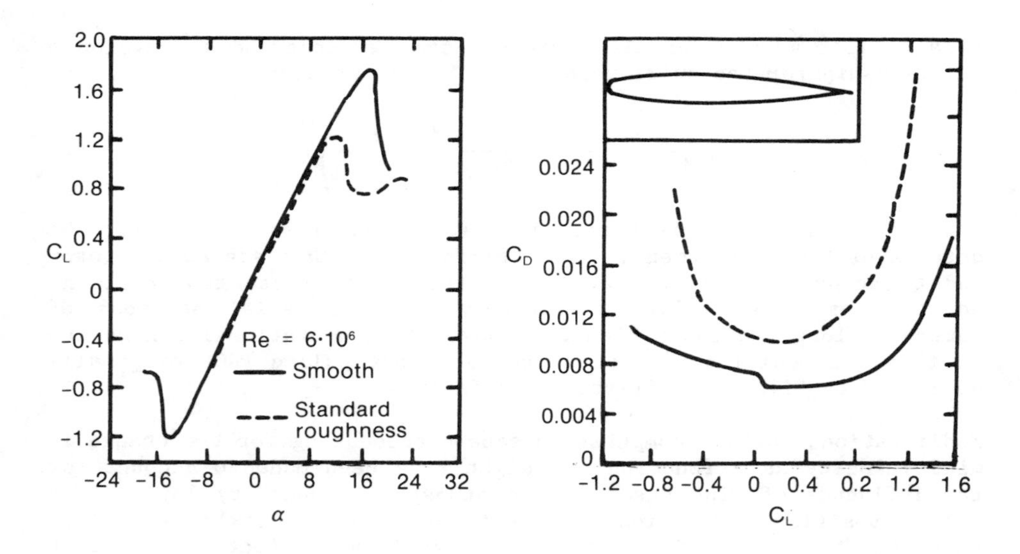

Figure 6. Lift and Drag Curves for a NACA 23012 Airfoil

Using computer codes, like PROP, the power output for typical classes of wind turbines can be computed. The conversion efficiency of a wind turbine is defined by its power coefficient, given by

$$C_p = \frac{(\text{power output})}{\frac{1}{2}\rho A V_w^3} , \quad (8)$$

where ρ is the air density, A is the swept area, and V_w is the wind speed.

Figure 7 illustrates the power coefficient, C_p, versus tip speed ratio, which is the blade tip speed divided by the wind speed. Both the modern horizontal-axis wind turbines and the vertical-axis wind turbines have measured power coefficients that exceed 0.4. In general, the greater the number of blades the lower will be the tip speed ratio at which peak C_p occurs. The reader should also note that no machine has a C_p greater than $16/27 \simeq 0.6$. This is the maximum possible extraction efficiency, called the Betz Limit, which is based on a fairly reasonable set of constraining assumptions concerning the flow field. Power coefficient, while important to the aerodynamicist, is only one of a number of important parameters associated with wind power. The wind power neophyte should note that the power output is a function of the wind velocity cubed, assuming a constant C_p in Eq. 8. In addition, the total thrust force on the wind turbine can be shown to be a function of the wind velocity squared.

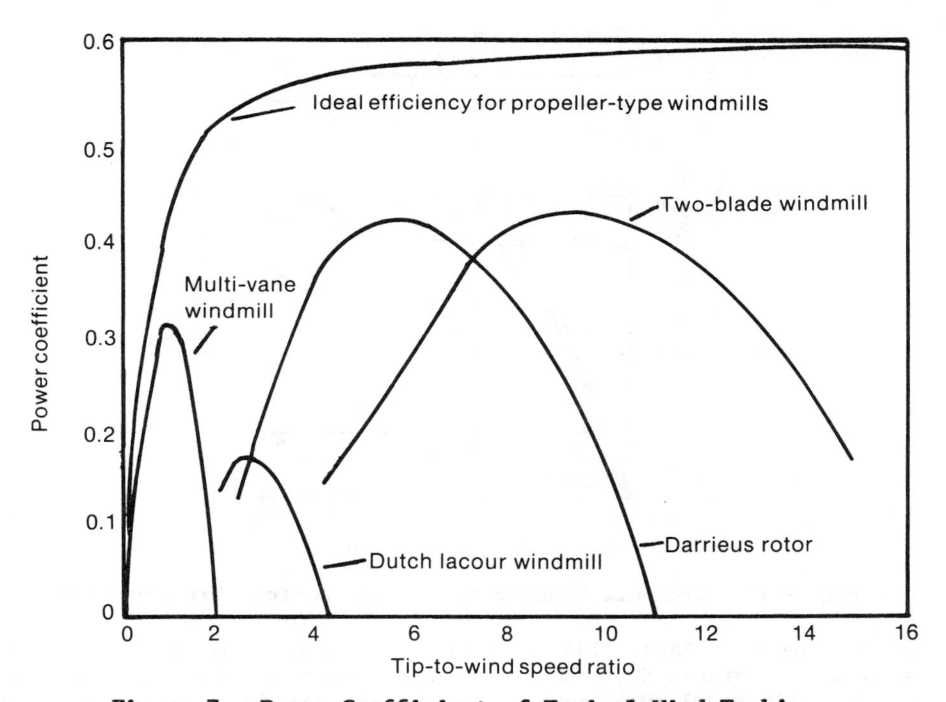

Figure 7. Power Coefficient of Typical Wind Turbines

RESONANT VIBRATION ANALYSIS

Resonance occurs in a mechanical system when a disturbing force excites the system with a frequency approximately equal to one of its natural frequencies. The resulting system motion usually has an amplitude much larger than if the system were forced at some other frequency. In wind turbines as with other rotating machinery, the exciting forces are generated by the rotating structure and are then transmitted to the fixed structure at frequencies that are integer multiples of the rotation rate. A common way to present natural frequency data and to look for possible resonances is to plot a Campbell diagram. The Campbell diagram consists of a plot of the natural frequencies of the system versus the rotor speed, and a second set of star-like straight lines that pass through the origin and express the relationship between the possible exciting frequencies and the rotor speed. The slopes for these straight lines are selected as an integer number of oscillations per revolution of the rotor. Because it is expected that the excitation frequencies will always be integer multiples of the rotor speed, the intersection of one of the straight lines with one of the natural frequency curves indicates a potential for resonant vibration near the rotor speed of the intersection point. Figure 8 is a Campbell diagram for a hypothetical wind turbine.

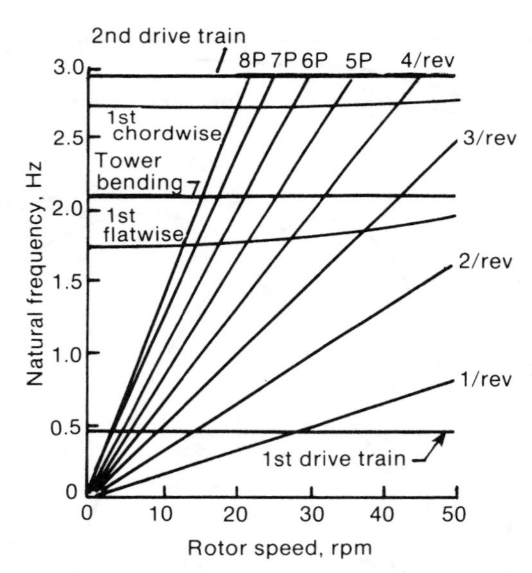

Figure 8. Campbell Diagram for a Hypothetical Wind Turbine

Another way to present this same information is to make a tabulation as shown in Figure 9. In this presentation of the Department of Energy (DOE) Mod-1 wind turbine system natural frequencies, the per revolution frequencies of various important motions are tabulated in columns. Regions of possible resonance near integer multiples of the rotation speed have been designated as regions to be avoided. It is expected that a two-bladed rotor will only transmit forces that are even multiples of the rotor speed, which is the reason that the odd multiples have not been designated as avoidance areas for the fixed structure. However, experimental measurements on both large and small field test turbines have clearly demonstrated that the rotor does, in some cases, transmit forces at the odd harmonics of rotor speed. Although the reasons for this have not been fully explained, it has generally been assumed that either the rotor was not perfectly balanced or the aerodynamic properties of the two blades are different. Until the reasons for these force inputs are explained, it is probably prudent to avoid all of the integer harmonics of the rotor speed.

Figure 9 presents both the calculated system frequencies and the experimentally measured frequencies, which were obtained after the experimental Mod-1 was built by DOE. In general, the agreement is quite good, and where differences do occur they are attributed to modeling assumptions that were not met in the hardware. For a complete discussion of the results presented in Figure 9, the interested reader is referred to Sullivan (4).

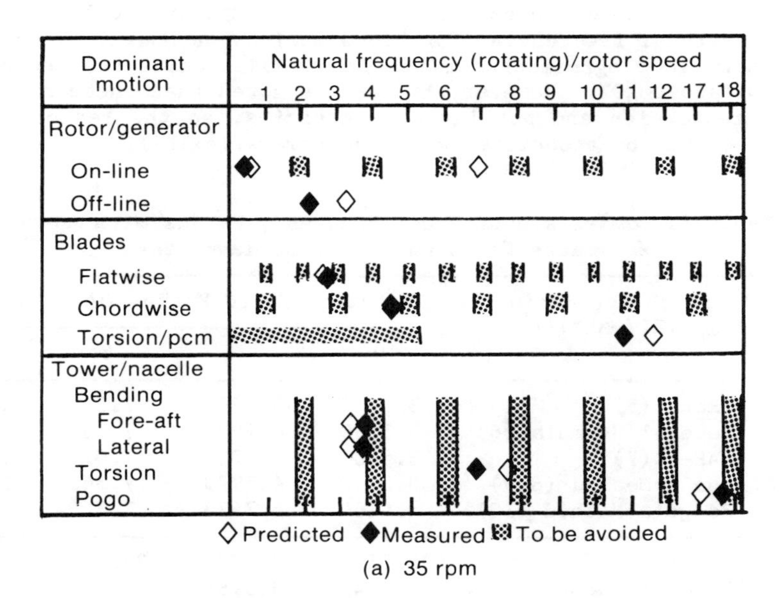

Figure 9. MOD-1 System Natural Frequencies

To estimate the natural frequencies for the rotor blades, the influence of the centrifugal forces and the resulting stiffening of the blades must be considered. The design of helicopter rotors as well as other long, flexible rotating blades has motivated a number of studies on the vibration of rotating beams. Analysis leads to a variable-coefficient differential equation for the mode shapes, which is generally solved using an approximate technique. Recently,

Wright (5) developed a power series solution, which gives precise values for the natural frequencies and the mode shapes of centrifugally stiffened beams. Although Wright's solution is not practical for design work, it does provide criteria with which other approximate solutions can be evaluated. Table 1, taken from Wright's paper, provides a comparison for a number of approximate methods with the exact power series solution for the simple case of a uniform cantilever beam without hub offset at several rotation speeds. Included in this comparison are results obtained using SAP V where the beam was modeled using 10 beam elements and the geometric stiffness option to account for rotational effects. Although SAP V is not highly accurate, it does give reasonable results. It is widely available at a nominal cost and can be used to model highly complicated stiffness and mass distributions. Perhaps the most notable conclusion from this comparison is that all of these approximations provide fairly accurate results over a wide range of rotation speeds, even for the fourth mode. At this time there are very little field test data with which to make an experimental comparison; however, Figure 9 does have three data points comparing calculated and measured natural frequencies for blade motions on Mod-1. These three points show good agreement for the first mode, while the second mode is off by about 20%. However, it must be noted that these are really system frequencies and not blade frequencies, so the reason for the differences may be associated with the tower flexibility.

Table 1. Comparison of Exact Frequency Ratios with Approximate Estimates for a Uniform Cantilever Beam

Mode	Frequency Ratio $\omega_n/\{EI/\rho AL^4\}^{1/2}$	Rotation Speed Ratio $\Omega/\{EI/\rho AL^4\}^{1/2}$			
		0	3	6	12
1	Exact (5)	3.5160	4.7973	7.3604	13.1702
	Peters' Formula (6)	3.5160	4.7964	7.3561	13.1634
	SAP-V (7)	3.500	4.908	7.629	13.65
	Hoa's Method (8,9)	3.5160	4.7973	7.3604	13.1706
	Hodge's Method (10)	3.5160	4.7973	7.3604	13.1702
2	Exact	22.0345	23.3203	26.8091	37.6031
	Peters' Formula	22.0345	23.3188	26.7933	37.5155
	SAP-V	21.69	23.18	27.17	39.27
	Hoa's Method	22.0352	23.3210	26.8098	37.6050
	Hodges' Method	22.0345	23.3203	26.8091	37.6031
3	Exact	61.6972	62.9850	66.6840	79.6145
	Peters' Formula	61.6973	62.9720	66.5421	78.7188
	SAP-V	60.12	61.55	65.65	79.88
4	Exact	120.902	122.236	126.140	140.534
	Peters' Formula	120.901	122.194	125.759	138.322
	SAP-V	116.6	118.0	122.2	137.6

In general, it seems that natural frequency computations are performed using a variety of techniques, with good computational accuracy when compared to exact results, but these results are only good for the lower natural frequencies when compared with field test data from actual hardware.

FORCED DYNAMIC ANALYSIS

The primary analysis tool used to determine the forced response of wind turbine systems is any one of a number of dynamic analysis computer codes. These codes have their analytical origins in the field of helicopter dynamics. They can be separated into two categories, rotor codes and system codes. Rotor codes are used to investigate blade dynamics and usually assume that the blade hub is fixed in space. System codes account for the flexibility of the tower, and often model the important actions of the electrical and control system. The system code is much more complex and expensive to use. It would take about two years to become proficient in the use of a system dynamics code. The smaller turbines, like those shown in Figure 3, are generally designed using rotor codes, while the large multimegawatt turbines have been designed using system codes. In this paper, we consider only the simpler rotor codes. They provide the essentials of the analysis method, without the complexity of all the subsystems that are elements of the system codes. The reader interested in looking into systems codes further should see References 11,12,13, and 14.

The rotor code to be discussed here is called FLAP (Force and Loads Analysis Program). The code allows only flapping motion of an individual wind turbine blade. It accounts for the blade-bending deformation about the smallest blade inertial axis. The rotor is assumed to rotate at a constant speed, but the hub is allowed to move in a prescribed yawing motion. Rotors that are tilted and yawed relative to the mean wind direction can be accommodated in a straightforward manner.

The model and the code are designed to operate with aerodynamic models of varying sophistication. Currently the model is configured to include the effects of mean wind, wind shear, and tower shadow. The code is structured such that time-dependent turbulent wind fluctuations can be added in the future. The rotor blade flapping motion is represented by a set of coordinate shape functions that are simple polynomials. Four functions are included in the computer code, but any number of the functions can be used, from only one up to the maximum of four. At present, cantilever blade attachment conditions and a teeter hinge attachment have been implemented in the code.

The current version for the aerodynamic model uses a quasi-steady linear aerodynamic model to compute the blade aerodynamic forces. However, the code has been designed to use more sophisticated aerodynamic force models, including time-dependent aerodynamics such as those involved in dynamic stall computations.

The model operates in the time domain, and the blade acceleration equation is integrated via a modified Euler trapezoidal predictor-corrector method. The method involves the use of a set of low order relations, is self-starting and stable, and allows frequent step size changes. The procedure is entirely automated within the computer program. Results of the blade loads analysis are printed in tabular form, and include the deflection, slope, and velocity, the flapwise shear and moment, edgewise shear and moment, blade tension, and blade twist moment for any point along the blade axis.

The program, written in FORTRAN V, is in the public domain and was developed for easy end-user modification and customization. A substantial effort has been made to make the actual code contain its own documentation through extensive use of comments within the program.

In order to understand the type and level of analysis that is involved in the FLAP code a brief description of the model follows. Figure 10 shows the orientation of the turbine blade under analysis with all the intermediate coordinates required to represent the blade motion. The capital X, Y, Z coordinates are the fixed reference system. The mean wind velocity at the hub, V_{hub}, and its fluctuating components, δV_X, δV_Y, and δV_Z, are given in this system. The rotor spin axis is allowed to tilt through a fixed angle χ and the rotor is allowed to have a prescribed time-dependent yawing motion given as $\phi(t)$, where ϕ is the yaw angle. The yaw axis is coincident with the Z coordinate axis. The hub, located a distance "a" from the yaw axis, is considered to be rigid and to have some radius h. The flexible portion of the blade begins at the outer hub radius, h. The airfoil shape may begin at h or at some position farther out along the blade z axis. The blade is coned at some angle β_o as shown in the figure.

The x,y,z coordinates are located in the surface of revolution that a rigid blade would trace in space, with the y axis normal to this surface. The x_p, y_p, z_p are the blade principal bending coordinates, where the z_p axis is coincident with the elastic axis of the undeformed blade. Bending takes place about the x_p coordinate. It is further assumed that the blade principal axes of area inertia do not change along the z_p axis. The influence of blade twist on bending displacment is neglected. The orientation used to set the angle θ_p for computations is the principal axis near the blade tip, because the deformation is largest there. The final coordinate system is the η, ζ, ξ system, which is on the principal axes of the deformed blade at some point along the elastic axis.

The blade is assumed to be a long, slender beam so that the normal strength of material assumptions concerning the bending deformation are valid. It is assumed that the blade bends only about the weakest principal axis of inertia; in the figure this is the x_p axis. The one-dimensional strength of materials bending assumption then gives the following moment curvature relationship:

$$M_{x_p} = - EI_{x_p} \frac{d^2v}{dz_p^2} , \tag{9}$$

where M_{x_p} is the bending moment about the blade principal axis of inertia x_p, E is the elastic modulus, I_{x_p} is the area moment of inertia about the x_p axis, and v is the bending displacement in the y_p direction.

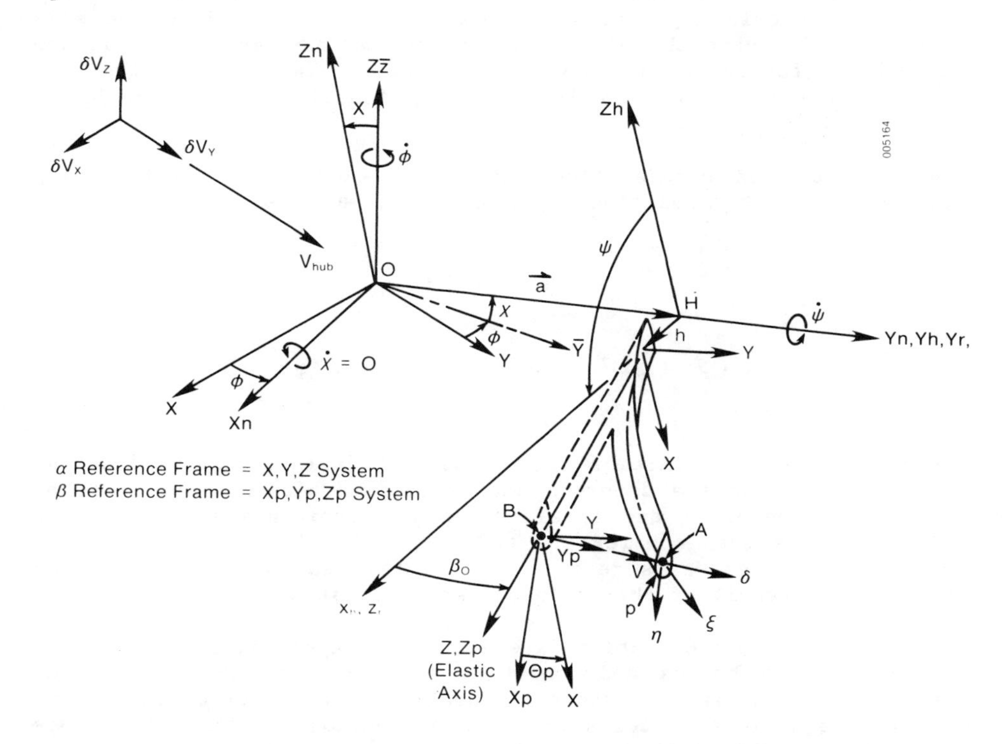

Figure 10. Illustration of the Rotor System Coordinates with Positive Displacements and Rotations Shown

The combined wind effects of wind shear and tower shadow are written in the form

$$V_{uu}\,(r,\Psi) \simeq V_{hub}\,\{1 + W_s\,(r,\Psi) - T_s(\Psi)\}\ , \tag{10}$$

where $W_s(r,\Psi)$ is a function that accounts for the variation in mean wind across the rotor disk (wind shear), and $T_s(\Psi)$ is a function that accounts for the wind velocity deficit caused by the interference of the tower.

Although the figure shows fluctuating wind components, given as $\delta V_x, \delta V_y, \delta V_z$, these turbulence excitations are not modeled in the code. It is clear from recent field studies of experimental turbines that wind turbulence plays an important part in the input. However, no validated approaches for simulating this input are yet available.

The flow at the rotor disk plane is further modified by the action of the aerodynamic lift force. This tends to slow the flow as discussed in the section on aerodynamic analysis. The effect of the aerodynamic forces is determined by computing the interference factors a and a' as described previously. Often these computations are simplified by ignoring the a' factor, which does not greatly influence the bending forces acting on the rotor blade. In addition, the lift coefficient is assumed to be a linear function of the angle of attack.

The velocity and acceleration analyses make use of the following classical kinematic relationships, written symbolically as

$$_\alpha\vec{V}_A = {}_\alpha\vec{V}_B + {}_\beta\vec{V}_A + \vec{\Omega} \times \vec{R}_{A/B} \tag{11}$$

and

$$_\alpha\vec{a}_A = {}_\alpha\vec{a}_B + \dot{\vec{\Omega}} \times \vec{r}_{A/B} + 2\vec{\Omega} \times {}_\beta\vec{V}_A + \vec{\Omega} \times \left(\vec{\Omega} \times \vec{r}_{A/B}\right) + {}_\beta\vec{a}_A \; . \tag{12}$$

The α and β reference frames are defined in Figure 10, as well as the points A and B on the deformed and undeformed blades. The computation of the velocity and acceleration for arbitrary points on the rotor blade is both complex and tedious, and will not be discussed. The interested reader is referred to Thresher and Hershberg (15) for the detailed result of these linearized computations.

The blade equations of motion are derived by applying the principles of Newtonian mechanics and employing a Galerkin analysis to reduce the resulting partial differential equation to a set of ordinary differential equations in terms of model deflections at the rotor blade tip. This results in the following set of expressions for the blade tip modal flapping acceleration, $\ddot{s}_k$, as follows:

$$\ddot{s}_k M_{k\ell} = -s_k K^B_{k\ell} - s_k \left\{ \Omega^2 K^\Omega_{k\ell} + 2\Omega s\theta_p \dot{s}_n K^c_{nk\ell} - c\phi g K^g_{k\ell} \right\}$$

(Bending) (Tension Stiffening)

$$- c\theta_p \left\{ \Omega^2 \beta_o + 2\Omega\dot{\phi} c\phi + \ddot{\phi}\, s\Psi \right\} M^R_\ell$$

(Rigid Body Motion)

$$+ \quad s_k \Omega^2 s\theta_p{}^2 K^q_{k\ell}$$

(inertia Moment Stiffening)

$$+ \quad \Omega^2 s\theta_p c\theta_p M_\ell^B \quad + \quad F_\ell^a$$
(Blade Imbalance) (Aero Force)

$$+ \quad s_k \Omega^2 s\theta_p^2 M_{k\ell}$$
(Inertia Force Stiffening)

$$+ \; g\{-\chi c\theta_p + s\theta_p s\psi + \beta_o c\theta_p c\psi\} M_\ell^g \tag{13}$$
(Gravity Loads)

In these equations, the s_k's are blade modal deflections at the tip. The capital K's are stiffness coefficients, while the capital M's are mass coefficients, and F_ℓ^a's are the generalized aerodynamic forces. In addition, a short-hand notation for the sine and cosine of the blade azimuth angle and pitch angle has been used. The various coefficients are given by

$$\sin \psi = S\psi \qquad \cos \psi = C\psi$$

$$\sin \theta_p = S\theta_p \qquad \cos \theta_p = C\theta_p$$

$$M_{k\ell} = \int_o^L m\gamma_k\gamma_\ell dz \qquad K_{k\ell}^\Omega = \int_o^L T^\Omega(z_p)\gamma_k'\gamma_\ell' dz_p$$

$$M_\ell^R = \int_o^L m(h + z_p)\gamma_\ell dz_p \qquad K_{k\ell}^q = I_{\eta\eta}^m(L)\gamma_k'(L)\gamma_\ell'(L) - \int_o^L I_{\eta\eta}^m(z_p)\gamma_k'\gamma_\ell' dz_p$$

$$M_\ell^B = \int_o^L e_\eta m\gamma_\ell dz_p \qquad K_{nk\ell}^c = \int_o^L T_n^c(z_p)\gamma_k'\gamma_\ell' dz_p$$

$$M_\ell^g = \int_o^L m\gamma_\ell dz_p \qquad F_\ell^a = \int_o^L dA_\zeta\gamma_\ell dz_p$$

$$K_{k\ell}^B = \int_o^L EI_{x_p}\gamma_k''\gamma_\ell'' dz_p \qquad T^\Omega(z_p) = \int_{z_p}^L m(\xi)(h + \xi)d\xi$$

$$T_n^c(z_p) = \int_{z_p}^L m(\xi)\gamma_n(\xi)d\xi \qquad T^g(z_p) = \int_{z_p}^L m(\xi)d\xi \tag{14}$$

The γ_k are the assumed coordinate shape functions for the rotor blade deformation, which must at a minimum match the kinematic boundary conditions at the hub attachment. In addition, m is the mass per unit length of the blade; h is the hub radius, which is assumed to be rigid; e_η is the location of the mass center of the blade with respect to the elastic axis; E is the elastic modulus and I_{x_p} is the bending inertia about the x_p axis; and $I^m_{\eta\eta}$ is the mass moment of inertia per unit of the blade about the x_p axis; dA_ζ is the aerodynamic force per unit length normal to the blade's weakest bending axis.

Referring to Figure 10, Ω is $\dot{\psi}$, β_o is the blade precone angle, and χ is the tilt of the rotor shaft, while ϕ is the yaw angle. The descriptive phrase under each term of the equation is intended to give the reader a general feel for the forces acting on the spinning blade. A complete development of the above equations is given in Reference 15.

The computer solution of the equations of motion and the computation of the resulting displacments and loads require a sophisticated interactive program capable of performing a variety of tasks, including input and output of data and results, matrix inversion, time domain analysis, and the computation of spatially dependent blade properties and aerodynamic factors. To compare the predicted results with experimental data, a particular turbine system was modeled. The turbine modeled was a three-bladed, downwind system with a rotor diameter of 33 ft. The rotor rotation speed was 72 rpm and the blade first natural frequency was about 3.95 Hz, or about 3.3 times the rotor passage frequency. The blade has a constant chord of 18 in., no twist, and is coned at 3.5^o. The results for one particular case are shown in Figure 11.

The comparison results in this case are for a turbine with a relatively stiff blade. The flapping motions were small, and neither control system effects nor tower motion effects were significant. Until the FLAP code is verified against other more flexible rotor systems it cannot be considered a validated computation procedure. Additional comparison cases are given in Reference 16.

RESEARCH ON ADVANCED ANALYSIS METHODS

The insightful reader may have recognized that the analysis methods discussed have been quasi-steady in nature. We have considered dynamic processes, but the wind inputs, as well as the aerodynamic responses, have been assumed to be independent of time. The only

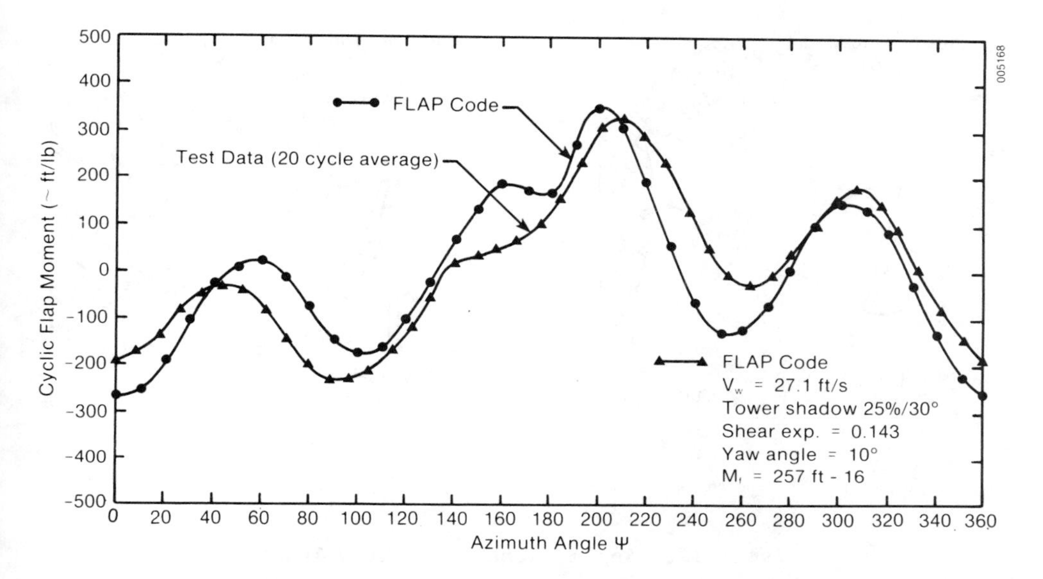

Figure 11. Best Prediction of Flap Moment at 20% Span at 18 mph

dynamics involved was due to the rotation of the rotor blade through the steady spatially varying atmospheric boundary layer, and the disturbed flow near the tower. This causes a rotor blade flapping response that is the same for every rotor revolution, after the start-up transients decay. In particular, the influence of turbulence and dynamic stall have not been considered. Note that the comparison with experimental data for the FLAP code in Figure 11 is for 20 revolutions averaged together, which removes the cycle to cycle variations.

Figure 12 from reference 1 illustrates the flow that a wind turbine rotor would see under steady flow condition and for turbulent wind inputs. During turbulent inflow condition, the wind field varies spatially and is time dependent. The wind speed can change, causing aerodynamic stall over portions of the rotor disk. The stall regions would be both a function of time and blade position. The stall could appear and disappear more or less randomly. Figure 13 shows how the lift coefficient changes under conditions of static stall and dynamic stall. For static stall, when the angle of attack becomes large, the flow over the airfoil separates and the lift coefficient begins to decrease with increasing angle of attack. In addition, the drag coefficent grows rapidly. In dynamic stall, when the angle of attack is changed rapidly, the separation is delayed and the lift coefficient continues to increase with increasing angle of attack, so that when the flow does stall the decrease in lift coefficient is abrupt. The influence of this type of behavior on structural loads is obviously damaging. It could greatly increase fatigue damage rates.

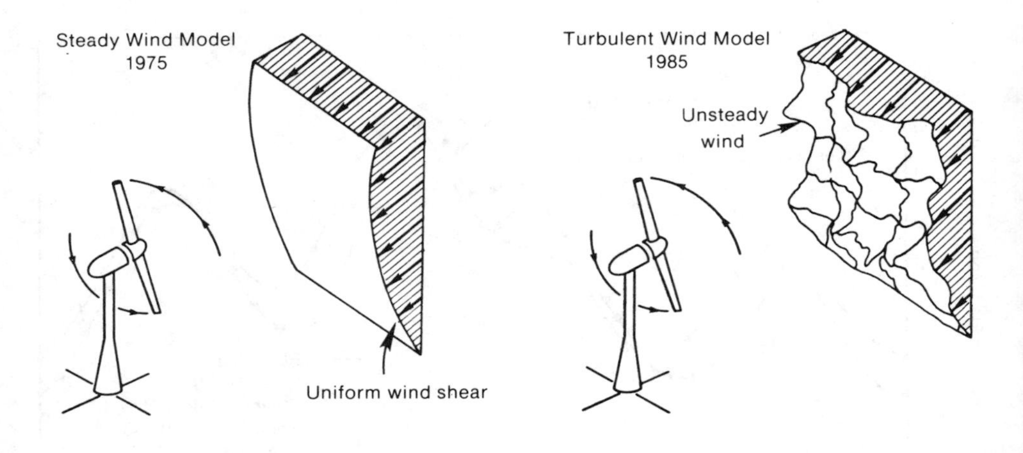

Figure 12. Representation of Wind Inputs

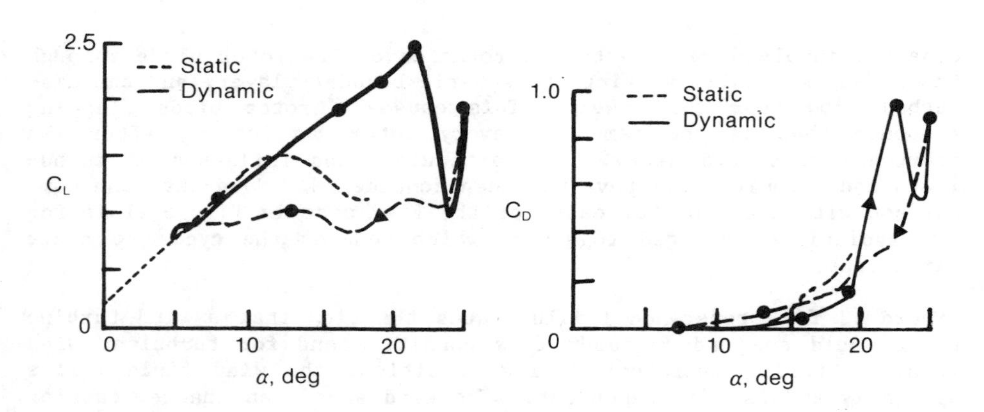

Figure 13. Effect of Airfoil Section Oscillation on Section Lift and Drag Coefficients; VR - 7 airfoil; α = 15 + 10 sin $\omega\tau$ degrees

There has been only limited study of the influence of turbulence on wind turbine dynamic respones. The modeling is quite complex due to the need to stochastically represent the wind field in both time and space. The interested reader may wish to review the studies by

Connell (17,18,19), and the modeling of Sundar and Sullivan (20) and Thresher and Holley (21,22). Dynamic stall has been studied by McCroskey (23) for helicopter application, but investigations for wind turbines in the presence of turbulent in-flows has not been accomplished.

Early field experience with the large multimegawatt wind turbine sponsored by DOE, and with the smaller wind farm turbines indicates that the fatigue damage rates have been larger than expected. The field test experience with the large MOD-2 wind turbines is documented in references (24,25). In general, the cyclic fatigue loads have not been well predicted by any dynamics codes, and investigators have generally attributed this short-coming to neglecting the turbulence inputs. For the smaller wind farm turbines, the situation is not well documented, but subjectively judged to be quite similar. Issues that are currently under investigation include load phasing (the timing and magnitude of loads on different elements of the structure), load magnitude (the incremental change in a load caused by a turbulent wind fluctuation including dynamic stall), cycle counting (the development of methods for properly accounting for load cycles and their magnitude), and damage laws (determining how much fatigue damage result from a particular sequence of load cycles). With an improved understanding of these dynamic processes it is felt that lighter, more cost-effective wind turbines can soon be designed.

ADVANCED WIND TURBINES

Cost is the key to wind turbine design, but, as in the past, reducing dynamic loads will be an important aspect to improved turbine designs. Although there will never be complete agreement on the best advanced configuration, Table 2 presents some of the more promising options. From this list, it is clear that greater flexibility is viewed as desirable.

Table 2. Possible Configuration Changes for Advanced Turbines

Configuration Change	Reasons
Special airfoils	To control dynamic stall, reduce loads and and noise, as well as improve performance
Tetered hub	To reduce dynamic loads
Passive pitch control	To simplify the machine and reduce costs
Upwind soft yaw	To reduce loads and noise, and improve heading control
Variable speed rotor	To increase energy capture and reduce rotor and tower loads
Soft tower and rotor	To reduce cost and reduce loads

Although soft, flexible structures are less costly to build because they contain less material, it is much more difficult to predict their dynamic response due to the greater possibility of encountering aeroelastic interactions. The computer analysis capabilities previously outlined will need to be improved substantially to account for the more flexible structures. In addition, the natural frequencies are lower for softer systems and turbulence excitations become more energetic at lower frequencies, adding to the complexity of the dynamics problem.

THE OUTLOOK FOR WIND ENERGY

The future of wind energy as a commercial source of electricity in a wide market depends upon greatly reducing cost and simultaneously achieving gains in performance and reliability. Achieving these goals is a significant technical challenge. It will require an improved understanding of the atmospheric excitations and the aerodynamic and structural dynamic response, as well as the resulting fatigue damage. In addition, this improved understanding of the physical processes must result in advanced wind turbines that are not only cheaper, but more cleverly designed so that much of the structural loading is avoided or negated before it contributes to fatigue damage. With these accomplishments, wind energy could become economical on a significant scale, but without them it may forever remain a marginal energy source, that occasionally enjoys economic favor depending on the pricing for competing sources.

REFERENCES

1. This section condensed from the Department of Energy, Wind Energy Technology Division, "Five-Year Research Plan," DOE/CE-T11, U.S Department of Energy, Washington, DC, January 185.

2. Nelson, V., "A History of the SWECS Industry in the U.S.", Alternative Sources of Energy, No. 66, March/April, 1984.

3. Wilson, R. E., and Lissaman, P. B. S., Applied Aerodynamics of Wind Power Machines, Oregon State University, May 1974, Available from NTIS.

4. Sullivan, T. L., "A Review of Resonance Response in Large Horizontal-Axis Wind Turbines", Proceedings for the Wind Turbine Dynamics Workshop, NASA Conference Publications 2185/DOE CONF-810226, Feb. 24-26, 1981.

5. Wright, A. D., et al., "Vibration Modes of Centrifugally-Stiffened Beams", ASME J. of Appl. Mech., March 1980.

6. Peters, D. A., An Approximate Solution for the Free Vibration of Rotating Uniform Cantilever Beams, NASA N78-33289, formerly NASA TMX-62,299, Revised 8-28-78.

7. SAP-V, A Structural Analysis Program for Static and Dynamic Response of Linear Systems, University of Southern California, Department of Civil Engineering, Report No. EERC 73-11.

8. Hoa, S. V., "Vibration of a Rotating Beam with Tip Mass", Journal of Sound and Vibration, Vol. 67, No. 3, pp. 369-381.

9. Hodges, D. H. and Rutkowski, M. J, "Comments on Vibration of a Rotating Beam with Tip Mass", Journal of Sound and Vibration, Vol 72, No. 4, pp. 547-549.

10. Hodges, D. H., "Vibration and Response of Nonuniform Rotating Beams with Discontinuities", Journal of the American Helicopter Society, Vol. 24, No. 5, pp. 43-50.

11. Spera, D. A., "Comparison of Computer Codes for Calculating Dynamic Loads in Wind Turbines", NASA TM-73773, presented at the Third Biennial Wind Workshop, Washington, D.C., Sept., 1977.

12. Kaza, K. R. V., Janetzke, D. C., and Sullivan, T. L., Evaluation of MOSTAS Computer Code for Predicting Dynamic Loads in Two-Bladed Wind Turbines, NASA TM-79101, 1979.

13. Dugundji, J., and Wendell, J. H., General Review of the MOSTAS Computer Code for Wind Turbines, Report DOE/NASA/3303-1 or NASA CR-165385, June 1981.

14. Thresher, R. W., "Structural Dynamic Analysis of Wind Turbine Systems", ASME Journal of Solar Energy Engineering, March 1982.

15. Thresher, R. W. and Hershberg, E. L., Development of an Analytical Model and Code for the Flapping Response of a HAWT Rotor Blade, SERI Subcontract Report, SERI/STR-217-2629, February 1985.

16. Thresher, R. W., Hershberg, E. L., and Wright, A., "A Computer Analysis of Wind Turbine Blade Dynamic Loads", SERI/TP-214-2563, November 1984; ASME Wind Symposium IV Proceedings 1985, presented in Dallas, Texas, February 1985.

17. Connell, J. R., The Spectrum of Wind Speed Fluctuations Seen by a Rotating Blade of a Wind Energy Conversion System; Observations and Theory, Report PNL-4083, Pacific Northwest Laboratory, Richland, WA, 1981.

18. Connell, J. R., George, R. L., and Sandborn, V. R., Rotationally Sampled Wind and Wind Turbine Response at Goodnoe Hills Mod-2, Unit No. 2, Measured in July-August 1983: A Preliminary Analysis, Electric Power Research Institute Report, Palo Alto, California 1983.

19. George, R. L. and Connell, J. R., Rotationally Sampled Wind Characteristics and Correlation with MOD-OA Wind Turbine Response, Report PNL-5238, Pacific Northwest Laboratory, Richland, WA 1984.

20. Sundor, R. M. and Sullivan, J. P., "Performance of Wind Turbines n a Turbulent Atmosphere, Proceedings of the Workshop on Wind Turbine Dynamics, February 24-26, 1981, NASA Conference Publication 2185 as DOE Publication CONF-810226.

21. Thresher, R. W., et al., Modeling the Response of Wind Turbine to Atmospheric Turbulence, Oregon State University Report, RLO/2227-81/2, 1981.

22. Thresher, R. W. and Holley, W. E., The Response Sensitivity of Wind Turbines to Atmoshperic Turbulence, Oregon State University Report, RLO/2227-81, 1981.

23. McCroskey, W. J., The Phenomenon of Dynamic Stall, NASA TM 81264, March 1981.

24. Bovarnik, M. L., Shipley, S. A., and Finger, R. W., Goodnoe Hills Mod-2 Cluster Test Program, Volume 2, Final Report, May 1985.

25. Bovarnik, M. L. and Miller, R. D., Goodnoe Hills Mod-2 Cluster Test Program, Volume 3: Rotational Wind Sampling Tests and Analyses, EPRI AP-4060, Volume 3, Final Report, May 1985.

CHAPTER 25

HYDRO-POWER DEVELOPMENT IN REMOTE LOCATIONS OF DEVELOPING COUNTRIES

Granville J. Smith II
STS Energenics Ltd., Washington, D.C. 20006

ABSTRACT

In many developing countries hydropower can be used to replace the consumption of imported oil. The economic advantage of using hydropower increases if a low cost, locally manufactured turbine, called the cross flow turbine, can be used. This paper discusses the technical design and use of the cross flow turbine in the context of an hydroelectric development project in Africa.

INTRODUCTION

The question before us was always the same, How can the subsistance level of life in developing countries be raised? Sometimes we thought about it in cosmic terms and tried to envision global strategies, but most often the question arose in a microscopic context in which a small village or tribe in a remote area of a developing country had a specific problem or a special potential for improving their quality of life but lacked an essential ingrediant. Often that essential ingrediant was energy- energy to grind grain, energy to cut lumber, energy to run health facilities, energy to pump water, energy to communicate with the outside world.

United Nations' estimates of Africa's energy resources are enormous: 341 billion tons of coal, 57.8 billion barrels of crude oil, 189.4 billion cubic feet of natural gas, 1.35 million tons of uranium, and 200,000 megawatts of hydroelectic power. Even so, most African countries are developing these resources only very slowly, if at all. Moreover, the oil importing countries in Africa are spending 30% or more of their total export earnings on oil and this percentage is increasing. Energy imports to African countries create a heavy financial burden, but are generally viewed as an essential commodity. Thus, a difficult economic cycle is set up in which development depends on imported oil. Consumption of oil creates large debts which can not be paid because

0094-243X/85/1350497-25$3.00 Copyright 1985 American Institute of Physics

economic development is too slow and which preclude borrowing for additional oil. Economic development may stop or worsen. It is a vicious cycle which many countries can escape only after a very long period during which the internal resources of the countries are finally developed.

During the early development of Europe and the U.S. small hydropower plants, both mechanical and electrical, played an important role. It would seem possible that hydropower could play a similar development role in Africa, and that it might even help certain countries replace oil consumption by a local renewable resource. The World Bank completed in 1984 a survey of hydroelectic power in 100 developing countries.[1] In Africa the report identified large hydro potential relative to the size of the current and forecasted power market. In Table 1 one notes, however, the disparity between the technically feasible potential and the probable construction of hydropower plants. The difference is largely a matter of economics. The investment needed to realize even a small fraction of the potential is enormous. Just the preparation costs, that is, the costs of studies, institutional strengthening and project engineering, are estimated to be $994 million ('82 $'s) through 1990 for West and East Africa. The investment for a 100 megawatt plant would be most likely in the range from $100 million to $300 million dollars. Few developing countries can justify such large expenditures.

Table 1: Hydroelectric Power in West and East Africa.

	West Africa	East Africa
1980 Installed Electric Capacity (GW)		
Thermal	3.0	2.0
Hydro	2.8	7.5
All Others	0.0	0.0
1980 Electrical Production (Twh)		
Oil	4	4
Gas	5	5
Coal	negl.	1
Hydro	11	23
Hydro Additions through 1995		
Technical Potential (MW)	85,545	216,480
Probable Additions (MW)	6,910	10,523

It is not necessary to develop power in such large increments. Initial development of power in the United States consisted of many dispersed, small units. Driven by the need for energy, the availability of hydropower, heavy constraints on capital expenditures, many African countries have seriously investigated the construction of small hydroelectric powerplants.

This paper is about hydropower in the developing world and, more specifically, it is a hypothetical case history of the development of an hydropower plant in a remote area of West Africa.[2,3] Even more specifically, it focusses on technical questions associated with determining the right type and size of turbine to be used at a specific hydro site. The technical questions associated with choosing the correct turbine are conceptually interesting and require some physics and engineering not encountered very often in the class room. The paper gives some background, examples and references to these technical questions.

It is important to understand that the answers to development questions in developing countries come truly from the domain of liberal arts and require an appreciation of anthropology, economics, farming, politics, and a variety of other disciplines as well as physics and engineering. There are many forces which may or may not allow any one project to become a reality. In this regard, the author has chosen remote Africa as the location for the hypothetical case history, because he has worked there and, as a result of his own experience, has some hope of keeping the wider issues in the reader's mind while discussing the narrower technical issues. The particular case history is interesting, because technically one can argue, as we will, that a certain type of equipment is best for the project, whereas, as we will discover, for good reasons evolving out of the broader nature of the project, such equipment is not always practicable.

DIBAGUIL

The location of the hydropower project is at a series of water falls on the Dibaguil River, a small tributary of the Ngoko River near the village of Sembe, Republic of the Congo. As shown on the map in Fig. 1, Sembe is a real village as are the falls on the Dibaguil River. The Dibaguil River flows into the Ngoko River which forms part of the boundary between Cameroon and the Congo and is a tributary of the Sangha River which in turn is a major tributary of the Congo River.[4]

Fig. 1: Map of the People's Republic of the Congo locating project site in its northwest corner near the village of Sembe.

Near Sembe and across the Ngoko River live pygmees much as they have for centuries. An elephant was killed just before our first visit to the site by two Belgian hunters. They have the tusks; the pygmees have smoked the meat for later consumption. The trail to the site is cut by local natives with machetes. The Dibaguil falls are in dense forest and jungle.

Having located the site, the difficult task of determining its feasibility as a location for a hydropower plant must be undertaken. In Sembe a survey of power needs has already been completed. The 3,000 inhabitants want and need electricity. A local cocoa industry needs electricity. The survey shows a total electrical demand of about 100 kw while the need for mechanical power at the site does not exist. Therefore, the first critical decision is made. The hydropower plant will produce electricity.

Table 2: Current and future electrical demand at Sembe, Congo.

	On Peak Demand	Average Demand
Current Demand (1985)	75-125 kw	30-50 kw
Future Demand (1995)	150-250 kw	50-80 kw

HEAD

The site is surveyed so that an accurate head can be determined. The "head" is the distance through which water falls as it is conveyed to the turbine in the powerhouse. If we locate the powerhouse at the bottom of the falls, the head will be, according to our survey, 180 feet. A long pipe, called the penstock, will convey the water from the top of the falls to the powerhouse. Some frictional energy is lost as the water travels through the penstock and into the turbine effectively decreasing the head which the turbine "sees". Thus, hydropower engineers say the gross head is 180 feet and the net head is the gross head minus the head losses due to friction.

The head loss can be calculated using a variety of empirical formulae; one of which is the Manning equation:

$$H = V^2 n^2 L \;/\; (1.49)^2 R^{4/3}, \qquad (1)$$

where H is the head loss due to friction (ft), V is the velocity of flow in the penstock (ft/sec), n is a roughness coefficient which varies from 0.010 for smooth pipes to 0.017 for rough pipes, L is the length of the penstock (ft), and R is the hydraulic radius of the pipe (ft). For Dibaguil the penstock is 246 feet long and the penstock will be smooth and 2.00 feet in diameter, therefore the head loss is 0.27 feet. There are additional frictional losses in the bends in the penstock and across gates and other devices used to control the flow of water in the penstock. All of these losses must be carefully assessed in a final design.

POWER

The power of the turbine is linearly dependent on the head and the quantity of water, the flow, flowing through the turbine. From elementary physics the power equation is easily derived by considering the work done by a volume of water falling through a known distance at a known velocity. The equation is:

$$P(in) = \gamma \cdot g \cdot Q \cdot H, \tag{2}$$

where P is power, Q is the flow in cubic meters per second or cubic feet per second, γ is the specific density of water and g is the gravitational acceleration. P(in) is the theoretical power available just before the water enters the turbine. The actual power produced is somewhat less due to inefficiencies in the machinery. Thus, one normally writes the power equation with another factor, η, which accounts for the efficiency of the turbine:

$$P(out) = \eta \cdot \gamma \cdot g \cdot Q \cdot H. \tag{3}$$

Another way of writing the efficiency factor is the ratio of Equations (2) and (3) with Eq. (2) expressing the power input to the turbine and Eq. (3) expressing the power output from the turbine:

$$\eta = P(out)/ P(in). \tag{4}$$

The efficiency factor turns out not to be a constant or even a simple function and will require some further investigation.

FLOW

First, however, the flow must be considered. What is the flow which should be used in Eq. (3)? Sitting on a rock in the middle of the Dibaguil River, one can guess from past experience that the flow at that moment is about 3/10 cubic meters/sec (10 cubic feet/sec).[5] Since the flow is equal to the cross sectional area of the river times the velocity of the water, in a few minutes one can measure the flow using a simple velocity flow meter and a tape measure. There are some problems with measuring the velocity since it varies across the cross section, but an average is taken and, as a first approximation, the measurement is a good one. See Fig. 2.[4] The flow is about 4/10 m^3/s (15 cfs).

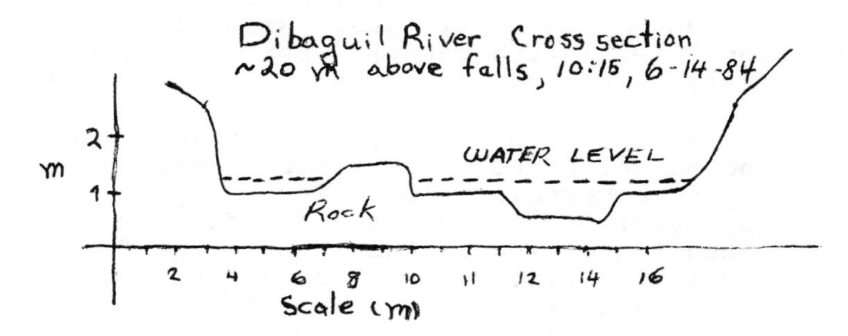

Fig. 2: Cross section of Dibaguil River taken from field notes of 6-14-85.

The determination of flow presents an interesting conceptual problem for now it is February and the dry season in the northern Congo. The flow during the wet season will undoubtably be greater. The problem is that not only does the flow vary by season but also by day and even by hour. This point is frequently demonstrated during our site visit, because rainfall in the Congo tends to be heavy for short durations and we have noted and even measured on several occasions the change in flow after a short, intense rain.

Eq. 3 says that the power from the turbine will vary in time as the flow changes, but there are some obvious limits imposed by the geometry of the machinery. Imagine, for example, that the turbine is shaped like a fan and is located in a pipe so that water flowing through the pipe turns the turbine blades or, as the rotating part of the turbine is normally called, the turbine runner. The quantity of flow is limited by the cross sectional area of the pipe and the velocity of the flow. For example, if the diameter of the pipe is 2.00 feet and the maximum velocity is $(2gH)^{1/2}$ where g is the acceleration due to gravity, then the maximum flow is equal to 3 m^3/s (107 cfs). One can equally well imagine that for very low flows the friction in the turbine would overwhelm the force on the runner and that the power produced would be zero. Thus, from the point of view of making a turbine operate, the flow through a given turbine has a maximum and minimum value.

It is possible to chose the turbine to be any size which is desirable. Clearly, because the objective is to produce as much electricity as possible, we must size the turbine so that it runs as much as possible at as high a power as possible. (There is also a serious question of economics which arises, because large turbines cost more than small ones. We will return to this question later in the paper). For what flow should the turbine be designed in order to maximize output?

An important step in the right direction is to determine the number of hours per year on average that the flow exceeds a certain value. For example, how many hours per year on average does the flow exceed 4/10 m^3/s? Such a question is easily answered for most of the rivers in the U.S., whose flow has been recorded on a continuous basis for many years. The flow on the Nile River has been recorded for 4000 years! Such flow information can be conveniently summarized in a "flow duration curve" like the one shown in Fig. 3 for the Dibaguil River.

The curve was drawn from the table of associated data shown in Fig. 3. Each flow value in the table is accompanied by a percentage which represents the percentage of time the flow at the site exceeds the value shown. Thus, 22% of the time the flow equals or exceeds 1600 liters/second (57 cfs) and 100% of the time the flow equals or exceeds 300 l/s (10 cfs). Normally, flow duration curves are constructed from average daily data. That is, numbers which represent the average flow during a 24 hour period. Since there exist seasonal flow variations and yearly flow variations, the more years of daily flow data one has, the closer will be the representation of an average year as summarized in the flow duration curve.[6]

For Dibaguil the problem is much more complex, because we have only one flow measurement! We do have daily rainfall data from the nearby weather station in Sembe. If sufficient information can be generated about the Dibaguil water basin, then it is possible to

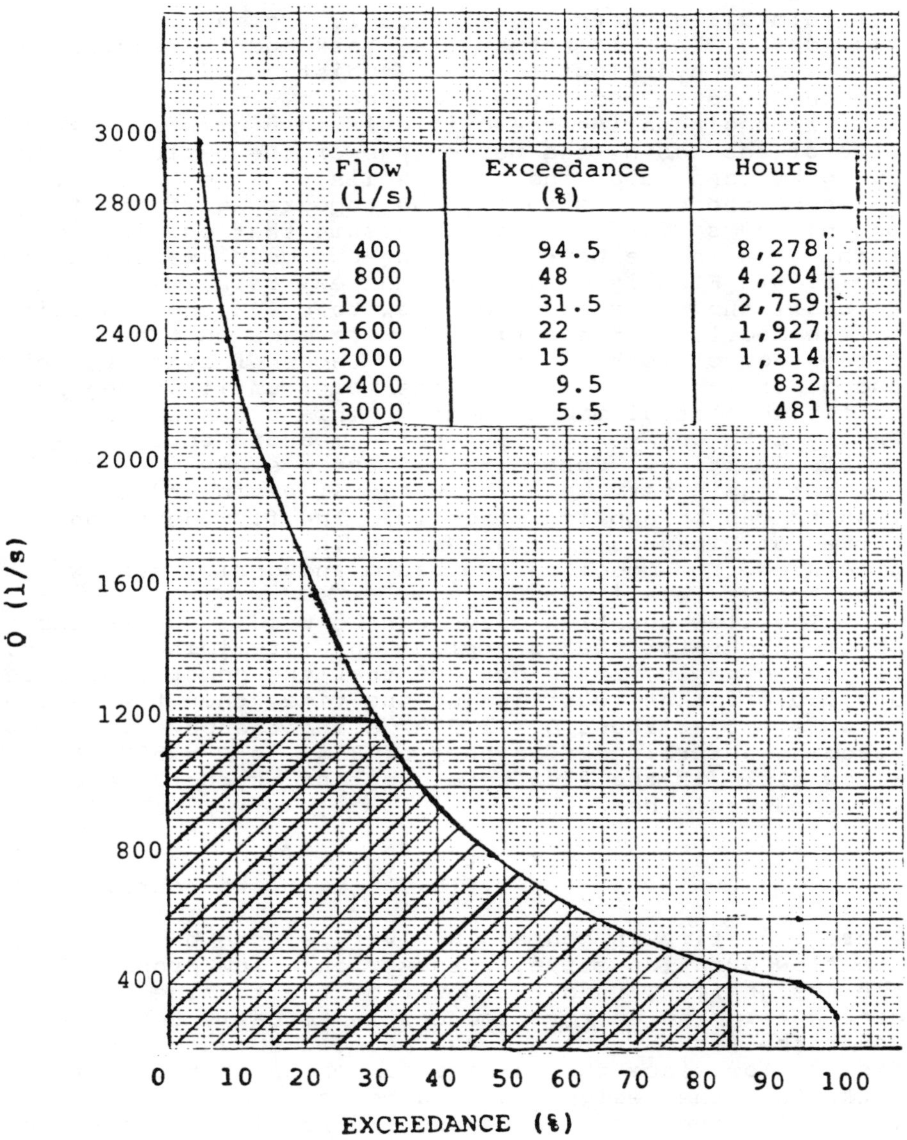

Flow (l/s)	Exceedance (%)	Hours
400	94.5	8,278
800	48	4,204
1200	31.5	2,759
1600	22	1,927
2000	15	1,314
2400	9.5	832
3000	5.5	481

Fig. 3: Energy production (shaded area) by a 550 kilowatt turbine with a design flow of $Q_D = 1.2\ m^3/s$.

use rainfall data to generate daily and even hourly flow data. This is not an easy task and it requires an experienced hydrologist and a sophisticated computer program used to model the water basin.

We begin making the critical measurements: exact size of the basin, the percentage of the basin covered by rock, sand, clay and other soils, the permeability rates of the basin's soils, the topography of the basin, and so forth. Next we assume that one inch of rain falls in one hour and using our model we calculate the flow. Finally we wait for a real rain, measure the rainfall and the associated flow to check the viability of the model. Since estimates were made on the percentage of rock and clay, the rates of evaporation, and other factors we adjust the model until it begins to fit the actual measurements. This process can easily take one year of intensive measurements and model adjusting. Fortunately, such a study was done on a water basin very near the Dibaguil basin and the results can be applied directly. The flow duration curve for Dibaguil, the one shown in Fig. 3, was generated from modeling the basin and deriving flow data from rainfall data.

The flow duration curve can be used to chose the power output for the powerplant on a preliminary basis. If, for example, it is decided that the power should be available for at least 30% of the time, then the design flow is set at about $Q = 1.2\ m^3/s$ and the power, assuming a turbine efficiency of 85%, is $P = 550$ kilowatts. If the flow were constant, then the energy produced would be:

$$E = P \cdot t \qquad (5)$$
$$= 4{,}818{,}000 \text{ kilowatt-hours,}$$

where t is time in hours. However, as can be seen from the flow duration curve, whenever the flow drops below the design flow, then the power of the turbine decreases. The energy produced is actually the area under the flow duration curve bounded above by the design flow since the turbine runner can not pass more water than the design flow and bounded below by the least flow the runner can pass and still operate. The energy calculated for this particular turbine for flows satisfying $Q \geqslant 0.35Q_D$ is shown as the shaded area under the flow duration curve. We have assumed that the turbine will not operate if the flow is below 35% of it design flow, Q_D.

POWER FOR SEMBE

During the first visit to Sembe the survey of power needs, as shown in Table 2, totaled 100 kw. We have just seen from the above calculation that a turbine designed at a 30% exceedance would produce about 550 kw when the flow is 1.2 m^3/s. There is nothing absolute about choosing the 30% on the flow duration curve. In dry years and dry seasons the probability is that the turbine will be operating substantially less than 30% of the time at full power. During the wet seasons the average will be higher. In any case, it would seem highly inappropriate to spend a large sum of money for a 550 plant when the local demand is only 100 kw.

Studying the development of other small **electric** power projects in developing countries one notes often that the power demand has rapidly expanded to meet available power. It is possible that if a 550 kw power plant is constructed, the demand will increase to this power level. If the power plant promotes significant additional demand it will be most likely from industrial customers. Therefore, it is very important to consider which industries may develop and what the power demand for those industries will be. For Sembe it appears that within ten years the peak demand could grow to 250 kw. Clearly it is not necessary to size the plant as large as 550 kw. In fact, a very serious mistake would have been made, because such a large turbine could not run below about 190 kw and, therefore, could not provide power efficiently to Sembe for a very long time! On the otherhand, if the plant is sized to produce 250 kw, then it can still efficiently produce the current demand of 100 kw.

We should also recall that the flow duration curve reflects the hydrology well only when there are many years of average daily flow data. Our flow duration curve is based on almost no measured data and, therefore, must be viewed with some skepticism. It would be a mistake to size the power plant at an exceedance flow of 1.2 m^3/s (30%) and discover after a few years that a more accurate 30% exceedance flow is 0.8 m^3/s. Hydroelectric power plants do not run so well when the flow is 0.8 m^3/s and the turbine is designed for 1.2 m^3/s as we shall discover under the section called TURBINES.

Finally, it is appropriate to think about what can be done for the industry which depends on the power from the plant, but can not get the power during the dry season or during a very dry year.

This problem is minimized if the turbine is designed to run well at the flow rates expected during the dry season. That is 10-15 cfs for the Dibaguil project. Even so, if the demand has grown sufficiently large, the industry may not have power when it needs it. What can be done to improve the situation? One possibility is to build a small dam at the top of the falls. The dam would impound some water which could be sent through the turbine for short periods of time to boost power output. When the power plant is first built, it would be difficult to accurately assess the need for and the potential energy contributions that a dam would contribute. However, after 10 years of operation one would have a good data base from which to make such a judgment.

TURBINES

The efficiency of the turbine mentioned above was chosen to be 85%. As a first approximation, 85% is a good choice, but efficiencies of turbines can vary widely depending on the geometry of the turbine, and the pressure and flow of water through the turbine. Turbine efficiency curves as a function of rated capacity (or flow) are shown in Fig. 4 for a typical reaction turbine and a typical impulse turbine.

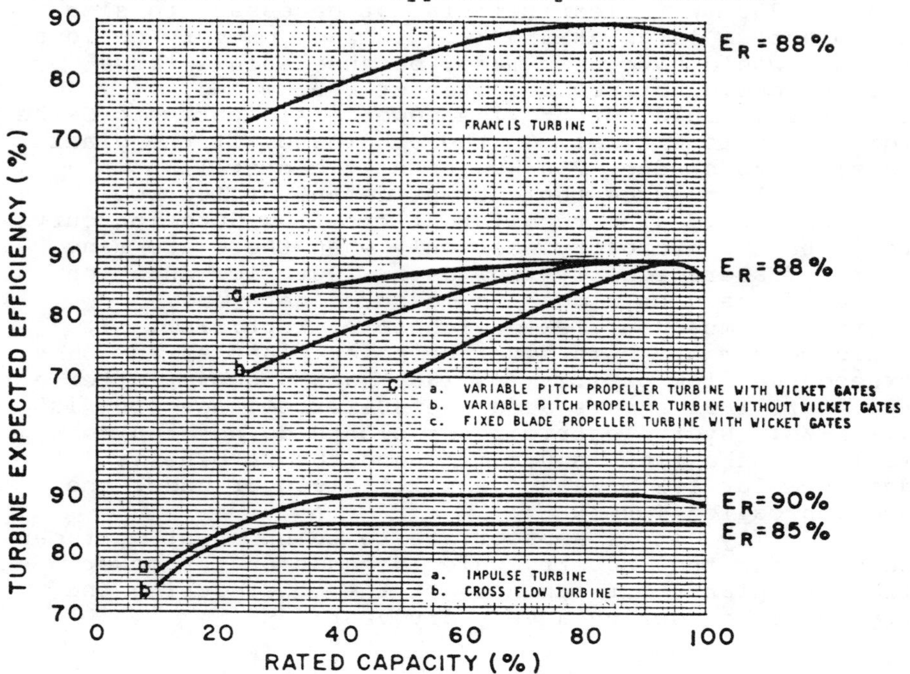

Fig. 4: Turbine efficiency curves as shown in Ref. 6c.

Turbines are normally classified as either reactive or impulse in reference to the type of force on the runner. Fig. 4 shows two impulse turbines and two reactive turbines. These turbines are further classified with respect to how the water is controlled as it flows through the turbine. That is, sometimes the pitch of the blade can be varied (the Kaplan turbine), sometimes the water flows into the runner radially and departs axially (the Francis turbine), and so forth.[7]

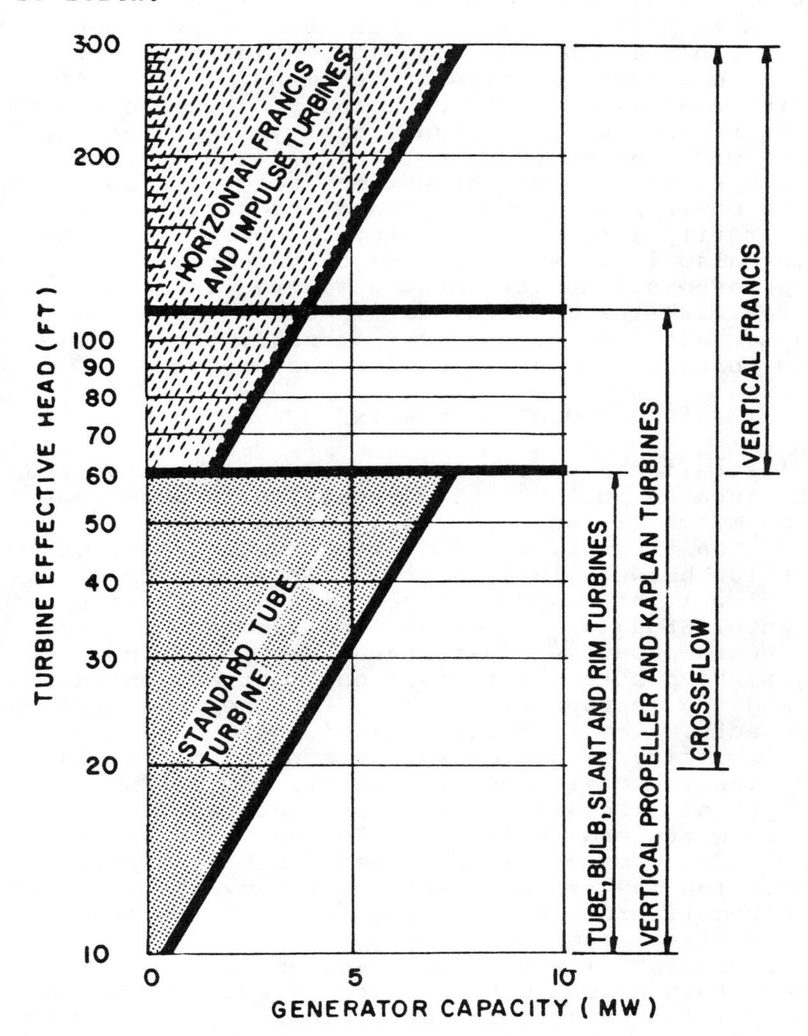

Fig. 5: Head ranges for different turbine types as shown in Ref. 6c.

Each one of these designs has special features which must be taken into account when making a final turbine selection.

One such special design feature limits the use of turbines by head range. Impulse turbines operate most efficiently at very high heads, while a standard tube turbine operates better at lower head ranges. Fig. 5 shows the head ranges for several turbine types.

Several very special constraints exist at our hypothetical site. The first constraint involves the observation that there exist in the Congo many sites similar to Dibaguil and the Congolese government may like to chose a turbine which it can manufacture. If the turbine can be manufactured locally, then the costs of the power plant will be significantly lower and the government will better be able to afford the cost of several projects. The government may also like to design the power plant in a way that the civil works can be primarily carried out by local contractors. This constraint also impacts the choice of turbine.

These external constraints considered in conjunction with the constraints imposed by head and flow and technical design suggest strongly that a crossflow turbine be considered for the project.

CROSSFLOW TURBINE

An Australian engineer A.G.M. Michell patented the crossflow turbine in 1903. The crossflow is sometimes called the Mitchell-Banki or Banki turbine after the Hungarian, Donat Banki, who did extensive research on the crossflow between 1917 and 1919. The history of the crossflow is nicely summarized in a dissertation, "Experimental Study of the Cross-flow Turbine, by Shahram Khosrowpanah.[8,9] Khosrowpanah provides nine advantages which strongly attract one's attention when considering the appropriate turbine for the Congolese hydro projects. The most critical of these advantages are its simplicity of design and construction, its low cost, and its low civil cost. However, the maximum power output of a locally manufactured crossflow is a few 100 kilowatts and it is possible that many of the Congolese hydro projects will exceed 1,000 kilowatts. This limitation may become a serious disadvantage, but, for the Dibaguil project, the crossflow turbine appears to be an excellent choice.

An "exploded" view of the crossflow is shown in Fig. 6 next to a cross section of the turbine in which its principal components are identified.[7]

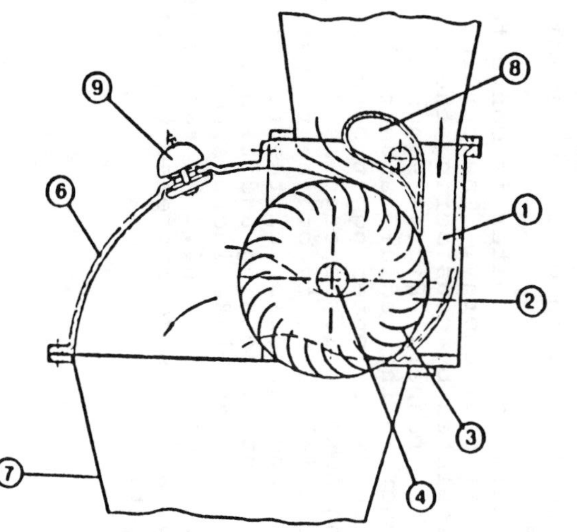

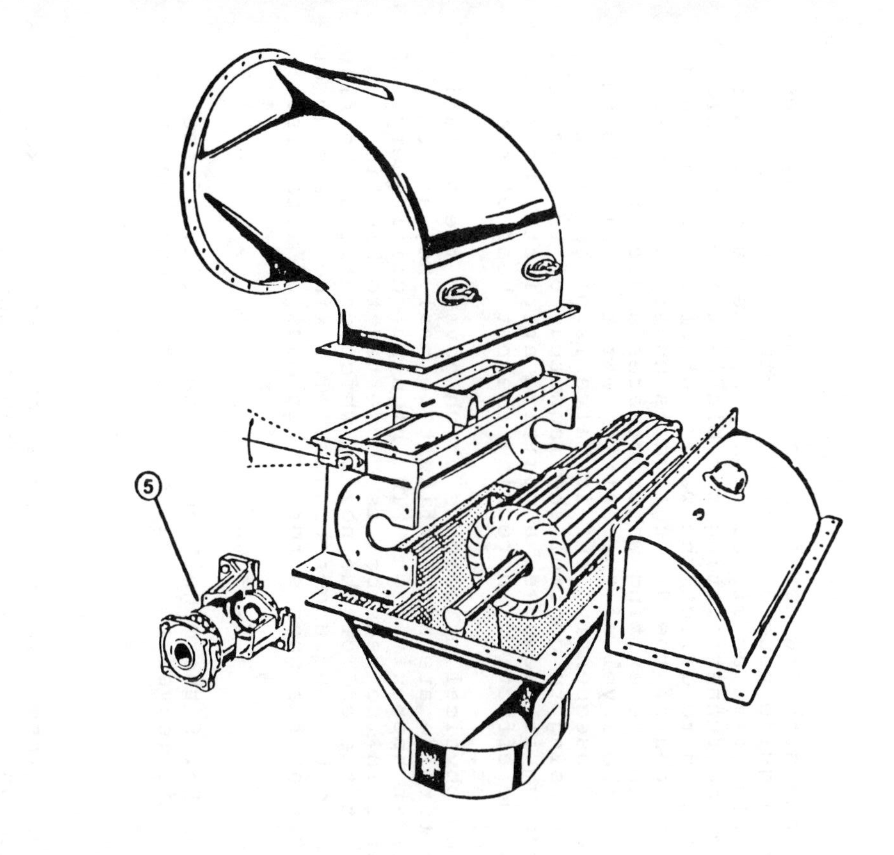

1. Nozzle
2. Runner
3. Blade
4. Shaft
5. Bearing
6. Casing
7. Draft tube
8. Guide-vane
9. Air valve

Fig. 6 : Components of the crossflow turbine as shown in Refs. 7 & 8.

Water enters the runner from the penstock through the nozzle. A guide vane directs the water to hit the runner blades at an efficient angle. The water passes through the runner hitting the blades twice. If a draft tube is used to direct the water out of the runner and the lower end of the draft tube is under water at all times, then the exiting water creates a vacuum which can pull the level of the exiting water as high as the runner. Consequently, an air valve is placed in the runner housing to control the magnitude of this vacuum. Some vacuum is viewed as a positive attribute, because it has the effect of increasing the net head of the power plant.

The physical principals which govern the operation of the turbine are not difficult to analyze, at least, to a first approximation. The high potential energy of the water in the penstock is converted to kinetic energy at the end of the nozzle which controls the flow of water onto the runner blades. The kinetic energy of the water at the nozzle for a unit mass of water is:

$$K.E. = gH \qquad (6)$$

and from Eq. 6 the magnitude of the velocity of the water as it leaves the nozzle and enters the runner is given by:

$$V = (2gH)^{1/2} \qquad (7)$$

The energy losses in the nozzle have not been taken into account in Eq. 7. The design of the nozzle must be very carefully considered in order to minimize nozzle energy losses.

The kinetic energy of the water is converted by the runner into rotational energy. The turbine power, P(out), can be calculated by considering the forces which act to turn the runner. Fig. 7 shows a cross section of the runner in the radial-tangential plane. One can visualize the water entering at point A with a velocity V_1, losing some of its energy as it acts to increase the peripherial velocity of the runner from U1 to U2, crossing the open interior of the runner, entering again at point C, lossing additional energy as it acts against the runner blade increasing its peripherial velocity from U3 to U4. At two stages the water losses energy and acts to increase the rotational energy of the runner.

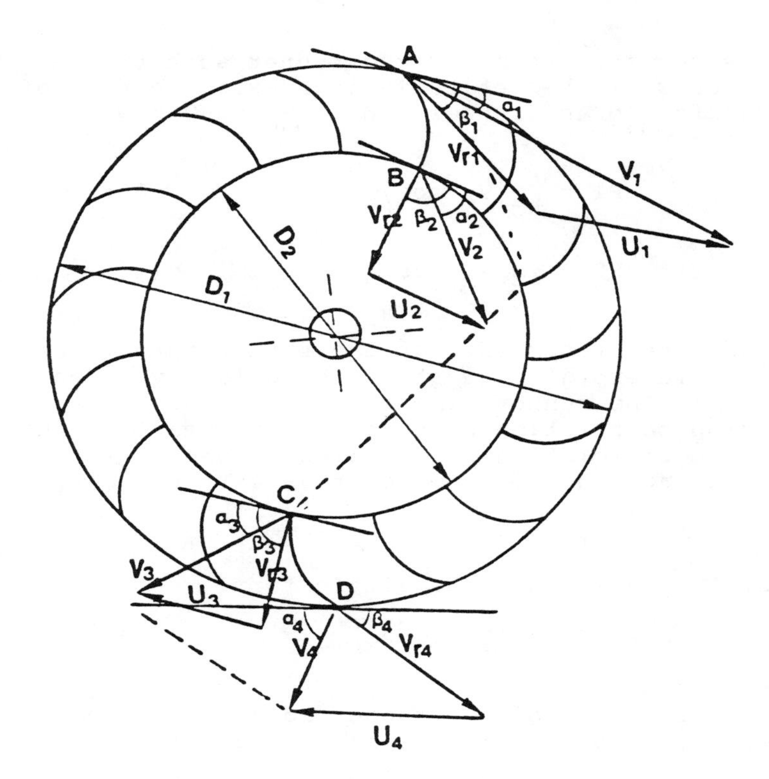

Fig. 7: Velocity diagram of the cross-flow turbine in the radial-tangential plane as shown in Ref. 8.

Only tangential forces act to rotate the runner and it is the mass flow, Q (lbs/sec), across the blades which provides the pressure force normal to the flow. The forces on the blades, to a first approximation, are:

$$F1 = \frac{\gamma Q}{g} (V_{r_1} \cos\beta_1 - V_{r_2} \cos\beta_2) \tag{8}$$

$$F2 = \frac{\gamma Q}{g} (V_{r_3} \cos\beta_3 - V_{r_4} \cos (\pi - \beta_4) \tag{9}$$

where F1 and F2 are the tangential forces at the first and second stages, respectively; V_r is the velocity of the water flow relative to the blades; β is the angle between the peripheral velocity (tangent to the runner) and the relative velocities, V . β is called the blade angle and the runner is designed such that $\beta_1 = \beta_4$, and $\beta_2 = \beta_3$.

Also, it is possible to build the runner such that $\beta_2 = \pi/2$ which one can show theoretically increases the rate of energy transfer to the runner. Eqs. 8 and 9 now simplify to:

$$F1 = \frac{\gamma Q}{g} V_{r_1} \cos\beta_1 \tag{10}$$

$$F2 = \frac{\gamma Q}{g} V_{r_4} \cos\beta_1 \tag{11}$$

As the water flows through the runner turbulance occurs and some energy is lost which could have been transferred to the runner blades. In a first approximation calculation such losses are small so that the magnitude of relative entry velocity can be set equal to the magnitude of the relative exit velocity:

$$V_{r_1} = V_{r_4} \tag{12}$$

Thus, Eqs. 10 and 11 have the simple form:

$$F1 = F2 = \frac{\gamma Q}{g} V_{r_1} \cos\beta_1 \tag{13}$$

The power output of the runner equals the product of the torque, T, on the runner and its angular velocity, :

$$\begin{aligned} P(out) &= T \cdot \omega \\ &= (F1 + F2) \cdot r_1 \frac{U_1}{r_1} \\ &= 2 \frac{\gamma Q}{g} (V_{r_1} \cos\beta_1) U_1 , \end{aligned} \tag{14}$$

where the pheripheral runner velocity $U_1 = \omega \cdot r_1$.

Eq. 14 can easily be rewritten in terms of the angle of attack of the water from the nozzle and the velocity of the water as it leaves the nozzle:

$$P(out) = 2 \frac{\gamma Q}{g} (V_1 \cos\alpha_1 - U_1) U_1 , \tag{15}$$

and the efficiency of the runner as given in Eq. 4 becomes:

$$\eta = 4 \frac{U_1}{V_1} \left(\cos\alpha_1 - \frac{U_1}{V_1} \right) \tag{16}$$

Eq. 16 contains some interesting information. Note that the efficiency maximizes when $\cos\alpha_1 = 1$ which is not possible, but it is possible to construct a runner with a small attack angle. According to Khosrowpanah, $\alpha_1 = 15°-16°$ is normally used. The maximum efficiency obtained by differentiating with respect to U_1/V_1 and setting the result equal to zero yields a maximum of efficiency of 92.4% for $\alpha_1 = 16°$. See Ref. 7 for additional information of the efficiency of the crossflow.

DESIGN OF THE RUNNER

One can calculate the overall dimensions of the runner using specifications provided by Banki in terms of the radius of curvature of the inside surface of the blades as shown in Fig. 8:

$$D_1 = 2 \cdot r_1 = 6.1236\,\rho, \quad (ft) \qquad (17)$$

$$r_2 = D_2/2 = 0.659\ r_1 = 2.0177\rho, \quad (ft) \qquad (18)$$

$$n = INT\left(\frac{19.24}{0.67 + t} + 0.5\right), \qquad (19)$$

where D and D are the outer and inner diameters of the runner as shown in Fig. 8, is the blade radius n is the number of blades and t is the blade thickness.

The thickness of the blade must be sufficient to withstand the bending stress which results when the water impacts them. The bending stress depends on the magnitude of the head and the length of the runner blades. P. Verhaart has analyzed the problems associated with broken blades resulting from bending stress and has written a short program for a hand held calculator which calculates the strength of the blades in terms of their inner radius and thickness. Using the program the thickness and length of blade can be chosen so that the blades will not break.[10]

The blades can be cut from mild steel pipe of an oppropriate inner radius, . The inner radius can be set by choosing the outer diameter of runner, D_1. D_1 and the runner length, L, must be sufficiently large to accommodate the design flow. Khosrowpanah has calculated the product of the diameter and the length, $D_1 \cdot L$, in terms of the flow, turbine efficiency, head and the nozzle entry angle, λ:

$$D_1 \cdot L = Q\ /\ (1.159) \cdot (H \cdot \eta)^{1/2} \lambda \ (ft^2) \qquad (20)$$

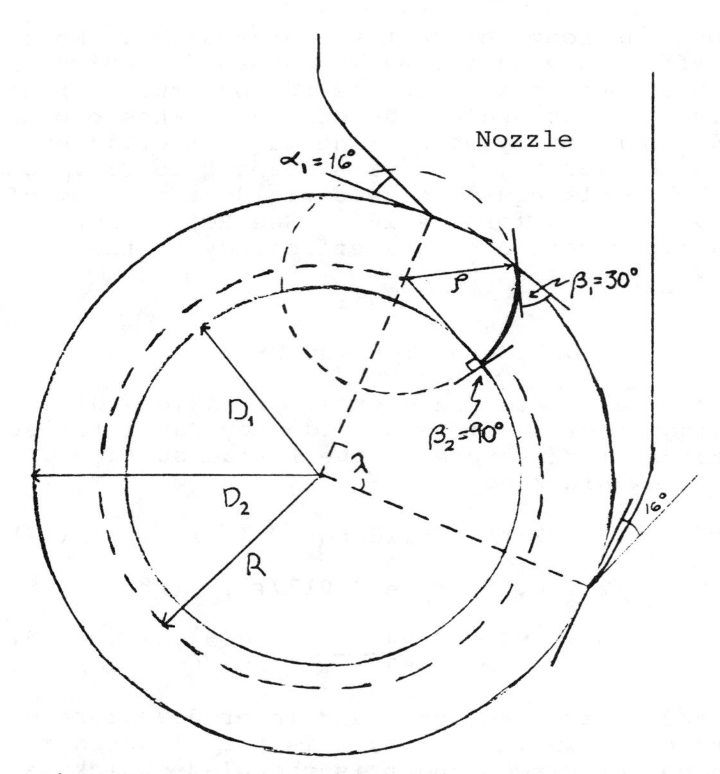

Fig. 8 : Cross section of the cross-flow turbine showing the nozzle location and dimensions relative to the runner radii.

Fig. 8 shows a typical configuration for the nozzle and defines the angle, λ, and the arc length, $D_1 \cdot \lambda/2$, over which water enters the runner. Note that it is possible to choose $\lambda = \pi/2$. Also, note from Fig. 7 that .80 is a conservative value for η.

The only remaining quantities needed to complete the design of the turbine runner are the angles which position the blades efficiently. Banki has chosen the angle α_1 = 16° and angle β_1 = 30°. The efficiency of the runner increases as decreases toward zero (Eq. 16), but it is not practically possible to weld the blades to the runner end plates if α_1 is less than 16°.

Banki has computed a pitch radius which can be used to position the blades. It can be derived directly from the angles already specified and the geometry shown in Fig. 8.

The pitch radius, R, is given by:

$$R = 2.2516\rho \qquad (21)$$

Eqs. 17 - 21 allow us to specify the dimensions of the crossflow runner for Dibaguil. Recall that the power has been chosen to be 250 kw, because the present and future demand in Sembe could be adequately served by 250 kw. Also, such a power level fit nicely with the head and flow constraints imposed by the topography and hydrology of Dibaguil falls. Using Eq. 3 one calculates the flow to be about 0.58 m^3/s (20.5 cfs). Eq. 20 yields $D_1 \cdot L = 0.94$ ft . This result is very interesting, because it says that if we choose the diameter of the end plates to be twelve inches, then the length of the runner is just 11.28 inches. The turbine runner is quite small. In many projects the head will be a much smaller quantity and, consequently, the dimensions of the runner will be much larger. (A general rule used by designers of all types of turbines is: "The higher the head, the lower the cost.")

Calculating the inner radius of the blades yields:

$$\rho = 1.96",$$

which is not a very convenient quantity, because we would like to construct the turbine from already available pipe. Therefore, choose $\rho = 2.00"$ and recalculate the other dimensions:

$$D_1 = 12.25",$$
$$D_2 = 8.07",$$
$$R = 4.50",$$
$$L = 11.05".$$

The fabrication of the runner can now be accomplished using these dimensions.

The only remaining unspecified quantities are the number of blades and the blade thickness. The blade thickness can be determined from Ref. 10. Let us assume that 1/4" thick pipe is satisfactory. Then, the number of blades is 25.

One must make a number of additional decisions during the actual construction of a crossflow. Design drawings and other fabrication information is available from the German Appropriate Technology Exchange (GATE). GATE has been very actively involved in promoting the use of crossflow turbines in developing countries.[11]

CONCLUSIONS

Having developed on a preliminary basis the design features of the crossflow turbine, it is appropriate to conclude with a review of the order of events which normally take place before, during and after the development of a small hydro project like the one at Dibaguil falls.

We start by assuming that we are working with governments, federal and local, which want to develop the hydropower site under investigation. It is also assumed that the investigator has had detailed discussions with the federal government, the local government and any other group that will be directly involved with making decisions about the final design, construction, financing and operation of the project. While at the project site, it is necessary to determine the power demands. For Dibaguil the demand was for electricity, but for other projects the demand could be for mechanical power or a mix of mechanical and electrical power. The type and magnitude of demand, as determined in the present and estimated for the future, set important conditions on the design of the project.

The hydro project can not produce more power, than allowed by the available water flow and head. We have used a flow duration curve to help us understand the flow and to choose an appropriate turbine power. The head for Dibaguil falls was measured and an assumption was made that the head did not vary radically as a function of flow. When investigating a site with a low head and high flow, it often happens that the net head for the plant changes considerably as the flow changes.

U.S. publications such as Ref. 6a,b, and c can be used to learn much more about turbines and other hydro equipment. However, they must be used very carefully for all projects in developing countries, because the technical and economic discussion are based on many assumptions valid only for the U.S.

A manual, which will be published soon, includes material similar to Ref. 6c, but refocusses the entire decision making process to account for the differences in available data and the differences in local capabilities.[12] Of equal importance are the differences in local management, politics, and economic perspective. In short, one must be in a proper frame of reference when making technical decisions.

Consider the decisions which were made with respect to the development of Dibaguil project and the decision to consider the cross flow turbine. The following critical events occurred:

1. The Congolese government had identified a large number of hydroelectric sites which could be developed.

2. The development of the hydroelectric power would replace or supplement the production of electricity in remote areas by diesel oil fired generator sets. Even though the feasibility studies done on hydro sites indicated a comparatively high initial capital investment, the existing generator sets were expensive to run and maintain. Thus, in many cases hydropower was economically advantageous compared to diesel power.

3. The Congolese government was promoting the development of new industrial enterprises in remote areas which would need dependable power sources. The development of hydropower would provide a direct economic benefit to the Congolese economy by stimulating the growth of local industry.

4. On the technical side, the head and flow at Dibaguil falls limited the choice of turbines to two or three types. The uncertainty in flow resulting from our poor knowledge of the hydrology of the water basin promoted the idea of using a turbine with the flexibility to operate efficiently over a wide range of flows. This condition favored the cross flow turbine.

5. The cost of the project could be decreased by building the turbine locally. Only the cross flow turbine could be reasonably be considered for local construction.

6. The best system over the long term would be one which could be locally maintained and repaired. Since the cross flow turbine would be locally built, it could also be locally repaired.

The various ministries of the Congolese government had to understand and fully support all aspects of the Dibaguil project before it was funded for development. It is interesting to note that for very good reason it is unlikely that a cross flow turbine will be used in the project.

Why, given the above discussion, should that be the case? Naturally, in the U.S. as well as in the Congo economic considerations guide technical decisions. A major reason for considering the cross flow turbine over a Francis turbine, which would work as well, was the potential for locally manufacturing the cross flow turbine at low cost. Associated with the construction of a new product like the cross flow turbine for the first time is considerable risk. Almost certainly the cross flow turbine would spend considerably more time in repair than if it were manufactured by a experienced manufacturer. The trade off, then, is the risk of not having a dependable project at lower cost against a more dependable project at higher cost. The Congolese government favored the latter approach. In other countries where the higher cost precluded development, the cross flow has been used extensively; and, perhaps in time, the Congolese government will acquire the necessary experience to build the cross flow turbines which clearly have a wide range of applicability in their country.

REFERENCES AND NOTES

1. "A Survey of the Future Role of Hydroelectric Power in 100 Developing Countries", World Bank Energy Department, Paper No. 17, August 1984. Table 1 is based on information in this reference.

2. There are many articles on the development of hydropower in developing countries. See for example:
   a. NRECA, 1981. "Small Hydropower for Asian Rural Development." Proceeding of NRECA Workshop on Small Hydropower for Asian Rural Development, Asian Institute of Technology, Bangkok, Thailand, June 8-11, 1981. National Rural Electric Cooperative Association, Washington, D.C.

   b. Jack J. Fritz, "The Potential for Small-Scale Hydro in Developing Countries," CIVIL ENGINEERING, December 1983.

   c. A.R. Inversin, "A Case Study, Micro-Hydropower Schemes in Pakistan," 1981, International Programs Division, National Rural Electric Cooperative Association, 1800 Massachusetts Ave., NW, Washington D.C.

3. NRECA, 1982. "Small-Scale Hydropower in Africa." Proceeding of NRECA Workshop on Small-Scale Hydropower in Africa, Abidjan, Ivory Coast, March 1-5, 1982.

4. STS ENERGENICS, Ltd. "Site Inspection Report for Dibaguil Hydroelectric Site." June 20, 1984.

5. Metric units are not often used in the U.S. for water measurements. The conversion is 35.3 cubic feet/cubic meter.

6. See the following references for additional information on flow duration curves:
   a. C.C. Warnick. "Hydropower Engineering." Prentice-Hall, 1984.
   b. "Reconnaissance Evaluation of Small, Low-Head Hydroelectric Installations." TUDOR Engineering Company, July 1, 1980.
   c. "Feasibility Studies for Small Scale Hydropower Additions." U.S. Army Corps of Engineers, July 1979.

7. See, also, F.W.E. Stapenhorst Inc., "Ossberger Turbine Generating Sets," Point Claire, Que., February 1983.

8. Sharam Khosrowparrah, "Experimental Study of the Cross-flow Turbine," Dissertation, Colorado State University, 1984.

9. D. Banki, "Nene Wasserturbine," Zeitschrift fuer GeSainte Turbinenwesen, Vol. 15 NR 21 (30 July 1918) R. Oldenborng Verlag., Berlin, Munick.

10. P. Verhaart, "Blade Calculations for Water Turbines of the Banki Type," Report WPS 3-83.03. R351, Eindhoven University of Technology. March 1983.

11. Helmut Seheurer, et al., "Small Water Turbine," Instruction Manual for the Construction of a Crossflowturbine, GATE, Postfach 5180, D-6236 Eschborn 1, Federal Republic of Germany, September 1980.

12. A.R. Inversin, Micro-Hydropower Sourcebook, to be published.

CHAPTER 26

LIQUID AND GASEOUS FUELS FROM BIOMASS

Thomas E. Bull
Office of Technology Assessment
U.S. Congress
Washington, D.C. 20510

ABSTRACT

An analysis of energy from biomass (plant products and animal wastes) in the U.S. indicates that about 95% of the potential resource is in the form of lignocellulosic material (wood and plant herbage). The reasons for this conclusion and the major processes for converting lignocellulose to liquid and gaseous fuels are described, together with a brief consideration of end uses for the fuels.

INTRODUCTION

The Energy Information Administration of the Department of Energy has estimated[1] that in 1983, the U.S. consumed about 2.6 EJ (10^{18} J) of wood for energy, which was about 3.6% of total U.S. energy consumption that year (73.3 EJ[2]). By comparison, hydroelectric energy consumption was 3.9 EJ (1.3 EJ of electric energy) and nuclear consumption was 3.2 EJ (1.1 EJ of electric energy). These and other major types of U.S. primary energy consumption are shown in Figure 1.

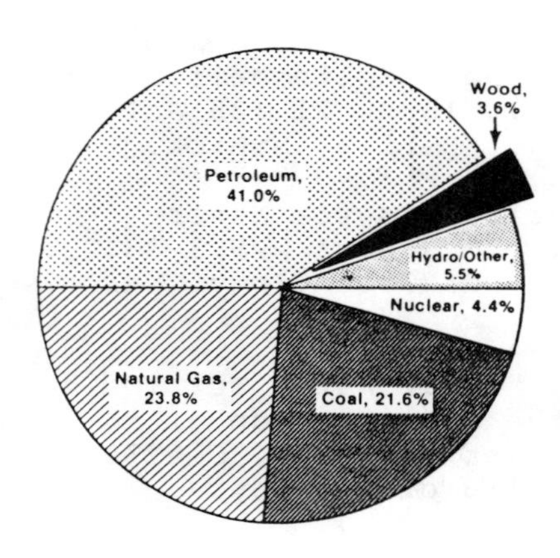

Figure 1 U.S. Energy Consumption, 1983

Source: EIA, "Monthly Energy Review," DOE/EIA-0035(84/09), (1984).

In addition to wood energy, which was by far the largest source of biomass energy, about 2 billion liters of fuel ethanol (0.04 EJ) were produced from corn and other grains in 1983 and small amounts of methane were produced from landfills and animal manure. While the production of liquid and gaseous fuels from grains, landfills and animal manure is likely to grow somewhat in the future, the major potential for fluid biomass fuels lies elsewhere[3].

0094-243X/85/1350522-26$3.00 Copyright 1985 American Institute of Physics

In order to characterize this potential, we will first consider conventional and unconventional sources of biomass feedstocks and then describe the major types of technologies for converting the feedstocks to fluid fuels. End uses for the fuels are also considered briefly.

MORE OR LESS CONVENTIONAL SOURCES OF BIOMASS

As mentioned above, wood is currently the largest source of biomass energy. The majority of this (1.7 EJ in 1983[4]) was consumed by industry, mostly as part of the process of recovering chemicals from spent paper pulping liquor and for steam to run mechanical equipment in the forest products industry. The residential sector consumed most of the remaining wood energy (0.9 EJ[4]) for home heating. In addition, the forest products industry delivered products such as lumber, paper, and furniture containing wood with an energy content of about 2 EJ.

Despite this large level of consumption, there is still considerable potential to increase the production of wood from the nation's forests. In addition, large quantities of biomass could be gotten from increased production of grass from pasturelands and harvesting of crop residues (the plant material left after harvesting conventional crops). Smaller quantities of energy can be obtained from converting grains to ethanol and using animal manure and food processing wastes for energy feedstocks. Each of these categories of conventional biomass sources is considered below.

WOOD

The U.S. has about about 740 million acres (300 million hectares) of forestland, with about half in the East (North plus South regions in Fig. 2). Of this, about 490 million acres is commercial, i.e. the soil type and climate enable the forestland to produce at least 0.3 dry tons/acre yr of commercial timber and the land has not been withdrawn from timber production. About three quarters of the commercial forestland is in the East. About 205 million acres of forestland are classified as non-commercial because of their low productive potential, while the remaining 45 million acres are reserved for parks, wilderness areas and other uses or are under consideration for such uses.

U.S. Department of Agriculture estimates of the productive potential of forestlands[5], together with actual measurements of biomass growth, indicate that commercial forestlands could produce up to 36 EJ/yr of timber with full stocking of productive, existing tree species on the land[6]. (This estimate does not include foilage, stumps or roots.) While the majority of this growth potential probably cannot be accessed for a variety of reasons[7], knowledgable observers believe it reasonable to expect that 40% of the growth

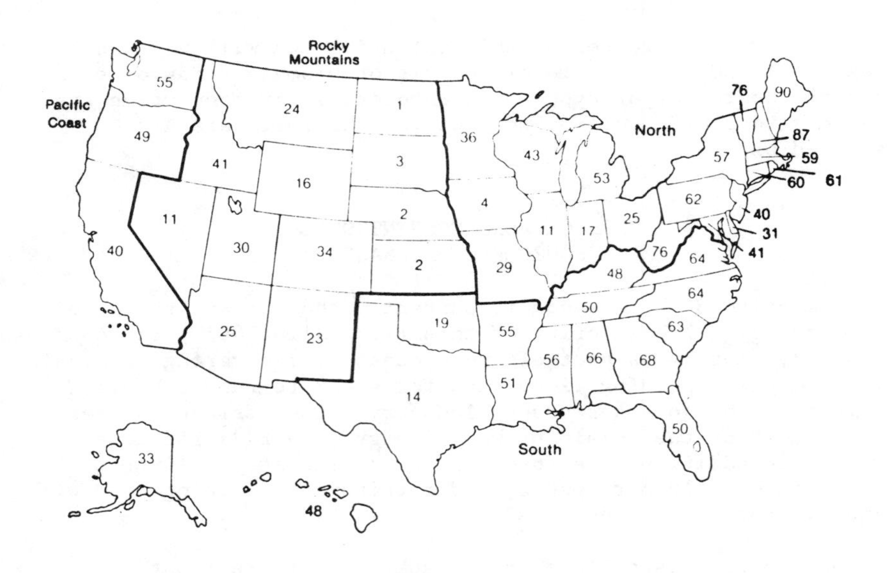

Figure 2 Forestland as a percentage of total land area

Source: Forest Service, U.S. Department of Agriculture

potential could be tapped. Assuming the 40% figure, one arrives at the potential for woody biomass production shown in Table I.

If the output of forest products not used for energy were to double (to 4 EJ energy equivalent/yr), then there could be up to 10 EJ/yr of wood available for energy. Achieving this potential would require replanting after harvest, periodic (e.g., every ten years) thinning of forests, and removal of non-productive or diseased trees, brush and vines. This type of forest management or less extensive management is currently practiced on less than 2 million acres of the roughly 90 million acres of forestland owned by the forest products industry. However, increased forest management would increase both the quality and quantity of commercial timber produced. And cost estimates indicate that the cost of the thinning and other operations could be paid for by selling the residues, delivered to the end user, for \$20-\$60/dry ton (depending on the location), with the majority being avialable for \$30-\$40/dry ton (\$2.00-\$2.50/GJ)[8].

Table I Potential wood availability by region

Forest region	Area of commercial forestland (10^6 acre)	Percent federally owned	Potential wood production[a] (EJ/yr)
South	188	8	3.0-6.0
North	171	7	2.3-4.6
Pacific coast	71	51	1.4-2.8
Rocky mountains	58	66	0.6-1.3
Total	488	20	7.3-14.6

[a]Assumes 40% of total growth potential. Following relative productivity factors are assumed: Pacific coast = 1, South = 0.78, North = 0.66, and Rocky mountains = 0.58. These were derived from data in Ref. 5

Source: Ref. 8

GRASS

Currently about 125 million acres of pasture and hayland in the eastern half of the U.S. have sufficient rainfall to support increased grass production. About 100 million acres of this could be harvested. By fertilizing this land and harvesting the grass more often, grass production could be increased by an average of about 2 dry ton/acre yr for a cost of about $40/dry ton ($3/GJ)[3]. This could result in about 2.7 EJ/yr of biomass for energy, with the distribution by state shown in Table II. However, some or most of this land could be converted to row crop production over the next 20 years (if the demand exists), and it would then be unavailable for the production of grass for biomass energy. Conversely, with low demand for this land for food production and modest development of more productive grass hybrids, the biomass energy output could reach 5 EJ/yr or more.

CROP RESIDUES

Another potential source of biomass energy is crop residues, or the plant matter left in the field after crop (mostly grain) harvests. While about 80% of the residues should be left in the field to protect the soil from erosion, the most productive cropland yeilds residues in excess of that needed for this purpose. Using standard soil loss equations for wind and water erosion, it has been estimated[3] that about 1 EJ/yr of crop residues could be removed and used for energy without exceeding the U.S. Department of Agriculture Soil Conservation Service standards for soil erosion. The resultant potential supply of crop residues is summarized by state in Table III. Cost estimates indicate that most of these residues could be supplied for less than $40/dry ton ($3/GJ)[3].

Table II Potential excess grass production, assuming 2 ton/acre annual production increase

State	Quantity[a] (EJ/yr)	State	Quantity[a] (EJ/yr)
Missouri.......	0.34	Ohio...........	0.10
Iowa...........	0.18	Virginia.......	0.09
Wisconsin......	0.18	Pennsylvania...	0.09
Kentucky.......	0.18	Indiana........	0.08
Minnesota......	0.17	Louisiana......	0.08
Tennessee......	0.15	Georgia........	0.08
Mississippi....	0.11	Michigan.......	0.07
Arkansas.......	0.11	North Carolina.	0.05
Illinois.......	0.11	West Virginia..	0.04
Florida........	0.11	South Carolina.	0.04
New York.......	0.11	Vermont........	0.02
Alabama........	0.11	Maryland.......	0.02
		Other..........	0.03
		Total	2.7

[a] Assumes additional production on all hayland, cropland pasture, and one-half of noncropland pasture in areas with sufficient rainfall to support the increased production. Also assumes 13 GJ/dry ton of grass. Uncertainty in total estimated at 30%.

Source: Ref. 8

Table III Average crop residue quantities usable for energy

State	Quantity[a] (EJ/yr)	State	Quantity[a] (EJ/yr)
Minnesota......	0.13	Texas..........	0.03
Illinois.......	0.12	Arkansas.......	0.03
Iowa...........	0.11	South Dakota...	0.03
Indiana........	0.08	Idaho..........	0.03
Ohio...........	0.05	Michigan.......	0.02
Wisconsin......	0.05	Missouri.......	0.02
California.....	0.04	Oregon.........	0.02
Washington.....	0.04	North Dakota...	0.02
Kansas.........	0.03	Other..........	0.14
Nebraska.......	0.03	Total..........	1.0

[a]Assumes 13 GJ/dry ton of residue.

Source: Ref. 8

ETHANOL FROM GRAIN

As mentioned above, current production of ethanol for blending with gasoline is about 2 billion l/yr (with an energy content of 0.04 EJ/yr). The long term potential for the production of fuel ethanol from grains is primarily a matter of economics, involving the cost of other liquid fuels, government subsidies and competition between food production and fuel production for the available cropland. The fuel replacement potential, on the other hand, involves technical questions related to what is often call the "net energy balance." Both of these aspects raise a number of important issues, which have been dealt with extensively elsewhere[8,9]. For the purposes of this section, however, the discussion will be limited to two major points, one about fuel displacement and one about food/fuel competition. And for the sake of brevity, ethanol from grains will be emphasized, although similar points can be made about ethanol from sugar crops.

The sum of the heats of combustion of the fuels consumed in cultivating, harvesting, and drying grain and converting it to ethanol is about the same as the heat of combustion of the resultant ethanol (about 22 MJ/l). Nevertheless, a little over half of this fuel is consumed in the distillery, which can be fueled with coal or solid biomass. In addition, when blended 10% in gasoline, the octane of a gasoline-ethanol mixture is about 1-4 octane numbers higher than that of the gasoline alone. (The exact boost in octane depends on the octane and composition of the gasoline.) Some oil refiners can take advantage of this chemical property of ethanol to reduce their refining energy requirements[10] by lowering the octane of the gasoline they produce. Also ethanol production consumes only the grain's starches, leaving the protein. This protein rich byproduct (usually gluten or distillers' dried grain, DDG) can be used as a substitute for other protein concentrates in animal feed, such as soybean meal, thereby eliminating the fuel consumption needed to produce that concentrate. When all of these factors are favorable for the replacement of fluid fuels (oil and natural gas), each liter of ethanol produced and blended in gasoline reduces the consumption of other fluid fuels by an amount that has roughly the same heat of combustion as that of the ethanol. (See Table IV.)

As production levels rise, however, several things happen to change this picture. First, the average refinery energy savings would drop because more ethanol would be blended[11] with gasoline from refineries that cannot take advantage of ethanol's octane boosting properties to lower their fuel requirements. Second, as farmers move to less productive land to satisfy the demand for grain, the farming energy requirements per unit of grain (and starch) produced would increase and the types of crops grown on specific parcels of land would change. In an unfavorable shift (e.g., corn replacing grain sorghum in Nebraska, grain sorghum production in Texas increasing), the consumption of fluid fuels for agriculture could increase by 35 MJ/l of ethanol produced[12].

Table IV Fluid fuel balance for the production of ethanol from corn

Source	Fluid fuel consumption (+) or replacement (-) (MJ/l ethanol)	
	Favorable	Unfavorable
Farming	+13	+35
Distillery byproduct	-3	--
Automobile	-22	-22
Oil refinery	-10	--
Total	-22	+13

Source: Ref. 8 and 12

Third, at ethanol production levels of about 4 times the current level, the market for the protein rich animal feed byproduct begins to be saturated, thereby eliminating this (modest) energy credit.

Taken together, these effects could change the fluid fuel "balance" for ethanol production from a net positive displacement of about the energy content of the ethanol itself to a net increase in fluid fuel consumption of about 13 MJ/l of ethanol produced, as shown in Table IV. As the level of ethanol production increases, the balance for each additional increment of ethanol production will shift from the net displacement level existing now to a net deficit at perhaps 2-4 times the present production level.

The other point to be made about ethanol production is its relationship to food prices, or its indirect costs. One acre of average cornland can produce a quantity of corn that, when converted to ethanol, yields enough protein concentrate (as a coproduct) to replace the soybeans that can be grown on 0.8 acre of average soybean land[13]. Since corn and soybeans are grown on similar types of land, this replacement effect reduces the increase in acreage under cultivation needed to produce fuel ethanol. As mentioned above, however, the animal feed market would begin to saturate if ethanol production levels quadruple; and the additional acreage under cultivation needed for each increment of ethanol production would also increase.

Current ethanol production levels probably have little effect on farm commodity prices, but as production increases farmers would increasingly have to expand cultivation onto land which is more expensive to farm than average cropland and on which crop yields fluctuate from year to year more than they do on average cropland. This would raise average farming costs and food prices and tend to cause farm commodity prices to fluctuate more from year to year. At ethanol production levels of 10-20 billion l/yr (about 2.5-5% of current gasoline consumption), the food price rise has been estimated to be about $1/yr in higher food prices for each liter/yr of ethanol production[8,14]. Furthermore, in order to maintain the current level of stability for food prices, the size of government subsidized buffer stocks of farm commodities (used to stabilize farm commodity prices) would have to be increased.

It is extremely difficult, perhaps impossible, to predict how the above considerations, future Federal and state subsidy policies, future oil prices, and other factors will develop to influence future ethanol production levels. Nevertheless, it is plausible to assume that ethanol production will not exceed 20 billion l/yr by the year 2000. At the other extreme, expiration of the direct Federal subsidy of $0.13/l for fuel ethanol in 1992, possible loss of state subsidies which are up to twice as large, and moderate oil prices could lead to ethanol production levels in 2000 which are lower than they are today.

SUMMARY

In addition to the sources of biomass fuel mentioned above, manure from animals in confined animal operations (cattle, swine, chickens, and turkeys) could be converted to biogas (usually 60% methane and 40% CO_2) through anaerobic digestion (bacterial attack in the absence of oxygen). If all such manure were converted, the annual production of biogas would have a heat of combustion of about 0.3 EJ. And various agricultural wastes, including orchard prunings, cheese whey, tomato pumice, etc., could be burned or converted to various fuels with combined heats of combustion of up to 0.1 EJ/yr.

Combining all these sources, OTA[8] has derived two plausible scenarios, shown in Fig. 3, for the potential supplies of biomass energy in the year 2000. While the totals for the heats of combustion of the fuels in the two scenarios differ by a factor of about three, the sceanrios are quite similar in two other respects. First, wood from increased forest management is the dominant potential source of biomass energy. Second, in both scenarios about 95% of the biomass energy is in the form of lignocellulosic material (wood and plant herbage). Producing large quantities of liquid or gaseous fuels from more or less conventional sources of biomass, therefore, is primarily a matter of synthesizing them from lignocellulose. Before proceeding to the conversion processes for lignocellulose, however, we will briefly consider unconventional sources of biomass.

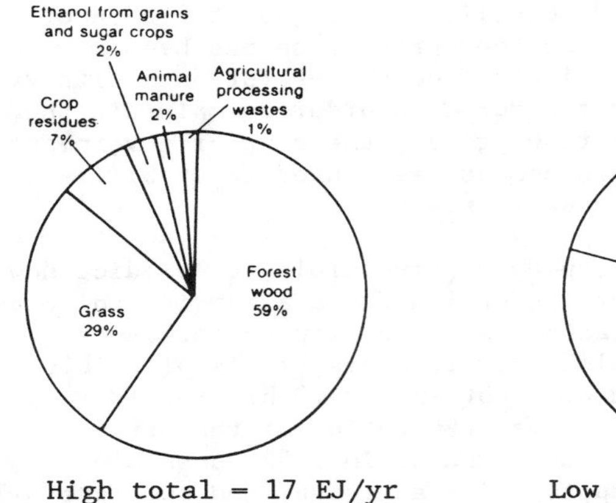

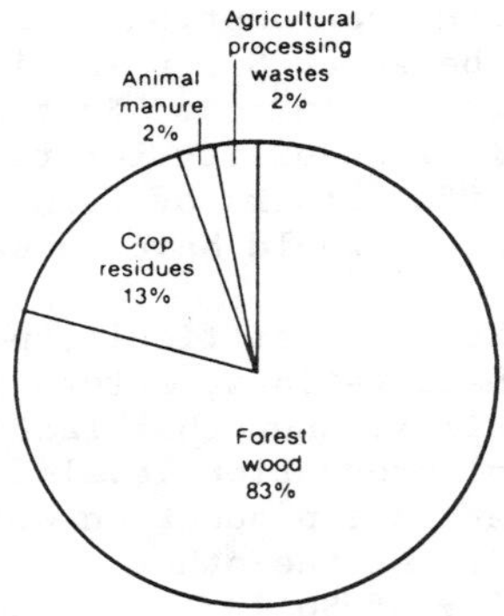

High total = 17 EJ/yr Low total = 6 EJ/yr

Figure 3 Potential bioenergy supplies (not including speculative sources or municipal wastes)

Source: Ref. 8

UNCONVENTIONAL BIOMASS SOURCES

A wide variety of plants, that have never been cultivated over large areas of the U.S., have been proposed as possible sources of biomass energy. These include crops that produce large quantities of oils, starches, and/or sugar, as well as fast growing trees and grasses. In addition, both fresh and salt water aquatic plants have been investigated. In order to assess the potential for these unconventional crops, it is necessary to consider some general aspects of photosynthetic efficiency and plant growth.

The maximum theoretical photosynthetic efficiency can be estimated as follows. About 10% of the light striking a leaf is reflected. About 43% of the energy contained in the light penetrating the leaf ends up as potential chemical energy in excited chlorophyll. And the basic chemical reactions (10 photon process) which use excited chlorophyll to convert CO_2 and water to glucose have an overall efficiency of 22.6%. Combining these factors yields an overall maximum photosynthetic efficiency of about 9%.

An efficiency approaching the theoretical maximum appears to have been achieved for a short time under laboratory conditions using an alga[15]. These results are controversial, however, and in practice several factors limit photsynthetic efficiency and plant growth. The most important of these factors, many of which are interdependent, are listed in Table V. For example, light saturation can be influenced by the CO_2 concentration, which is affected by other things. The key factors are light saturation, soil productivity (its ability to hold water, supply oxygen, release nutrients, and allow easy root development), weather (amount and timing of rainfall, absence of severe storms, length of growing season, temperature and insolation during growing season), and plant type (leaf canopy structure, longevity of photosynthetic system, sensitivity to various environmental stresses, etc.).

Table V Factors limiting plant growth

- o Water availability
- o Light saturation--a tendency for the photosynthetic efficiency to drop as the incident light intensity increases above values as low as about 10% of peak solar radiation itensity
- o Ambient temperature, especially wide fluctuations from ideal
- o Mismatch between plant growth cycle and annual weather cycle
- o Length of photoperiod (hours of significant illumination per day)
- o Plant respiration
- o Leaf area index--completeness of coverage of illuminated area by leaves or other photosensitive surfaces
- o Availability of primary nutrients--especially nitrogen, phosphorus, and potassium
- o Availability of trace chemicals necessary for growth
- o Physical characteristics of growth medium
- o Acidity of growth medium
- o Aging of photosynthetically active parts of plants
- o Wind speed
- o Exposure to heavy rains, hail, or icing conditions
- o Plant diseases and plant pests
- o Changes in light absorption by leaves due to accumulations of water film, dirt, or other absorbers or reflectors on surfaces of leaves or any glazing cover
- o Nonuniformity of maturity of plants in crop
- o Toxic chemicals in growth medium, air, or water, such as pollutants released by human activity
- o Availability of CO_2
- o Adjustment to rapid fluctuations in insolation or other environmental variables--i.e., "inertia" of plant response to changing conditions

Source: Ref. 3

Table VI provides a summary of some empirical photosynthetic efficiencies obtained in laboratory experiments and field cultivation. As can be seen, the sub-optimal conditions that exist in commercial farming lower the efficiency of corn, which is by far the most photosynthetically efficient crop grown over large areas of the U.S., to just under 1% as a national average[16]. Nevertheless, between 1950 and 1980 national average corn yields increased annually by about 2 bu/acre[17] and an extrapolation of past trends puts national average photosynthetic efficiency for corn at about 1.2% in the year 2000.

Table VI Photosynthetic efficiency summary

	Average PSE[a] during growth cycle (percent)
Maximum theoretical..............................	8.7
Highest laboratory short term PSE...............	about 9
Laboratory single leaves, high CO_2 or low O_2, C-3 plant, <7% full sunlight...........	6.3
Same as above, C-4 plant.........................	4.4
Corn canopy, single day, no respiration..........	5.0
Record U.S. corn (345 bu. of grain/acre, 120 day crop)..................................	3.0
Record sugar cane in continental U.S.............	3.0
Record Napier grass (El Salvador)................	2.5
Record U.S. state average corn, for states with more than 10^6 acres harvested (131 bu./acre, Illinois, 1982)..................	1.1
Record U.S. average corn (113 bu./acre, 1982)....	0.9

[a]Photosynthetic efficiency

Source: Ref. 3 and "Agricultural Statistics 1984," U.S.D.A.

For the purposes of estimating yields for land based crops grown over large areas of the U.S., the 1.2% efficiency that may be achieved for corn can serve as an optimistic, but plausible, goal for the growth of annual crops during their growing season. Perennial plants, however, are subject to cold weather at the beginning and end of their growing season; and an optimistic goal for their overall efficiencies probably is closer to 1%. Assuming that unconventional crops can be developed to achieve these average efficiencies, then the yields of various products shown in Table VII would be possible. These calculations also assume that the oil plants' synthesis of vegetable oil or hydrocarbons is 50-75% as efficient as the synthesis of lignocellulose and that the oil or hydrocarbon content of the plants could be increased to 25 weight percent, up from the 10-15 weight percent characteristic of the most promising unconventional oil and hydrocarbon producing plants. Obviously, these assumptions make oil and hydrocarbon yields more speculative than the other estimates.

In interpreting these estimates, it should be noted that yields from experimental plots and individual sites in the Midwest and Gulf Coast will exceed or already have exceeded the yields in Table VII. Projections based on estimated yields significantly larger than those in Table VII, however, either 1) are limited to the relatively small acreage of the best U.S. soils, 2) rely on technologies that do not now exist and are not anticipated in the forseeable future, or 3) require extensive management practices that are not likely to

Table VII Optimistic future average crop yields for plants under cultivation on millions of acres of average U.S. cropland

Region	Product	Plausible average yield (ton/acre yr)
Midwest	Dry plant matter	15
Gulf coast	Dry plant matter	21
Midwest	Sugar	4
Gulf coast	Sugar	6
Midwest	Starch (or grain)	4
Midwest	Vegetable oil or hydrocarbon	1.7-2.2 (1900-2500 1)
Area with 13 cm rainfall per year and no irrigation	Dry plant matter	0.6
Area with 13 cm rainfall per year and no irrigation	Vegetable oil or hydrocarbon	0.1 (110 1)

Source: Ref. 3

be cost effective unless there are dramatic increases in the prices for these farm commodities.

In addition to crops grown on land with adequate rainfall, Table VII shows estimates for yields of dry plant matter and vegetable oil or hydrocarbons from plants grown in arid regions without irrigation. In these cases, the plants' growth is limited by the amount of water. Experience suggests that, under these circumstances, plant growth would be about 1 kg of plant matter per 10^3 kg of water[18]. Dry land plants can survive with little water, but they grow only when water is available.

At the opposite extreme, there are several reasons to expect that aquatic plants may achieve higher average photosynthetic efficiencies than those for any land crop. Aquatic plants are never water stressed, nutrients and CO_2 levels can be held near the optimum, and the water's heat capacity prevents rapid temperature changes. All of these factors favor plant growth (see Table V).

Nevertheless, there are still significant problems to be solved before aquatic plants could be cultivated extensively for energy. For example, to achieve high photosynthetic efficiency, algae ponds must be stirred rapidly (which is energy intensive) to prevent light saturation; and harvesting of algae may also require energy intensive centrifuge filters. For kelp cultivation in ocean "farms," it is not known how to engineer the vast structures needed to support and protect the kelp and supply it with nutrients (from deep ocean waters) or what constitutes optimum growing conditions. Whether these and other problems can be solved in a way that makes aquatic plant cultivation competitive with other options is an open question. And until more experience has been gained and specific solutions to the problems have been developed and tested, no meaningful estimates of aquatic crop yields can be given.

This analysis of potential unconventional crop yields, while necessarily crude, does provide the basis for some guidance. First, using the yields in Table VII and the appropriate conversion efficiencies[19] for liquid fuel production from the primary products, it can be shown that oil crops might produce roughly twice as much liquid fuel per acre (in terms of its heat of combustion) and lignocellulose crops could produce three times as much liquid fuel per acre as could be gotten from sugar and starch crops. (Remember, however, that the estimates for oil crop yields are extremely speculative and achieving these levels of production will probably require extensive plant development.) Second, even if oil, sugar, or starch is the primary product, considerable lignocellulose will also be produced; and maximizing fluid fuel yields per acre cultivated will necessarily involve conversion of some of this lignocellulose. While neither of these points is conclusive, each reemphasizes the importance of lignocellulose for the production of fluid fuels from biomass, even in the area of unconventional crops.

Assuming that crops grown for their lignocellulose prove to be the most widely applicable, unconventional source of biomass energy, then an additional comparison is worth noting. The major candidates for lignocellulose crops are fast growing trees (e.g., poplar, red alder) and grasses (e.g., kenaf, tall fescue, Napier grass). The trees have the advantage that they produce a denser fuel (heat of combustion per unit volume), which is less expensive to transport and for which more commercial combustion technologies exist. The grasses, however, have several advantages of their own. Their higher leaf index (leaf area per unit land area), compared to trees, leads to lower light intensities on the leaves, which favors higher photosynthetic efficiency. Grasses can be harvested first a few months after initial planting, whereas trees require two years' or more of cultivation (and the associated costs) before the first harvest. In addition, if disease or pests destroy the crop, a grass farmer would loose only a fraction of a year's production, while a tree farmer could loose several years' investment in cultivation. Finally, if the energy farmer decides to convert the land to other crops, it is considerably easier to plow under a grass field than it would be to convert a field of fast growing trees with its extensive and tenacious root system.

LIQUID AND GASEOUS FUEL SYNTHESIS

In the previous two sections we have shown that lignocellulose is the most abundant source of biomass for energy and is likely to be so even when unconventional sources are considered. This is not to say that other types of energy crops may not serve an important role locally for certain soil types, climates, and economic conditions. But from a national perspective, by far the largest potential for biomass energy is with lignocellulose; and a consideration of liquid and gaseous fuels from biomass should therefore center around this material.

While lignocellulose can be pyrolyzed under pressure to form an acidic oil (which might yield useful products), our current understanding suggests two avenues for converting lignocellulose to more useful fuels. The two principal routes are 1) to convert the cellulose and hemicellulose to sugar and convert the sugar biochemically to ethanol or methane or 2) to convert the entire lignocellulose thermochemically to a low energy fuel gas for combustion or to synthesis gas (CO and H_2) for methanol or methane synthesis. Each of these categories of conversion processes is considered below.

COVERSIONS INVOLVING A BIOCHEMICAL STEP

Lignocellulose typically contains 50% cellulose, 15-25% hemicellulose, and 25-35% lignin. Cellulose is a polymer of six carbon sugars (like those in table sugar), while hemicellulose is a polymer of five carbon sugars. Lignin, on the other hand, is a polymer containing a hydrocarbon chain with methoxy-phenyl groups[20] attached.

The key to converting lignocellulose by the first route mentioned above is the inexpensive production of sugars by hydrolysis of the cellulose and hemicellulose. Once this is accomplished, the sugars can be readily fermented to ethanol with commercial technology or anaerobically digested to biogas using processes which are currently under development[21] (see Fig. 4).

Some types of biomass, such as Kentucky blue grass, water hyacinth, and ocean kelp, can be digested to biogas easily without special pretreatments, but the potential quantities of these feedstocks are relatively small with current technology. Municipal solid wastes are also converted to biogas in some landfills. The reactions are slow, however, and the biogas is more properly considered a fortunate and locally important byproduct of the landfill operation rather than an energy technology per se. Furthermore, due to weather conditions and other factors, the national total production of biogas from landfills is likely to be relatively small.

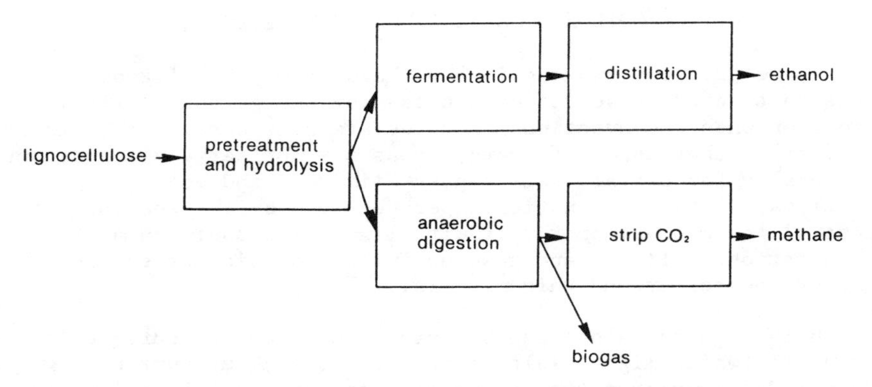

Figure 4 Block diagram for synthesis of ethanol or methane from lignocellulose using biochemical processes.

A third possibility, which is currently under development, is to produce biogas from potentially more abundant sources of biomass using anaerobic digesters without pretreatment. Jewell and coworkers[22] have recently conducted tests digesting untreated Napier grass, wheat straw, and sorghum in an anaerobic digester held at 55° C. Biogas yields were high (about 90% of the energy content of the sorghum was converted to biogas), but the reactions were slow, requiring 70 days to convert 70% of the biomass and an additional 200 days for the other 20%. With these slow reaction rates, the digesters would have to be quite larage (and thus expensive) relative to the rate of gas production.

While these studies should be continued to see if the reaction rates can be increased and to better understand the digestion process, current indications are that converting large amounts of biomass to methane will probably require development of inexpensive pretreatments to increase digestion rates from the more abundant sources of lignocellulose; and these pretreatments are likely to be similar to those needed for ethanol production.

While hemicellulose can easily be extracted from lignocellulose and hydrolyzed, cellulose is more of a problem. What was apparently the first acid catalyzed hydrolysis of the cellulose in wood was described in a German patent issued in 1880[23]. Variations on this process have been used to produce animal fodder and ethanol in various countries, but the processes are relatively expensive. One variation uses dilute acid at 150-200° C, but the high temperature causes decomposition of some of the sugar produced, resulting in low yields. This leads to relatively high feedstock costs per unit ethanol produced. Another variation uses concentrated acid at 40° C. The sugar yields are high, but the need to neutralize large quantities of acid results in high process chemical costs. A third possibility might be to use dilute acid or gaseous hydrofluoric acid[24] at elevated pressures, but we are not aware of any cost estimates for processes using this approach.

Partly because of the high costs for acid hydrolysis, considerable attention has been paid in recent years to biological hydrolysis of lignocellulose. Bacteria have been developed which can hydrolyze cellulose enzymatically to sugar with 80% of the theoretically maximum yield. However, in lignocellulose the lignin sterically protects the cellulose from enzymatic attack. Consequently, before the hydrolysis can take place the lignin matrix must be broken down or swollen. Various processes, involving extensive mechanical grinding or high pressure steam treatment and explosive decompression, have been developed to do this, but these processes tend to be expensive and energy intensive. Other processes have been proposed that preferentially dissolve the cellulose or lignin in various solvents to separate these components. And it may be possible to develop a micro-organism that consumes the lignin but not the cellulose.

Table VIII Plausible future cost of ethanol from wood in a 190 million liter/year facility (1984 $)

Fixed investment	$122 million
Working capital	$13 million
Total investment	$135 million
	$/liter
Labor, chemicals	0.10
Electricity (@ $0.052/kwh), (possibly) fuel	0.07
Feedstock ($30/ton, 415 l/ton)	0.07
Capital charges (20-40% of investment)	0.14-0.29
Total	0.38-0.44

Source: Adapted from G.H. Emert and R. Katzen, "Chemicals from Biomass by Improved Enzyme Technology," Presented at the symposium "Biomass as a Non-Fossil Fuel Source," April 1-6, 1979, ACS/CSJ Joint Chemical Congress, Honolulu, Hawaii.

If processes are successfully developed, the costs of producing ethanol from lignocellulose might be something like those shown in Table VIII. If the lignin in the feedstock could provide all of the processing fuel needs[25], the conversion efficiency (heat of combustion of ethanol produced divided by feedstock heat of combustion) could be over 50% (380 l ethanol/ton biomass). Probably, however, some additional fuel will be needed and the efficiency (heat of combustion of ethanol produced divided by feedstock plus fuel heat of combustion) is likely to be closer to 45% or less. (No comparable cost or efficiency estimates for biogas production are available).

Successful development of processes that produce cellulose or sugar inexpensively from lignocellulose could have significant effects beyond fuel production. For example, the cellulose extraction could serve as the basis for a less expensive route to paper production; and a cheap and abundant source of sugar would provide powerful competition for traditional sugar producers. Until and unless this occurs, however, the direct production cost for ethanol from sugar or starch crops is likely to be lower than that for ethanol from lignocellulose, except in very site specific and atypical circumstances[26]. But, as noted above, production of ethanol from sugar and starch crops can lead to high indirect costs associated with the competition between food/feed crops and fuel crops for the available farmland; and increased farm commodity prices could eventually raise sugar and starch prices to a level where lignocellulose is competitive as a feedstock.

THERMOCHEMICAL CONVERSION

The second route for conversion of lignocellulose to fluid fuels is the thermochemical route. The critical step in these processes is gasification of the biomass. Air blown gasifiers can be used to produce a low energy fuel gas; while the more expensive oxygen blown or pyrolysis gasifiers produce synthesis gas. If synthesis gas is produced, commercial processes can then be used to catalytically convert it to methanol or methane[27] (see Fig. 5). (In most commercial methanol synthesis, natural gas or residual fuel oil is used to produce the synthesis gas.)

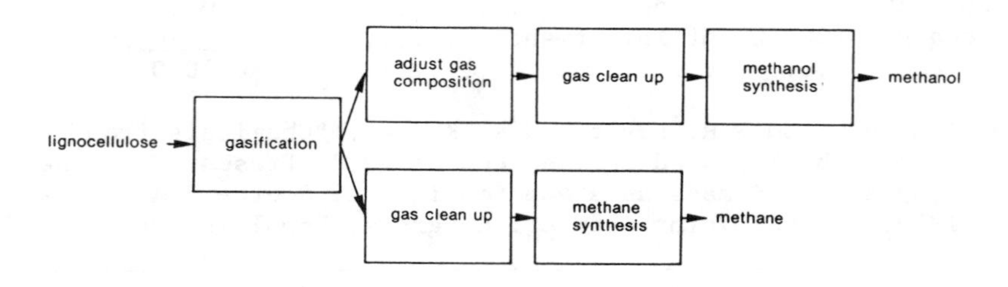

Figure 5 Block diagram for the synthesis of methanol or methane using thermochemical processes

While synthesis gas can be used directly as a boiler fuel or for process heat, in many cases a low energy fuel gas from the less expensive air blown gasifier is adequate for this purpose. This type of gasifier partially oxidizes the biomass with air to provide the heat necessary to decompose and gasify the biomass. In effect, the gasifier burns the biomass incompletely and the exiting gas can serve as a gaseous fuel.

Unlike synthesis gas, which has an energy density (heat of combustion per unit volume) of 10-15 MJ/m^3 at 25° C and 1 atmosphere, the gas from an air blown biomass gasifier is more typically 7 MJ/m^3 or less, due to dilution with nitrogen. Because of its low energy density, it is not economical to transport the fuel gas over large distances; and air blown gasifiers are typically operated at the site where the fuel gas is used.

Use of fuel gas with an energy density of around 7 MJ/m^3 does not present any serious problems in a boiler; but if poor design of the gasifier or operating difficulties lowers the density significantly below this level, the maximum power output from the boiler (pounds of steam per hour) may drop. While this may be of little consequence in some applications, in others it could place strict requirements on the gasifier design and reliability.

In a 1982 study[28], 14 air blown gasifiers then in commercial operation were surveyed. The gasifiers ranged in power from 0.4-20 kW. All operated at above 80% efficiency; and although none of the users had recorded operational costs, the authors of the study estimated the fuel gas being produced cost $2.75-$5.30/MJ (in 1984 dollars). By way of comparison, the average cost of natural gas for industrial customers in 1984 was about $4.10/MJ.

For methanol or methane synthesis, it is desirable to produce a synthesis gas that is free of nitrogen. Removal of the nitrogen from the gas prior to methanol or methane synthesis would be expensive. And inclusion of nitrogen in the gas during synthesis would also be expensive, due to the larger volumes of gas that would have to be heated, compressed and otherwise processed.

Consequently, the two types of gasifiers suitable for methanol or methane synthesis are the oxygen blown and pyrolysis gasifiers. The oxygen blown gasifier uses oxygen to partially oxidize the biomass to produce the heat needed to decompose and gasify the remaining feedstock. Pyrolysis gasifiers use only heat to decompose and gasify the moist biomass. It is not know which gasifier type will ultimately prove to be the most economic. The oxygen blown gasifier requires the energy intensive and expensive equipment needed to produce oxygen, but the gasifier size can be kept relatively small because of the high reaction rates. Pyrolysis gasifiers require no oxygen plant, but the gasifiers are relatively large in order to accomodate feedstock residence times of up to an hour or more. Both types of gasifiers have experienced problems with feedstock clogging and excessive wear on certain parts (typical problems for solids handling).

Since a pyrolysis gasifier is more energy efficient than an oxygen blown one, we will consider some of the chemistry and thermodynamics of methanol synthesis from wood using pyrolysis gasification in order to get an idea of what the overall conversion efficiencies may be. The data used here are taken from the experiments and development work conducted by Wright-Malta Corp. on their pyrolysis gasifier and reported by Coffman[29]. No comparable estimates are available for the efficiency of methane synthesis, although Coffman's analysis does provide some general indications of what these efficiencies might be.

The Wright-Malta gasifier operates at about 14 atmospheres. The green wood (45% moisture) is continuously fed into the gasifier through a special lock hopper. In a methanol plant, exit gases from downstream processes would be passed through the gasifier jacket to help heat the biomass as it gradually passes through the gasifier. The exit gas from the gasifier would be put through a so called shift reaction, to adjust its composition; and the resulting synthesis gas would be compressed to about 100 atmospheres in the presence of a catalyst to produce methanol.

Based on his experiments, Coffman concluded that the wood gasification reaction at 14 atm. and 600 C in his gasifier could be written as

$$8\ C_6H_9O_4 + 58\ H_2O \text{ (i.e. 45\% moisture wood) } \rightarrow$$

$$2\ C + 20\ CH_4 + 12\ H_2 + 4\ CO + 22\ CO_2 + 42\ H_2O \qquad (1)$$

This reaction is exothermic (about 1.2 MJ per mole of reaction as written) and it is self sustaining, provided preheated biomass is introduced continuously.

The next step is to increase the gas temperature to 800 C in the presence of an equilibrating catalyst. Based on a computer program which calculates the gas composition with the minimum Gibbs free energy, Coffman determined that the shift reaction would alter the gas composition as follows

$$20\ CH_4 + 12\ H_2 + 4\ CO + 22\ CO_2 + 42\ H_2O \rightarrow$$

$$5\ CH_4 + 51\ H_2 + 25\ CO + 16\ CO_2 + 33\ H_2O \qquad (2)$$

This reaction is endothermic and about 3.5 MJ of heat per mole of reaction as written would have to be supplied externally.

Exit gas from the shift reaction would then be fed through the gasifier jacket to help heat the incoming biomass and vaporize its moisture. After cooling this gas and removing the water and CO_2 in a Benfield potassium carbonate system, it would be compressed to 100 atm. and converted to methanol at 350 C in the presence of a catalyst via the following reaction

$$5\ CH_4 + 51\ H_2 + 25\ CO \rightarrow 5\ CH_4 + H_2 + 25\ CH_3OH \qquad (3)$$

This reaction is exothermic, releasing about 2.4 MJ of heat per mole of reaction as written. In order to attain good methanol yields, the reaction vessel would be cooled with water; and the resulting steam would be used to generate power for the gas compressor and other mechanical devices in the plant. Furthermore, the byproduct hydrogen and methane together have a heat of combustion of about 4.6 MJ per mole of reaction as written, or over 1 MJ more than is needed to drive the shift reaction.

Overall, no heat source beyond that provided by the feestock is needed and the theoretical energy efficiency of the methanol synthesis (heat of combustion of methanol product divided by heat of combustion of the feedstock wood, if it were dry) is over 77%. In practice, it is not uncommon for chemical plants to achieve 90% of theoretical yields, and an overall efficiency of 70% seems plausible. This compares quite favorably with typical coal to methanol processes, in which about 55% of the heat of combustion of the feedstock coal ends up as methanol[30].

Table IX Plausible future cost of methanol from wood in a 140 million liter/year facility (1984 $)

Fixed investment	$45 million
Working capital	$5 million
Total investment	$50 million
	$/liter
Labor, chemicals, etc	0.02
Feedstock ($30/ton, 750 l/ton)	0.04
Capital charges (20-40% of investment)	0.07-0.15
Total	0.13-0.21

Source: Adapted from Ref. 26

If one assumes a conversion efficiency of 70% and Coffman's estimates for plant and operating costs, the cost of methanol from wood might be something like that shown in Table IX. In practice, because of the cost and difficulty of collecting large amounts of biomass to central locations, more of the total biomass resource may be accessible if plants smaller than the one in Table IX are used. Normally this scale down involves an increase in the cost of the plant per unit of output. However, if smaller plants are prefabricated in a factory and simply assembled in the field (rather than being field erected like the one in Table IX), then the investment cost per unit output and the cost of the methanol could still be comparable to the figures shown in Table IX. If so, then methanol from wood and plant herbage would be competitive with methanol from coal.

If methane is the desired product, the shift reaction would probably be eliminated, since the products of gasification (Eq. (1)) include considerable amounts of methane and the ratio of 3:1 for the H_2 to CO concentrations is ideal for methane synthesis,

$$3\ H_2 + CO \ \text{-->}\ H_2O + CH_4. \tag{4}$$

Based on these stohiciometric yields, the energy efficiency of methane synthesis could be 88%, if the energy released in the exothermic gasification and methane synthesis steps is sufficient to run the compressors and other mechanical equipment. In all probability, it is not, however. Furthermore, some of the products in Eq. (1) reported as methane are actually larger hydrocarbons, which would have to be separated cryogenically from the methane in order to produce pipeline quality gas. These factors would lower the overall efficiency, but relatively high efficiencies for methane synthesis still seem plausible. No cost estimates for this synthesis are available, however.

END USE OF THE FUELS

If properly dried and cleaned, methane synthesized from biomass can be introduced into natural gas pipelines and subsequently used for space heating, hot water, and the various industrial processes that use natural gas. While this is practical for methane from some land fills and some other sources of biomass, most of the potential supplies of lignocellulose are located away from existing gas pipelines. Consequently, either the biomass would have to be transported long distances or numerous new feeder pipelines would have to be built; and it is not clear that either of these options will be economic.

Alternatively, the methane could be used at or near the place it is produced. This is often the best solution where wet biomass, such as animal manure, is the feedstock; but in most cases where lignocellulose is the feedstock, it would probably be less expensive to produce and use the low energy fuel gas from an air blown gasifier. In either case, however, the number of locations where these options prove viable may be limited.

Because of these limitations, ethanol or methanol may ultimately prove to be the fluid fuels from biomass that can serve the largest markets. As mentioned above, ethanol can be used as an octane boosting additive to gasoline. Due to the fact that ethanol requires less air for complete combustion than does an equal volume of gasoline, an engine with a fixed carburetor runs "leaner" on an ethanol/gasoline blend than it does on gasoline alone. This limits the amount of ethanol that can be added to about 10-20% of the gasoline in most cars with fixed carburetors. For engines with automatically adjusted carburetors (driven by oxygen sensors in the exhaust stream), the amount of ethanol that can be blended depends on how far the carburetor can be adjusted.

Methanol can also be used as an octane boosting additive to gasoline. Generally the methanol is blended together with cosolvents (usually higher alcohols) to increase the solubility of methanol in the gasoline and thereby help prevent phase separation when small amounts of water are present in the fuel (a common occurance). Commercial methanol blends are limited to about 3% methanol, because of the phase separation problem and the fact that methanol in higher concentrations can attack certain elastomers and metals commonly found in fuel systems. Replacing the components attacked by methanol with more methanol resistant materials and careful handling of the fuel to prevent introduction of water would enable higher concentrations of methanol in gasoline blends. Like ethanol, however, the maximum concentration would be determined by the leaning effect and how far the carburetor can be adjusted.

Both ethanol and methanol can also be used as stand alone fuels in vehicles, and methanol is the required fuel for high performance racing cars. However, due to the alcohols' high heat of vaporization and low vapor pressure at low temperature, the carburetor must be heated (typically with engine waste heat) and some provisions are necessary for starting the engine in cold weather (such as blending small amounts of hydrocarbons in the alcohol). But because of the high octane of the alcohols, ethanol fueled engines could be 10-15% more efficient and methanol fueled engines 15-20% more efficient than comparable gasoline engines, provided the engines' compression ratios are increased to take advantage of the higher octane fuels. In addition, it is possible to use engine waste heat to decompose methanol into CO and H_2 before it is burned. Since the heat of combustion of the CO/H_2 mixture is 20% greater than that of the methanol from which it came, this approach can lead to still higher engine efficiencies. Other modifications that take advantage of the alcohols' unique properties to increase engine efficiency and performance are also possible, but these examples serve to illustrate the potential advantages of alcohols over gasoline as an automobile fuel.

Before automakers will manufacture and people will buy vehicles designed to take advantage of these properties of alcohol, however, the alcohols must be readily avialble to the public. But service station owners are unlikely to make the alcohols available until there are enough alcohol fueled vehicles on the road to make the necessary investments economic. A possible interim strategy might be to manufacture vehicles capable of running on anything from pure gasoline to pure alcohol. While the manufacture of such vehicles is technically feasible, they probably could not take advantage of the alcohols' unique properties to increase fuel efficiency; and it is unclear whether consumers would be willing to bear the added cost (relative to a pure gasoline vehicle) in the hopes that the alcohols would become readily available at competitive costs within the lifetime of the vehicle. Because of this impasse, it may be easier initially to convert methanol to gasoline with a zeolite catalyst using a process developed by Mobil Oil Corp. But for the longer term future, the unique possibilities afforded by the alcohols used as stand alone fuels should not be overlooked.

CONCLUSION

The preceding analysis indicates that, for the forseeable future, lignocellulose is likely to continue to be the largest source of biomass used for energy, even when unconventional biomass sources are considered. Processes for converting lignocellulose to fluid fuels include both biochemical and thermochemical routes. Both types of processes need further development; but, for liquid fuel production, current indications are that the thermochemical route is more energy efficient and produces a less expensive fuel. (Less complete information is available for gaseous fuel production.) The principal liquid fuel from thermochemical conversion of lignocellulose is methanol. While methanol can be used as a stand alone fuel in vehicles, initially it may be more practical to convert it to gasoline. Once fuel production levels are high enough to accommodate substantial sales of vehicles specifically designed to run on methanol, however, the gasoline production step could be phased out.

REFERENCES

1. EIA, "Monthly Energy Review," Energy Information Administration, U.S. Department of Energy, DOE/EIA-0035(84/09), p. i, September, (1984).

2. Ref. 1, pg 4. This number includes 2.6 EJ of wood energy, which is not normally included in EIA summary tables.

3. OTA, "Energy from Biological Processes, Volume II--Technical and Environmental Analyses," OTA-E-128, September (1980), available from National Technical Information Service,, 5285 Port Royal Road, Springfield, VA 22161, order # PB 81-134 769.

4. Ref. 1, p. ii.

5. USDA, "Forest Statistics of the U.S., 1977," U.S. Forest Service, (1978).

6. See Ref. 3 and J.S. Bethel, et. al., "Energy From Biological Processes, Volume III Part A: Energy from Wood," (1981), NTIS order # PB 81-134 777.

7. For example, the timber may be too far from roads or on steep slopes where harvesting is difficult. In addition, the owners of some of the forestland may not want to disturb their land. In one survey, for example, about half of the forestland owners (owning 9% of the forestland) indicated that they would not harvest their forest because of its scenic value or because their tracts were too small. In the same survey, however, 10% of the owners (owning 53% of the forestland) intended to harvest within 10 years and one third of the owners (owning 87% of the land) intended to harvest "some day." (N.P. Kingsley and T.W. Birch, "The Forest-Land Owners of New Hampshire and Vermont," USDA Forest Service Resource Bulletin NE-51, (1977).)

8. OTA, "Energy from Biological Processes, Volume II--Technical and Environmental Analyses," OTA-E-128, September (1980), NTIS order # PB 81-134 769.

9. OTA, "U.S. Vulnerability to an Oil Import Curtailment: The Oil Replacement Capability," OTA-E-243, p. 62-3 and p. 73-5, September (1984), NTIS order # PB 85-127 785.

10. The actual savings depends on the configuration of the refinery and the relative amounts of various products being produced. The reforming step, which increases the octane of middle distillates to a level suitable for gasoline, produces hydrogen by creating more double bonded, ring, and aromatic compounds. In many refineries, this hydrogen is needed to upgrade other fuels; and the hydrogen lost by less severe reforming has to be replaced with hydrogen from other sources, such as

the reaction of residual oil with steam. In this case, there would be little incentive to reduce the severity of the reforming and little reduction in refinery energy requirements.

11. Use of neat ethanol in vehicles designed for this fuel would eliminate the refinery savings, but the potential 15% increased efficiency--J/mile--of these vehicles could save an additional 3.3 MJ/l of ethanol used. However, production of several hundred thousand vehicles per year would be needed before automobile manufacturers would redesign the motors to take full advantage of this potential. For lower production runs, engines designed for gasoline would be adapted to ethanol, thereby largely eliminating this potential savings.

12. Assuming marginal acreage is 70 percent as productive as average acreage in each state and using the data from "Energy and U.S. Agriculture: 1974 Data Base," Vol. II, U.S. Department of Agriculture, Federal Energy Administration.

13. Other grains are less productive, however, and a displacement of 0.3 acres of soybeans per acre of grain production is more typical of these other grains.

14. Schnittker Associates, "Ethanol: Farm and Fuel Issues," prepared for the U.S. National Alcohol Fuels Commission, August (1980).

15. V.C. Goedheer and J.W. Kleinen Hammans, Nature, 256, 333(1975).

16. By way of comparison, trees growing on average commercial forestland have an average photosynthetic efficiency of 0.07-0.15%.

17. Duvick estimated that 60% of the increase up to 1976 was due to genetic improvements and 40% to improved agricultural practices. (D.N. Duvick, Maydica, XXII 187 (1977).)

18. Ref. 3, p. 93

19. For methanol synthesis from lignocellulose, a 70% energy conversion efficiency is assumed (see section on conversion processes) with a feedstock energy densisty of 13 GJ/ton, appropriate for grass or crop residues. (If wood were the feedstock, the feedstock energy density would be 16-18 GJ/ton.) For sugar, the conversion is 510 l ethanol (22 MJ/l) per ton. Vegetable oils and hydrocarbons are taken to contain 39 GJ/ton. Although the oils may need additional refining, any conversion loss here has been ignored.

20. Benzene rings with OCH_3 groups attached.

21. See Gosse Schraa and William J. Jewell, J. Water Pollution Control Federation 56, 226 (1984).

22. W.J. Jewell, R.J. Cummings, B.K. Richards, and D.J. Rector, "Methane Production from the High Solids Digestion of Terrestrial Biomass,"

Presented at the Institute for Gas Technology's Ninth Annual Symposium on Energy from Biomass and Wastes, Orlando, Florida, January 31, 1985.

23. H.F.J. Wenzl, The Chemical Technology of Wood, translated by F.E. Brauns, Academic Press, New York, (1970).

24. C.M. Ostrowski, J. Aitken, and D. Free, "New Developments in Fuel Ethanol Production by Gaseous Anhydrous Hydrofluoric Acid (HF) Hydrolysis of Hardwood, Populus Tremuloides," Bioenergy '84 Proceedings, Vol. 3, Bio-Energy Council, P.O. Box 12807, Arlington, VA 22209-8807.

25. An energy efficient plant for converting corn into anhydrous ethanol consumes fuel with a heat of combusion of about 16 MJ/l of ethanol produced.

26. For example, acid hydrolysis to produce ethanol together with various chemical feedstock byproducts, such as phenol (from the lignin), could be economic today. The chemical feedstock market, however, would be saturated long before any significant effect on the fuel markets could be noted.

27. There is also the possibility of converting synthesis gas to fluid fuels with bacteria, but this concept is not well developed and will not be considered here. See J.G. Zeikus, R. Kerby, and J.A. Krycki, Science 227, 1167 (1985).

28. R. Hodam, R. Williams, and M. Lesser, "Engineering and Economic Characteristics of Commercial Wood Gasifiers in North America," SERI/TR-231-1459, contractor report prepared for Solar Energy Research Institute, November (1982).

29. J.A. Coffman, "Steam Gasification of Biomass," Wright-Malta Corp., Ballston Spa, N.Y. 12020. No date given.

30. OTA, "Increased Automobile Fuel Efficiency and Synthetic Fuels: Alternatives for Reducing Oil Imports," OTA-E-185, p. 171, September (1982), NTIS order #PB 83-126 094,

CHAPTER 27

ENVIRONMENTAL EFFECTS OF OBTAINING LIQUID FUELS FROM BIOMASS

Steven E. Plotkin
Office of Technology Assessment
U.S. Congress
Washington, D.C. 20510

INTRODUCTION

The use of biomass materials as an energy source has generally been viewed as highly attractive, especially in the 1970's when concerns about energy shortages and environmental problems seemed to peak simultaneously. Biomass energy is a form of solar energy, and thus can share the high regard with which solar energy is widely viewed. It is domestically available (indeed, the low energy density of biomass feedstocks virtually precludes biomass imports except in finished product form), it generally is renewable if properly managed, and the technology involved in its production usually is not exceedingly complex. Many biomass conversion technologies can be run at relatively small scale without major sacrifices in economies-of-scale, so biomass energy appeals to groups who favor decentralization of U.S. energy supply. Finally, biomass materials generally are relatively free of the minerals--especially sulfur--that cause some of the pollution problems in coal and oil combustion and synthetic fuel conversion. Thus, biomass energy often is characterized as a "safer, smaller-scale, and more environmentally benign alternative to coal and nuclear development."[1]

It probably is wise to be skeptical about environmental claims for technologies that have not yet been widely deployed. It is worth noting that one biomass technology that is widely deployed--residential wood-burning for space heat--has created significant air pollution problems where its application has been intensive. Thus, the potential environmental effects of producing large amounts of liquid fuels--alcohols--from biomass sources should be examined carefully. These effects will stem from fuel substitution -- reducing requirements for other fuels -- as well as from the massive changes in agriculture and forestry that will be required, the construction and operation of conversion facilities, and the actual use of the fuels, most likely in automobiles and other vehicles.

FUEL SUBSTITUTION EFFECTS

Alcohol fuels from biomass can substitute for domestic and imported petroleum (as gasoline or diesel fuel) or synthetic fuels from coal and oil shale with the "choice" dependent primarily on relative prices.

0094-243X/85/1350548-14$3.00 Copyright 1985 American Institute of Physics

A likely scenario in which biomass liquids production grows rapidly would be one where oil prices -- both domestic and imported -- escalate sharply. Because methanol from coal appears to be somewhat similar in production costs to wood-based methanol, given the latest technologies, biomass alcohols would likely substitute for domestic and imported petroleum, and most likely the latter (assuming domestic production proceeds at the maximum rate allowed by the high price). Because the importation of oil has few significant environmental effects on the U.S. aside from occasional spills, the fuel substitution-related environmental benefits of biomass alcohol use are likely to be relatively minor in this time period. In the longer time period, however, large scale alcohol fuel production will reduce the demand for coal and oil shale-based synthetic fuels. Substitution for such synfuel production should generate significant environmental benefits.

EFFECTS OF OBTAINING BIOMASS FEEDSTOCKS

Of the remaining potential avenues for environmental changes, the process of obtaining the massive amounts of feedstock materials necessary for large-scale production clearly will cause the most important effects. All of the fully credible biomass fuel cycles involving agricultural crops, grasses or trees require various degrees of ecological alteration, replacement or disruption on vast land areas. Table 1 summarizes the land requirements to replace 1000 gallons of gasoline equivalent per year using a variety of alcohol fuel cycle alternatives. Given the limitations in the market for distillery byproducts,[2] saving ten percent of current gasoline use would require a minimum of about 25 million acres with a combination of sugar/starch crops and grasses[3] unless grass or wood crops of higher productivity than today's are developed and exploited extensively as energy crops. If this savings were attempted strictly by the use of gasohol (with the ethanol made from corn), the land requirements probably would be at least 40 million acres.[4] If methanol from wood were the major source, upwards of 50 million acres of forestland (assuming average productivity) would have to be placed in intensive management. The extensiveness of the biomass base, cited as an environmental advantage by opponents of coal and nuclear power, obviously may carry with it some severe disadvantages.

Each of the alcohol feedstocks -- wood, grains and sugar crops, grasses and legume herbage, and crop residues -- will have distinct environmental effects created both by its physical nature and the institutions influencing its use. It is generally true of these feedstocks that, based on their physical attributes and the variety of management options available, their potential impacts may range widely, from minor (or even positive) to severe. It is the choice of management options, influenced by a complex set of behavioral incentives, that dictates whether or not the provision

TABLE 1. INCREASED ACREAGE REQUIREMENTS FOR GASOLINE DISPLACEMENT

A net premium fuel displacement equivalent to 1000 gallons per year of gasoline can be achieved from:

- About 1.4 acres of new cropland put into production with corn, distillery fueled with coal, byproducts used fully to replace soybean production and increase corn production further and ethanol used as octane-boosting additive to gasoline.
- About 2.1 acres grown in grass and converted to methanol used as an octane-boosting additive to gasoline.
- About 3.3 acres of new cropland grown with corn, distillery fueled with coal, byproducts used fully to replace soybean production and increase corn production further, ethanol used as stand-alone fuel.
- About 3.8 acres grown in grass and converted to ethanol used as an octane-boosting additive to gasoline.
- About 3.9 acres grown in grass, converted to methanol used as stand-alone fuel.
- About 6.4 acres grown in grass, converted to ethanol used as a stand-alone fuel.
- About 7.4 acres grown with corn, distillery fueled with coal, no byproduct utilization, ethanol used as octane-boosting additive to gasoline.
- About 25 acres grown with corn, distillery fueled with coal, no byproduct utilization, ethanol used as stand-alone fuel.
- About 330 acres in grain sorghum, distillery fueled with coal, ethanol used as stand-alone fuel.
- Infinite acres of corn or other grain if oil used as distillery boiler fuel and ethanol used as stand-alone fuel.

Source: Adapted from Office of Technology Assessment, Energy from Biological Processes, July, 1980.

of sufficient feedstock for large scale alcohol production will be viewed as environmentally acceptable.

Obtaining large quantities of wood for methanol production will involve fully utilizing the logging residues that now are left in the forest or burned, and increasing the scale and intensity of management (more acreage under intensive management, shorter times between thinnings, more complete removal of biomass, more conversion of low-quality "stands" of trees). It might involve an increased harvest of forestland with lower productive potential -- so-called "marginal lands" -- and it will almost certainly mean that lands not now subject to logging will be logged. All of these effects are intensifications of current trends resulting from increased demand for wood products and a decline in the availability of timber from old-growth stands.

If handled with care, there are substantial benefits to be gained from a "wood-for-energy" strategy. Forests that were severely degraded in the past by unsound forestry practices can be cleared and replanted. The removal of logging slash will, when managed carefully, lower the incidence of fires, remove the habitat of bark beetles and other forest pathogens, and promote reforestation. The added yields of high quality timber from increased management may relieve the pressure on the few remaining unprotected stands of scenic, old growth timber as well as on other stands of high quality timber that also have significant esthetic, recreational and ecological value. And although intensified management will change both the visual and ecological character of the affected forests, the changes need not be unattractive or damaging to the abundance of wildlife in the forests.

Nevertheless, there is substantial potential for damage to the forests if they are mismanaged. High rates of biomass removal coupled with short rotations may cause a depletion of nutrients and organic matter from the more vulnerable forest soils.[5] (A particular problem here is the current lack of knowledge about the long-term effects of the intensive management that might become the norm and the probability that some of the more important effects would be both hard-to-detect and difficult to ascribe to a specific cause.) The impacts of poor logging practices -- erosion, degraded water quality, esthetic damage, and damage to valuable ecosystems -- may be aggravated by the lessening of recovery time (because of the shorter rotations) and any lingering effects of soil depletion on the forests' ability to rebound. The intensified management may further degrade ecological values if it incorporates widespread use of mechanical and chemical brush controls, very large area clearcuts and elimination of "undesirable" tree species, and if it neglects to spare large pockets of forest to maintain diversity. Finally, the incentive to "mine" wood from marginal lands with nutrient deficiencies, thin soils and poor climatic conditions risks the destruction of forests that, although "poor" from the standpoint of commercial productivity, are rich in esthetic, recreational, and ecological values.

The balance between careful management and poor practices will be determined by the system of incentives represented by the costs and benefits of different practices and the set of regulations and enforcement capabilities available to Federal, state and local forestry agencies. Table 2 lists the present incentives for and against good management.[6] In general, the regulatory incentives for good management are weak, reflecting the difficulty that local, state, and Federal governments have had in controlling geographically extensive activities such as farming and forestry. Even where strong forest management laws are on the books, enforcement officials must wage a strong campaign to gather public support for their actions, which often must be taken against small landowners of long standing in their communities. The economic incentives are mixed, with clear indications that incentives for good management are fairly strong for managers with a long term financial stake in the land. Some of the positive economic incentives may be negated, however, by a failure to understand the environmental consequences of different management strategies. These may be caused by shortcomings in the general state-of-the-art of forest ecology as well as in the education of potential managers. Also, manager and landowner may be separate entities. The landowner's incentive for good management may not affect actual management practices unless he maintains careful supervision.

As a particular example of the complexity of the set of incentives influencing impacts, the specific factors that will affect the extent of "mining" of marginal forests include:

1. The direct cost of wood harvesting. Development of more versatile harvesting equipment can lower the cost of operating on steep slopes and thus promote harvesting on these lands, which are particularly vulnerable to soil erosion after cutting.

2. The stringency and enforcement of environmental standards. The stronger the controls, the more likely it is that loggers will avoid the more vulnerable stands.

3. The price of woodchips for energy. At a high enough price for the wood, the "value-added" to high quality land by clearing will become less important to the economics of logging than the value of the harvested wood, and poorer quality lands will become more attractive targets for harvesting.

4. The price of agricultural land and "high value" forestland. At high land prices, wood harvesting for energy would tend to gravitate to higher quality, less

TABLE 2. INCENTIVES FOR AND AGAINST GOOD FOREST MANAGEMENT

PRO

o Logging costs are lower on flatter -- and thus less erosive -- lands.

o Careful management, including controlled regeneration, leads to improved timber values.

o Careless management can damage recreational and aesthetic values that determine the future sales price for the land.

o Some environmental control measures, such as minimization of roadway length and short-term prevention of roadway erosion, can lower harvesting costs.

o Some states provide excellent management assistance.

CON

o Many benefits of good management (e.g., protection of stream quality) accrue to adjacent landowners or the general public rather than to the investor.

o Most benefits are long term while the costs are immediate.

o Most states lack strong forest management laws, or else do not have sufficient manpower for proper enforcement.

o Proper economic/environmental tradeoffs cannot always be made because of inadequate scientific knowledge -- especially in dealing with new wood-for-energy practices.

o High wood transport costs may increase pressure on marginal lands during local shortages.

o Federal controls on erosion impacts (based on Section 208 of the Clean Water Act) have been slow to implement.

o Multiplicity of small operators in residential wood market may be hard to control and educate.

SOURCE: Plotkin, S.E., "Biomass Energy and the Environment," Environment, November 1980.

erosion-prone/depletion-prone lands because clearing for agriculture or stand conversion will be more profitable.

5. The distribution of different soil/slope/rainfall conditions. These conditions will vary in forestland potentially available for cutting.

6. The attitude of private landowners. The private citizens who currently own much of the land available for clearing have diverse views about harvesting. Although most appear to be amenable to some harvesting, some are not, and others may object to some of the more high intensity forms of harvesting.

7. The cost of transporting wood. The higher this cost is, the more likely it is that local shortages could force harvesting onto vulnerable lands.

Obtaining alcohols through the more traditional "gasohol" route -- using surpluses and new supplies of corn and other grains and sugar crops -- may have a higher probability of causing substantial environmental damages. Although we can't be sure, much of the additional land that would be used for new crop production probably is now pasture or hayland -- land that may have been put into this use because it was erosive and would be protected by the perennial grasses. Forestland may also be vulnerable to conversion, especially because the clearing costs can now be offset in many cases by the value of the wood as fuel.

Increased erosion appears to be the major danger of the added production. Agriculture currently is the primary cause of soil erosion in the U.S., sending at least a billion tons of soil into the Nation's surface waters each year.[7] Much of the Nation's cropland is eroding at rates in excess of the Soil Conservation Service's guidelines,[8] and the land most available for new intensive production appears to be about 20 percent more erosive, on the average, than present cropland.[9] Environmental damages from the expected added erosion include the filling of reservoirs and lakes, degrading of aquatic habitats, and long term debilitation of land quality and productivity through the slow drain of topsoil (a net loss of 5 tons per acre per year leads to a loss of an inch of topsoil in 30 years). Other important damages from added crop production include potential loss of forestland -- between 10 and 30 million acres may be at risk[10] -- and substantial increases in the use of agricultural chemicals, especially pesticides (current U.S. useage: 1 billion lb/yr[11]).

As with forest management, the actual effects of an increase in agricultural production will be quite dependent on the managerial behavior of the farmers. A multiplicity of management mechanisms such as no-till and reduced tillage practices, appropriate land selection, soil analysis to minimize fertilizer useage, and integrated pest management procedures are available to reduce the adverse impacts of intensive crop production. Also as with forest management, the problem is getting the farmer to actually use these mechanisms. Neither existing government conservation and erosion and chemical control programs[12] nor the market[13] appear to be particularly successful at doing this. In the absence of strengthened regulatory and economic incentives for environmental protection, it appears likely that a strong ethanol-from-corn (or other food crop) program would have serious adverse impacts on water quality and other environmental factors that are vulnerable to large-scale intensive agriculture.

Some of these adverse environmental impacts -- as well as the potential food/fuel competition -- could be reduced or avoided by growing grasses and other herbage for alcohol production rather than food crops. Existing pasture and hayland can increase its grass yield by about 1 to 2 ton/acre-year through fertilization and more frequent harvesting,[14] and high yield grasses may yield even greater returns on these lands or on potential croplands. Although ethanol from corn is about twice as "land efficient" as methanol from grass if the corn byproduct is used to displace soybean production, saturation of the byproduct market reverses this advantage (see Table 1).

The environmental effects associated with obtaining substantial quantities of grass feedstocks could be relatively mild. The intensive production of grasses, in contrast to annual row crop production, is not expected to lead to significant increases in erosion because the root systems of grasses survive after harvest, grasses provide more coverage of the soil, and erosive cultivation is not needed. At the present time, pesticide use in grasslands is virtually nonexistent. Although it is possible that the added stress of multiple harvesting and fertilization could lead to greater vulnerability to pests and increased pesticide requirements, the lower level of runoff and erosion would reduce the loss of any agricultural chemicals to surface waters. Also, much of the intensive grass production will occur on land that is now in some sort of grass production, and thus major ecosystem changes will not occur (except for possible loss of wildlife that cannot cope with the several-times-a-year fertilization and harvest) unless production expands beyond current grasslands.

Crop residues represent the remaining major alcohol feedstock. About a 1 EF (10^{18} joules) of these residues, or 20 percent of the total produced, can be made available for conversion to methanol without violating Soil Conservation Service guidelines on erosion.[15]

The important role of residues as a protective cover (when

left in place, erosion on conventionally tilled land may be cut in half[16]) leads to the strong concern that a high price may tempt farmers to remove their residues indiscriminately. Substantial increases in both cropland erosion and its negative water quality consequences could occur if residue collection for energy is encouraged without providing strong incentives for farmers to follow erosion control guidelines. Also, there may be some potential for negative productivity effects on some lands because of the loss of soil organic matter. This effect appears not likely to be important on the medium-textured, limestone-based soils off of which much of the residues will be obtained.[17]

PRODUCTION EFFECTS

The production of ethanol and methanol from biomass will pose a number of air and water pollution problems that will require careful regulatory attention. None of these problems, however, is inherently different from problems posed by large numbers of present fuel combustion sources and chemical manufacturing facilities.

Ethanol distilleries use substantial amounts of fuel -- and therefore can create air pollution problems -- and produce very large amounts of sludge wastes, called "stillage," from the fermentation process. An efficient 50-million-gal/yr. distillery will consume slightly more fuel than a 30 Mw powerplant.[18] If government policy is to avoid the use of premium fuels at these plants, then coal and biomass may be likely fuels. Particulates and, secondarily, sulfur oxides would then be the air pollutants of major interest. Until, or unless, federal emission standards are created for these plants, emission control will be the responsibility of State and local governments.

The untreated stillage is very high in biological and chemical oxygen demand and must be kept out of surface waters. Although the stillage from grains is a valuable animal feed product and will often be recovered without the need for any further incentives, problems may arise if the market for the feed becomes saturated, or in cases when the protein is removed before ethanol production (in this latter case, the stillage would have little value). Also, the stillage from sugar crops is less valuable than grain stillage and will require strict regulation to avoid damage to aquatic ecosystems. EPA has had a history of problems with rum and other distilleries, and ethanol plants will be similar to these.

Additional problems may arise if substantial numbers of small on-farm distilleries go into operation. Aside from a higher potential for inefficient fuel combustion accompanied by high emissions of unburned particulate hydrocarbons, small facilities producing anhydrous (dry) alcohol could pose significant occupational hazards from high pressure steam and chemicals (such as cyclohexane and ether) used for the last distillation step. Also, a proliferation of small producers may be exceedingly difficult to properly monitor, and environmental enforcement may be inadequate.

Methanol synthesis plants pose somewhat different kinds of dangers, although the general problem of monitoring small scale plants continues with this technology as well. Little external fuel use is needed, so the standard fuel combustion emissions will not be a problem. However, the gasification process will generate a variety of toxic compounds that present both an air and water pollution problem. These compounds include a multitude of oxygenated organic compounds (organic acids, aldehydes, ketones, etc.), phenols, and particulate matter.

The major air pollution concerns are occupational. Although cleanup of the gas stream is required to prevent poisoning of the plant catalysts (and to avoid wasting resources on pressurizing gases that do not contribute to the synthesis reaction), raw gas leakage would pose a significant hazard to plant personnel. Similarly, although the tars and oils generated by the gasification probably will be recycled in most plants, these materials may be carcinogens and must be kept from human contact. Good plant housekeeping will be a special environmental requirement of methanol synthesis plants.

EFFECTS OF USING ALCOHOL FUELS

The final environmental impacts of major concern in the alcohol fuel cycle are the effects of using the fuels, as blending agents and in pure (neat) form, in automobiles and other vehicles, and in gas turbines (in neat form only).[19] Of particular interest are the potential air quality impacts. Used as blending agents in gasoline, both ethanol and methanol raise evaporative and aldehyde emissions from spark ignition engines and thus may pose some additional smog problems for urban areas. Because aldehyde concentrations are reduced by catalytic converters, the increase may be significant only in pre-catalyst vehicles. Also, the pollution effects are not all bad: emissions of polycyclic organic matter appear to decrease (demonstrated for methanol, speculative for ethanol), carbon monoxide decreases in older automobiles, and nitrogen oxide and hydrocarbon emissions either increase or decrease depending on the original state of tune of the automobile.

The use of pure alcohols in modified automobiles is likely to be judged as beneficial despite an increase in aldehyde emissions. This is because emissions of other reactive hydrocarbons as well as of nitrogen oxides will be substantially reduced, and particulate emissions should decrease virtually to zero. The particulate decrease occurs in diesels, also, and this is especially valuable because of the well-known control problems with particulates and the possibility that their elimination would also allow the use of oxidation catalysts for control of hydrocarbons.

Finally, the use of methanol in gas turbines has relatively limited experience. However, one likely air quality effect of substituting methanol for distillate oil is a substantial drop in nitrogen oxide emissions.

Aside from its effects on outdoor air quality, the widespread use of alcohol fuels will affect the health and safety of the public--and of specific occupations--in other ways. In particular, the exposure of the public and of workers in fuel distribution to gasoline will decrease, while exposure to the alcohols will increase. Short-term exposure to gasoline is considered more poisonous, tissue-disruptive, and irritative than methanol exposure when effects of eye contact, inhalation of fumes, skin penetration, skin irritation, or ingestion are considered.[20] Ethanol is considerably less toxic than either. Also, the risk of fire and explosion generally are lower with alcohol fuels than with gasoline, although there is some increased hazard in closed areas.[21] Thus, the use of alcohol fuels should ease occupational and public health and safety hazards associated with fuel distribution **if exposures are similar**. This caveat about exposures is especially crucial in considering the possibility of ingestion, because fuel ethanol may be a tempting target for diversion to illegal beverage use, and methanol also has a history of such diversion. Although high purity ethanol does not pose a special hazard beyond that of commercial high-proof alcoholic beverages, fuel-grade ethanol may be contaminated with toxic materials during its manufacture or transport. Special measures to combat illegal diversion will need to be considered if alcohol fuel use is to be safely promoted.

CONCLUSIONS

Although the environmental effects of all energy sources depend in large measure on the behavior of those that develop and use them, the effects of biomass energy development are especially dependent on this behavior. At one extreme, with poor management, biomass energy could be quite damaging to the environment. At the other, with careful management and efficient environmental controls on conversion plants, biomass energy could be an unusually benign source of energy.

For large-scale production of liquid fuels from biomass, several of the potential problems associated with very small scale conversion and haphazard harvesting of feedstocks appear unlikely. The present level of expertise of American farmers and forest managers should set a sharp limit on the level of adverse effects from growing and harvesting feedstocks. Similarly, the network of institutional controls on pollution that has been built up over the past two decades should work well on the moderate-to-large scale plants that would likely characterize such a production level. However, there are still some areas of concern, impacts that may be overlooked by our present system.

The major concern is the slow, often subtle impacts of large-scale intensive agriculture and forestry. These impacts include erosion, pesticide buildup in soils and sediment, ecosystem losses, and a possible gradual degrading of forest soils from nutrient depletion and other causes. Our current institutional framework has been relatively unsuccessful in controlling these impacts in the past, and it is difficult to be confident about improvements in the future, especially given the current political climate. Furthermore, there are important gaps in our scientific understanding of these impacts, especially with respect to intensive harvesting practices in the Nation's forests. These gaps greatly complicate the choice of appropriate development strategies aimed at protecting the environment--and, simultaneously, the "renewability" of the biomass resource base.

At the very least, the existence of these problems and information gaps implies that we should move with care in implementing any large scale development plans for biomass energy. This means coupling a significant research program and provisions for periodic reviews of environmental conditions with any development program. It probably also means reexamining the system of institutional controls and economic incentives that determines the environmental health of our agricultural and forestry resource bases. Without such a strategy, biomass energy development might disappoint some of its most fervent supporters.

REFERENCES

1. Plotkin, S.E., "Biomass Energy and the Environment," Environment, November 1980.

2. It should be noted that increased export demands for high protein products (for animal or human consumption) could raise the current expected saturation point for DG or gluten and this could legitimately be counted as an additional acreage credit if it were assumed that without the byproducts, the U.S. would place additional cropland into production in response to the new demands.

3. Assuming a mixture of ethanol and methanol. Note that some of the methanol might have to be used as a standalone fuel, at a higher "acreage requirement per unit of gasoline requirement."

4. In this case, the 1.4 acres per 1000 gallons does not apply because the byproduct market will saturate at ethanol production levels of about 2-3 billion gallons/yr, and the refinery energy savings/unit of ethanol decreases as ethanol production increases. See the chapter on Liquid and Gaseous Fuels from Biomass.

5. See C.J. High and S.E. Knight, "Environmental Impact of Harvesting Non-Commercial Wood for Energy Research Problems," Thayer School of Engineering, Dartmouth College Paper DSD No. 101, October 1977, also C.G. Wells and J.R. Jorgenson, "Effect of Intensive Harvesting on Nutrient Supply and Sustained Productivity," and E.H. White and A.E. Harvey, "Modification of Intensive Management Practices to Protect Forest Nutrient Cycles," in Proceedings: Impact of Intensive Harvesting on Forest Nutrient Cycling (Syracuse, N.Y., State of New York, August 13-16, 1979.

6. Plotkin, S.E., op. cit.

7. This is a conservative estimate. Many sources estimate a sedimentation rate of between 2 and 3 billion tons per year.

8. 1977 National Erosion Inventory, Soil Conservation Service, USDA.

9. Based on computer runs conducted for the Office of Technology Assessment by the Soil Conservation Service, from the 1977 Inventory.

10. "Tables of Potential Cropland," 1977 National Erosion Inventory, op. cit.

11. Draft Impact Analysis Statement: Rural Clean Water Program (Washington, D.C., U.S. Department of Agriculture, Soil Conservation Service, June 1978).

12. See General Accounting Office, "To Protect Tomorrow's Food Supply, Soil Conservation Needs Priority Attention," CED-77-30, February 14, 1977. Also, General Accounting Office, "National Water Goals Cannot be Attained Without More Attention to Pollution from Diffused or Nonpoint Sources," CED-78-6, 1977. Also, Council on Environmental Quality, Environmental Quality. Ninth Annual Report. December 1978.

13. The major problem is the separation in space and time of costs and benefits. Not only do many of the benefits tend to accrue to the public or downstream landowner and not to the farmer, but they are often slow in being realized because of the subtlety of long term erosion of chemical effects.

14. Barber, S., et.al., "The Potential of Producing Energy from Agriculture," Purdue University, Contractor Report to OTA, May 1979.

15. Based on the work of W.E. Larson and his colleagues for the SCS. For example, see W.E. Larson, et.al., "Residues for Soil Conservation," Paper No. 9818, Science Journal Series, ARS-USDA, 1978.

16. Ibid.

17. Several experiments have found no relationship between soil organic content and yields of corn and other large-seeded crops on such soils. For example, see Moldenhauer, W.C. and C.A. Onstad, "Achieving Specified Soil Loss Levels," Journal of Soil and Water Conservation, Vol. 30, No. 4, 166-168, 1975.

18. Assuming a total energy use of 55,000 Btu/gal. of ethanol for a modern new distiller, OTA, Energy from Biological Processes, Volume II, op. cit.

19. These effects are reported in detail in the OTA report (Volume II) and in H.R. Adelman, R.K. Pefley, et.al., The End Use of Fluids from Biomass as Energy Resources in Both Transportation and Non-Transportation Sectors, Contractor Report to OTA, January 1979.

20. Stear, N.V., ed., Handbook of Laboratory Safety (Cleveland, Ohio; Chemical Rubber Company, 1971).

21. Ibid.

CHAPTER 28

ICE PONDS

Theodore B Taylor
President
Nova, Inc.
10325 Bethesda Church Rd.
Damascus, MD 20872

INTRODUCTION

The cooling capacity of natural cold air is a huge natural resource. The thermal energy released when a mass of dry air cools by 10°C, for example, is more than 1,000 times the kinetic energy of that mass when it is moving at 15 km/hr.

"Ice ponds" capture this renewable resource by exposing water sprays to natural air when the air temperature is below freezing, and collecting the mixture of ice and water in a watertight reservoir. If the unfrozen water is allowed to drain from the ice pond, the remaining ice is porous, with a bulk density that is typically about 2/3 the density of water.

The accumulated ice can be used for process cooling or air conditioning by circulating meltwater from the pond to a cooling load and returning the warmed meltwater to the ice pond, where it melts some ice and cools back down to 0°C. If the ice is to be used for cooling during warm weather, it must be insulated to protect it from environmental melting.

Ice ponds can also be used for purifying water, since ice crystals, as they form, reject dissolved substances. If the sprayed water is only partly frozen, the dissolved impurities stay in solution in the excess water, which can either be drained back to the source of water to be purified (e.g. when desalinating seawater) or retained in another reservoir for futher concentration and ultimate disposal or recycling of the impurities.

For most applications some of the water to be frozen is placed in the reservoir before ice-making operations start in the winter. Excess water from the pond can then be recirculated through the sprays until more water is needed to make up for the water that has been frozen. This process is continued until the ice is piled as high as practicable above the reservoir, generally corresponding to several times more ice above ground than below it. The reservoir dimensions and spraying equipment capacity are generally chosen to allow piling the ice to full capacity in an average winter. Depending on the subsequent use for the ice, the reservoir may or may not be drained of excess water. Water would be held in the reservoir if the ice pond provides a source of chilled water for cooling, but not if the ice is to be marketed as drained bulk ice, for example. After gravity drainage for several days the typical quantity of liquid water that remains clinging to the ice particles is about 3% of the mass of the drained ice. This percentage is called the "wetness."

The fundamental attraction of ice ponds is their low energy consumption per unit of cooling or purified water supplied, compared with conventional

0094-243X/85/1350562-14$3.00 Copyright 1985 American Institute of Physics

alternatives. However, they can only be used in places that have cold winters and they require large reservoirs and land areas if they are used to supply cooling capacity or purified water throughout the warm season. The ice pond area required for summer air conditioning, for example, is typically about 1/4 the floor area that is cooled.

HISTORY

The basic idea of using natural winter ice for summer cooling goes back to antiquity, when ice collected from natural water bodies was stored under insulating blankets of organic refuse, such as sawdust or crop residues, and used for cool storage of food in summer. Ice houses that use this natural resource are still in use in many parts of the world. The idea of using water sprays to produce the ice where it is to be subsequently used for cooling was patented in the United States in 1836.

In the 1960's and early 1970's much field testing of the use of naturally frozen water sprays for purifying salty river water and groundwater was done in Wyoming and Saskatchewan (1,2). Although test results were positive, work on this process was discontinued, apparently because of a lack of continued funding by the governments of the United States and Canada.

Princeton Ice Pond Projects

The first ice pond for air conditioning was an experimental system built at Princeton University in late 1979, as part of an R&D program funded by the Prudential Insurance Company. About 1000 tons of ice were accumulated in a 20 x 20 meter plastic lined reservoir 5 meters deep, using a commercial snow-making machine to make the ice. The ice was covered with straw between two plastic sheets in the spring of 1980, and the ice was used for cooling a small adjoining building during the summer.

Success in this demonstration was followed by construction and testing of a second ice pond test facility at the same site. The pond was circular, 18 meters in diameter, and 3 meters deep. It was covered with a clear span steel structure for supporting a fabric cover, the sides of which could be raised in the winter to allow circulation of cold air to the interior. After about 600 tons of ice made by a snow-making machine had been piled up some 5 meters above grade level, the ice was covered with insulating blankets of plastic foam and the building's side flaps were secured. Most of the ice was used in the summer of 1981 for air conditioning the same building used for the 1980 tests. The pond was again filled with ice the winter of 1981/82. Although the ice was not used for air conditioning in the summer of 1982, it was again covered with insulation, and its rate of environmental melting observed. Some of the ice was preserved through March, 1983, for somewhat more than a year (3).

Following the successful testing of the two experimental ice ponds Prudential decided to build a much larger ice pond to serve the cooling needs of a 130,000 square foot office building, one of two buildings forming Prudential's ENERPLEX office building complex in Princeton, about a mile north of the site of the University's first two test ponds. Construction of the ENERPLEX ice pond started in 1982. System tests were performed in 1983 and 1984, and the ice pond was first used for air conditioning the occupied office building in 1985.

The ENERPLEX ice pond is approximately 40 by 50 meters across and 6 meters deep. It is covered by a clear span steel structure that supports a fabric cover and approximately 30 centimeters of fiberglas insulation suspended directly below the outer cover. The entire covering structure and insulation system can be raised about 6 meters during the ice making season, by electrically operating four tower mounted cable systems at each of the corners of the building. The ice is produced by three Highland snow-making machines made by Snow Machines, Incorporated. The ice pond's storage capacity is 7,000 tons when the ice is piled up to a maximum height about 6 meters above grade level. Cold water from the ice pond is circulated through fan coils in the office building, which is about 100 meters from the ice pond, and the warmed cooling water is returned to the ice pond for recooling. Further testing of the system is proceeding in the summer of 1985.

Natural Cycle Cooling System at Kutter's Cheese Factory

The first ice pond for year-round process cooling was designed and built by NOVA at Kutter's Cheese Factory, Inc. (KCFI), located about 25 miles east of Buffalo, N. Y., in the fall and winter of 1982. Funding was provided for this project by the New York State Energy Research and Development Authority (NYSERDA), with cost-sharing by KCFI. The project also included a statewide assessment of ice ponds for process cooling in New York State. The project continued as a combined R&D and commercial demonstration program through the fall of 1984, and the ice pond system is now operating routinely (4).

The new cooling system at KCFI is called a Natural Cycle Cooling System (NCCS), since it makes use of natural seasonal and daily cycles in the temperature, and therefore the cooling capacity of ambient air in a particular location. It consists of four separate ponds, with appropriate pumps, spray nozzles, and interconnecting plumbing between the ponds and the cheese factory.

The largest pond is an annual storage ice pond 18 by 18 meters across and 1.8 meters deep, surrounded by a 1.1 meter high vertical wall. When filled with ice to a maximum height about one meter above the top of the wall, it holds about 650 tons. Unlike the ice ponds at Princeton, the ice is made by spraying with inexpensive nozzles, rather than snow-making equipment. When the ice pond has been filled in March it is covered with a sheet of white, opaque plastic fabric that is supported several meters above the ice surface by low pressure air provided by a small blower. When the maximum height of the ice has dropped below the top of the wall in late April or early May, the outer cover is removed, the ice surface is covered by slabs of rigid foam insulation 10 centimeters thick, and the air supported cover is re-installed. The primary use of the ice pond is for cooling a cheese storage room held at 7°C by circulating 0°C water from the ice pond through a fan coil in the storage room. In 1984 this ice pond system met the storage room's entire cooling load until the end of September, when all the ice melted.

A second, smaller ice pond is used to supply 0°C cooling water only during the cold season, from mid-November through mid-April. It is 9 by 9 meters across and 1 meter deep, and is never covered. For two successive winters it has been used to accumulate about 150 tons of ice while also supplying ice water for cooling cheese storage space. The ice is produced by water sprayed from a single nozzle suspended over the center of the pond, and has reached a maximum height of about 8 meters. Since no attempt is made to protect the ice from environmental melting, this pond can be used only to save electrical energy that would be used by a conventional refrigeration system during the cold season. A conventional system is turned on as needed for the rest of the year. Such a system is called a "winter only" ice pond.

A pair of smaller ponds used for night-time spray cooling and cooling water storage complete the Natural Cycle Cooling System (NCCS) at the cheese factory. These are both 6 by 6 meters across and 1 meter deep. The system is used primarily to supply cool water for dehumidifying a cheese processing room during the summer, but can also supplement the two ice ponds during the winter. One of the spray cooling ponds is covered with a framework of waterproof insulating slabs with small spaces in between. Stored, warmed cooling water from the adjacent pond is sprayed on to the insulating cover during the early morning. The spraying cools this water to within about 1°C of the air's wet bulb temperature, and the cooled water is collected under the insulating cover. This cool water, which is typically at a temperature of about 15°C in July or August, is circulated the following day through a fan coil in the processing room, where a wet bulb temperature below 25°C is maintained even on the hottest days. The warmed cooling water is discharged to the cooling water storage pond, where it is held until the spray cooling cycle is repeated early the following morning. Use of this system has avoided the need for a conventional air conditioning system in a new cheese processing room.

Use of the ice ponds or spray cooling system for cooling pasteurized cream at the cheese factory is pending the establishment of health related standards for use of this type of system for direct process cooling in the food processing industries. One possibility is the use of an intermediate heat exchanger between the water from an ice pond or spray cooling system and recirculated "sweet water" that passes through heat exchangers for cooling the food product. This adds somewhat to the system cost and increases the minimum temperature that can be achieved in the cooled product, but is an economically viable option. Another possibility is to use the cooling water from the NCCS to improve the performance of commercial refrigeration systems by cooling the system's condensers. NCCS water would allow lower condensing temperatures than are available with ambient air or water from a conventional cooling tower. Both possibilities are now being investigated.

<u>Greenport, New York Water Purification Project</u>

One of the applications of ice ponds that was investigated in the statewide assessment of ice ponds for process cooling was the desalination of seawater in places in New York state close to the sea and facing serious municipal water supply problems. One such location that was identified in the study is Greenport, a community of about 4,500 people near the eastern tip of Long Island. Greenport's water supply, which is entirely local groundwater, is becoming contaminated with pesticides, nitrates, and intruding seawater. Preliminary studies and contacts with local officials indicated that Greenport would be an appropriate site for field testing of the ice pond desalination concept (5).

Accordingly, such a project was started in the fall of 1983. Funding was provided by NYSERDA and the New York Power Authority, with cost sharing by the Village of Greenport (6).

The first phase of the project was the establishment of the scientific feasibility of purifying seawater to potable levels by partially freezing spray water with the same concentration of sodium chloride as seawater, using a 2.5 x 2.5 x 3.5 meter refrigerated enclosure for the purpose. Measurements of the electrical conductivity of melted samples of drained ice made in the enclosure showed that the concentrations of total dissolved solids dropped from about 30,000 parts per million (ppm) to as little as 5 ppm, depending on how much of the accumulated ice was allowed to melt before the samples were taken.

Immediately following this laboratory scale demonstration and some preliminary estimates of costs of large scale seawater desalination in ice ponds, a field test program was started on property of the Village of Greenport on the shore of Long Island Sound. This test program, which extended from December 1983 through March 1985, demonstrated that the process could be used for desalinating large quantities of seawater to much better than potable quality, in terms of concentrations of dissolved solids--i. e. from about 30,000 ppm to less than 5 ppm. To reach potable quality (about 400 ppm) required melting of less than 15% of the accumulated ice.

Ice was accumulated both winters in a plastic lined pond 20 by 20 meters across and an average of one meter deep. The ice was produced by spraying seawater from a set of fire hose nozzles at the periphery of the pond. The seawater was drawn from Long Island Sound through an offshore pickup pipe, and excess saltwater drained from the pond was returned to the Sound until the drainwater reached potable quality. The drain was then closed, allowing the potable meltwater to accumulate in the pond. About 150 tons of ice were produced the first winter, and about 350 tons in February and March of 1985. Typically about 5% of the saltwater was frozen during each pass through the spray system, when the air temperature was $-7^{o}C$. (The temperature at which seawater starts to freeze is $-1.7^{o}C$.)

Tests of the performance of a commercial snow-making machine similar to the ones used at the ENERPLEX ice pond were also performed in the winter of 1984/85, using seawater drawn from Long Island Sound. The spray ice fractions, as expected, were several times those achieved with the simple spray nozzles, indicating the advantage of seeding the sprays with ice crystals. Attempts to increase the ice fractions of the fire nozzle sprays by mixing with the spray from the snow-making machine were unsuccessful, however.

Smaller scale field tests at Greenport and Damascus, Maryland early in 1985 also showed that the process sharply reduces the concentrations of many, and quite possibly all, dissolved inorganic or organic substances at initial concentrations as low as a few parts per billion.

This two year project included the preliminary design of a much larger pilot desalination facility for supplying 38,000 cubic meters (10 million gallons) of water for summer distribution in the Greenport municipal water system, along with engineering and economic analyses of ice pond systems for desalinating up to 1 billion gallons of seawater per year, and for removing contaminants from aquifers or industrial aqueous waste streams. Plans are now underway for detailed design, construction, operation, and monitoring of the 10 million gallon per year pilot facility at Greenport.

Fig. 1. Greenport, NY Desalination Ice Pond (Jan. 1983)

PRESENT STATUS OF ICE POND TECHNOLOGY

Overview

Most applications of ice ponds for cooling or water purification are still in the R&D phase. The only systems that could be built commercially now are "winter only" ice ponds designed to save energy at appropriate sites for serving substantial cold weather air conditioning or process cooling loads. These applications require considerably less capital investment per annual unit of energy cost savings than annual cycle ice ponds, simply because they do not require the large ice storage volumes or insulating covers. Although we see ways that further R&D could improve the performance of ice ponds designed for use only during the colder 3 to 5 months of the year, we have found numerous situations where present knowledge and experience would allow construction of such systems for capital costs that would be repaid by energy savings in 1 to 3 years.

The following sections briefly set forth some of the key findings, so far, related to ice pond technology.

Ice Production

Three favored ways for using natural cold air to accumulate ice in a watertight reservoir, at low cost, are the following:

1. Surface freezing of water in the reservoir, while spreading water on the ice surface at about the same rate as the water freezes. This is much more efficient than simply building up a layer of ice on a body of standing water that becomes more and more insulated from the cold air as the ice layer builds up.

2. Supercooling and subsequent partial freezing of water sprays that are not seeded with ice crystals. Compared with surface freezing, this greatly increases the area of exposed water above the ice pond reservoir. Even though the fraction of the sprayed water that freezes in each pass is typically only a few percent at air temperatures as low as -10°C, this process makes it possible to build up ice thicknesses more than an order of magnitude greater than by surface freezing.

3. Use of relatively fine sprays that are seeded with ice crystals to prevent supercooling and greatly increase the rate of heat transfer from the droplet surfaces to the cold air. This is the type of process used in the design of snow-making equipment. A common way to produce the ice crystals for seeding the main spray is by expansion of wet compressed air sufficiently to lower the temperature of the air below about -40°C, and joining the plume of resulting ice crystals with the main spray a short distance from the spray nozzles. Such equipment can produce fairly dry ice plumes, typically with a water content less than 50%. The penalty, compared with unseeded sprays or surface freezing techniques, is higher capital cost for the same annual rate of production of ice. Currently available commercial snow-making machines also generally consume more energy per unit mass of ice produced, but that would probably not be the case for seeded spray systems designed specifically for ice pond applications, for which the deposited ice can be much wetter than on ski slopes.

The average energy consumption per unit mass of ice produced in a winter depends on the local winter climate, the method for using the ice, and, in some cases, the average total thickness of ice produced (when the thickness is greater than about 30 meters, and the energy required to lift sprayed water against gravity becomes significant).

A rough rule of thumb is that an average of at least 500 hours per winter of below freezing air temperatures is sufficient to consider using ice ponds. This condition is met in more than 3/4 the area of the United States, all of Canada and the Soviet Union, most of Europe, and parts of Asia at high latitudes or altitudes (such as northern China or Japan). Ice ponds are generally inappropriate in Africa, Latin America, or Australia, except at altitudes greater than about 3,000 meters.

Most of the energy consumed in making the ice is used by the pumps that must raise the water pressure sufficiently to meet the requirements set by spray nozzle designs. This pressure is generally in the range from 3.5 to 10

kg/cm^2 (50 - 150 psi). Much lower pressures, of course, are required to spread water on a surface if the surface freezing technique is used. Seeded spray systems that used compressed air can consume significant amounts of energy in the air compressors.

The total energy consumption per unit mass of ice produced per winter is generally lower in colder climates because the energy consumption per unit rate of ice production decreases with lower temperatures, typically in inverse proportion to the difference between the air temperature and the freezing point of the sprayed water. (Note that the freezing point of water with large concentrations of dissolved solids can be considerably lower than 0^oC.)

The total energy consumption per metric ton of ice produced by various ice pond systems is presented in Table 1.

TABLE 1

ICE POND ENERGY CONSUMPTION
PER UNIT MASS OF ICE

Surface freezing only	Less than 0.1 kw hr/ton
Sprays in very cold climate (e.g. Minnesota)	0.5 "
Sprays in Washington, D. C. area	2 "
Very large water purification ice ponds in coastal Northeastern U. S.	2-6 "

In spite of the relatively small energy consumption required for the surface freezing technique, spray freezing is the preferred method in most cases. This is because the surface freezing process requires much larger reservoir areas, and correspondingly higher capital costs for a particular annual rate of ice production. Furthermore, surface freezing tends to form relatively impermeable bulk ice that provides a much smaller heat transfer area for recooling water returned from a cooling load than the granular, porous ice generally produced by water sprays.

The low energy cost of producing ice, by any of these methods, is evident from Table 1. At a cost of electric power of, say, $0.07/kw hr, the indicated range of energy consumption corresponds to an energy cost range from less than 0.7 cents to a maximum of 42 cents per ton of ice, where the latter figure is for ice that has been piled up to a height of about 200 meters!

These ranges of energy consumption and costs are much lower than for conventional refrigeration systems. The input electrical energy per metric ton of ice equivalent cooling capacity ranges from a low of 16 kw hr/ton, for high performance, water cooled compressor systems for supplying chilled water at 7°C, to as much as 50 kw hr/ton for air cooled refrigeration systems for producing ice. At $.07/kw hr the corresponding range of energy costs per equivalent ton of ice is $1.10 to $3.50.

Ice Preservation

Operationally and economically acceptable means for reducing the environmental melting of the ice in annual storage ice ponds to a small fraction of the initial inventory have yet to be demonstrated. The insulating systems used so far are too labor intensive or involve capital costs that are too high, or both. The problem is the control of surface melting of the ice, not control of heat flow from the ground, which is strongly suppressed by the insulating value of the ground itself.

NOVA's present preferred solution to this problem, which we hope to test in the winter of 1985/86, is to enclose the entire ice pond, including the sprays, inside an air supported structure to which the necessary flow of cold air for making ice is supplied by a set of fans. The outer cover thus encloses the entire ice pond system at all times. Insulation is provided by a layer of stagnant air held inside a double walled blanket of reinforced plastic sheeting attached to the under side of the cover. Radiation flow downward is suppressed by low emissivity, high reflectivity surfaces on the plastic sheeting. Total energy consumption by the high flow ventilating system used during ice-making operations is smaller than the energy required for the spray pumping system. Energy consumption by the air supporting blowers is much smaller.

There is the special case of very large ice ponds that require no insulation because the annual environmental melting, even in a location with very warm summers, accounts for only a fraction of the mass of ice at the end of the winter. Such an ice pile might be a substantial source of desalinated seawater, while also providing ice for serving a large cooling load. We have called such ice ponds "dual purpose ice ponds." The thickness of ice (at a bulk density 2/3 the density of water) that would be environmentally melted during the warm season at a Long Island site, for example, is about 35 meters. To serve a significant annual cooling load an uncovered ice pond should therefore have an average ice thickness significantly greater than 35 meters. For the ice pile to have a reasonable shape, it should cover at least several hectares. An example of possible use of such a dual purpose ice pond is to handle the air conditioning load at Kennedy Airport, while also providing up to a 4 million cubic meters (1 billion gallons) of desalted seawater.

Water Purification

There are several important differences between ice pond systems for desalinating seawater and those for purification of contaminated water, such as groundwater or industrial waste water. Special requirements for decontamination systems include the following:

o Their function is not only to generate water with greatly reduced concentrations of contaminants, but also to extract the contaminants for further processing or disposal.

o The decontamination process generally requires a set of ice ponds operating in series, to allow concentration of contaminants to the greatest possible extent in the residual water. The maximum achievable concentration of contaminants is generally set by depression of the freezing point as the concentration increases, along with the minimum air temperatures that occur reasonably frequently in the region.

o The initial and allowable final concentrations of dissolved contaminants may be orders of magnitude smaller than in seawater. Examples are highly toxic organic compounds or heavy metals.

NOVA has constructed two simplified mathematical models relating to the ice pond purification process. The first applies to the process of building up the ice pile in each ice pond stage, and is used for calculations of contaminant concentrations after gravity drainage of the ice, but before appreciable surface melting. The second provides estimates of the contaminant concentrations in the bulk ice and drainwater as functions of the amount of surface melting.

A major simplifying assumption in both models is that sprayed water or surface meltwater mixes completely with any water held in the ice as the water percolates downward through the ice mass. This assumption is borne out by drainage experiments performed at the Greenport test facility.

The input parameters for the ice production and gravity drainage models are as follows:

M_o = initial mass of water in pond, before recirculation starts

t = time that sprays have been turned on

F = spray flow rate

w = wetness of ice after gravity drainage to equilibrium

f_i = average spray ice fraction during spraying

f_e = average spray evaporation fraction

f_o = average overspray fraction

M_i = mass of ice in pond at time t, with no melting

C_o = initial concentration of dissolved solids in water

The model's output parameters and their relations to the input parameters are:

M_i = mass of ice at time t

$$= M_o f_i (1-f_e)(1-f_o) \quad (1)$$

C_w = concentration in drainwater at time t

$$= \frac{C_o}{\{1-[f_o+f_e+f_i(1-f_e)(1-f_o)][Ft/M_o]\}^{\frac{f_e+f_i}{f_o+f_e+f_i}}} \quad (2)$$

(Note that Ft/M_o is the number of times the initial water inventory in the pond has been recycled through the spray system.)

C_i = bulk concentration in ice at time t

$$= wC_w \quad (3)$$

$(Ft)_d$ = total spray water flow when ice pond "runs dry"

$$= M_o/[f_i(1-f_e)(1-f_o)(1+w) + f_e + f_o] \quad (4)$$

The input parameters for the ice washing model are:

C_{io} = concentration in bulk ice after gravity drainage, before melting or washing with potable water source starts

C_{wo} = concentration of drainwater before melting and washing

C_{ip} = concentration in bulk ice after a period of melting and washing

C_{wp} = concentration of drainwater after melting and washing

w = wetness of ice after gravity drainage to equilibrium

The model's output parameter is generally the fraction of the mass of ice that needs to melt at the surface (or needs to be added from another source of potable water) to achieve a prescribed concentration of contaminants in the bulk ice after gravity drainage and washing. Sometimes the melting fraction is prescribed, and the concentrations after washing and drainage are output parameters. The relations between the parameters are:

f_m = fraction of ice melted

$$= w(\ln C_{io} - \ln C_{ip}) = w(\ln C_{wo} - \ln C_{wp}) \quad (5)$$

$$C_{ip} = C_{io}\exp(-f_m/w) \quad (6)$$

$$C_{wp} = C_{wo}\exp(-f_m/w) \quad (7)$$

Relations 5, 6, and 7 indicate why the wetness, w, is such an important parameter. As part of the exponent in the concentration formulae its greatest effect is in governing the amount of ice that must melt to achieve a desired water purity level. It also directly determines the amount of contaminants persistently clinging inside the ice mass and therefore the ultimate concentrations of impurities in the melted ice accumulation. Since small values of w are generally beneficial, optimum performance of the process is achieved when gravity drainage is allowed to proceed as long as possible before substantial melting occurs. This can happen during periods when air temperatures are low enough to avoid excessive melting while the drainage proceeds.

Relation (2) shows that the contaminant concentrations that can be achieved in the drainwater at the end of the ice-making phase, before washing, can be lowered considerably by excessive overspray that causes the exponent in the equation to be significantly less than one.

Parameter studies using these models have shown that three ice pond purification stages are generally sufficient for achieving potable quality in the melted, washed ice, and for concentrating contaminants by more than a factor of 100 in the retained, concentrated drain water, for a wide range of initial concentrations of dissolved solids. The storage volumes of successive ice ponds for each stage of purification and concentration are typically between 10% and 20% of the volume of the preceding stage, and the volume of residual water at high concentrations is in the vicinity of 1% of the total volume of water processed.

SUMMARY OF APPLICATIONS OF ICE PONDS

Possible uses of ice ponds are listed in Table 2. Most of these have been discussed above. Production of bulk ice for shipment to such users as the food packing, fishing, and concrete production industries can be considered a cooling application, but is distinguished here because the cooling capacity is not delivered as chilled water. Water with extremely low concentrations of dissolved solids (e. g. less than 1 ppm) is sometimes required by such industries as producers of beverages or pharmaceuticals. Ice ponds may be useful in reducing costs of recovery of minerals from seawater or recycling waste water contaminants by greatly reducing the volume of water that needs to be processed. "Structural ice" means large piles of ice deposited by water sprays in places where massive barriers are useful, such as breakwaters or the equivalent of earth filled dams in places with very cold winters. All these applications are under study or being actively developed. In all cases the fundamental attraction of ice ponds is the very low cost of producing the ice in places with cold winters.

ECONOMIC SUMMARY

The competitive market price for ice pond products ranges over at least two orders of magnitude, from about \$0.30/ton for purified water for municipal uses to more than \$30/ton for bulk ice, and possibly still higher prices for extremely pure water for special uses. The value of the cooling capacity of a ton of stored ice depends strongly on local electric power costs (including considerations of time-of-day and peak demand pricing) and the energy consumption performance of competing refrigeration equipment.

TABLE 2

USES OF ICE PONDS

- o AIR CONDITIONING AND PROCESS COOLING
  - "WINTER ONLY"
  - ANNUAL STORAGE, INSULATED OR VERY LARGE UNINSULATED
- o BULK ICE PRODUCTION
- o WATER PURIFICATION
  - SEAWATER DESALINATION
  - CONTAMINATED WATER PURIFICATION (e.g. AQUIFERS, INDUSTRIAL WASTES)
  - ULTRA-PURE WATER PRODUCTION
- o RECOVERY OF DISSOLVED SOLIDS IN WATER
  - SEAWATER MINERALS
  - RECYCLABLE WASTE WATER CONTAMINANTS
- o STRUCTURAL ICE

This "value" ranges from about $1/ton when comparing ice ponds with high performance air conditioning equipment used in low power cost locations up to about $5/ton for comparison with process cooling refrigeration systems not designed to minimize energy consumption, and used in high power cost locations.

The economic incentives for using ice ponds also depend on whether they are added to an existing system, to save energy costs, or completely replace refrigeration equipment. In the former case the capital cost of the ice pond system must be paid back through savings in operating expenses in reasonably short times--generally less than three years for most commercial establishments. In most cases the capital cost of an annual storage ice pond is greater than the cost of refrigeration equipment it might replace. In this case the added cost of the ice pond typically needs to be paid back in less than 3 years, by savings in operating costs, to be economically attractive to private enterprises.

All ice pond systems exhibit substantial economies of scale, since large fractions of their capital costs are roughly proportional to the ice pond area, and the accumulated mass of ice per unit area tends to increase with size. The size threshold for economic viability depends on the type of ice pond and the specific setting and application. It is smallest for "winter only" ice ponds--in some cases corresponding to an ice pond area as small as 100 square meters. Enclosed annual storage ice ponds can begin to look economically attractive when their area is in the vicinity of 3,000 square meters, sufficient for air conditioning 10 to 20 thousand square meters of conditioned space. Seawater desalination plants begin to look economically attractive for municipal water systems at sizes corresponding to several hectares, sufficient for supplying at least 500,000 cubic meters (roughly 120 million gallons) of water per year, at costs in the vicinity of $0.50/ton.

Present estimates of total capital costs of fully enclosed, annual storage ice ponds range from about $25 per ton of stored ice, at a total capacity of 10,000 tons, down to about $10/ton at a capacity of 200,000 tons. The costs of the insulated enclosures, which are equipped to supply cold outside air to mix with the water sprays, account for more than 2/3 the total cost of the system.

The cost of land for ice ponds can range from trivial to prohibitive, depending on the specific setting. Land costs are generally prohibitive in heavily built up urban areas, but trivial in many rural industrial locations. In some cases, of which the ENERPLEX ice pond is an example, zoning laws are such that the land needed for an ice pond can be used at no additional cost because it can be included in the minimum area that must be allocated to "green space." As a rough rule of thumb land costs up to the vicinity of $20 per square meter (roughly $100,000 per acre) are not sufficient to price ice ponds out of the market.

REFERENCES

1. Stinson, Donald L., "Atmospheric Freezing for Water Desalination," AICHE Symposium Series No 166, Vol. 73.

2. Spyker, J. W., "Desalination of Brackish Water by Spray Freezing," Saskatchewan Research Council Technical Report N. 106, May 1980.

3. Kirkpatrick, D. L., Masoero, M., Rable, A., Roedder, C. E., Socolow, R. H., Taylor, T. B., "The Ice Pond--Production and Seasonal Storage of Ice for Cooling," Solar Energy, in Press, 1985.

4. "Investigation of Ice Pond Technology for Process Cooling in New York State," Final Report by Nova, Inc., under contract with the New York State Energy Research and Development Authority, October, 1984.

5. Ibid.

6. "Advanced Development of Ice Pond Technology for Desalination and Water Purification," Final Report by Nova, Inc., under contract with the New York State Energy Research and Development Authority, June 1985.

APPENDIX A

A CHRONOLOGY OF ENERGY CONSERVATION AND PRODUCTION

D.W. HAFEMEISTER
PHYSICS DEPARTMENT
CALIFORNIA POLYTECHNIC UNIVERSITY

1945
--August 15; Office of Price Administration (OPA) lifted gasoline rationing.

1946
--May 6; Division of Oil and Gas established in the Department of Interior.
--May 21; President Truman ordered the U.S. Government to take possession of bituminous coal mines due to miners strike.
--June 18; National Petroleum Council established.

1947
--January 1; Atomic Energy Commission began operation.
--March 25; Coal mine disastor killed 11 in Centralia, Illinois.
--June 16; Federal Power Commission authority extended to all natural gas producers.

1952; Severe air pollution (0.7 ppm SO_X and particulates) in London kills 2500 people in three days.

1953
--August 7; Congress gave U.S. Government jurisdiction to the ocean floors beyond the 3 mile boundary.
--December 8; President Eisenhower delivered "Atoms for Peace" speech before the United Nations.

1954
--August 30; Atomic Energy Act of 1954 encouraged peaceful uses of nuclear energy.

1957
--King Hubbert correctly predicted that U.S. petroleum production would peak between 1966 and 1971; U.S. oil production for the lower forty-eight states peaked at 9.1 million barrels/day in 1970. Hubbert used the Verhulst differential equation to describe the consumption of a finite resource, but he did not consider the economics effects of higher prices.

1959
--March 10; President Eisenhower limited oil imports to stimulate development of domestic production and refining capacity.

1962
--October 11; Congress authorized the President to impose mandatory

0094-243X/85/1350576-4$3.00 Copyright 1985 American Institute of Physics

oil import quotas.

1963
--December 17; Clean Air Act provided assistance to states for air polution research; major amendments made to the act in 1970.

1965
--October 2; Water Quality Act established the Water Control Administration in Department of Health, Eduction and Welfare (HEW).
--October 20; Solid Waste Disposal Act provided assistance for study, collection, and disposal of solid wastes.
--November 9; First major power blackout covered the northeast U.S.

1967
--November 21; Clean Air strengthened to give authority to the Secretary of HEW to set auto emission standards.

1969
--January 1; National Environmental Policy Act (NEPA) established the Council of Enrvironmental Quality (CEQ) and the framework for Environmental Impact Statements.
--January-February; Major oil spill from offshore drilling near Santa Barbara, Calilifornia.
--December 30; Oil depletion allowance reduced from 27.5% to 22%.

1970
--March 5; President Nixon issued an Executive Order to require that federal agencies evaluate their activities for the proctection of the environment under NEPA.
--July 9; President Nixon requested Congress to create the Environmental Protection Agency (EPA).
--October 23; Merchant Marine Act Ammendment provided subsidies for oil and liquified natural gas tankers.
--December 24; Geothermal Steam Act authorized leases for geothermal steam.

1971
--July 23; Calvert Cliffs decision required the Atomic Energy Commission to comply with the National Enviornmental Policy Act.

1973
--June 29; Energy Policy Office created with former Governor Love as director.
--October 17, 1973 to March 17, 1974; Orangization of Arab Petroleum Exporting Countries (OPEC) embargoed the U.S. and the Netherlands because of their support for Isreal.
--November 7; President Nixon created Project Independence to end oil imports by 1980.
--November 27; Emergency Petroleum Allocation Act provided authority for the allocation of oil.
--December 4; Federal Energy Office in the Excecutive Office of the President created with William Simon as director.

--December 15; Congress mandated daylight savings to save energy.

1974
--June 22; Congress authorized the Federal Energy Administration (FEA) to order electrical utilities and industry to convert from burning oil and natural gas to coal.
--September 3; Congress authorized funds for research on geothermal energy and on solar heating and cooling demonstrations.
--October 5; Congress repealed manditory daylight saving to save energy.
--October 11; Energy Reorganization Act abolished the AEC and created the Energy Research and Deveolopment Administration (ERDA) and the Nuclear Regulatory Commisison (NRC).
--Ocotber 26; Congress transferred solar energy research to ERDA and authorized the Solar Energy Research Institute (SERI) which was started on July 5, 1977 in Golden, Colorado.
--December 31; Congress required ERDA to submit an annual comprehensive plan on energy R&D.

1975
--January 4; Congress established the 55 mph speed limit to save energy.
--March 17; Supreme Court ruled that the states do not have jurisdiction over the outer continental shelf.
--October 29; ERDA dedicated their first wind power system at Sandusky, Ohio.
--December 22; Energy Policy and Conservation Act (EPCA) established a forumula to set the price of U.S. crude oil, the Strategic Petroleum Reserve, emergency powers for the president on energy matters, and average automobile fuel economy stadards of 27.5 mpg by 1985.

1976
--April 5; Congress authorized the future production of existing naval petroleum reserves.
--August 14; Energy Conservation and Production Act (ECPA) created incentives for conservation and renewables, funded weatherization for low income homes, and established a program to establish energy conservation standards for new buildings.

1977
--April 7; President Carter indefinitely deferred the reprocessing of spent nuclear fuel and delayed the construction of the Clinch River Breeder Reactor.
--October 1; Department of Energy created from ERDA and the FEA.

1978
--November 9; National Energy Act established weatherization grants for low income families, conservation programs for local governments, energy standards for consumer products, programs to convert utilities to coal, residential energy tax credits, and adjusted controls on natural gas prices.

1979

--March 28; the nuclear reactor accident at Three Mile Island power plant.

--spring; gasoline shortages in several areas of the U.S.

--August 17; President Carter began to gradually lift price controls on domestic crude oil.

--November 3; U.S. embassy in Iran siezed by revolutionaries; President Carter suspends oil imports from Iran on November 14.

1980

--April 2; Windfall profit tax established on crude oil along with assistance for the weatherization of homes of low income people.

--June 30; Energy Security Act created the Synthetic Fuels Corporation, an Energy Conservation Bank, and funding for solar, biomass, and geothermal projects.

1981

--January 28; President Reagan completed the decontrol of prices for crude oil.

--1981 to 1985; President Reagan and the Congress debate the funding levels for various conservation programs.

1982

--May 24; President Reagan proposed to the Congress to transfer most of the responsibilities of the Department of Energy to the Department of Commerce.

1985

--February 4; supplies of natural gas and oil appear plentiful in the near term, but the Department of the Interior sharply reduced the estimates for offshore oil (from 27 to 12 Bb) and gas (from 163 to 91 TCF).

--June 27; EPA modifies mileage test, lowering 27.5 mpg by about 2.

--June 28; EPA curbs tall smokestacks to avoid distant pollution.

--July 16; Appellate Court confirms EPCA's appliance standards by voiding DOE's "no-standard" standard.

ACKNOWLEGEMENT

This list has been compiled from a variety of sources. In particular we would like to thank Prentice Dean[1] of the Department of Energy and Gary Pagliano and Langdon Crane of the Congressional Research Center[2].

1. P. Dean, ENERGY HISTORY CHRONOLOGY FROM WORLD WAR II TO THE PRESENT, DOE/ES-0002, Department of Energy, Washington, DC, 1982.

2. ISSUE BRIEFS ON ENERGY, Congressional Research Service, Library of Congress, Washington, DC.

APPENDIX B

ENERGY DATA

DATA SOURCES:

AER: ANNUAL ENERGY REVIEW 1984, Energy Information Administration, Washington, DC, April, 1985.

ECI: ENERGY CONSERVATION INDICATORS 1983, Energy Information Administration, Washington, DC, October, 1984.

DOE: ENERGY PROJECTIONS TO THE YEAR 2010, Department of Energy, October 1983.

LBL: L. Schipper, A. Ketoff and A. Kahane, Annual Review of Energy 10 (1985).

CEC: RELATIVE COST OF ELECTRICITY PRODUCTION, July 1984, and ENERGY EFFICIENCY STANDARDS, 1985, California Energy Comm.

Diagram 1. Total Energy Flow, 1984 (AER)
(Quadrillion Btu)

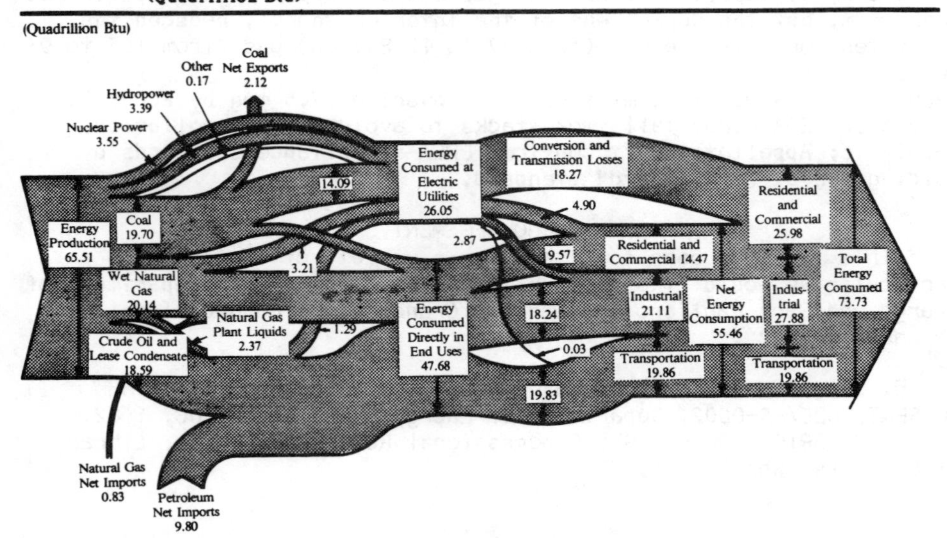

*Total Energy Consumption with conversion and transmission lossess allocated to end-use sectors in proportion to the sectors' use of electricity.
Note: Sum of components does not equal total due to independent rounding; the use of preliminary conversion factors; and the exclusion of changes in stocks, miscellaneous supply and disposition, and unaccounted for quantities.

0094-243X/85/1350580-20$3.00 Copyright 1985 American Institute of Physics

Figure 3. Consumption of Energy by Source, 1949-1984 (AER)

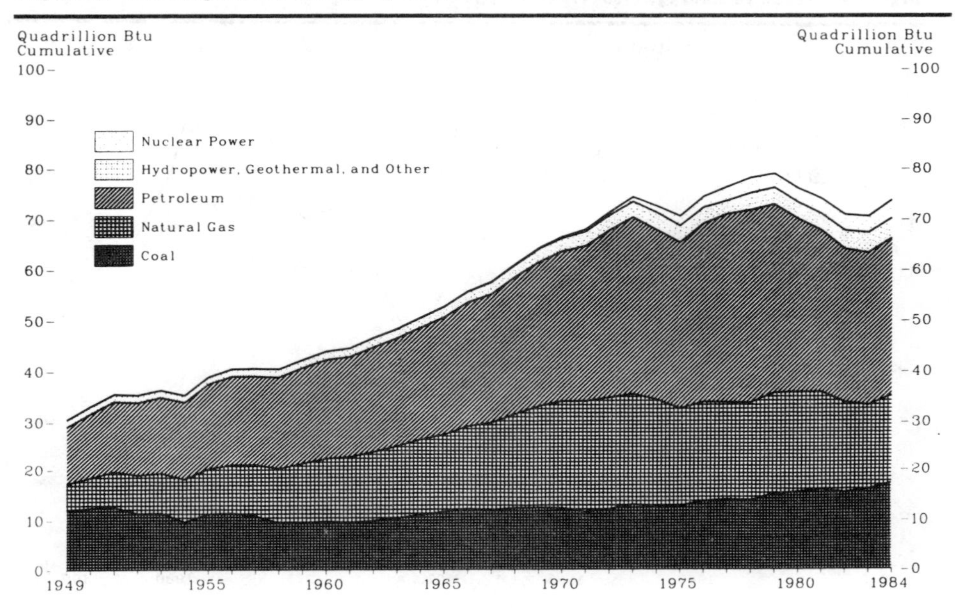

Figure 4. Consumption of Energy by End-Use Sector, 1949-1984 (AER)

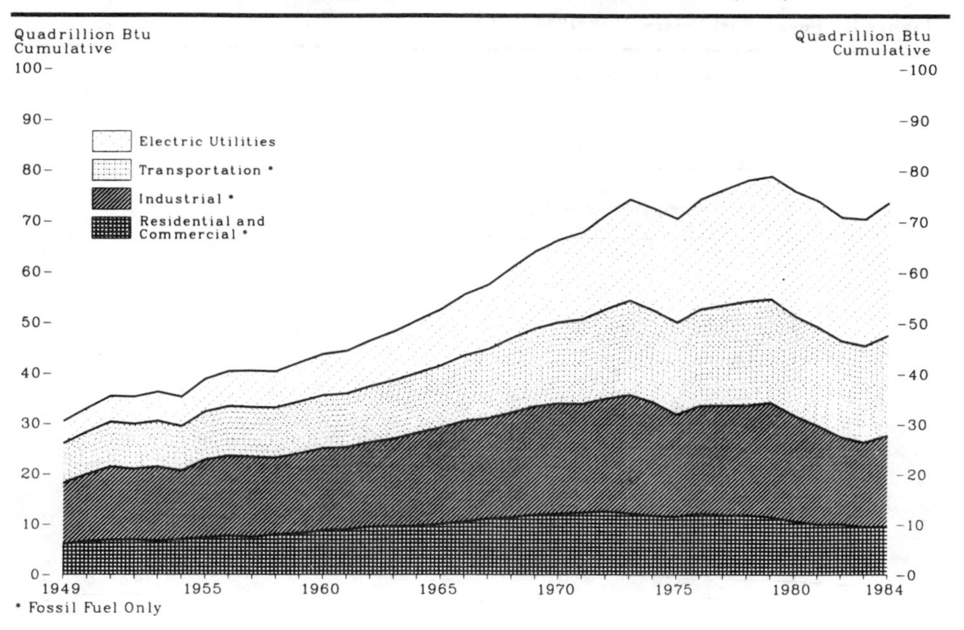

Figure 6. Trade in Energy, 1953–1984 (AER)

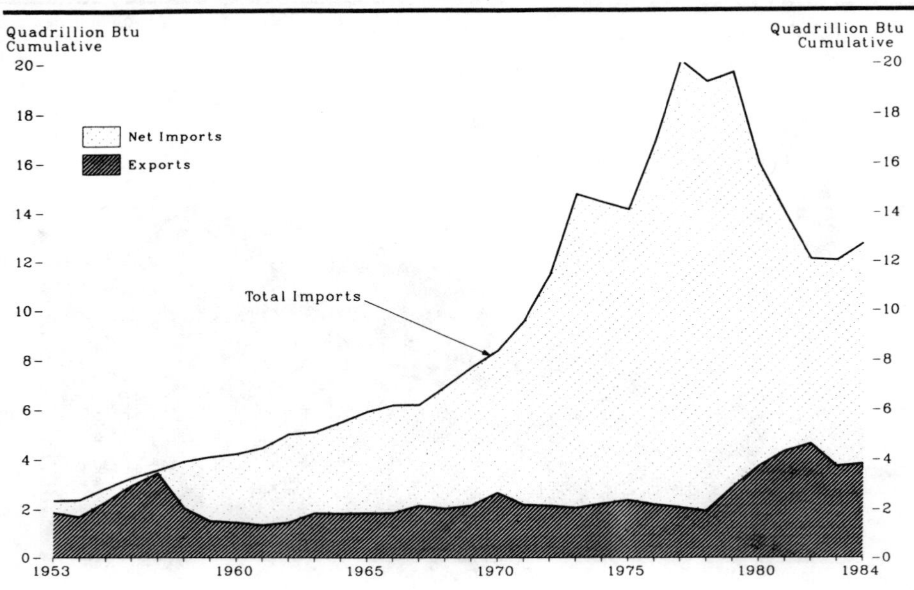

Figure 11. Fossil Fuel Prices, 1949–1984 (AER)

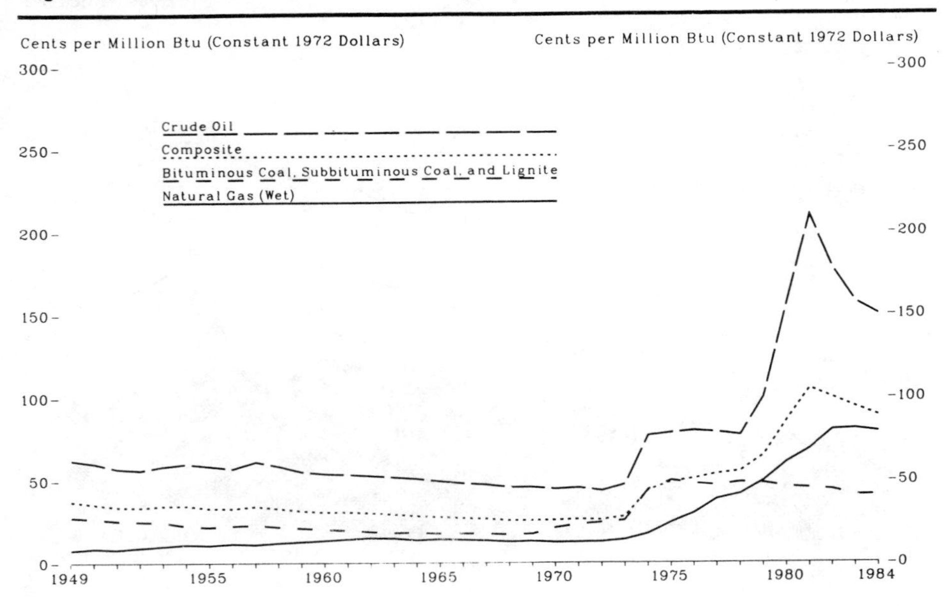

Figure 16. Value of Net Trade in Fossil Fuels, 1958–1984 (AER)

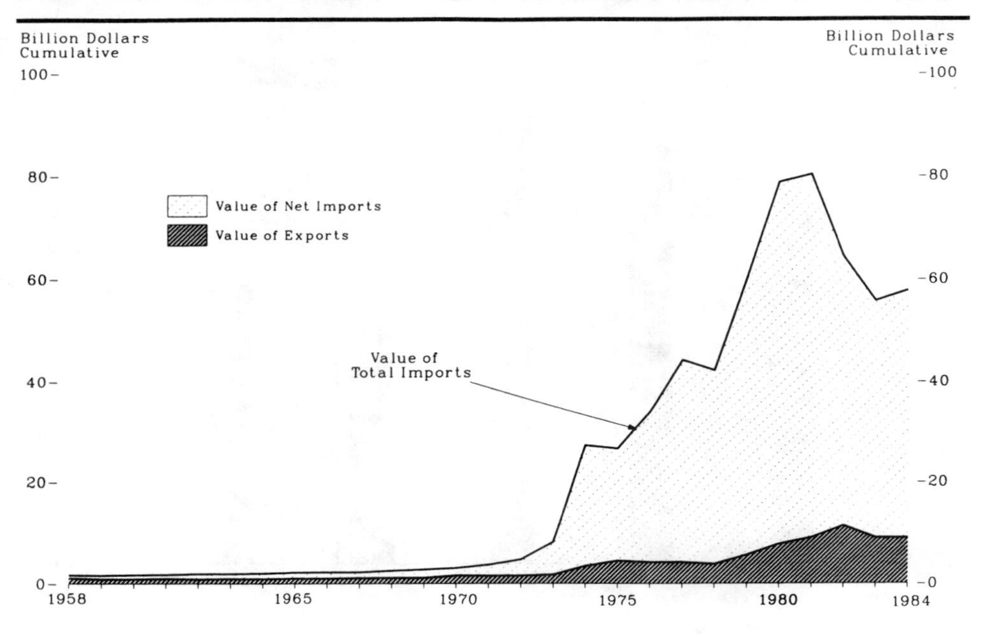

Figure 21. Energy Consumption per Capita and per Constant Dollar of Gross National Product, 1949–1984 (AER)

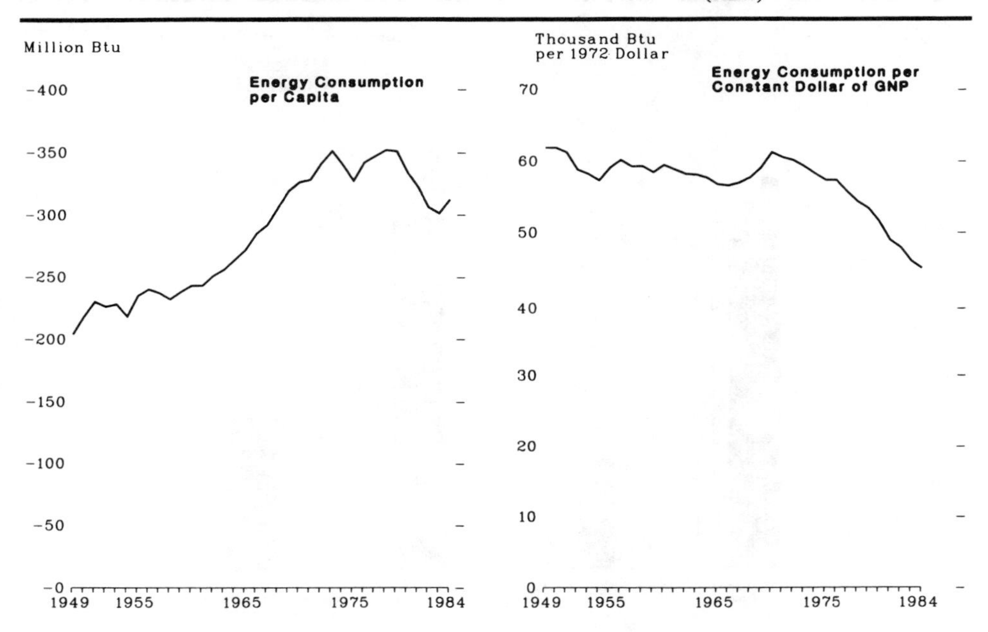

Figure 22. Energy Consumed by Households, 1978–1982 (AER)

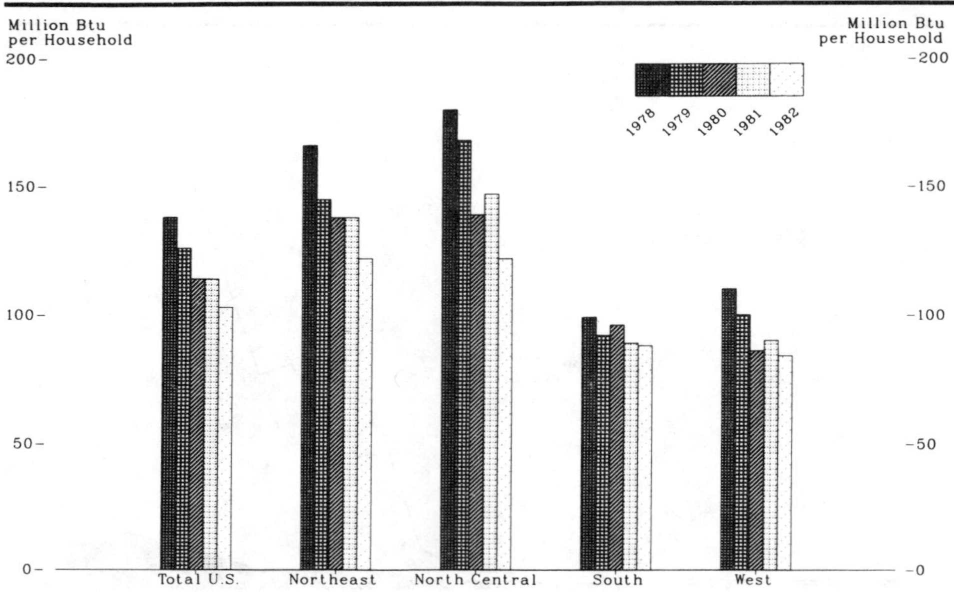

Note: See Appendix 1 for Census Regions.

Figure 23. Household Energy Consumption Indicators, 1978–1982 (AER)

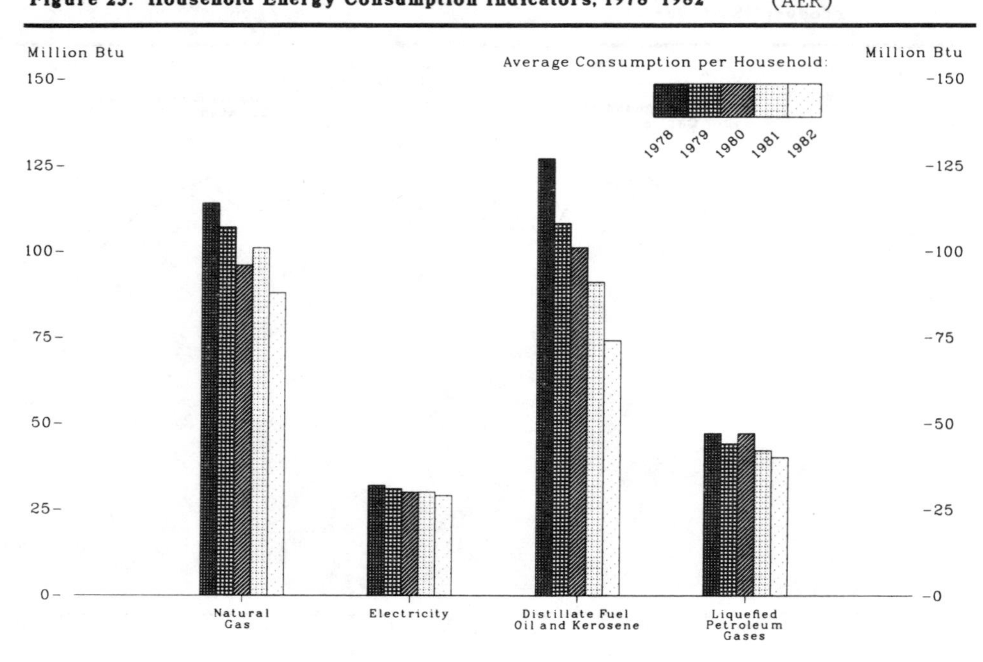

Figure 25. Household Appliance Data, 1980–1982 (AER)

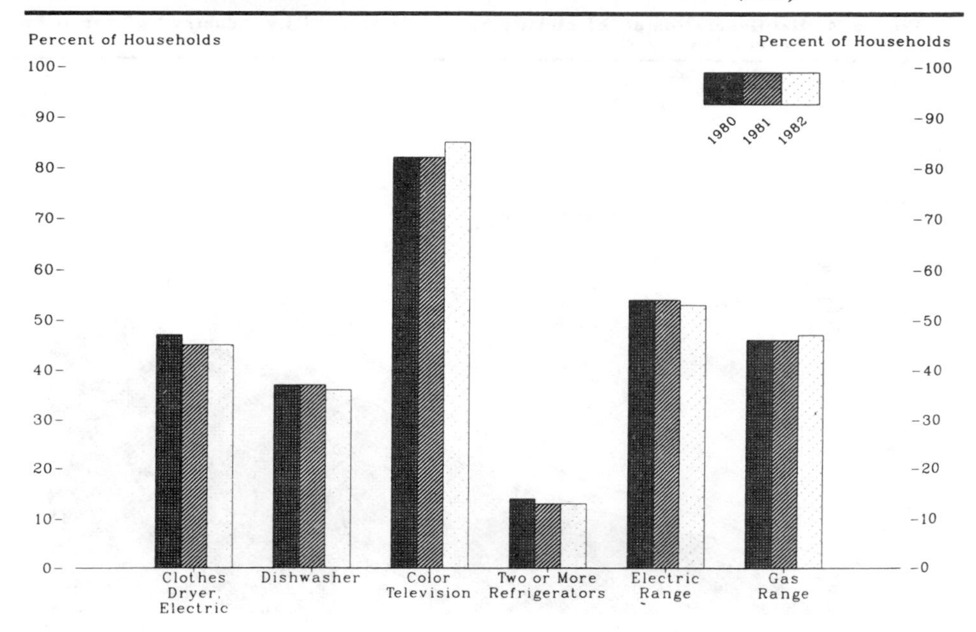

Figure 36. Proved Reserves of Liquid and Gaseous Hydrocarbons, Yearend, 1949–1983

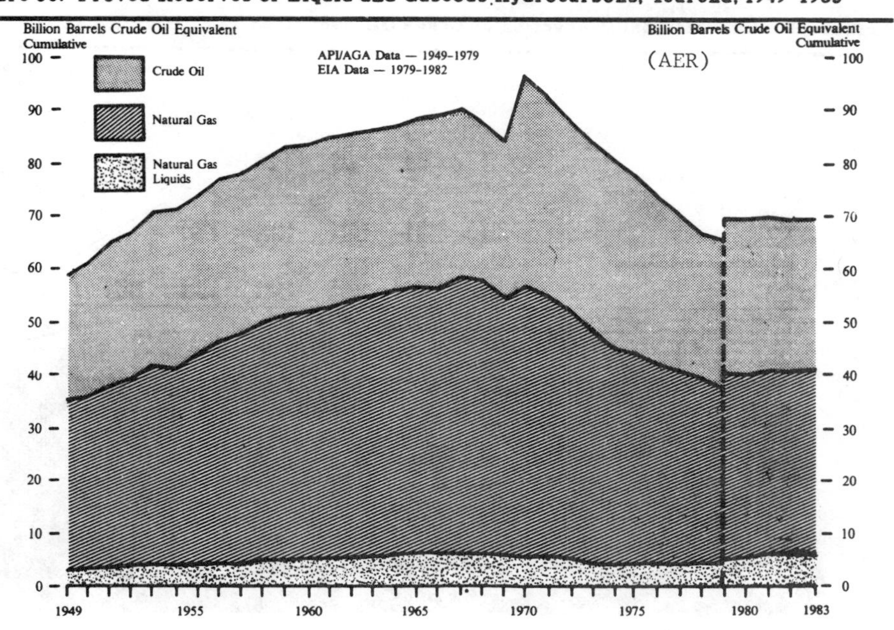

Figure 74. Net Generation of Electricity by the Electric Utility Industry by Type of Energy Source, 1949-1984

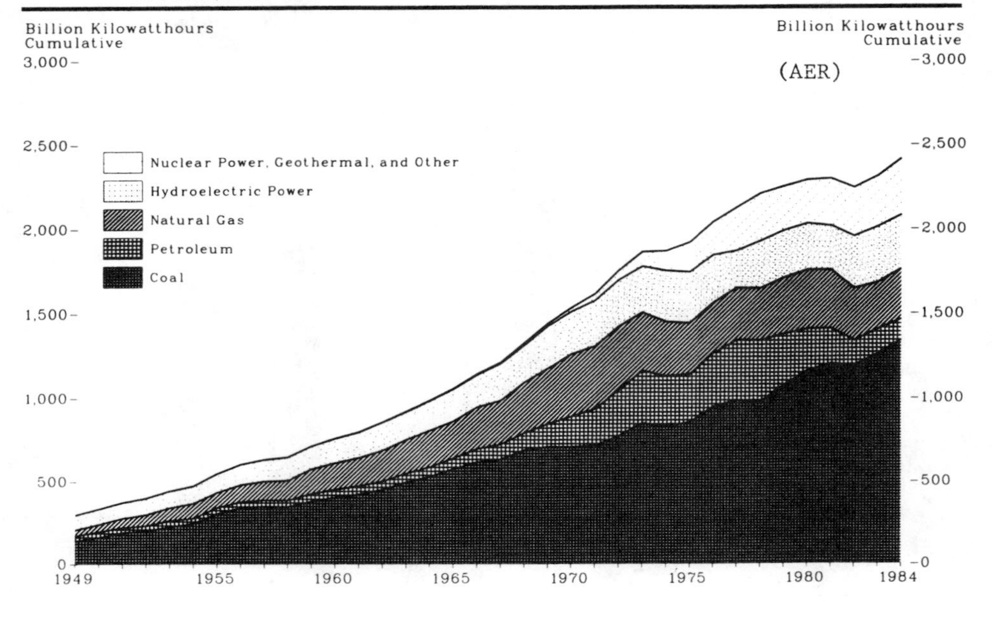

Figure 95. International Production of Crude Oil, 1960-1984 (AER)

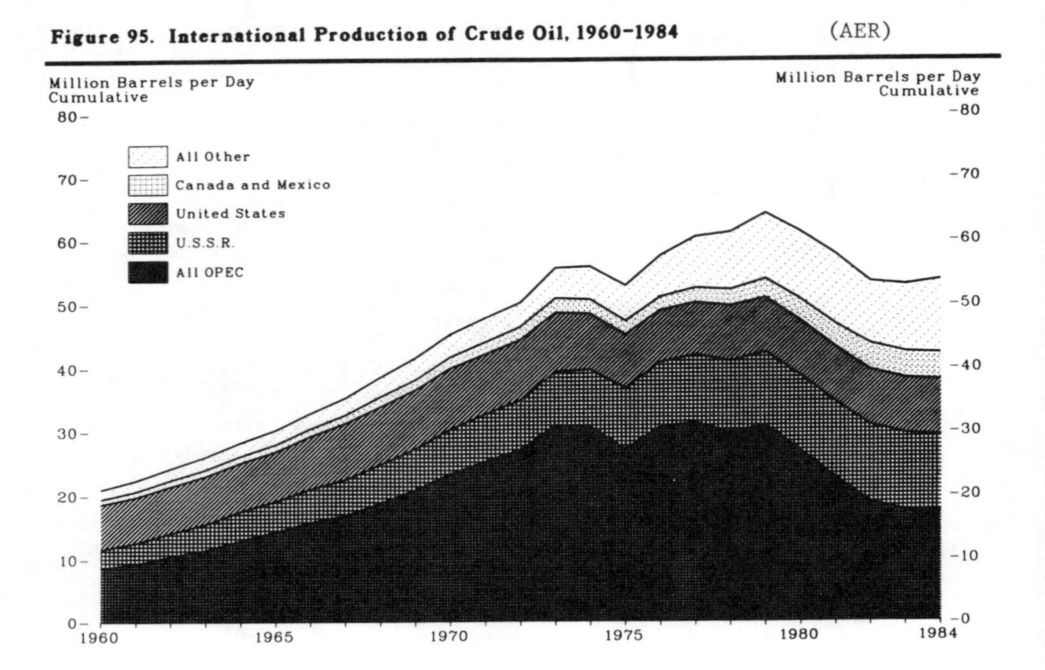

(AER)

Figure 96. International Crude Oil Flow, 1982
(Thousand Barrels per Day)

Arrows Indicate Origins and Destinations but Not Necessarily Specific Routes, Several Minor Routes and Quantities Are Not Displayed.

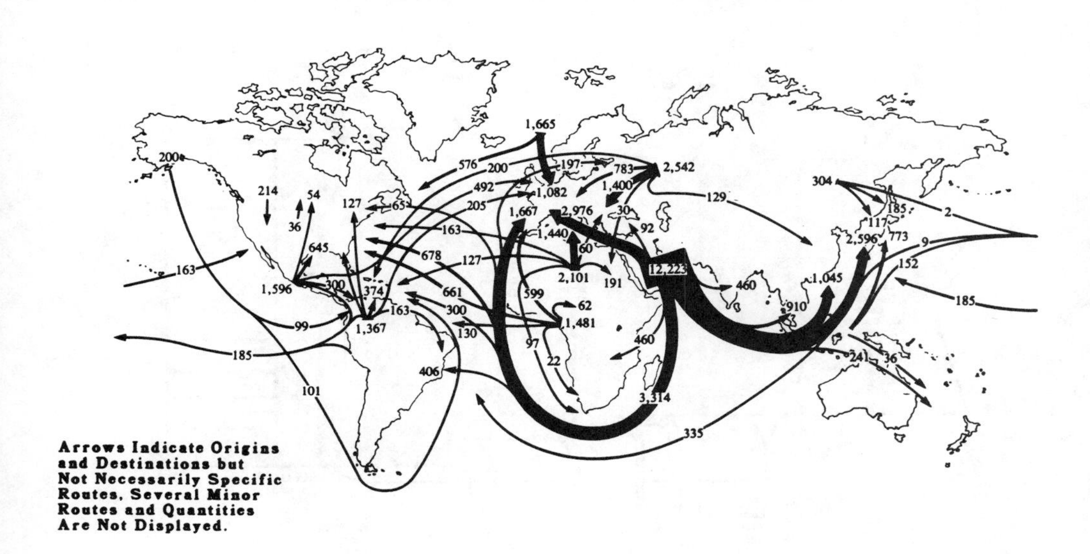

Figure 100. International Natural Gas Flow, 1982
(Billion Cubic Feet)

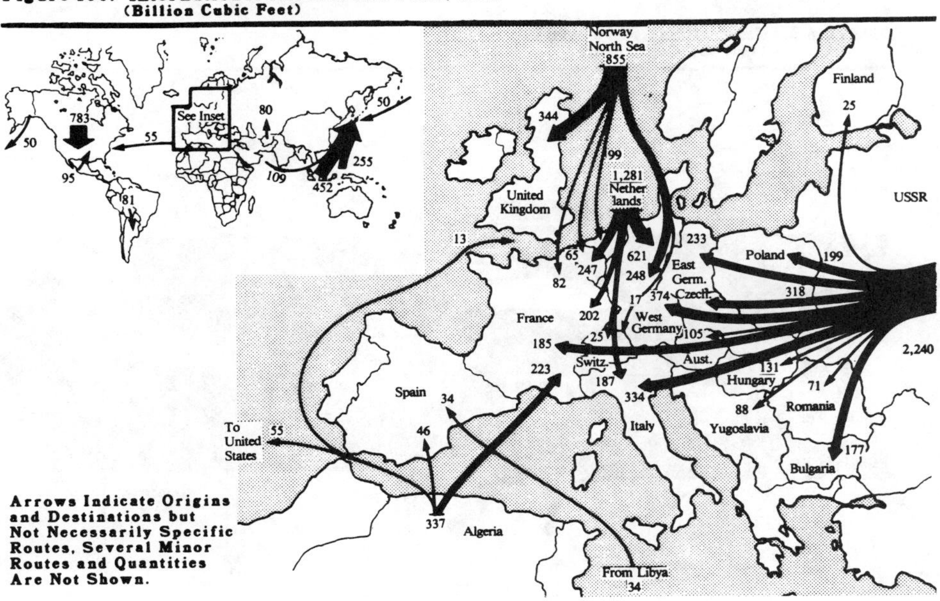

(AER)

Arrows Indicate Origins and Destinations but Not Necessarily Specific Routes, Several Minor Routes and Quantities Are Not Shown.

Diagram 2. Petroleum Flow, 1984 (AER)
(Million Barrels per Day)

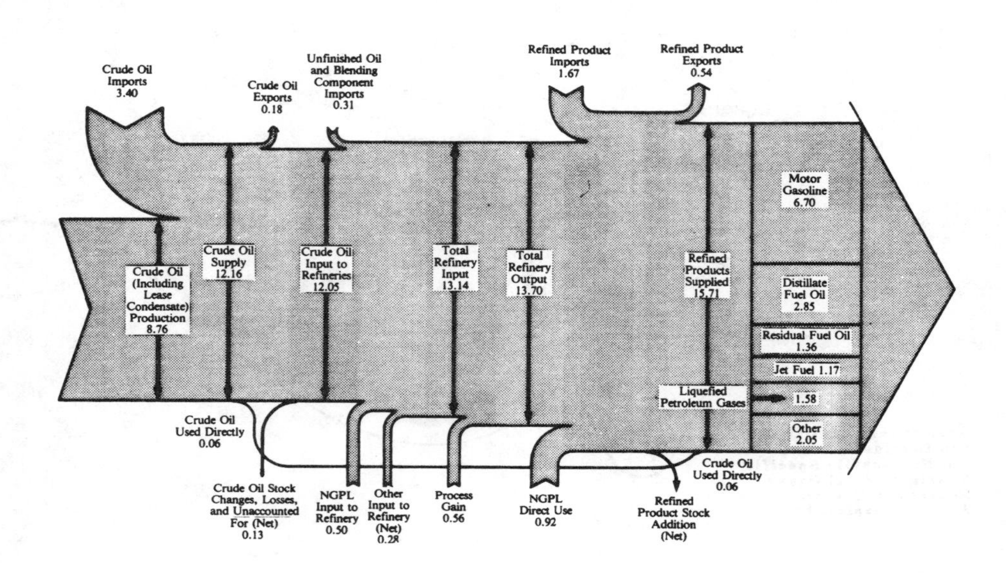

Diagram 3. Natural Gas Flow, 1984 (AER)
(Trillion Cubic Feet)

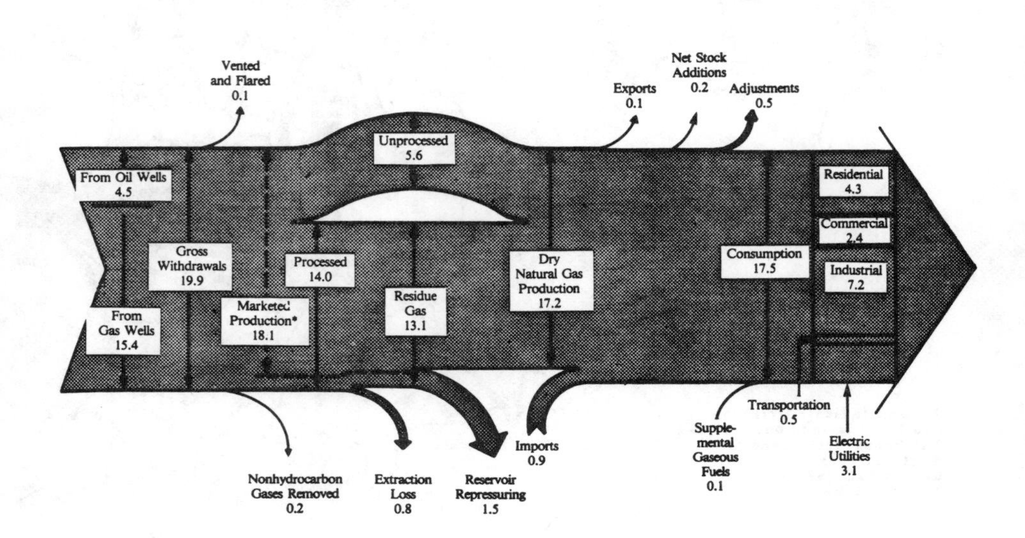

Diagram 4. Coal Flow, 1984 (AER)
(Million Short Tons)

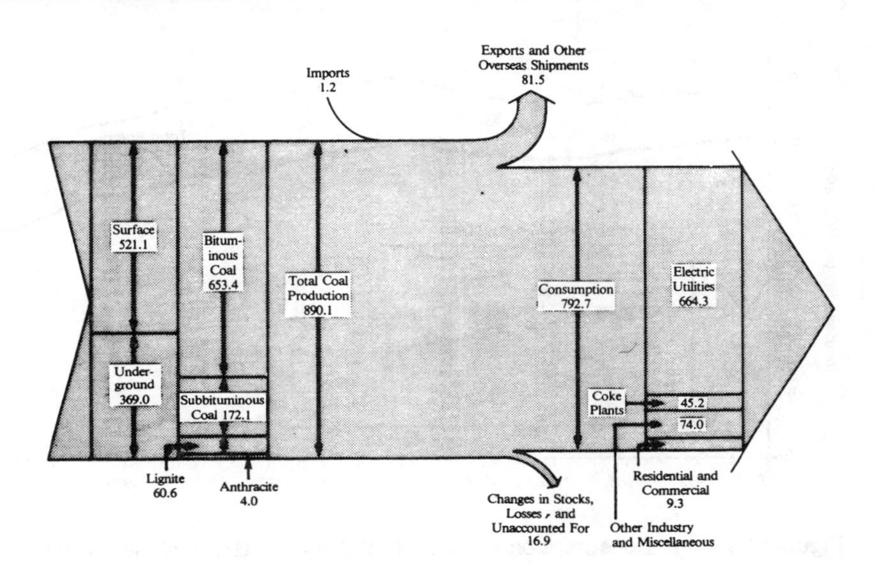

Diagram 5. Electric Utility Electricity Flow, 1984 (AER)
(Billion Kilowatthours)

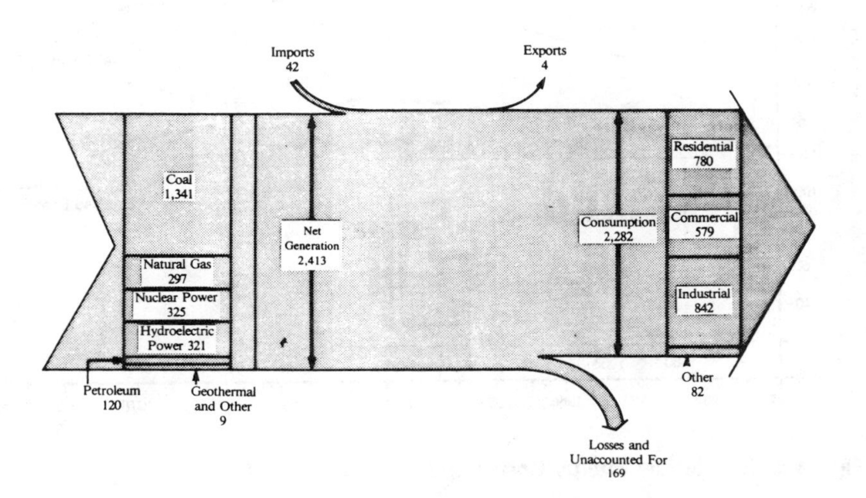

Aggregate Indicators

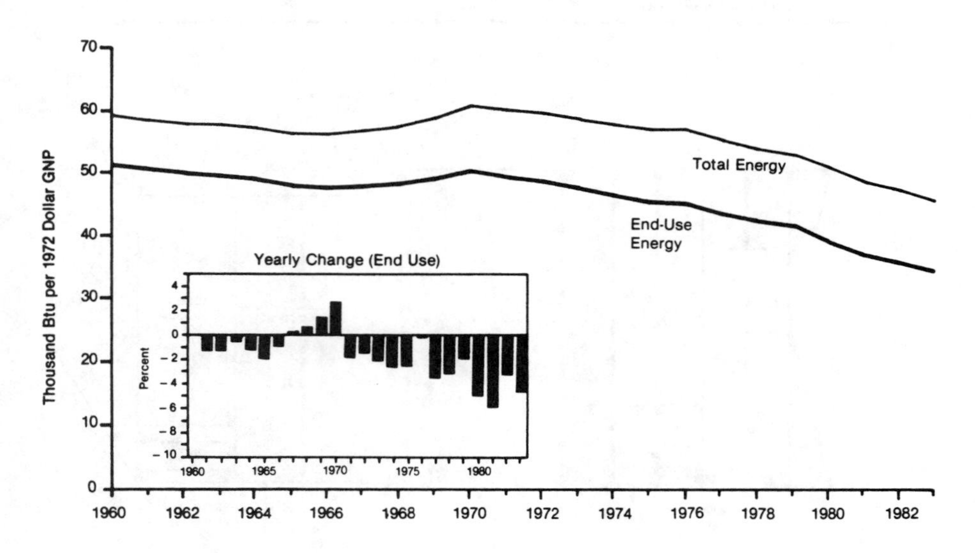

Figure 1. Energy Consumption per Constant Dollar of Gross National Product (ECI)

Residential Indicators

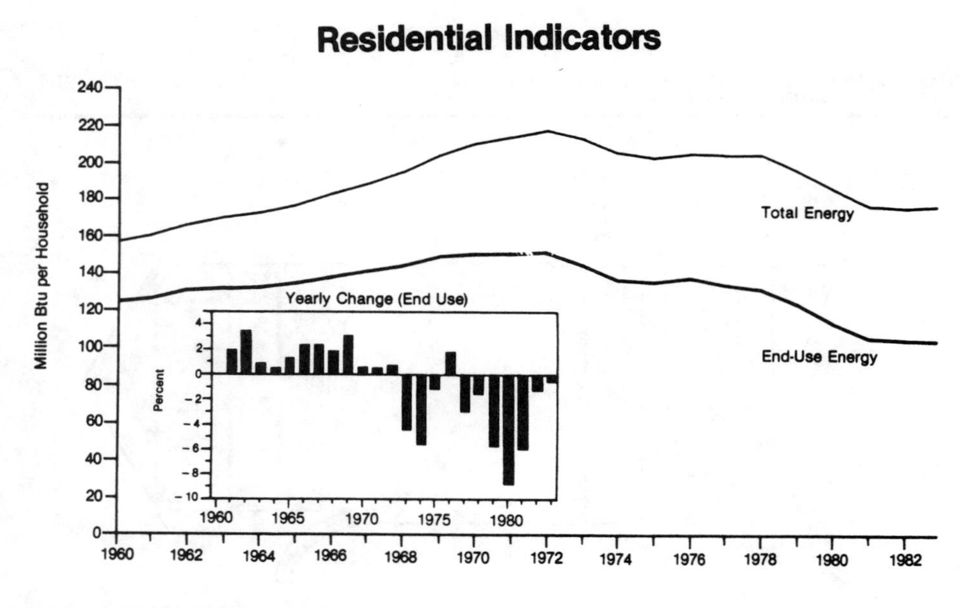

Figure 8. Residential Energy Consumption per Household (ECI)

Residential Indicators

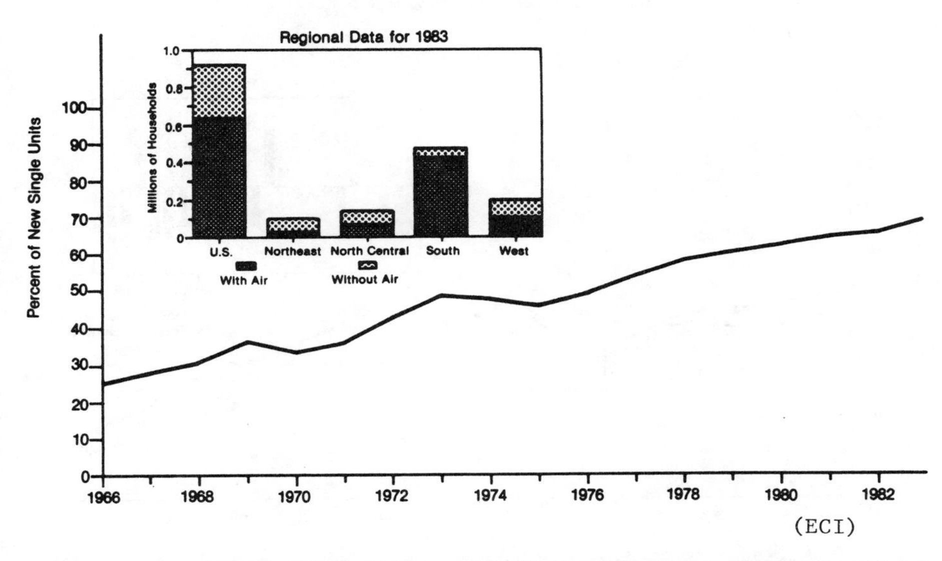

(ECI)

Figure 15. New Single-Family Houses with Air Conditioning, by Region and Total

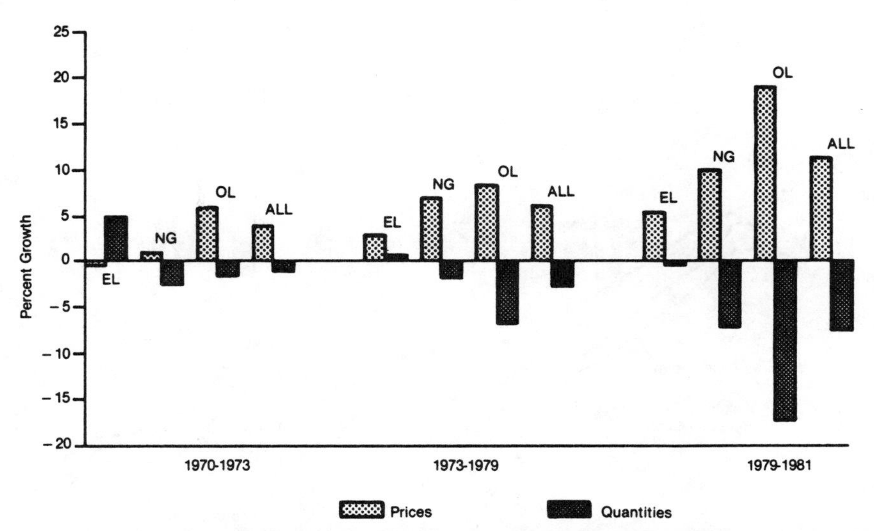

Figure 21. Residential Energy Prices and Consumption per Household, by Energy Source

EL = Electricity
NG = Natural Gas
OL = Oil
ALL = Average of all fuels

(ECI)

Industrial Indicators

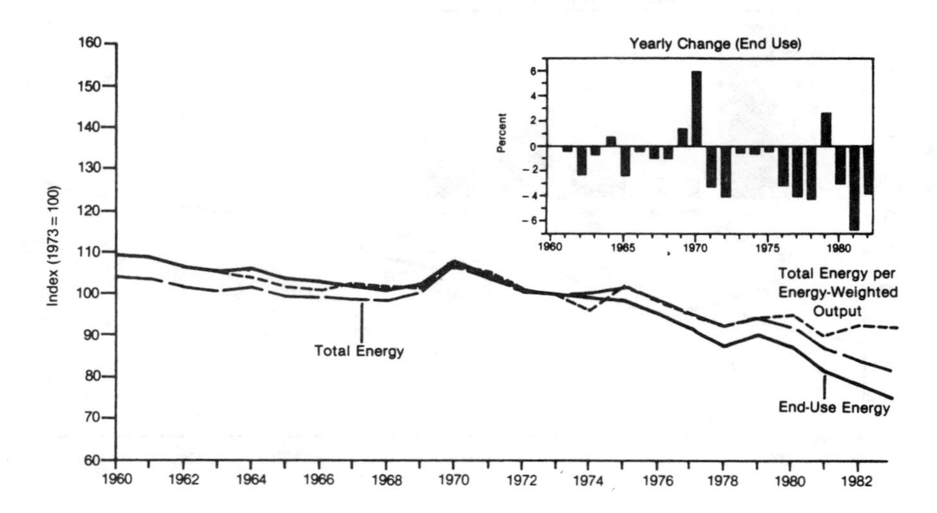

Figure 32. Industrial Energy Consumption per Unit of Industrial Output (ECI)

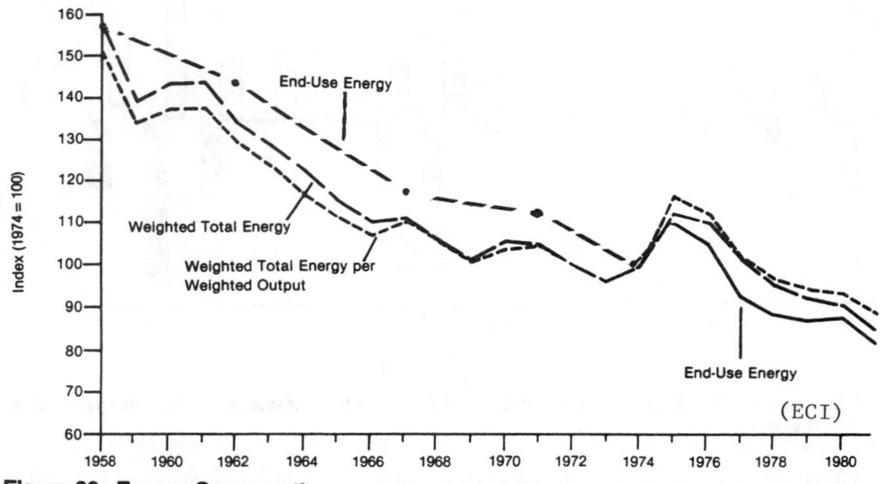

Figure 36. Energy Consumption per Unit of Output in Chemicals, Rubber, and Plastic

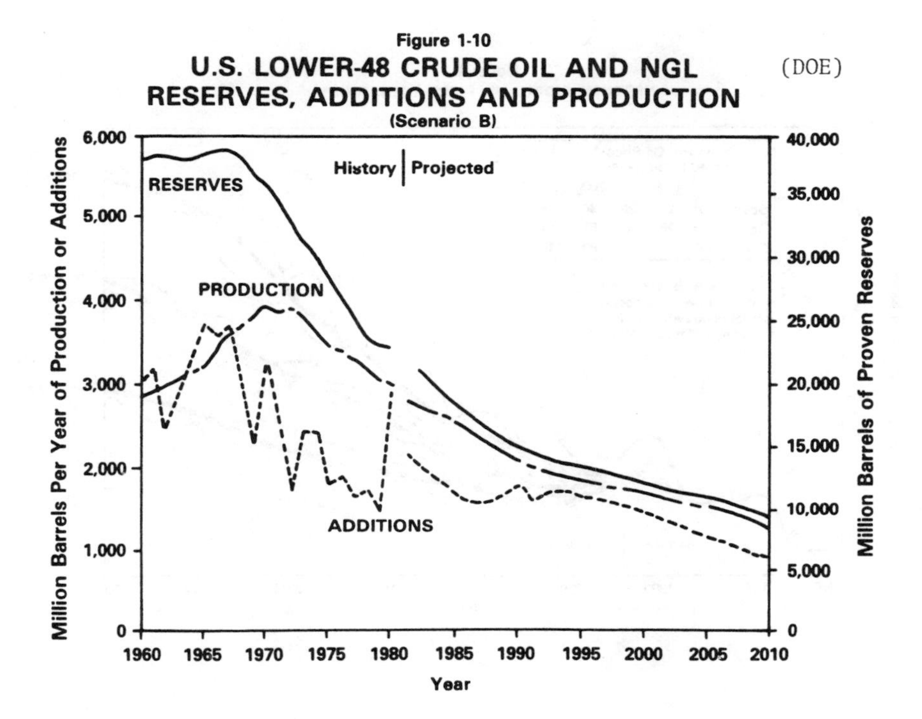

Figure 1-10
U.S. LOWER-48 CRUDE OIL AND NGL RESERVES, ADDITIONS AND PRODUCTION
(Scenario B)
(DOE)

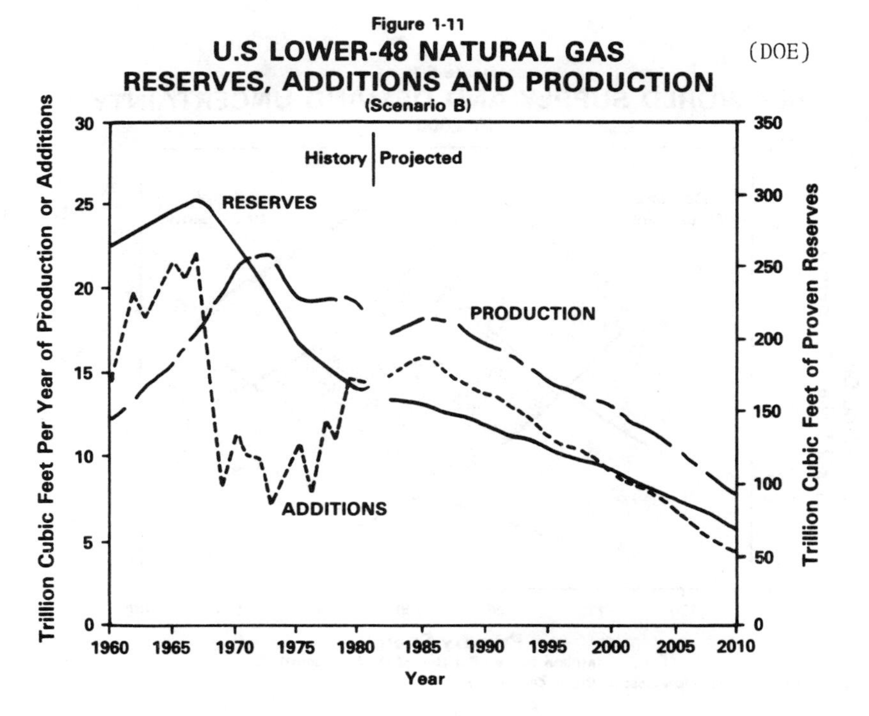

Figure 1-11
U.S LOWER-48 NATURAL GAS RESERVES, ADDITIONS AND PRODUCTION
(Scenario B)
(DOE)

Figure 3
NEPP-1983 WORLD OIL PRICE SCENARIOS* (DOE)

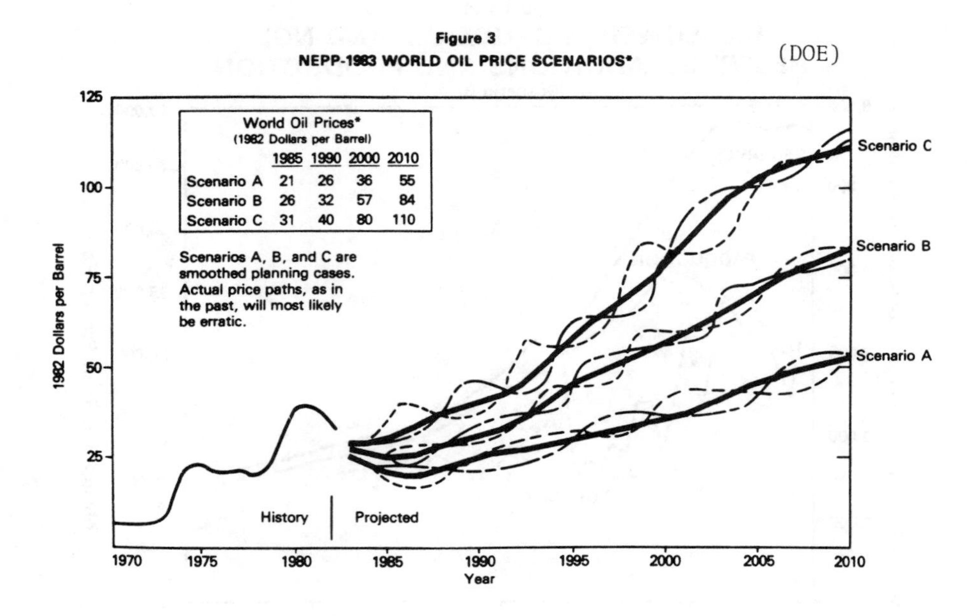

Figure 5-1 (DOE)
FREE-WORLD SUPPLY AND DEMAND UNCERTAINTY IN 2000

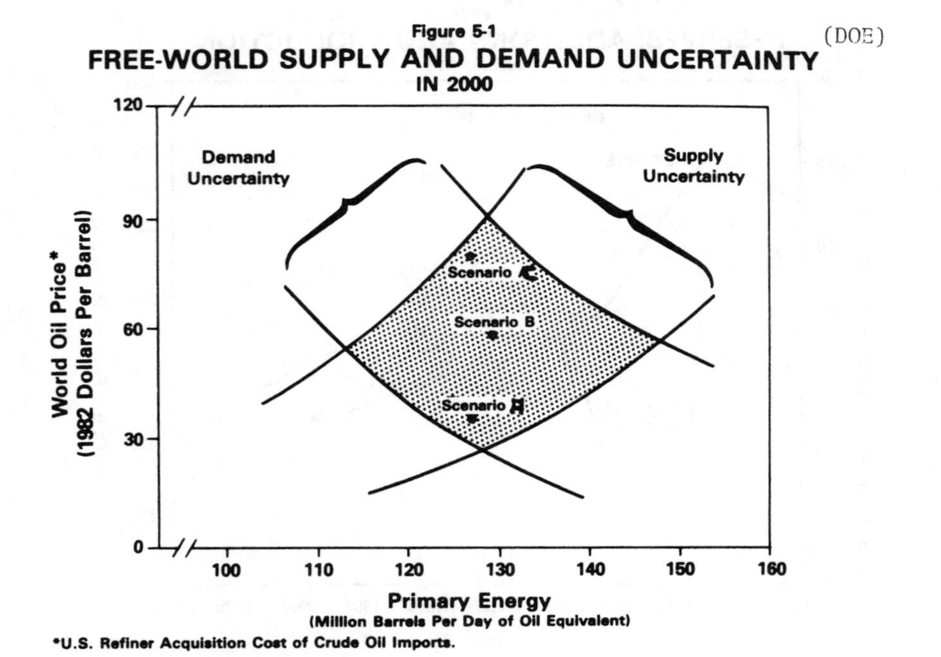

*U.S. Refiner Acquisition Cost of Crude Oil Imports.

Figure 7-2 (DOE)

PROJECTIONS FOR U.S. PRIMARY ENERGY CONSUMPTION FOR 1980 AND 1985 VERSUS REAL PRIMARY ENERGY CONSUMPTION

(Quadrillion Btu per Year)

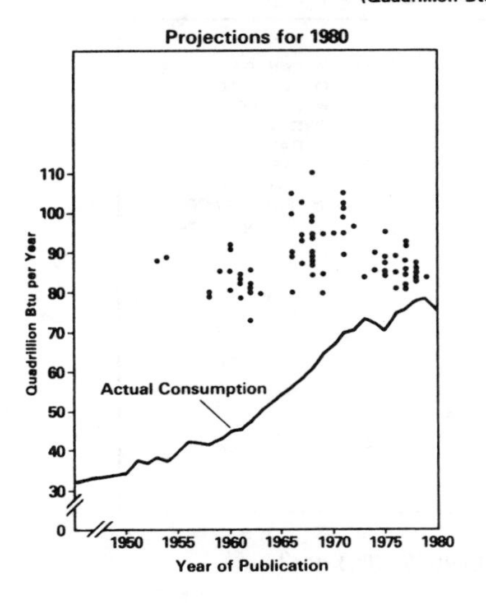

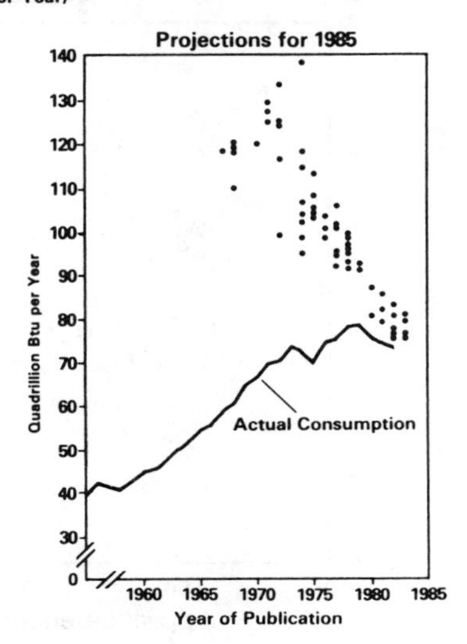

Figure 7-3 (DOE)

PROJECTIONS OF U.S. PRIMARY ENERGY CONSUMPTION FOR THE YEAR 2000

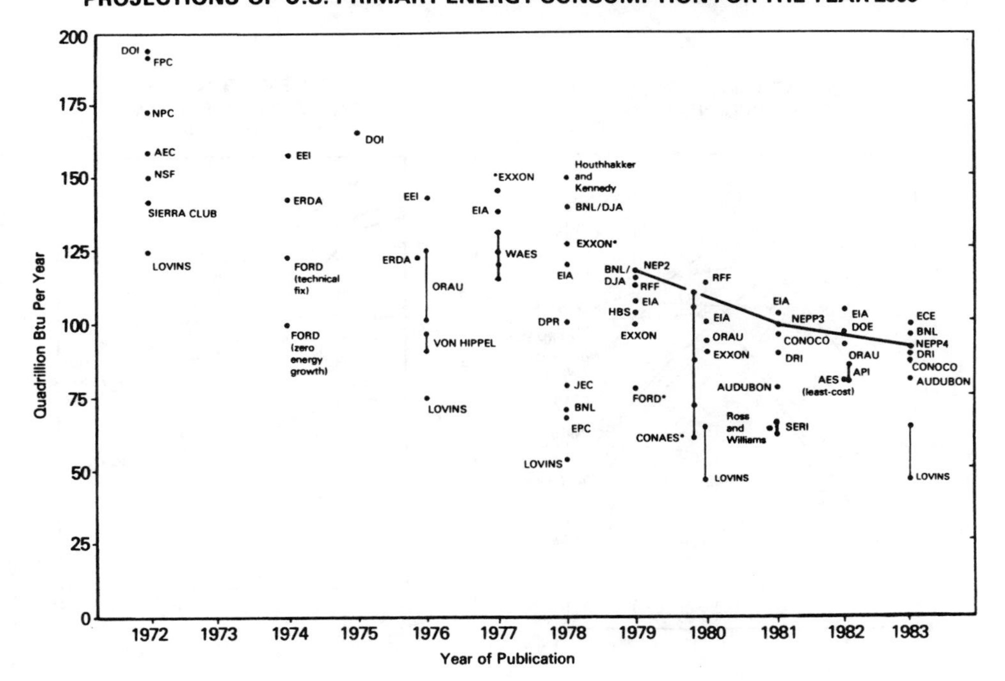

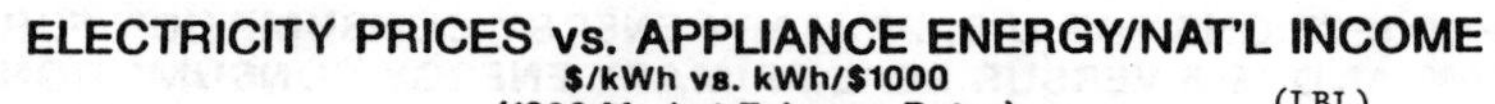
ELECTRICITY PRICES vs. APPLIANCE ENERGY/NAT'L INCOME

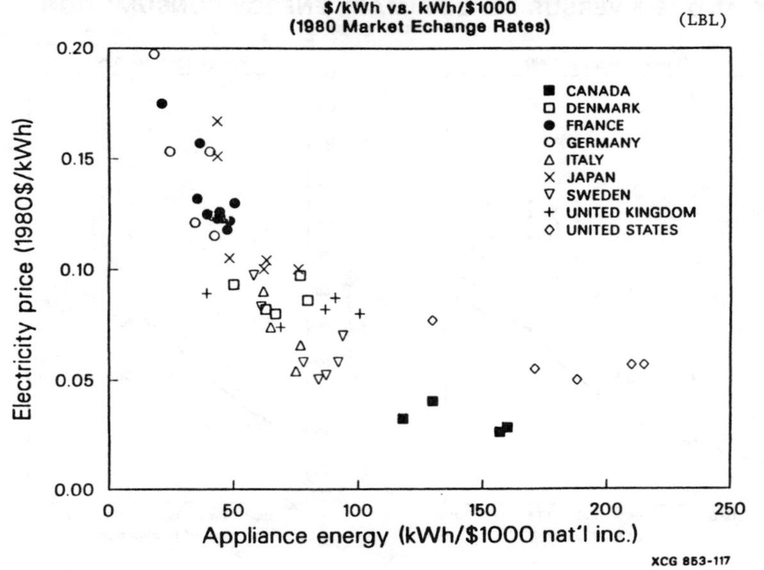
$/kWh vs. kWh/$1000
(1980 Market Echange Rates)
(LBL)
CANADA
DENMARK
FRANCE
GERMANY
ITALY
JAPAN
SWEDEN
UNITED KINGDOM
UNITED STATES
Electricity price (1980$/kWh)
0.00
0.05
0.10
0.15
0.20
0
50
100
150
200
250
Appliance energy (kWh/$1000 nat'l inc.)
XCG 853-117

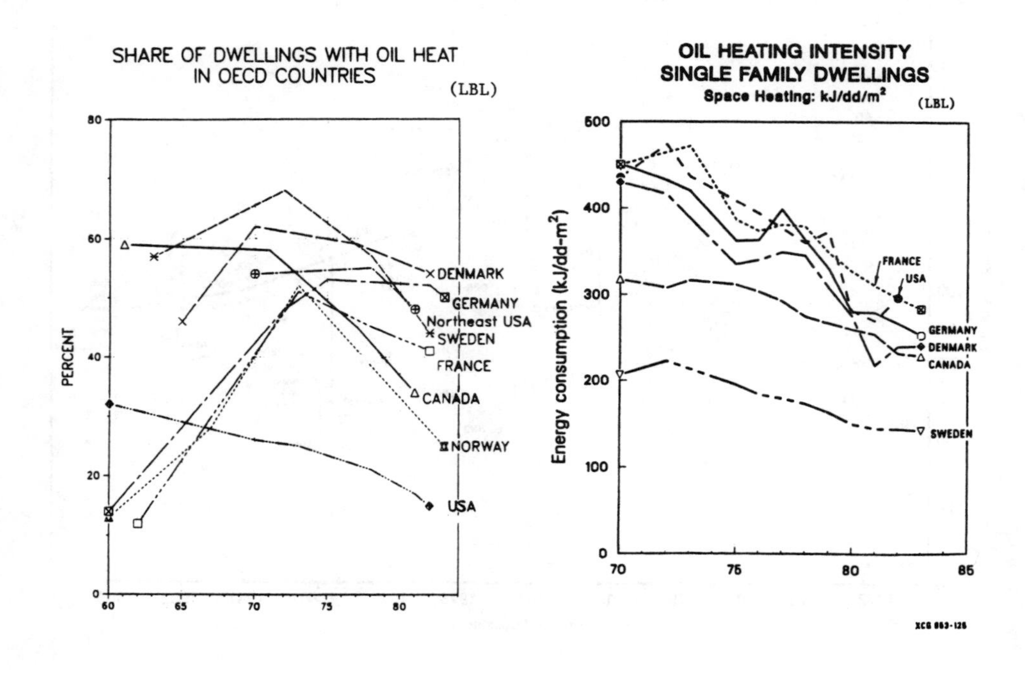
SHARE OF DWELLINGS WITH OIL HEAT
IN OECD COUNTRIES
(LBL)
PERCENT
DENMARK
GERMANY
Northeast USA
SWEDEN
FRANCE
CANADA
NORWAY
USA
OIL HEATING INTENSITY
SINGLE FAMILY DWELLINGS
Space Heating: kJ/dd/m²
(LBL)
Energy consumption (kJ/dd-m²)
FRANCE
USA
GERMANY
DENMARK
CANADA
SWEDEN
XCG 853-126

Table 4. Estimates by the California Energy Commission for new power plants in California during 1990-2020. (CEC) Levalized cost data for investor-owned utilities at the bus bar in 1984 constant dollars. The cost of delivered electricity is approximately obtained by dividing the levelized cost by 0.9 to take into account losses, and then adding 2-5 ¢/kWh for distribution and delivery. RELATIVE COST OF ELECTRICITY PRODUCTION, July 1984.

Technology (Fuel)	Instant Capital Cost[2] 1984 $/kW	In-service Capital Cost[18] 1990	Capital Escalation Real %/Yr.	Fixed Charge Rate[5] Real %/Yr.	Fuel Cost 1984 $/10⁶ Btu	Heat Rate Btu/kWh	Capacity Factor %	Fixed Charges Level ¢/kWh	Fuel[19] Level ¢/kWh	O&M[7,19,20] Level ¢/kWh	Total Levelized Cost
Conventional Steam											
Residual Oil	817	924	1.5	6.8	5.57	9,400	80	0.9	5.2-9.9	0.2	6.3-8.7-11.0
Natural Gas/SNG	817	924	1.5	6.8	5.29	9,400	80	0.9	4.6-9.3	0.2	5.7-8.0-10.4
Methanol (coal-based)	769	870	1.5	6.8	8.05[17]	9,400	80	0.8	7.1-8.6[26]	0.2	8.1-8.6-9.6
Coal	1916	2,231	1.5	7.0	2.14	10,125	65	2.7	3.3-5.0	0.8	6.8-7.5-8.5
Nuclear PWR	1840-2773	2187-3912	2.0-14.0	6.8	1.00	10,500	55	3.1-5.5	1.2	0.4	4.8-6.0-7.2[21]
Combustion Turbine											
Distillate	399	428	1.0	6.3	6.30	11,000	10	3.1	6.8-13.1	0.6	10.5-13.7-16.8
Natural Gas/SNG	407	437	1.0	6.3	5.29	11,000	10	3.1	5.3-10.9	0.6	9.0-11.8-14.6
Methanol (coal-based)	321	345	1.0	6.3	8.05[17]	11,000	10	2.5	8.3-10.0[26]	0.6	11.4-12.0-13.1
Combined Cycle											
Distillate	752	845	1.5	6.7	6.30	8,370	70	0.9	5.2-10.0	0.2	6.3-8.8-11.1
Natural Gas/SNG	770	866	1.5	6.7	5.29	8,370	70	0.9	4.1-8.3	0.2	5.2-7.3-9.4
Methanol (coal-based)	681	766	1.5	6.7	8.05[17]	8,370	70	0.8	6.3-7.6[26]	0.2	7.3-7.7-8.6
Int. Gasifier-Coal	1610	1826	1.5	6.8	2.14	9,460	80	1.8	3.0-4.7	0.7	5.5-6.2-7.2
Geothermal											
Dry Steam	642-944[10]	765-1125	3.0	6.8	1.66	22,800	80	0.7-1.1	4.7	0.1	5.5-5.7-5.9
Hydroelectric											
Conventional	570-1920	634-2136	1.0	5.6	--	--	35	1.2-3.9	--	0.3	1.5-2.8-4.2
Peaking	722-1036	803-1152	1.0	5.6	--	--	10	5.1-7.4	--	0.7	5.8-7.0-8.1
Wood-fired Boiler	1563	1718	1.5	6.5	1.26	16,200	80	1.6	3.1-4.7	0.1	4.8-5.5-6.4
Utility Wind	1084-1597[8]	1104-1626	0[12]	6.3	--	--	35	2.3-3.3	--	0.2	2.5-3.0-3.5
Solar											
Photovoltaics	2453-3879[8]	2493-3942	0[12]	6.3	--	--	22	8.1-12.9	--	0.4	8.5-12.5-13.3
Thermal (3 hr. stor.)	2613-3355[8]	2746-3526	0[12]	6.3	--	--	30	6.6-8.4	--	1.5	8.1-8.9-9.9
Thermal (9 hr. stor.)	3640-5226[8]	3825-5491	0[12]	6.3	--	--	50	5.5-6.5	--	0.9	6.4-6.9-7.4
Repowering (nonreheat)											
Distillate	485[9]	521	1.0[13]	6.5	6.30	8,700[15]	70	0.6	5.4-10.4	0.2	6.2-8.8-11.2
Natural Gas/SNG	485[9]	521	1.0[13]	6.5	5.29	8,700[15]	70	0.6	4.2-8.6	0.2	5.0-7.2-9.4
Methanol (coal-based)	414[11]	445	1.0[13]	6.5	8.05[17]	8,700[15]	70	0.5	6.6-7.9[26]	0.2	7.3-7.7-8.6
Fuel Cell - Utility											
Natural Gas/SNG	776-1232[8]	825-1309	0[12]	6.7	5.29	8,300	85	0.7-1.2	4.0-8.2	0.3	5.0-7.4-9.7
Methanol (coal-based)	776-1232[8]	825-1309	0[12]	6.7	8.05[17]	8,300	85	0.7-1.2	6.3-7.6[26]	0.3	7.3-8.0-9.1

TABLE 2-53Z3. ALTERNATIVE COMPONENT PACKAGES FOR CLIMATE ZONE 3 (SF-Bay) (CEC-Houses) 1985

Component	Package A	B	C	D	E
BUILDING ENVELOPE					
Insulation Minimums					
Ceiling	R 30	R 30	R 30	R 30	R 30
Wall[1]	R 11	R 19	R 11	R 11	R 11
"Heavy" Walls	(R 4.5)	(R 3.5)	(R 2.5)	N/A	N/A
"Light Mass" Walls	[R 5.0]	[R 5.0]	[R 4.5]	N/A	N/A
Slab Floor Perimeter	R 7	R 7	R 7	NR	N/A
Raised Floor	R 11	R 19	R 11	N/A	R 19
Attic[2]	NR	NR	NR	REQ	REQ
GLAZING					
Maximum U Value	1.10	0.65	1.10	0.65	0.65
Maximum Total Area	NR	16%	14%	20%	20%
Maximum Total Nonsouth Facing Area	9.6%	N/A	N/A	N/A	N/A
Minimum South Facing Area	6.4%	NR	NR	NR	NR
SHADING COEFFICIENT					
South Facing Glazing	NR	NR	NR	NR	NR
West Facing Glazing	NR	NR	NR	NR	NR
East Facing Glazing	NR	NR	NR	NR	NR
North Facing Glazing	NR	NR	NR	NR	NR
THERMAL MASS[3]	REQ	NR	NR	25%	NR
INFILTRATION CONTROL					
Continuous	NR	NR	NR	NR	NR
Electrical Outlet Plate Gaskets	NR	NR	NR	NR	REQ
Air-to-Air Heat Exchanger	NR	NR	NR	NR	NR
SPACE HEATING SYSTEM					
If Gas, Seasonal Efficiency=	MIN	MIN	MIN	MIN	79%
If Heat Pump[4], ACOP=	MIN	MIN	MIN	2.5	2.8
SPACE COOLING SYSTEMS					
If Air Conditioner, SEER=	MIN	MIN	MIN	8.0	9.5
DOMESTIC WATER HEATING TYPE					
Gas, heat pump, or solar with any type backup	YES	YES	Solar w/ Gas Backup	YES	YES

LEGEND: NR = Not Required; N/A = Not Applicable; REQ = Required; ACOP = Adjusted Coefficient of Performance; MIN = minimum efficiencies required by Section 2-5306.

1. The value in parentheses is the minimum R-value for the entire wall assembly if the wall weight exceeds 40 pounds per square foot. The value in brackets is the minimum R-value for the entire assembly if the heat capacity of the wall meets or exceeds the result of multiplying the bracketed minimum R-value by 0.65. The insulation must be integral with or installed on the outside of the exterior mass. The inside surface of the thermal mass, including plaster or gypsum board in direct contact with the masonry wall, shall be exposed to the room air. The exterior wall used to meet the R-value in parentheses cannot also be used to meet the above thermal mass requirement.

2. Where an attic is required, it must be installed under not less than 75 percent of the roof that is over any conditioned space.

3. To calculate the amount of thermal mass required for Package A (an option with passive solar design), use the method set forth in Section 2-5351(c)4.

   Package D (an option for buildings with concrete slab floors) requires 25 percent of the floor area directly exposed to the conditioned space. To determine the floor area, count only the first floor in conditioned areas. Uncarpeted (e.g., linoleum or tiled) first floor areas, such as entry ways, kitchens, bathrooms, and conditioned utility rooms or closets may all be counted towards this requirement.

TABLE 2-53W3. ALTERNATIVE COMPONENT PACKAGES FOR CLIMATE ZONE #03 (SF-Bay) FOR HIGH RISE OFFICE BUILDINGS (CEC) - 1987

Component	PACKAGE A	PACKAGE B	PACKAGE C
OPAQUE ENVELOPE			
Minimum Roof Total R-Value (R_t)	14.9	14.9	22.9
Minimum Opaque Wall Total R-Value (R_t) (one of the following):			
Heat Capacity [Btu/°F/ft^2]			
0.0-3.99	7.4	7.4	3.1
4.0-9.99	6.5	6.5	2.6
10.0-14.99	4.4	4.4	1.7
15.0-19.99	3.0	3.0	1.4
20.0 or more	2.0	2.0	1.3
Minimum Suspended Exterior Floor Total R-value (R_t)	9.5	9.5	9.5
GLAZING			
Maximum Allowed Total & West-Facing Vertical Glazing (one of the following):			
Shading Coefficient			
1.00-0.72	29%	42%	29%
0.71-0.61	35%	49%	35%
0.60-0.46	38%	56%	38%
0.45-0.01	45%	61%	45%
See Section 2-5342(b)5. for adjustment for overhangs.			
Maximum Allowed Horizontal Glazing (one of the following):			
Shading Coefficient			
1.00-0.51	Not Allowed	9%	Not Allowed
0.50-0.01	Not Allowed	16%	Not Allowed
LIGHTING (Either:)			
Maximum Adjusted Lighting Power Density, watts per square foot	1.50	1.50 Daylighting Controls Required	1.40
Maximum Adjusted Connected Lighting Load, watts	See Section 2-5342(d)2	See Section 2-5342(d)2	Not Allowed
SPACE CONDITIONING SYSTEM (Both:)			
General Requirements	See Section 2-5342(e)1	See Section 2-5342(e)1	See Section 2-5342(e)1
Performance Criteria (applicable to any Alternative Component Package)			

Fan Wattage Index	1.10
Source Heating Power Index	34.9
Source Cooling Power Index	24.0

APPENDIX C

SUMMARIES OF REPORTS FROM THE CONGRESSIONAL OFFICE OF TECHNOLOGY ASSESSMENT

1. RESIDENTIAL ENERGY CONSERVATION, 1979.
2. ENERGY EFFICIENCY OF BUILDINGS IN CITIES, 1982.
3. INDUSTRIAL ENERGY USE, 1983.
4. INCREASED AUTOMOBILE FUEL EFFICIENCIES AND SYNTHETIC FUELS, 1982.
5. U.S. VULNERABILITY TO AN OIL IMPORT CURTAILMENT: THE OIL REPLACEMENT CAPABILITY, 1984.
6. OIL AND GAS TECHNOLOGIES FOR THE ARCTIC AND DEEPWATER, 1985.
7. ACID RAIN AND TRANSPORTED AIR POLLUTANTS: IMPLICATIONS FOR PUBLIC POLICY, 1984.

1. RESIDENTIAL ENERGY CONSERVATION, 1979.

Americans are responding to a changed energy situation by rapidly curtailing the direct use of energy in their homes. The patterns of energy use established by households in the 1960's have changed dramatically. Residential energy use, which grew at a rate of 4.6 percent per year during the 1960's, has grown at an average annual rate of 2.6 percent since 1970. In 1977, Americans used 17 quadrillion Btu* (Quads) of energy in their homes, 22 percent of the total national energy use. Had the growth rate of the 1960's continued, the Nation would have used an additional 2.5 Quads—equivalent to 430 million barrels of oil—in 1977.

As impressive as these figures are, they can be better. Savings of more than 50 percent in average use by households, compared to the early 1970's, are already being achieved in some new homes, and experiments with existing homes indicate that similar reductions in heating requirements can be realized through retrofit. These savings can be achieved with existing technology, with no change in lifestyle or comfort—and with substantial dollar savings to homeowners. However, more sophisticated design, quality construction, and careful home operation and maintenance will be required.

For the residential sector as a whole, the potential energy savings can be seen in another way. If the trend of the 1970's were to continue for the balance of the century, the residential sector would use about 31 Quads of energy in 2000. But if investments were made in home energy conservation technologies up to the point where each investor received the highest possible dollar savings (in fuel costs) over the investment's life, energy use in the year 2000 would be reduced to between 15 and 22 Quads, depending on the price of energy. The cumulative savings between now and 2000 compared to the 1970's trend would be equivalent to between 19 billion and 29 billion barrels of oil. Despite the sound economic reasons for achieving these savings, there are reasons why they may not be reached. This report examines the underlying problems and what to do about them.

0094-243X/85/1350600-28$3.00 Copyright 1985 American Institute of Physics

Following this section, the study's major findings are presented. They lead to these conclusions, among others:

1. Analysis of data on price and consumption, combined with research on consumer motivation, indicates that the desire to save money is the principal motivation for changes in energy habits (turning down the thermostat at night) and investment in conservation (purchasing insulation or having the furnace improved). This report outlines the approximate level of energy savings that might result from investments up to the point where dollar savings over the life of the investment are greatest. If it is national policy to encourage energy savings beyond this point—for example, to the point where investments in energy savings provide smaller economic return but greater energy savings—additional incentives would be required. The difference between these two points is substantial in energy terms, because once a dwelling is efficient, costs of operation are relatively insensitive to energy prices. Such a shift would be analogous to the standards set in 1975 to improve energy performance of new cars. In addition to price or economic incentives, regulation could also increase energy savings.
2. One of the principal ways to improve energy use lies in the area of information and technology transfer. Those who actually implement policy need more training. Policy may be made in Washington, but is carried out by tradespersons, builders, local code inspectors, loan officers, appraisers, energy auditors, heating technicians, State and local officials, do-it-yourselfers—literally thousands of individuals. The essentially human nature of the effort is both a strength and a weakness—many are willing to take some action, but there are many obstacles to perfect performance.
3. The diversity of the housing stock, number of persons involved, requirements for technology transfer, and product availability all argue for careful pacing of Federal policy, based on setting goals over at least a decade. For example, short-term programs, aimed at one particular solution, appear to constrain the market and may not encourage optimal solutions. This is particularly true of programs aimed at the existing housing stock. Anticipation of the tax credit for insulation caused increased prices and spot shortages and may not have produced substantial insulation beyond what would have occurred in any event. Another reason for deliberate policymaking is that knowledge of the nature of a house as an energy system is imperfect. Although a good deal is already known about saving energy, more remains to be learned. Because choices will vary with climate, local resources need to be developed; these resources will include both trained personnel and improved data.

Policy choices will reflect the goals for savings and costs. If the current trajectory is appropriate, present programs appear to be adequate in number and range. A lower growth rate can probably be accomplished by vigorous congressional oversight, some administrative adjustments, review and fine-tuning of program operation, and improved information efforts. If the sector is already moving fast enough, less emphasis could be placed on residential energy use. In order to move much more rapidly, stronger measures will be required. A great deal of energy could be saved in homes above present levels; these savings would still be cost-effective to the consumer. A stronger program approach might reflect national security goals and a high return on the housing dollar.

The following sections consider the trends illustrated by this volume and the major factors affecting residential energy use and conservation: price, consumer attitudes, the poor, existing housing stock, building industry response, design opportunities, the role of States and localities, the utilities, and Government programs.

Trends in Residential Energy Consumption

The decade of the 1970's has brought significant changes in the historical patterns of growth in energy consumption in the residential sector. Earlier, Americans as a group were increasing their use of energy in the home at an average rate of 4.7 percent per year; in the 1970's, the annual growth rate has averaged 2.6 percent. Moreover, the remaining growth is attributable primarily to a growth in the number of households; the amount of energy used in each household has remained almost constant between 1970 and 1977. In 1970, 63.5 million households collectively used 14 quadrillion Btu of energy (Quads) or about 230 million Btu apiece. (A Quad is equivalent to 500,000 barrels of oil per day for 1 year—or the annual energy required for the operation of eighteen

1,000-MW powerplants—or 50 million tons of coal.)

In 1977, residential use of energy accounted for 22 percent of total energy consumption, totaling 17 Quads. By comparison, the commercial sector in 1977 used 11 Quads (14.5 percent of total), transportation accounted for 20 Quads (26 percent), and industry used 28 Quads (37 percent). Total 1977 U.S. energy use was 76 Quads.

Many factors have contributed to the slowed growth in residential energy use in this decade. Among them are energy price increases, economic fluctuations, demographic trends, the OPEC embargo, and consumers' responses to rising awareness of energy. Demonstrating a precise cause-and-effect relationship between any one of these factors and the lower growth rate is statistically impossible. Fortunately, isolating and quantifying the contribution of each factor is probably of limited utility to policymaking.

The rapidity of the slowdown suggests that actions taken to reduce consumption so far are primarily changes in the ways people use their existing energy equipment—e.g., turning down thermostats and insulating. A longer time frame is normally required to bring about widespread replacement or improvement of capital stock, including heating equipment and housing units.

No one can say with certainty whether the residential energy growth rate will stabilize at today's rate, drop still further, or creep back up toward earlier trends. Countervailing forces could work in either direction. The current demographic trend toward slower population growth is expected to continue for the near term, but household formation rates are likely to exceed population growth rates. Energy use in the residential sector can be expected to grow faster than population as long as new households are forming at a higher rate, although construction of highly efficient new housing would alter that presumption.

On the other hand, if energy prices continue to rise, greater investments in conservation (energy productivity) measures will become cost-effective for consumers. Moreover, while there will be more households, each is likely to be smaller; having fewer people at home generally means smaller dwelling units and lower levels of energy consumption in each home. Very few experts believe that residential energy growth rates will ever again approach the very high pre-1970 rates.

If residential energy use were to continue growing by 2.6 percent annually until the year 2000, total residential consumption in that year would approximate 31 Quads. This is considerably lower than the 48 Quads American homes would consume in 2000 if growth patterns of the 1960's had continued. Yet actual consumption in 2000 might be even lower than 31 Quads, driven down by rising prices and a number of other factors, including improved design and technology as well as evolving consumer awareness of the economic benefits of conservation.

If residential energy growth were to match the rate of household formation—that is, if the energy consumption per household were to remain constant between now and 2000—total residential sector consumption in that year would be 24 Quads. This trend would represent an annual growth rate of 1.6 percent, which is the household formation rate projected by the Oak Ridge National Laboratory housing model. This modest decline from 1970-77 trends would appear to be relatively easy to achieve under current laws and programs (with improvements in their implementation in some cases) and without sacrificing personal comfort, freedom, or social goals that require increases in energy consumption for those at the low end of the economic spectrum. Much of the decline could be accomplished through replacement of capital stock and construction of smaller, more efficient housing units to accommodate new households.

An even lower consumption level in 2000 could be achieved through an optimal economic response—one in which all residential consumers made the maximum investment in conservation technologies that they could justify through paybacks in reduced energy costs over the remaining lives of their dwelling units. Such responses would depend on the levels of energy prices over the next two decades. Using a range of plausible energy prices, possible residential energy consumption levels were projected to be between 15 and 22 Quads in 2000, based on optimal economic response. Few observers expect the lower end of the range to be achieved even using the highest price assumptions, because of imperfections in the marketplace. Circumstances requiring especially vigorous public policies could create additional incentives to consumers to approach this level of savings.

The middle ground between the 1970's trend and the optimal economic response trend is seen by many as a reasonable public policy target. Measuring our progress toward this conservative goal would be relatively easy; each year, the goal would be to maintain constant national average energy consumption per household by keeping the growth in residential energy use to a rate determined by the household formation rate. This target appears to be manageable within our current social, political, and economic situation. This option would not involve sacrifice, because it would allow for a constantly improved level of residential amenities that can be achieved by means of improved energy productivity (less energy per unit of amenity provided). Some critics will view this goal as too easy, too modest; considering depletion of nonrenewable resources, maximum return on housing dollars, environmental quality, and the national security implications of our oil imports. (Comparative energy use projections showing these Quad levels appear graphically in figure 1.)

Figure 1.—Comparative Energy Use Projections (Residential sector)

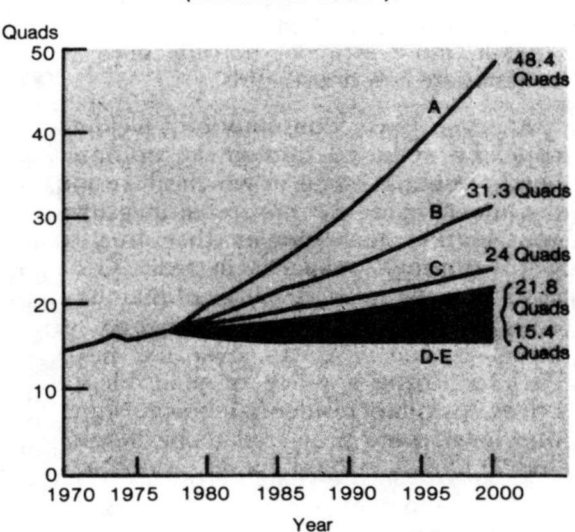

B — Residential consumption based on simple extrapolation of 1970-77 trend.

A — Residential consumption based on simple extrapolation of 1960-70 trend.

C — Residential consumption based on constant level of energy use per household; growth results from increase in number of households.

D-E — Range of "optimal economic response" based on assumption that energy saving devices are installed as they become cost-effective. Range is formed by price; upper boundary represents response to lowest projected price, lower boundary represents response to highest projected price.

NOTES: These curves are not given as predictions of the future, but as points of comparison for discussion. See chapter I for detailed information.

For SI users. Quads can be substituted using exajoule (EJ) on this figure within the accuracy of the calculations. One Quad ≈ 1 EJ.

Residential Energy Prices

Rising energy prices appear primarily responsible for reduced residential consumption in recent years. Rapid growth in the 1960's accompanied a decline in real energy prices, while the growth slowdown of the 1970's has concurred with a rise in real prices. The increase in energy prices has been especially marked since 1974, when the embargo reached its peak and the Arab oil cartel began a quintupling of oil prices. The OPEC nations' recent decision to raise oil prices in 1979 and other Middle East developments can be expected to affect U.S. energy consumption patterns further.

For the residential consumer, the 1970's have already brought a 65-percent rise in home oil-heating bills, a 37-percent increase in the natural gas bill, and a 25-percent rise in the electricity bill (in constant 1976 dollars). In current dollars, the increases have been far more dramatic. Even so, price controls on oil, average costing of electricity, and Government regulation of natural gas prices at the wellhead have resulted in subsidized retail prices that fail to reflect the full replacement cost of oil, gas, and electricity generated from either nuclear or fossil fuels.

It is important that energy prices represent true replacement costs whether this is higher or lower than current energy prices. It is only under this circumstance that consumers have a correct signal to use in determining how much to invest in conservation if they are to achieve maximum dollar savings. Furthermore, if society decides that information on items such as environmental damage, resource depletion, and reliance on foreign oil would not be accurately given by normal market forces, than it is possible to adjust the replacement cost accordingly or to provide equivalent financial incentives. In any case, since dollar savings are the principal motivation for energy conservation, it is important that conservation policy be concerned with energy prices.

Price increases clearly mean less disposable income for consumers. Stretching the available resources through higher productivity of energy use is a less costly approach than developing new supplies. Improving energy productivity in household use helps to counter the inflationary impact of rising costs. A number of policy responses are possible between holding prices steady or allowing them to rise directly in response to costs; these include matching technological solutions to energy consumption—such as heat exchangers, "smart" thermostats, and draft-excluding devices—can be easily encouraged by Government action assisting the market.

Improved data collection is needed on homes that use little purchased energy. Construction of such homes on a demonstration basis, perhaps one in every county, could provide the type of direct learning experience most valuable and influential for builders and buyers.

Technologies now in the development or commercialization stage will offer opportunities for energy savings well beyond the options now available. More efficient furnaces, new approaches for the design and construction of walls and windows, and electronic systems to monitor and control the operation of homes are now being tested and used experimentally. As these devices become more reliable and lower in cost, the options for reducing home energy use will increase dramatically.

States and Localities

States and localities bear the major responsibility for implementation of federally authorized residential conservation programs. Building code revision and enforcement, information and education efforts, quality control, and regulation of utilities all come within the jurisdiction of States, counties, and towns. The priority assigned to conservation goals by these levels of government will directly influence the level of effort and thus the resources available to consumers and builders.

Current Federal policies both help and hinder State and local efforts. Central difficulties include rapid pacing of Federal initiatives that may not match the capabilities of the locality; failing to involve States and localities in preparing guidelines and regulations; placing responsibility for administering a large number of complicated programs on State energy offices that are frequently small, understaffed, and underfunded; and imposing Federal priorities that may not match local needs. Programs designed with the needs and capabilities of the States in mind are most likely to take root and remain effective as Federal priorities change and Federal funding fluctuates.

Localities work most closely with new construction through the building permit process. Local code inspection offices may require special help, both technical and financial, to improve their level of activity. This will certainly be the case if Federal actions to mandate energy changes in building codes continue. While the needs of localities may press a State energy office beyond its capabilities, these offices must recognize the importance of providing resources to localities.

Transfer of information and technology from the Federal Government can be improved. Trained personnel, either from Washington offices or regional offices, could greatly assist States in working out technical problems and establishing ground rules for program operation.

Utilities

The ways in which gas and electric utilities can most effectively stimulate energy conservation in the residential sector are just beginning to be understood and exercised. As experience with utility-based conservation activities is gained, early concerns about utility involvement in nontraditional activities (such as insulation financing) and uncertainty about the impacts of innovative pricing and service delivery options (particularly time-of-use pricing and load management) are being replaced with encouraging empirical data.

Utilities can encourage residential energy conservation through information programs and home energy audits; financing and/or marketing insulation and other conservation devices; altering the rate structures to reflect costs that vary with time of use; and instituting programs of load management in the residential sector. Relatively few utilities have carried out aggressive conservation programs to date, although most electric and gas companies have undergone some adjustments in their management and planning functions as a result of changing circumstances. While economic and social criteria encouraged rapid energy growth in the years before 1973, more recent phenomena—including rising fuel costs, massive increases in capital requirements for new capacity, uncertainty about future demand, and changing regulatory requirements—have all caused utilities to expect and even encourage diminished growth.

Activities authorized by the National Energy Conservation Policy Act of 1978 should yield useful data over the next few years. The effects of audit programs, cost-based rates, load management, and time-of-use pricing should be carefully analyzed and the information widely shared. Following evaluation, Congress may wish to consider removal of the prohibition against utility involvement in sale or installation of residential conservation measures.

2. ENERGY EFFICIENCY OF BUILDINGS IN CITIES, 1982

This study is concerned with the prospects for and barriers to increased energy efficiency in the building stock of the Nation's cities. From the perspective of the energy specialist, the report assesses the specific capital investments that might be made to make existing buildings more energy efficient. Then from the perspective of the real estate specialist, the report identifies which of these investments in energy efficiency are likely to take place and which are not, and why.

Overall, OTA estimates that about 7 Quads per year of energy savings is technically possible by 2000, through feasible investments in the improved energy efficiency of building types covered in this report (see table 1). Nearly 3 Quads of these potential energy savings are likely to come about because of investments in energy efficiency made by building owners who have personal or business reasons to invest money in improved energy efficiency of their buildings.

The other 4 Quads of potential energy savings, on the other hand, may not occur because building owners fail to make investments in the energy efficiency of their buildings. Part of the failure to retrofit is due to the difficulty and costliness of improvements in energy efficiency to some building types. Part of the failure is due to building owners' stringent requirements for return on investments in energy efficiency. The diversity of buildings and owners and their implications for national energy use is described below.

Technical Description

The national potential (estimated in table 1) for increased energy efficiency of the building stock is the result of physical changes to improve the energy efficiency of millions of buildings. For convenience, these physical changes are referred to as *energy retrofits* in this report. While recognizing that each building is to some extent a unique problem, OTA did identify the major characteristics of buildings which influence the types of energy retrofits that are likely to be most effective. These are:

- **Size.**—Energy retrofits which improve the energy efficiency of the building envelope (walls, windows, and roof) are more important for small buildings than for large buildings. On the other hand, certain kinds of retrofits which bring about similar savings in small buildings and large buildings will cost relatively less per unit of energy saved in large buildings because of economies of scale.
- **Wall and roof type.**—Masonry or curtain walls and flat roofs without attics or with very small crawl spaces are much more difficult to insulate than are wood frame walls and roofs with attics and ample crawl spaces.
- **Mechanical system (HVAC) type.**—Physical changes to the way space heating or cooling is produced and circulated can provide significant increases in building efficiency but vary with the type of heating ventilation and air conditioning (HVAC) system used by the building.
- **Building use.**—Most *commercial buildings* are used from 9 to 5 on weekdays (offices) or 9 to 9 daily (shopping centers) and are unoccupied outside these hours. This provides opportunities for improved energy efficiency by careful control of temperature and lighting between operating and nonoperating hours. Opportunities also exist for more efficient and task-specific lighting in commercial buildings. Finally, retrofits to the hot water system of *multifamily buildings* can usually save considerable energy.

Table 1.—The Gap Between Likely Energy Savings Through Retrofit and Technically Feasible Savings by the Year 2000: Building Types Covered in This Report (quadrillion Btus of primary energy)

	Projected energy use[a]	Technical savings potential[b]	Likely savings[c]	Gap: technical savings potential not realized
Multifamily buildings (all)	2.4	1.0	0.3	0.7
Commercial buildings (all)	6.3	3.5	1.3	2.2
Low income single family (all)	1.6	0.8	0.2	0.6
Moderate and upper income single-family homes in cities	3.5	1.8	0.9	0.9
Total buildings covered in this report	13.8	7.1	2.7	4.4

[a]Projected energy use in 2000 assumes no reduction from current energy use by these buildings and is based on a set of assumptions, that are described in the appendix to ch. 2, about demolition of existing buildings and construction of new buildings needing retrofit. A quadrillion Btu equals approximately 5000,000 barrels of oil per day for a year.
[b]The technical savings potential is defined as that resulting from all retrofits to these building types which as of 1981, are technically feasible and which would be cost effective over a 20-year lifetime, assuming no real increases in energy prices and a 3-percent real return on investment.
[c]Likely savings are those which are likely to come about from investments by building owners under current conditions of availability of capital, retrofit information, and public programs.

Capital Costs

OTA reduced 43 potential combinations of the four building characteristics described above to 13 building types for which the lists of appropriate retrofit options are distinct (although there may be considerable overlap among them). The 13 building types are shown in table 2. **OTA identified no major category of building typically found in cities for which substantial savings were not available from retrofits of low or moderate capital cost compared to savings.**

For some of the building types, a major part of the potential savings are likely to come from retrofits of low capital cost compared to savings (see table 3) in the sense that they will pay for themselves in energy savings in 2 years or less and will earn real rates of return over the life of the retrofit (20 years on average) of more than 50 percent per year

Table 2.—Thirteen Types of Buildings With Significantly Different Retrofit Options

Building type and wall type	Mechanical system type	More energy savings from: Low capital cost[a] retrofit package[a]	More energy savings from: Moderate capital cost[a] retrofit package[a]
Small house with frame walls (single family or 2-4 units)	Central air system	X	
Same	Central water system[b]	X	
Same	Decentralized system	X	
Small rowhouse with masonry walls (single family or 2-4 units)	Central air system		X
Same	Central water system		X
Same	Decentralized system		X
Moderate or large multifamily building (masonry or clad walls)	Central air system	X	
Same	Central water system	X	
Same	Decentralized system		X
Moderate or large commercial building (masonry or clad walls)	Central air system	X	
Same	Central water system		X
Same	Complex reheat system	X	
Same	Decentralized system	X	

[a]See table 3 for a definition.
[b]OTA's assumption is that this building type has a central water system and window air-conditioners.

assuming no increase in the real cost of energy. These building types include all small frame houses, moderate or large multifamily buildings with central air or water mechanical systems, and all commercial buildings except the usually older commercial buildings with water or steam heating systems and window air-conditioners. Clearly the problem of financing retrofits for these buildings should be minimized by the fast payback (and high return) of their retrofit options. Some of these fast payback retrofit options include wall insulation in frame buildings, economizer cycles which make greater use of outside air for air-conditioning in commercial buildings and hot water flow restrictors in multifamily buildings.

For all of the remaining building types, on the other hand, substantial savings are more likely to come from retrofit options of moderate capital cost compared to savings, which will payback in 2 to 7 years and whose real rate of return can range from as high as 50 percent to as low as 13 percent per year over a 20-year retrofit life (also see table 3). These building types include all small masonry rowhouses, moderate or large multifamily buildings with decentralized heating and cooling systems, and older commercial buildings with water or steam systems and window air-conditioners. For owners of such buildings there may be significant problems of financing substantial energy retrofits. Some examples of effective retrofits with moderate capital cost include: roof insulation and storm windows for masonry rowhouses, hot water heat pumps for multifamily buildings with decentralized systems, and replacing low efficiency window air-conditioners with more efficient models.

For most of the building types there are also retrofit options of high capital cost compared to savings with paybacks of longer than 7 years and annual real rates of return of less than 13 percent per year (over 20 years). If lifecycle costing is used, such retrofits may in fact be less expensive over the full life of the measure of the cost of the energy they would save. However, their very slow payback and low annual rate of return create serious financing obstacles. **For most of the building types OTA examined, such high cost retrofits would save no more than 20 percent of the full technical savings potential.** The three exceptions and the estimated percentage of total savings from high cost retrofits are:

- Masonry rowhouse with a heating system using air (40 percent).
- Masonry rowhouse with a water or steam system (25 percent).
- Large multifamily building with an air system (30 percent).

Table 3.—Three Ways to Express the Relative Cost Effectiveness of Energy Retrofits

Relative capital cost[a]	Simple payback[b] (in years)	Annual real return on investment[c] (percent)
Low capital cost[d]	Less than 2 years	More than 50% per year
Moderate capital cost[d]	2 to 7 years	13 to 50% per year
High capital cost[d]	7 to 15 years	3 to 13% per year
Cost of retrofit exceeds savings[e]	More than 15 years	Less than 3% per year

[a] See ch. 3 for a full definition. Low capital cost is defined as less than $14.00 per annual million Btu saved. Moderate capital cost is defined as $14.00 to $49.00 per annual million Btu saved. High capital cost is defined as $49.00 to $105.00 per million Btu saved. In all OTA's calculations in ch. 3, all electricity savings are multiplied by 2.46 to reflect the higher cost of electricity.
[b] Annual real discount rate that equates costs and savings over a 20-year measure lifetime. This assumes that fuel savings escalate at the same rate as inflation.
[c] Number of years for value of first year's energy savings to equal retrofit costs. Assumes value of energy savings is $7.00 per million Btu (approximately equal to the average price of distillate fuel oil in 1980).
[d] Compared to savings.
[e] Not cost effective.

3. INDUSTRIAL ENERGY USE, 1983

For many years to come, energy need not be a constraint to economic growth in the United States. OTA projects that in the next two decades investments in new processes, changes in product mix, and technological innovation can lead to improved industrial productivity and energy efficiency. As a result, the rate of industrial production can grow three times faster than the rate of energy use needed for that production.

Because the investments needed to improve energy efficiency are long term, a reduction in energy use growth rates resulting from investments begun now will continue through the 1980's and 1990's. Furthermore, this improvement will continue beyond 2000 as the proportion of new, energy-efficient capital stock increases. Improvements in energy efficiency for the next several years will be largely a result of housekeeping measures and investments that began during the 1970's.

In 1981, the industrial sector used 23 Quads of direct fuel, electricity, and fossil fuel feedstock, of which petroleum and natural gas constituted 73 percent. Four industries—paper, petroleum refining, chemicals, and steel—accounted for almost half of all industrial energy used. Over the past decade, soaring energy prices have led to significant changes in the absolute amount and mix of energy used in industry. Energy used per unit of product in the industrial sector decreased by almost 20 percent. This improvement was accomplished by housekeeping measures, equipment retrofits, and new process technologies that produce existing products and new product lines.

In addition to reducing the energy use growth rate, industry will continue its shift away from premium fuel use. For the next two decades, industrial coal use—particularly in boilers and in some large, direct heat units—will increase substantially because coal is cheaper than oil and natural gas. Moreover, the demand for purchased electricity will probably grow faster than the total industrial energy demand if the price difference between natural gas and electricity continues to decrease.

While industry has made significant strides in reducing energy use, **opportunities for further gains in energy efficiency from technical innovation are substantial.** Because capital stock has not turned over as quickly in recent years as it did in the 1960's, there is a large backlog of retrofit improvements to be made. Furthermore, high capital costs and the limited capital pool have kept many new process technologies from penetrating product markets. **OTA projects that new processes or process technologies would save more energy than would retrofits and housekeeping measures, and would reduce overall costs by improving productivity and product quality. However, such process shifts will entail large capital outlays, which in turn, will require general economic growth over many years.** Without economic growth, there will not be enough product demand or capital to support these productivity improvements.

A product mix shift away from energy-intensive products will also continue to contribute to the decline in energy use growth rates. Product mix shift will occur within specific industries (e.g., a shift from basic chemical production to agricultural/specialty chemical manufacture) as well as from one industry to another (e.g., a shift away from steel to aluminum and plastics in auto manufacture). These shifts are driven by changing demand patterns and international competition, as well as by increasing energy prices.

OTA found that corporations have a strategic planning process that evaluates and ranks investments according to a variety of factors: product demand, competition, cost of capital, cost of labor, energy and materials, and Government policy. In analyzing energy-related investment behavior, **OTA found no case in which a company accorded energy projects independent status.** Although energy costs are high in each of the four industries examined by OTA, costs of labor, materials, and capital financing are also high. **Thus, energy-related projects are only part of a general strategy to improve profitability and enhance a corporation's competitive position.**

Most firms regard energy efficiency as one more item in which to invest and not as a series of projects that are different from other potential investments. This view differs significantly from the view of firms that produce energy or energy-generating equipment where the **entire investment** is focused on increasing energy production. This difference has important policy implications because incentives aimed at reducing energy demand growth must compete with other strategic factors and are therefore diluted. Energy incentives directed at increasing energy supply suffer no such competition.

Of the four most energy-intensive industries, chemicals and paper will show the largest growth in production over the next two decades and will also show a substantial increase in energy efficiency. In the paper industry, energy use has risen slightly since 1972, but the industry is now more energy self-sufficient. In 1981 the paper industry generated half of all its energy needs through the use of wood residues as fuel. By 2000 self-generation of energy could result in the paper industry meeting over 60 percent of its needs internally. The limitation on the percentage of self-sufficiency is the value of the product foregone by using feedstock (wood) as fuel. Also, the paper industry's use of oil will decline as residual oil is displaced in boilers by coal and biomass fuels. OTA projects that over the next two decades, energy use per ton of paper will decline, owing to specific process changes in papermaking steps, such as oxygen-based bleaching, computerized process controls, and new methods of making paper.

The petroleum refining industry will show a decline in overall product output but will continue to improve its energy efficiency, although only slightly. Energy efficiency gains from retrofit and housekeeping measures will be merely offset by a shift to heavier, high-sulfur, crude oil feedstocks and by increased use of energy in refining because of market requirements for high-octane, unleaded motor fuels. Of the four industries, this is the only one in which product or process shifts are not projected to lead to less energy use. Nonetheless, overall efficiency can be expected to improve as a result of a number of anticipated technological changes in refinery operations, such as the extensive use of vapor recompression and waste heat boilers in the distillation and cracking processes and the use of computerized process controllers to optimize plant operations.

In the chemicals industry, energy efficiency improvements will result from a combination of retrofits to existing processes and technical innovation in new processes and products. For example, vapor recompression, process controls, and heat recuperators and exchangers will be added to existing processes to improve thermal efficiencies. In addition, there is a trend toward increased use of electricity and coal and away from premium fossil fuels, especially natural gas. OTA projects that by 2000, coal use will account for almost one-third of the fuel used in the chemicals industry. An important source of energy efficiency improvement in the chemicals industry is a shift in product mix. Because of higher profit margins and less foreign competition, the industry will increase production of less energy-intensive, higher value chemicals, such as pharmaceuticals and pesticides, relative to more energy-intensive chemicals such as ethylene and ammonia.

As the steel industry retools to meet foreign competition, there will be a large reduction in energy intensity. The major source of this decline will be investments in new processes—i.e., 1) the replacement of ingot casting by continuous casting, and 2) the substitution of electric arc furnace or minimills for the blast furnace/basic oxygen furnace combination. With continuous casting, significant energy will be saved by not having to reheat cooled metal ingots before shaping. Electric arc minimills will save energy by substituting scrap metal feedstocks for iron ore, thus reducing coke demand. This trend will also result in the substitution of steam coal for metallurgical coal since the former will most often be used to generate electricity.

Over the years, Congress has passed a number of measures that affect the industrial use of energy. In general, the goals of these measures have been to reduce oil imports, encourage domestic production of fossil fuels, and reduce energy demand through efficiency improvements. **OTA found that legislation directed specifically at improving energy efficiency in industry has little influence on investment decisions.** At the highest levels of corporate financial decisionmaking, there is an awareness of Government tax and industrial policies. However, OTA found that technical decisions and energy project evaluation tend to be separate from and subservient to corporate financial decisions. Moreover, the decision to invest depends not only on an individual project's return on investment, but also on such corporatewide parameters as debt-equity ratio, debt service load and bond rating, and, most importantly, the aforementioned strategic considerations of corporate decisionmaking. **Because energy must compete with other factors of production when investment choices are made, policy incentives directed at energy demand alone will be just one of a number of considerations in making these**

choices. Unless such incentives are substantial, they are unlikely to alter a decision that would have been made in the absence of such incentives.

To assess the effects of a range of incentives on energy use in industry, OTA selected a set of policy initiatives directed at energy specifically or at corporate investment in general. The latter include the accelerated cost recovery system (ACRS) provisions of the Economic Recovery Tax Act of 1981 (ERTA) and increased capital availability for investment, while the former include broadened and expanded tax credits for energy investments and the imposition of energy taxes on premium fuels. These policies are compared to a reference case consisting of current economic conditions and the tax code as amended by ERTA.

The effect of the ACRS on increasing energy efficiency depends on the ability of the ACRS to increase investment. **OTA found that the ACRS is a positive stimulus to investment when the industry is profitable and growing.** Under these conditions, total investment and energy efficiency improvements would be accelerated by the ACRS. **As long as conditions of high interest rates, low-to-moderate demand growth, and the like exist, however, the ACRS will do little to increase energy efficiency.**

Energy investment tax credits at a 10-percent level have little direct influence on capital allocation decisions in large American firms, and thus have little or no influence on energy conservation. These tax credits appear to be too small to exert any change in the return on investment of a company when the only factor they affect is energy. **However, energy investment tax credits directed at energy production, such as cogeneration by third parties, would be effective.** In this case, the entire investment would be covered by the tax credit, and energy would be the principal product being produced by the investment. Regarding investments in technologies that improve the energy efficiency of industrial process technologies, however, OTA could find no case where decisions to undertake a project depended on gaining a 10-percent energy investment tax credit.

Taxes at a rate of $1 per million Btu on premium fuels—natural gas and petroleum—would change the fuel mix and cause energy efficiency to improve, although not by more than a few percent. Because of the already large cost differential between premium fuels and coal, the increase in costs as a result of the tax would not significantly change the economic incentive to switch to coal. The effect of the tax would be more significant for electricity, but there the availability of industrial production technologies that use electricity instead of petroleum or natural gas would be the limiting factor. Consequently, imposition of the tax would cause only a slight increase in conversion to coal and electricity from natural gas and petroleum. Investments in energy efficiency through retrofits and new process technology would still primarily be limited by capital availability and growth in product demand.

The fuel tax would have different consequences for each of the energy-intensive industries investigated. OTA found that a premium fuels tax would accelerate energy self-sufficiency and decrease natural gas consumption in the paper industry. The petroleum refining industry might be affected by a premium fuels tax in two ways: 1) some energy-related projects would be given a higher priority, and 2) earnings would decline because of a general decrease in product demand. In the chemicals industry, the domestic impact of a premium fuels tax is potentially detrimental. The greatest impact would likely be on the ability of the industry to export products as well as to make the domestic market more vulnerable to imports. Finally, a premium fuels tax would be least detrimental to the steel industry because only a small percentage of the industry's energy is derived from petroleum sources.

The best way to improve energy efficiency is to promote general corporate investment by reducing the cost of capital. Corporations that believe energy prices will continue to rise have a strong impetus to use capital for more energy-efficient equipment. Low interest rates affect energy efficiency to the extent that lower rates may allow a company's cash flow to go further, its debt service to be less burdensome, or its ability to take on more debt to increase. Lowering interest rates would increase capital availability and therefore allow more projects to be undertaken. Improvement in capital availability would magnify the effect of the ACRS because the ability to make use of the latter depends on the investment climate. **At the same time, however, it should be recognized that growth in product demand is essential if investment is to take place, even with lower interest rates.**

4. INCREASED AUTOMOBILE FUEL EFFICIENCIES AND SYNTHETIC FUELS, 1982

In 1981, U.S. oil imports averaged 5.4 million barrels per day (MMB/D) —approximately 34 percent of its oil consumption and 15 percent of its total energy use. This is potentially a serious risk to the economy and security of the United States. Furthermore, recovery from the current recession will increase demand for oil and, although currently stable, domestic oil production is likely to resume a steady decline in the near future.

Several options exist for reducing imports. However, even with moderate increases in automobile fuel efficiency, moderate success at developing a synthetic fuels industry and expected reductions in stationary use of fuel oil, U.S. oil imports could still be over 4 MMB/D by 2000, if the U.S. economy is healthy and has not undergone unforeseen structural changes that might reduce oil demand well below projected levels.

Only with vigorous promotion of all three options and technological success of a full range of oil import reduction options could the Nation hope to eliminate oil imports before 2010.

Congress faces several decisions on how to reduce the U.S. dependence on imported petroleum. Two options, increased automobile fuel efficiency and synthetic fuels, are particularly likely to be subjects of congressional debates. First, Congress may want to consider new incentives to increase auto fuel efficiency beyond that mandated by the 1985 CAFE (Corporate Average Fuel Efficiency) standards. Second, Congress will have to decide whether to continue into the second phase of the program to accelerate synfuels development under the Synthetic Fuel Corp. (SFC). The purpose of this report is to assist Congress in making these decisions and comparing these two options by exploring in detail the major public and private costs and benefits of increased automobile fuel efficiency and synthetic fuels production. A third option for reducing imports—increased efficiency and fuel switching in stationary (nontransportation) oil uses—is examined briefly to allow an assessment of potential future levels of oil imports. Finally, electric-powered automobiles are examined.

Import Reductions

In the judgment of the Office of Technology Assessment, increased automobile fuel efficiency, synthetic fuels production, and reduced stationary (nontransportation) use of oil can significantly decrease U.S. dependence on oil imports during the next two to three decades. Indeed, reducing oil imports as quickly as possible requires that all three options be pursued. Electric cars are unlikely to play a significant role, however.

Although a precise forecast of the future contributions of the import reduction options is not feasible now, it **is** possible to draw some general conclusions about their likely importance and to estimate what their contributions **could be** under specific circumstances (see fig. 1).

First, **increases in auto fuel efficiency will continue, driven by market demand and foreign competition.** OTA believes that, with strong and consistent demand for high fuel efficiency, there is a good chance that actual

average new-car fuel efficiencies would be greater than OTA's low scenario in which average new-car fuel economy* was projected to be:

30 miles per gallon (mpg) .... in 1985
38 mpg .... in 1990
43 mpg .... in 1995
51 mpg .... in 200C

Figure 1.—Potential Oil Savings Possible by the Year 2000[a]
(relative to 1980 demand)

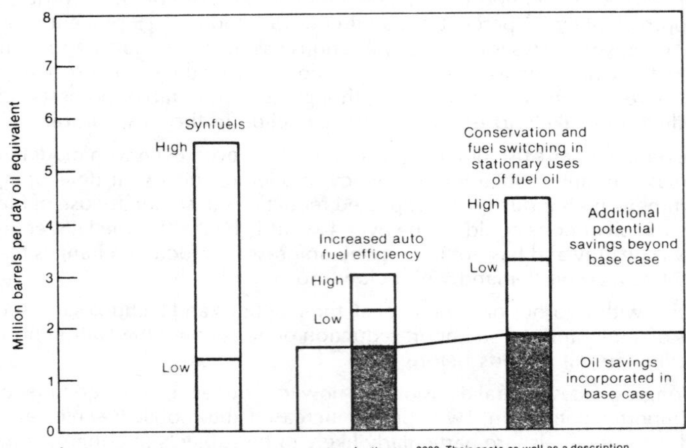

[a]These potential oil savings (or synfuels production) are for the year 2000. Their costs as well as a description of the technologies and other characteristics of the high and low scenarios are given in this summary.

with a moderate shift in demand to smaller cars. Although this scenario is based on modest technical expectations, it is dependent on favorable market conditions. Domestic automakers are unlikely to commit the capital necessary to continue the current rapid rate of increase in efficiency unless they improve their sales and profits.

If the industry is able to attain the fuel efficiency levels shown above, the United States would save 800,000 barrels per day (bbl/d) of oil by 2000 compared with the case where post-1985 new-car efficiency remained at 30 mpg. The savings would increase to at least 1.1 MMB/D by 2010 because of continued replacement of older, less fuel-efficient automobiles.

With a poorer economic picture and weaker demand for high fuel efficiency, new-car efficiencies could be 40 mpg or less by 2000, with correspondingly lower savings. Achieving 60 to 80 mpg by 2000 would require favorable economic conditions, strong demand for fuel efficiency, and relatively successful technical development.

Second, **substantial contributions to oil import reductions from production of synthetic fuels appears to be less certain than substantial contributions from the other options.** Potential synfuels producers are likely to proceed cautiously for the following reasons: 1) investment costs are very high (even with loan guarantees covering 75 percent of project costs); 2) there is a fairly small differential between the most optimistic of OTA's projected synfuels production costs and the current price of oil; 3) investors are now uncertain about future increases in the real price of oil; and 4) there are high technological risks with the first round of synfuels plants (possibly exacerbated by the cancellation of the Department of Energy's (DOE) demonstration program).

OTA projects that, even under favorable circumstances, fossil-based production of synthetic *transportation fuels* could at best be 0.3 to 0.7 MMB/D by 1990 and 1 to 5 MMB/D by 2000. Biomass synfuels could add 0.1 to 1 MMB/D to this total by 2000. In less favorable conditions, for example if SFC financial incentives were withdrawn, it appears unlikely that even the lower fossil synfuels estimate for 1990, and perhaps 2000, could be achieved unless oil prices increase much faster than they are currently expected to.

Achieving much more than 1 MMB/D of synfuels production by 2000 would require fortuitous technical success and either: 1) unambiguous economic profitability or 2) continued financial incentives requiring authorizations considerably larger than those currently assigned to SFC. Achieving production levels near the upper limits for 2000 are likely to be delayed, perhaps by as much as a decade, unless there is virtually a "war mobilization"-type effort.

Third, **there are likely to be large reductions in the stationary use of fuel oil (currently 4.4 MMB/D) in the next few decades.** With just cost-effective conservation measures, stationary fuel oil use could be reduced significantly. Additional conservation measures by users of electricity and natural gas could make enough of these fuels available to replace the remaining stationary fuel oil use by 2000. Total elimination of stationary fuel oil use by 2000 is unlikely, however, because site-specific factors and differing investor payback requirements will mean that a significant fraction of the numerous investments needed for elimination will not be made.

Fourth, **even a 20-percent electrification of the auto fleet—a market penetration that must be considered improbable within the next several decades—is unlikely to save more than about 0.2 MMB/D.** Electric cars are most likely to replace small, low-powered—and thus fuel-efficient—conventional automobiles, minimizing potential oil savings.

Plausible projections of domestic oil production—expected by OTA to drop from 10.2 MMB/D in 1980 to 7 MMB/D or lower by 2000—suggest that oil imports could still be as high as 4 to 5 MMB/D or more by 2000 unless imports are reduced by a stagnant U.S. economy or by a resumption of rapidly rising oil prices. Achieving low levels of imports—to perhaps less than 2 MMB/D within 20 to 25 years—is likely to require a degree of success in the three major options that is greater than can be expected as a result of current policies.

Costs

Except for stationary fuel oil reductions, economic analysis of the options for reducing oil imports involves a comparison of **tentative** cost estimates for mostly **unproven** technologies that will **not be deployed for 5 to 10 years or more.** Even if costs were perfectly estimated for today's market (and the estimates are far from perfect), different rates of inflation in the different economic sectors affecting the options could dramatically shift the comparative costs by the time technologies are actually deployed. Figure 2 presents OTA's estimates for the investment costs for all options except electric cars. The costs are expressed in dollars per barrel per day, which is the amount of investment needed to reduce petroleum use at a rate of 1 bbl/d. In OTA's judgment, **the estimated investment costs (in dollars per barrel per day) during the 1990's of automobile efficiency increases, synthetic fuels production, and reduction of stationary uses of oil are essentially the same, within reasonable error bounds. If Congress wishes to channel national investments preferentially into one of** these options, differentials in estimated investment costs cannot provide a compelling basis for choice.

On the other hand, **investments during the 1980's to reduce stationary oil use** (from the current 4.4 to 3 MMB/D or less by 1990) **and increase automobile fuel efficiency** (to a 35 to 45 mpg new-car fleet average by

Figure 2.—Estimated Investment Costs for the Oil Import Reduction Options

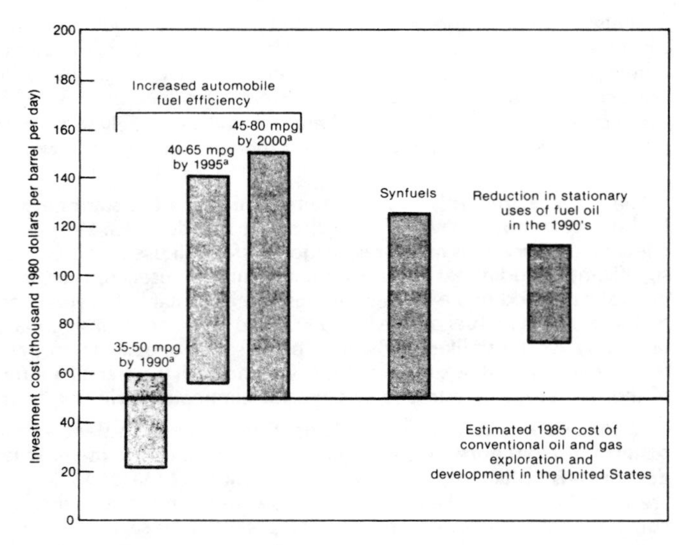

ªAverage new-car fuel efficiency (55%/45% city/highway EPA fuel efficiencies). Fuel savings calculated by comparing fuel efficiency of cars produced at end of each 5-year period to the average new-car fuel efficiency of cars produced at end of previous 5-year period.

1990) **are likely to cost less than the 1990-2000 investments in any of the options.**

Electric vehicles are likely to be very expensive to the consumer—costing perhaps $3,000 more per vehicle than similar, conventional automobiles or $300,000 to $400,000/bbl/d of oil saved. (The latter is not strictly comparable to investment costs for the other options.) If batteries must be replaced at moderate intervals, which is necessary today, the total costs of electric cars would escalate.

WHAT ARE THE INVESTMENT COSTS FOR REDUCING U.S. OIL CONSUMPTION?

Introduction

Investment costs are an important consideration when comparing alternatives for reducing U.S. oil consumption. OTA's analysis indicates that synfuels production, increased fuel efficiency in automobiles, and conservation and fuel switching in stationary uses of oil all will require investments of the same order of magnitude for comparable reductions in oil consumption in the 1990's; whereas, synfuels production appears to require larger investments than the other alternatives for the 1980's. Uncertainties in the cost estimates as well as the fundamental differences in the nature of the investments are too large, however, to allow a choice between approaches on this basis alone.

In order to compare investment costs, they have been expressed as the investment needed to either produce or save 1 barrel per day oil equivalent[1] of petroleum products. This method was chosen in order to avoid problems that arise when comparing investments in projects with different lifetimes and for which future oil savings may be discounted at different rates. In addition, from a national perspective the per unit investment cost is important in that it is the parameter used in the aggregate to make choices among competing investments. Conventional oil and gas exploration are considered first to provide a reference point. Following this, OTA's estimates for the investment costs for increased automobile fuel efficiency, EVs, synfuels, and increased efficiency and fuel switching in stationary uses are discussed briefly.

Conventional Oil and Gas Production

Two estimates of recent investment costs for conventional oil and gas exploration and development in the United States are shown in table 11. The data in this table were developed from estimates of the annual investments in oil, gas, and natural gas liquids exploration and development per barrel of increased proven reserves of these fuels (corrected for depletion). These latter estimates were then converted to investments for an increase of 1 barrel per day (bbl/d) of production (corrected for depletion) using the 1980 ratio of crude oil reserves to crude-oil production and assuming an 8 percent refining loss. The ratio of reserves to production for natural gas was not used because price controls on natural gas tend to inflate this ratio and thus the estimated costs; and investments for oil exploraton and development were not separated from those for natural gas because there is no practical way to do so.

Table 11.—Estimated Investment Costs for Conventional Oil and Natural Exploration and Development

Year	Estimated investment cost (thousand 1980 dollars per barrel per day of petroleum production[a])	
	Estimate A[b]	Estimate B[c]
1974	13	15
1975	17	19
1976	20	17
1977	22	20
1978	29	18
1979	31	57[d]
1980	Not available	39[e]
Extrapolated to 1985[f]	53	49

There are significant uncertainties in these estimates due to numerous anomalies in the data, some of which are detailed in footnotes to table 11, and because the ratio of reserves to production changes with market prices, production techniques (e.g., enhanced oil recovery), and the nature and quantity of reserves. Nevertheless, these data do indicate that it is reasonable to expect costs of $50,000/bbl/d or more for conventional petroleum exploration and development by the mid-1980's if recent cost trends continue.

Automobile Fuel Efficiency

OTA's estimates of the investment plus associated product development costs for increased automobile fuel efficiency are shown in table 12. There are notable technical, accounting, and market uncertainties associated with this type of cost analysis, however.

The estimates in table 12 were derived by first estimating the efficiency gains that can reasonably be expected over time from various changes in the automobile system. They are based on both published estimates and OTA's analysis. The rates at which these technologies may be incorporated into new cars were then estimated and resultant schedules for capital turnover derived. Next, the investment cost calculations were based on published estimates for the cost of replacing the applicable capital equipment (e.g., facilities for producing a new engine or transmission, etc.). The actual investment cost and resultant fuel efficiency increases, however, will depend on a number of factors specific to individual production plants (and their future evolution), the way various production tradeoffs are resolved, and the results of future product development programs.

In addition to capital investment, development costs have been included as part of the investment necessary to produce modified vehicles. During the 1970's, domestic auto manufacturers' R&D (mostly development) costs averaged from 40 to 60 percent of their capital investments.[6] In table 12, development costs are assumed to be 40 percent of the capital investment allocated to fuel efficiency (see below), but the actual costs of developing the technologies for producing more efficient cars at minimum cost are highly uncertain.*

Beyond the uncertainties in the investment and development costs, there is the problem of determining what fraction of the investments should be ascribed to fuel efficiency. This arises because some of the investments can be used not only to increase fuel efficiency, but also to make other changes in the car. The cost allocation problem associated with multipurpose investments is well known in accounting theory, and there is no fully satisfactory solution to it.[7]

For table 12, it was assumed that 50 percent of the cost of engine and body redesign, 75 percent of the cost of most transmission changes, and 100 percent of the cost of advanced materials substitution and energy storage and automatic engine cutoff devices should be allocated to fuel efficiency. This results in between 55 and 80 percent of the investments being allocated to fuel efficiency, depending on the time period and scenario chosen. For further details on how this and other problems in estimating the cost of fuel efficiency were resolved, see chapter 5.

During the period 1985-2000, total capital investments in changes associated with increasing fuel efficiency (i.e., allocating 100 percent of the multipurpose investments to fuel efficiency) could average $2 billion to $5 billion per year, depending on the number of new cars sold and the rate at which fuel efficiency is increased. However, if one deducts the cost of changes that would have been made under "normal" circumstances, the added capital investment needed to achieve the lower mpg numbers in table 12 would be $0.3 billion to $0.7 billion per year. The higher mpg numbers in table 12 would require added capital investments (above "normal") of $0.6 billion to $1.5 billion per year. Adding 40 percent of the capital investment for development costs results in added outlays of $0.4 billion to $0.9 billion per year and $0.8 billion to $2 billion per year for the low and high scenarios, respectively.

A detailed examination of the scenarios presented in chapter 5 shows that a 1990 new-car average fuel efficiency of 35 to 45 mpg (depending on the proportion of small, medium, and large cars sold) probably can be achieved with what is termed here "normal" rates of capital turnover. However, the validity of this conclusion and of the above incremental investment and development cost estimates will depend on market demand for fuel efficiency, and, in OTA's judgment, there is no credible way to predict future market demand for fuel efficiency.

Electric Vehicles

Use of EVs more nearly approximates synfuels than increased automobile fuel efficiency, in that EVs involve switching from conventional oil to another energy source rather than reducing energy consumption. Consequently, the costs (per barrel per day of oil replaced) for EVs are included in table 13 with synfuels. As shown in table 13, the costs for EVś appear to be significantly higher than for the various synfuels options, due to the high purchase price of the vehicle (relative to a comparable gasoline-fueled car) and the fact that EVs would be replacements for relatively fuel-efficient cars (because of an EVs limited size and acceleration). Furthermore, if batteries must be replaced at regular intervals and the cost of this is included as an investment cost, the total investment per barrel per day rises dramatically.

Synfuels

The best available estimates for the investment costs for various liquid synthetic transportation fuels are shown in table 13. Because of uncertainties in the cost estimates, no meaningful intercomparison among synfuels on the basis of cost is currently possible. In addition, as discussed in chapter 6, the final investment in synfuels is likely to be different from these estimates. As the processes approach commercial production, they will be revised as costs to overcome problems encountered in demonstration units are determined. Construction costs will inflate at an unknown rate relative to general inflation. And delays during construction due to such possibilities as lawsuits, strikes, late delivery of construction materials, or other causes can increase the investment cost. In sum, current investment estimates provide a very tentative guide to what synfuels plants constructed in the 1990's will cost.

Table 12.—Capital Investment Allocated to Fuel Efficiency Plus Associated Development Costs

			Average capital investment plus associated development costs[b]	
Time of investment	Mix shift	New-car fuel efficiency at end of time period[a] (mpg)	Thousand 1980 dollars per barrel per day oil equivalent of fuel saved[c]	1980 dollars per car produced[d]
1985-1990	Moderate[e]	38-48	20-60[g]	50-190[g]
	Large[f]	43-53		
1990-1995	Moderate[e]	43-59	60-130[g]	70-180[g]
	Large[f]	49-65		
1995-2000	Moderate[e]	51-70	50-150[g]	50-150[g]
	Large[f]	58-78		

Table 13.—Investment Cost for Various Transportation Synfuels and Electric Vehicles

	Thousand 1980 dollars per barrel per day oil equivalent to end users				
	Shale oil	Methanol from coal	Coal to methanol and Mobil methanol to gasoline	Direct liquefaction	Electric vehicle
Mining	(Included in conversion plant)	4-15	4-15	4-15	5-19
Conversion plant	49-73[a]	47-93[a]	53-110[a]	67-100[a]	0-69[b]
Refinery	0-10[c]	0	0	4-22[d]	0
Distribution system	0	0-2[e]	0	0	0
End use	0	0-11[f]	0	0	320-390[g]
Total	49-83	51-121	57-125	75-137	325-478

5. U.S. VULNERABILITY TO AN OIL IMPORT CURTAILMENT: THE OIL REPLACEMENT CAPABILITY, 1984.

Since 1973, the United States has experienced two major oil supply disruptions and shortfalls which resulted in large and enduring increases in oil prices. Although the Nation has made great strides in reducing oil consumption in response to those price increases, another disruption and shortfall could still have significant negative consequences for the U.S. economy.

Much of the continuing debate on how to deal with another disruption and shortfall has centered around emergency response mechanisms such as oil stockpiling and standby fuel allocation schemes. Little attention has been paid to ways of responding to a shortfall of indefinite duration because it has always been assumed that any oil cutoff would end after a period of 1 or 2 years. An indefinite shortfall is not implausible, however. Indeed, the lasting increases in oil prices that resulted from events in the 1970s are the economic equivalent of lasting supply shortfalls. And as a result of the most recent shortfall, the period 1978-83 saw a 60-percent increase in the real price of oil and an unadjusted decline in oil demand of nearly 4 million barrels per day (MMB/D).

Judging from this historical experience, therefore, an important aspect of the United States' vulnerability to a future oil import curtailment is the Nation's ability to adjust to a lasting oil supply shortfall and price rise. As demonstrated by the Nation's response to the most recent price shock, a lasting shortfall would require technological and economic adjustments that go well beyond short-term emergency responses, although those responses would also certainly be necessary.

At the request of the Senate Committee on Foreign Relations, OTA addressed the possibility of a lasting shortfall by asking the following questions: how could the United States respond to a large and protracted oil supply shortfall by technical means alone? and how do the economic consequences of a shortfall depend on the deployment rate of oil replacement technologies?

As a starting point for its analysis, OTA made a number of assumptions:

1. Acceptance of the International Energy Program[1] IEA agreements results in a 3 MMB/D[2] shortfall in the United States (compared to pre-shortfall demand of 16 MMB/D).
2. The shortfall is assumed to be of indefinite duration (i.e., to last at least 5 years) at the outset and to begin in the mid-1980s.
3. The economy would not undergo major structural changes, such as major shifts in output mix or behavior during the 5-year period.
4. The Strategic Petroleum Reserve, as well as private oil stockpiles, would be used to reduce the immediate effects of the shortfall, but they would be depleted within 3 years, dropping from a drawdown rate of 1.5 MMB/D the first year of the shortfall to zero by the end of the third year.

MAJOR FINDINGS

At the onset of an oil supply shortfall, emergency measures such as reductions in private and public oil stocks can cushion the immediate effects of the oil shortfall. After 5 to 10 years, long leadtime technologies such as enhanced oil recovery and synthetic fuels production can begin to provide liquid fuels, which are essentially indistinguishable from the lost oil. In the period of about 1 to 5 years after onset, however, oil consumers will either have to forgo certain energy services or invest in nonoil energy technologies.

OTA has examined each sector in the U.S. economy and identified the technologies that, based on **technical** considerations, are likely to be able to replace the largest quantities of oil, at the least cost, for each sector. The rate that each oil replacement technology (fuel switching and increased efficiency of use) could be deployed was then estimated from existing capacities to produce and install the necessary equipment, historical peak rates of installation and various end-user constraints. Based on this analysis, **OTA has concluded that the United States has the technical and manufacturing capability to replace up to 3.6 MMB/D of oil use within 5 years after the onset of an oil supply shortfall** (see fig. 1 and table 1).

Figure 1.—Potential Replacement of Oil Through Fuel Switching and Increased Efficiency

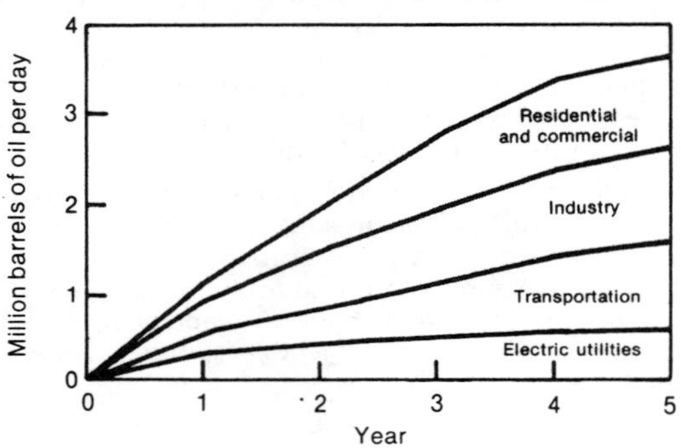

Table 1.—Major Oil Replacement Options

Sector	Oil replacement potential after 5 years (MMB/D)[a]
Electric utilities:	
Switching to coal and completion of new powerplants currently under construction	0.5
Increased use of natural gas	0.1
Subtotal	0.6
Industry:	
Switch to natural gas	0.45
Switch to coal	0.2
Increased efficiency	0.15
Reduced refinery throughput	0.2
Subtotal	1.0
Residential and commercial (heat and hot water in buildings):	
Switch to natural gas	0.45
Switch to electricity	0.4
Increased efficiency and switch to other fuels	0.15
Subtotal	1.0
Transportation:	
Increased efficiency of cars and light trucks	0.7
Increased efficiency in other transportation modes	0.1
Increased production and use of ethanol	0.1
Switch to other alternative transportation fuels	0.1
Subtotal	1.0
Total	3.6

The criteria used to select the most promising technologies for each major end use of oil were: 1) the technology must be commercial now or is likely to be commercial by mid-1985, 2) individual units can be installed or built in less than 2 to 3 years, 3) the technology has sufficiently broad applicability to be capable of replacing a significant fraction of the oil consumed for that end use, and 4) the technology is currently among the lowest cost alternatives to oil for that end use. In other words, OTA selected those technologies that—based on current engineering cost estimates and technical judgments—could replace large quantities of oil in a relatively short time at costs below OTA's estimate of the probable post-shortfall price of oil ($50 to $70 per barrel in 1983 dollars).

The options satisfying these criteria that can replace the largest amounts of oil within 5 years after the start of a shortfall are:

1. increased efficiency and switching to alternative fuels to reduce oil use for space and water heating in buildings and for steam in industry and electric utilities, and
2. increased average efficiency of automobiles and light trucks on the road.

Smaller, additional amounts of oil can be replaced in transportation and materials uses of oil (e.g., petrochemicals) using a variety of other technologies, but the near- to mid-term opportunities are more limited because of longer leadtimes and/or higher costs.

At the end of 5 years, deployment of the major oil replacement technologies would leave transportation fuels and materials production as the predominant remaining uses for oil. Aside from refinery use of oil for fuel (8 to 10 percent of refinery throughputs), less than 5 percent of the remaining oil consumption would be for space and water heating and steam, mostly in residential and commercial buildings in the northeastern United States and in small industrial boilers throughout the country.

This oil replacement would require about 2 trillion cubic feet (TCF) of natural gas (11 percent of 1982 consumption) and 115 million tons of solid fuels (coal and wood) per year (13 percent of 1982 production) as substitutes for oil. Nearly all of the increment of natural gas, however, could be made available through investments in increased efficiency of natural gas use.

End-user investment costs for the major oil replacement technologies can vary from $0 to $5,000 per barrel per day (B/D) of oil replaced (for conversion of an industrial boiler to natural gas) up to $35,000 to $60,000 per B/D of oil replaced (for installation of a central electric heat pump for residential space heating and hot water) (see table 2). However, with residential electricity costing 8 cents per kilowatthour (kWh) (1983 average was 7.2 cents/kWh), even the cost of installing a heat pump in an average oil-heated residential building could be recovered in 2 to 6 years, depending on the price of oil following the shortfall and on the actual investment cost. The payback period for the other options considered would be shorter, unless there were rapid inflation in equipment costs and/or natural gas prices. (Although some inflation in the price of equipment would be expected, as mentioned above, there is no fundamental reason why these prices should become prohibitively high. Furthermore, natural gas price rises could be moderated by investments in increased efficiency of natural gas use, with investment costs similar to those for increased efficiency of oil use.)

Total investment would amount to $30 billion to $40 billion per year,[3] on average, or about 7 to 9 percent of recent annual investments in producer durables and residential structures.

6. OIL AND GAS TECHNOLOGIES FOR THE ARCTIC AND DEEPWATER, 1985.

This assessment addresses the technologies, the economics, and the operational and environmental factors affecting the exploration and development of energy resources in the deepwater and Arctic regions of the U.S. Outer Continental Shelf (OCS) and the 200-mile Exclusive Economic Zone (EEZ) established in March 1983. For the purposes of this study, OTA defined "deepwater" as those offshore areas where water depths exceed 400 meters or 1,320 feet. The "Arctic" is defined as the Beaufort, Chukchi, and Bering Seas north of the Aleutian Islands.

Leasing submerged coastal lands for oil and gas development began with State programs in California, Louisiana, and Texas years before there was a Federal offshore leasing program. Leasing in Federal offshore lands began in 1954 after the OCS Lands Act of 1953 provided the Secretary of the Interior guidance and authority for such activity. The industry leased, explored, and developed OCS oil and gas under the provisions of the 1953 Act for 25 years. Most of the offshore activity during that period was in the Gulf of Mexico and the Pacific Ocean off southern California. Then, in 1978, an emerging national awareness of the environment coupled with the Arab oil embargo and increased concern about energy supplies led to enactment of the OCS Lands Act Amendments.

Congress included in the 1978 amendments a directive that the Secretary of the Interior seek a balance in the OCS leasing program that would accommodate "expeditious" development while protecting the environment and the interests of the coastal States. The amendments established procedures for considering environmental and State concerns in leasing decisions, required the orderly formulation of future leasing schedules, and ordered experimentation with a variety of alternative bidding systems. In seeking to balance energy development and other values, the offshore leasing program has been the target of criticism from coastal States, environmentalists, and the industry. These criticisms have sharpened in the 1980s as offshore activities have expanded into the deepwater and Arctic frontier areas.

The revised leasing system mandated by the 1978 amendments has been in place slightly more than 6 years. In this time, there have been numerous legal challenges to the leasing program, changes in leadership and reorganizations at the Department of the Interior, and a shift in leasing from nearshore areas to offshore frontier regions. In spite of the fact that it has proven to be one of the government's most controversial natural resource programs, the offshore leasing program has generally performed well in achieving the objectives set by Congress. It is unlikely that any statutory framework devised to expand and expedite exploration for oil and gas on Federal lands, while giving equal weight to protecting the environment and honoring the sovereign goals of the States, can be anything but adversarial and contentious. Despite the conflicts which have arisen, leasing of offshore oil and gas has worked more smoothly and efficiently than other Federal energy leasing programs.

The existing OCS Lands Act appears to provide Congress and the executive branch sufficient latitude to guide the leasing program in any direction that public policy may dictate. In general, the OCS Lands Act allows the administrative flexibility needed to adjust leasing terms and conditions to deepwater and Arctic frontier areas.

OFFSHORE RESOURCES AND FUTURE ENERGY NEEDS

The current abundance of oil and gas in the world market is considered to be a temporary anomaly. As economic recovery and cheaper energy stimulates consumption in the United States, the Nation will be faced with increasing petroleum imports unless substantial new domestic reserves are discovered. Oil imports, which have declined in recent years, are expected to gradually rise and may again reach the high levels of the 1970s.

Forecasts by the Department of Energy and the Gas Research Institute indicate domestic energy shortfalls may necessitate oil imports of about 7 million barrels per day and natural gas imports of about 3 trillion cubic feet per day by the end of the century. Forecasts by OTA and the Congressional Research Service anticipate higher oil import rates in the 1990s, perhaps again reaching the historic 1977 high of 9.3 million barrels per day. Predictions of declining real oil prices in the short term, which would reduce incentives for exploration and production of domestic resources, make even these forecasts optimistic. Oil imports of the magnitude expected in the 1990s would make the country vulnerable to supply interruptions and would increase the trade deficit.

Where will new domestic oil and gas resources be found to meet future U.S. energy needs? The onshore areas of the lower 48 States are the most densely explored and developed oil provinces in the world. But—with the exception of Prudhoe Bay, the largest field in North America—few sizable onshore discoveries have come on line in the past decade. Domestic reserves continue to dwindle. It is unlikely—but not impossible—that a giant field similar to Prudhoe Bay will be found onshore in the lower 48 States.

Most of the undiscovered oil and gas in the United States is expected to be in offshore areas or onshore Alaska. But resource estimates of undiscovered oil and gas, while useful as indicators of relative potential, are little more than educated guesses. Experts agree that prospects for oil and gas offshore are good, but they also admit there is a chance that only an insignificant amount of oil and gas may be found. In fact, only one major offshore field of a size needed to significantly increase reserves—the Point Arguello Field off southern California—has been discovered since offshore exploration was accelerated in the 1970s.

Exploration in the offshore frontier regions during the last 5 years has yielded some information—most of it negative—about potential oil and gas resources. In 1981, the U.S. Geological Survey (USGS) estimated that between 26 and 41 percent of the future oil and between 25 and 30 percent of the future natural gas was offshore. However, the Minerals Management Service (MMS) recently lowered the estimates of undiscovered recoverable offshore oil by half and of natural gas by 44 percent as a result of unsuccessful exploration efforts in Alaska and the Atlantic (see table 1).

Much of the 1.9 billion acres within the offshore jurisdiction of the United States is still unexplored. Only actual exploratory drilling can determine the presence of hydrocarbons. The offshore oil and gas industry will drill the most promising geological structures as exploration expands in the Arctic and deepwater frontiers. If significant reserves are not discovered in the first round of drilling, the government may need to consider a "second-round" leasing strategy to induce the industry to drill second-level prospective structures.

If Congress wishes to pursue the objectives of the OCS Lands Act, it is important that the oil and gas industry have access to Federal offshore lands to more accurately determine the resource potential of frontier areas. A "second-round" leasing strategy may also be needed to assess the extent of smaller offshore reservoirs that could cumulatively contribute to the Nation's energy security.

Table 1.—Revised Offshore Oil and Gas Resource Estimates

	Oil (billion barrels)			Gas (trillion cubic fee)		
Planning area	1981	1985	% change	1981	1985	% change
Alaska:						
Beaufort Sea	7.8	0.89		39.3	3.93	
Navarin Basin	1.0	1.30		1.30	1.58	
Chukchi Seal	1.6	0.54		13.8	3.02	
St. George Basin	0.4	0.37		2.5	3.47	
Norton Basin	0.2	0.09		1.2	0.43	
Other	1.2	0.11		2.2	1.42	
Total Alaska	12.2	3.30	−73	64.6	13.85	−78
Atlantic:						
North Atlantic	1.4	3.30		5.6	2.14	
Mid-Atlantic	3.1	0.35		14.2	6.02	
South Atlantic	0.9	0.22		3.6	4.04	
Other	—	—		0.3	0.11	
Total Atlantic	5.4	0.68	−87	23.7	12.31	−48
Gulf of Mexico:						
Western Gulf	5.2	1.90		65.7	26.76	
Central Gulf	—	3.72		—	30.69	
Easter Gulf	1.0	0.41		2.8	2.19	
Total Gulf of Mexico	6.2	6.03	− 3	68.2	59.64	−13
Pacific:						
Northern California	0.5	0.25		0.9	1.12	
Southern California	2.4	1.54		3.9	2.42	
Central California	—	0.36		—	0.51	
Washington and Oregon	0.3	0.04		1.4	0.65	
Total Pacific	3.2	2.19	−31	6.2	4.70	−24
Total offshore	27.0	12.2	−55	162.7	90.5	−44

SOURCE: U.S. Geological Survey, Circular 860, *Estimates of Undiscovered Recoverable Conventional Resources of Oil and Gas in the United States* (1981). Minerals Management Service, *Estimates of Undiscovered Oil and Gas Resources for the Outer Continental Shelf* (Personal correspondence, Feb. 4, 1985).

7. ACID RAIN AND TRANSPORTED AIR POLLUTANTS: IMPLICATIONS FOR PUBLIC POLICY, 1984.

Until recently, air pollution was considered a local problem. Now it is known that winds can carry air pollutants hundreds of miles from their points of origin. These *transported* air pollutants can damage aquatic ecosystems, crops, and manmade materials, and pose risks to forests and even to human health. Throughout this report we discuss three of these pollutants: acid deposition (commonly called acid rain), atmospheric ozone, and airborne fine particles.

The Clean Air Act—the major piece of Federal legislation governing air quality in the United States—addresses local air pollution problems but does not directly apply to pollutants that travel many miles from their sources. However, reports of natural resource damage in this country, Canada, Scandinavia, and West Germany have made transported air pollutants—particularly acid rain—a focus of scientific and political controversy. Many individuals and groups, pointing to the risk of irreversible damage to resources, are calling for more stringent Federal pollution controls. Others, emphasizing scientific uncertainties about transported air pollutants and drawing different conclusions about how to balance risks against costs, contend that further pollution controls are premature, may waste money, and would impose unreasonable burdens on industry and the public.

OTA's analysis of acid deposition and other transported air pollutants concludes that these substances pose substantial risks to American resources. Thousands of lakes and tens of thousands of stream miles in the Eastern United States and Canada are vulnerable to the effects of acid deposition. Some of these have already been harmed. Elevated levels of atmospheric ozone have reduced crop yields on American farms by hundreds of millions of bushels each year. Acid deposition may be adversely affecting a significant fraction of Eastern U.S. forests; it, along with such other stresses as ozone and natural factors such as drought, may account for declining forest productivity observed in parts of the East. Both sulfur oxides and ozone can damage a wide range of manmade materials. Airborne fine particles such as sulfate reduce visibility and have been linked to increased human mortality in regions with elevated levels of air pollution.

The costs of reducing pollutant emissions are likewise substantial. Most current legislative proposals to control acid deposition would cost about $3 billion to $6 billion per year. Adding these new emissions control proposals to our Nation's current environmental laws would increase the total costs of environmental regulation by about 5 to 10 percent. Average electricity costs would rise by several percent—as high as 10 to 15 percent in a few Midwestern States under the most stringent proposals. Additional emissions controls could also have important indirect effects, such as job dislocations among coal miners and financial burdens on some utilities and electricity-intensive industries.

Any program to reduce emissions significantly would require about 7 to 10 years to implement. If *no* further action is taken to control emissions, 30 to 45 years will elapse before most existing pollution sources are retired and replaced with facilities more stringently regulated by the Clean Air Act. **The effective time frame of most proposals to control acid deposition, therefore, is the intervening period of about 20 to 40 years—long enough to be significant to natural ecosystems.**

If all the *risks* posed by transported air pollutants were realized over this time period, resulting resource damage would outweigh control costs. The risks discussed throughout this report, however, are *potential* consequences, not necessarily the consequences that will, in fact, occur.

One of the most difficult questions facing Congress, therefore, is whether to act *now* to control acid deposition or *wait* for results from ongoing multimillion-dollar research programs. Both involve risks. *Delaying action* for 5 or 10 years will allow emissions to remain high for at least a decade or two, with the risk of further ecological damage. But predicting the magnitude and geographic extent of additional resource damage while waiting is not possible. *Acting now* involves the risk that the control program will be less cost effective or efficient than one designed 5 or 10 years from now. Significant advances in scientific understanding over this time period, however, are by no means assured.

The distributional aspects of transported air pollutants further complicate the congressional dilemma. Because these pollutants cross State and even international boundaries, they can harm regions far downwind of those benefiting from the activities that produce pollution. Moreover, different economic sectors bear the risks of waiting or acting now to control.

The policy decision to control or not to control transported air pollutants, therefore, must be based on **the risks of resource damage, the risks of unwarranted control expenditures, and the distribution of these risks among different groups and regions of the country.** This study describes the tradeoffs implicit in this choice by characterizing the *extent* of the risks, their *regional distribution*, and the *economic sectors* that bear them. It concludes with a list of policy options available to Congress for addressing transported air pollutants.

THE POLLUTANTS OF CONCERN

The transported air pollutants considered in this study result from the emission of three *primary* pollutants: sulfur dioxide, nitrogen oxides, and hydrocarbons. As these pollutants are carried away from their sources, they can be transformed through complex chemical processes into *secondary* pollutants: ozone and airborne fine particles such as sulfate and nitrate. Acid deposition results when sulfur and nitrogen oxides and their transformation products return from the atmosphere to the Earth's surface. Elevated levels of ozone are produced through the chemical interaction of nitrogen oxides and hydrocarbons (see fig. 1). Numerous chemical reactions —not all of which are completely understood—and prevailing weather patterns affect the overall distribution of acid deposition and ozone concentrations.

Current levels of precipitation acidity and ozone concentrations are shown in figures 2 and 3, respectively. Peak levels of acid deposition—as measured by precipitation acidity, or pH*—center around Ohio, West Virginia, and western Pennsylvania. High levels of acid deposition are found throughout the United States and southeastern Canada. Peak values for ozone are found further south than for acid deposition, centering around the Carolinas. A band of elevated ozone concentrations extends from the mid-Great Plains States to the east coast.

During 1980 some 27 million tons of sulfur dioxide and 21 million tons of nitrogen oxides were emitted in the United States. Figure 4 displays the regional pattern of emissions. About 80 percent of the sulfur dioxide and 65 percent of the nitrogen oxides came from within the 31 States bordering or east of the Mississippi River. Figure 5 shows how these emissions have varied since 1900. Since 1940, sulfur dioxide emissions increased by about 50 percent and nitrogen oxides emissions about tripled. Throughout the same period, taller emission stacks became common, allowing pollutants to travel farther.

Figure 5 also shows a range of *projected* future emissions of these pollutants, assuming that current air pollution laws and regulations remain unchanged. *Actual* future emissions will depend on such factors as the demand for energy, the type of energy used, and the rate at which existing sources of pollution are replaced by newer, cleaner facilities. By 2030—the end of the projection period—most existing facilities will have been retired. Despite relatively strict pollution controls mandated for new sources by the Clean Air Act, **emissions of both sulfur and nitrogen oxides are likely to remain high for at least the next half century.**

The pollutants responsible for acid deposition can return to Earth in rain, snow, fog, or dew—collectively known as acid precipitation—or as dry particles and gases. **Averaged over the Eastern United States, about equal amounts of sulfur compounds are deposited in wet and dry forms.** However, in areas remote from pollution sources, such as the Adirondacks, wet deposition may account for up to 80 percent of the sulfur deposited; in urban areas near many sources of pollution the situation is reversed.[1]

Pollutants emitted into the atmosphere can return to Earth almost immediately or remain aloft for longer than a week, depending on weather patterns and the pollutants' chemical interactions. During this time they move with prevailing winds, which in the Eastern United States tend to move from west to east and from south to north.

Figure 2.—Precipitation Acidity—Annual Average pH for 1980

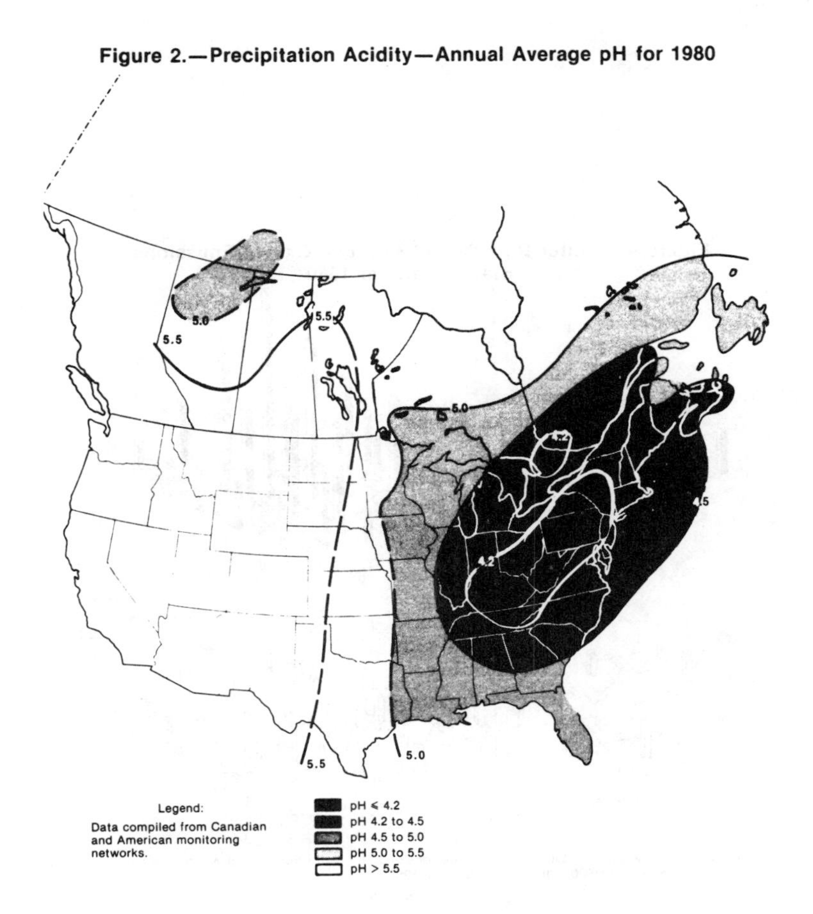

Figure 3.—Ozone Concentration—Daytime Average for Summer 1978

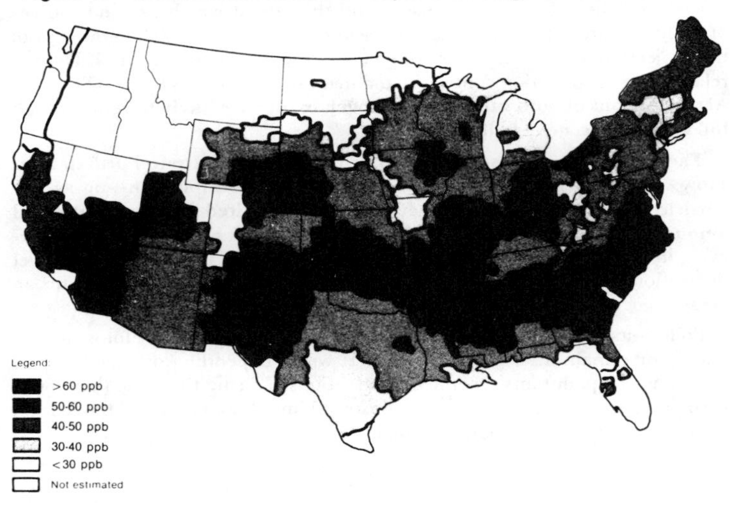

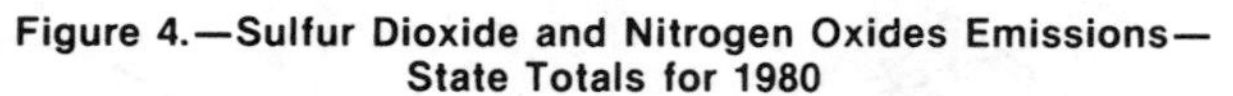

Figure 4.—Sulfur Dioxide and Nitrogen Oxides Emissions—State Totals for 1980

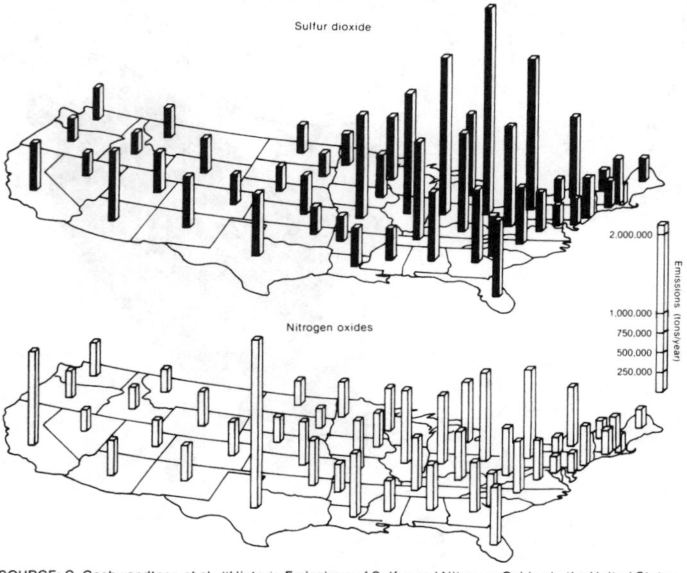

SOURCE: G. Gschwandtner, et al., "Historic Emissions of Sulfur and Nitrogen Oxides in the United States From 1900 to 1980," draft report to EPA, 1983.

Preliminary analyses suggest that about one-third of the total amount of sulfur compounds deposited over the Eastern United States as a whole originates from sources over 500 kilometers (km) (300 miles) away from the region in which they are deposited. Another one-third comes from sources between 200 and 500 km (120 to 300 miles) away, and the remaining one-third comes from sources within 200 km (120 miles).[2]

Because pollution sources are unevenly distributed across the Eastern United States and Canada, the *relative* contribution of emissions from local, midrange, and distant sources varies by region. As shown in figure 6, sulfur deposition in the Midwest—a region with high emissions—is dominated by emissions from sources within 300 km (180 miles). The sulfur compounds that reach the less-industrialized New England region typically have traveled farther. The "average" distance—considering the contribution from both local and distant sources—is about 500 to 1,000 km (300 to 600 miles).

Figure 5.—Sulfur Dioxide and Nitrogen Oxides Emissions Trends—National Totals, 1900-2030

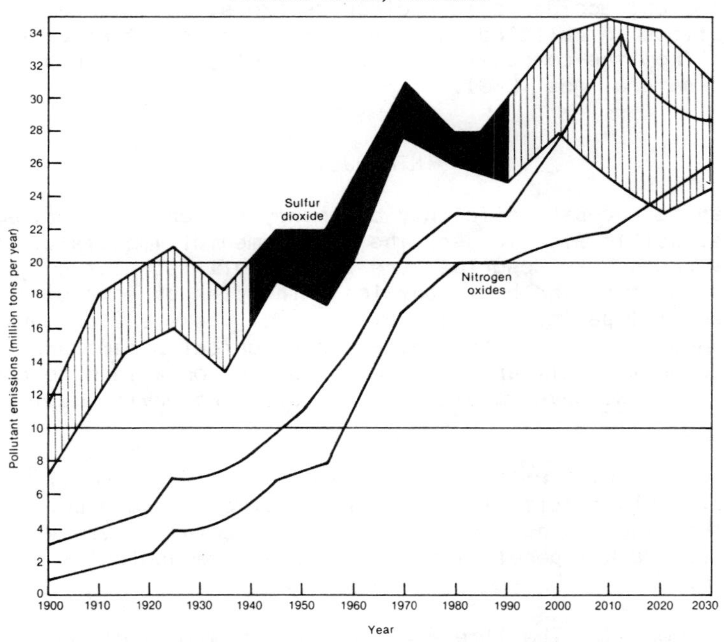

The graph above displays estimates of historical emissions, and projections of future emissions of sulfur dioxide and nitrogen oxides. Pre-1940 estimates and post-1990 projections are subject to considerable uncertainty. Projections of future emissions incorporate a wide range of assumptions about future economic growth, energy mix, and retirement of existing facilities; they assume no change in current air pollution laws and regulations.

SOURCES: Office of Technology Assessment. Composite from: U.S. Environmental Protection Agency, "National Air Pollution Emission Estimates, 1940-1980," 1982; G. Gschwandtner, et al., "Historic Emissions of Sulfur and Nitrogen Oxides in the United States From 1900 to 1980," draft report to the U.S. Environmental Protection Agency, 1983; *Emissions, Costs and Engineering Assessment*, Work Group 3B, United States-Canada Memorandum of Intent on Transboundary Air Pollution, 1982; and "Summary of Forecasted Emissions of Sulfur Dioxide and Nitrogen Oxides in the United States Over the 1980 to 2010 Period," ICF Inc. and NERA for the Edison Electric Institute, 1982; forecasts 1980 to 2030 by E. H. Pechan & Associates, Inc., for the Office of Technology Assessment, 1984.

APPENDIX D

PHYSICS OF SOME ENVIRONMENTAL ASPECTS OF ENERGY

DAVID HAFEMEISTER
PHYSICS DEPARTMENT
CALIFORNIA POLYTECHNIC UNIVERSITY
SAN LUIS OBISPO, CA 93407

ABSTRACT

Approximate numerical estimates are carried out on the following environmental effects from energy production and conservation: (1) The greenhouse effect caused by increased CO_2 in the atmosphere; (2) Loss of coolant accidents in nuclear reactors; (3) Increased radon concentrations in buildings with very low air infiltration rates; (4) Acid rain from the combustion of fossil fuels; and (5) Explosions of liquified natural gas (LNG).

INTRODUCTION

Enhanced end-use efficiency of energy not only conserves natural resources, but it also lessens the environmental impacts of obtaining those resources. We describe here some simplified models by which one can understand the basic physics principles of some of these environmental impacts. Our calculations of these environmental impacts have used only widely accepted numerical parameters, and the results agree with either direct observations or with more complex[1] calculations. We have considered the following environmental effects:

I. CO_2. The present rate of increase of CO_2 (ppm/y) will be estimated. This result will be exptrapolated to the middle of the last century and to the middle of the next century under various conditions. Market penetration of noncarbon technologies will be considered.

II. NUCLEAR. The time available after a loss of coolant accident for an emergency core cooling system to react will be estimated for a light water reactor (LWR) and for a high temperature gas reactor (HTGR). The case of a total loss of electrical power will be considered.

III. RADON. The energy available from reducing infiltration leaks will be estimated as well as the associated health effects.

IV. ACID RAIN. The approximate pH of the nation's rain will be estimated. Plumes from power plants, scrubbing, and pollution scaling laws will be considered.

0094-243X/85/1350628-14$3.00 Copyright 1985 American Institute of Physics

V. LNG. The thermal energy and power from an LNG explosion will be calculated.

I: ATMOSPHERIC CO_2 (THE "GREENHOUSE" EFFECT).

The concentration of CO_2 in the atmosphere[2,3] has risen from about 295 ppm in 1860 to about 345 ppm in 1984. Many scientists have predicted that a doubling of the CO_2 concentration will raise the average temperature of the earth by about 3^{0}C (1.5^{0}C at the equator and 4.5^{0}C at the poles). (This temperature rise is not strictly caused by a "greenhouse" effect since actual greenhouses block convective heat transfer and the atmosphere does not do this.) The planet Venus is an extreme example of this effect; its thick CO_2 atmosphere (about 90 times the earth's atmospheric pressure with a CO_2 concentration of about 96%) causes a surface temperature of about 482^0 C. This logarithmical rise in temperature could drastically effect the earth's food supplies and flood the low regions of the earth. It is generally believed that the cause of the increase in CO_2 is primarily due to the burning of fossil fuels. On a heating basis, coal produces 24% more CO_2 than oil, and 76% more CO_2 than natural gas. Deforestation, thus far, seems to have exacerbated the problem, but has not been the main cause of the increase in CO_2.

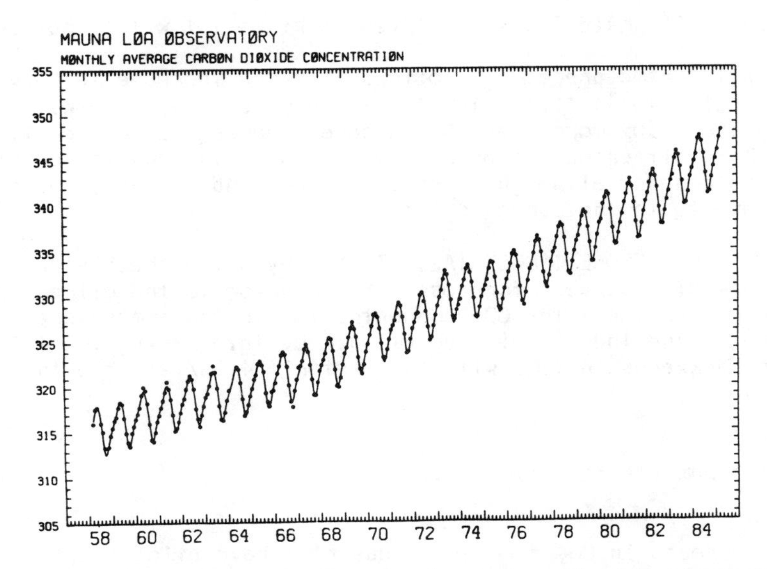

FIGURE 1. Concentration of atmospheric carbon dioxide observed with a continuously recording, non-dispersive infrared gas analyzer at Mauna Loa Observatory, Hawaii. The smooth curve represents a fit of the data to a four harmonic annual cycle which increases linearly with time, and a spline fit of the interannual component of the variation. The dots indicate monthly average concentrations. Data courtesey of C.D. Keeling, R.B. Bacastow, and T.P. Whorf, reference 3.

I.A. CO_2 GROWTH. Let us estimate the approximate deposition rate of CO_2 into the atomosphere and compare it to the current (1984) annual rise of about 1.5 ppm/year (by volume). The U.S. consumes about 35 million barrels per day (Mb/d) of petroleum equivalent which is about 30% of the world's consumption. The mass of fossil fuel burned each year on the earth is about

$$(35/0.3 \times 10^6 \text{ Mb/d})(365 \text{ d/y})(125 \text{ kg/b}) = 5.3 \times 10^{12} \text{ kg/y}.$$

About 80% of this amount is carbon, corresponding to

$$(0.8)(5.3 \times 10^{15} \text{ g})/(12 \text{ MolWt}) = 0.35 \times 10^{15} \text{ moles/y},$$

of carbon, or about

$$(0.35 \times 10^{15})(6.023 \times 10^{23}) = 2.1 \times 10^{38} \text{ molecules/y}$$

of CO_2. Since the atmospheric pressure is about 10^5 Pascals, the total mass of the atmosphere is about

$$F/g = (10^5 \text{ Pascals})(4\pi)(6.4 \times 10^6 \text{ m})^2/(9.8 \text{ m/s}^2) = 5.3 \times 10^{18} \text{ kg}$$

which corresponds to

$$(5.3 \times 10^{21} \text{ g})(6.023 \times 10^{23})/(29 \text{ MolWt}) = 1.1 \times 10^{44} \text{ molecules}$$

in the entire atmosphere. It follows that the increase in CO_2 per year is $(2.1 \times 10^{38})/(1.1 \times 10^{44}) = 1.9$ ppm/year (by volume). This value is about 30% more than the measured increase of CO_2 of 1.5 ppm/y. This difference is often attributed to the absorption of the CO_2 by the oceans, although there is a large and, as yet, uncertain sink somewhere in the carbon cycle.

I.B. CO_2 BEFORE INDUSTRIALIZATION. By using the historic growth rate of 4.3%/year for fossil fuels prior to the oil embargo of 1973, we can estimate the CO_2 concentration in the atmosphere at the beginning of the industrial revolution. By integrating the 1.5 ppm/year backwards in time with the historic 4.3%/year growth rate, we obtain

$$\int_{-\infty}^{0} (1.5 \text{ ppm/y})(e^{0.043t})dt = 35 \text{ ppm}$$

for the increase in CO_2 from the industrial revolution. By subtracting this value from the value of 345 ppm in 1984, we obtain 310 ppm for the CO_2 level prior to the industrial revolution. This is in good agreement with the often quoted value of 295 ppm.

I.C. THE CO_2 LEVEL IN THE YEAR 2050. By integrating forward in time from 1984 to 2050 we can obtain the CO_2 level in the year 2050. By assuming a 2%/year growth rate, we obtain

$$\int_0^{66} (1.5\ \text{ppm/y})(e^{0.02t})dt + 345\ \text{ppm} = 551\ \text{ppm}.$$

For other growth rates we obtain 444 ppm (0%/y), 485 ppm (1%), 657 ppm (3%), and 833 ppm (4%) respectively. Thus, it seems clear that the CO_2 level will double during the middle of the next century if carbon consumption continues to grow at 2 to 4%/year. The lower growth rates of 0 and 1%/y would markedly reduce the "greenhouse" effect.

I.D. THE RISE OF THE OCEANS. If the temperature increase caused by the increase in CO_2 is amplified in the polar regions to about a 5^0C increase, the polar icecaps would partially disintegrate. It might take from decades to centuries for an ice shelf to disintegrate. Let us estimate how much the oceans would rise above if we initially assume that only the West Antarctic Ice Shelf is dislodged? The West Antarctica Ice Sheet, formed only 10 million years ago, has an area of about 1.5 million km^2 with an average thickness of about 1.5 km. The volume of water from the ice shelf is $(1.5 \times 10^6\ km^2)(1.5\ km)(0.9\ \text{water/ice}) = 2.0 \times 10^6\ km^3$. Spreading this volume over the area of the oceans (70% of the earth) gives a rise in the height of the oceans of

$$(2.0 \times 10^6\ km^3)/(0.7)(4\pi)(6400\ km)^2 = 6\ \text{meters}.$$

This rise in the level of the oceans would cover about 2% of the U.S., and about 30% of Florida and Louisiana. Since it is generally believed that the temperature difference between glacial and interglacial periods is about 5 °C, the rising of the oceans to flood the world's low lands is a distinct possibility. The complete disintegration of the Antarctica and Greenland ice would take much longer and is much less likely, but that event would raise the world's oceans by about 100 meters.

I.E. MARKET PENETRATION BY ALTERNATIVE TECHNOLOGIES. Assume that the fraction, f, of the energy market based on noncarbon technologies (sun, wind, nuclear, geothermal, conservation, the capturing of CO_2 and releasing it deep in the ocean, etc.) is able to increase linearly according to the relation $f = t/P$ where P is the time period for total market penetration. Let us estimate the ultimate CO_2 concentration in the atmosphere by the year 2050 (P = 66 y) assuming growth rates of 0 to 4%/y. In order to simplify the calculation assume that the lifetime of the additional CO_2 in the atmosphere is considerably longer than 66 years. The CO_2 concentration in the atmosphere is

$$\int_0^P (1.5\ \text{ppm})(e^{at})(1 - t/P)\ dt\ +\ 345\ \text{ppm}$$

where a is the energy growth rate. This integrates to

$$(1.5\ \text{ppm}/Pa^2)(e^{aP} - 1 - aP) + 345 \quad .$$

Using P = 66 years, we obtain CO_2 levels of 395 ppm (0%), 407 ppm (1%), 426 ppm (2%), 453 ppm (3%), and 492 (4%). These values are considerably lower than those quoted in Section I.C and they are consistent with those obtained by Laurmann[4] who used a more complicated function to describe market penetration.

I.F. THE EFFECTS OF SYNFUELS AND NUCLEAR ENERGY. At one point the synfuel industry might have produced 5 Mb/d by the turn of the century. On the basis of delivered energy, synfuels will produce about 40% more CO_2 than burning coal, 70% more CO_2 than burning oil, and about 130% more CO_2 than burning natural gas. The use of energy in the world would rise to 135 to 150 Mb/d at the turn of the century with growth rates of 1 to 2%/y. Since this is considerably larger than the additional contributions of CO_2 from a large synfuels program (perhaps 5 Mb/d), the near term synfuels program wouldn't have mattered too much. However, a very large, world-wide synfuel industry in the future would complicate matters in the long term. At present nuclear power contributes about 2.5 Mb/d on a world wide basis, and it might contribute 5 to 10 Mb/d in the future. However, since nuclear power presently is only useful for (base-loaded) electricity which is 25% of the world's primary energy budget, it is not clear that nuclear power alone can save us from CO_2.

II: LOSS OF COOLANT ACCIDENTS.

If a nuclear reactor loses its coolant and the emergency core coolant system (ECCS) fails to operate in the worst possible combination of circumstances, the reactor core could melt through the steel reactor vessel and then into the earth. The heat energy for a melt down comes mainly from the beta decay of the fission fragments in the core. It is also possible to melt the core of a reactor if all the electrical power systems for the reactor fail.

II.A. THERMAL RISE TIME IN A PWR AFTER A LOCA. Consider the case of a loss of coolant accident (LOCA) in a pressurized water reactor[5] (PWR). Assume that the large amounts of water in the accumulator do not enter the reactor vessel until after the fuel rods have become quite hot. Let us estimate how long will it take the fuel of a PWR to rise to a temperature of 1370°C (2500°F) at which rapid damage to the core will begin to take place from the exothermic reaction of zircalloy and the water that comes from the accumulator. (UO_2 melts at 2200°C.) Assume the following: (1) The mass of UO_2 in a 1 GWe reactor is about 10^5 kg; (2) The thermal efficiency of a PWR is about 1/3; (3) The average temperature of the fuel is about 400°C before the LOCA; and (4) The thermal power after the LOCA is[6]

$$P = P_0(0.0766)(t^{-0.181}) \quad \text{for } t < 150 \text{ sec},$$

and

$$P = P_0(0.130)(t^{-0.283}) \quad \text{for } 150 \text{ sec} < t < 4 \times 10^6 \text{ sec},$$

where P_0 is the thermal power of the reactor before the LOCA, and t is the time in seconds after the LOCA. The rise time of the temperature after the LOCA is obtained by equating the heat necessary to heat the core to 1370 °C to the integral over time of the thermal power P. The heat needed to raise the core is $Q = NC(\Delta T)$ where N is the number of moles, C is the molal specific heat of UO_2, and ΔT is the temperature rise of the core (1370 °C - 400 °C = 970 °C). The number of moles of UO_2 in the core is about

$$N = (10^8 \text{ g})/(238 + 32)(\text{g/mole}) = 3.7 \times 10^5 \text{ moles}.$$

Since 600 °C is above the Debye temperature of UO_2, we can use the high temperature specific heat, C = 3R = 24.9 Joules/mole-°C. It follows that the necessary heat to raise zircalloy to 1370 C

$$Q = NC(\Delta T) = (3.7 \times 10^5)(24.9)(970) = 8.9 \times 10^9 \text{ J}.$$

Since a 1 GWe (electric) reactor has a thermal power of 3 GWt (thermal), the rise time of the LOCA is obtained from

$$Q = \int_0^t P\, dt = \int_0^t (0.0766)(3 \times 10^9)(t^{-0.181})\, dt$$

$$= (2.8 \times 10^8)(t^{0.819}) \text{ J} = 8.9 \times 10^9 \text{ J}.$$

Solving this equation we obtain t = 68 seconds which is very close to the published values of about one minute from more sophisticated calculations using the parabolic heat equation.

II.B. THERMAL RISE TIME IN A HTGR AFTER A LOCA. The high temperature gas reactor[7] (HTGR) uses graphite as a moderator and helium gas as a coolant in contrast to the PWR which uses light water as both a moderator and a coolant. As an interesting comparison let us calculate the thermal rise time for a HTGR after a LOCA in which the ECCS fails, using the following assumptions: (1) The heat capacitiy of the core of a HTGR is determined mainly by its 500,000 kg of graphite; the graphite reflectors and the nuclear fuel may be ignored; (2) The thermal effeciency of an HTGR is about 39%; (3) The average temperature of the core is about 750 °C; and (4) The core of an HTGR should be kept below about 1700 °C.

The HTGR will have a considerably longer thermal rise time because its core has about 100 times greater thermal mass (the number of moles times the specific heat) than the core of a PWR. This follows because the mass of the HTGR core is about 5 times greater (5×10^5 kg/10^5 kg), and because the molecular weight of graphite is about 20 times smaller than UO_2 (270/12 = 22). In addition the rise time of the HTGR will be further lengthened because the HTGR can withstand higher temperatures (1700 °C vs. 1370 °C for the PWR). The heat necessary to heat the HTGR core to 1700 °C is

$$Q = NC(\Delta T) = (5 \times 10^8/12)(24.9)(1700 - 750) = 9.9 \times 10^{11}\ J \quad .$$

This value of Q is 140 times larger than the PWR value because of the large heat capacity (mass times specific heat) of the core of the HTGR. By integrating the radioactive heat (before and after 150 sec), we obtain

$$Q = (10^9\ W/0.39)\{\int_0^{150}(0.0766)(t^{-0.181})dt + \int_{150}^{t}(0.13)(t^{-0.283})\ dt\}$$

$$= (10^9\ J)\{(0.46)(t^{0.717}) - 2.3\} = 9.9 \times 10^{11}\ J \quad .$$

From this we obtain t = 12 hours which closely agrees with the value of about 10 hours obtained from more sophisticated calculations.

II.C LOCA FROM THE LOSS OF ALL ELECTRICAL POWER. The fire at the Brown's Ferry boiling water reactor (BWR) in Alabama shut down all the electrical power necessary for cooling the core. This could have resulted in a LOCA since the cooling water was evaporated by the heating from the residual radioactive fission fragments. Let us estimate how long would it take for the core to become uncovered under the following assumptions: (1) About 700,000 kg of cooling water must be evaporated for the core to become uncovered; and (2) Assume that only the usual heat of vaporization of water must be considered, 2270 J/g.

The amount of heat needed to evaporate the water is about

$$Q = (7 \times 10^8\ g)(2270\ J/g) = 1.6 \times 10^{12}\ J \quad .$$

Setting this value of Q equal to the integrated radioactive heat for the PWR (which is about the same for the BWR) from II.B (with an efficiency of 0.33), we obtain t = 19 hours which was similar to the amount of time, about 13 hours, for the operator at Brown's Ferry to recover the situation and turn on the back-up pump.

III: INDOOR RADON.

The average level of radioactive radon in buildings[8] is about 1 picoCurie/liter (1 pCi/l = 1 nCi/m^3). This value is about 5 times the corresponding outdoor background level of about 0.2 pCi/l. Considerably higher levels of radon as high as 25 pCi/l have been measured in houses. Radon-222 results from the decay of radium-226 which is part of the uranium-238 decay chain; the radon enters the building through the foundations, and from the building materials and the water supply. The level of radon in these buildings is directly affected by the infiltration rate of fresh air from the outside since the fresh air replaces the inside air which contains radon-222 ($T_{1/2}$ = 3.8 days). The principal health risk from radon-222 arises from the fact that the four radioactive daughters are not chemically inert and can attach themselves to airborne particulates. Since the lifetimes of these daughters are all less than 30 minutes, the level of radon should be affected by the infiltation rate of clean air. Because of this, the EPA has recommended guidelines to the state of Florida to consider remedial action to lower the radon level below about 2 to 3 pCi/l.

The air filters into a building because of temeprature differences and because of the Bernouli effect caused by wind velocity. Because the infiltration of outside air replaces the air in the buildings causes a loss of about 25% of the energy to heat and cool our buildings, many "house doctor" groups are considering ways to reduce this needless loss of energy. Because of the infiltation pathways, the typical house has about one air change per hour (ach). There is a much wider variation (about 2 to 3 orders of magnitude) in the source term (the rate of radon infiltration) in the nation's housing stock than there is in the the rate of exchanges of outside air (about 1 order of magnitude in ach). The source term varies so much that one should consider the health effects of very tight housing on a case-by-case basis by measuring the indoor air quality. If the leakage pathways were reduced to 0.33 ach on new construction, the energy loss by infiltration would be reduced to 33% of its former value. On existing buildings "house doctors" can use blower doors to find the leakage pathways and reduce infiltation to about 0.5 ach. Of course, the "house doctors" should save the energy, but what will be the health effects of the increased radon levels and increased indoor air pollution in our buildings? Can we have both conservation and clean air?

III.A ENERGY SAVINGS FROM REDUCED INFILTATION. Let us estimate how much energy could be saved if the rate of air exchange was reduced from 1 ach to 0.5 ach? Assume the following: (1) The 70 million American living units have an average of 130 m^2 (1500 ft^2) floor space and 2.5 m (8.5 ft) ceilings; (2) The mass density of air is 1.2 kg/m^3 (0.0735 lb/ft^3) and its specific heat is C = 1.0 J/g^0C (0.238 BTU/lb ^0F); (3) The average heating season for the U.S. is about 2670 ^0C (4800 ^0F) degree days/year (dd/y); and (4) The average efficiency of a furnace is $\eta = 2/3$.

The rate of energy lost by infiltration is

$$dQ/dt = (dm/dt)(C)(\Delta T)/(\eta)$$

where dm/dt is the infiltraton rate of air mass, and ΔT is the temperature difference between the outside and the inside, 18.3^oC (65° F) - T(outside). Integrating this loss rate over the year, we obtain the energy lost by infiltation from a building over a year

$$Q = (dm/dt)(C)(dd/y)(24\ h/d)/(\ )$$

where dd/y is the number of degree days per year, $(1/24) \int (\ T)\ dt$ where t is in hours. It follows that the energy saved by closing the nation's infiltration pathways could be as high as

$$(1\ ach - 0.5\ ach)(130 \times 2.5\ m^3)(1.2\ kg/m^3)(24\ h/d)$$
$$(1000\ J/kg^oC)(2670\ dd/y)(7 \times 10^7\ homes)(3/2)$$
$$= 1.3 \times 10^{18}\ J/y = 1.2 \times 10^{15}\ BTU/y = 1.2\ quads/y.$$

This value is equivalent to 0.6 Mb/d of oil (13% of the 4 to 5 Mb/d that is used to heat the nation's homes), or 1.2 trillion cubic feet of natural gas (6% of the annual consumption of natural gas). Additional energy savings would come from reduced air conditioning as well as from commercial and industrial buildings.

III.B RADON FROM REDUCED INFILTRATION. Let us estimate very approximately the increase in the U.S. cancer rate if the infiltration rate of all houses was reduced from 1 ach to 0.5 ach. We will assume that the hypothesis that accepts a linear relationship between low-dose radiation and increased probability of getting lung cancer is correct; a doubling of the low-dose radiation level will double the probability of getting cancer from that particular source of radiation. For the purposes of this calculation we will use analysis of the excess cancer rate of uranium miners that was carried out by the United Nations Committee on Radiation (UNSCEAR). From their review of the relevant data on uranium miners (which appear to be approximately linear) they concluded that about 100 annual additional cases of lung cancer would be caused if one million persons spent all of their time in an environment with 1 pCi/l of radon-222. Some observers consider that the UNSCEAR number to be a factor of two too large, and that the other uncertainties of the radon problem cause a total uncertainty of about a factor of 10. We will assume that persons will spend about 50% of their time inside their homes.

If the air change rate is reduced by a factor of two from 1 ach to 0.5 ach, the radon level in the buildings would be increased by a factor of two from about 1 to 2 pCi/l. The number of additional cases of lung cancer could be

$$(2 - 1)(pCi/l)(10^{-4}/pCi\text{-}y/l)(2.3 \times 10^8/US)/2 = 10{,}000/y$$

within a range of 2,000 to 20,000/y. Since this result is about 10% of the present number of new lung cancer victims per year in the U.S. (100,000/y), indoor radon appears to be a significant contribution. In fact, the U.S. Environmental Protection Agency, Canada, and the Scandinavian countries have recommended guidelines to minimize the radon problem; for example the Sweden has recommended a minimum ventilation rate standard of 0.5 ach and 2 pCi/l for newly constructed homes.

Fortuitously, there is a technical fix so that we can have both conservation and health; Japan and the European nations are already marketing a \$400 air-to-air heat exchanger that transfers about 75% of the heat energy from the exhaust air to the clean incoming air. The payback period will depend on a variety of factors such as climate, new/old construction, desired number of ACH, costs, etc. These heat exchangers would also allow the possibility of reducing the present radon concentration in our current untightened buildings which cause about 10,000 lung cancers per year. It is clear that there is a point of diminishing return in tightening a house too much when considering a house with an air-to-air heat exchanger. Since the energy loss rate is proportional to the number of ACH, and the harmful health effects are inversely proportional to the number of ACH, there is an optimal spread of ACH values, depending on the quality of the indoor air in a particular house.

IV: ACID RAIN

The increased burning of coal contains sulfur has exacerbated the problem of acid rain[9] in the United States and in Europe. The most extreme example of acid rain with a pH of 2.4 (equivalent to vinegar) was recorded in Scotland in 1974. The average pH of the rainfall in some regions of the eastern U.S. has fallen to about 4.1. The pH of the rain in Pasadena, California has ranged between 2.7 and 5.4 with an average of 3.9. (Because of the CO_2 in the atmosphere, the pH of normal rain is about 5.6.) Even the rain in the Rocky Mountains has become considerably more acidic (from increased NO_x emissions); the pH of the rain in Colorado dropped from 5.4 to 4.6 in the three year time span of 1975 to 1978. Since water with an excess acidity (pH below 4 or 5) interferes with reproduction and spawning of fish, it has become difficult to support fish life in some lakes.

IV.A EASTERN AND WESTERN COAL. In 1979 the EPA relaxed the SO_2 emissions standards so that western coal (subbituminous) would not have to be scrubbed to the former standard of 90% sulfur removal on new electrical power plants; in the future, plants burning western coal would only have to have 70% of its SO_2 removed. This new formulation of the standard would result in approximately the same rate of sulfur emissions on an energy basis (lbs/BTU) from both kinds of coal. Eastern coal (bituminous) contains about 2.5% sulfur and has about 12,500 BTU/lb whie Western coal has about 8,500 BTU/lb. Since the rate of sulfur emissions from eastern (2.5%) and western coal will be approximately the same, we can determine the sulfur content of western coal (W%)

$$(2.5\%)(100\% - 90\%)/(12{,}500\ \text{BTU/lb}) = (W\%)(100\% - 70\%)/(8500\ \frac{\text{BTU}}{\text{lb}})$$

From this we obtain W = 0.57% which is quite close to the sulfur content of 0.5% for western coal. The Western coal has less sulfur because it was formed under fresh water while the eastern coal was formed under salt water.

IV.B. ACID RAIN. Let us calculate very approximately the average pH of rain in the U.S. We assume the following: (1) Most of the sulfur comes from the burning of about 600 million tons/year of coal which has an average sulfur content of about 2% (by weight); (2) the U.S. has an area of about 3 million square miles (7.7×10^6 km^2) with an average rainfall of about 25 inches (0.63 m); and (3) The nitrogen oxide compounds contriubte about one-third of the total acidity. (In the western U.S. NOx can be the predominant cause of the acidity.)

The number of gram moles of H_2SO_4 produced each year is about

$$(0.02)(600 \times 10^6 \text{ tons})(9.1 \times 10^5 \text{ g/ton})/(32 \text{ MolWt of S})$$

$$= 3.4 \times 10^{11} \text{ gram-moles/y.}$$

The number of gram moles of hydrogen ions will be three times this figure because H_2SO_4 contributes two ions and the NO_X compounds one third of the total acidity. The volume of rain that falls on the U.S. each year is about

$$(7.7 \times 10^{12} \text{ m}^2)(0.63 \text{ m})(10^3 \text{ l/m}^3) = 4.8 \times 10^{15} \text{ liters/y.}$$

The maximum possible average pH of the rain in the U.S is obtained by taking the logarithm of the ratio of

$$(1.0 \times 10^{12} \text{ moles}/(4.8 \times 10^{15} \text{ liters}) = 2.1 \times 10^{-4}$$

which gives a pH of about 3.7. This value is considerably lower than the average value of about 4.5 for the eastern U.S. for at least two reasons, tall smoke stacks and "dry" acid rain. The tall 300 meter smoke stacks tend to disperse the acid rain to other places such as Canada, the Atlantic Ocean, and Europe. In addition, the acidity of the acid rain will be further reduced since about 20% of the sulfur is deposited onto the earth as a "dry" acid rain made up of small particulates.

IV.C POWER PLANT PLUMES. The calculation of the dispersal of SO_2 from a power plant is difficult because turbulance and thermal eddies in the air are considerably more important than classical molecular diffusion. In addition the local geography can severely modify the air currents so that any simple formulation of the problem can be incorrect by an order of magnitude. However, an approximate[10] solution to this difficult problem is meaningful since it forces us to focus on the basic science of SO_2 plumes from a power plant. Let us assume that the steady state diffusion equation is valid for the large distances in the direction of the wind, x, such that $x^2 \gg (y^2 + z^2)$;

$$U \frac{\partial C}{\partial x} = D_y \frac{\partial^2 C}{\partial y^2} + D_z \frac{\partial^2 C}{\partial z^2}$$

where U is the wind velocity in the x direction, C is the concentration of the impurity (SO_2 in this case), and D_y and D_z are the macro diffusion constants which are about 6 orders of mangnitude larger than the molecular diffusion constants in completely still air. It can be easily shown that the solution to this equation is

$$C = \frac{S}{2\pi U\sigma_y\sigma_z} \exp(-(y^2/\sigma_y^2 + z^2/\sigma_z^2)/2)$$

where S is the emission rate of SO_2 and the standard deviation widths of the Gaussian solutions are given by $\sigma_y = \sqrt{2D_y/U}$. Let us determine the SO_2 concentrations in the direction of the wind (y = 0) at distances of x = 1 km and 10 km from the power plant. We will also determine the SO_2 concentrations at the center of the plume that has risen from a tall smokestack (300 m above the ground) and at ground level. Assume the following: (1) A 1 GWe power plant burns about 10^4 tons of coal per day with a 2% sulfur content; (2) the wind velocity is 5 m/s; (3) the diffusion constants (enhanced by turbulance) are $D_y = D_z = 25\ m^2/s$ (for slightly unstable (Pasquill-Gifford) stability condition); (4) the SO reflected from the earth can be treated as an image source 300 m below the earth; and (5) a concentration of 2.6 mg/m^3 of SO_2 corresponds to 1 ppm. Compare your answer to the threshold for increased hospital admissions when SO_2 concentrations exceed 0.1 ppm for 4 days or when SO_2 exceeds 1.0 ppm for 5 minutes. These pollution levels can be compared to the extreme case of London in 1952 when 3 days of 0.7 ppm of SO_2 (and particulates) caused about 2500 excess deaths.

The standard deviations for the SO_2 distribution is

$$\sigma = \sqrt{(2)(25)(1000)/(5)} = 100 \text{ m}$$

at a distance of 1 km and σ = 316 m at 10 km. The power plant emits sulfur at the rate of

$$S = (0.02 \times 10^4 \text{ tons/d})(910 \text{ kg/ton})/(8.6 \times 10^4 \text{ s/d}) = 2.1 \text{ kg/s}.$$

Since the molecular weight of SO_2 is twice that of sulfur, this corresponds to 4.2 kg/s of SO_2. Inserting these values into the solutions of the diffusion equation for the sources at z = 300 m and z = - 300 m, we obtain C(x,y,z) for the following places (in km): C(1,0,0.3) = 5.1 ppm, C(1,0,0) = 0.11 ppm, C(10,0,0.3) = 0.60 ppm, C(10,0,0) = 0.66 ppm. From these values we see that the SO_2 concentration has decreased by almost a factor of ten within the plume (y = 0, z = 0.3) as the distance from the plant was increased from 1 to 10 km. However, we note that the SO_2 concentration at ground level (z = 0) has increased over this distance from 0.11 ppm to 0.60 ppm. The tall smokestack has decreased the severity of the problem, but it has shifted the highest dose rates to more distant neighbors. Since these values of SO_2 exceed the conditions for increased hospital admittance, it is clear that one should not remain in the plume.

IV.D POLLUTION SCALING LAW. City A has a size of L km by L km and it has a SO_2 level of 0.015 ppm. The SO_2 is emitted at a constant rate per unit area from many small sources. Let us determine the SO_2 level in city B that has exactly the same weather conditions, and pollution production density, but B has an area of 10L by 10L?

Since B is 100 times larger it will produce 100 times as much pollution. Since the wind will blow the pollution away through a cross-sectional area of LH (where H is the inversion layer) that is only 10 times larger for city B, the pollution level in city B will be 10 times larger than city A. The SO_2 level of 0.15 ppm in city B exceeds the Primary (Public Health) Air Quality Standard of 0.14 ppm for a maximum 24 hour concentration.

V: LNG EXPLOSIONS

The amount of liquified natural gas (LNG) that will be transported in the future may increase because some regions of the world need natural gas and other regions of the world are faced with the choice of either flaring the gas in the atmosphere or selling it. Most of the LNG will be transported in ships which contain up to five spherical tanks with a diameter of 35 m that will hold about 25,000 m^3 of LNG. If one of the spherical tanks should rupture, the LNG vapors would spread horizontally, rather than rise, because the density of the cold vapors (-160^{0}C) is greater than the density of air. Let us estimate the equivalent explosive energy and power if one of the spherical tanks should rupture and explode.[11] Assume the following: (1) The energy content of LNG is about 3.3×10^{10} J/m^3; (2) One kiloton of explosives has an energy equivalence of 10^{12} calories; and (3) The type of LNG explosion will depend on atmospheric conditions and on the time of ignition as to whether it would be a horizontal fire storm or a "fireball", but for this example we will assume it takes five minutes to the burn all the LNG.

The explosive energy available from one of the sperhical tanks is

$$E = (2.5 \times 10^4 \text{ m}^3)(3.3 \times 10^{10} \text{ J/m}^3) = 8.2 \times 10^{14} \text{ J} = 200 \text{ kilotons}$$

which is about 15 times that of the Hiroshima bomb. The average explosive power during the 5 minute burn is

$$P = E/t = (8.2 \times 10^{14} \text{ J})/(300 \text{ s}) = 2.7 \times 10^{12} \text{ W} = 2700 \text{ GWt}$$

which is about the same as the thermal power of the U.S. (2500 GWt = 78×10^{15} BTU/y). If the duration of the fireball was 30 seconds, the explosive power would have been 10 times higher. If all five spherical containers in a large LNG tanker exploded, the energy and power values would be five times higher.

REFERENCES

1. A good general reference on the methodology of environmental risk is by W. Lowrance, OF ACCEPTABLE RISK, Kaufmann, Los Altos, California, 1976. More specific estimates on risks from energy production can be found in: J. Holdren, G. Morris, and I. Mintzer, Ann. Rev. Energy 5, 241 (1980). B. Cohen and I. Lee, Health Physics 36, 707 (1979). R. Wilson and E. Crouch, RISK BENEFIT ANALYSIS, Ballinger, Cambridge, MA, 1983. D. Hafemeister, Am. J. Phys. 50, 713 (1982).
2. S. Schneider and R. Chen, Ann. Rev. Energy 5, 107 (1980). J. Williams, CARBON DIOXIDE AND SOCIETY, Pergamon Press, Oxford, 1978. CHANGING CLIMATE, National Academy of Sciences, Washington, DC, 1983.
3. C.D. Keeling, R.B. Bacastow, and T.P. Whorf, "Measurements of Concentration of Carbon Dioxide at Mauna Loa Observatory, Hawaii," in CARBON DIOXIDE REVIEW: 1982, W.C. Clark, editor, Oxford Univ. Press, p. 377-385 (1982), and unpublished data courtesy of C.D. Keeling. The measurements were obtained in a cooperative program of the U.S. National Oceanic and Atmospheric Administration, and the Scripps Institution of Oceanography.
4. J. Laurmann, Science 205, 896 (1979).
5. A. Nero, A GUIDEBOOK TO NUCLEAR REACTORS, Univ. California Press, Berkeley, 1979.
6. Reports to the American Physical Society: REACTOR SAFETY STUDY, Rev. Mod. Phys. 47, S1, S95 (1975), and RADIONUCLIDE RESEASE FROM SEVERE ACCIDENTS AT NUCLEAR POWER PLANTS, Rev. Mod. Physics 57 (1985).
7. H. Agnew, Sci. Amer. 244, 55 (June, 1981).
8. R. Budnitz, J. Berk, C. Hollowell, W. Nazaroff, A. Nero, and A. Rosenfeld, Energy and Buildings 2, 209 (1979). J. Spengler and K. Sexton, Science 221, 9 (1983). Entire issue of Health Physics 42, January, 1983. Chapter by R. Sextro, et al in this book.
9. G. Likens, R. Wright, J. Galloway and T. Butler, Sci. Amer. 241, 43 (Oct. 1979). ACID RAIN AND TRANSPORTED AIR POLLUTANTS, Off. Technology Assessment, Washington, DC, 1984. ACID DEPOSITION: ATMOSPHERIC PROCESSES IN EASTERN NORTH AMERICA, National Academy of Sciences, Washington, DC, 1983. K. Rahn and D. Lowenthal, Science 228 (1985).
10. S. Williamson, FUNDAMENTALS OF AIR POLLUTION, Addison-Wesley, Reading, Mass., 1973. J. Seinfeld, AIR POLLUTION, McGraw Hill, New York, 1975.
11. TRANSPORTATION OF LIQUIFIED NATURAL GAS, Office of Technology Assessment, U.S. Congress, Washington, 1977. J. Fay, Ann. Rev. Energy 5, 89 (1980).

APPENDIX E

THE DOE-2 COMPUTER PROGRAM FOR THERMAL SIMULATION OF BUILDINGS

B. E. Birdsall, W. F. Buhl, R. B. Curtis, A. E. Erdem,
J. H. Eto, J. J. Hirsch, K. H. Olson, F. C. Winkelmann

Applied Science Division
Lawrence Berkeley Laboratory
Berkeley, California 94720

Summary

The DOE-2 building energy analysis computer program was designed to allow engineers and architects to perform design studies of whole-building energy use under actual weather conditions. Its development was guided by several objectives: (1) that the description of the building entered by the user be readily understood by non-computer scientists, (2) that, when available, the thermal calculations be based upon well established algorithms, (3) that it permit the simulation of commonly available heating, ventilating, and air-conditioning (HVAC) equipment, and (4) that the predicted energy use of a building be acceptably close to measured values. Applications of DOE-2 include design of energy-efficient buildings, analysis of the performance of innovative building components and systems, and energy standards compliance. It is also widely used as an educational tool in the building sciences.

DOE-2 Program Structure

A building, examined thermodynamically, involves non-linear flows of heat through and among all of its surfaces and enclosed volumes, driven by a variety of heat sources. Mathematically, this corresponds to a set of coupled integral-differential equations with complex boundary and initial conditions. The function of a program like DOE-2 is to simulate the thermodynamic behavior of the building by approximately solving the mathematical equations.

DOE-2 performs its energy use analysis of buildings in four principal steps.

First is the calculation of the hour-by-hour heat loss and gain to the building spaces and the heating and cooling loads imposed upon the building HVAC systems. This calculation is carried out for a space temperature fixed in time and is commonly called the LOADS calculation. It answers the question: how much heat addition or extraction is required to maintain the space at a constant temperature as the outside weather conditions and internal activity vary in time and the building mass absorbs and releases heat?

Second is the calculation of the energy addition and extraction actually to be supplied at the heating and cooling coils by the HVAC system in order to meet the possibly varying temperature set-points and humidity criteria subject to the schedules of fans, boilers and chillers, and to outside air requirements. This calculation results in the demand for energy that is made on the primary energy sources of the building. This step, called the SYSTEMS

This work was supported by the Assistant Secretary for Conservation and Renewable Energy, Office of Buildings Energy Research and Development, Building Systems Division of the U. S. Department of Energy under Contract No. DE-AC03-76SF00098.

0094-243X/85/1350642-8$3.00 Copyright 1985 American Institute of Physics

calculation, answers the question: what are the heat extraction and addition rates when the characteristics of the HVAC system, the time-varying temperature set-points, and the heating, cooling and fan schedules are all taken into account?

Third is the determination of the fuel and electricity requirements of primary equipment such as boilers and chillers, and the electric generators, etc., in the attempt to supply the energy demand of the HVAC systems. This PLANT calculation answers the question: how much fuel and electrical input is required by the HVAC system given the efficiency and operating characteristics of the plant equipment and components?

The *fourth* step, ECONOMICS, evaluates the costs of equipment, fuel, electricity, labor and building envelope components. It answers the question: Is the expenditure of funds for energy conserving materials and systems cost effective, when compared with alternative systems?

A detailed description of the simulation algorithms used in DOE-2 is given in Ref. [3].

LOADS

General Considerations

The LOADS program computes the hourly cooling and heating loads for each space of the building. A *load* is defined as the rate at which energy must be added to or removed from a space to maintain a constant air temperature in the space. A *space* is a user-defined subsection of the building. It can correspond to an actual room, or it may be much larger or smaller, depending upon the level of detail appropriate to the simulation.

The space loads are obtained by a two-step process. First, the heat gains (or losses) are calculated; then the space loads are obtained from the space heat gains, taking into account the storage of heat in the thermal mass of the space. A *space heat gain* is defined as the rate at which energy enters or is generated within a space in a given moment. The space heat gain is divided into radiative and convective components, depending on the manner in which the energy is transported into or generated within the space. The components are:

1. solar heat gain from radiation through windows and skylights,
2. heat conduction through walls, roofs, windows, and doors in contact with the outside air,
3. infiltration air (unintended ventilation),
4. heat conduction through walls and floors in contact with the ground,
5. heat conduction through interior walls, floors, ceilings, and partitions,
6. heat gain from occupants,
7. heat gain from lights,
8. heat gain from appliances or office equipment.

The calculation of heat conduction through walls involves solving the one dimensional diffusion equation

$$\frac{\partial^2 T}{\partial x^2} = \frac{1}{\alpha}\frac{\partial T}{\partial t}$$

each hour, where T is the temperature and α the thermal diffusivity. In DOE-2 the equation is presolved for each wall or roof using triangular temperature pulses as excitation functions. The resulting solutions, called "response factors" are then used in the hourly simulation modulated by the actual indoor and outdoor temperatures. This approach assumes that the wall properties, including inside film coefficients, do not change during the simulation.

The solar gain calculation starts with the direct and diffuse solar radiation components, which are obtained from measured data or computed from a cloud cover model, taking into account the actual position of the sun each hour. The radiation is projected onto glass surfaces, after taking into account the shading (for the direct component) of exterior shading surfaces, and is transmitted, absorbed, and reflected in accordance with the properties of the glass in the window. As with the conduction through walls, the problem is presolved for a finite class of window properties.

Heat flow through interior walls and through surfaces in contact with the soil is treated as steady state, i.e., the capacitive effects of the walls are ignored in the hourly calculation, although they are taken into account in the calculation of the weighting factors. For interior walls that are light and not load bearing, this is a reasonable assumption. Interior walls between a sunspace and an interior space, on the the other hand, can be massive and delayed conduction through such walls can be modelled.

The internal heat gains from people, lights, and equipment are basically fixed by the user's input of peak values multiplied by hourly values in the schedules for these gains.

In general, space heat gains are not equal to space cooling loads. An increase of radiant energy in a space does not immediately cause a rise in the space air temperature. The radiation must first be absorbed by the walls, cause a rise in the wall surface temperature, and then (by convective coupling between the wall and the air) cause an air temperature rise. This is handled in DOE-2 through weighting factors. The weighting factors are determined from a detailed heat balance which gives the response in time of a zone (with all its mass and walls and fenestration) to a unit pulse of each of the zone heat gains. The weighting factor method is based on the assumption that the superposition principle applies to heat transfer processes occurring in a space. In effect, this means that the heat gain due to solar radiation, for example, can be calculated independent of the other heat gains.

Fig. 1 shows DOE-2 predictions for solar heat gain and the resulting cooling load for a passive solar home in Chicago. Note that the weighting factors attenuate and delay the effects of the heat gain, so that a cooling load from solar radiation extends through the nighttime hours long after the sun has set.

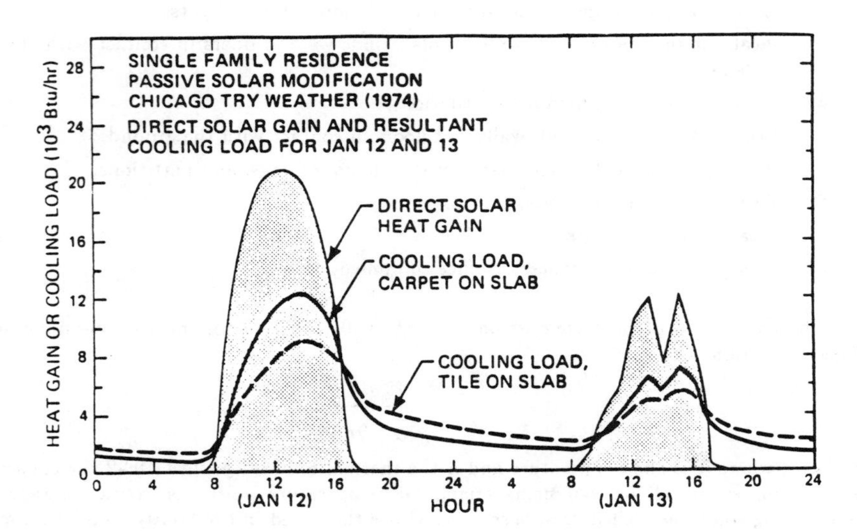

Fig. 1: DOE-2 predictions for solar heat gain and resultant cooling load for tile floor vs. carpetted floor.

Special LOADS Features

In DOE-2.1C, the latest version of the program, there are several additional features that greatly extend its usefulness. These include in the LOADS program the ability to take advantage of credit for daylighting, the ability to model sunspaces and the transmission of solar radiation through interior windows, and a mechanism by which users can substitute their algorithms for those used by the program.

- **Daylighting Credit**

The daylighting simulation in DOE-2, coupled with the thermal loads and HVAC analysis, allows users to evaluate the energy- and cost-related consequences of daylighting strategies. The program takes into account the availability of daylight from sun and sky, window management in response to solar gain and glare, and various electric lighting control schemes.

The daylight illuminance calculation in DOE-2.1C considers such factors as: window size and orientation, glass transmittance, inside surface reflectances of the space, sun-control devices such as blinds and overhangs, luminance distribution of the sky, and discomfort glare.

For each daylit space, a preprocessor calculates and stores a set of daylight factors for a series of sun positions covering the annual ranges of solar altitude and azimuth at the specified building latitude. These factors relate interior illuminance and glare levels to outdoor daylight levels.

In the hourly daylighting calculation, the illuminance from each window or skylight is found by interpolating the stored daylight factors using the current-hour sun position and cloud cover, then multiplying by the current-hour exterior horizontal illuminance. If the glare-control option has been specified, the program will automatically close window blinds or drapes in order to decrease glare below a pre-defined comfort level. Adding the illuminance contributions from all the windows then gives the total number of footcandles at each reference point.

The program then simulates the lighting control system to determine the artificial lighting electrical energy needed to make up the difference, if any, between the daylighting level and the required illuminance. Finally, the lighting electrical requirements are passed to the thermal calculation which determines hourly heating and cooling requirements for each space.

- **Sunspace Model**

DOE-2.1C allows the user to model the different forms of heat transfer that can occur between a sunspace (or atrium) and adjacent spaces. These include

1. direct and diffuse solar gain through interior glazing,
2. forced or natural convection through vents or an open doorway,
3. delayed conduction through an interior wall, taking into account the solar radiation absorbed on the sunspace side of the wall,
4. conduction through interior glazing.

The model also simulates the venting of the sunspace with outside air to prevent overheating, and, for residential applications, the use of a sunspace to preheat outside ventilation air.

- **Functional Approach**

DOE-2.1C allows the user to modify the way that DOE-2 does its calculations in LOADS without having to recompile the code. This "Functional Approach" involves writing FORTRAN-like functions in the LOADS input that compute the program variables as desired by the user. The possibilities of this feature are many and include, for example, deploying window shades to block the sun if a space has a cooling load, making the outside air film conductance dependent upon the wind direction, entering measured values of daylight factors, printing user designed reports, and modelling innovative products and control strategies.

SYSTEMS

General Considerations

The SYSTEMS program simulates the equipment that provides heating, ventilating and/or air conditioning to the thermal zones and the interaction of this equipment with the building envelope. This simulation comprises two major parts:

1. Since the LOADS program calculates the "load" at constant space temperature, it is necessary to correct these calculations to account for equipment operation.
2. Once the net sensible exchange between the thermal zones and the equipment is solved, the heat and moisture exchange between equipment, heat exchangers, and the heating and cooling coil loads can be passed to the primary energy conversion equipment or utility.

The dynamics of the interaction between the equipment and the envelope are calculated by the simultaneous solution of the room air-temperature weighting factor equation with the equipment controller relation. The former relates the "load" from LOADS and the heat extraction rate (the sensible coil load) to the zone temperature. The latter relates the heat extraction rate to the controlling zone temperature. Once the supply and thermal zone temperatures are known, the return air temperature can be calculated and the outside air system and other controls can be simulated. Thus the sensible exchange across all coils are calculated.

The moisture content of the air is calculated at three points in the system: the supply air leaving the cooling coil, the return air from the spaces, and the return air after being mixed with outside ventilation air. These values are calculated assuming that a steady state solution of the moisture balance equations each hour will closely approximate the real world. The return air humidity ratio is used as the input to the controller activating a humidifier in the supply airflow or resetting the cooling coil controller to maintain maximum space humidity set points. The moisture condensation on the cooling coils is simulated by solving the coil leaving air temperature and humidity ratio simultaneously with the system moisture balance.

Once the above sequence is complete, all sensible and latent coil loads are known. These values are then either passed to the PLANT program as heating and cooling water circuit loads or, in the case of direct-expansion cooling equipment, the energy conversion is simulated in SYSTEMS.

System Types

The DOE-2 program provides the user with 22 generic system types with many sizing and control options, depending upon the type chosen. Among the systems which can be modelled are: variable air volume, powered induction unit, dual duct, fan coil, water-to-air heat pump, air and air-to-air heat pump, packaged air conditioning, and residential furnace.

Special SYSTEMS Features

The DOE-2 program allows the simulation of a large number of special features within the generic system types. These features include (1) baseboard heaters, (2) humidity control, (3) "economizer cycle", i.e., use of outside air for cooling when its temperature or enthalpy is low enough, (4) natural ventilation through open windows, (5) optional start-up time for fans, (6) night temperature setback, (7) nighttime pre-cooling of building mass, and (8) recovery of sensible heat from the exhaust air stream.

PLANT

The PLANT program simulates primary HVAC equipment, i.e., central boilers, chillers, cooling towers, electrical generators, pumps, heat exchangers, and storage tanks. In addition, it also simulates domestic or process water heaters, and residential furnaces. Each type of equipment is modelled using empirically determined performance curves which characterize thermal performance as a function of part load, entering fluid temperatures, and other variables.

The purpose of the primary equipment is to supply the energy needed by the fans, heating coils, cooling coils, or baseboards, and the electricity needed by the building's lights and office equipment. Building loads can be satisfied by using the user-defined plant equipment or by the use of utilities: electricity, purchased steam, and/or chilled water.

Plant Management

The management of a plant is determined by setting up time schedules and/or load ranges under which specified equipment will operate. An example of such management is scheduling chillers to run at night to produce cold water or ice, which is stored for use the following day to cool the building; a strategy of this kind can avoid the high peak electrical demand charges incurred by running chillers during the day.

ECONOMICS

The ECONOMICS portion of the program computes the costs of energy for the various fuels or utilities used by the equipment. A wide variety of tariff schedules can be modelled. In addition, sale to the utility of electricity generated at the building can be simulated.

Rate Schedules

DOE-2 allows the following energy resources to be used: chilled water, steam, electricity, natural gas, fuel oil, coal, diesel oil, methanol, LPG, and biomass. For each of these resources that is used by a building the user may specify uniform cost rates, escalation rates, fixed monthly charges by season, various block charges by season, whether there are demand charges and how much, time-of-day charges, and, for electricity only, details about ratchet periods and types and conditions of sale to utilities. Not all of these apply to every fuel or resource, of course, and defaults exist for the simplest tariffs. On the other hand, most of the existing tariff structures can be simulated.

Investment Statistics

In addition to the possibility of treating the costs of energy, DOE-2 allows the user to simulate the life cycle costs of a building and to compare the costs between two configurations of the building. Assuming one is the base case and the other is a retrofit or an alternative design, investment statistics such as pay back period, savings to investment ratio, etc., are computed over the life cycle of the building.

Validation

DOE-2 has been verified against field measurements on existing buildings [4]. Examples of DOE-2 comparisons with measured data are shown in Figs. 2 and 3. The general conclusion of the verification studies is that total energy usage for a well-modelled building can be predicted within 15%.

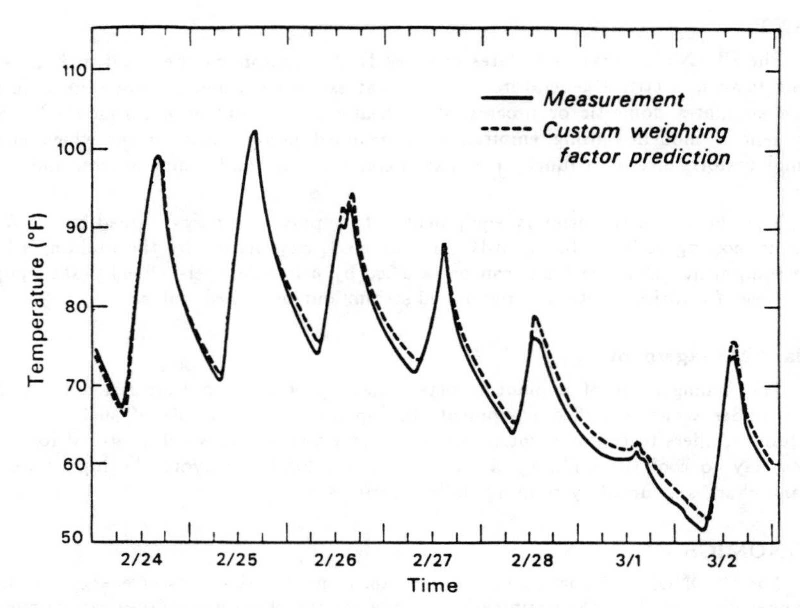

Fig. 2: Comparison of DOE-2 predictions and temperature measurements in the Los Alamos National Laboratory direct-gain passive solar test cell.

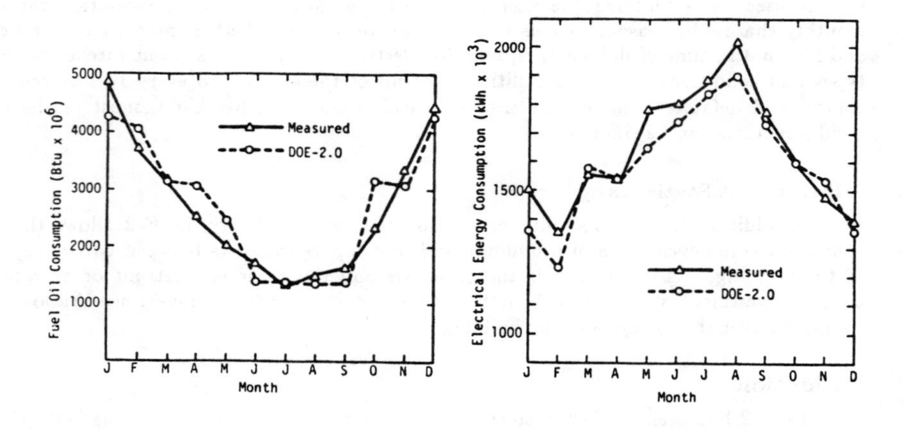

Fig. 3: Comparison of DOE-2 and measured data on fuel oil and electricity consumption for the Marriott Hotel in Boston (from Ref. 4).

Applications

DOE-2 is now in wide use throughout the world. Its applications span a broad spectrum from research to education to building design. As a research tool, DOE-2 is commonly used to study the thermal behavior of individual building components and the interaction of these components with the rest of the building. As an example of this type of research application, the Windows and Daylighting Group at Lawrence Berkeley Laboratory (LBL) is using DOE-2 to study the annual performance of advanced glazing materials (e.g., electrochromic or photochromic glass) whose transmittance can be actively or passively controlled based on temperature, solar radiation, or other environmental variables [5].

The program has also been extensively used to perform parametric analyses to determine the energy use response of a building as a function of one or more variables. The resulting database of DOE-2 results can be quite large, so that statistical approaches, such as multiple regression analysis, are valuable in summarizing the data. For example, in the recent upgrade of the ASHRAE 90 standard for energy conservation in non-residential buildings, thousands of DOE-2 runs were made varying important building envelope parameters for different U. S. climates [6]. For each climate, a regression analysis yielded coefficients in a linear fit which gave cooling energy, heating energy, and cooling peak for office buildings as a function of envelope conductance, window aperture, and electrical lighting power density.

A similar statistical approach was used by the Energy Analysis Group at LBL for houses [7]. A data base of residential energy use was generated from 10,000 DOE-2 runs for seven prototype houses in 45 U. S. locations, covering a wide range of typical conservation measures. Linear regression was then used to relate heating and cooling loads to key building components, such as wall conductance or infiltration air change rate. From the regression equations, two simplified energy prediction tools were developed -- an Energy Sliderule and the PEAR microcomputer program, both of which allow energy estimates to be made in a fraction of the time required by a full DOE-2 simulation.

For more information on DOE-2, contact Karen H. Olson, Building Energy Simulation Group, Lawrence Berkeley Laboratory, Building 90, Room 3147, Berkeley, CA 94720.

REFERENCES

1. "DOE-2 REFERENCE MANUAL", Lawrence Berkeley Laboratory, Report. No. LBL-8706 Rev. 2, 1981.
2. "DOE-2 SUPPLEMENT", Lawrence Berkeley Laboratory, Report No. LBL-8607 Rev. 4, Suppl., 1984.
3. "DOE-2 ENGINEERS MANUAL", Lawrence Berkeley Laboratory, Report No. LBL-11353, 1982.
4. "DOE-2 Verification Project, Phase I, Interim Report" Los Alamos National Laboratory, Report No. LA-8295-MS, 1981.
5. R. Johnson, D. Connell, S. Selkowitz, and D. Arasteh, "Advanced Optical Materials for Daylighting in Office Buildings", Lawrence Berkeley Laboratory, Report No. LBL-20080, 1985.
6. R. Johnson, R. Sullivan, S. Nozaki, S. Selkowitz, C. Conner, and D. Arasteh, "Building Envelope Thermal and Daylighting Analysis in Support of Recommendations to Upgrade ASHRAE/IES Standard 90", in "Recommendations for Energy Conservation Standards and Guidelines for New Commercial Buildings", Pacific Northwest Laboratory, Report No. PNL-4870-4, 1983.
7. Huang, Y. J., Bull, J. and Ritschard, R., "Simplified Calculations for Residential Energy Use Using a Large DOE-2 Data Base", Lawrence Berkeley Laboratory, Report No. LBL-20107, 1985.

APPENDIX F

THE ELECTRICAL ANALOG: RC NETWORKS FOR HEAT TRANSFER CALCULATIONS

Frederick C. Winkelmann
Applied Science Division
Lawrence Berkeley Laboratory
Berkeley, CA 94720

Electrical network theory has proven to be a powerful tool in the solution of the complex heat transfer problems encountered in the analysis of heating and cooling of buildings. For those not familiar with the use of RC networks in heat transfer, we give here a brief introduction to the analogy between electricity flow and heat flow, and show some simple examples of RC networks applied to heat conduction through building walls.

We begin with Ohm's law, which states that a potential difference ΔV (volts) across a resistance R_e (ohms) produces a current I given by

$$I = \frac{\Delta V}{R_e} \quad , \tag{1}$$

with the current flowing from higher to lower potential.

The corresponding statement for one-dimensional steady-state heat conduction through a wall is that a temperature difference ΔT (°K) across a thermal resistance $R_t (m^2 \,{}^\circ K/W)$ produces a heat flux $Q(W/m^2)$ given by

$$Q = \frac{\Delta T}{R_t} \quad , \tag{2}$$

where the direction of heat flow is from higher to lower temperature.

For problems involving non-steady flow, capacitive effects may be important. A time-varying voltage dV/dt (volts/s) across a capacitance C_e (farads) produces a current I into the capacitor given by

$$I = C_e \frac{dV}{dt} \quad . \tag{3}$$

Analagously, the heat flow $Q(W)$ into a thermal capacitance $C_t (J/ \,{}^\circ K)$ subject to a time-varying temperature $T(\,{}^\circ K)$ is given by

$$Q = C_t \frac{dT}{dt} \quad . \tag{4}$$

This work is supported by the Assistant Secretary for Conservation and Renewable Energy, Office of Buildings Energy Research and Development, Building Systems Division of the U. S. Department of Energy under Contract No. DE-AC03-76SF00098.

0094-243X/85/1350650-5$3.00 Copyright 1985 American Institute of Physics

The origin of the analogy between (1) and (2) and between (3) and (4) can be demonstrated by examining the basic differential equations for the flow of electricity and heat (for simplicity we restrict ourselves to one-dimensional flow although all of the arguments apply for two- or three-dimensional flow as well). For electricity, the ***diffusion equation*** relates the space and time variation of the potential, V:

$$\frac{\partial^2 V}{\partial x^2} = \frac{\rho c_e}{k_e}\frac{\partial V}{\partial t} , \tag{5}$$

where ρ is the density of the medium (kg/m^3), c_e is the specific capacitance (farads/kg), and k_e is the conductivity $[(ohm-m)^{-1}]$. (The inverse of k_e is the electrical resistivity, $r_e[ohm-m]$.) The quantity $\alpha_e \equiv k_e/\rho c_e$ is called the ***electrical diffusivity.***

Ohm's law establishes the relationship between current density $I[amp/(m^2$ *normal to flow direction*$)]$ and voltage gradient:

$$I = -k_e \frac{\partial V}{\partial x} . \tag{6}$$

For heat conduction, the heat diffusion equation relating the space and time variation of temperature is

$$\frac{\partial^2 T}{\partial x^2} = \frac{\rho c_t}{k_t}\frac{\partial T}{\partial t} , \tag{7}$$

where ρ is the density of the medium (kg/m^3), c_t is the specific thermal capacitance $(J/kg-{}^{\circ}K)$, and k_t is the thermal conductivity $[W/{}^{\circ}K-m]$. (The inverse of k_t is the thermal resistivity, $r_e[{}^{\circ}K-m/W]$.) The quantity $\alpha_t \equiv k_t/\rho c_t$ is the ***thermal diffusivity.***

The equivalent of Ohm's law is the law of Biot-Fourier which relates the heat flux $Q[W/(m^2$ ***normal to flow direction***$)]$ and the temperature gradient:

$$Q = -k_t \frac{\partial T}{\partial x} . \tag{8}$$

Comparison of eqns. (5) and (7), and eqns. (6) and (8) establishes the following analogy:

Electrical Quantity	Thermal Quantity
V	T
I	Q
c_e	c_t
k_e	k_t
r_e	r_t

A simple problem illustrating the use of the analogy is the following: determine the steady-state heat flux through a homogeneous wall of thickness L_w and thermal conductivity k_w if the outside air temperature is T_o and the inside air temperature is T_i. Figure 1 shows the network for this problem. The wall has a lumped resistance $R_w = L_w / k_w [m^2 \,°K/W]$. The resistance of the outside surface air film has been decomposed into a radiative part $R_{o,r}$ and a convective part $R_{o,c}$. Because the radiative and convective heat exchange at the wall surface are independent, their resistances are taken in parallel. A similar situation occurs at the inside wall surface.

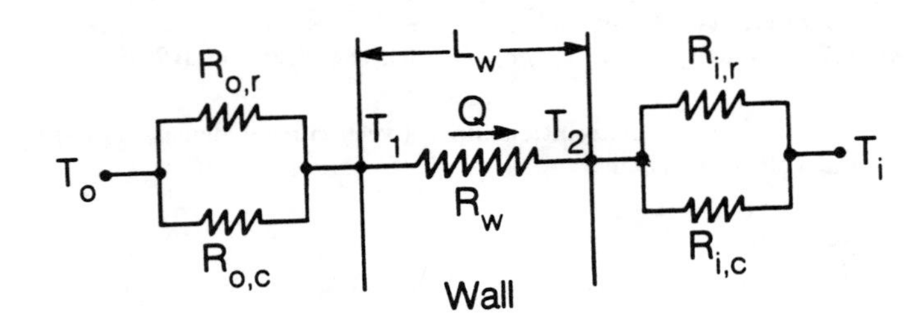

Fig. 1: Resistance network for steady-state heat conduction through a wall.

A temperature difference $T_o - T_i$ occurs across a total resistance R given by $[R_{o,r}$ in parallel with $R_{o,c}]$ in series with R_w in series with $[R_{i,r}$ in parallel with $R_{i,c}]$. Thus,

$$R = R_o + R_w + R_i \tag{9}$$

where

$$R_o = \frac{R_{o,r}\, R_{o,c}}{R_{o,r} + R_{o,c}} \tag{10}$$

and

$$R_i = \frac{R_{i,r}\, R_{i,c}}{R_{i,r} + R_{i,c}} \tag{11}$$

"Ohm's law" then gives the heat flow

$$Q = \frac{T_o - T_i}{R} \; [W/m^2] \,. \tag{12}$$

Given Q, one may easily calculate the wall surface temperatures T_1 and T_2. Using "Ohm's law" across R_o and R_i, and using the fact that the flow through R_o, R_w, and R_i is the same and equal to Q, we have

$$Q = \frac{T_o - T_1}{R_o} = \frac{T_2 - T_i}{R_i} \,. \tag{13}$$

Using eqn. (12) for Q then gives

$$T_1 = T_o \left(1 - \frac{R_o}{R}\right) + T_i \frac{R_o}{R} \tag{14}$$

$$T_2 = T_i \left(1 - \frac{R_i}{R}\right) + T_o \frac{R_i}{R} \quad . \tag{15}$$

For non-steady-state conduction $\left(\partial T/\partial t \neq 0\right)$, the right-hand side of eqn. (7) is non-zero and capacitive effects in the wall must be considered. The network of Fig. 1 becomes that shown in Fig. 2. Here C_w is the lumped thermal capacitance of the wall. Figure 2 also shows a "current source" I_s representing solar radiation $[W/m^2]$ absorbed by the outside surface of the wall.

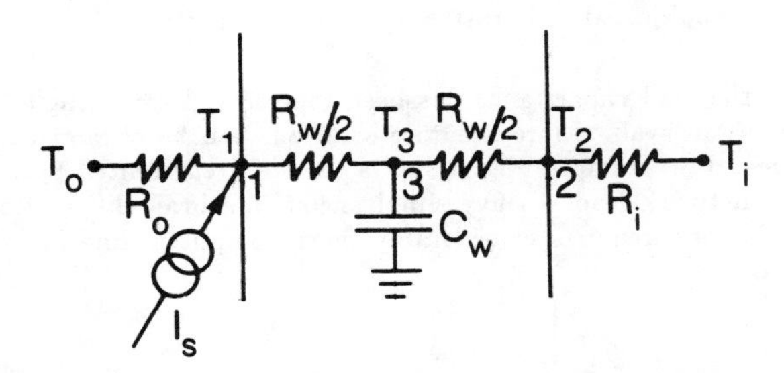

Fig. 2: RC network for transient heat conduction through a wall.

We now want to find T_1 and T_2 as a function of time given time-dependent values of T_o, T_i and I_s. One solution approach is to apply Kirchoff's Law (which, for electrical networks, states that the sum of the currents entering a node is zero) to nodes 1, 2, and 3. This yields three equations in the three unknowns, T_1, T_2, and T_3:

$$\textit{at node 1:} \quad \frac{T_o - T_1}{R_o} + \frac{T_3 - T_1}{R_w/2} + I_s = 0 \tag{16}$$

$$\textit{at node 2:} \quad \frac{T_3 - T_2}{R_w/2} + \frac{T_i - T_2}{R_i} = 0 \tag{17}$$

$$\textit{at node 3:} \quad \frac{T_1 - T_3}{R_w/2} + \frac{T_2 - T_3}{R_w/2} - C_w \frac{dT_3}{dt} = 0 \tag{18}$$

These can be written in matrix form as

$$[A]\,[T] + [B]\,[\dot{T}] = [D] \tag{19}$$

$$with\ [T] = \begin{bmatrix} T_1 \\ T_2 \\ T_3 \end{bmatrix} \quad and \quad [\dot{T}] = \begin{bmatrix} dT_1/dt \\ dT_2/dt \\ dT_3/dt \end{bmatrix} .$$

Coupled algebraic and differential equations like eqns. (17)-(19) are quite amenable to computer solution using a number of different techniques, such as Laplace Transform, Z-transform, frequency analysis (Fourier series), transfer functions, or finite-differences. Although most networks satisfy linear differential equations, some physical processes, such as free convection, give resistances which are temperature dependent. (For example, the convective air film resistance $R_{i,c}$ in Fig. 1 has a weak dependence on temperature of the form $| T_2 - T_i |^n$ where $n \approx 0.3$.) These cases produce non-linear equations, which can either be linearized about an operating point or subjected to iterative solution techniques.

In Fig. 2 the wall capacitance has been represented by a single lumped value located at the mid-point of the wall. More accurate solutions can be obtained by sub-dividing the wall into more nodes. For example, Fig. 3 shows the wall represented by three capacitors and four resistors. This network leads to five simultaneous algebraic-differential equations in the unknowns T_1 to T_5, and requires considerably more computer time for a solution than the one-capacitor network.

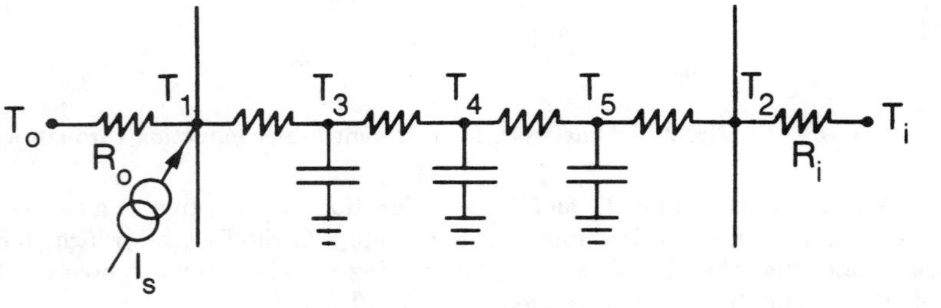

Fig. 3: Five-node RC network representation for transient heat conduction through a wall.

The inside air node (T_i) in Figs. 1, 2, and 3 is, of course, coupled not only to the wall in question, but to all other walls (and floor and ceiling) in the room as well, and each of these walls may have from two to ten or more nodes depending on wall thickness and capacitance. Or it may be desired to simulate two- or three-dimensional heat flow, in which case nodes in the **y** and **z** directions, as well as in the **x** direction, are required in each wall. Furthermore, additional resistors may be needed to represent convective heat transfer between rooms, infiltration of outside air, and ventilation through open windows. Thus, in actual buildings the total number of nodes, and therefore the number of simultaneous equations can be in the hundreds or even thousands. This means that very efficient solution techniques (usually involving partitioning of large, sparse matrices) are required. For a discussion of such techniques, and for examples of the RC network approach applied to simulation of heating and cooling equipment, the reader is referred to J. A. Clarke, *Energy Simulation in Building Design*, Adam Hilger Ltd., Boston 1985.

APPENDIX G

AIR INFILTRATION IN BUILDINGS

Max Sherman

Applied Science Division
Lawrence Berkeley Laboratory
University of California
Berkeley, CA 94720

INTRODUCTION

Infiltration, the naturally-induced air flow through the building envelope, typically accounts for one-third of the energy load while at the same time acting as the primary mechanism for removing internally-generated pollutants and thus assuring adequate indoor air quality. To minimize energy use one may wish to minimize infiltration, but to insure adequate indoor air quality one may wish to increase infiltration. This apparent contradiction implies that the prediction of infiltration for a new or existing structure can be one of the more critical calculations an engineer or architect can make.

CALCULATION OF INFILTRATION USING THE LBL MODEL

To predict air infiltration in buildings it is necessary to have a model of the process that is simple enough to be practical, but one that captures the essential processes in effect. The LBL infiltration model is based on physical simplifications of the many effects that enter into the process of air infiltration. The considerations behind these simplifications have been examined in great detail in previous works and will be summarized in the sections to follow. The original work[1] was a Ph.D. thesis in physics; the first complete presentation[2] in the professional literature led to its inclusion in the *1981 ASHRAE Handbook of Fundamentals*, and was followed closely by its first international presentation.[3] The most current version[4] should be consulted for practical uses.

Infiltration is primarily a flow phenonemon; it is the (flow) response of the building to the pressures induced by the driving forces, which may either be mechanically-induced[‡] or weather-induced. There are two primary weather mechanisms which induce infiltration: wind force and stack force (i.e. temperature-induced density differences). The approach taken in the LBL model is to calculate the

This work was supported by the Assistant Secretary for Conservation and Renewable Energy, Office of Building Energy Research and Development, Building Systems Division, of the U.S. Department of Energy under Contract No. DE-AC03-76SF00098.

0094-243X/85/1350655-8$3.00 Copyright 1985 American Institute of Physics

infiltration induced by each driving force *assuming the other driving forces are absent* and then combine them.

Stack-Induced Infiltration

The stack effect is caused by a difference in temperature between the air inside and the air outside the building. This temperature difference causes a density difference and this buoyancy creates a pressure gradient along any vertical boundary.

The leaks in a building are distributed over the entire envelope, thus a detailed summation would be required to determine the flow at each point on the envelope. To avoid this level of detail, we have grouped the envelope leakage into three categories: floor, wall, and ceiling leakage area. Within each area we assume that the leaks are evenly distributed.

$$Q_s = L\ f_s\ \sqrt{|\Delta T|} \tag{1}$$

where:

Q_s	is the stack-induced infiltration $[m^3/s]$
L	is the effective leakage area $[m^2]$
f_s	is the stack parameter $[m/s\text{-}K^{1/2}]$
ΔT	is the inside-outside temperature difference $[K]$

Wind-Induced Infiltration

When wind flows around a building, it induces pressure differences across the external faces of the envelope. These pressure differences are proportional to the local wind speed and the shielding coefficient for the building environment. (We have calculated the generalized shielding coefficient for five degrees of obstruction around the building.)

Most wind data is taken from a weather tower not necessarily at the height of the building. The measured wind speed must be converted from a weather station into a local wind speed for our model. We use a method that uses two terrain parameters to describe the wind profile and do the necessary conversion; these terrain parameters are included in the description and definition of the wind parameter.

$$Q_w = L\ f_w\ v \tag{2}$$

where:

Q_w	is the wind-induced infiltration $[m^3/s]$
L	is the effective leakage area $[m^2]$
f_w	is the wind parameter $[-]$
v	is the unobstructed speed of the wind $[m/s]$

‡ Although the references discuss how to include mechanically-induced infiltration in the total, this summary will not.

Superposition

Although we may be able to calculate the pressures induced across the shell of the building from each of the two driving forces, it is not a simple matter to sum the point pressures on the surface of the building caused by both acting in concert. To simplify the point pressure problem, we can calculate the wind-induced (Q_w) and stack-induced (Q_s) infiltration independently and then combine them. We cannot, however, simply add them to find the total infiltration, rather we assume the leaks to be simple orifice flow and infer that we approximate the combined effect by adding the flows in quadrature:

$$Q_{weather} = \sqrt{Q_w^2 + Q_s^2} \tag{3}$$

where

$Q_{weather}$ is the combined infiltration $[m^3/s]$

Like the effective leakage area, L, the parameters f_s and f_w are both time and weather-independent, but do depend on the the building and its environment. While the effective leakage area can vary by over an order of magnitude from house the house, the wind and stack parameters tend to vary less than a factor of two.

SPECIFIC INFILTRATION

If we combine the stack and wind induced terms together, the total (weather-induced) infiltration can be expressed as follows:

$$Q_{weather} = L \; s \tag{4.1}$$

$$s \equiv \sqrt{f_w^2 v^2 + f_s^2 \, | \Delta T \, |} \tag{4.2}$$

where

s is the specific infiltration $[m/s]$.

To the extent that we can ignore variations in the wind and stack parameters, the specific infiltration is dependent only upon weather and not on the leakage of the building. As such the specific infiltration is a good indicator of the severity of climate relative to infiltration. Accordingly, we have calculated the heating season average of the specific infiltration for 60 cities across the country and then interpolated the results to generate a contour map. Figure 1 (note unit change) presents these results. Once the leakage area has been determined (as described below) a reasonable estimate of the seasonal infiltration can be made by multiplying the specific infiltration by the leakage area (to convert to a volume flow rate) and then, if desired, dividing by the building volume (to convert to air changes per hour‡).

‡ Air changes per hour, ACH, is a commonly used unit. It is calculated by dividing the total infiltration by the building volume (i.e. $ACH = L \; s / V$)

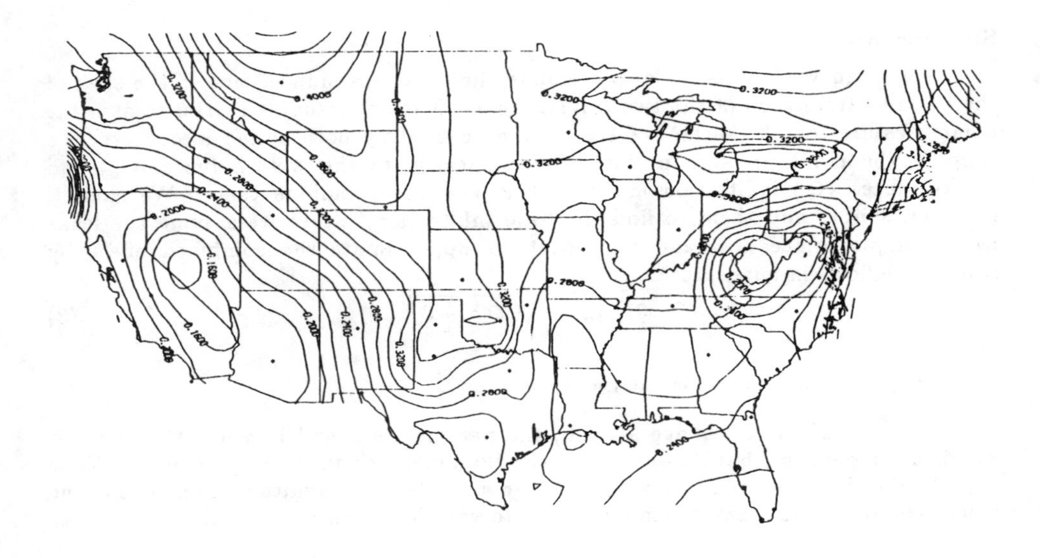

Heating Season Infiltration $(m^3/hr\text{-}cm^2)$

XBL 8212-12076

Figure 1: A contour map of the specific infiltration(s) over the heating season throughout the U.S. varying from lows below 0.2 in the southwest to highs near 0.4 in the northern tier; contour steps are 0.02.

DETERMINATION OF EFFECTIVE LEAKAGE AREA

Effective leakage area provides a simplified description of the process of air leakage through a building envelope under specified pressure; it is a quantity conceptually equivalent to the sum of the areas of all the cracks and holes in the building envelope through which air is able to pass. Air flow through the building envelope is a combination of viscous (i.e. laminar) flow and turbulent flow. The former is proportional to the outside to inside pressure difference (ΔP) while the latter is proportional to the square root of ΔP. Empirically it has been found that air flow through the envelope can be characterized by the equation:

$$Q = K\ (\Delta P)^n \tag{5}$$

where:

Q	is the air flow induced through the envelope $[m^3/s]$
K	is the leakage coefficient $[m^3/s - Pa^n]$
ΔP	is the applied pressure difference $[Pa]$
n	is the leakage exponent [-].

The effective leakage area is defined assuming that in the pressure range characteristic of natural infiltration (-10 to +10 Pa) the flow versus pressure behavior of a building more closely resembles square-root (turbulent) than viscous flow. Because four Pascals is representative of the typical pressures inducing infiltration, it is common to use it as the reference pressure and assume orifice (turbulent) flow (n=1/2). Therefore, the leakage area of the envelope is defined at four pascals:

$$L \equiv \frac{Q\,(\Delta P_r\,)}{\sqrt{\frac{2}{\rho}\Delta P_r}} \tag{6}$$

where:

L	is the effective leakage area $[m^2]$
$Q\,(\Delta P_r\,)$	is the induced air flow at the $(4\ Pa\,)$ reference pressure $[m^3/s\,]$
ρ	is the density of air $[1.2\ kg\ /m^3]$
ΔP_r	is the reference pressure difference $[4\ Pa\,]$

Effective leakage area can be defined for pressurization of the building (i.e. higher internal than external pressure) or for depressurization (i.e. evacuation); depending on circumstances, one or the other or the average is used. For most U.S. housing stock the effective leakage area will run between 300 cm^2 and 1000 cm^2; super tight houses have been measured as low as 50 cm^2 and old, leaky houses have been measured as high as 3000 cm^2.

Fan Pressurization

The most common method for measuring the effective leakage area of a building is called ***fan pressurization***. This method uses a blower door, a door-mounted, variable speed fan capable of moving large volumes (up to 7000 m^3/hr) of air into or out of a structure, and a differential pressure gauge such as an inclined manometer. By supplying a constant air flow with the fan, a pressure difference across the building envelope can be maintained. When the differential pressure is held constant, all air flowing through the fan must also be flowing through the building envelope.

To make the fan pressurization measurement, the blower door is sealed into an exterior doorway, and the pressure gauge is set up with one pressure tap placed outside in a location protected from the wind. The inside pressure tap should be placed out of the direct flow path of the fan. All exterior doors and windows should be closed, and if a fireplace and/or a wood-burning stove are present, their dampers are closed to prevent soot from entering the building during depressurization. All interior doors, except closet doors, should be open. The blower door is then used to blow air into (pressurize) and to suck air out of (depressurize) the building at a series of fixed pressure differentials (e.g., from 10 to 70 Pa at 10 Pa intervals). The air flow (or some quantity directly related to it) and the pressure difference are measured at each point. The resulting pressure versus air flow curves for both pressurization and depressurization are used to find the effective leakage area of the building by fitting the data to the air flow equation using a regression technique.

AC Pressurization

Another technique for measuring the effective leakage area, called AC pressurization[5], differs from fan pressurization (DC pressurization) by creating a periodic pressure difference across the building envelope that can be distinguished from naturally occurring pressure fluctuations. Assuming that there are no leaks in the building envelope and that the structure is rigid, the change in pressure can be precisely determined from the structure's volume and the piston's displacement. The airtightness of a building affects the pressure change from the periodic volume change, including both the amplitude and phase of the pressure change. Therefore, any deviation from the pressure change for a sealed building can be attributed to leakage through the envelope. The measured volume change and pressure response can thus be used to calculate the airflow through the envelope. The AC pressurization apparatus includes components that perform four basic functions: (1) volume change, (2) displacement monitoring, (3) pressure measurement, and (4) analysis/control. Several options for accomplishing each of these functions have been investigated.

The pressure-flow relationships used for AC pressurization measurements are substantially more complex than those for DC pressurization. Because the drive component provides a periodic volume change, and thereby induces a periodic pressure response, the flow through the envelope must be determined from the continuity equation for a compressible medium:

$$Q + \dot{V}_d + c\ \dot{P} = 0 \tag{7}$$

where:

Q	is the air flow through the envelope $[m^3/s]$
$\dot{V}_d$	is the induced rate of change in interior volume $[m^3/s]$
c	is the effective capacity of the envelope $[m^3/Pa]$
$\dot{P}$	is the resultant rate of change of the internal pressure $[Pa/s]$

Theoretically, this expression could be used to calculate the instantaneous air flow (Q) directly from the measured volume and pressure changes. In practice, this is not possible because of the accuracies required for both the estimation of the capacity, c, and the measurement of the pressure (especially its time derivative). However, because all the terms are periodic (i.e., AC), we can use *synchronous detection* (i.e., phase-sensitive detection) to analyze the data and increase its accuracy. Specifically, we lower our precision requirements by extracting the component that is in phase with the pressure signal:

$$\left\{ Q\,\Delta P \right\} + \left\{ \dot{V}_d\,\Delta P \right\} + c \left\{ \dot{P}\,\Delta P \right\} = 0 \tag{8}$$

where:

ΔP	is the inside-outside pressure difference $[Pa]$
$\left\{ \cdots \right\}$	indicates a cycle average.

Because the pressure signal is periodic and the outside pressure is independent of our drive signal, the term in $\dot{P}$ vanishes. Inserting the power-law definition of the air flow, simplifying, and solving for the leakage area yields the following:

$$L = -\sqrt{\frac{\rho}{2P_r}}\ \frac{\left\{ \dot{V}_d \dfrac{\Delta P}{P_r} \right\}}{\left\{ \left| \dfrac{\Delta P}{P_r} \right|^{n+1} \right\}} \tag{9}$$

This basic equation of AC pressurization is used to determine the leakage area directly from the measured rate of volume change (from the piston velocity) and pressure response.

This technique holds great promise but it is still in the development stages. Comparisons with conventional fan pressurization show reasonable agreement, but indicate a tendency for AC pressurization to treat large leaks (e.g. open flues or chimneys) differently from the more common leaks. This fact suggests that the technique may be suitable as a probe to determine the character of leaks in a building.

EXAMPLE

In this section we will go through an example of how fan pressurization and the LBL model may be used to predict infiltration. The example house is a single-story, ranch-style, (tightened) California house with a ventilated crawl-space, whose vital statistics are as follows:‡ Floor area = 100 m^2; Volume = 240 m^3; s = 0.25 $m^3/hr - cm^2$; f_s = 0.13 $m/s - K^{1/2}$; and f_w = 0.12.

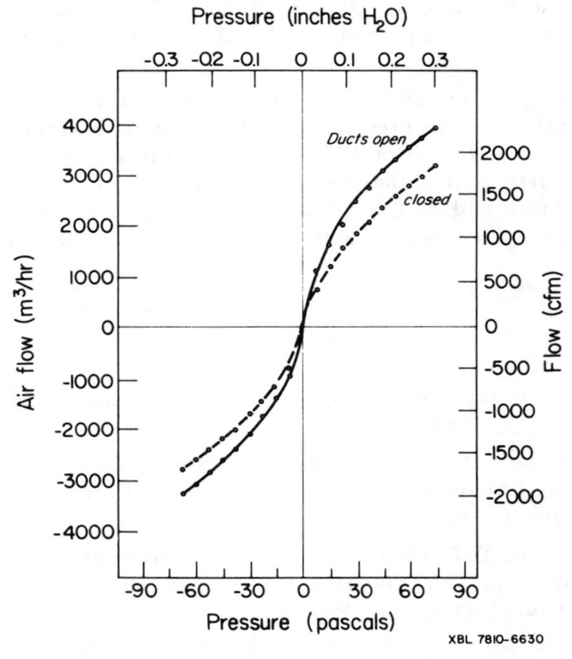

Figure 2: A plot of the flow rate through the building envelope vs. applied pressure for the example house.

To determine the leakage area, a fan pressurization test was done on the example house. The analysis of the results (for the ducts closed condition) yields a value of 400 cm^2 of leakage area. Combining the seasonal specific infiltration (0.25 $m^3/hr - cm^2$) with the effective leakage area (400 cm^2) yields a seasonal average infiltration of 100 m^3/hr or a seasonal average air change rate of 0.42 per hour.

The equations described above can also be used to find the instantaneous infiltration for a particular wind speed and temperature difference. We take for example a cool, blustery California day (i.e. 7 m/s wind speed and a 16K inside-outside temperature difference). The stack-induced infiltration (in cubic meters per hour) is the product of the leakage area (400 cm^2), the stack parameter (0.047 $m^3/hr - cm^2 - K^{1/2}$) and the square root of the temperature difference (16K) yielding 75 m^3/hr. Similarly, the wind-induced infiltration (in cubic meters per hour) is the product of the leakage area (400 cm^2), the wind parameter (0.043 $m^2 - s/cm^2 - hr$) and the wind speed (7 m/s) yielding 120 m^3/hr. Combining the stack-induced and wind-induced infiltration together with quadrature yields 142 m^3/hr or 0.59 air changes per hour.

CONCLUSION

Prior to the introduction of the LBL model, prediction of infiltration was at best ad hoc. Engineering estimates were crude and generally only used in sizing equipment for which accuracy and reproducibility were not important. The LBL model begins with the fundamental physics of air flow and then uses simplifying assumptions to create a simple, physical model of infiltration, which is accurate to 20%.

The effective leakage area and local weather data can be used as inputs to the LBL infiltration model. The fan pressurization technique is a simple method that is used to measure the air tightness of a building. AC pressurization is a potentially superior method for determining the effective leakage area of a building. Both the pressurization techniques and LBL infiltration model can be used in a variety of ways: energy estimates, air quality calculations, consensus standards, and code requirements.

REFERENCES

(1) M.H. Sherman, Air Infiltration in Buildings. Ph.D Thesis, University of California, Berkeley, CA, 1980.

(2) M.H. Sherman and D.T. Grimsrud, "Infiltration-Pressurization Correlation: Simplified Physical Modeling." *ASHRAE Trans.* 86 (II), pp. 778-807, 1980. Lawrence Berkeley Laboratory Report, LBL-10163.

(3) M.H. Sherman and D.T. Grimsrud, "The Measurement of Infiltration Using Fan Pressurization and Weather Data." Lawrence Berkeley Laboratory Report, LBL-10852, EEB-epb-81-9. In *Proceedings of the First International Air Infiltration Centre Conference*, London, UK, October 1980.

(4) M.H. Sherman, and M.P. Modera, *Comparison of Predicted and Measured Infiltration Using the LBL Infiltration Model*, accepted for publication in ASTM STP-72, Measured Air Leakage, April 1984, Lawrence Berkeley Laboratory Report, LBL-17001.

(5) M.P. Modera, and M.H. Sherman, *AC Pressurization: A New Technique for Leakage Area Measurement*, accepted for publication in *ASHRAE Trans.* 91 (II), June 1985, Lawrence Berkeley Laboratory Report, LBL-18395.

APPENDIX H
RESIDENTIAL VENTILATION AND HEAT RECOVERY WITH AIR-TO-AIR HEAT EXCHANGERS

William Fisk
Lawrence Berkeley Laboratory
Berkeley, CA 94720

A major disadvantage of using ventilation to control indoor pollutant concentrations is the energy required to heat or cool the ventilation air. To overcome this disadvantage, techniques of ventilation have been developed in which energy is recovered from the exhausted ventilation air. One such technique is to provide balanced mechanical ventilation and recover energy in an air-to-air heat exchanger. Using fans, roughly equal amounts of air are supplied to and exhausted from the structure. In an air-to-air heat exchanger, heat is transferred from the warmer to the cooler airstream, ideally without any mixing between the two streams. Due to the transfer of heat, the incoming air is preheated in the winter and precooled in the summer, thus reducing the load on the heating or air conditioning system. In some air-to-air heat exchangers, only heat is transferred between airstreams; these are often referred to as "sensible exchangers". In other exchangers, often called "enthalpy exchangers", both heat and moisture are transferred between airstreams. Air-to-air heat exchangers have been used for some time in industry and commercial buildings and a large variety of designs are available. In recent years, the use of mechanical ventilation systems with air-to-air heat exchangers has become increasingly common in residences.

A residential heat exchanger (Fig. 1) generally consists of a core, where heat is transferred between the two airstreams, a pair of fans, and a case that contains the core, fans, and fittings for attachment to ductwork. Some models also contain coarse air filters within the case. Most available models of residential air-to-air heat exchangers are designed to be attached to a ductwork system. Air is withdrawn from various locations throughout the residence and supplied to other locations via the duct system. Other models can be mounted in window frames or through an opening in the wall and utilize no ductwork.

0094-243X/85/1350663-3$3.00 Copyright 1985 American Institute of Physics

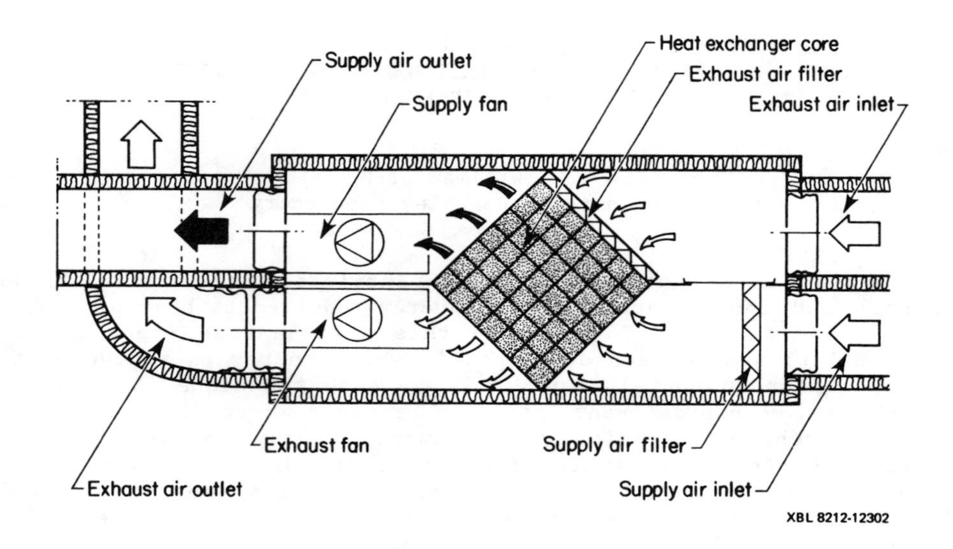

Figure 1. Schematic Diagram of an Air-to-Air Heat Exchanger.

The capability of various residential air-to-air heat exchangers to transfer (recover) heat has been investigated in both laboratory and field studies.[1-4] The most commonly quoted performance parameter is the temperature efficiency which is defined by the equation

$$\epsilon = \frac{T_{c,o} - T_{c,i}}{T_{h,i} - T_{c,i}} \times 100\% \quad (1)$$

where T is an airstream temperature, and subscripts c, h, i, and o refer to the cold airstream, warm airstream, into the heat exchanger, and out of the heat exchanger, respectively. The temperature efficiency is a measure of the heat exchanger's ability to preheat or precool the incoming air and it is desirable for it to be as high as possible.

The temperature efficiency of various residential air-to-air heat exchangers generally falls within the range of 45 to 85% depending primarily on the physical characteristics of the heat exchanger and, to a lesser degree, the operating conditions. Heat exchanger performance can be greatly reduced by freezing of moisture within the core. Freezing is initiated[5] when outdoor temperatures are as high as -3 to -12 °C (27 to 10 °F), therefore, a freeze protection system[6] that periodically defrosts the heat exchanger or preheats the outdoor air is required in many climates.

The economic benefit or cost of providing ventilation with air-to-air heat exchangers, in contrast to relying on the same time-averaged amount of natural infiltration, depends on numerous factors including the following: 1) the initial cost of the heat exchanger and its installation; 2) the amount of ventilation provided; 3) the temperature efficiency and fan power requirements of the heat exchanger; 4) climate, 5) the cost of fuel or electricity used for heating; 6) the lifetime of the heat exchanger; and 7) periodic maintenance costs. In previous economic analyses,[7,8] the payback periods have ranged from five to greater than thirty years, however, these analyses have not accounted for any benefits that result from better control of ventilation rate when ventilation is provided mechanically.

Increasing the ventilation rate of a residence by using an air-to-air heat exchanger will reduce indoor pollutant concentrations to a highly variable degree. A common assumption, in buildings where the indoor pollutant concentration greatly exceeds the outdoor concentration, is that indoor pollutant concentrations are inversely proportional to the total ventilation rate. However, it has been shown that the relationship between ventilation rate and indoor pollutant concentration is often much more complex than the case of simple inverse dependence.[3,9] The presence of significant pollutant removal processes other than ventilation and couplings between ventilation rate and indoor pollutant source strengths make it necessary to use more complex models to relate ventilation rate and indoor pollutant concentrations.[9]

In summary, the energy performance, economics, and air quality impacts of providing ventilation with air-to-air heat exchangers all vary widely and depend on numerous factors. The suitability of this technique of ventilation should be assessed on a case-by-case basis. Builders and homeowners are using air-to-air heat exchangers with increasing frequency.

REFERENCES

1. W.J. Fisk, G.D. Roseme, and C.D. Hollowell, Lawrence Berkeley Laboratory Report, LBL-11793, Berkeley, CA, 1980.
2. W.J. Fisk, et al., Lawrence Berkeley Laboratory Report, LBL-12559, Berkeley, CA, 1981.
3. F.J. Offermann, et al., Lawrence Berkeley Laboratory Report, LBL-13100, 1982.
4. J.R. Hughes, Solar Age 10(1), 1985, p. 88.
5 W.J. Fisk, et al., ASHRAE Trans. 91, Pt 1B, 1985, p. 145.
6. W.J. Fisk, et al., ASHRAE Trans. 91, Pt 1B, 1985, p. 159.
7. W.J. Fisk and I. Turiel, Energy and Buildings, 5, 1983, p. 197.
8. I. Turiel, W.J. Fisk, and M. Seedall, Energy 8 (5), p. 323.
9. W.J. Fisk, Lawrence Berkeley Laboratory Report, LBL-18875, 1985, submitted to Building and Environment.

APPENDIX I

DISTRICT HEATING - SOME SWEDISH EXPERIENCES

Enno Abel
Chalmers University of Technology
S-41296 Gothenburg, SWEDEN

By building up a district heating system in a city, a number of relatively small individual heat loads can be joined together in one big centralized load. By that the number of heat source alternatives which are costeffective is increased and a supply system can be built up that is both thermodynamically efficient and gives so low energy costs that the capital investments in the district heating system cam be repaid. In general, district heating of buildings can be considered a serious alternative if the outdoor temperature is at or below freezing more than about 1,000 hours a year.

The local conditions and the size of the district heating plant determine the applicability of district heating. In general however at least some of the advantages below can be exploited.

1. Less expensive fuels can be used.
2. The maintenance and the operation can be made more efficient and less costly.
3. Waste heat from industrial processes can be utilized.
4. Heat from waste incineration plants can be used.
5. Large heatpumps might be used, if the cost of electricity is not too high.
6. The cogeneration of electricity and heat can be utilized.

For waste heat from industry and heat from waste incineration plants to be costeffective, it is necessary that the utilization times be very high. Due to the shape of the utilization curve of the heat load, the thermal power of the waste heat source or the waste incineration plant must be small compared with the connected heat load. Usually the ratio these loads ought to be 1:5 or less.

The cogeneration of electricity and heat would be one of the most efficient applications of district heating systems. However the district heating plant must be quite large to enable the cost effective cogeneration of heat and electricity. Due to the shape of the heat load, the power plant cannot possibly be dimensioned for a condenser heat flow bigger than one half of the peak load of the heating system. Furthermore, the thermal efficiency of the steam turbine process will only be about 30% due to the relatively high condensing temperature of up to about 120°C. Thus, the electricity output cannot be much more than about 25% of the peak heat load of the district heating plant.

0094-243X/85/1350666-5$3.00 Copyright 1985 American Institute of Physics

A cost effective power plant can hardly have an electricity output lower than 15 - 20 MW which dictates a minimum peak load of the district heating system of 60 - 80 MW thermal.

A general idea about the minimum size of a district heating system with cogeneration, can be obtained from Swedish statistics. Sweden, along with Denmark, has the most experience with district heating. Sweden has more than 80 district heating plants in operation with the connected thermal loads ranging between 25 MW and 2,000 MW. In all the systems hot water is used for heat distribution. In Fig.1 the total length of the double pipe coundit (outlet and return) is plotted versus the connected thermal power for a number of smaller Swedish district heating plants.

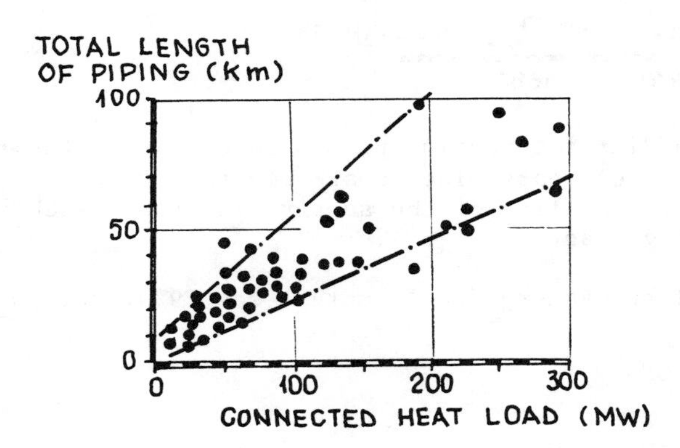

Fig.1 Total length of piping for a sample of existing Swedish district heating systems.

If the 60 - 80 MW load mentioned above is used as a basis, it can be seen from the diagram that the district heating system usually has to include about 10 - 30 km (5 - 20 miles) of piping before it becomes possible to consider cogeneration. The initial stage of developing a district heating system represents a major investment that needs careful planning. It could be preferable to start with the building of a number of local centralized plants in areas of high energy density. Subsequently these local systems can be integrated in a common centralized system, big enough for cogeneration.

From an economic point of view, district heating reduces the total energy cost by replacing a part of the fuel and operation costs with capital costs. These capital costs have to be repaid even in time of decreasing energy use. After a building has been connected to the district heating system, further measures of heat conservation must be very cost effective as the capital costs have to be repaid even if the use of energy decreases.

Fig.2 shows the development of district heating systems in Sweden. The drop in electricity production by cogeneration in the early 80's resulted from good availability of hydro and nuclear power. The actual capacity of the cogeneration plants is about 7 BkWh (2.3 GW), but this capacity is only partly used for the time being.

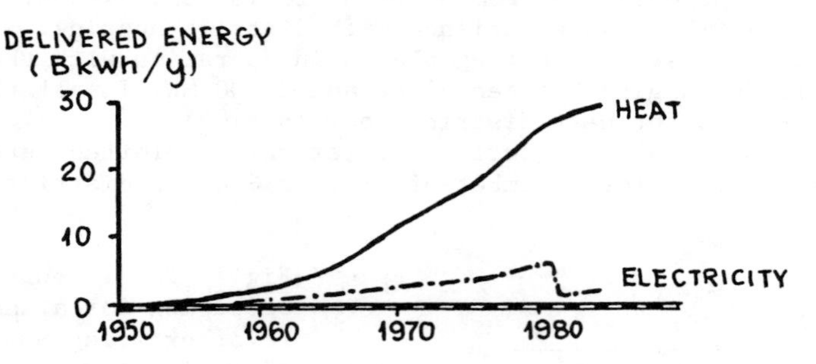

Fig.2 The growth of district heating in Sweden

Fig.3 shows how the Swedish district heating plants have changed their fuel usage during the last three years. The remarkable reduction in the use of oil illustrates the flexibility of the supply side that is characteristic for district heating plants.

The average efficency (delivered energy/fuel energy) was 79% 1982, 80% 1983 and 81% 1984.

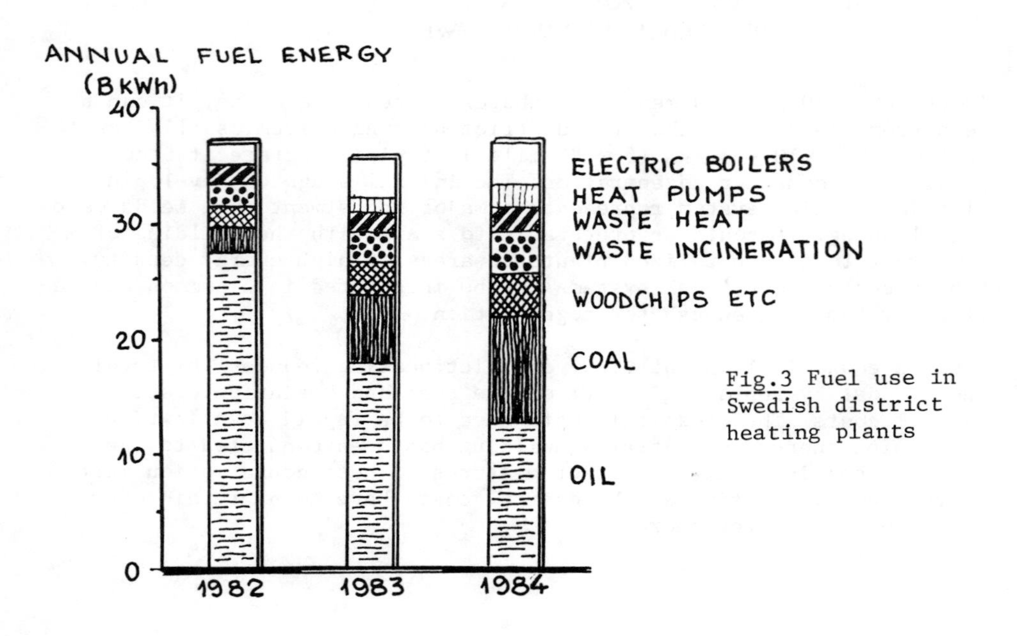

Fig.3 Fuel use in Swedish district heating plants

The way of doing it practiced by the utilities is exemplified below. Fig.4-6 show the changes and the planned changes of the heat supply systems for the city of Gothenburg, a city of 400,000 inhabitants. In 1984 the amount of heat delivered into the 300 km piping system was 2.5 BkWh. The total heat loss in the distribution system was 7%.

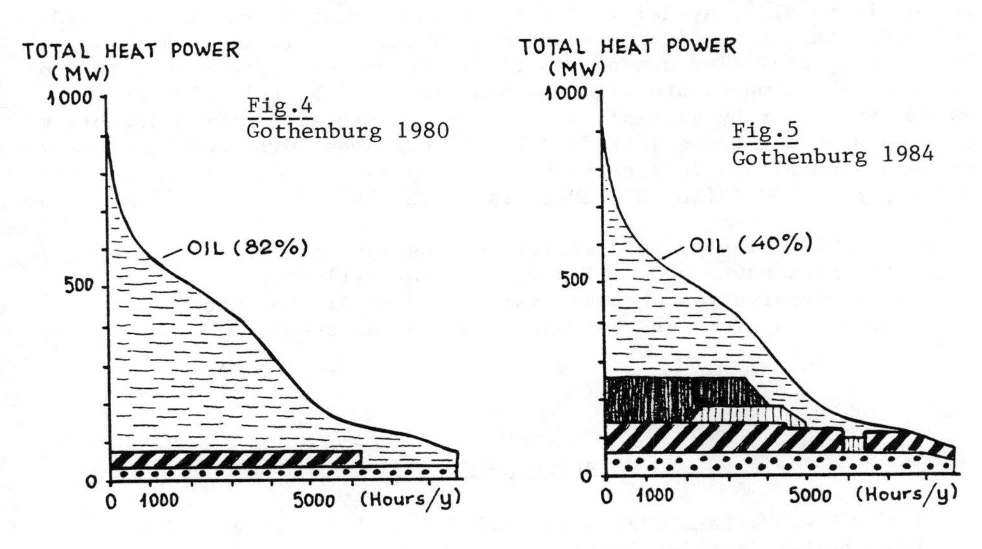

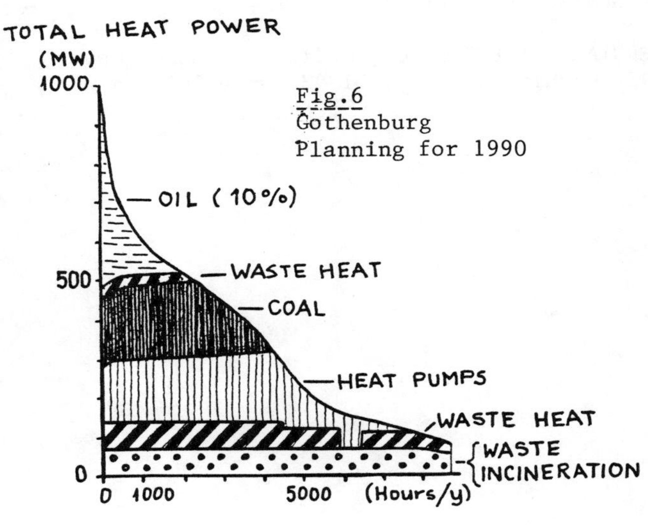

By adopting new heat sources, the annual amount of oil needed is decreased from 200,000 tons 1980 to 100,000 tons 1984 and is to be decreased down to about 30,000 tons by 1990.

One of the major advantages with district heating is that it facilitates the use of low temperature waste heat, heat from cogeneration plants and heat from other sources with a limited temperature output. With regard to this and considering that the heat losses in the distribution system increase with the temperature of the heat carrying fluid, it is important to keep the temperature levels low. During winter design conditions the outlet temperature of the distribution system ought not to exceed 100°C, yet up to 120°C can be unavoidable in older systems. The temperature must be reset in accordance with the change in the heat load so that it always is kept as low as possible. The demands to limit the maximum temperature and to decrease the temperature with the heat load can be fulfilled only by water systems or by variable vacuum steam systems. In practice steam systems seem not to be justified for larger areas with moderate thermal load densities. This could be illustrated by a quotation from the OTA report, INDUSTRIAL AND COMMERCIAL COGENERATION:

> "... U.S. city-scale district heating systems owned by utilities have, up until now, enjoyed little success when compared to European systems, primarily because European systems use hot water instead of steam."

REFERENCES

1. SVENSKA VARMEVERKSFORENINGEN. STATISTIK 1984. (Statistics of the Swedish District Heating Utilities)

2. INDUSTRIAL AND COMMERCIAL CONGENERATION, Office of Technology Assessment, OTA-E-192, Washington, D.C., 1983, page 165.

APPENDIX J

BIOGRAPHICAL NOTES ON THE AUTHORS

Enno Abel: Professor Building Services Engineering, Chalmers Univ., Sweden. Former Director, HVAC Department, BSK Consulting Company.

Elliot Aronson: Prof. Psychology, U. Cal-Santa Cruz. Former Chairman, Social Science Aspects of Energy Use, Nat. Acad. Sciences.

Van Baxter: Director, Oak Ridge Test Faciltiy, Oak Ridge National Laboratory, Recipient of the 1982 ASHRAE Willis Carrier Award.

Sam Berman: Leader, Lighting Research Group, Lawrence Berkeley Lab (LBL). Former Professor of Physics, Stanford University.

Thomas Bull: Senior Analyst and Project Director, Office of Technology Assessment, U.S. Congress.

Douglas Bulleit: President, Integrated Communication Systems Inc. Former Vice-President of Corporate Development, Herry International.

David Claridge: Associate Prof. Civil-Env. & Arch. Engineering, Univ. Colorado. Former Group Manager, SERI and Analyst, OTA.

Ben Cooper: Professional Staff, Energy Committee, U.S. Senate. Former Assistant Professor Physics, Iowa State University.

Paul Craig: Professor of Applied Science, Univ. California-Davis. Former Advisor to Presidental Science Advisor, Energy R&D, NSF.

Gautam Dutt: Research Scientist, Center for Energy and Environmental Studies, Princeton Unviersity.

Margaret Fels: Research Scientist, Center for Energy and Environmental Studies, Princeton University.

William Fisk: Staff Scientist and Project Leader, Building Ventilation and Indoor Air Qualtiy Group, LBL.

Howard Geller: Associate Director and Principal Resarcher, American Council for an Energy-Efficient Economy.

John Gibbons: Director, Office of Technology Assessment. Former Director, Office of Energy Conservation, Federal Energy Admin.

0094-243X/85/1350671-3$3.00 Copyright 1985 American Institute of Physics

Baruch Givoni: Professor, Graduate School of Architecture, UCLA. Former Senior Scientist, U.S. Army Instit. Environ. Medicine.

Jose Goldenberg: Companhia Energetica de Sao Paulo, Sao Poulo, Brazil.

David Grimsrud: Co-Leader, Building Ventilation and Indoord Air Quality Group, LBL. Former Prof. Physics, St. Olaf College.

David Hafemeister: Professor of Physics, California Polytechnic Univ. Former Special Assistant, U.S. Senate and State Department.

Daniel Hamblin: Leader, Demand Analysis Group, ORNL. Economist, Power Forecasting Division, Bonneville Power Administration.

John Ingersoll: Senior Staff Scientist, Hughes Aircraft Co. Member, Energy Technolgoy Group, Graduate School of Architecture, UCLA.

Henry Kelly: Senior Associate, Office of Technology Assessment. Former Associate Director, Solar Energy Research Institute.

Jonathan Koomey: Graduate Student, Energy/Resources Program, Univ. California, Berkeley, and Energy and Economic Systems Group, LBL.

Eric Larson: Research Scientist, Center for Energy and Environmental Studies, Princeton University.

Barbara Levi: Research Scientist, Center for Energy/Enviornmental Studies, Princeton Univ. Consulting Editor, Physics Today.

Mark Levine: Group Leader, Energy and Economic Systems, LBL. Former staff member, Ford Energy Policy Project.

Paul Maycock: President, Photovoltaic Energy Systems, Inc. Former Director of Photovoltaics Division, Department of Energy.

Robert Mowris: Graduate Student, Building Energy Engineering, Univ. Colorodo. Research Assistant, Lawrence Berkeley Laboratory.

Tony Nero: Co-Leader, Building Ventilation and Indoor Air Quality Group, LBL. Former Assistant Prof. Physics, Princeton Univ.

Joe O'Gallagher: Senior Reserach Associate, Enrico Fermi Instit., University Chicago. Former Assistant Prof. Physics, U. Maryland.

Robert Peddie: Former Chairman and Chief Executive, South Eastern Electricity Board, Board Member, Cent. Electricity Gov. Board, U.K.

Steven Plotkin: Senior Analyst and Project Director, Office of Technology Assessment. Former Senior Enviornmental Engineer, EPA.

John Reitz: Manager, Physics Department, Research Staff, Ford Motor Company. Former Professor of Physics, Case Western Reserve Univ.

Arthur Rosenfeld: Professor of Physics, Univ. Cal-Berkeley and Leader, Energy-Efficient Buildings Research Program, LBL.

Marc Ross: Professor Physics, Univ. Michigan. Senior Scientist, Energy/Enviornment Systems Div., Argonne National Lab.

Stephen Selkowitz: Group Leader, Windows and Daylighting, LBL. Executive Editor, ENERGY AND BUILDINGS journal.

Richard Sextro: Staff Scientist and roject eader, Building Ventilation and Indoor Air Qualty Group, L .

Max Sherman: Group Leader, Energy Performance of Buildings, LBL. U.S. Representative, Air Infiltration, International Energy Agency.

Granville Smith II: Vice President, STS Energenics, Ltd. Former Professional Staff, Energy Committee, U.S. Senate.

Robert Socolow: Director, Center for Energy and Environmental Studies and Professor of Mechanical Engineering, Princeton Univ.

Theodore Taylor: President, NOVA Incorporated. Co-author of NUCLEAR THEFT: RISK AND SAFEGURADS.

Robert Thresher: Principal Scientist, Wind Energy Research Center, Solar Energy Research Institute.

Teresa Vineyard: Research Associate, Energy Division, Oak Ridge National Laboratory.

Robert Williams: Senior Research Scientist, Center Energy/Env. Studies, Princeton. Former Scientist, Ford Energy Policy Project.

Fred Winkelmann: Staff Scientist, Buildings Energy Simulation Group, Lawrence Berkeley Laboratory.

Roland Winston, Professor Physics, University of Chicago. Inventor of the Compound Parabolic Concentrator.

AIP Conference Proceedings

		L.C. Number	ISBN
No. 1	Feedback and Dynamic Control of Plasmas – 1970	70-141596	0-88318-100-2
No. 2	Particles and Fields – 1971 (Rochester)	71-184662	0-88318-101-0
No. 3	Thermal Expansion – 1971 (Corning)	72-76970	0-88318-102-9
No. 4	Superconductivity in *d*- and *f*-Band Metals (Rochester, 1971)	74-18879	0-88318-103-7
No. 5	Magnetism and Magnetic Materials – 1971 (2 parts) (Chicago)	59-2468	0-88318-104-5
No. 6	Particle Physics (Irvine, 1971)	72-81239	0-88318-105-3
No. 7	Exploring the History of Nuclear Physics – 1972	72-81883	0-88318-106-1
No. 8	Experimental Meson Spectroscopy –1972	72-88226	0-88318-107-X
No. 9	Cyclotrons – 1972 (Vancouver)	72-92798	0-88318-108-8
No. 10	Magnetism and Magnetic Materials – 1972	72-623469	0-88318-109-6
No. 11	Transport Phenomena – 1973 (Brown University Conference)	73-80682	0-88318-110-X
No. 12	Experiments on High Energy Particle Collisions – 1973 (Vanderbilt Conference)	73-81705	0-88318-111–8
No. 13	π-π Scattering – 1973 (Tallahassee Conference)	73-81704	0-88318-112-6
No. 14	Particles and Fields – 1973 (APS/DPF Berkeley)	73-91923	0-88318-113-4
No. 15	High Energy Collisions – 1973 (Stony Brook)	73-92324	0-88318-114-2
No. 16	Causality and Physical Theories (Wayne State University, 1973)	73-93420	0-88318-115-0
No. 17	Thermal Expansion – 1973 (Lake of the Ozarks)	73-94415	0-88318-116-9
No. 18	Magnetism and Magnetic Materials – 1973 (2 parts) (Boston)	59-2468	0-88318-117-7
No. 19	Physics and the Energy Problem – 1974 (APS Chicago)	73-94416	0-88318-118-5
No. 20	Tetrahedrally Bonded Amorphous Semiconductors (Yorktown Heights, 1974)	74-80145	0-88318-119-3
No. 21	Experimental Meson Spectroscopy – 1974 (Boston)	74-82628	0-88318-120-7
No. 22	Neutrinos – 1974 (Philadelphia)	74-82413	0-88318-121-5
No. 23	Particles and Fields – 1974 (APS/DPF Williamsburg)	74-27575	0-88318-122-3
No. 24	Magnetism and Magnetic Materials – 1974 (20th Annual Conference, San Francisco)	75-2647	0-88318-123-1

No. 25	Efficient Use of Energy (The APS Studies on the Technical Aspects of the More Efficient Use of Energy)	75-18227	0-88318-124-X
No. 26	High-Energy Physics and Nuclear Structure – 1975 (Santa Fe and Los Alamos)	75-26411	0-88318-125-8
No. 27	Topics in Statistical Mechanics and Biophysics: A Memorial to Julius L. Jackson (Wayne State University, 1975)	75-36309	0-88318-126-6
No. 28	Physics and Our World: A Symposium in Honor of Victor F. Weisskopf (M.I.T., 1974)	76-7207	0-88318-127-4
No. 29	Magnetism and Magnetic Materials – 1975 (21st Annual Conference, Philadelphia)	76-10931	0-88318-128-2
No. 30	Particle Searches and Discoveries – 1976 (Vanderbilt Conference)	76-19949	0-88318-129-0
No. 31	Structure and Excitations of Amorphous Solids (Williamsburg, VA, 1976)	76-22279	0-88318-130-4
No. 32	Materials Technology – 1976 (APS New York Meeting)	76-27967	0-88318-131-2
No. 33	Meson-Nuclear Physics – 1976 (Carnegie-Mellon Conference)	76-26811	0-88318-132-0
No. 34	Magnetism and Magnetic Materials – 1976 (Joint MMM-Intermag Conference, Pittsburgh)	76-47106	0-88318-133-9
No. 35	High Energy Physics with Polarized Beams and Targets (Argonne, 1976)	76-50181	0-88318-134-7
No. 36	Momentum Wave Functions – 1976 (Indiana University)	77-82145	0-88318-135-5
No. 37	Weak Interaction Physics – 1977 (Indiana University)	77-83344	0-88318-136-3
No. 38	Workshop on New Directions in Mossbauer Spectroscopy (Argonne, 1977)	77-90635	0-88318-137-1
No. 39	Physics Careers, Employment and Education (Penn State, 1977)	77-94053	0-88318-138-X
No. 40	Electrical Transport and Optical Properties of Inhomogeneous Media (Ohio State University, 1977)	78-54319	0-88318-139-8
No. 41	Nucleon-Nucleon Interactions – 1977 (Vancouver)	78-54249	0-88318-140-1
No. 42	Higher Energy Polarized Proton Beams (Ann Arbor, 1977)	78-55682	0-88318-141-X
No. 43	Particles and Fields – 1977 (APS/DPF, Argonne)	78-55683	0-88318-142-8
No. 44	Future Trends in Superconductive Electronics (Charlottesville, 1978)	77-9240	0-88318-143-6
No. 45	New Results in High Energy Physics – 1978 (Vanderbilt Conference)	78-67196	0-88318-144-4
No. 46	Topics in Nonlinear Dynamics (La Jolla Institute)	78-57870	0-88318-145-2

No. 47	Clustering Aspects of Nuclear Structure and Nuclear Reactions (Winnepeg, 1978)	78-64942	0-88318-146-0
No. 48	Current Trends in the Theory of Fields (Tallahassee, 1978)	78-72948	0-88318-147-9
No. 49	Cosmic Rays and Particle Physics – 1978 (Bartol Conference)	79-50489	0-88318-148-7
No. 50	Laser-Solid Interactions and Laser Processing – 1978 (Boston)	79-51564	0-88318-149-5
No. 51	High Energy Physics with Polarized Beams and Polarized Targets (Argonne, 1978)	79-64565	0-88318-150-9
No. 52	Long-Distance Neutrino Detection – 1978 (C.L. Cowan Memorial Symposium)	79-52078	0-88318-151-7
No. 53	Modulated Structures – 1979 (Kailua Kona, Hawaii)	79-53846	0-88318-152-5
No. 54	Meson-Nuclear Physics – 1979 (Houston)	79-53978	0-88318-153-3
No. 55	Quantum Chromodynamics (La Jolla, 1978)	79-54969	0-88318-154-1
No. 56	Particle Acceleration Mechanisms in Astrophysics (La Jolla, 1979)	79-55844	0-88318-155-X
No. 57	Nonlinear Dynamics and the Beam-Beam Interaction (Brookhaven, 1979)	79-57341	0-88318-156-8
No. 58	Inhomogeneous Superconductors – 1979 (Berkeley Springs, W.V.)	79-57620	0-88318-157-6
No. 59	Particles and Fields – 1979 (APS/DPF Montreal)	80-66631	0-88318-158-4
No. 60	History of the ZGS (Argonne, 1979)	80-67694	0-88318-159-2
No. 61	Aspects of the Kinetics and Dynamics of Surface Reactions (La Jolla Institute, 1979)	80-68004	0-88318-160-6
No. 62	High Energy e^+e^- Interactions (Vanderbilt, 1980)	80-53377	0-88318-161-4
No. 63	Supernovae Spectra (La Jolla, 1980)	80-70019	0-88318-162-2
No. 64	Laboratory EXAFS Facilities – 1980 (Univ. of Washington)	80-70579	0-88318-163-0
No. 65	Optics in Four Dimensions – 1980 (ICO, Ensenada)	80-70771	0-88318-164-9
No. 66	Physics in the Automotive Industry – 1980 (APS/AAPT Topical Conference)	80-70987	0-88318-165-7
No. 67	Experimental Meson Spectroscopy – 1980 (Sixth International Conference, Brookhaven)	80-71123	0-88318-166-5
No. 68	High Energy Physics – 1980 (XX International Conference, Madison)	81-65032	0-88318-167-3
No. 69	Polarization Phenomena in Nuclear Physics – 1980 (Fifth International Symposium, Santa Fe)	81-65107	0-88318-168-1
No. 70	Chemistry and Physics of Coal Utilization – 1980 (APS, Morgantown)	81-65106	0-88318-169-X

No. 71	Group Theory and its Applications in Physics – 1980 (Latin American School of Physics, Mexico City)	81-66132	0-88318-170-3
No. 72	Weak Interactions as a Probe of Unification (Virginia Polytechnic Institute – 1980)	81-67184	0-88318-171-1
No. 73	Tetrahedrally Bonded Amorphous Semiconductors (Carefree, Arizona, 1981)	81-67419	0-88318-172-X
No. 74	Perturbative Quantum Chromodynamics (Tallahassee, 1981)	81-70372	0-88318-173-8
No. 75	Low Energy X-Ray Diagnostics – 1981 (Monterey)	81-69841	0-88318-174-6
No. 76	Nonlinear Properties of Internal Waves (La Jolla Institute, 1981)	81-71062	0-88318-175-4
No. 77	Gamma Ray Transients and Related Astrophysical Phenomena (La Jolla Institute, 1981)	81-71543	0-88318-176-2
No. 78	Shock Waves in Condensed Mater – 1981 (Menlo Park)	82-70014	0-88318-177-0
No. 79	Pion Production and Absorption in Nuclei – 1981 (Indiana University Cyclotron Facility)	82-70678	0-88318-178-9
No. 80	Polarized Proton Ion Sources (Ann Arbor, 1981)	82-71025	0-88318-179-7
No. 81	Particles and Fields –1981: Testing the Standard Model (APS/DPF, Santa Cruz)	82-71156	0-88318-180-0
No. 82	Interpretation of Climate and Photochemical Models, Ozone and Temperature Measurements (La Jolla Institute, 1981)	82-71345	0-88318-181-9
No. 83	The Galactic Center (Cal. Inst. of Tech., 1982)	82-71635	0-88318-182-7
No. 84	Physics in the Steel Industry (APS/AISI, Lehigh University, 1981)	82-72033	0-88318-183-5
No. 85	Proton-Antiproton Collider Physics –1981 (Madison, Wisconsin)	82-72141	0-88318-184-3
No. 86	Momentum Wave Functions – 1982 (Adelaide, Australia)	82-72375	0-88318-185-1
No. 87	Physics of High Energy Particle Accelerators (Fermilab Summer School, 1981)	82-72421	0-88318-186-X
No. 88	Mathematical Methods in Hydrodynamics and Integrability in Dynamical Systems (La Jolla Institute, 1981)	82-72462	0-88318-187-8
No. 89	Neutron Scattering – 1981 (Argonne National Laboratory)	82-73094	0-88318-188-6
No. 90	Laser Techniques for Extreme Ultraviolt Spectroscopy (Boulder, 1982)	82-73205	0-88318-189-4
No. 91	Laser Acceleration of Particles (Los Alamos, 1982)	82-73361	0-88318-190-8
No. 92	The State of Particle Accelerators and High Energy Physics (Fermilab, 1981)	82-73861	0-88318-191-6

No. 93	Novel Results in Particle Physics (Vanderbilt, 1982)	82-73954	0-88318-192-4
No. 94	X-Ray and Atomic Inner-Shell Physics – 1982 (International Conference, U. of Oregon)	82-74075	0-88318-193-2
No. 95	High Energy Spin Physics – 1982 (Brookhaven National Laboratory)	83-70154	0-88318-194-0
No. 96	Science Underground (Los Alamos, 1982)	83-70377	0-88318-195-9
No. 97	The Interaction Between Medium Energy Nucleons in Nuclei – 1982 (Indiana University)	83-70649	0-88318-196-7
No. 98	Particles and Fields – 1982 (APS/DPF University of Maryland)	83-70807	0-88318-197-5
No. 99	Neutrino Mass and Gauge Structure of Weak Interactions (Telemark, 1982)	83-71072	0-88318-198-3
No. 100	Excimer Lasers – 1983 (OSA, Lake Tahoe, Nevada)	83-71437	0-88318-199-1
No. 101	Positron-Electron Pairs in Astrophysics (Goddard Space Flight Center, 1983)	83-71926	0-88318-200-9
No. 102	Intense Medium Energy Sources of Strangeness (UC-Sant Cruz, 1983)	83-72261	0-88318-201-7
No. 103	Quantum Fluids and Solids – 1983 (Sanibel Island, Florida)	83-72440	0-88318-202-5
No. 104	Physics, Technology and the Nuclear Arms Race (APS Baltimore –1983)	83-72533	0-88318-203-3
No. 105	Physics of High Energy Particle Accelerators (SLAC Summer School, 1982)	83-72986	0-88318-304-8
No. 106	Predictability of Fluid Motions (La Jolla Institute, 1983)	83-73641	0-88318-305-6
No. 107	Physics and Chemistry of Porous Media (Schlumberger-Doll Research, 1983)	83-73640	0-88318-306-4
No. 108	The Time Projection Chamber (TRIUMF, Vancouver, 1983)	83-83445	0-88318-307-2
No. 109	Random Walks and Their Applications in the Physical and Biological Sciences (NBS/La Jolla Institute, 1982)	84-70208	0-88318-308-0
No. 110	Hadron Substructure in Nuclear Physics (Indiana University, 1983)	84-70165	0-88318-309-9
No. 111	Production and Neutralization of Negative Ions and Beams (3rd Int'l Symposium, Brookhaven, 1983)	84-70379	0-88318-310-2
No. 112	Particles and Fields – 1983 (APS/DPF, Blacksburg, VA)	84-70378	0-88318-311-0
No. 113	Experimental Meson Spectroscopy – 1983 (Seventh International Conference, Brookhaven)	84-70910	0-88318-312-9

No. 114	Low Energy Tests of Conservation Laws in Particle Physics (Blacksburg, VA, 1983)	84-71157	0-88318-313-7
No. 115	High Energy Transients in Astrophysics (Santa Cruz, CA, 1983)	84-71205	0-88318-314-5
No. 116	Problems in Unification and Supergravity (La Jolla Institute, 1983)	84-71246	0-88318-315-3
No. 117	Polarized Proton Ion Sources (TRIUMF, Vancouver, 1983)	84-71235	0-88318-316-1
No. 118	Free Electron Generation of Extreme Ultraviolet Coherent Radiation (Brookhaven/OSA, 1983)	84-71539	0-88318-317-X
No. 119	Laser Techniques in the Extreme Ultraviolet (OSA, Boulder, Colorado, 1984)	84-72128	0-88318-318-8
No. 120	Optical Effects in Amorphous Semiconductors (Snowbird, Utah, 1984)	84-72419	0-88318-319-6
No. 121	High Energy e^+e^- Interactions (Vanderbilt, 1984)	84-72632	0-88318-320-X
No. 122	The Physics of VLSI (Xerox, Palo Alto, 1984)	84-72729	0-88318-321-8
No. 123	Intersections Between Particle and Nuclear Physics (Steamboat Springs, 1984)	84-72790	0-88318-322-6
No. 124	Neutron-Nucleus Collisions – A Probe of Nuclear Structure (Burr Oak State Park - 1984)	84-73216	0-88318-323-4
No. 125	Capture Gamma-Ray Spectroscopy and Related Topics – 1984 (Internat. Symposium, Knoxville)	84-73303	0-88318-324-2
No. 126	Solar Neutrinos and Neutrino Astronomy (Homestake, 1984)	84-63143	0-88318-325-0
No. 127	Physics of High Energy Particle Accelerators (BNL/SUNY Summer School, 1983)	85-70057	0-88318-326-9
No. 128	Nuclear Physics with Stored, Cooled Beams (McCormick's Creek State Park, Indiana, 1984)	85-71167	0-88318-327-7
No. 129	Radiofrequency Plasma Heating (Sixth Topical Conference, Callaway Gardens, GÁ, 1985)	85-48027	0-88318-328-5
No. 130	Laser Acceleration of Particles (Malibu, California, 1985)	85-48028	0-88318-329-3
No. 131	Workshop on Polarized ^{3}He Beams and Targets (Princeton, New Jersey, 1984)	85-48026	0-88318-330-7
No. 132	Hadron Spectroscopy–1985 (International Conference, Univ. of Maryland)	85-72537	0-88318-331-5
No. 133	Hadronic Probes and Nuclear Interactions (Arizona State University, 1985)	85-72638	0-88318-332-3
No. 134	The State of High Energy Physics (BNL/SUNY Summer School, 1983)	85-73170	0-88318-333-1